Handbook of Ecosystem Theories and Management

ENVIRONMENTAL *and* ECOLOGICAL MODELING

Series Editor: **S.E. Jørgensen**

PUBLISHED TITLES

Handbook of Environmental and Ecological Modeling

S.E. Jørgensen, *The Royal Danish School of Pharmacy, Section of Environmental Chemistry, Copenhagen, Denmark*

B. Halling-Sørensen, *Institute of Environmental Science and Engineering, The Technical University of Denmark, Copenhagen, Denmark*

S.N. Nielsen, *The Royal Danish School of Pharmacy, Section of Environmental Chemistry, Copenhagen, Denmark*

Handbook of Estimation Methods in Ecotoxicology and Environmental Chemistry

S.E. Jørgensen, *The Royal Danish School of Pharmacy, Section of Environmental Chemistry, Copenhagen, Denmark*

B. Halling-Sørensen, *Institute of Environmental Science and Engineering, The Technical University of Denmark, Copenhagen, Denmark*

H. Mahler, *Software Developer, Copenhagen, Denmark*

Handbook of Ecosystem Theories and Management

S.E. Jørgensen, *The Royal Danish School of Pharmacy, Section of Environmental Chemistry, Copenhagen, Denmark*

F. Müller, *Kiel University, Kiel, Germany*

FORTHCOMING TITLES

Thermodynamics and Ecological Modeling

S.E. Jørgensen, *The Royal Danish School of Pharmacy, Section of Environmental Chemistry, Copenhagen, Denmark*

James Kay, *University of Waterloo, Waterloo, Ontario, Canada*

Handbook of Ecosystem Theories and Management

Edited by

S.E. Jørgensen
F. Müller

CRC Press is an imprint of the
Taylor & Francis Group, an informa business

CRC Press
Taylor & Francis Group
6000 Broken Sound Parkway NW, Suite 300
Boca Raton, FL 33487-2742

First issued in paperback 2019

ISBN-13: 978-1-56670-253-9 (hbk)
ISBN-13: 978-0-367-39891-0 (pbk)

Library of Congress Card Number 99-053502

Library of Congress Cataloging-in-Publication Data

Handbook of ecosystem theories and management / [edited by] S.E. Jørgensen, F. Müller..
p. cm. — (Environmental and ecological modeling)
Includes bibliographical references.
ISBN 1-56670-253-4 (alk. paper)
1. Biotic communities. 2. Ecosystem management. I. Jørgensen, Sven Erik. 1934–
II. Müller, Felix, 1954– III. Series.
OH541 . H244 2000
577—dc21

99-053502
CIP

Visit the Taylor & Francis Web site at
http://www.taylorandfrancis.com

and the CRC Press Web site at
http://www.crcpress.com

Preface

During the last four decades an increasing interest for ecosystem theory has emerged as a consequence of our increasing concern for the environment. The ecological and environmental literature from the last four decades contain many papers which focus on systems ecology, i.e., the system properties of ecosystems, and the question, how an ecosystem functions as a system. As a consequence of these trends in system ecology and ecosystem theory, Otto Fränzle, Winfried Schröder and Felix Müller decided to publish a German handbook in ecosystem analysis and its application in environmental science and management. It was published in 1997 under the title 'Handbuch der Umweltwissenschaften'. A number of German speaking authors covered various themes not only about systems ecology, but also about environmental science and management, mainly from a holistic view point. The volume was published as a folder which will be supplied with new chapters every 6 months to cover an increasing spectrum of themes. As English is the dominant language in science, it was, therefore, not difficult for us, the editors of this volume, to decide in 1996 that a publication of an English edition with a much wider selection of authors from all over the world would be a natural step to take. It was, however, decided to limit the focus only to ecosystem theory and its immediate application in management.

The ecological literature from the last two decades presents many different ecological theories. It is our ambition to present all or at least most of these theories in this volume to allow the readers a comparison of the different viewpoints. We attempt to illustrate that these theories, to a certain extent, are consistent and form a pattern (see Chapter II.10) which is an excellent basis for a more profound understanding of ecosystems. Ecosystems are incredibly complex, and it is therefore not surprising, that many different viewpoints are needed to obtain a more complete image of an ecosystem. A pluralistic approach is a 'must' if we want to cover most of the different aspects of ecosystem characteristics.

The volume encompasses three Sections including an introduction, a presentation of the different theories and general outlines for the immediate application of ecosystem theory in environmental management.

Section I, ECOLOGICAL AND EMPIRICAL FOUNDATIONS, deals with the following questions: What are ecosystems? Which tools can we use to study them and to uncover their characteristic properties? It consists of chapters with the following topics: Ecosystems as Complex Systems, Ecosystems as Structural Entities, Ecosystems as Functional Entities, Uncertainty in Ecology and Ecological Modelling, Fundamentals of Ecosystem Theory from General Systems Analysis, Methods of Ecosystem Theory, stemming from Ecosystem Research, Ecological Statistics and Ecological Modelling. As seen, these first chapters deal with an introduction of typical ecosystem properties, while the last chapters cover the tools applicable to ecosystem research, namely system analysis, ecological research, statistics and modelling.

Section II is entitled DIFFERENT APPROACHES IN RECENT ECOSYSTEM THEORY. It covers a wide spectrum of different theoretical approaches in system ecology, presented by authors who have published relevant topics in scientific papers. The first chapter in this Section focuses on the use of thermodynamics in ecosystem theory. It consists of three subchapters that cover the application of the three basic thermodynamic laws in ecosystem theory, a recent theory based on an extended version of the second law of thermodynamics and on a new tentative fourth law of thermodynamics. The second chapter in Section II deals with the self-organisation of ecosystems, while the third chapter looks into the application of information theories in order to obtain a better understanding of ecosystems.

The fourth chapter describes cybernetics and the fifth chapter applies hierarchical theory in our effort to develop ecosystem theory. The sixth chapter is a presentation of the concept emergy, used to facilitate a better understanding of ecosystem behaviour. The seventh chapter covers several facets considering ecosystems as networks where various components are dependent on each other through the particular webs of the considered elements. The chapter presents the ascendancy as a central concept in quantification of the network, the indirect effects which can be shown to exceed the direct effects in many cases and utility theory which looks into the cycling of matter and energy for instance by expressing how many times we are using the same energy and matter.

Chapter eight analyses ecosystem dynamics. Are ecosystems stable in mathematical terms? Are ecosystems resilient systems? Is catastrophe theory applicable to describe ecosystem behaviour under certain circumstances? Do ecosystems show chaotic behaviour? Are they operating on the edge of the chaos as self-organised critical systems? How do the dynamics of ecosystems operate in space and time?

The ninth chapter combines ecosystem theory and landscape ecology. This is followed by the formation of a pattern up to that point described from different approaches.

Section III: APPLICATIONS OF ECOSYSTEM THEORETICAL ASPECTS overviews the direct application of ecosystem theories and their potential contributions to environmental practice. In addition, there are several indirect applications. For instance, ecosystem theory is applied in ecological modelling. We can most probably develop better models by using our knowledge about ecosystems to improve the models and have them become better tools in environmental management. A better knowledge about ecosystem reactions has also a general indirect application in corners of environmental management. Section III is however entirely devoted to the direct applications of ecosystem theory in conservation biology, in assessment of ecosystem health, integrity and sustainability and in the two recently developed ecological disciplines ecological engineering and ecological economics. The direct applications use various theoretically based orientors to a high extent. Section III is therefore terminated by a chapter that serves as an overview of the applicable orientor approach, entitled as 'Ecological Orientors - A Path to Environmental Applications of Ecosystem Theories'.

It has been our ambition to collect different contributions about ecosystem theories in a useful book. We do, therefore, hope that it has been possible to produce a handbook which may be of general applicability for environmental managers and systems ecologists who want to have a quick access to more knowledge about ecosystems. As it can be seen from the list of authors and the titles of the chapters, this book covers a very wide range of differing viewpoints and concepts in ecosystem theory which has made it difficult to avoid a certain overlap. We acknowledge this as a weak point of the present handbook, but we have deliberately made this choice rather than omit a facet of the ecosystem theoretical complex.

We would like to express our sincere gratitude to Otto Fränzle and Søren Nors Nielsen who gave us good advice during the development of this book project and supported us in our effort to find the most appropriate content for a handbook of ecosystem theory. We also owe the many excellent authors a special debt of gratitude for their excellent contributions without which the book never would have been realised. The expert assistance of Gerti Rosenfeld and Paulette Clowes in converting the manuscripts to the present book form is also gratefully acknowledged.

Contents

List of Contributors

Barkmann, Jan
Ecology Center, Schauenburger Strasse 112, University of Kiel, 24118 Kiel, Germany

Bendoricchio, Giuseppe.
Department of Process Chemistry and Engineering, University of Padova, Via Marzola 9, 35131 Padova, Italy

Bossel, Hartmut
Am Galgenköppel 6, 34289 Zierenberg, Germany

Breckling, Broder
Ecology Center, Schauenburger Strasse 112, University of Kiel, 24118 Kiel, Germany

Brown, M.T.
Environmental Engineering Sciences and Center for Wetlands, University of Florida Gainesville, FL 32611, USA

Cleveland, Cutler
Center for Energy and Environmental Studies, Boston University, Boston, USA

Costanza, Robert
Institute for Ecological Economics, University of Maryland, P.O. Box 38, Solomons, Maryland 20688, USA

Dierssen, Klaus
Institute of Botany, University of Kiel, Germany

Dong, Quang
South East Environmental Research Program, Florida International University, Miami, Florida B 33199, USA

Fath, Brian
Institute of Ecology, University of Georgia, Athens, GA 30602, USA

Fränzle, Otto
Institute of Geography, University of Kiel, Germany

Gnauck, Albrecht
Institute of Ecosystems and Environmental Informatics, Technical University of Cottbus P.O. Box 10 13 44, 03013 Cottbus, Germany

Golley, Frank
Institute of Ecology, University of Georgia, Athens, GA 30602, USA

Grant, W.E.
Dept. Of Wildlife and Fishery Services, Texas A&M University, College Station, TX 7784-32258 USA

Gunderson, L.H.
Department of Zoology, University of Florida, Gainesville, FL 32611 USA

Hári, Stefanie
Ecology Center, Schauenburger Strasse 112, University of Kiel, 24118 Kiel, Germany

Holling, C.S.
Department of Zoology, University of Florida, Gainesville, FL32611 USA

Jørgensen, Sven. E.
Royal Danish School for Pharmacy, Department of Environmental Chemistry, University Park 2, 2100 Copenhagen Ø, Denmark

Kay, James
Environmental and Resource Studies, University of Waterloo, Ontario, N2L 3GI, Canada

Marin, S.L.
Dept. Of Wildlife and Fishery Services, Texas A&M University, College Station, TX 7784-32258 USA

Mitsch, W.
School of Natural Resources, Ohio State University, 2021 Correy Road, Columbus, OH 43210, USA

Moll, R.H.H.
Statistics Canada, Ottawa, Ontario, Canada

Müller, Felix
Ecology Center, Schauenburger Strasse 112, University of Kiel, 24118 Kiel, Germany

Nielsen, Søren N.
Royal Danish School for Pharmacy, Department of Environmental Chemistry, University Park 2, 2100 Copenhagen Ø, Denmark

Odum, H.T.
Environmental Engineering Sciences, University of Florida, PO Box 116450, Gainesville, Fl 32611, USA

O'Neill, Robert, V.
Environmental Sciences Division, Oak Ridge National Laboratory, Oak Ridge, TN 37831

Pahl-Wostl, C.
EAWAG, Ueberlandstr. 133 8600 Duebendorf, Switzerland

Patten, Bernard C.
Department of Zoology and Institute of Ecology, University of Georgia, Athens, GA 30602, USA

Pedersen, E.K.
Dept. Of Wildlife and Fishery Services, Texas A&M University, College Station, TX 7784-32258 USA

Perrings, Charles
Environmental Economics and Environmental Economics, University of York, York Y01 5DD, UK

Peterson, G.D.
Department of Zoology and Institute of Ecology, University of Georgia, Athens, GA 30602, USA

Rapport, D. J.
Faculty of Environmental Sciences, University of Guelph, Guelph, Ontario, Canada

Straskraba, M.
Biomathematical Laboratory, Biological Research Centre, Czech Academy of Science, Branisovska 31, 37005 Budejovice, The Czech Republic

Svirezhev, Yuri
PIK – Potsdam Institute for Climate Impact Research, P.O. Box 60 12 03, 14412 Potsdam, Germany

Ulgiati, S.
Department of Chemistry, University of Sienna, Pian die Mantellini 44, 53100 Siena, Italy

Ulanowicz, Robert E.
Chesapeake Biological Laboratory, University of Maryland, P.O. Box 38, Solomons, Maryland 20688, USA

Windhorst, Wilhelm
Ecology Center, Schauenburger Strasse 112, University of Kiel, 24118 Kiel, Germany

Fundamental and Empirical Foundations

Introduction

The following section covers the fundamental and basic information needed to understand the subsequent Parts II and III of the handbook. The following questions are discussed: 1) What are ecosystems? 2) Which structural and functional characteristics do ecosystems have? 3) Which tools can be applied to study the properties of ecosystems, including the use of system analysis and empirical tools: data collection, statistics and models.

The handbook can be understood as a bouquet of 'different ecosystem flowers' which we hope that the readers will find is nice and attractive or at least complementary. In this perspective, the papers in Section 1 will deliver the foundations. They will discuss ecosystem definitions and the general problems of ecosystem theory. They will describe the fundamental subsystems of ecosystem structures, functions and organisations, and they will analyse the integration of ecological units and general systems theory. They will illustrate the general problems of ecological uncertainties and will elucidate them from a modelling point of view. Finally, the basic methodologies are illustrated which potentially provide the data and techniques to test ecosystem theoretical hypotheses. These are e.g. ecosystem research, ecological statistics, and environmental modelling.

When reading the following chapters it may become obvious that many of the concepts are still in an early developmental stage. The terminology of different approaches is not totally harmonised: Some notions consequently are not compatible throughout all papers of this book. We accepted these differences because they indicate the recent state of ecosystems theory. It is a pattern of different approaches which originate in different conceptions. Actually we prefer the combination of these ideas as they are all together fundamentals for a holistic synthesis, which will not be one 'unified' theory, but rather an emergent sum of different concepts.

I.1 Ecosystems as Complex Systems

Sven E. Jørgensen and Felix Müller

1. Introduction

1.1 The scope of system ecology

System ecology was initiated in the fifties by the scientific community and the author E.P. Odum (1953), but the real 'take off' with tail wind coming from society started, however, around the mid sixties. This came as a result of the environmental discussion, triggered by Rachel Carson's book 'Silent Spring' (1962), while Arthur Koestler's book 'The Ghost in the Machine' (1967) contributed to the initial development of ecosystem theory. 'Limits to Growth' by Meadows et al. (1972) also has to be mentioned in this context as an important initiative of environmental discussion, which provoked a significant need for a new and more holistic ecology.

The interest for ecosystems comprehended as complex systems from an holistic view point, meaning that the whole system is more than the sum of its parts because it is providing characteristic emergent properties, has increased enormously during the last couple of decades due to our concern for the environment. We want to understand the reaction of nature to our influence and we have realised that reductionistic methods alone cannot cope with the environmental problems, mainly due to –

1) unexpected effects and rare events, unexpected in time and space may occur at any time due to the high amount of linkages and indirect effects in ecosystems, and because
2) ecosystems are extremely complex which renders them impossible to analyse and thereby unable know all the details.

Consequently, we cannot attempt to find the understanding of the whole by adding all the reductionistic details. We have to use another approach, which is system ecology, where the system analytical focus is put on the properties of entire ecosystems.

Ecosystems are what is called medium number systems (O'Neill et al., 1986). They have many components and the components as well as the systems are all unique and different. In a generalising estimation, a typical ecosystem would have in the order of 10^{15}-10^{20} components. Some of them may belong to the same species or the same type of non-living components, for instance suspended clay particles. The number of species in ecosystems may vary considerably, but will in most ecosystems be of the order of 1000 - 100,000. Individuals belonging to the same species may have some characteristic properties in common, but each individual is still different from all the other individuals – for instance they have their own genetic code, which implies that we are dealing with systems of 10^{15}-10^{20} components with clearly different properties. This variability is indispensable, as it is the basis for the Darwinian selection and thus a prerequisite for evolution. The conclusion is therefore that it would be an impossible task to analyse and know all the components of an ecosystem in detail, and even if we could gather the detailed knowledge about all the components, it may soon be useless, because the biological components are currently changing as they adapt to the steadily changing conditions.

The enormous complexity of ecosystems may be illustrated by the following example: the molecules in a gas make up a system with even a higher number of components; compare with Avogadro's number, 6.02×10^{23}, but in this case we are dealing with components that can be classified as approximately 10 or 20 different types with only 10-20 different sets of properties, which are not changeable, but fixed. Therefore we can apply statistical methods and get a fairly good description on the basis of a relatively few measurements. Due to the high number of elements, subsystems and interrelations, ecosystems behave as complex adaptive systems and have several common features:

- They are composed of numerous components of many different kinds as presented above.
- The components interact and react non-linearly and on different temporal and spatial scales (see Chapter II.8).
- The systems organise themselves to produce complex structures and behaviours (see Chapters II.2 and II.7).
- The systems maintain thermodynamically unlikely states, made possible by their openness (see Chapter II.1.1).
- Some form of heritable information allows the systems to respond adaptively to environmental changes (see Chapter II.3).
- The structure and dynamics of these systems are effectively irreversible, and there is always a legacy of history (see Chapters II.1.1 and II.7.3).
- They are hierarchically organised (see Chapter II.5).

Consequently, we need a specific science of ecosystems, which can cope with these features and assist us in our effort to understand their reactions to our impacts on them. This science should be based on ecosystem theories, that try to understand the reactions of ecosystems without knowing all the details, like we also try to understand human reactions without knowing all the properties of all human cells and their biochemical reactions. Systems ecology must emphasise the system properties and ecosystem theories must be based on our possibilities to draw general conclusions from our knowledge about these properties. This scientific approach implies that we will not be able to give full descriptions of ecosystems covering all aspects. We can only give a partial description covering some of the aspects – *therefore we need a pluralistic approach.*

1.2 The scope of ecosystem theory

Theories are extremely interesting spheres of science. Therefore a multitude of different connotations and attitudes of 'theory' can be found in the scientific literature. Sadly, the most common idea about theories is that they are very complicated and hardly intelligible descriptions of actually less complicated facts. Or, they simplify the situation too much and offer just 'vague speculations', 'unclear descriptions' and are less helpful and unusable than describing the 'real' situation. However, it is not all negative. Popper (1989), for example, characterises theories metaphorically as *"a fishing net that scientists cast to catch the world, to explain it and to control it"*. And, to add another positive and encouraging anecdote, in classical Greek philosophy, theorising was considered as one of the highest forms of intellectual activity.

In a first approximation 'theory' can be understood as a set of hypotheses posting novel entities and properties which frequently are not accessible to direct observation (Audi, 1995). Theories, furthermore, are empirically or deductively based, aggregating and integrating representations of the proved and reliable knowledge of a branch of science. They function as fundamental elements and frameworks of a scientific system within which all single phenomena of the objective should be explainable by coherent, theory based laws. Theories are used for the description, explanation, and organisation of scientific facts, laws and relationships. We can classify theories as follows (after Müller, 1997):

- Theories consist of sets of hypotheses which have been proven by underlying empirical facts.
- Theories have to be non-evaluative descriptions of a context of casualties.

- Theories must be capable of being falsified (Popper, 1989).
- Theories must be prognostic.
- Theories must be applicable to single cases.
- Theories provide a directing and motivating function for the development, orientation and structure, in the empirical research activities of the respective scientific discipline.

Ecosystems theory has to utilise empirical data, from various scientific areas for the integration, organisation and improvisation of the understanding of the general features, patterns, and casualties of the interactions in and between ecosystems. Ecosystem theory is built up by a hierarchical set of comprehensive ecological hypotheses, which have to be integrated into an abstract system of coherent contexts and principles. These are often rather complex hypotheses about biotic, abiotic and anthropogenic interrelations which frequently are formulated as ecological models. These models can be proven through model verification and validation procedures against empirically measured ecosystem data sets. While doing this we should be aware of the fact that the first hand only functional hypotheses, which are theory based simplifications of reality. Therefore validation at the same time is a test of both the model and the hypotheses. In this situation there is a 'double doubt'. Consequently, it would be required to perform validation on a wide range of cases (as wide as possible) i.e. more ecosystem research. Applying such models, the hypotheses of ecosystem theory can be approximately falsified within the extents of the models' validates, and through model utilisation they are prognostic as well as applicable to single cases. Thus, models are important tools for the explanation of ecosystem phenomena.

1.3 The scope of ecosystem analysis

The focal object of this book has been defined for the first time by A.G. Tansley (1935). His ecosystem comprehension was the following:

> " *But the more fundamental conception is, as it seems to me, the whole system (in the sense of physics), including not only the organism-complex, but also the whole complex of physical factors forming what we call the environment of the biome – the habitat factors in the widest sense. It is the system so formed which, from the point of view of the ecologist, are the basic units of nature on the face of the earth. These ecosystems, as we may call them, are of the most various kinds and sizes. They form one category of the multitudinous physical systems of the universe, which range from the universe as a whole down to the atom.* "

Thus, the ecosystem is a fundamental unit of ecological research (Odum, 1980), which can be characterised by the multiple interactions between and within the biotope and the biocenois which represents the community of living organisms in the investigated space. There are many modern definitions of ecosystems (see Breckling and Müller, 1997) which can be distinguished into two groups. The first class of definitions refer to structural features and comprehends ecosystems as really existing, 'touchable' segments of space. In the second group of definitions, the functional system of interactions is being stressed. Ecosystems from this viewpoint are derived models of the processes which operate between the structural sub-units of the entity. In an optimal case, both viewpoints are combined. Then ecosystems are models of networks of biotic and abiotic interactions in a certain area. Some authors (e.g., Ellenberg, 1983; Stugren, 1986; Klötzli, 1993; Odum, 1980 or Jax et al., 1993) argue that certain minimal requirements have to be fulfilled to call a natural or near-nature entity 'ecosystem'. They are

- internal cycles of matter
- an emergent behaviour of the system which exceeds the reactions of the parts,
- an energy flow ion foodwebs which induces biodiversity and nutrient cycles

3. Quantum mechanical uncertainty caused by ecosystem complexity

How can we describe complex ecosystems in detail? The answer is that it is impossible, if the description must include all details, including all interactions between all the components in the entire hierarchy and containing all details on feedbacks, adaptations, regulations and the entire evolution process. To describe the corresponding problems, Jørgensen (1988 and 1994b) has introduced the application of the uncertainty principles of quantum mechanics in ecology rooted in the complexity of ecosystems. In nuclear physics the uncertainty is caused by the observer of the incredibly small nuclear particles, while the uncertainty in ecology is caused by the enormous complexity of ecosystems.

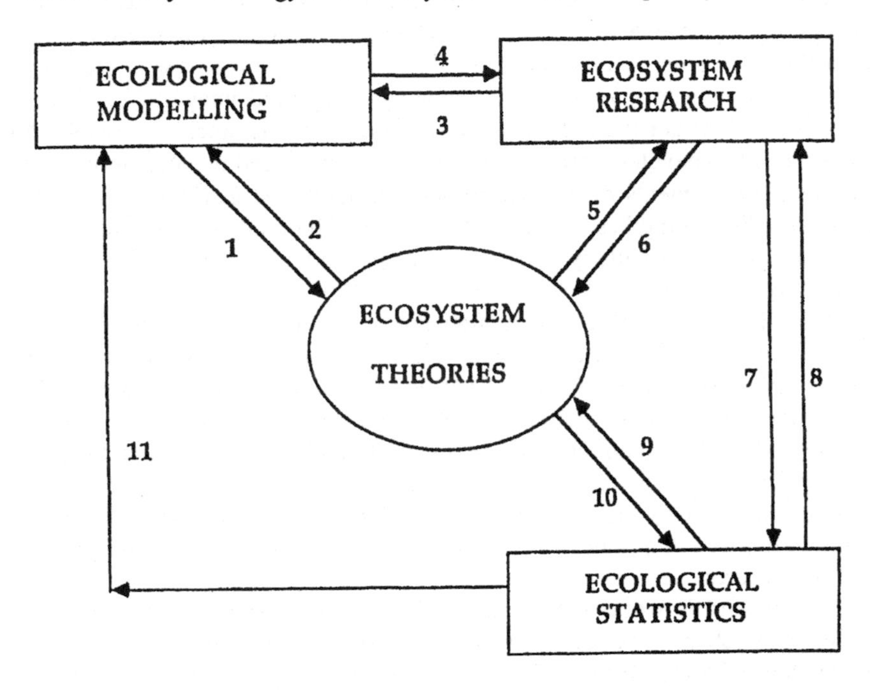

Figure 1: The figure illustrates the interactions between ecosystem theories, ecological modelling, ecological research and ecological statistics. The arrows represent the interactions: 1) the use of models to test ecological theoretical hypotheses or to provide components of ecosystem theories directly, 2) new ecological theoretical developments should be used to make our models more ecological. 3) the use of data to test ecological models 4) models can provide research priorities 5) questions raised in ecological theoretical context should be answered by ecological research 6) interpretation of ecological research results may lead to elements in ecosystem theories 7) data from ecological research should be treated statistically 8) requirement of statistical significance may inspire to additional ecological research 9) results of ecological statistics are used as elements in ecological theories 10) ecosystem theories may be tested by ecological statistics 11) submodels or black box models, resulting from ecological statistics are used in ecological modelling.

For instance, if we take two components and want to know all the relationships between them, we would need at least three observations to show whether the relations are linear or non-linear. Correspondingly, the relationships among three components will require 3*3 observations for the shape of the plane. If we have 18 components we would correspondingly need 3^{17} or approximately 10^8 observations. At present this is probably an approximate, practical upper limit to the number of observations which can be invested in one project aimed for one ecosystem. This could be used to formulate a practical uncertainty relationship in ecology according to Jørgensen (1997):

$$10^5 * \Delta x / \sqrt{3^{n-1}} \leq 1 \qquad (1)$$

where Δx is the relative accuracy of one relation, and n is the number of components examined or included in the model.

The 100 million observations could, of course, also be used to give a very exact picture of one relation. Costanza and Sklar (1985) talk about the choice between the two extremes: knowing 'everything' about 'nothing' or 'nothing' about 'everything'. The first refers to the use of all the observations on one relation to obtain a high accuracy and certainty, while the latter refers to the use of all observations on as many relations as possible in an ecosystem.

How can we obtain a balanced complexity in the description? This question will be discussed in the next chapter. Equation (1) formulates a practical uncertainty relation, but, of course, the possibility that the practical number of observations may be increased in the future cannot be excluded. Ever more automatic analytical equipment is steadily emerging on the market. This means that the number of observations that can be invested in one project may be one, two, three or even several magnitudes larger in one or more decades. However, a theoretical uncertainty relation can be developed. If we go to the limits given by quantum mechanics, the number of variables will still be low, compared to the number of components in an ecosystem.

One of Heisenberg's uncertainty relations is formulated as follows:

$$\Delta s * \Delta p \geq h/2\pi \qquad (2)$$

where Δs is the uncertainty in determination of the place, and Δp is the uncertainty of the momentum. According to this relation, Δx of equation (2) should be in the order of 10^{-17} if Δs and Δp are about the same order of magnitude. Another of Heisenberg's uncertainty relations may now be used to give the upper limit of the number of observations:

$$\Delta t * \Delta E \geq h/2\pi \qquad (3)$$

where Δt is the uncertainty in time and ΔE in energy.

If we use all the energy that the Earth has received during its lifetime of 4.5 billion years we get:

$$173 * 10^{15} * 4.5 * 10^{9} * 365.3 * 24 * 3600 = 2.5 * 10^{34} \text{ J}, \qquad (4)$$

where $173 * 10^{15}$ W is the energy flow of solar radiation. Δt would, therefore, be in the order of 10^{-69} sec. Consequently, an observation will take at least 10^{-69} sec., even if we use all the energy that has been available on Earth as ΔE, which must be considered the most extreme case. The hypothetical number of observations possible during the lifetime of the earth would therefore be:

$$4.5 * 10^{9} * 365.3 * 3600/10^{-69} = \sim \text{of } 10^{85} \qquad (5)$$

This implies that we can replace 10^5 in equation (I.1.1) with 10^{60} since

$$10^{-17}/\sqrt{10^{85}} = \sim 10^{-60}$$

If we use $\Delta x = 1$ in equation (I.1.1.) we get:

$$\sqrt{3^{n-1}} \leq 10^{60}$$
$$\text{Or} \quad n \leq 253. \tag{6}$$

From these very theoretical considerations we can clearly conclude that we shall never be able to get a sufficient number of observations to describe even one ecosystem in all detail. These results are completely in harmony with Niels Bohr's complementarity theory. He expressed it as follows: *"It is not possible to make one unambiguous picture (model) of reality, as uncertainty limits our knowledge."* The uncertainty in nuclear physics is caused by the inevitable influence of the observer on the nuclear particles; in ecology it is caused by the enormous complexity and variability.

Quantum theory may have an even wider application in ecological theories. Schrödinger (1944) suggests, that the 'jump like changes' you observe in the properties of species are comparable to the jump-like changes in energy by nuclear particles. Schrödinger was inclined to call De Vries' mutation theory (published in 1902), the quantum theory of biology, because the mutations are comparable to quantum jumps in the gene molecule.

As an extrapolation of quantum mechanics to ecology Patten (1982) defines an elementary 'particle' of the environment, called an environ – previously he used the word holon – as a unit which is able to transfer an input to an output. Patten suggests, that a characteristic feature of ecosystems is the connectance. Input signals are received by the ecosystem components, they are transformed and translated into output signals. Such a 'translator unit' is an environmental quantum according to Patten. The concept is borrowed from Koestler (1967), who introduced the word 'holon' to designate the unit on a hierarchic tree; see Allen and Starr (1982), O'Neill et al. (1986) and Müller (1992). The term comes from Greek 'holos' = whole, with the suffix 'on' as in proton, electron and neutron to suggest a particle or part.

Stonier (1990) introduces the term infon for the elementary particle of information. He envisages an infon as a photon, whose wavelength has been stretched to infinity. At velocities other than c, its wavelength appears infinite, its frequency zero. Once an infon is accelerated to the speed of light, it crosses a threshold, which allows it to be perceived as having energy. When that happens, the energy becomes a function of its frequency. Conversely at velocities other than c, the particle exhibits neither energy nor momentum – yet it could retain at least two information properties: its speed and its direction. In other words at velocities other than c, a quantum of energy becomes converted into a quantum of information – an infon.

4. Pluralistic view of complex ecosystems

The use of maps in geography is a good parallel to the complex and diversity-influenced situation in ecology. As we have road maps, aeroplane maps, geological maps, maps in different scales for different purposes, we have in ecology many models of the same ecosystem. We need them all, if we want to get a comprehensive, pluralistic view of the ecosystem. A map can furthermore never provide a complete picture. We can always make the scale larger and larger and include more and more details, but we cannot get all the details, for instance where all the cars of an area are situated just now, and if we could, the image would be invalid a few seconds later, because we want to map too many dynamic details at the same time. An ecosystem also consists of too many dynamic components to enable us to model all the components simultaneously and even if we could, the model would be invalid a few seconds later, because the dynamics of the system have changed the situation.

In nuclear physics we need many different pictures of the same phenomena to be able to describe our observations. We say that we need a pluralistic view to cover our observations completely. Our observations of light for instance require us to consider light as waves as well as particles. The situation in ecology is similar. Because of the immense complexity we need a pluralistic view to cover a description of the ecosystems according to our observations. We need many models covering different viewpoints. It is consistent with Gödel's Theorem from 1931 (see Gödel, 1986), that the infinite truth can never be condensed in a finite theory. There are limits to our insight, or, we cannot produce a map of the world with all possible details, because that would be the world itself.

Furthermore, ecosystems must be considered as irreducible systems (Wolfram, 1984 a and b) in the sense that it is not possible to make observations and then reduce the observations to more or less simple laws of nature, as it is true of mechanics for instance. Too many interacting components force us to consider ecosystems as irreducible systems. The entire ecological network plays a role for all the processes in an ecosystem. If we isolate a few components and their interacting processes by a laboratory or an *in situ* experiment, we will exclude the indirect effects of the components interacting through the entire network. As the indirect effects often are more dominant than the direct ones (see Chapter II. 7.3), our experiment will not be capable to uncover the results of the relations as they are observed in nature. The same problem is found today in nuclear physics, where the picture of the atoms is now 'a chaos' of many interacting elementary particles. Assumptions on how the particles are interacting are formulated as models, which are tested toward observations. We draw upon exactly the same solution to the problem of complexity in ecology. It is necessary to use what is called experimental mathematics or modelling to cope with such irreducible systems. Today, this is the tool in nuclear physics, and the same tool is increasingly used in ecology.

5. Holism versus reductionism

Holism and reductionism are two different approaches to reveal the secrecies of nature.

Holism attempts to reveal the properties of complex systems such as ecosystems by studying the systems as a whole. According to this approach the system properties cannot be found by a study of the components separately due to the high complexity and due to the presence of emergent properties. Therefore – although it is far more difficult – it is required that the study be on the system level. This does not imply that a good knowledge of the components and their properties is redundant. The more we know about the system on all levels, the better we are able to extract the system properties. But it does imply, that a study of the components of ecosystems will never be sufficient, because such a study will never reveal the system properties and the functional patterns of interactions. The components of ecosystems have been co-ordinated co-evolutionary to such an extent, that ecosystems operate as indivisible unities.

Reductionism attempts to reveal the properties of nature by separating the components from their wholeness to simplify the study and to facilitate the interpretation of the scientific results. This scientific method has been the core approach since Newton. It is, of course, always very useful to find governing relationships in nature – for instance primary production versus radiation intensity, mortality versus concentrations of toxic substances, etc. But the reductionistic methods have obvious shortcomings, when the functions of entire ecosystems are to be investigated. A human being cannot be described on the basis of the properties of all the cells of the body. The function of a church cannot be found through studies of the bricks, the columns, etc. There are numerous examples which illustrate the additional need for holistic approaches.

It is far easier to take a complex system apart than it is to reassemble the parts and restore the important functions. Similarly, no science has succeeded in understanding the structure and dynamics of a complex system from a reductionistic approach alone. During the last few decades population ecologists have tried to describe and predict the fluctuations in the local abundances of species and community ecologists have tried to describe and predict changes in the composition of locally coexisting species; but progress has been slow and generality has been limited. It is not because the causes of past changes in abundance or species composition are inherently unknowable or because prediction of the future trajectory is impossible. It is because even small differences can be amplified (see Chapter II.8.4) by non-linear processes to produce divergent outcomes.

The conclusion from these considerations is clear: we need both approaches, but because it is much easier to apply the reductionistic method, analytical work has been the overwhelming synthetic work in science, particularly during the period from 1945 to 1975. The last twenty to thirty years of ecological research have shown with increasing clarity that the need for the holistic approach is urgent. Many ecologists feel that a holistic ecosystem theory is a necessary basis for a more comprehensive understanding of the ecosphere and the ecosystems and for a solution to all threatening global problems.

The need for a more holistic approach increases with the complexity, integration, number of interactions, feedbacks and regulation mechanisms. A mechanical system – a watch for instance – is divisible, while an ecosystem is indivisible, because of the well developed interdependence. The ecosystem has developed this interdependence during billions of years. All species have evolved, step by step, their properties determined by the prevailing conditions – i.e., all external factors and all other species. All species are influenced by all other biological and non-biological components of the ecosystem to a certain extent. All species are therefore confronted with the question: 'which of the possible combinations of properties will give the best chance for survival and growth, considering all possible factors, i.e., all forcing functions and all other components of the ecosystem?' That combination will be selected and will offer most benefit in the long run to the entire system, where all components try to optimise the answer to the same central question, mentioned above. This game has continued for billions of years. A steady refinement of the properties has taken place, and it has been possible through this evolution to consider ever more factors, which means that the species have become increasingly integrated with the system and ever more interactions have developed.

Patten (1991 and 1992) expresses the direct and the indirect effect numerically. The direct effect between two components in an ecosystem is the effect of the direct link between them; see also Chapter II.7.2. The link between phytoplankton and zooplankton for instance is the grazing process. The indirect effect is the effect caused by all relationships between two components except the direct one. The grazing of zooplankton for instance has also a beneficial effect on phytoplankton, because the grazing will accelerate the turnover rate of the nutrients. It is difficult mathematically to consider the total indirect effect and to be able to compare it with the direct effect. This problem is treated in Chapter II.7.2, but it can already be revealed in this context that Patten has found that the indirect effect often may be larger than the direct one. It implies that a separation of two related components in an ecosystem for examination of the link between them will not be capable to account for a significant part of the total effect of the relationship. The conclusion from Patten's work is clearly that it is not possible to study an ecosystem on the system level, taking all interrelations into account by studying the direct links only. An ecosystem is more than the sum of its parts.

Lovelock (1979) has taken a full step in the holistic direction, as he considers our planet as one co-operative unit. The properties of the Earth cannot be understood according to his opinion without an assumption of a co-ordinated co-evolution of approximately ten billion species on earth. Lovelock (1988) was struck by the unusual composition of the atmosphere. How could methane and oxygen be present simultaneously? Under normal circumstances these two gases would react readily to produce carbon dioxide and water. Looking further he found that the concentration of carbon dioxide was much smaller on Earth than it would have been if the atmospheric gases had been allowed to reach an equilibrium. The same is true for the salt concentration in the sea. Lovelock concluded that the planet's persistent state of non-equilibrium was a clear proof of the life activities and that the regulations of the composition of the spheres on earth have coevolved over time. Particularly the cycling of essential elements has been regulated to the benefit of life on earth. Lovelock believes that innumerable regulating biomechanisms are responsible for the homeostatis or steady-state far-from-equilibrium of the planet. Three examples will be mentioned here to illustrate this challenging idea.

Ocean plankton emits a sulphurous gas into the atmosphere. A physical-chemical reaction transforms the gas into aerosols on which water vapour condenses, setting the stage for cloud formation. The clouds then reflect a part of the sunlight back into space. If the Earth becomes too cool, the number of plankton is cut back by the chill. The cloud formation is thereby reduced and the temperature rises. The plankton operate like a thermostat to keep the Earth's temperature within a certain range.

The silica concentration of the sea is controlled by the diatoms. Less than 1% of the silica transported to the sea is maintained at the surface. Diatoms take up the silica and when they die they settle and remove the silica from the water to the sediment. The composition of the sea is maintained far from the equilibrium known from salt lakes without life, due to the presence of diatoms. Life – the diatoms – controls that life conditions are maintained in the sea.

Sulphur is transported from the lithosphere into the sea causing an imbalance. If there were no regulations, the sulphur concentration of the sea would be too high and sulphur would be lacking in the lithosphere as an essential element. However, many aquatic organisms are able to get rid of undesired elements by methylation processes. Methyl compounds of mercury, arsenic and sulphur are very volatile, which implies that these elements are transported from the hydrosphere to the atmosphere by methylation processes. *Polysiphonia fastigiata*, a marine alga, is capable of producing a huge amount of dimethylsulphide (Lovelock, 1979). This biological methylation of sulphur can be used to explain that the delicate balance of essential elements between the spheres is maintained.

The Gaia hypothesis presumes that the components of the ecosphere and therefore also the components of the ecosystems co-operate more than they compete, when we contemplate the effects from a system's view point. This point is illustrated here by an example which shows how symbiosis can develop and lead to new species. The example is described by Barlow (1991) and the event was witnessed and described by Kwang Jeon. Kwang Jeon had been raising amoebas for years, when he received a new batch for his experiments. The new batch disseminated a severe illness and the amoebas refused to eat and failed to reproduce. Many amoebas died and the few that grew and divided did so reluctantly. A close inspection revealed that about 100,000 rod-shaped bacteria, brought in by the new amoebas, were present in each amoeba. The surviving bacteriolized amoebas were fragile. They were easily killed by antibiotics and oversensitive to heat and starvation. For some five years, Jeon nurtured the infected amoebas back to health by continuously selecting the tougher ones. Although they were still infected, they started to divide again at the normal rate. They had not got rid of their bacteria, but they adapted and were cured of their disease. Each recovered amoeba contained about 40,000 bacteria and adjusted their destructive tendencies in order to live and survive inside other living cells.

From friends Jeon reclaimed some of the amoebas that he had sent off before the epidemic. With a hooked glass needle he removed the nuclei from the infected and uninfected organisms and exchanged them. The infected amoebas with new nuclei survived, while the uninfected amoebas supplied with nuclei from cells that had been infected for years struggled for about four days and then died. The nuclei were unable to cope with an uninfected cell. To test this hypothesis, Jeon injected uninfected cells with nuclei from infected amoebas with a few bacteria, just before they died. The bacteria rapidly increased to 40,000 per cell and the amoebas returned to health. Obviously, a symbiosis had been developed.

The amoeba experiment shows that co-operation is an extremely important element in evolution. An ultimate co-operation of all components in the ecosystems leads inevitably to a Gaia perception of ecosystems and the entire ecosphere.

It is interesting that Axelrod (1984) demonstrates through the use of game theory that co-operation is a beneficial long-term strategy. The game anticipates a trade situation between you and a dealer. At mutual co-operation, both parties earn two points, while at mutual defection both earn zero points. Co-operating while the other part defects stings: you get minus one point while the 'rat' gets something for nothing and earns four points. Should you happen to be a rat, while the dealer is co-operative, you get four points and the dealer loses one. Which strategy should you follow to achieve most? Two computer tournaments have given the result that the following so-called 'tit-for-tat' strategy seems to be winning: starts with a co-operative choice and thereafter does what the other player did in the previous move. In other words, be open for co-operation unless the dealer is not. But only defect one time after the dealer has defected.

It may be possible to conclude that the acceptance of the Gaia hypothesis does not involve mysterious, unknown, global forces to be able to explain these observations of homeostasis. It seems to be possible to explain the hypotheses by five factors:

1. Selection (steadily ongoing tests of which properties give the highest chance of survival and growth) from a range of properties, offered by the existing species.
2. Interactions of randomness (new mutations and sexual genetic recombinations are steadily produced) and necessity, i.e., to have the right properties for survival under the prevailing conditions, resulting from all external factors and all other components of the ecosystem.
3. A very long time has been available for this ongoing 'trial and error' process, which has developed the ecosphere step-wise toward the present, ingenious complexity, where all components have unique and integrated properties.
4. The ability of the biological components to maintain the results already achieved (by means of genes) and to build upon these results in the effort to develop further.
5. As the complexity of the ecosystems and thereby the complexity of the entire ecosphere develops, the indirect effect becomes more and more important; see Chapter II.7.2 for further explanations. It implies that the selection based upon the 'effects' on the considered component will be determined by the entire ecosystem and that this selection process will assure that all components of the ecosystem will evolve toward being better and better fitted to the functional and structural entity. It means that the system will evolve toward operating more and more as a whole – as an integrated system – and that the result of the selection will be more and more beneficial for the entire system.

6. Toward theories of complex ecosystems

It is the aim of this book to give an overview of present ecosystem theories. The presentation will concentrate on the developments in systems ecology, but to understand all the presented theories, it is also necessary to understand all the components, that have contributed to the state of the art. Therefore the book will touch on quantum mechanics, non-equilibrium thermodynamics, catastrophe theory, chaos theory, fractal theory and network theory and it will illustrate these theories by examples and applications which have been carried out to understand the reactions and processes of ecosystems.

A comprehensive ecological theory is not available, yet, but many elements of such a theory have already been presented. At first glance, they are very different and a unification seems impossible. However, complex systems such as ecosystems cannot be described by the use of only one viewpoint. Many aspects are needed simultaneously to give a full coverage of all the many and different concepts of ecosystems and their system properties. Therefore a pluralistic view is the only possible foundation for a comprehensive ecosystem theory and the scientific problem seems more to be able to present and compare the presented elements of an ecosystem theory than to demonstrate which of the different views offers the most correct description of nature; see Chapter II.10.

It is the aim of this book to attempt to draw parallels between two or more ecological theories and thereby show that the various ecological theories are different entrances to the same matter to a certain extent. They are as complementary as ying and yang.

It is the goal of ecological system research to resolve the complex entities and processes that confront us in living nature by dividing them into elementary units in order to explain them by means of summation and integration of these elementary units and processes. Biochemistry and classical cell biology can present many good examples of these ideas. However, the actual whole shows properties that are absent from its isolated parts. The problem of life is the problem of organisation of the components which create emerging properties and it is the problem of synergism in contrast to simple algebraic additions.

During the last decades, ecology has not only turned from components to systems, but has also expanded the scale of the scientific research. Maurer and Brown (1988) and Brown et al. (1995) use the term 'macro-ecology' to emphasise the need for statistical pattern analysis and for expanding the scale of ecological research due to the implications of important advances in other disciplines such as biogeography, paleobiology and earth sciences. Characterising an additional aspect, Schneider (1994) defines quantitative ecology as the use of scaled quantities in understanding ecological patterns and processes.

The basic philosophy or thinking of sciences is currently changing along with other facets of our culture such as arts and fashions. The driving forces behind such developments are often very complex and very difficult to explain in detail, but it will be attempted to indicate at least some tendencies which are consistent with the discussion of the controversy and parallelism between ecology and physics mentioned above:

1. The sciences have realised that the world is more complex than we thought some decades ago. In nuclear physics we have found several new particles and are faced with environmental problems. We have realised how complex nature is and how much more difficult it is to cope with problems in nature than in laboratories. Computations in sciences were often based on the assumption of so many simplifications, that they became unrealistic.
2. Systems ecology – we may call it the science of (the very complex) ecosystems – has developed very rapidly during the last decades and has evidently shown the need for systems sciences – also for interpretations, understandings and implications of the results obtained in other sciences, including physics. If we sacrifice important properties of the whole by separating systems into parts, we cannot understand systems (Jørgensen et al., 1992). We need therefore 'to see the forest, not the trees.'
3. It has been realised in sciences that many systems are so complex that it will not be possible to know all their details. In nuclear physics there is always an uncertainty in observations, expressed by Heisenberg's uncertainty relations. The uncertainty is caused by the influence of our observations on the nuclear particles. We have a similar uncertainty relation in ecology and environmental sciences caused by the complexity of the systems (see Chapter I.4 in this volume). The Heisenberg dilemma extended to

ecology asserts that a science of parts cannot explain the multi-scale reality of wholes (Jørgensen et al., 1992). In addition, many relatively simple physical systems such as the atmosphere, show chaotic behaviour, which makes long term predictions impossible, see Chapter II.8.4. The conclusion is unambiguous: we cannot and will never be able to know the world with complete accuracy. Our description of nature can never be fully deterministic. We can only consider propensities (Ulanowicz, 1997). We have to acknowledge that these are the conditions for modern sciences.

4. It has been realised, that most systems in nature are irreducible systems (Wolfram 1984 a and b), i.e., it is not possible to reduce observations on system behaviour to a law of nature, because the system has so many interacting elements that the reaction cannot be surveyed without the use of models. From cells to biosphere the essence of systemisation is interconnection – all things acting together in a tangle of complexity that may be partly charted, as for instance in molecular biology, but can never be fully unravelled without sacrificing the essence (Jørgensen et al., 1992). For such systems other experimental methods must be applied. It is necessary to construct models and compare the reactions of the models with our observations to test the reliability of the model and get ideas for model improvements, construct an improved model, and compare its reactions with the observations again to get new ideas for further improvements and so forth. By such an iterative method we may be able to develop a satisfactory description of nature which is consistent with our observation.
5. As a result of the tendencies 1-4, modelling as a tool in science and research has developed. Ecological or environmental modelling has become a scientific discipline of its own – a discipline which has experienced rapid growth during the last decade. Developments in computer science and ecology have of course favoured this rapid growth in modelling as they are the components on which modelling is founded.
6. The scientific analytical method has always been a very powerful tool in research. However, there has been an increasing need for scientific synthesis, i.e., for putting the analytical results together to form a holistic picture of natural systems. Due to their extremely high complexity, it is not possible to obtain a complete and comprehensive picture of natural systems by analysis alone, but it is necessary to synthesise important analytical results to explain system-properties. The synthesis and the analysis must work hand in hand. The synthesis (for instance in the form of a model) will indicate which analytical results are needed to improve it and on this basis new analytical results can then be used as components in the synthesis. There has been a clear tendency in science to give the synthesis a higher priority than it has been previously. This does not imply that analysis should be given a lower priority. Analytical results are needed to provide components for the synthesis, and the synthesis must be used to give priorities for the needed analytical results.

For some ecologists ecosystems are either biotic assemblages or functional systems. The two views are separated. It is, however, important in the context of ecosystem theory to adopt both views and to integrate them. Because an ecosystem cannot be described in detail, it cannot be defined according to Morowitz's definition, before the objectives of our study are presented. With this in mind the definition of an ecosystem used in the context of ecosystem theory as presented in this volume, becomes:

An ecosystem is a biotic and functional system or unit, which is able to sustain life and includes all biological and non-biological variables in that unit. Spatial and temporal scales are not specified a priori, but are entirely based upon the objectives of the ecosystem study.

Ecosystem studies are widely using the notions of order, complexity, randomness and organisation. They are sometimes used interchangeably in the literature, which causes much confusion. As the terms are used in relation to ecosystems throughout the volume, it is necessary to give a clear definition of these concepts in this introductory chapter.

According to Wicken (1988, p. 357) randomness and order are each other's antithesis and may be considered as relative terms. Randomness measures the amount of information required to describe a system. The more information is required to describe the system, the more random it is.

Organised systems are to be carefully distinguished from ordered systems. Neither kind of system is random, but whereas ordered systems are generated according to simple algorithms and may therefore lack complexity, organised systems must be assembled element by element according to an external wiring diagram with a high level of information. Organisation is functional complexity and carries functional information. It is non-random by design or by selection, rather than by a priori necessity. Ecosystems evolve towards a higher level of organisation rather than towards a higher complexity (Odum 1969).

Saunders and Ho (1981) claim that complexity is a relative concept dependent on the observer. We will adopt Kay's definition (Kay, 1984, p.57), which distinguishes between *structural complexity* , defined as the number of interconnections between components in the system and *functional complexity*, defined as the number of distinct functions carried out by the system. The structural and functional complexity is in contrast to what could be named random complexity, i.e., the complexity caused by disorder. A system of thermodynamic equilibrium has for instance a random complexity as one would need a lot of information to know where all the molecules are, their rates and directions. This is expressed thermodynamically by the concept of entropy. A random system has a very high entropy which may be translated in informatic terms as a need for an enormous amount of detailed information. A structured and functional system has less entropy and less information is needed to describe the system as it is sufficient to describe the principle of the ordered structure and function.

REFERENCES

Allen, T.F.H. and T. B. Starr, 1982. Hierarchy: Perspectives for Ecological Complexity. University of Chicago Press, Chicago, IL.

Audi, R (Ed), 1995. The Cambridge Dictionary of Philosophy. Cambridge.

Axelrod, R., 1984. The Evolution of Cooperation. Basic Books.

Barlow, C. (Ed), 1991. From Gaia to Selfish Genes. Selected Writings in the Life Sciences. MIT Press, Cambridge, MA.

Breckling, B. and F. Müller, 1997, Der Ökosystembegriff aus heutiger Sicht - Grundstrukturen und Funktionen von Ökosystemen. In: Fränzle, O., F. Müller and W. Schröder (Hrsg.): *Handbuch der Ökosystemforschung*. Landsberg.

Brown, W., 1973. Heat-flux transitions at low Rayleigh number. *J. Fluid Mech.* 69: 539-559.

Carson, R., 1962. Silent Spring. New American Library, New York.

Costanza, R. and H. E. Daly, 1987. Toward an ecological economics. Ecological Modelling 38: 1–7.

Costanza, R. and F.H. Sklar. 1985. Articulation, accuracy, and effectiveness of mathematical models: a review of freshwater wetland applications. *Ecological Modeling* 27: 45-69.

Ellenberg, H. 1973. Ökosystemforschung. Ziele und Stand der Ökosystemforschung. Berlin.

Gödel, K. 1986. Collected Works (editied by Solomon Feferman). Oxford University Press, Oxford.

Jax, K., E. Vareschi and G.-P. Zauke 1993. Entwicklung eines theoretischen Konzepts zur Ökosystemforschung Wattenmeer. *UBA-Texte 47*. Berlin.

Jørgensen, S.E., 1988. Fundamentals of ecological modeling (2nd. ed.). Amsterdam (Elsevier).

Jørgensen, S.E., 1994a. Fundamentals of Ecological Modelling (second edition) In: *Developments in Environmental Modelling, 19*. Elsevier, Amsterdam, 628.

Environmental Modelling, 19. Elsevier, Amsterdam, 628.
Jørgensen, S.E., 1994b. Models as instruments for combination of ecological theory and environmental practice. *Ecol. Modelling* 75/76: 5-20.
Jørgensen, S.E., 1997. Integration of Ecosystem Theories: A Pattern (second edition). Kluwer Academic Publishers, Dordrecht, Boston, London. 386.
Jørgensen, S.E., B. Halling-Sørensen, and S.N. Nielsen (Eds), 1995. Handbook of Environmental and Ecological Modelling. CRC Press, Boca Raton, FL.
Jørgensen, S.E., B. Patten, and M. Straskraba ,1992. Ecosystems emerging. *Ecol. Modelling* 62: 1-28.
Kay, J., 1984. Self Organization in Living System [Thesis]. Systems Design. Engineering, University of Waterloo, Ontario, Canada.
Klötzli, F. 1993a. Ökosysteme: Aufbau, Funktionen, Störungen. Stuttgart.
Klötzli, F. 1993b. Ökosysteme. In: Kuttler, W. (Hrsg.): *Handbuch zur Ökologie*. Berlin.288-295
Koestler, A., 1967. The Ghost in the Machine. Macmillan, New York.
Lovelock, J.E., 1979. Gaia: a New Look at Natural History. Oxford University Press, Oxford.
Lovelock, J.E., 1988. The Ages of Gaia. Oxford University Press, Oxford.
Maurer, B.A. and J.H. Brown, 1988. Distribution of biomass and energy use among species of North American terrestrial birds. *Ecology* 69: 1923-1932.
Meadows, D.H., D.L. Meadows, J. Randers, and W.W. Behrens, 1972. The Limits to Growth: A Report for the Club of Rome's Project on the Predicament of Mankind. Earth Island, London.
Müller, F., 1992. Hierarchical approaches to ecosystem. *Ecol. Modelling* 63: 215-242.
Müller, F., 1997. State of the Art in Ecosystem Theory. *Ecological Modelling* 100:165-161.
Odum, E.P., 1953. Fundamentals of Ecology. Saunders, Philadelphia, PA.
Odum, E.P., 1969. The strategy of ecosystem development. *Science* 164: 262-270.
Odum, E. P. 1980. Grundlagen der Ökologie. Stuttgart.
O'Neill, R.V., D.L. DeAngelis, J.B. Waide, and T.F.H. Allen, 1986. A Hierarchical Concept of Ecosystems. Princeton University Press, Princeton, NJ.
Patten, B.C., 1982. Environs: relativistic elementary particles for ecology. *Am. Nat.* 119: 179-219.
Patten, B.C., 1991. Network ecology: indirect determination of the life-environment relationship in ecosystems. In: M. Higashi and T.P. Burns (Eds). *Theoretical Studies of Ecosystems: The Network Perspecive*. Cambridge University Press, 288-351.
Patten, B.C., 1992. Energy, emergy and environs. *Ecol. Modelling* 62: 29-70.
Popper, K., 1989. *Logik der Forschung,* Tübingen.
Saunders, P.T. and M.W. Ho, 1981. On the increase in complexity in evolution, II: The relativity of complexity and the principle of minimum increase. *J. Theor. Biol.* 90: 515-530.
Schneider, D.C., 1994. Quantitative Ecology: Spatial and Temporal Scaling. Academic Press, San Diego, CA.
Schrödinger, E., 1944. What is Life? Cambridge University Press, Cambridge.
Stonier, T., 1990. Information and the Internal Structure of the Universe. Springer-Verlag, London.
Stugren, B.,1986, Grundlagen der Allgemeinen Ökologie.Stuttgart.
Tansley, A.G., 1935, The use and abuse of vegetational concepts and terms. Ecology 16: 284-307.
Ulanowicz, R.E., 1997. Ecology, the Ascendant Perspective. Columbia University Press, New York, 201.
Wicken, J.S., 1988. Thermodynamics, evolution, and emergence: ingredients for a new synthesis. In: B.H. Weber, D.J. Depew, and J.D. Smith (Eds). *Entropy, Information, and Evolution: New Perspectives on Physical and Biological Evolution*. MIT Press, Cambridge, MA, 139.
Wolfram, S., 1984a. Cellular automata as models of complexity. *Nature* 311: 419-424.
Wolfram, S., 1984b. Computer software in science and mathematics. *Sci. Am.* 251: 140-151.

I.2 Ecosystem Structure

Frank B. Golley

1. Introduction

Let us begin an examination of ecosystem structure with a definition of the word 'structure'. The Websters dictionary defines 'structure' as the construction or the composition of a thing. Implicit in the dictionary definition is the sense that structure is constructed by human hands, guided by the human mind which conceives a plan showing how the various structural parts fit together into the construction as a whole.

In modern science, which is grounded in a materialistic and mechanical model of reality, structure takes on a different meaning which stresses the second part of the definition, that of composition. Natural science studies the composition of natural systems because it provides fundamental information that underlies, or comes before, the study of the functions of the system components. The full examination of a system is often described in ecology, at least as a study of 'structure and function'. A probable reason for this bias in point of view is that we are visual animals. We give significance to the visual pattern through which dynamic action takes place. We begin biology with the study of anatomy and then with that background we move to transmission along nerves, digestion and reproduction. In history, science begins with structure, i.e., with anatomy, land forms, lists of species and so on and then moves to function,

This is not to say that science is not interested in construction because it is. The goal of research is to discover the plans that guide the construction and maintenance of nature. In ecology one way this objective is expressed is by the phrase 'ecological engineering' which defines a field of work which utilises ecological principles to achieve specific results in the environment. The concept is even extended to organisms such as the beaver, which are termed ecological engineers. Ecological engineering would presumably use the word 'structure' in the same way as would civil engineering.

However, that natural world is fundamentally different from the built or constructed world. The structure we observe in the natural world is the consequence of the interaction between the genetic process in the organism and the selective forces in the environment. Because the organism is continually maintaining itself in the dynamic environment, its composition is responsive and dynamic, although often there is a degree of fixicity which is what makes structure what it is. Adaptation may not be obtained through change in structure; therefore the flexibility of structure is essential for survival.

This contradiction between a static, fixed and a dynamic, changing natural structure can be confusing, If the ecologist is a materialist, he or she might seek to describe structure in mechanical terms, as in systems. This point of view is opposed to the naturalistic ecologist who thinks of structure as changing and flexible and describes it through lists of species. The cause of the confusion is that the ecologist may have different definitions of structure than those given in the dictionary.

In this chapter I will discuss the structure of ecosystems. I will begin at a broad level by considering the nature of entities in ecology. I will then consider ecosystems as one kind of ecological entity and will treat development of the ecosystem concepts from a structural perspective. I will end with a consideration of several structural approaches in ecology and a suggestion for research.

2. The nature of entities in ecology

In a recent book manuscript (Golley and Keller, 1998), we have defined an entity as an object that is bounded in space/time and is internally coherent. Boundedness does not imply closure. Ecological systems are open systems, which means that energy, matter and information can flow into or out of the entity to and from its environment. The discussion below is paraphrased from the discussion of entities in Golley and Keller (1998).

Fredrick Ferré is one of the few philosophers who has considered the metaphysical character of ecological entities. He approaches the reality of entities from our use of the concept in decision making: i.e. *"what is 'essential' or 'accidental' for entities is a matter of interest and purpose interwoven with facts (Ferré, 1996). The actual attributes, relations and functions of something are not irrelevant to the decisions we make. They provide the basis for our decisions. Entities 'are' the joint product of what we find and what we make".*

In these comments, Ferré is recalling an old argument in the philosophy of science. This argument may be expressed simply through the perspectives of John Stuart Mill and William Whewell, philosophers of the nineteenth century. Mills argues that all knowledge is sensory in origin. Facts map the features of an entity on our human sense. Whewell argues that while we sense facts, some part of the knowledge is added by the knower. Ferré emphasises the joint product of sense and context in his choice of explanation and in his classification of entities. He identifies six kinds of entities. Aggregate entities, such as mountains and lakes, are physical bodies with boundaries. They are strongly dependent upon the knowledge that is added to them. Aggregate entities seldom enter into ecological relationships; rather they form a background for the ecological play.

Systematic, formal and organic entities are closer to the concerns of ecology. Systematic entities include ecosystems. The use of the word 'system' refers to the ecosystem's ability to retain structure and function under continually changing environmental conditions. Formal entities are those which are most strongly based on the knowledge added to them. A species is an example of a formal entity. Living organisms are examples of an organic entity. Organic entities are creative because they generate, through the genetic process, unique, new forms of life.

Ferré completes his taxonomy of entities with two final categories: compound and fundamental entities. Compound entities are those which have strong internal relationships but are without an apparent internal system dynamics. Ferré uses water or sodium chloride as examples of compound entities. Fundamental entities are those which provide a deep structure beneath entities in general. Ferré comments that metaphysics needs such a category, even though it might be of little use to ecologists. If entities are recognised by our sense, interpreted through our experience and knowledge, then entity is a tautology. It is likely that we recognise what we already know. Fundamental entities stand separate from our interpretation. They exist before we observe them and they are exclusively factual. Ferré states "qualitative differences among fundamental entities may be great but these differences reside in the patterns achieved, not in the process that achieves the patterns."

Ferré seeks to bring order out of a chaos of usage by using Whewell's perspective that we know through sense observation, extended by past experience and knowledge. This is a familiar method for ecologists. Ecologists recognise entities that are largely defined through sense observation, such as individual organisms, and entities that are largely defined through the addition of knowledge to sensed experience, such as ecosystems. Between these two familiar cases lie other ecological entities. A classification of entities, such as that of Frederick Ferré, is helpful at the outset of this chapter because the structure of ecosystems depends, in part on our definition of ecological entities.

2.1 Boundaries

The term 'entity' implies boundedness. The boundary is the envelope around the entity which converts it from an amorphous tendency into concrete reality. Nevertheless, the boundary is permeable. Flows of energy, matter and information pass into and out of the entity, from and to its environment, across its boundary.

If we begin with the individual human organism, we can identify multiple boundaries. First, there is the surface of our skin, lung, reproductive and digestive tract. These surfaces protect the organism and alter the rates and characteristic of the flows of energy and matter from the environment which enter the organism. Second, the region around us that is interpretable by our senses is bounded by the capacity of our senses. For example, we can hear and identify birds in the canopies of large trees at a distance much greater than our vision would allow. Third our personality extends us into the community through kinship, friendship and interactions. For example, the conservationist who cares about his or her environment can create conditions which cool, protect and improve the environment of an entire neighbourhood. The boundary of the organism is defined differently for each process or feature in the environment. If we speak about boundaries, we must specify the process and features we are concerned with.

If we turn to the ecosystem, once again we face a set of possibilities about kinds of boundaries. Since the ecosystem is more diffuse and less clearly bounded than the organism, there is more room for argument over the presence of boundaries. For some ecosystems, such as a lake or a patch of forest in a field, the boundary is clear and one can step across the boundary from land to water or from meadow to forest. Country people recognise such ecosystems and even have special names for them. But, in other cases, the boundaries may be indistinct or very large in scale. Species may be distributed in gradients across large areas, as demonstrated by Robert Whittaker for plant species in the Smoky Mountains (Whittaker, 1953). Location for the boundary may be arbitrary in this case. Ecologists have given the name of 'ecotone' to this type of boundary. Ecotones are important habitats because there are species especially adapted to the conditions in ecotones. Ecotones also may be extensive in area.

There is another sense in which the environment of the organism and the ecological system affects how we view boundaries of systems. We have suggested that the environment of the system is made up of those elements that are not part of the system yet interact with it. For example, water, which is part of the natural environment of the terrestrial organism, enters the organism through the mouth and crosses the cells of digestive tract and enters the body. The boundary to the movement of water could be the mouth or the gut wall. But in another sense the organism is part of a population of organisms. Influence can occur from other organisms directly or indirectly on the organism of interest. The other organisms in the population are part of the environment of the individual. Similarly, the environment of the ecosystem is the higher level systems in the nested hierarchy of which the ecosystem is a part. Thus, the environment of a prairie pot hole is the prairie in which it is nested. This second meaning of environment makes the concept of boundaries more complex.

The differences in the meaning of boundary and environment have led to much mischievous argument in the ecological sciences because the difference between organism and ecosystem influences our understanding of the entity itself. The organism, with a boundary that can be sensed so that we can, without much ambiguity, determine what is inside and outside of the organism, is appealing to the common observer. The ecosystem, in which the boundary is sometimes defined as a change in the rate or direction of flow, is much less easily recognised. The boundary is a property of the ecosystems as an entity and is not the sum of the boundaries of the individual components.

Landscape ecology is concerned with boundaries of ecosystems. At the scale of a landscape the boundary is usually an arbitrary point where the flow rate changes as it moves from one kind of system to another, at the edge of a watershed, for example, or where the components differ in some fundamental way, or where field becomes forest. The boundaries of the ecosystem are not fixed or rigid. As a consequence, we say that ecosystem boundaries are fuzzy, that is, they are imprecise, changing and dynamic. Thus, the boundaries of organisms and ecosystems are different to a fundamental degree.

2.2 The Systems Approach

Now that we have defined ecosystem and have discussed that nature of the boundaries of ecosystems, we have a rudimentary concept about this type of entity. This is a struggle for many ecologists who have been trained within the organismic paradigm. They must learn to think abstractly from 'concrete' organisms to abstract systems. As they do this, they carry concepts of reality and interaction from the organism to the ecosystem in an appropriate way. These two kinds of entities are deeply, if not fundamentally, different and they require a different way of thinking.

This apparent difference is not a case of either-or; rather, it is a case of addition. The ecosystem contains most of the features that are characteristic of individual organisms plus other features characteristic of ecosystems. Thus, it is perfectly possible to find properties of individuals within systems. And while individuals can be considered to be systems too, it is not necessary to take this viewpoint to understand modern biology. It might be helpful, for example in understanding the response of an individual to stress, but it is usually considered not essential in biological thinking.

A further element in this confused set of definitions and viewpoints involves the development of ecosystem science. Application of the ecosystem to a new area or in a new project, for example, as in the IBP, required us to use the limited information that already existed on organisms and environment and begin constructing models of system structure and function. Often this was not an especially helpful way to begin because data was not realistic. Today, after 30 years of research, we still find that we understand some parts of ecosystems very poorly. For example, the breakdown of organic compounds by micro-organisms is one area where new discoveries are frequent. Slowly we have been moving from the view of systems as black boxes to identification of specific control points, transformations and storages. That is, we have moved from a static parts perspective to a dynamic network perspective, which is, as yet, incomplete.

This conclusion raised yet another complication in ecosystem structure, the concepts of chaos and non-linear systems . Fritjof Capra (1996) wrote:

"In the 1920's, however quantum theory forced (the physicists) to accept the fact that the solid material objects of classical physics dissolve at the sub-atomic level into wavelike pattern of probabilities. These patterns, moreover, do not represent probabilities of things, but rather probabilities of interconnections. The sub-atomic particles have no meaning as isolated entities but can be understood only as interconnections or correlation's, among various processes of observation and measurement. In other words, sub-atomic particles are not 'things' but interconnections among things, and these, in turn, are interconnections among other things and so on. In quantum theory we never end up with any things, we always deal with interconnections."

Quantum theory gave us a view different from that of classical Newtonian physics. The quantum physicist recognised that we cannot predict any specific event occurring in space/time, we can only speculate on the probability of the event happening. Matter and energy combine into shifting patterns of waves that combine, dissolve and recombine. As complex systems are built up, new properties appear that were not foreshadowed in parts alone. These patterns of order appear on the edge of chaos; as Stuart Kaufmann (1995) would put, between chaos and fixed, static order. It is on the edge of chaos that creativity can emerge.

These words resonate in my mind with echoes from the 1960's. They connect with environmentalistic thought, indigenous native American statements and practices, and New Age thinking. To put them into practice requires not only a compatible mind but a fundamental shift in thinking. This is the reason I introduced this chapter with a discussion of entities. We are accustomed to organising nature into entities which we see, enjoy and manipulate. When we translate from ecological theory to practice, we must communicate with a public that thinks in these ways. Practical people can do very little with discussions of interconnected wholes. Thus, there is a problem of translation in moving from the admittedly static concept of entities to the more current discussion of chaos and non-linearity. While I can offer a mature and well thought out translation here, such a statement would link change, dynamic behaviour and evolution of entities with non-linear feedback, dynamic connections and the role of chance. The key point is that in either form of expression we are emphasising connectedness and dynamic behaviour.

2.3 Hierarchies

I have alluded to hierarchical patterns when I considered the environment of ecosystems. I consider hierarchical theory, as developed by Allen and Starr (1982) and O'Neill et al, (1985) as a fundamental concept of ecosystem structure. This theory allows us to organise systems of widely different scale and character into coherent patterns that fit what we observe on the landscape. Ecosystem hierarchies are nested. By nested we mean that we establish a scale from large to small sized and from complex to simple. All the systems of the same scale are members of a hierarchical class. All systems of a small scale are components or subsystems of the higher order systems. In this process of reduction we ultimately reach the smallest unit of interest. Ecologists in European countries call this smallest unit an 'ecotope'. Ecotope should be used by all ecosystem scientists for the smallest system that is homogeneous for the properties of interest. If one is interested in global change, then the earth is the ecosystem of interest. If one is studying a patch of remnant forest in south-eastern U.S. soybean fields then the ecosystem is a forest ecotope. The ecologist can scale from the ecotope to the earth system or from the earth to the forest ecotope.

The structure of the hierarchy is made up of subsystems, sub subsystems and so on down to the finest scale needed in the analysis. It is only at the finest scale that the ecologist deals with the familiar interactions of biota and environments that were at one time the only subject of ecology. The student of global change may be interested in the ocean subsystem in a global model and treat the oceans as a whole, comparing its behaviour to that of the atmosphere. Anther student of global change may be interested in the way a particular microbe breaks apart a lignin molecule in an offshore environment and be considering nature at the finest scale.

As has been stated by many authors, these nested systems have little in common with social hierarchies, with a purpose of allocating power. In a military system for example, power flows from the top and each level is distinct and separate. In contrast, in the nested hierarchy control passes both from below and from above. The highest levels of scale contain, by definition, all the sub components of the other hierarchical levels. This is what nested means.

It is important to emphasise that the principle of self similarity holds for the construction of nested hierarchies. Self similarity means that the criteria used to identify and define a system is applied at all levels of scale. If texture of soil is the defining property in the hierarchy of interest, all levels of the hierarchy are constructed on soil texture. In this way one might have a hierarchy of systems in which proportions of sand and clay create different patterns. This means that in the sense of hierarchy theory the familiar ecological scale of organism, population, community, ecosystem is incorrect. Organism, population and community form a nested hierarchy in which communities are defined as systems of interactions between the members of the biota and the environment. As a consequence, ecosystems form different kinds of nested hierarchies.

3. Construction of the ecosystem concept

In this section of the chapter, I will summarise material presented in Golley (1995). It is helpful in understanding the concept of ecosystem structure to view how the founders of the ecosystem concept used structure in their statements. The word 'ecosystem' was coined by Sir Arthur George Tansley in a 1935 paper, titled "The Use and Abuse of Vegetation Concepts and Terms." Tansley was a plant ecologist, concerned about the classification of vegetation is space and time. In 1904 he called for a survey of the vegetation of the British Isles and in 1912, when this survey lost its momentum, Tansley worked to form the world's first ecological society, the British Ecological Society, which he served as the first president.

In the paper on vegetation concepts and terms Tansley attacked the Clementsian idea that vegetation was a superorganism, that went through a process of development, and achieved a condition of equilibrium called the climatic climax. Tansley accepted the idea that vegetation was like an organism and might be termed a quasi-organism, but he protested vigorously the extension of the term 'organism' to include both individual organisms and vegetation units together. He preferred to think of vegetation as a system and he coined the term 'ecosystem' in place of Clement's organism.

Tansley was obviously in touch with the modern trends in science in the middle of the century. Systems thinking emerged for many parts of science, including Lotka's physical biology. Ludwig von Bertalanffy system theory of life, and so on. Tansley's term fit the time, it was simple, easy to remember, interpretable by the layman and was modern. It replaces other terms that referred to a complex of organisms and environmental conditions interacting together. The point is that Tansley supplied the word; the underlying reality to which the word referred had been known for a long time.

Tansely did not do much more than coin the term. His definition makes a variety of important points that are not relevant to this story and will not be discussed. He did not apply the ecosystem concept in his research or writing nor did he expand upon the concept in a word model or conceptual model. Thus, Tansley (1935) did not address the nature of ecosystem structure except tangentially in the vegetation concepts and terms paper when he stated that (page 291)

"The relatively stable climax community is a complex whole with more or less definite structure, i.e., inter-relation of parts adjusted to exist in the given habitat and to co-exist with one another. It has come into being through a series of stages which have approximated more and more to dynamic equilibra in these relations."

It is not exactly clear to me what Tansly meant by 'inter-relation of parts'. If he was referring to the linkages of the connections between components or the web or chains of connections, then his comment on structure has a modern ring to it.

The ecosystem concept was operationalised by Raymond Lindamen on Cedar Bog Lake, Minnesota and by G. Evelyn Hutchinson on Lindsay Pond, Connecticut. Lindamen became Hutchinson's postdoctoral fellow on a Sterling Fellowship at Yale university where he died prematurely in 1942. Lindamen and his wife, Eleanor Hall Lindamen, sampled the biota and environmental factors of a small lake for about two years. They organised the resulting large and complex set of data and records by fitting the individual species populations into feeding groups, which, in turn, made up a feeding cycle. Lindamen's conceptual model of the cedar Bog Lake ecosystem is shown in Figure 1. The structure included phytoplankters, zooplankters, plankton predators, pondweeds, browsers, benthic predators, swimming predators, and bacteria. Ooze is placed at the centre of the diagram, because it is the sink for organic matter.

Gradually a lake of Cedar Bog type, which is a senescent body of water formed by a melting ice block after retreat of glaciation, fills in with organic matter and becomes a bog or wet meadow and eventually a forest. From the ooze there are internal cycles of elements among the system components and an out flow from the system to the larger environment in which it is nested.

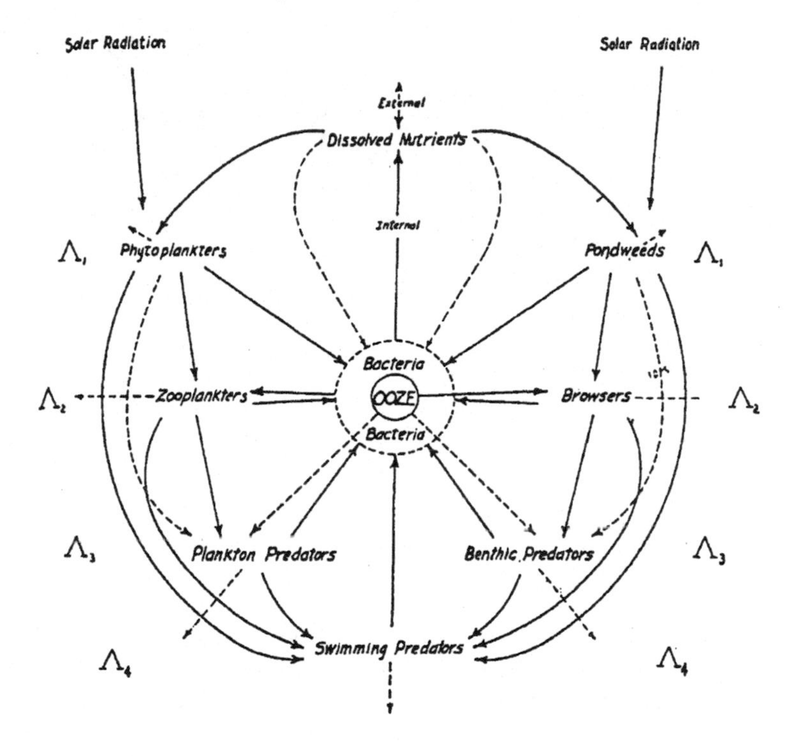

Figure 1: A food cycle diagram based on Cedar Bog Lake, Minnesota, from Lindamen (1941). Note the place of ooze surrounded by bacteria in the centre of the diagram.

Lindamen's diagram has several sources, as described is his doctoral dissertation. The subsystems are similar to those in Chancey Juday's 1940 paper on the energy budget of Lake Mendota. Juday's subsystems are those familiar to lake ecologists, such as August Thienemann, and also Charles Elton's trophic divisions that make up his pyramid of biomass (Elton, 1928). Raymond Lindamen (1941) illustrated diagrams of trophic patterns in lakes, beginning with Shelford (1918), Alsterberg (1922 and 1925), Thienmann (1926), Strom (1928), Rawson (1939), Wasmund (1939) and others. Several of these diagrams placed detritus or ooze in the centre, as did Lindamen. Clearly, Lindamen's structural diagram had a long history.

Lindamen's circular diagram stressing nutrient cycles was replaced a little more than 10 years later by a new diagram developed by Howard T. Odum (Figure 2) to represent his understanding of the energetics of Silver Springs, Florida (Odum, 1957). This diagram was introduced to students in the second edition of Eugene Odum's (1959) 'Fundamentals of Ecology' and became an icon for energy dynamic studies for at least a decade. The H.T. Odum model had its precursors in the Shelford Model (1918), in Elton's diagram of the nitrogen cycle at Spitzbergan Island and in the block and arrow models used by Lotka (1925) to illustrate flows of energy and matter in his 'Elements of Mathematical Biology'.

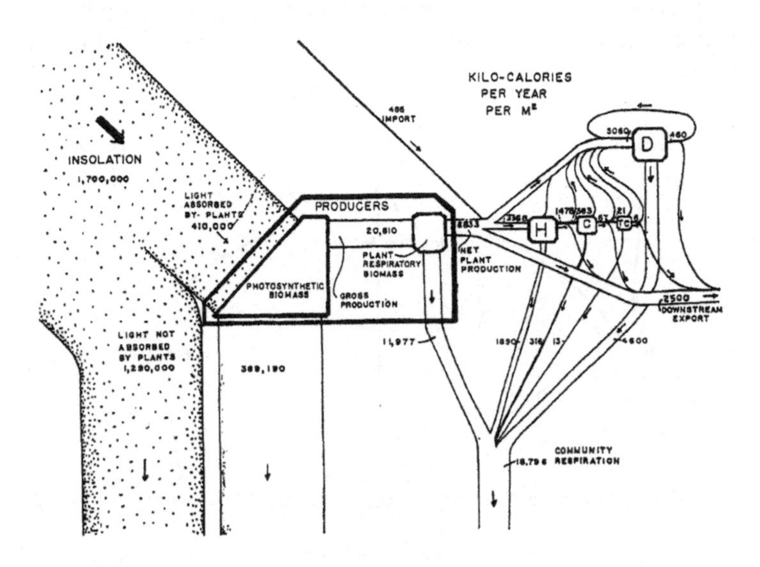

Figure 2: The energy flow diagram, with units in kilocalories per square meter per year in the Silver Springs community, Florida, based on Odum (1957). The letters H indicate herbivores, C carnivores ad D decomposers.

Comparison of the Lindamen and H.T. Odum models illustrates two different approaches to ecosystem structure. Lindamen and his precursors used biologically defined components to represent supracomponents of the system. For example, in Lindamen's diagram these are phytoplankton, zooplankton, bacteria and so on. In contrast, in H.T. Odum's diagram these subsystems are generalised to primary producers, herbivores, carnivores and decomposers. This difference, while subtle, is fundamental to understanding a shift in perception characteristic of the late 1950's and the 1960's in ecology. The Odum categories required one to focus on food habits and food relationships exclusively. Further, one had to sum across taxa by deciding the main type of food intake of an organism or partition the organisms energy into two or more components. I recall teaching this approach in 1958 at a NSF science teachers meeting, operated by George Baxter at the University of Wyoming. Many of the students with in-depth knowledge of taxa and their food habits were uncomfortable with placement of these organisms into functional categories. Later, in the mid 1960's George Van Dyne had a similar problem in trying to implement a model of the grassland biome. Variation in biological behaviours created serious difficulties with ecosystem classification. Eventually the ultimate in abstraction was reached when ecosystem modelers abandoned any attempt to retain biological reality and the categories in their models solely represented nodes in transfer networks.

A second important point derived from a comparison of these models is that the Lindamen model forces the ecologist to stay within the biological paradigm where the individual organism and the species play a dominant role. If one is dealing with a category called zooplankton, then *Daphnia* spp is a real category that carries with it all the biological attributes of *Daphnia* as a genus. The arrows connecting the biological components may indicate any relationship, although feeding relationships are most widely used. But, the dominant structure is the complex of biological species named in the diagram. In contrast, the more abstract H.T. Odum model permits one to manipulate the component, reduce it to a node and emphasise the connections between nodes. The process of abstraction can result in a network diagram in which relationships, not biological properties, dominate. Bruce Hannon (1973) emphasised the *"energy flow interdependence, both directly and indirectly, of each species upon the other, 'at any prescribed time' represents the structure of the ecosystem"*. The shift from biology to interdependence allowed a connection to be made to system models of all types and ecological modelling became recognised as an active area of systems science. Ideally, the ecologist would want to retain biological reality in the system model and emphasise interdependence. The problem of implementing this desirable approach is that most ecotopes contain hundreds, if not thousands, of biological species. Key questions – such as – what is the role of biological diversity? How much can this diversity be reduced without changing the function of the ecosystem? What are the functions of an ecosystem? How can we model species with many functions, acting at different times, and being nested within complexes of other species? How independent are the subsystems of ecosystems? How does interdependence maintain system stability? – have not been answered satisfactorily.

In the 1960's a seven year long study of the biological basis of natural productivity was launched. This global study was called the International Biological Program or IBP. The IBP had several parts, one of these concerned the study of ecosystems. The ecosystem study program, called the biome program, was most active in the United States where ecologists attempted to initiate a project in all the major geographical regions of the continent. The IBP tried to use ecosystem modelling as an organising structure and as a predictive tool. IBP became controversial because it tries to solve the problem of lack of information about the functional biology of species and interdependence of species and the environment by moving from the Lindamen type model to the H.T. Odum type emphasising abstract ecological categories. The solution of this controversy indicated the state of the subject at that time. The tundra working group rejected the notion of a conceptual or mathematical model in favour of a word model. This group developed, through several versions, a statement about the functional dynamics of the tundra ecosystems. It was a reasonably accurate statement about the state of the knowledge of the tundra ecosystem at that time and was useful for that reason. But it was a model made up solely of words and sentences. In other biomes ecosystem models were mainly used for organising information. Any attempt to implement as a full model was abandoned by most of the biomes early in their development. Only the grassland biome carried through the original plan and produced an advanced ecosystem model called Elm (Innis 1978).

At the same time as the IBP another ecosystem project was initiated at the Hubbard Brook Experimental Watershed in New Hampshire by F.H. Bormann, Gene Likens, Noye Johnson, Robert Poerce and John Eaton. The concept of this study was presented for the first time in a science paper by Bormann and Likens (1967). In this case the focus was on nutrient cycling. The diagram (Figure 3) invented by Bormann and Likens to illustrate their approach had no direct ancestors. It was most similar to the word model of the IBP tundra ecologist. Their ecosystem consisted of storages of chemical elements and transfer flows, called intrasystem cycle, between compartments, with the ecosystem linked to the biosphere (Earth ecosphere) by inputs and outputs. This diagram is expanded in their book (Likens et al., 1977) on biogeochemistry of a forested ecosystem.

We are left then with a variety of approaches to the description of ecosystem structure. There may be an emphasis on the components, the interactions between components or both. There may be an emphasis on biological or geological reality or on various levels of abstraction. The ultimate of abstraction is the model in which components and interactions are represented solely by symbols and quantitative values. These various approaches have different purposes; there is no one correct or most useful approach for all ecosystems.

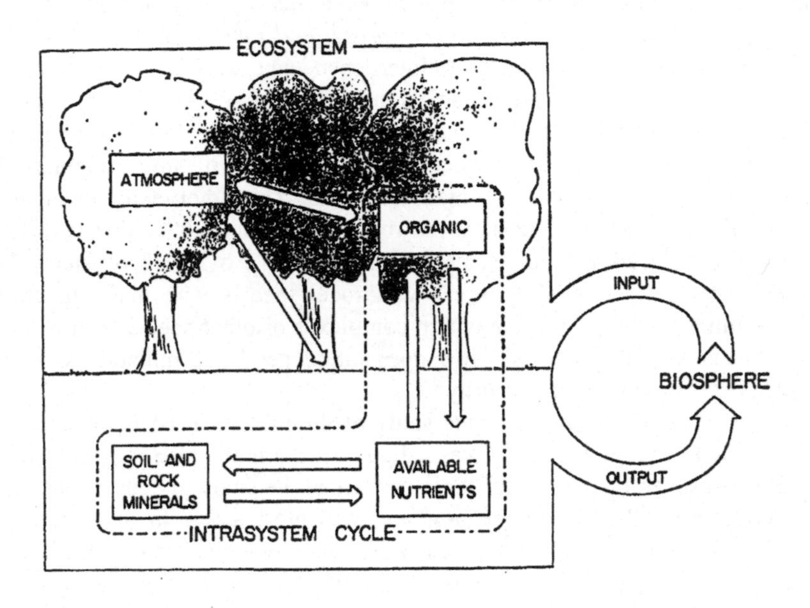

Figure 3: Diagram of Bormann and Likens (1967) showing the relationships of the geological, meteorological and biological components within the ecosystem, connected through inputs and outputs to the biosphere. This diagram is based on research at Hubbard Brook Experimental Watershed in New Hampshire.

This historical survey has given us a broad picture of the ways ecologists have developed ecosystem structure. The remaining task is to compare these several approaches with an ideal approach which would capture all of the relevant interactions among organism and environmental factors. Firstly, it is not clear to me what the sum total of all the relevant interactions would be. We aim at a total description of behaviours and we claim that even if we obtained such a goal, still it would be inadequate because we have neither a parts list of all the elements in an ecosystem nor a blueprint that shows us how to assemble the parts into a whole. Ecological engineering speaks about assembling ecosystems and Akira Miyawaki, in Japan (Miyawaki and Golley, 1993), has assembled tree species to reconstruct forests successfully at over 500 sites. But, Miyawaki and other ecologists who use this approach let nature provide the other members of the biota by natural processes of invasion. Allowing nature to take its course is not the same thing as understanding the structure of ecosystems and using this knowledge to rebuild a system from its parts.

Secondly, it is clearly not so that 'every thing is connected to everything else.' In fact the measured connectivity in parts of systems suggests that connectivity is far below 100%. Indeed, it may be less than 10%. This means that there is a structure which groups organism into collectiveness of some sort in most ecosystems. Within a collective, there may be tight connections but there seems to be only links between collectives. At the first Congress of Ecology in Den Hague, I analysed data collected over several years at the Savannah River Ecology Laboratory in old fields resulting from land abandonment (Golley, 1974). Organisms appeared to be connected to key plant species. For example, *Lespedeza cuneata* had its fauna including herbivores, carnivores, top carnivores, predators and parasites, while *Andropogon virginica* had a different fauna associated with it. There appeared to be a great deal of redundancy of species in the middle of the network but less at the beginning and the end of the food webs. If this pattern is general, it would explain why trophic levels often have not provided adequate structures for research. They are too simple.

4. Conclusion

In this chapter I have shown that ecosystem structure is the network of interactions between components of the system. Ordinarily the components are populations of organisms and the interactions involved, exchange of energy, matter and information. However, the term 'ecosystem' is generic and can be applied to all members of nested spatial hierarchies, ranging from the ecosphere (the Earth) to the smallest of ecotopes. For this reason, components of systems might be landscape units, such as lakes, rivers, forest or other geographical features of ecoregions and landscapes, but usually ecosystem components are species.

In description of ecosystem structure it is usual to compile a list of species present in the system and then a list of interactions between species. From these lists it is possible to construct maps or diagrams of the interaction network of the system. If the data are quantitative, then the diagram will describe storage of matter or energy in the system and the rates of flow. In this way a quantitative measure of importance of components or flow pathways can be obtained.

If the species approach is not possible, then it is common practice to aggregate species into groups and to treat the group as a component. This approach does not allow for identification of sources of natural variation or selection, exchange if status of species illustrating the principle of redundancy, or the source of emigration. In other words, the simplicity of the analysis has serious cast of loss of reality associated with it.

The process of abstraction can be carried further for purposes of mathematical analysis of structure arrangements of networks. These experiments have no or little relationship to real ecosystems and while ecologists may accord them respect and even use them to guide research, they are dependent upon the initial conditions, the assumptions of the model, the chosen structure and selected rate processes. In other words, they represent the dynamics of machines. Machines are not ecosystems and actually are poor representations of ecosystems.

The consequence of this analysis is that the ecosystem scientist is caught in a dilemma. If they select for reality, they must deal with enormous numbers of components, which represent unique properties of interest, and relatively poor information on many components. If they select for abstraction, depending upon how far down this road they travel, they lose reality and the model more resembles the fundamental properties of the machine. How does the ecologist solve this dilemma?

In my opinion, the solution to this structural problem is to focus on local ecotopes which are important for environmental reasons. The practical work may attract funding because of its relevance and if not, still local people will be interested and will benefit from the research. Given time, the ecosystem investigator will get to understand how the system is constructed, where instability comes from, how the system changes in time and space and other essential information to understand and manage it. This sort of in depth understanding of ecosystems is the basis of ecosystems management, Given time, it also solves the dilemma of understanding structure and function in the real world.

REFERENCES

Allen, T.F.H. and T. B. Starr. 1982. Hierarchy: Perspectives for Ecological Complexity. Chicago, University of Chicago Press.

Alsterberg, G. 1922. Die respiratorischen Mechanismen der Tubificiden. Lunds University. *Arrskrift* 11 (18): 1-16.

Alsterberg, G. 1925. Die Nahrungszirkulation einiger Binnenseetypen. Arch. Hydrobiol. 15:291-228.

Bertalanffy, L. von. 1952. Problems of Life: An evaluation of Modern Biological and Scientific Thought. Harper Torchbooks, New York.

Bormann, F.H. and G.E. Likens. 1967. Nutrient cycling. *Science* 155 (3761): 424-429.

Capra, F. 1996. The Web of Life: A New Scientific Understanding of Living Systems. Anchor Books, New York.

Elton, C. 1928. Animal Ecology. Sidgwick and Jackson, London.

Ferré, F. 1996. Being and Value: Toward a Constructive Postmodern Metaphysics. State University of New York Press, Albany.

Golley, F. B. 1974. Structural and functional properties as they influence ecosystem stability. Proceedings First International Congress of Ecology, pp97-102.

Golley, F.B. and D. Keller. 1998. Science of Synthesis. University of Georgia Press, Athens, in press.

Hannon, B. 1973. The structure of ecosystems. *J. Theoretical Biology* 41:535-546.

Hutchinson, G.E. 1964. The lacustrine microcosm reconsidered. *American Scientist* 52:334-341.

Innis, G.S. 1978. Objectives and structures for a grassland simulation model. In: Innis, G. (Ed) *Grassland Simulation Model* pp1-21. Springer- verlag, New York.

Juday, C. 1940. The annual energy budget of an inland lake. *Ecology* 21 (4):438-450.

Kaufmann, S. 1995. At Home in the Universe: the Search for Laws of Self-organization and Complexity. Oxford University Press, New York.

Likens, G., F.H. Bormann, R.S. Pierce, J.S. Eaton and N.M. Johnson. 1977. Biogeochemistry of a Forested System. Springer Verlag, New York.

Lindamen, R. 1941. Energy Dynamics of a Senescent Lake. PhD dissertation. University of Minnesota.

Lindamen, R. 1942. The trophic-dynamic aspect of ecology. *Ecology* 23 (4):339-418.

Lotka, A.J. 1925 (1956). Elements in Mathematical Biology. Dover, New York.

Miyawaki, A. and F.B. Golley. 1993. Forest reconstruction as ecological engineering. Ecological Engineering 2:333-345.

Odum, E.P. 1959. Fundamental of Ecology. Second Edition. W.B. Saunders and Co. Philadelphia.

Odum, H.T. 1957. Trophic structure and productivity of Silver Springs, Florida. *Ecol. Monogr.* 27:55-112.

O'Neill, R.V., D.L. DeAngelis, J.B. Waide and T.F.H. Allen. 1985. A hierarchical Concept of Ecosystems. Princeton University Press, Princeton.

Rawson, D.S. 1939. The Bottom Fauna of Lake simcoe and its Role in the Ecology of the Lake. Univ. Toronto Stud. Publ. Ontario Fish Res. Lab., no 40.

Shelford, E. 1918. Conditions of Existence. In: Watrd, H and G. Whipple (Eds) *Freshwater Biology*. Pp 21-60. Wiley, New York.

Strom, K.M. 1928. Recent advances in limnology. *Proceedings of the Linnean Society London*, 1928:96-110.

Tansley, G.A. 1935. The use and abuse of vegetation concepts and terms. *Ecology* 16(3): 284-307.

Thienmann, A. 1926. Der Nahrunskreislauf im Wasser. *Verh. Deutsch. Zool. Ges.* 31:29-79.

Wasmund, E. 1939. Lakustrische Unterwasserboden. *Handbuch der Bodenlehre* 5:97-190.

Whittaker, R.H. 1953. A consideration of the climax theory. The climax theory as a population and pattern. *Ecol. Monographs* 23:41-78.

I.3 Ecosystems as Functional Entities

Felix Müller and Wilhelm Windhorst

1. Introduction

In ecosystem science strict divisions are very often made between research and inventories on the spatial distribution of abiotic and biotic system elements on one hand and investigations about the flows, storages, and balances of energy or matter on the other. Although the central task of ecosystem science is the integration of those structural and functional aspects into an encroaching organisational concept it is necessary to describe the fundamental viewpoints and the corresponding methodologies. This will be done in the following paper with respect to ecosystem functions, which are the characteristics of ecosystemic processes and process networks. As the comprehension of 'function' is a widespread field, different attitudes will be taken into account, including the anthropocentric view of natural functions and their contributions to sustainable landscape management strategies.

2. The Multiple comprehensions of functions

The term 'function' can be characterised by an enormous number of definitions and understandings. As many disciplines are working with this notion, its multiple utilisations make up a very colourful picture. This can for example be seen when a dictionary is used to define 'function': We will find explanations like the 'general activity or performance of an object', 'specific task within a broader context', the 'role of an object', or the 'specific contribution of a unit' which is similar to the 'purpose of an entity' or to the 'functioning in the sense of successfully following a specific objective'. Important connotations for the principle understanding of 'functions' are

- the active interactions of structural patterns,
- effective interrelationships of processes,
- the roles of parts for wholes e.g. organs for organisms,
- purposive processes and activities,
- the contributions of parts to the performance of the whole, or
- the potentials of biological systems for human utilisation.

Apart from the uncertainty of definitions in the biological context of 'functions', problems arise from the implicit teleology which arises if the term is used without a prefix or an attribute. In this case 'function' imputes a certain non-human expediency to biological entities. Thus the corresponding question is: what is the 'intrinsic' purpose of natural systems? And this is a very dangerous issue. If we ignore all warning shots from the philosophy of science and if we carefully apply Ulanowicz's interpretation of Popper's propensities (Ulanowicz, 1998), self-organisation, self-conservation, and autocatalysis seem to be the outcome of the interrelationships between the ecosystem-physiological (functional) biological processes. From this point of view, 'functions' create orientors (specific regularities in the systems' dynamics which are characterised by certain attractor states, see Müller and Leupelt, 1998 or Müller and Jørgensen in this volume) which may be used to indicate the state of the respective ecosystem. Therefore, the introduction of functions which could be interpreted as the imputation of purposes may be allowed as a heuristic tool in systems analysis, but only if we take a teleonomic approach instead of a teleological one. Doing this, we have to be aware that functions as purposive actions are always accompanied with normative implications. The carrier of 'functions' is covered with positive connotations. Realising these traps, we furthermore have to keep in mind that 'function' is not causality. Functional approaches (e.g.: why is that organ good for the organism?) can be used to assist causal concepts, but they cannot replace them.

The bundling of these fields of problems and arguments will be carried out from three aspects in the following text, (i) 'functions' in systems analysis, (ii) 'functions' in systems ecology, and (iii) 'functions' in ecological economy. These three approaches will be integrated in the final part of this chapter which will try to elucidate the consequences of the functional systems ecology for the conception of sustainable development.

2.1 Systems analytical and mathematical aspects

Approximating 'functions' from a generalising perspective, an advisable initial question is: What is the role and understanding of 'functions' in mathematics and systems analysis? Mathematical functions are rules for classifications. They are dynamic terms which are based upon the relations between a dependent and an independent variable. Mathematical functions can be formulated as equations, and these equations can be introduced into simulation models. Therefore, ecosystem modelling and ecosystem theory actually build upon mathematical functions.

From a systems analytical aspect, Bahg (1990) describes 'function' as the inherent order within the interactions and processes of a system, similar to Jantsch (1988) who understands 'function' as the total characteristics of all acting processes in a system. Therefore, 'function' is often used to characterise an ordered operation, an ordered development or an ordered organisation. Herewith we can classify 'function' as an holistic feature from the system analytical point of view. In that context, the total amount of relations in an ecosystem becomes its 'function' if we observe it from a teleonomic point of view, which can for example focus on the contribution of the relations for the self-organisation of the entity, but which can also be the benefit that the whole gains for long-term human utilisation from the internal pattern of interactions.

This connection leads to another, definitory systems theoretical aspect. Systems consist of elements, which can be classified into subsystems, and relations, that are the interactions between the elements and subsystems. The distribution of the elements forms the structure of the entity (see Golley in this volume). That can be described as the spatial structure, the temporal structure, or it can be designed with respect of specific aspects, like matter flows or energy dynamics. If the set of elements is called structure, the set of relations is the system's function. As a general feature it includes all interrelations that are all flows of energy, matter, water or information; thus 'function' also includes all changes because they are connected with modifications of the processes, in any time. The integral of structures and functions can be described as the system's organisation, and if we look at the dynamics of open, biological system, the temporal development of the organisational criteria is the basis of self-organisation.

2.2 Ecological aspects

It is not surprising that the compression of the term 'function' also produces many different ideas. They are reaching from determinations about the role of an ecological process (Jax et al., 1993) over interdependencies between two terms (Leser, 1995) or the meaning of organisms for the ecosystemic process performance (Leser, 1994) to the total characteristics of all active processes (Jantsch, 1988). Odum (1983) states that functions consist of flows and changes in state variables which are operated due to modifications in the energy budgets of the ecosystems. Similarly, Schwerdtfeger (1975) describes functioning as the sum of interactions in an ecological structure. Schaefer (1992) offers 4 potential definitions: function is (i) the course of all living processes in an ecological entity, (ii) the activities of an organism, (iii) the significance of an organism for the ecological processes, and (iv) the maintenance of a steady state. Gertberg (1994) defines functions as purposes (see section 2.3) and thus as potentials for specific natural services. This viewpoint is one of the basic elements in modern landscape planning which exceeds the limitations of structural aspects, species, and abundances.

If we accept the systemic definition of functions (as ecosystem physiological functions), that comprehends them as patterns of interacting ecosystem processes, we can find some examples and characteristics in the Tables 1, 2 and 3. In these cases it is assumed that ecosystems provide many different relations between their numerous elements. A function can be a subset of the total set of ecosystemic relations. This subset of course can include the whole set in the sense of Bahg's definition, but it can also be a part of it, only. Furthermore, it can be selected based upon the question of whether it fulfils any purpose for a specific target which is defined by the observer. Thus, all potential interpretations of the notion can be applied within this definition.

In Table 1, environmental (purposive) arguments have been listed. The interactions that are classified here can be taken as examples for (ecophysiological) functions in ecological systems. If the target of the observer is an intersubjective systems description he will include the listed processes of energy budgets, water balances, matter flows and storages and community interrelations.

Table 1: Some exemplary types of functional, ecological interactions

Energetic interactions
- elements and relations of the ecosystem radiation budgets
- elements and relations of the ecosystem heat balances
- energy transfer in food webs, storage in organisms, detritus, or soils
- energy export by evapotranspiration or respiration
Hydrological interactions
- water transport through plants in ecosystems
- water interception, transpiration, evaporation
- water transport in soils, leaching, interflow, seepage, storage
- water transport in the ground water or in the surface waters
Matter interactions
- soluted transport with water flows
- gaseous matter transports into the atmosphere or from the atmosphere
- substance metabolism and storage in organisms
- soil nutrient and contaminant processes (e.g. adsorption, desorption, solution)
Biocenotical interactions
- competition
- commensalism
- parasitism
- mutualism

These interrelations can be individually characterised by the variables from Table 2. As an ecosystem analysis always has to take into account many single functions, complex combinations of the characteristics will appear and the whole extent of the depicted pairs in Table 2 will be filled in. The listed characters can also be used to characterise human inputs into ecological systems. One of the characteristics in Table 2 refers to the hierarchical features of a process. In the paper of Hári and Müller (this volume) the implications of the respective hierarchical approach to system ecology are documented.

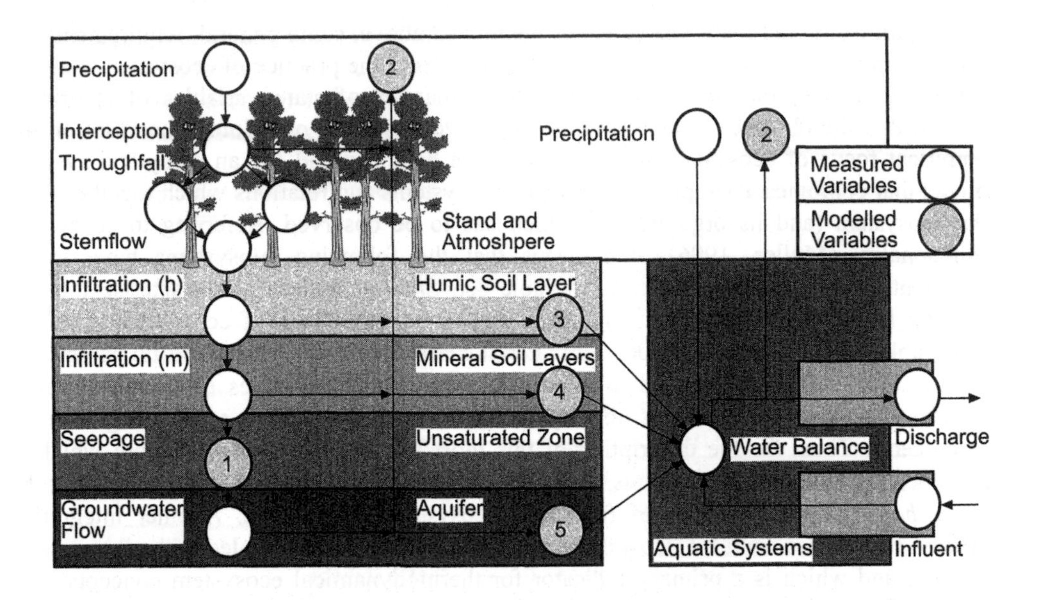

Figure 1: Hydrological flows and storages as parts of ecosystem functions. Abbreviations: (1) = seepage rate model data, (2) = evapotranspiration model data, (3) = erosion model data, (4) = interflow model data, (5) = groundwater flow model data.

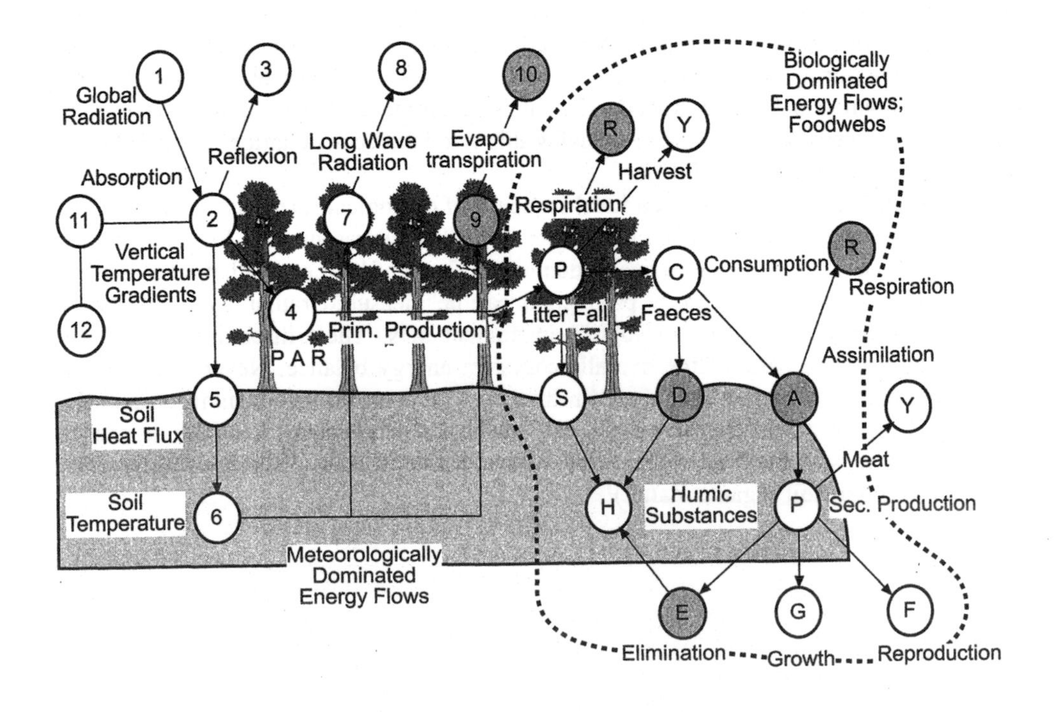

Figure 2: Energetic flows and storages as parts of ecosystem function.

The inputs, stores, fluxes and losses of *matter (nutrients and environmental chemicals)* have a tight functional link to the processes of the water and energy budgets in ecosystems. Besides the transportations through water, matter can also be translocated by gaseous flows, transformed by biochemical processes, and retained by the adsorptive processes at soil surfaces. Furthermore, all substances are transferred in the food-web, and consequently the detritus compartment is a very important store for energy, nutrients, and environmental chemicals. Three different spectra of eco-chemical features should be investigated for a general functionality analysis:

- Essential nutrients (e.g. C, N, P, Ca, Na, K, Mg),
- Important trace elements (e.g. Fe, Cu, Mo, Co, Cl, Si, Mn), and
- Basic characteristics of the matter budget (e.g. pH, Eh).

These variables can be analysed or modelled based on samples from the subsystems which are listed in the general flow scheme of Figure 3. The total amount of dry deposition, gaseous emissions, and matter balances have to be calculated by simulation models.

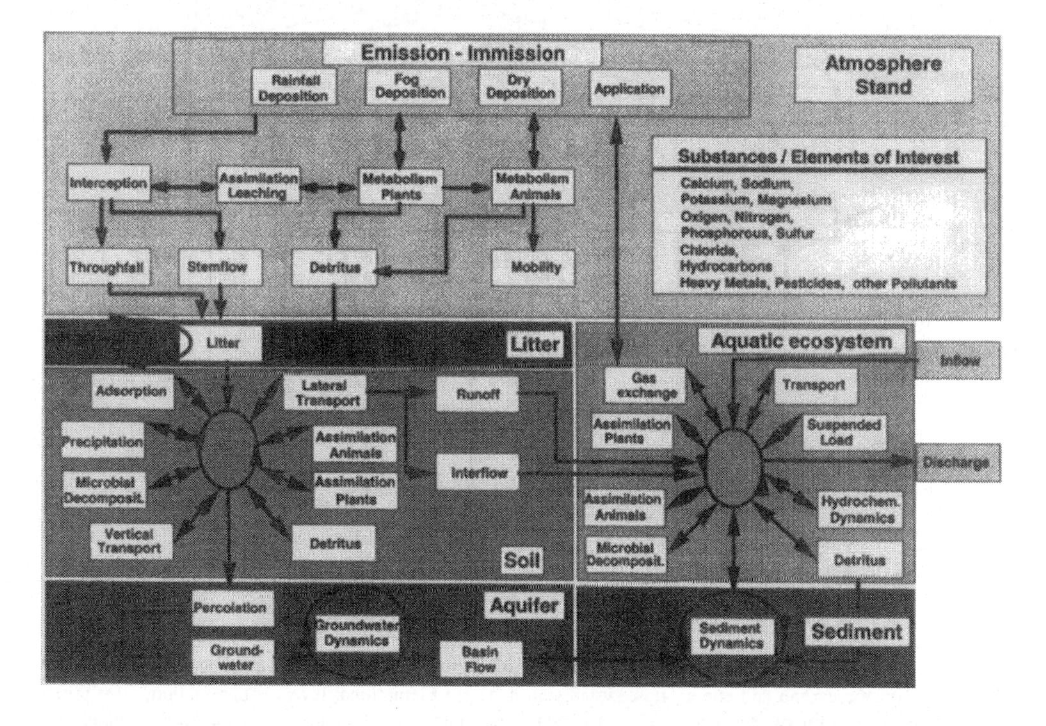

Figure 3: A conceptual model of the fluxes, storages and transformations of substances through ecosystems, watersheds and landscapes. Within these areas the relevant compartments are the atmosphere at the stand level, the litter layer and the other soil horizons, the aquifer, sediments and the interrelated aquatic ecosystems which are connected with the surrounding landscapes by matter inflow and discharge.

The *biotic community* is the active, living part of ecosystems; plants, animals, and micro-organisms are the driving forces for the fluxes of water, energy and matter, and they form the fundamental objects of all self-organising processes. A functional idea of the community structure is given in Figure 4. It is based on the non-equilibrium principle (Schneider and Kay, 1994; Jørgensen this volume; Müller and Nielsen, this volume), whereby different spatial and temporal scales are distinguished. To evaluate the integrity of these process networks, bioceonotical parameters have to be quantified on long-term sampling sites. Important indicators for the functionality of the community are – in addition to all other variables presented in this paper – the following variables:

- species abundances and diversity measures;
- phenology of selected species;
- leaf area index of the stands as model input data;
- selected organism groups to determine the carbon (and energy) contents, nutrient flows and stores, and concentrations of toxic substances.

A special problem in regard to the community variables is the high diversity of organisms. Therefore, certain species have to be selected to define the biocenotical spectrum that has to be investigated. For this step, the following criteria can be used: turnover rates, dominance structure, scarcity of the species, regional positions in global distribution patterns of species, representation of all trophic levels, potential utilisation as bioindicators.

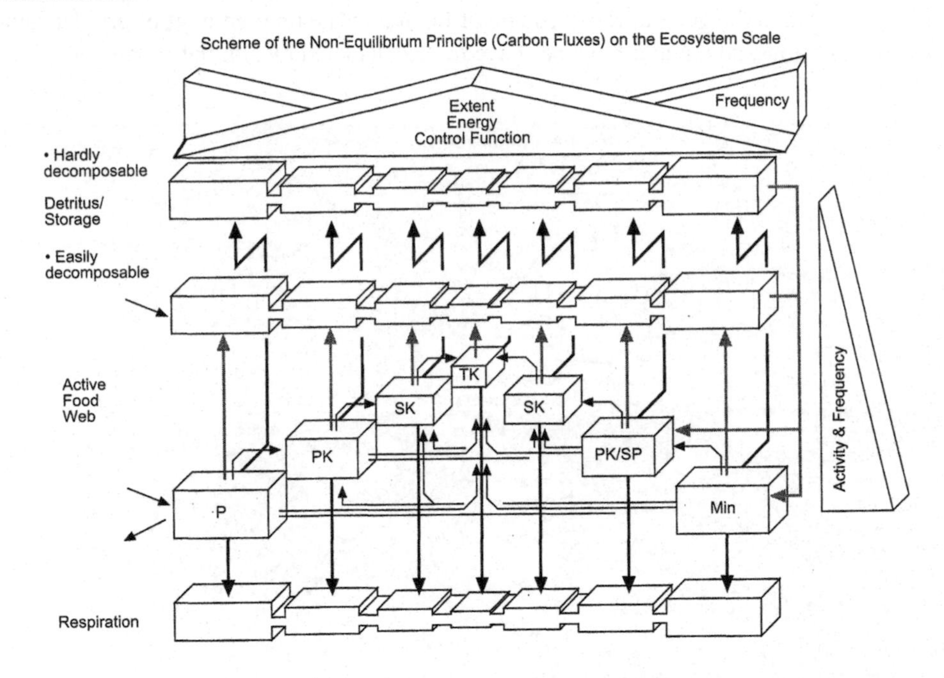

Figure 4: The basic functional concept of the community biological variables as parts of ecosystem functions. It is oriented on the carbon and energy flow systems on different hierarchical levels. Abbreviations: P = primary producers, PK = primary consumers, SK = secondary consumers, TK = top carnivores, MIN = mineralising organisms, PK,SP = saprophageous organisms. The flows are originating in two food webs (phytophageous web, saprophageous web) that are connected via a hierarchy of storages in the detritus compartment.

Looking at Figure 4 it might be suggested that the biocoenotical interactions can be reduced to fluxes of energy or matter. That presumption is not correct. There are many more interactions which as a whole form the community activities of ecosystems. One classification scheme of these processes is summarised in Table 4, illustrating functional attributes such as antibiosis, parabiosis, and symbiosis.

Within these short overviews, the classical ecological functions, the set of ecosystemic interactions, have been described briefly. The flows of water, energy, matter, and information and the additional community interactions form an extensive ensemble of ecological relations. The selection of respective sets of functions has been carried out here for the scientific purpose of an holistic ecosystem description (see also Fränzle in this volume). In the following chapters some further approaches to ecosystemic functions will be presented, focussing on human ecosystem utilisation and its long-term optimisation.

Table 4: Different types of biotic interactions, after Breckling and Müller (1997), (-) symbolises an inhibition, (+) stands for a support, and (0) stands for no effect; 'p1' means partner 1, 'p2' addresses the second partner

Type of interaction	p1	p2	Characteristics
Neutralism			no mutual influence
Antibiosis			one partner is inhibited
- competition	-	-	contest for the same resources
- predation	+	-	biomass transfer to the predator
- parasitism	+	-	parasite uses and harms the host
- amensalism	-	0	inhibition of one partner without profit for the other
- allelopathy	0	-	inhibition by metabolic products
Parabiosis			one partner is supported
- Parökie	+	0	neighbourship
- Synökie	+	0	living in their species' nests
- Epökie	+	0	living on other organisms
- Entökie	+	0	living in other organisms
- fabric interactions	+	0	using other organisms or species for constructions
- phoresis	+	0	using other organisms or species for transports
- metabiosis	+	0	using basic requirements that are provided by other organisms
Symbiosis			both partners are supported
- protocooperation	+	+	co-operation is not obligatory
- mutualism	+	+	mutual support
- alliance	+	+	loose partnership
- eusymbiosis	+	+	co-operation is obligatory

2.3 Socio- economic aspects

At this position, we can neglect all the teleological and all the normative reservations which have been rumbling in the background of the section 2.1 and 2.2. From an anthropocentric viewpoint, no discussions about purposes and their scientific implications are necessary. Nature is looked at here as a source for human benefits, it functions to optimise the human contentment, it provides societal services, and it makes available a high diversity of resources for the growth of human economies.

In the following sections, these aspects of ecosystem function will be illuminated from the aspect of sustainability which aims at a fair and just distribution and development of natural goods, their maintenance and their protection for a long-term benefit of mankind. All these essential items can be understood as purposes that – as an ensemble – form the anthropocentric aspects of ecosystem functions. Therefore, sustainable development is an interdisciplinary concept which includes a very comprehensive contribution from ecology and environmental science. Bossel (1998a) accentuates this relation when he states that in searching for principles to guide sustainable development, it is only natural to have a closer look at the global ecosystem, which has demonstrated sustainability over a few billion years. Ecosystems are working models of sustainable complex systems, and it is reasonable to study them for clues to the sustainable management of the human enterprise. Apart from this strategy of nature as a model for human dynamics, the ecological constraints alone determine the *real* potentials of human life. They provide the general living conditions of humanity, they support all human nutrition systems, they define the long-term limits of human behaviour strictly, and thus ecosystemic processes in fact represent the general constraints for all human activities and decision processes. Regrettably, this restriction of human dependence from the natural integrity, as well as the relations between human health and ecological health, have not been taken into account sufficiently in the past. Therefore, the human interactions with nature (which also form a class of ecosystem functions) actually have to be modified in a process of gradual qualitative change.

Whenever anthropogenic modifications of ecological structures are carried out not only the ecological functions are changed but also the utility of the system is modified. To document the respective economic consequences, different strategies of landscape evaluation have been proposed in the last few years. They are all based on the demands of basic functions of human existence, as they have been very generally modelled with Bossel's basic orientors (Bossel, 1992 or Müller and Jørgensen, this volume) or as they have been discriminated as the socio-geographic functions (basic functions of human existence) which are transport, education, logistics, accommodation, recovery, work, and communication (see Figure 5).

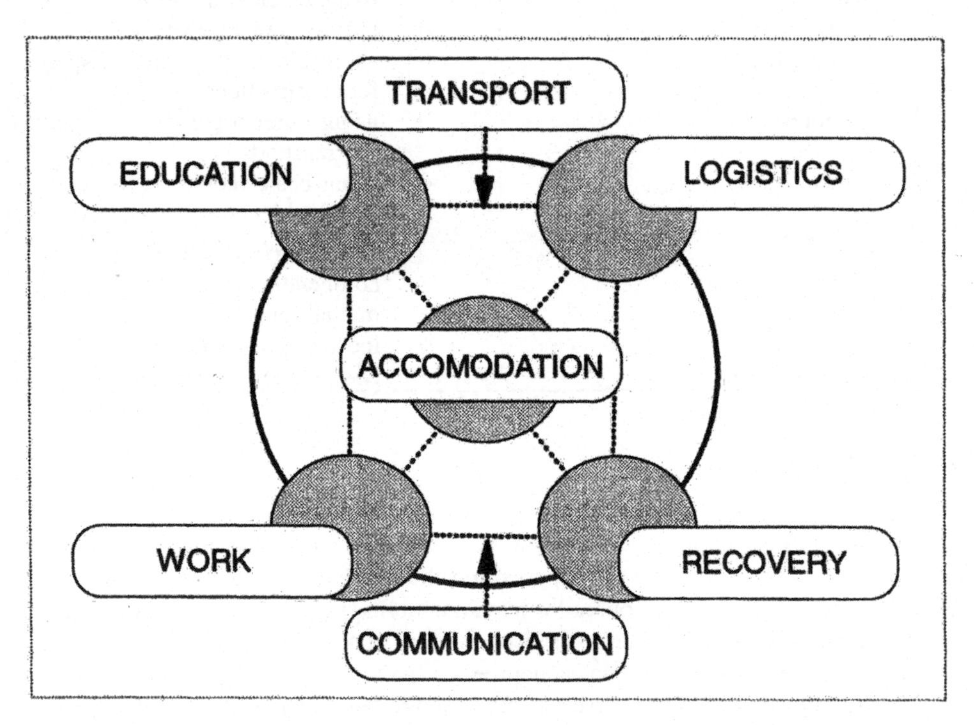

Figure 5: Basic functions of human existence, after Maier et al. (1977). These features have been identified as the most important spatio-temporally relevant scopes and needs of human life by geographers.

Table 5: Different perspectives of 'natural functions' in the context of environmental planning and management. After Bouma (1972), Maarel and Dauverlier (1978), Finke (1987), Marks et al. (1989), and Haase (1991), de Groot (1992); see also Müller (1998a)

Economic perspective (e.g. Siebert)	**Ecological perspective (e.g. Bouma)**	**Planning perspective (e.g. Maarel)**	**Lake utility (e.g. Haase)**	**Landscape perspective (e.g.Marks)**
Public consumption goods - Air - Amenity of landscapes - Recreation	**Production functions** - Food - Raw materials - Hydro power - New elements	**Production functions** - Abiotic components - Agriculture	**Production functions** - Energy - Water - Biomass	- Erosion resistance - Filter potential - Buffer potential - Transformator potential
Resource supply - Water - Solar radiation - Minerals - Oxygen	**Welfare functions** - Education - Health - Recreation - Aesthetics - Arts	**Carrier functions** - Urban activities - Industrial activities - Agricultural activities -Waste absorption	**Landscape functions** - Temperature regulation - Hydrological regulation	- Groundwater protection pot. - Groundwater storage potential - Discharge regulation potent.
Waste reception - CO_2 - SO_2 - Garbage	**Scientific functions** - Information storage - Research	**Information functions** - Research - Education - Storage	**Aesthetic functions** - Habitat functions - Psychological effects - Diversity	- Immision protection potential - Bioclimatic potential - Nature protection poetent
Space supply - Industry - Accommodation - Agriculture - Infrastructure	**Ecological functions** - Gene provision - Oxygen production - Recycling	**Regulation functions** - Purification - Stabilisation	**Human ecological functions** - Biometeorological functions - Sports	- Recovery potential - Biotic yield - Landscape potent

Within these socio-economic functions, the ecological subsystems provide 'natural services', such as production, control, early warning, recycling, cleaning, filtering, buffering, transport, discovery, recreation and safety. Additional functions have been defined in the literature. Some of the models that are recently discussed are demonstrated in Table 5 where the 'natural functions' are elucidated from very different points-of-view. All of these features can be used to indicate the influence that a measure on the benefits which nature provides will have for human land-use. Generally, in the literature, the following four classes of natural goods and services are distinguished:

- Regulation functions are related to the capacity of ecosystems to regulate essential ecological processes and life support systems. They are the direct results of the ecological interactions which are prerequisites for all living systems on Earth.
- Carrier functions are provisions of ecosystems of space and suitable substrates for human activities. The 'Rat Von Sachverständigen Für Umweltfragen' - SRU (1987) stresses carrier functions for residing, maintenance, waste disposal, transport, traffic,

communication, recreation and leisure. With all these activities, the system's integrity will be modified.

- Production functions provide natural resources. Their object is the supply of the society with goods and products from ecosystems. The SRU (1987) distinguishes goods from (i) natural resources, (ii) from wild living resources, and (iii) from agricultural resources. Human nutrition in all variances is based upon these functions as well as all industrial processes that are dependent on raw materials.
- Information functions are cultural attributes. They describe the flow of information between society and nature and its effects: *"Natural ecosystems contribute to the maintenance of mental health by providing opportunities for reflection, spiritual enrichment, cognitive development and aesthetic experience"* (de Groot 1992, p. 9).

A comprehensive analysis of these functions has been carried out by de Groot (1992). He has distinguished 37 natural functions which are listed in Table 6. These functions are additionally characterised by the dominant scales that they operate on. This comprehensive list draws a powerful picture of the enormous values nature can offer mankind, and therefore it is well-suited for the development of sustainable landscape management strategies. On the other hand, the parameters listed in Table 6 also elucidate that it is nearly impossible to evaluate the natural goods and services with a monetary concept. Therefore, the decision makers have to consider their plans carefully on the base of this qualitative, functional information.

Table 6: Functions of the natural environment; after de Groot (1992); The columns R, L, and G symbolise the most significant scales of the respective functions: L- predominantly local interactions, R - regional interactions, G - global interactions.

No.	Function	L	R	G	No.	Function	L	R	G
	Regulation functions					**Production functions**			
1	Protection cosmic influences			■	1	Oxygen			■
2	Regulation energy balance	■	■	■	2	Water for human use	■	■	
3	Regul. chemistry atmosphere			■	3	Food	■	■	■
4	Regul. chemistry oceans			■	4	Genetic resources	■	■	■
5	Regulation of the climate	■	■	■	5	Medicinal resources	■		
6	Regul. runoff, flood prevention	■	■	■	6	Raw materials for clothing	■		
7	Catchment, groundw. recharge		■		7	Raw materials for building etc.	■		
8	Erosion and sediment control	■	■		8	Biochemicals	■		
9	Regul. soil fertility	■	■	■	9	Fuel and energy	■		
10	Regul. biomass production	■	■	■	10	Fodder and fertiliser	■		
11	Regul. organic matter	■	■	■	11	Ornamental resources	■		
12	Regulation nutrient budgets	■	■	■					
13	Storage human waste	■							
14	Regul. biological control	■	■	■					
15	Habitat maintenance	■	■	■					
16	Diversity maintenance	■	■	■					
	Information functions					**Carrier functions**			
1	Aesthetic information	■	■		1	Human habitation, settlements	■	■	■
2	Spiritual information	■	■		2	Cultivation	■	■	■
3	Historic information	■			3	Energy conversion	■	■	■
4	Cultural inspiration	■	■		4	Recreation, tourism	■	■	■
5	Educational inform., Science	■	■	■	5	Nature protection	■	■	■

3. Ecosystem functions and sustainable development

What we can learn from the socio-economic aspects of 'function' is that sustainable development is a preferentially anthropocentric strategy. To fulfil the focal objectives of sustainability, central goals have to be the protection of natural resources, which means the protection of future life support systems for mankind, and the avoidance of risks to human welfare and health. If we invest a closer look at these aims – e.g. by introducing de Groot's functions or the potentials of Marks et al., 1990 – it will turn out that these basic requirements are purely ecological demands. As these items have to be fulfilled for sustainable landscape management, a number of prerequisites have to be met to apply functional approaches into really sustainable strategies. Such concepts therefore should be:

- Long term strategies: They have to take into account that the time scale of the assessed anthropogenic process generally should be adapted to the time scales of the respective natural processes. The speed of natural adaptation therefore usually is much smaller than the speed of man's technical changes.

- Multi scale strategies: Sustainable landscape management should realise that natural processes and functions are organised and regulated in accordance with their spatial and temporal characteristics (Allen and Starr, 1982; O'Neill et al., 1986; Müller 1992). Therefore, at least three different scales have to be analysed if the management result shall meet the sustainability demands: the focal scale of the actual problem, the constraining superior hierarchical level, and the inferior level which determines the biotic potential of the system (O'Neill et al., 1989).
- Interdisciplinary strategies: To derive measures for a sustainable management of the ecosphere, ecological arguments have to be put into a framework that also includes cultural, social and economic aspects. To develop respective objectives and methods the competent sciences have to co-operate intensively (see Jüdes, 1998).
- Holistic strategies: Sustainable strategies have to perceive the relationships between man and nature as functional entities which follow the general laws of open systems. To deal with these functional units, the techniques of systems analysis should be applied. This means that the enormous significance of indirect effects is noted and that the approach is capable of coping with the high complexity of the investigated systems.
- Realistic strategies: It is important to inform the responsible decision makers about the (in)validity of the ecological data used and the prognoses (Breckling, 1992). All ecological information should be escorted by descriptions of potential uncertainties, methodological mistakes, and statistical insecurities.
- Nature oriented strategies: As has been mentioned at the beginning of this chapter, in search of sustainable strategies, nature and its functional features seem to be a good model (Bossel, 1998b). Therefore, the potential reaction patterns of natural systems in many cases can be taken as guidelines for human reactions and management activities.
- Synergetic and theory based strategies: The correctness of the assumptions which are made throughout decision processes will be higher, the deeper the theoretical fundament of the applied concept is. Thus, approaches of systems science and synergetics are very helpful to understand and foresee the behaviour of complex entities. Applying these sciences, it turns out that one of the most important functional features of such systems is their potential for self-organisation (see Müller et al., 1997 a,b).
- Limiting and hierarchical strategies: All sustainable activities have to accept the natural system of constraints in which the investigated entity operates. Ecological assessments thus have to look for the system's carrying and assimilation capacities. Also,

As was shown in section 3, these items can be integrated to the concept of ecosystem functionality which delineates a holistic, ecological attitude to the concept of sustainable development. The corresponding inclusion of the functional approach of ecological integrity into environmental management has been in a conceptual stage thus far. Therefore, not only from the viewpoint of ecosystem theory, a consequent application in sustainable landscape management is a most important and promising step towards a successful future environmental policy and science.

REFERENCES

Allen, T.H.F. and T.B. Starr 1982. Hierarchy - Perspectives for ecological complexity. The University of Chicago Press.

Bahg, C.G. 1990. Major systems theories throughout the world. *Behav. Sci.* 1990, 35: 79-101.

Bossel, H. 1992. Real-structure process description as the basis of understanding ecosystems and their development. *Ecological Modelling* 63: 261-276.

Bossel, H. 1998a. Ecological orientors: Emergence of basic orientors in evolutionary self-organization. In: Müller, F. and Leupelt, M. (eds). *Eco targets, goal functions and orienters.* Springer. Berlin, Heidelberg, New York, 19-33.

Bossel, H. 1998b. Ecosystems and society: Orientation for sustainable development. In: Müller, F. and Leupelt, M. (eds.) *Eco targets, goal functions and orienters.* Springer. Berlin, Heidelberg, New York, 366-380.

Bouma, F. 1972. Evaluatie van natuurfuncties. Verkenningen van het I.v.M.-VU, Series A, No.3, Amsterdam .

Breckling, B. 1992. Uniqueness of ecosystems versus generalizabilty and predictability in ecology. *Ecological* Modelling 63, 1-4: 13-28.

de Groot, R.S. 1992. Functions of nature. Wolters-Noorhoff.

Finke, L. 1987. Ökologische Potentiale als Element der Flächenhaushaltspolitik. ARL Forschungs- und Sitzungsberichte Bd. 173, 203-228.

Gertberg, W. 1994. Umwelt und menschliches Handeln, eine fluxbezogene, systemische Betrachtung anhand graphischer Modelle. Diss. Freising-Weihenstephan.

Haber, W. 1995. Naturhaushalt. In: Handwörterbuch der Raumplanung. Hannover, 661-663.

Haase, G. 1991. Theoretische und methodische Grundzüge der Interpretation. In: Haase, G. (Hrsg.) *Naturraumerkundung und Landnutzung.* Leipzig.

Jantsch, E. 1988. Die Selbstorganisation des Universums. München.

Jax, K., Vareschi, E. and Zauke, G.P. 1993. Entwicklung eines theoretischen Konzepts zur Ökosystemforschung Wattenmeer. *UBA-Texte* 47/93. Berlin.

Jüdes, U. 1998. Human orientors: A system approach for transdisciplinary communication of sustainable development by using goal functions. In: Müller, F. and Leupelt, M. (eds). *Eco targets, goal functions and orientors.* Springer. Berlin, Heidelberg, New York, 381-394.

Leser, H. 1994. Westermann-Lexikon Ökologie und Umwelt. Braunschweig.

Leser, H. 1995. Landschaftsökologie. Stuttgart.

Maarel, E. van der 1978. Naar een Globaal Ecologisch Model (GEM) voor de Ruimtelijke Ontwikkling van Nederland. Min. van Volkshuisv. en Ruimt. Ord., Den Haag.

Maier, J., R. Paesler, K. Ruppert and F. Schaffer 1977. Sozialgeographie. Braunschweig.

Marks, R.; Müller, M.J.; Leser, H. and Klinik, H.-J. 1989. Anleitung zur Bewertung des Leistungsvermögens des Landschaftshaushalts. *Forschungen zur deutschen Landeskunde,* Band 229, Trier.

Müller, F. 1992. Hierarchical approaches to ecosystem theory. *Ecological Modelling* 63: 215-242.

Müller, F. 1996. Ableitung von integrativen Indikatoren der Funktionalität von Ökosystemen und Ökosystemkomplexen für die Beschreibung des Umweltzustandes im Rahmen der Umweltökonomischen Gesamtrechnungen (UGR). In: Statistisches Bundesamt (Hrsg.) *Beiträge zu den Umweltökonomischen Gesamtrechnungen, Bd. 2.* Wiesbaden.

Müller, F. 1998. Gradients in ecological systems. Proceedings of the Eco-Summit 1996 in Copenhagen, *Ecological Modelling* 108: 3-21.

Müller, F., B. Breckling, M. Bredemeier, V. Grimm, H. Malchow, S.N. Nielsen and E.W. Reiche 1997a. Ökosystemare Selbstorganisation. In: Fränzle, O., F. Müller and W. Schröder (Hrsg.): *Handbuch der Ökosystemforschung.* Landsberg, Chapter III-2.4.

Müller, F., B. Breckling, M. Bredemeier, V. Grimm, H. Malchow, S.N. Nielsen and E.W. Reiche 1997b. Emergente Ökosystemeigenschaften. In: Fränzle, O., F. Müller and W. Schröder (Hrsg.): *Handbuch der Ökosystemforschung.* Landsberg, Chapter III-2.5.

Müller, F. and Leupelt, M. (eds.) 1998. Eco targets, goal functions and orienters. Springer. Berlin, Heidelberg, New York.

Odum, H.T. 1983. Systems ecology. An introduction. New York.

O'Neill R.V., D.L. de Angelis, J.B. Waide and T.H.F. Allen 1986. A hierarchical concept of ecosystems. *Monographs in Population Ecology* 23, Princeton University Press.

O'Neill, R.V., A.R. Johnson and A.W. King 1989. A hierarchical framework for the analysis of scale. *Landscape Ecology* 3, 2/4: 193-206.

Rat der Sachverständigen für Umweltfragen (SRU 1987): Umweltgutachten 1987. Stuttgart und Mainz

Schaefer, M. 1992. Wörterbücher der Biologie. Ökologie. Jena.

Schneider, E.D. and J. Kay 1994. Life as a manifestation of the second law of thermodynamics . *Math. Comput. Modelling,* 19: 25-48.

Schwerdtfeger, F. 1975. Synökologie. *Ökologie der Tiere*, Bd. 3. Hamburg, Berlin.

Ulanowicz, R.E. 1998. Network orientors: Theoretical and philosophical considerations why ecosystems may exhibit a propensity to increase ascendency. In: Müller, F. and Leupelt, M. (eds.) *Eco targets, goal functions and orienters.* Springer. Berlin, Heidelberg, New York, 193-208.

I.4 Uncertainty in Ecology and Ecological Modelling

Broder Breckling and Quan Dong

1. Introduction

Ecology contributes much to our knowledge about nature and is essential for rational decision making in environmental and conservation issues (Roughgarden et al. 1989, Lubchenco et al. 1991, Dong 1996). Despite continuous efforts, ecology has not been able to offer universal laws or precise ubiquitous principles. A large amount of ecological observations usually only characterise a specific site at a specific time. Even some basic ecological concepts are ambiguous in definition. In many applications, it fails to make precise predictions. As a consequence, ecology is considered as more a postdictive rather than a predicitive science and is regarded as a 'soft science' (Peters 1992). In short, ecology still deals with a high level of uncertainty (Ludwig et al. 1993). This status limits ecology in its successful application to environmental management.

Ecology is developing into a quantitative science, following the success of quantitative models in physics. Physics, especially classical mechanics, represents the paradigm of a 'hard science'. It elaborates general theories and quantitative models. These models describe mechanical processes precisely and produce reliable predictions in their field. Nevertheless, physics and ecology study different types of subjects. The authors believe that the kinds of uncertainty ecology deals with are inherent to the biotic processes and therefore partially irreducible: we cannot that scientific progress will reduce uncertainty to any intended level. This inherent and irreducible uncertainty will prevent ecology from achieving the generality and predictability of mechanics. Therefore, strategies and techniques to deal with ecological uncertainty are needed and desirable. For administration and management, a general purpose of science is to reduce uncertainty in decision making. We need to know the potential sources and features of different types of uncertainty, and the limitations of current ecology. Mathematical models are major quantitative tools to produce ecological inferences (Jørgensen 1988). Many different kinds of quantitative models exist, including analytical models, simulation models, and statistical models (Loehle 1983, Haefner 1996, Hilborn and Mangel 1997). Each of them have specific merits and drawbacks in handling uncertainties. Ecologists often favour one of them over another. Uncertainty trade-offs exist in ecological modelling and have rarely been addressed. Environmental decision makers and mangers need practical strategies to handle uncertainty in ecological knowledge and ecological modelling and to be prepared for unexpected events. This paper will show the sources and features of uncertainty in ecology and particularly in ecological modelling and suggest strategies to handle ecological uncertainty. The authors suggest, we should use plural, multiple and linked approaches in integrative frameworks to address uncertainty problems.

2. The concept of uncertainty

In general, uncertainty refers to a degree of dissatisfaction of some knowledge and expectations. A subject is uncertain when it is either not exactly known, determined, or not reliably predicted (Webster 1983, Prigogine and Stengers 1984, Lemons 1996). The specific meaning of uncertainty may vary in different situations and depend on context. In science, predictability and reproducibility are two important measures of satisfaction. Predictability allows a certain derivation and satisfying characterisation of a future state from a preceding state based on the available knowledge and information about preceding states and determining processes. We consider a phenomenon as reproducible if its characteristics of particular interest can occur reliably and repeatedly under our control or manipulation. An irreproducible event is unique or singular and occurs only once. Science cannot test hypotheses about a subject if it occurs once and never again. Uncertainty is an outcome of unpredictability and irreproducibility. One main interest of science is to find regularity. Regular recurrences allow scientists to specify rules. Understanding and description of underlying rules is a condition to achieve predictability. Irregularity indicates the absence of detectable phenomenological rules. Many studies only expect to address the central tendency of a series of events. Variability refers to a magnitude of deviation from the central tendency, which can be taken as a measure of regularity and is often an inherent character of the system .In these cases, irregularity and variability measure uncertainty.

In ecology, irregularity, variability, irreproducibility, and unpredictability are remarkable. Some uncertainties are reducible while some are irreducible but estimable. Others are neither reducible nor estimable (Hilborn and Ludwig 1993). In order to discuss the issue of uncertainty in ecology, we inspect the paradigm of classical physics, which emphasises causality and mechanics, compare objects in ecology and mechanics, and argue that ecology falls only partially under that paradigm. We discuss model examples as case studies on uncertainty and finally make some recommendation on how to deal with it in ecological applications.

3. The paradigm of causality and mechanics in ecology

Mechanics is based on a strict and explicit form of causality: any event or phenomenon results from particular, antecedent causes. Identical causes produce identical outcomes. An object remains in its particular state unless an (external) cause moves it away. This is what Decartes described as *res extensa*, which makes the physical world (with only the exception of the soul and the will of human beings) as the basis of the mechanical world view (Prigogine and Stengers 1984). With this kind of internal constancy, mechanics makes a strict formalisation, which occurs in two major forms:

– the static form is implied in logical terms: 'if - then' operations describe the effect and its precedent causes. These statements can be used for qualitative descriptions.
– the dynamic form maps temporal changes to an equation structure. This is what differential equation models do: From state x, state y can be derived at a time later by applying a particular calculation. Linear systems, which can be integrated, allow a direct access of any later state directly from a given initial state. Non-integrable and most of the non-linear systems require a successive numerical step-by-step approximation (an iterative application of the equation system).

Decartes (1596-1650) first applied the mechanical view to lives, any form of plants and animals. Immediately following this time, a long historical debate started as to whether it is adequate and useful to resume the organismic processes in these mechanical forms. Spinoza (1632-1677) was one of the first prominent opponents, pointing to the web-structure of mutual causation, which hinders a strict separation of cause and effect in this field. Centuries passed before quantitative ecology emerged as a natural science and explicitly adopted the quantitative classical mechanical forms to describe population dynamics, energy and matter flows, and network interactions in ecological systems. In the beginning only simple mathematical models were created (Lotka, 1924; Volterra, 1926). Later, Patten (1971, 1972, 1975), Odum (1983), and others applied principles of physics to larger, complex ecological systems. Today, the static and dynamic approaches of physics are widely used in ecology, in the population – and community approach as well as in the ecosystems approach (Jones and Lawton 1995).

4. Ecological systems fall under the mechanics paradigm only partially

Causal mechanical projections require sufficient information about the given state, as well as underlying processes. In most of the practical system dynamic applications, the process structure is assumed to be constant (see Patten 1971, 1972, 1975; Odum 1983), or at least, deterministic rules that govern structural changes must be available. Major sources of unpredictability and uncertainty can then be distinguished according to the states, processes, structural characteristics and boundary conditions (Table 1).

Table 1: Reasons for unpredictability of a system state in the future

- If structural constancy and invariant conditions can be assumed, so that *a mechanical description is applicable,* then unpredictability can be caused by:
  - insufficient information about present states of considered system
  - insufficient information about underlying processes (assumed to be constant)
  - insufficient information about boundary conditions (external influences from the systems environment)
  - predictable unpredictability (i.e. chaotic system dynamics) – the system interaction itself increases microscopic disturbances to macroscopic order of magnitude. Predictability would thus require infinite precision about given states, which is unachievable for any physical as well as biological system).
- If there are heterogeneous and varying conditions, so that *a mechanical description is not applicable* then *additional unpredictability* can result from
  - changes of the systems composition (appearance of new elements, fading of present ones)
  - changes of the underlying process characteristics (connection of elements and mutual influences will modify)
  - changes of the boundary conditions (external influences from the systems environment alter)

Ecological uncertainty is inevitable because organisms, their structures and behaviours, and their ecological contexts are constantly changing. Organisms as living systems defy the mechanical assumption of inner constancy of an object. Due to the activity of organisms, the properties of any ecological context are not completely constant. This holds regardless of whether particular (groups of) organisms are investigated or whether they are resumed to certain functions and only particular effects or aggregated results of organismic activity are focused. The main cause for this is that organisms have the following particular properties: (1) they can reproduce and die, i.e. the size of populations can change considerably even under constant environmental conditions, and (2) organisms can modify their structural and functional attributes considerably over time because of ontogenetic development as well as (phylo-) genetic changes. All this happens at the level of ecological interaction and involves physical and chemical transitions as an underlying level (Ekschmitt et al. 1996). As a consequence, physical and chemical conservation laws, especially the law of mass and energy conservation, and thermodynamics, only entail boundary conditions; they are valid for the level of physical and chemical processes (Ekschmitt et al. 1996). A wide variety of different possible and equally probable ecological states exists under these conservation laws. For example, the primary production of an ecosystem cannot exceed an optimal use of the available irradiance as an energetic boundary. But it can be at any lower level, caused by antagonistic activities of the organisms themselves. There is no conservation law for the number of

parameters often represents the central tendency. The actual value of the parameter may occur differently due to stochasticity. (A true probability for outcomes of coin tossing can be 0.5, but the outcome of each toss is either 0 or 1.) If the changes in a parameter are limited and independent to other variables, statistics are often used for its estimation. Uncertainty then is handled as variance – the deviation from the central tendency. A general statistical formation of simple models can be written as,

$$N_t = f(N_{t-k}) + z \tag{3}$$

where z denotes a random error. In the case of the logistic model, we simply replace $f(N_{t-k})$ with equation 2. Finally, the system could be intrinsically unstable. May (1974, 1979) examined the stability of the discrete version of the logistic model. He found that chaos could occur in this system, given a certain parameter range. When chaos occurs, even a high level of accuracy and precision in measuring state variables and parameters will not improve long-term prediction. The magnitude of uncertainty in prediction approaches the magnitude of the overall variability of the system quickly through time.

Our categorisation of uncertainty differs slightly from the one suggested by O'Neill and Gardner (1979). They suggested three major sources of uncertainty in biological modelling: 1) model formulation, 2) parameter estimation, and 3) stochasticity. To reduce uncertainties from different sources, trade-offs exist between simple analytical models and complex realistic simulation models. For example, a common way to reduce uncertainty caused by ignorance is to include additional variables into models, making them more complex. This is a critical step in model formulation. Complex models contain more parameters and other specific details. Capable of representing more specific biological information, these models are likely to abstract less and ignore less than simple models. Thus, they are more realistic. Meanwhile, the data requirement grows very fast, potential bias in parameter estimation, potential errors in identification of functional formula, and interplay of these errors increase quickly with the level of complexity. Such kinds of errors and biases may bring about a large uncertainty in model outcomes. Gardner et al. (1980) found that complex ecological models usually amplify parameter errors. Furthermore, with simple models, analytical solutions exist, which provide a convenient base for statistics. The statistics can provide measures of stochasticity. It is very difficult to find analytical solutions for complex models. Thus, it is more difficult to design a general statistical formula for complicated simulation models. This statistical problem leaves room for uncertainty about and caused by stochasticity. Complex models usually postdict better but may not necessarily predict better (Gauch 1993). Heated debates occurred among ecological theoreticians in favour of simple vs. complex models (e.g., Hall and DeAngelis 1985). Some modellers believe there is likely an optimal level of complexity (Costanza and Sklar 1985, Hilborn and Mangel 1997). We suggest that modellers do not have to take a single position in the dichotomy. We can combine simple and complex models as well as analytical and simulation approaches to reduce uncertainty in our models. While simple models can capture and represent some of the dominant trends, complex models may help to focus on sensitive points.

5.2 A theoretical population model on the interaction of different integration levels

Ecological models usually focus on one particular integration level and implicitly assume the non-represented external context as constant, as invariant or irrelevant. Some of the impacts from upper levels may be considered as forcing functions, while others and also the lower levels usually fall under the so-called *ceteris paribus* (everything else constant) condition. They ignore structural uncertainty, which results from the organisational context. It is a very difficult task to estimate whether model results also hold under a range of changing context conditions. Only if heterogeneity, variability and different levels of organisation are explicitly represented in a model, can we study how this kind of uncertainties on one level may crucially affect other organisation levels. As these type of model studies have been difficult to perform previously, we want to point to this aspect as a very important and still neglected field of investigation (Müller 1996). We exemplify this source of uncertainty by the following model.

In a previous study (Breckling 1991, 1992), an individual based model has been developed which depicts the interaction of one prey- and two predator populations in a structured environment in abstract terms. Individual properties consisting of dispersal, colonising and foraging abilities are specified. The organisms interact in widely separated habitat patches, where extinctions can frequently occur, which have to be balanced by successful (re-) colonisation. The interference of individual properties, the local population dynamics and extinction-recolonisation patterns on the metapopulation level cause a multi-level interaction, which cannot be reduced to one level, if we want to estimate the persistences of the populations. Here, we present a related situation based only on the population-metapopulation interactions. We focus on the core aspect, which is the interference of foraging efficiency and colonisation ability. The population dynamics in each single habitat patch follows a conventional Lotka-Volterra equation with capacity limited prey growth. Colonisation is modelled as diffusion between neighbouring habitats. In the Appendix , the model equations and parameters are specified. Figures 1 – 3, show some model results. We find, that in a spatially structured environment it is not possible to draw conclusions directly from a basic interaction level (Lotka-Volterra competition of the organisms) to a higher integration level (persistence of the populations). Ignoring the spatial structure and basing expectations only on a representation of the local interactions we would always find the weaker forager out-competed. On the other hand, including the spatial structure it turns out that the more efficient forager population eventually may fail on the larger scale even though it is still able to out-compete the less efficient one regularly when they share a common habitat patch. At first glance this appears counter-intuitive. But if the different levels are explicitly represented, the interaction can be analysed precisely: More efficient predators deplete the prey resource faster. If the local interaction is unstable (i.e. the predators can extinguish the prey population and subsequently die of starvation), the duration of the average time before extinction limits the time span to colonise neighbouring habitats. Considered at the metapopulation level, the weaker forager gains a longer time span to colonise before the resource is depleted. So the less efficient forager may persist on the metapopulation level, while the more efficient one fails. The model assumes the foraging efficiency as the only difference between the predators. Detailed model evaluations show that small uncertainties or small parameter changes on the lower level can reverse the results on higher level completely. We found a parameter set, for which changes of the colonisation ability of the prey determines, which one of the two predators persists in the habitat complex and which one goes extinct. The example shows, that ecological uncertainty is not only a matter of detailed knowledge but also affected by the interaction of different integration levels.

5.3 Identifying and reducing ignorance in simple phenomenological models with complex mechanistic models

A useful technique to detect and examine the potential extent and consequences of ignorance is to examine the implications of the available biological knowledge which is not used by a simple abstract model. For example, rich information is available about physiological processes and behaviours at the individual level. This type of information can be synthesised and used to examine their consequences at population level dynamics (DeAngelis and Gross 1992).

In fisheries studies, the stock-recruitment relationship has been a focal subject for decades. As a pioneer in this field, Ricker (1954) suggested a simple abstract model to describe the relationship between the densities of the stock and the recruitment. Ricker's model includes a negative feedback from the stock to the recruitment, which could produce overcompensation (Figure 4). At the same period, Beverton and Holt (1957) proposed another simple model with a different form of negative feedback (Figure 4). In the Beverton-Holt model, the negative feedback produces exact compensation. These two models have been widely used in studies of population dynamics for over 40 years (Frank and Leggett 1994). The patterns of feedback in the Ricker and the Beverton-Holt models lead to qualitatively different dynamics. Thus, knowledge of density dependence is essential for fishery management (Frank and Leggett 1994). However, for a specific real population in a fishery, which pattern applies is often unclear. Time series data often fail to reveal the pattern (Shepherd and Cushing 1990). Great irregular variability exits and obscures the pattern detection. Two questions appeared for fisheries scientists: 1) Is the variability merely due to stochasticity? and 2) Can we identify the ignorance and improve these models by incorporating some important processes?

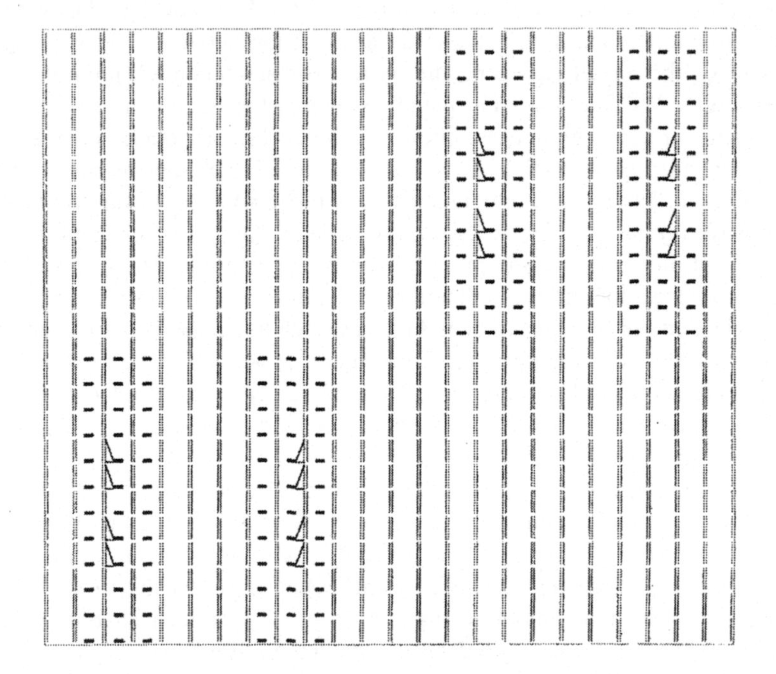

Figure 1: Initial conditions for a grid-based metapopulation model, identical for all model runs. Prey population size: black bar in the centre of a cell (habitat patch). Maximum height is according to habitat capacity. Predator 1 population size (Pred1, high foraging efficiency) is indicated downwards in the left third of the cell together with a backslash. Predator 2 population size (Pred2, low foraging efficiency) is drawn on the right together with a slash for an easier distinction. The model equations and parameter are given in the appendix.

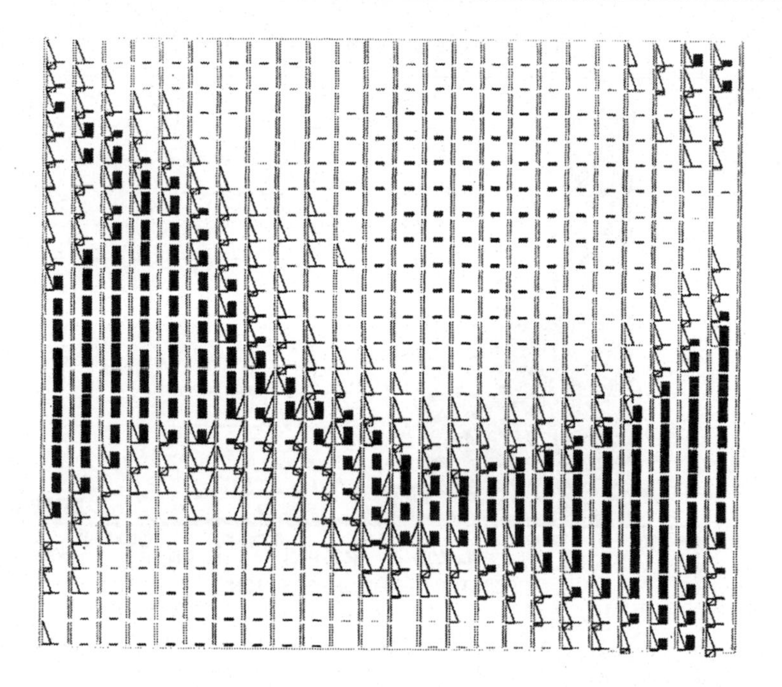

Figure 2: Scenario with a high colonisation rate of prey:
A: The grid after 220 iterations. Pred 2 has only half the foraging efficiency of Pred 1. As Pred 1 thus is faster in gaining biomass and colonising, Pred 2 is successively out-competed on the grid level and finally goes extinct.

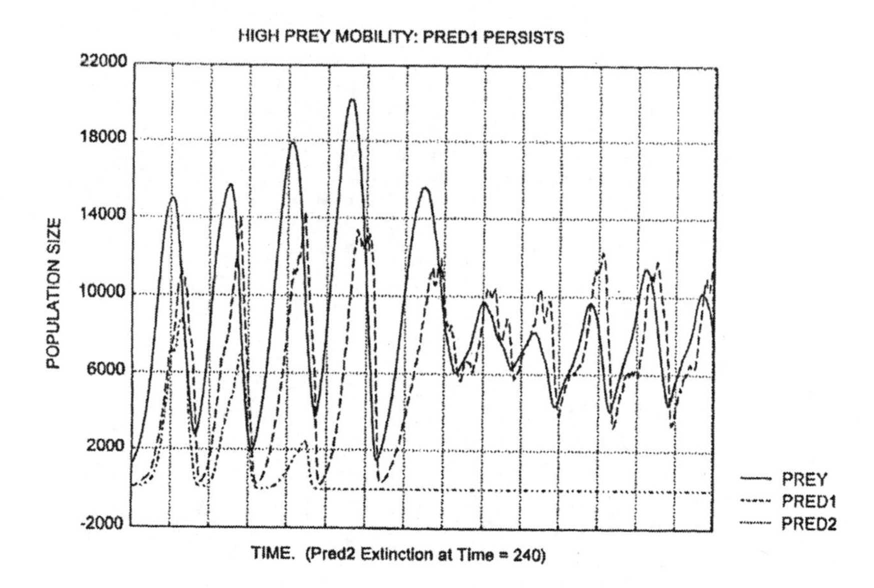

Figure 2b: The overall population density of one example run displayed over time. Pred 1 persists after the extinction of Pred 2.

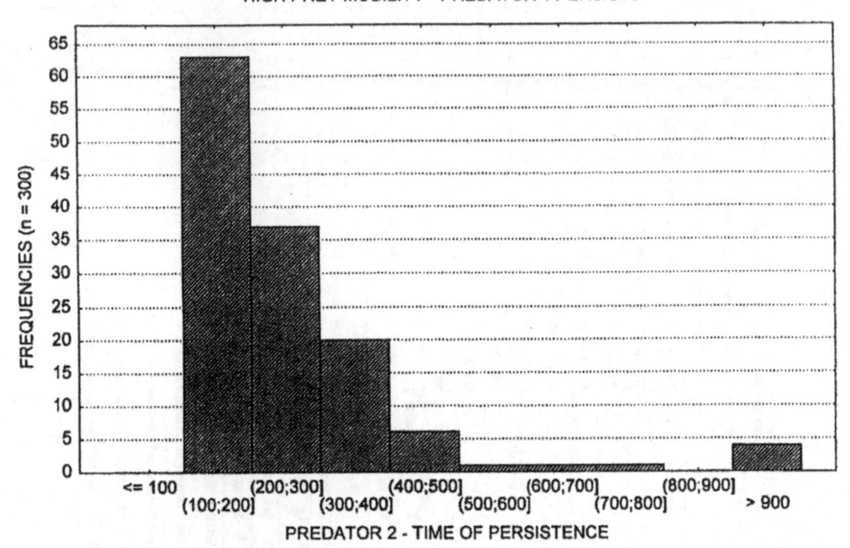

Figure 2c: The updating sequence of the grid cells is chosen at random. For different random seed we find a range of different persistence times of Pred 2 with a peak between 100 < Time < 200. The cumulationof extinction times of pred 2 is shown here for 300 re-runs of the model with different random seeds. Pred 1 usually persists over the whole simulation with only a few exceptions.

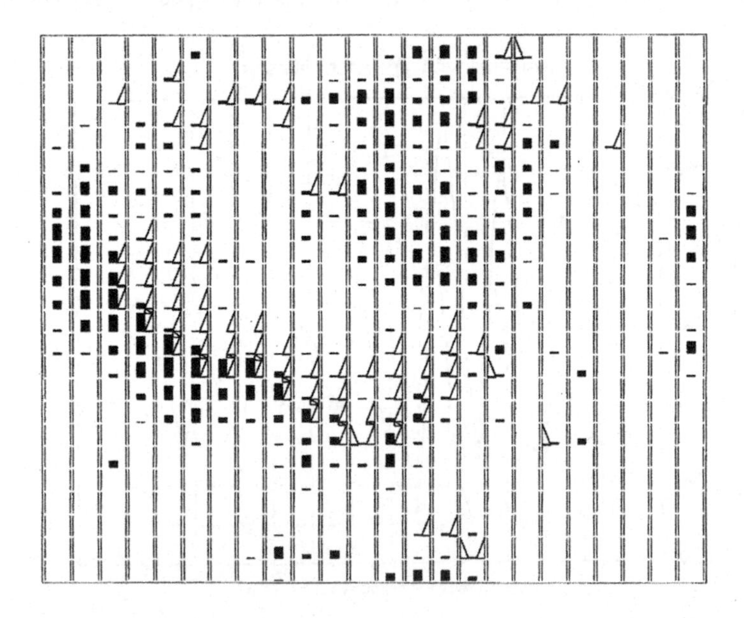

Figure 3: Scenario with a low mobility parameter for prey. Changing only one model parameter, the outcome of the Predator 1 - Predator 2 - competition is reversed.

a: The grid after 700 iterations, shortly before Pred1 goes extinct. The prey distribution pattern is largely different due to a lower colonisation rate. Because the foraging efficiency of Pred2 is only half of Pred1, the weaker predator leaves enough time for the prey to disperse. Pred1 on the other hand as the more efficient forager depletes the prey faster than the prey can spread. Local prey extinctions are not replaced rapidly enough by prey re-colonisation events to allow long term persistence of Pred1. Pred2 can survive in grid regions, which were not depleted by Pred1.

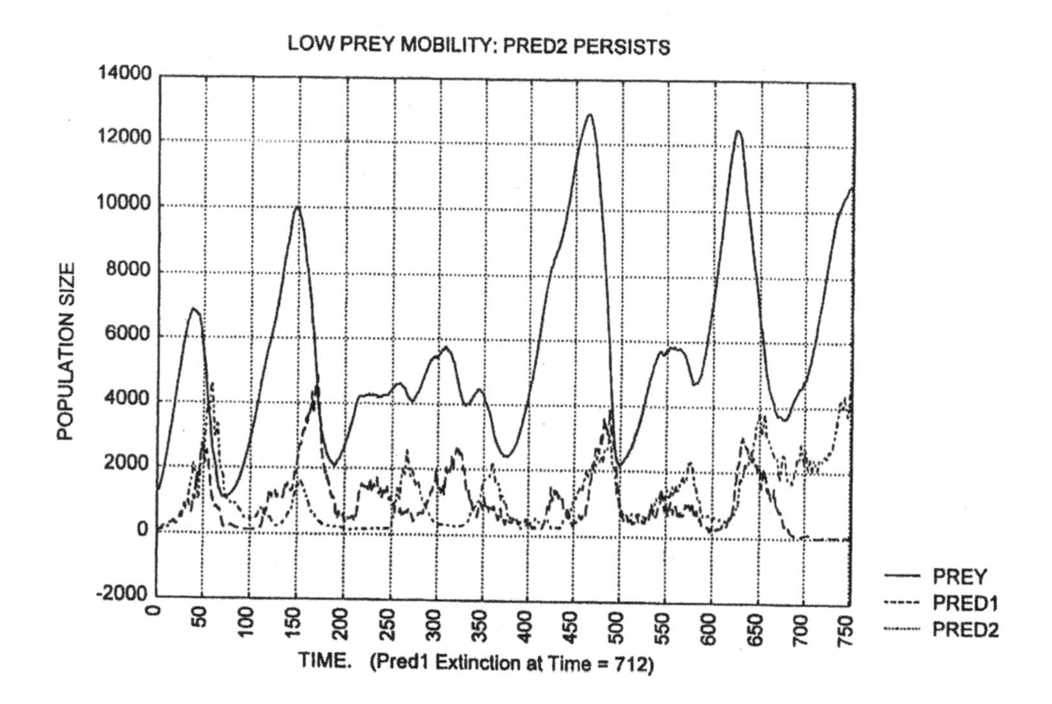

Figure 3b: The overall population density of one example run displayed over time. The extinction of Pred 1 occurs at Time = 712 because of local prey-overexploitation.

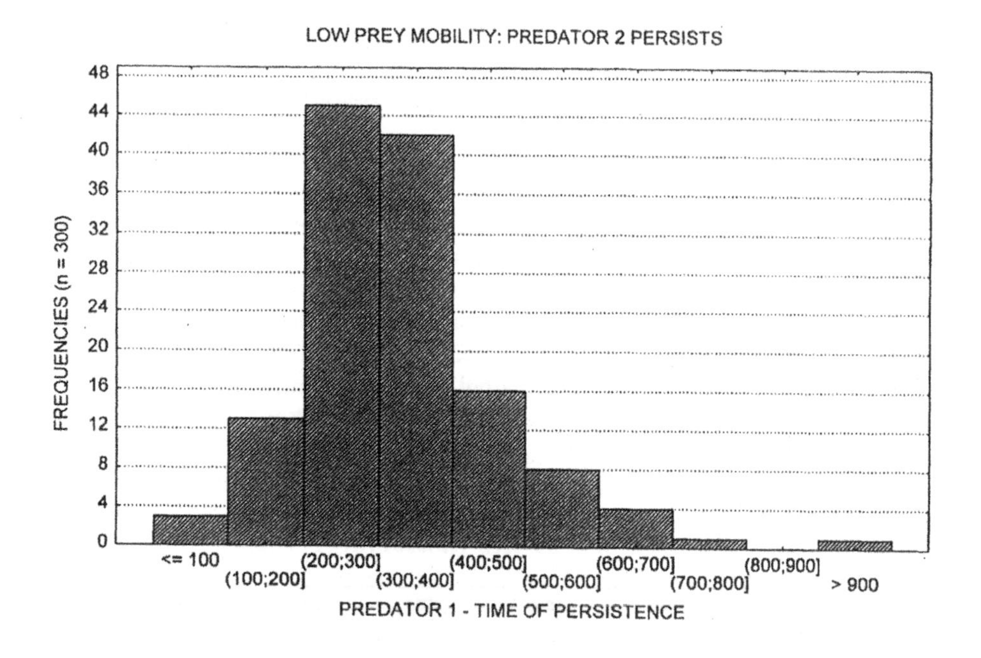

Figure 3 c: The updating sequence of the grid cells is chosen at random. For different random seed, we find a range of different persistence times of Pred1 with a peak between 200 < Time < 300. The cumulation of extinction times of Pred1 is shown here for 300 re-runs of the model with different random seed. Pred2 usually persists over the whole simulation with only a few exceptions.

which type of density dependence will occur. With a low density of the juveniles, the stock-recruitment curve follows the Beverton-Holt pattern. Intra-class interactions dominate and produce this pattern. High juvenile density leads to a Ricker type of stock-recruitment curve. Inter-class competition and cannibalism can bring about much variability in recruitment. After all, the stage-specific densities, instead of overall density, determine survival, growth and future densities. Thus, repeated samples based on an unstructured population model that ignores juveniles and their interaction with age-0 fish would have limited success in reducing uncertainty. Data on stage-specific density and interaction are needed. Our mechanistic modelling approach detected potentially important ignorance on the juvenile fish, identified critical inter-class processes, estimated the inter-class interaction quantitatively, and suggested that a stage-structured approach would be more appropriate for these kinds of populations. Thus, this modelling approach may be a great help to improve our understanding and prediction, by integrating information in diverse research fields, using advanced computational technology (Dong and Polis 1992).

5.4 Using simple models as a base to develop complex realistic models: An integrated modelling approach to Periphyton Dynamics

Ecological systems are complex and quite often demand a complex model to represent it. Complex models describe the system more realistically, summarise more information, investigate more ecological processes, evaluate scenarios of management and engineering plans and examine plans of data collections in more details. However, the marginal value of adding model complexity may decline rapidly with respect to the scientific hypothesis tests and decision-making supports. When the model complexity continues to increase, the ignorance reduction and improvement of mechanistic understanding subside, meanwhile, errors propagate, intangibility of interplay among parameters and between parameters and function forms grows, and handling cost and difficulties increase. Ignorance may still exist in complex models. Data availability often does not match model requisites. It is tempting to include trivial details of our knowledge in models due to their abundance and instructive description (Hilborn and Mangel 1997). Complex models that are based mainly on available data are often biased and may miss some critical processes, if the data distribution is biased. Thus, predictive power of complex models can be lower than the one of simpler models (Ludwig and Walters 1985, Gauch 1993, Hilborn and Mangel 1997).

Building a complex, realistic model is a progressive process. It often starts from a simple one and then grows in complexity. In this process, selection of new components to add in is critical. Careful component selection and structure design can help to reduce the uncertainty caused by both ignorance and errors. Stability analysis and sensitivity analysis are useful for this purpose. These analyses can help to identify the most critical parameters and dominant relations, which contribute more than others to the system behaviours. Below is an example to show how a complex model can benefit from the analysis of a simple model. In a currently ongoing project, Dong and colleagues (in preparation) developed a meta-frame modelling approach to study the periphyton communities and their response to and effect on phosphorus dynamics across the Everglades landscape in south Florida, USA. Periphyton is an attached assemblage of a variety of algal taxa. Periphyton is an important primary producer and serves as a key component of the food web base in this large subtropical freshwater wetland. Periphyton algae provide not only food materials but also a habitat for grazers and detritivores, which in turn support populations of the fish fauna that are the prey of alligators and large flocks of wading birds (Browder et al. 1994). During this century, the Everglades has been degraded by human activities. Phosphorus enrichment is one of the major contributors to this degradation and has led to dramatic changes in the standing crop, community structure, production, and decomposition of periphyton. Currently, the major goals of restoration efforts are to reduce excessive phosphorus loads and to re-establish a water control system that approximates the historically natural condition. Periphyton are also important indicators of phosphorus – thresholds in the Everglades. A good ecological understanding of the periphyton's response and effect can provide critical support to decision making for the Everglades restoration and management (McCormick and Scinto in press).

The first modelling step is to construct a model with a limited structural complexity. We started with only two state variables (periphyton standing crop and phosphorus concentration) and two differential equations. The growth and death of periphyton, phosphorus uptake from water, phosphorus release back to water, and phosphorus deposition by periphyton were simulated. With this model, equilibrium analysis and stability analysis reveal general behaviours of this system. The result suggests that multiple equilibria are likely to exist, and structural instability (catastrophe) may exist in the system (Figure 6). Small gradual changes in phosphorus supply in a certain range may lead to a dramatic change in the standing crop of periphyton. When a simple model demonstrates an unstable situation including a possibility of chaos or catastrophe, the critical components of instability become a major concern in model development. The stability analysis indicates that the detrimental effect of phosphorus on the maximum standing crop of periphyton is a critical destabilising process. Thus, information about the effect of phosphorus on the standing crop of periphyton is needed to reduce uncertainty. Further examination of data from field and mesocosm studies suggested that the standing crop declines as the periphyton community switches from calcareous to non-calcareous, when phosphorus supply increases. The phosphorus uptake and deposition differ with shifts between periphyton communities.

Thus, a more complex model is needed, and has been built on the basis of the simple model, to capture more realism and to understand the interaction. This model explicitly described two types of periphyton communities, their responses to phosphorus status and the feedback of their responses on phosphorus. In addition to capture more realism, inclusion of community switches makes the model more applicable to restoration issues. The community status serves as a criterion of restoration success. This model is still under development. Model complexity continues to increase while the modelling process proceeds.

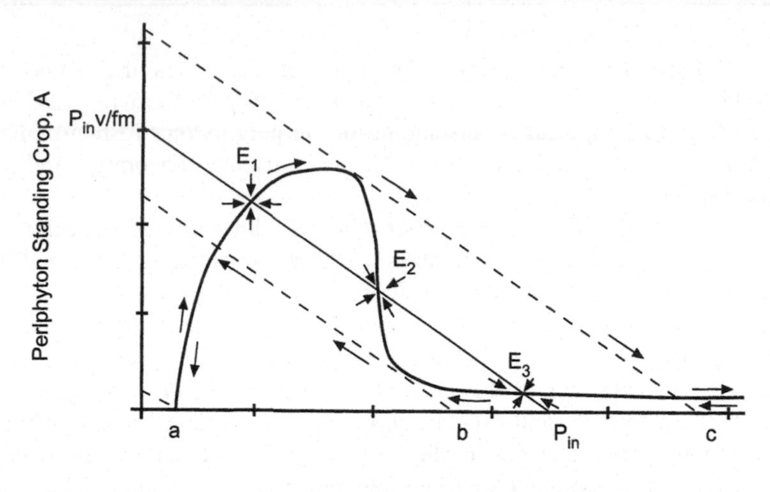

Figure 6: Potential catastrophe in a periphyton dominated slough ecosystem in the Everglades. The interactions of solid lines are equilibria. Stable equilibria are surrounded by small inward arrows from all directions. Outward arrows from an equilibrium to any direction indicate instability. P_{in} is the in-flow phosphorus concentration. Large broken arrows indicate the direction of change as a consequence of the P_{in} increase, and large solid arrows indicate the direction of change with the reduction of P_{in}. Given the P_{in}, the plane is divided into two basins of attraction, one containing all trajectories that move towards the stable node E_1 and one containing all trajectories that move towards another stable node E_3, separated by a saddle point E_2. When P_{in} increases from a very low level, the periphyton standing crop, A, raises and approaches a maximum value. Further increase of P_{in} , even very small beyond the point b, causes a dramatic drop of A, representing the dissolution of the periphyton mat. The state plane suddenly becomes a basin of attraction for a stable equilibrium at low A. Now suppose that efforts are made to reduce P_{in} from a high level. A remains low, as P_{in} is decreased, even after passing b. In fact, one must decrease P_{in} all the way to the point a. Below this P_{in} level, the whole state plane then becomes a basin of attraction for high A, and the system returns to the attractor. In this system, P_{in} is a driving force. A small difference in P_{in} or in the initial conditions could lead to a large difference in the equilibrium state. This behaviour is an example of a structural instability or a 'fold catastrophe', and may have significant

Such a suite of models with a general core is a meta-frame model. We have been developing this modelling approach to answer different questions about the same ecological system. A specific model – a frame – is constructed to addresses a few specific questions. Management or scientific inquiries have different requirements and tolerance about the level and type of uncertainty. These concerns dedicate the uncertainty trade-off selection and thus the model/frame specification. The core concisely summarises our best ecological knowledge about the focal process. Complexity increases in new frames. The high degree of model complexity eventually prevents analytical solutions. Simulations are performed to link ecological processes at different temporal and spatial scales and reconstruct system dynamics. This hierarchical multi-frame modelling design allows us to answer questions progressively along a spectrum of temporal and spatial scales and degree of details and specifications. We actually combined simple analytical models and complex simulation models into one integrative framework to handle trade-offs among ignorance and errors. A general scheme on the iterative process of empirical information and model development is given in Figure 7. Viewing the problem from different angles, using different approaches and different levels of complexity in representation helps to evaluate the different aspects of uncertainty.

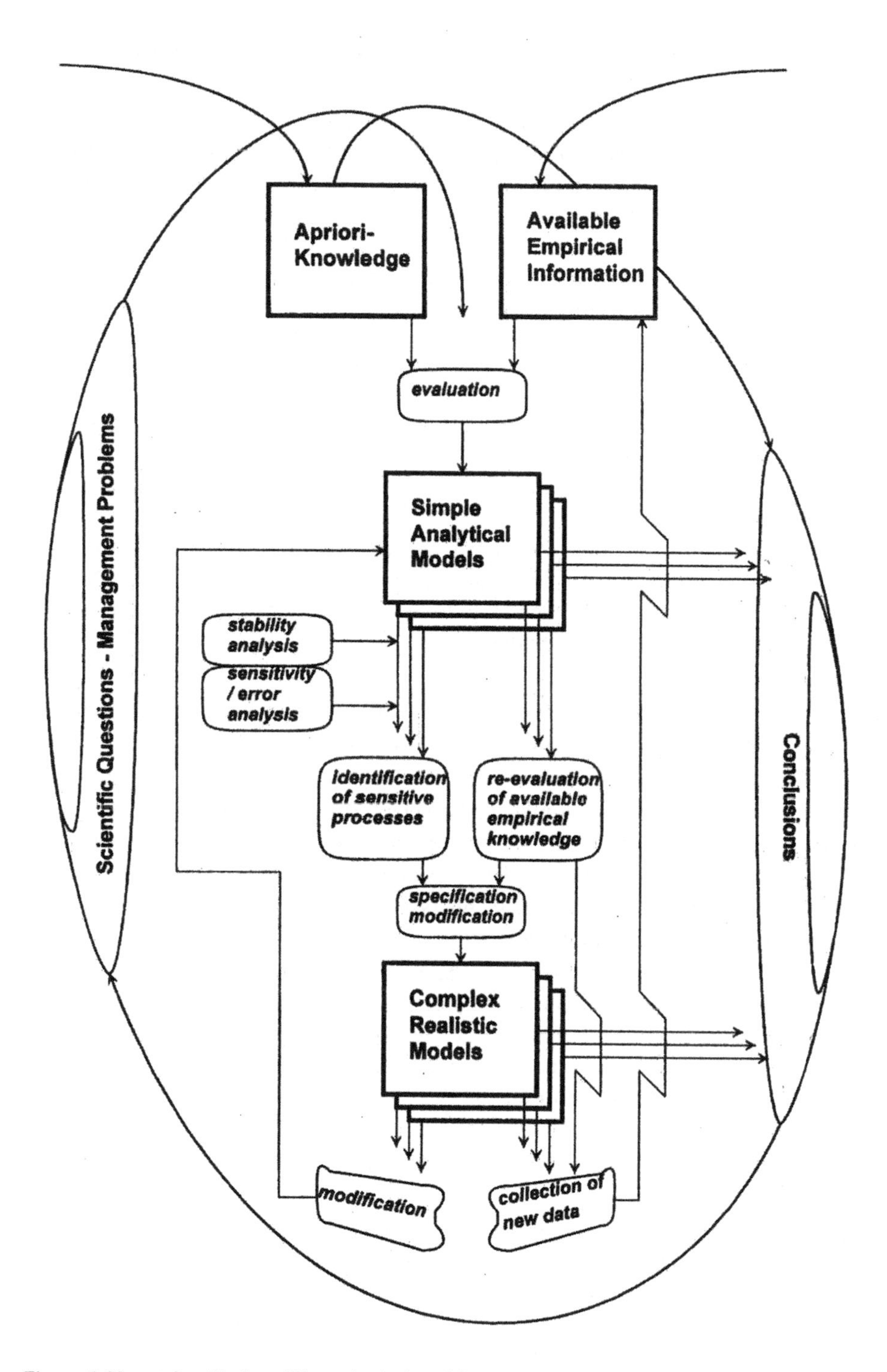

Figure 7: Uncertainty Trade - off in ecological models.
Simple analytical models derived from a priori knowledge and available empirical information can help to specify foci of special attention for more complex, realistic models, where stability analysis, re-evaluation of available knowledge etc. contribute to a more detailed understanding. Complex models can give new hints to capture dominant processes in modified simple models. As a whole, the modelling process can give hints for the acquisition of additional empirical information. Mutual improvements of different model approaches in this iterative cycle can help to reduce the remaining uncertainty. This means, to view an ecological problem from different angles using different approaches may lead to a better understanding than searching for one single model representation

0.05 Pred1_Starv
pred1 death rate without prey available
7.0 Pred1_Mini
pred1 minimum viable population (if pred1 is decreasing)

Parameter for Pred2 alike Pred1, except
0.006 Pred2_Fora pred2 foraging efficiency

REFERENCES

Allen, T.H.F and T.B. Starr, 1982. Hierarchy - perspectives for ecological complexity. Chicago (University of Chicago Press).

Breckling, B., 1992. Uniqueness of ecosystems versus generalizability and predictability in ecology. *Ecological Modeling* 63: 13-27.

Breckling, B., 1991. Variabilität, Kontext-Spezifität und Vorhersagbarkeit im Individuenorientierten Modell. *Verhandlungen der Gesellschaft für Ökologie* 20: 803-813 (Variability, context-specifity and predictability in individual based modeling; in German).

Breckling, B. and H. Reuter, 1996. The use of individual based models to study the interaction of different levels of organization in ecological systems. *Senckenbergiana maritima* 27 (3/6): 195-205.

Beverton, R.J.H. and S.J Holt, 1957. On the dynamics of exploited fish populations. Ministry of agriculture, fisheries and food (London) Fisheries investigating series 2 (19).

Browder, J.A., P.J. Gleason, and D. R. Swift, 1994. Periphyton in the Everglades: spatial variation, environmental correlates, and ecological implications. In: S.M. Davis and J.C. Odgen, (Eds.). *Everglades: the ecosystem and its restoration.* Boca Raton (St. Lucie Press).

Carpenter, S.R., 1996. Microcosm experiments have limited relevance for community and ecosystem ecology. *Ecology* 77: 67-680.

Costanza, R. and F.H. Sklar, 1985. Articulation, accuracy, and effectiveness of mathematical models: a review of freshwater wetland applications. *Ecological Modeling* 27: 45-69.

Crawley, M.J., 1987. What makes a community invasible? In: A.J. Gray, M.J. Crawley, and P.J. Edwards, (Eds.). *Colonization, Succession and Stability.* 26th Symposium of the British Ecological Society p. 429-453.

DeAngelis, D.L. and L. Gross, 1992. Individual-Based Models and Approaches in Ecology - Populations, Communities and Ecosystems. London, New York (Chapman and Hall).

DeAngelis, D.L., 1992. Mathematics: A bookkeeping tool or a means of deeper understanding of ecological systems. *Verhandlungen der Gesellschaft für Ökologie* 21: 9-13.

Dong, Q. and D.L. DeAngelis, 1998. Population consequences of cannibalism and competition for food in a smallmouth bass population: an individual-based modeling study. *Transactions of the American Fisheries Society* 127(2): 174-191.

Dong, Q. and G.A. Polis, 1992. The Dynamics of Cannibalistic Populations: A Foraging Perspective. 1992. In: M.A. Elgar and B.J. Crespi, (Eds.). *The Ecology and Evolution of Cannibalism.* Oxford (Oxford University Press) p. 13-37.

Dong, Q., 1996. Current state and trend of ecological studies in the western countries. *Acta Ecologica Sinica* 16(3): 314-324 (Chinese with English abstract).

Ekschmitt, K., B. Breckling, and Karin Mathes, 1996. Unsicherheit und Ungewissheit bei der Erfassung und Prognose von Ökosystemen. *Verhandlungen der Gesellschaft für Ökologie* 25: 495-500 (Sources of uncertainty and ignorance in the record and prognosis of ecosystem development, in German).

Frank, K. T. and W. C. Leggett, 1994. Fisheries ecology in the context of ecological and evolutionary theory. *Annual Review of Ecology and Systematics* 25: 401-422.

Gardner, R.H., R.V. O'Neill, J.B. Mankin, and D. Kumar, 1980. Comparative error analysis of six predator-prey models. *Ecology* 61(2): 323-332.

Gauch, Jr. H.G., 1993. Prediction, parsimony and noise. *American Scientist* 81: 468-478.

Haefner, J.W., 1996. Modeling biological systems, principles and applications. New York (Chapman and Hall).

Hall, C.A.S., 1988. An assessment of several of the historically most influential theoretical models used in ecology and of the data provided in their support. *Ecological Modeling* 43: 5-31.

Hall, C.A.S. and D.L. DeAngelis, 1985. Models in ecology: paradigm found or paradigm lost? *Bulletin of Ecological Society of America* 66: 339-346.

Hilborn, R. and D. Ludwig, 1993. The limits of applied ecological research. *Ecological Applications* 3(4): 550-552.

Hilborn, R. and M. Mangel, 1997. The ecological detective, confronting models with data. Princeton, New Jersey (Princeton University Press).

Hurlbert, S.H., 1984. Pseudoreplication and the design of ecological field experiments. *Ecological Monographs* 54: 187 - 211

Jørgensen, S.E., 1988. Fundamentals of ecological modeling (2nd. ed.). Amsterdam (Elsevier).

Jones, C.G. and J.H. Lawton (Eds.), 1995. Linking species and ecosystems. New York (Chapman and Hall).

Krebs, C. J., 1978. Ecology: the experimental analysis of distribution and abundance (2nd ed.). New York (Harper and Row).

Lemons, J. (Ed.), 1996. Scientific Uncertainty and Environmental Problem Solving. Cambridge (Mass) (Blackwell).

Loehle, C., 1983. Evaluation of theories and calculation tools in ecology. *Ecological Modelling* 19: 239-247.

Lotka, A.J., 1924. Elements of Physical Biology. Baltimore (Williams and Wilkins).

Lubchenco, J., 1998. Entering the century of the environment: a new social contract for science. *Science* 279: 491-497.

Lubchenco, J., A.M. Olson, L.B. Brubaker, S.R. Carpenter, M.M. Holland, S.P. Hubbell, S.A. Levin, J.A. MacMahon, P.A. Matson, J.M. Melillo, H.A. Mooney, C.H. Peterson, H.R. Pulliam, L.A. Real, P.J. Regal, and P.G. Risser, 1991. The sustainable biosphere initiative: an ecological research agenda. *Ecology* 72(2): 371-412.

Ludwig, D. and C.J. Walters, 1985. Are age structured models appropriate for catch-effort data? *Canadian Journal of Fisheries and Aquatic Sciences* 42: 1066-1072.

Ludwig, D., R. Hilborn, and C. Walters, 1993. Uncertainty, resource exploitation, and conservation: Lessons from history. *Science* 260: 17-36 (2. Apr. 1993).

May, R., 1974. Biological populations with non-overlapping generations: Stable points, stable cycles, and chaos. *Science* 186: 645-647.

May, R., 1979. Theoretical ecology, principles and applications. Cambridge (Mass.) (Blackwell).

McCormick, P.V. and L.J. Scinto (In press). Influence of phosphorus loading on wetlands periphyton assemblages: a case study from the Everglades. In: K.R. Reddy, (Ed.). *Phosphorus Biogeochemistry in Florida Ecosystems.*

McCullough, D.R., 1992. Concepts of large herbivore population dynamics. In: D.R McCullough and R.H. Barrett, (Ed.). *Wildlife 2001: populations.* London (Elsevier Applied Science). p. 967-984.

Müller, F., 1996. Emergent properties of ecosystems - consequences of self-organizing processes? *Senckenbergiana maritima* 27 (3/6): 151-168.

O'Neill, R.V., D.L. DeAngelis, T.F.H. Allen, and J.B. Waide, 1986. A hierachical concept of ecosystems. Monographs in Population Biology 23. Princeton, NJ (Princeton University Press).

O'Neill, R.V. and R.H. Gardner, 1979. Sources of uncertainty in ecological models. In: B.P. Zeigler, M.S. Elizas, G.J. Klir, and H.I. Oren, (Eds.). *Methodology in systems modeling and simulation.* Amsterdam (North-Holland Publishing Co.). p. 447-463.

Odum, H.T., 1983. Systems ecology: An introduction. New York, Wiley

Pahl-Wostl, Claudia, 1995. The dynamic nature of ecosystems - Chaos and order entwined. New York, Wiley.

Patten, B.C. (Ed.), 1971. Systems analysis and simulation in ecology. Vol. 1. New York.

Patten, B.C. (Ed.), 1972. Systems analysis and simulation in ecology. Vol. 2. New York.

Patten, B.C. (Ed.), 1975. Systems analysis and simulation in ecology. Vol. 3. New York.

Peters, R.H., 1992. A Critique for Ecology. Cambridge (Cambridge University Press).

Prigogine, I., 1982. Vom Sein zum Werden - Zeit und Komplexität in den Naturwissenschaften (3rd ed.). München (Piper). Orig.: From being to becoming - Time and complexity in Physical sciences.

Prigogine, I. and Isabelle Stengers, 1984. Order out of Chaos, Man's new dialogue with nature. New York (Bantam).

Ricker, W.E., 1954. Stock and recruitment. *Journal of the Fisheries Research Board of Canada* 11: 559-623.

Roughgarden, J., R.M. May, and S.A. Levin (Eds.), 1989. Perspectives in Ecological Theory. Princeton, NJ (Princeton University Press).

Shepherd, J.G. and D.H. Cushing, 1990. Regulation in fish populations: myth or mirage? *Philosophical Transactions of the Royal Society of London* B 330: 151-164.

Toffoli, T. and N. Margolus, 1987. Cellular automata machines. Cambridge (Mass.), London (MIT Press).

Volterra, V., 1926. Variations and fluctuations of the numbers of individuals in animal species living together. Reprinted in: R.N. Chapman, 1931. Animal Ecology. New York (McGraw Hill).

Webster, N., 1983. Webster's New Twentieth Century Dictionary (2nd ed). New York (Simon and Schuster).

Werner, E.E. and J.F. Gilliam, 1984. The ontogenetic niche and species interactions in size-structured populations. *Ann. Rev. Ecol. System* 15: 393-425.

I.5 Fundamentals of Ecosystem Theory from General Systems Analysis

Albrecht Gnauck

1. Introduction

Within the historical specialisation and interpretations of ecology, different conceptions of systems theory play an important role (Milsum, 1966; Bertalanffy, 1968; Morowitz, 1968; Rosen, 1970; Patten, 1971; Peschel and Wunsch, 1972; Jeffers, 1978; May, 1981; Straškraba and Gnauck, 1985; Jørgensen, 1992; Pahl-Wostl, 1995; Patten and Jørgensen, 1995; Mathes et al., 1996). Fleishman (1982) referred to this phenomenon as the system period of science. Within ecology this process led to new disciplines including systems ecology (Patten 1971, Halfon 1979, Innis and O'Neill 1979, Odum 1983), geographical ecology (MacArthur 1972), theoretical ecology (May 1974), dynamic ecology (Collier et al. 1974), cybernetic ecology (Straškraba 1980), network ecology (Patten, 1991; Allen and Hoekstra,1992; Pahl-Wostl, 1995), hierarchical ecology (Wiegleb, 1996; Müller, 1997) and others. Two hundred years ago Kant (1781) stated that the scientific degree of a special theory of nature depends on the usage of mathematics within this theory. The analysis of general or abstract systems is based on mathematical system theory. Such types of systems will be understood as invariants of special classes of concrete systems or will be equated with these classes. The mathematical analysis of an abstract system leads to knowledge which will be valid for all concrete systems of the same class. For complex systems (e.g. ecosystems), mathematical systems theory is a suitable tool for generally describing the interrelations between the system and its environment and within the system itself (Peschel, 1965; Klir, 1969; Mesarovic et al., 1970; Patten, 1971).

In the first phase, mathematical system theory applied to ecology was understood as dynamic statistical ecology (Thomann, 1967; Pielou, 1969; Patil et al., 1971). Data analysis and ecosystems modelling were carried out by use of statistics and time series analysis methods. Later on, in a second phase, mathematical ecology shifted more to systems analysis. This was facilitated by the understanding of ecosystems as parts of higher units, thus allowing a more generalised theoretical description (Patten, 1971, 1972, 1974, 1976; Innis and O'Neill, 1979; Halfon, 1979; Odum,1983; Allen and Hoekstra, 1992; Jørgensen, 1992 and Patten and Jørgensen, 1995). The notions of ecosystems as linear or non-linear, time-independent or time-dependent, spatial, multidimensional, static or dynamic, distributed or lumped, digital or analogous, and other types of mathematical descriptions are theoretically founded, developed and used independently from each other in different ecological disciplines. However, in most cases these special spatio-temporal systems exhibit the same mathematical structural behaviour, they overlap each other, or they are special cases of the same theoretical system. From the present point of view there are some differences within the formal structures of general systems descriptions. But there is some evidence that they may be reconciled in the future.

The application of linear or linearised general system concepts to control information processes in ecology led to a better understanding of ecological theory (Bossermann and Ragade, 1982 Duckstein and Klimontovich, 1984). A general system description used in state space theory (Zadeh and Desoer, 1963), dynamical system theory (Mesarovic and Takahara, 1975) and automata theory (Gill, 1962; Gluschkow, 1963) is given by the input transfer function **F** and the output function **G**:

$$\mathbf{F}: X \times U \rightarrow X \text{ with } x = \mathbf{F}(x, \mu) \quad \text{and} \quad \mathbf{G}: X \times U \rightarrow Y \text{ with } y = \mathbf{G}(x, \mu)$$

The ecosystem behaviour is given qualitatively by non-linear and non-stationary multiple operators **F** (input operator, state transition operator) and **G** (transfer operator, output operator, evolution operator) where **F** transforms the changes of inputs into changes of states, and **G** transforms the changes of states into output changes under consideration of internal and external disturbances which are connected with the general coupling structure of the system. The disturbances are given implicitly and/or explicitly. Such a transfer system exhibits different working states. Each state is responsible for a certain transformation of the set of input signals into the set of output signals. Because of changing states during signal processing within the system, a state description will represent a compressed recording of the system history.

The quantitative ecosystem behaviour is represented by mathematical models which can be simple algorithms (e.g. linear regression equations) or complicated systems of non-linear equations (e.g. systems of non-stationary partial differential equations) or Boolean equations (e.g. binary decision trees). From the point of view of systems theory the following modelling procedure is valid (cf. Jørgensen 1986):

1. Problem analysis (definition of the ecosystem to be modelled, definition of ecosystems environment)
2. Conceptualisation (objectives and goals of modelling)
3. Model structure formulation (choice of essential ecosystem variables, expected variances of inputs, states, disturbances, outputs)
4. Model calibration and parameter estimation (comparison with model structure, changes of the set of variables, a priori parameters, changes of system constants, changes of model structure)
5. Model validation (checking the general validity of the model)
6. Usage of the model for the same type of ecosystem.

Looking at ecosystems as controllable transfer systems (Reinisch, 1974, 1982), the general notions of automata and optimality are of basic interest (Stugren, 1978). In general, automata represent transfer systems. Their dynamic behaviour is determined in each time stroke and spatial co-ordinate by relations between the elements of the sets of input variables $\{U(\mathbf{R}, \rho, t)\}$, output variables $\{Y(\mathbf{R}, \rho, t)\}$ and state variables $\{X(\mathbf{R}, \rho, t)\}$ where $\mathbf{R}$ indicates a non-linear relationship, $\rho = \rho(\underline{u}, \underline{x}, \underline{y})$ is the spatial co-ordinate and t is the time co-ordinate (Gluschkow, 1963; Arbib, 1968; Starke, 1969). Locke (1984) formulated five classes of systems which are important for general system theory:

1. Space-time-systems (cellular systems, cellular automata)
2. Automata (linear automata, sequential automata, superponable automata, combinatorical automata)
3. Infinite dimensional systems (systems with distributed parameters, systems with dead time behaviour, steady state systems)
4. Linear time-dependent systems (linear differential systems, finite systems, discrete linear systems)
5. Non-linear differential systems (analytical systems, superponable systems, bilinear systems)

Within applied ecology three main lines of generalisation coming from general system theory can be distinguished. The first line is given by set theory. This type of generalisation is widespread elaborated and mostly used in ecology (Bertalanffy, 1968; Patten, 1971; Menshutkin, 1974; Nisbet and Gurney, 1982; Conrad, 1983; Jørgensen, 1986; Bossel 1992). An ecosystem $\boldsymbol{E}$ is defined by some weakly structured sets A, B, C, ..., and by some functions f, g, h,... where A, B, C,... are sets of signals, sets of biocoenotic components, sets of states, sets of matter, energy and information storage and others. The functions represent the couplings between the elements of these sets. The formal ecosystem behaviour is described by mapping these sets or cross-products of these sets into the set $E = (A, B, C, ..., f, g, h, ...)$.

The second type of generalisation is given by theory of mathematical algebra. An ecosystem E is represented by some general algebras A, B, C,... and by morphisms ϕ, γ, ψ,...which carry over some algebras of the system to other algebras without changes to their structural properties. The formal description is given by $E = \{A, B, C, ..., \phi, \gamma, \psi, ...\}$. This approach takes into account that the ecosystem defining sets are structured by couplings of the elements between different ecosystem compartments. Applications are well known in the fields of theoretical biology and ecology (Milsum, 1966; Morowitz, 1968; Laue, 1970; Rosen, 1971; Jones, 1973; Florkin, 1977; Lewis, 1977; Jeffers, 1978; Gibbons, 1985).

The third type of generalisation of ecosystems deals with the mathematical theory of categories (Gnauck and Straškraba, 1980; Takahara, 1981; Klir, 1985). An ecosystem E is defined as a category C with sets as objects. The morphisms of C are functions with structure keeping behaviour.

The formal description is given by $C = \langle S_1, S_2, ..., Mor(S_1, S_2), Mor(S_1, S_3), ...\rangle$, where S_i are subsystems with some common properties and $Mor(S_i, S_j)$,... are sets of morphisms. The subsystems of C are also categories which are imbedded in C. The interrelations between categories are realised by functors Φ. If C is a stochastic category, then the morphisms are stochastic functions with discrete measures of probability as values. Levich (1982) formulated a category concept of population biology, while Gnauck (1987) developed a description of limnological ecosystems on the base of a stochastic category.

2. Classification of ecosystem abstractions

Generally, dynamic systems (or models) can be divided into the classes of axiomatic systems and empirical systems. Axiomatic systems are always derived from the equations of systems motion. In a wider sense they may be interpreted as generalisations of Newtonian laws (Jørgensen 1992). The systems belonging to the second class are representations of system knowledge from experiments which produce data of the input-output-behaviour (black box models) or on the cause-effect relationships of ecological processes (analytical or causal models).

When applying methods of systems theory to ecological problems two main tasks arise:

1. Abstraction and quantification of ecological processes and ecosystems by means of mathematical systems theory.
2. Generalisation of the modelling results and elaboration of new system theoretical structures to obtain a theoretical framework of ecology.

The solution of these tasks is connected with the analysis of dynamic system characteristics. Specifically, the theoretical work of Kalaman et al. (1967) on such dynamic characteristics like observability, controllability, reachability, reconstructability and disturbability, plays a central role in ecosystem dynamics. Developments in ecosystem theory, as well as practical problems of ecosystem health and ecotechnology, are influenced by these characteristics. They led to the development of different mathematical concepts in ecology (May 1974, 1981, Svirezhev und Logofet 1978, Dubois 1981, Pahl-Wostl 1995). They differ in the classes of disturbances, in the set of admissible initial states and in the set of system responses related to each time interval. Table 1 gives an overview on essential dynamic characteristics referred to in systems theory.

The motion of an ecosystem, ie. the time variations of its state variables, may be determined by the initial state $\underline{x}(0)$ and by solutions of the time dependent systems equations. In the case of constant system parameter values, only stationary motions are described mathematically by linear and non-linear models. Because of internal interrelationships between some system variables, auto-oscillations will arise. They correspond to periodic solutions of the system equations (Nicolis und Prigogine 1977, Nisbet and Gurney 1982). External influences on ecosystems lead to changes of frequencies and amplitudes of the system signals. An evaluation of the motions of the system by special system theoretical measures, for example degree of stability, damping factor or quadratic error integral, is valid only for linear or linearised systems.

Table 1: Dynamic characteristics of ecosystems

characteristic	comment
observability	analysis of an ecosystem state $\underline{x}(t)$ for time $t = t_0$ by measurement of output variables $\underline{y}(t)$ and input variables $\underline{u}(t)$ for all time strokes $t_i > t_0$.
controllability	investigation of the actual ecosystem state $\underline{x}(t_0)$ to control it in a new state $x(t_i)$ for all $i > 0$.
reachability	investigation of the ecosystem state $\underline{x}(t)$ for time t_0 if it can be reached from a stationary state $\underline{x}(t_0)$ by a suitable control series of inputs $\underline{u}(t)$
reconstructability	investigation if the ecosystem state at time t_0 can be determined or reconstructed by measurements of input and output variables
disturbability	investigation if a stationary ecosystem state $\underline{x}(t)$ at time t_0 can be shifted by a suitable series of disturbances (inputs) $\underline{r}_i$ (t) to a new state $\underline{x}(t_0)$ and to answer the question if the new state is a stationary one
stability	investigation if the motions of a nonlinear system will return asymptotically to the stationary initial state after disturbances stopped
sensitivity	investigation of motions of state variables by changing of input and disturbance variables

The same is true for auto-oscillations where exact measures of stability cannot be given. For linear systems, evaluations of dynamic characteristics can be presented by rank numbers of system matrices (Ludyk, 1981; Reinisch, 1982). The algebraic criteria formulated by Kalman et al. (1967) allow only *yes* or *no* answers. When calculating the rank orders of complex systems some numerical difficulties arise (Müller and Lückel, 1984). Consistent measures for evaluating dynamic characteristics of non-linear time-variant systems are still unsolved problems (Göldner and Kubik, 1978; Litz, 1983; Schwarz, 1991). The extended graph theoretical evaluation methods developed by Wassel (1978) for controllability problems, and by Söte (1980) for observability problems will not substitute numerical investigations of ecosystem dynamic characteristics. State descriptions of complex systems require classifications related to dynamic properties of the system under investigation, to the transfer behaviour, and to the coupling structure of subsystems within the ecosystem. The non-linear feedbacks within an ecosystem are causes that change the characteristics of input signals and ecosystems states during signal transfer processes by modulation and/or by quantifiabilisation.

In Table 2 some mathematical descriptions from general system theory applied in ecology are listed.

Therefore, a general ecosystem classification is possible not only by its mathematical abstraction, but also by its characteristics of signals and the type of change in dynamic ecological properties (Table 3).

Table 2: General classes of ecosystem models

class of systems	state equation
Space-time-systems	
cellular systems	$u(\underline{x}, t) = f(\varphi_t, \psi_t, t)$
cellular automata	$u(\underline{x}t+1) = f(\varphi(t), \psi(t), t)$
Automata	
linear automata	$u(t+1) = A(u(t) + Bx(t))$
sequential automata	$u(t+1) = f(u(t), x(t))$
superposable automata	$u(t+1) = Au(t) \bullet Bx(t)$
Infinite-dimensional systems	
linear systems with distributed parameters	$\partial u(t) = Au(t) + Bx(t))\partial t$
time delay systems	$\partial u(t) = Au(t) + \Sigma a_i\, u(t-t_i) + Bx(t)^i$
steady-state system	*special partial differential equations*
Linear time-dependent system	
linear differential systems	$\partial u(t) = Au(t) + Bx(t)$
finite systems	$u(t) = G(x(t))$
discrete linear systems	$u(t+1) = Au(t) + Bx(t)$
Nonlinear differential systems	
analytical systems	$du(t) = f(u(t) + x(t)h(u(t))$
superposable systems	$du(t) = Au(t) \bullet Bx(t)$
bilinear systems	$du/t) = Au(t) + x(t) \cdot Bx(t)$

3. General dynamic systems

Bossel (1992) defined a system as a set of interrelated objects that have certain general properties given by its function, by its elements and structure, and by its non divisible identity. When generalising these outlines in general dynamic systems theory, three basic types of dynamic systems may be distinguished: The general input-output system, the general state system and the abstract automaton (Gluschkow, 1963; Kalman et al., 1967; Starke, 1969; Windeknecht, 1971; Klir, 1972; Mesarovic and Takahara, 1975; Brauer, 1979).

The general dynamic input-output system is defined as a relation $\boldsymbol{S}$ between the set $U(T)$ of input variables and the set $Y(T)$ of output variables where U, Y, T are non-empty sets and $\boldsymbol{S} \subseteq (U(T) \times Y(T))$. U is called input alphabet and Y is called output alphabet. $T \in \{ \Re \}$ is the set of all time strokes which is ordered by the < - relation with $R : T = \{ t \mid t_0 \; R \; t \} \cup \{ t_0 \}$.

The relation R is characterised by the following rules:

1. $\sim (t R t)$	$\forall t \in T$
2. $t R t' \wedge t' R t'' \Rightarrow t R t''$	for any $t, t', t'' \in T$
3. $t \neq t' \Rightarrow t R t' \vee t' R t$	for any $t, t' \in T$.

The sets $U(T)$ and $Y(T)$ denote the sets of all representations of the set T into the sets U and Y with $U(T) = \{ u \mid u: T \rightarrow U \}$ and $Y(T) = \{ y \mid y: T \rightarrow Y \}$. $u(t)$ and $y(t)$ are called time functions. $\boldsymbol{S}$ is defined on the set of all admissible input functions:
$D(\boldsymbol{S}) = \{u \mid \exists \, y \, (y \in Y(T) \wedge (u, y) \in \boldsymbol{S})\}$ with values defined on $V(\boldsymbol{S})$ with
$V(\boldsymbol{S}) = \{ y \mid \exists \, x \, (x \in U(T) \wedge (u, y) \in \boldsymbol{S})\}$. In general holds $D(\boldsymbol{S}) \subseteq U(T)$ and $V(\boldsymbol{S}) \subseteq Y(T)$.

Table 3: Classification of ecosystems by the type of dynamical characteristics

classification	**comment**
Characteristics of signals	
modulation of ecological signals	change of amplitudes, frequencies and phases of signals
quantifiability	time domain discretisation of the amplitudes and of the whole time courses of ecological signals
Adaptability of system	
adaptive	change of ecosystems state in dependence of available internal and external information on system variables, change of inputs and disturbances, change of parameters, change of ecosystem structure
non-adaptive	fixed parameters, no change of the ecosystem structure

A general dynamic state system $\boldsymbol{S}$ is defined as an algebraic structure $\boldsymbol{S} = (U, Y, X, U(T), Y(T), T, F, G)$ with the following characteristics:

1. U, Y, X are non-empty sets of inputs, outputs and states,
2. $T \in \{\Re\}$ a non-empty set of time strokes
3. $U(T)$ and $Y(T)$ the sets of admissible input and output functions,
4. F the set of transfer functions which changes the states of the system by input functions,

5. *G* the set of output functions which gives out the system response due to the action of the input functions.

The introduction of special system characteristics into the system definition leads to special classes of dynamic systems used in ecology.

1. ***S*** is called time-invariant, if $\forall t_0, t_1 \in T: f(t_0) = f(t_1)$ and $g(t_0) = g(t_1)$.
2. ***S*** is called a continuous system if $T = \Re$.
3. ***S*** is called a discrete system if $T = \Gamma$ (set of integers).
4. ***S*** is called a linear system if the sets *U*, *Y*, *X* are based on linear spaces.

The abstract (stochastic) automaton is defined as a special case of an algebraic structure. The quintupel $A = (U, Y, X, f, g)$ is an abstract determined automaton where $u \in U$ is the set of inputs, $y \in Y$ is the set of outputs and $x \in X$ is the set of (inner) states. *f* is a definite representation of $X \times U$ into *X* and g is a definite representation of $X \times U$ into *Y*. The automaton ***A*** works in a discrete time scale with innumerable-infinite time strokes t = 1, 2, 3,.... This automaton works in the following manner: If $x(t)$ is the state of ***A*** at time *t* and $u(t)$ is the input signal at time *t*, then the automaton ***A*** gives out the signal $y(t) = g(x(t), u(t))$ at time *t* and gets the state $x(t+1) = f(u(t+1))$. This type of automaton is called Mealy-automaton. When the output signal of A depends only on the state it is called a Moore-automaton. Starke (1969) summarised the representations *f* and *g* to a function γ defined on $\{X \times U\}$ with values in $\{Y \times U\}$. The transfer function and the output function of ***A*** are interpreted as an ordered pair $\gamma(x, u) = [\lambda(x, u), \delta(x, u)]$.

The analysis and control of complex systems are in the most cases difficult to survey (Casti, 1979; Ludyk, 1979). While statements on dynamic characteristics of ecosystems have to consider a lot of working states, the quality of mathematical system analysis depends on the flexibility of models used. Therefore, stochastic descriptions of the ecosystem structure and its transfer behaviour form a base of possible generalisations. Ecological processes and biocoenotic-structural components of ecosystems can be characterised by different time parameters, for example time delay, threshold values, altering, physiological parameters and others. Because of the random changes of internal ecosystem variables and of fluctuations of environmental variables with often high frequencies, the starting time strokes of changes of state variables will vary. This means that within the time courses of input variables switching processes take place at different time strokes (Pospelov, 1973). The resulting state transitions will then take place on intervals $(a_i(t), b_i(t))$, with probability densities $w_i(t)$ of time delays of ecosystem variables, and probabilities $p_i(t)$ for each realisation of a state transition:

$$p_i(t) = \int_{a_i(t)}^{b_i(t)} w_i(t)dt$$

Time delays in the courses of action of system components lead to retardations in the changes of ecosystem states. State transitions can be observed after a certain time interval only. A state transition within an ecosystem is then characterised by a quadrupel $\Theta_i(t) = \{a_i(t), b_i(t), w_i(t), p_i(t)\}$. Interpreting an ecosystem as a complex switching network (Gnauck 1987) changes of ecosystem states may be formulated by stochastic input and internal feedback variables. This is a general interpretation of the stochastic system transfer function. The stochastic output function of the ecosystem is then given by stochastic internal feedback and output variables which can be formulated as recurrent Boolean time functions. An evaluation of such stochastic processes is possible by specification of probabilities of any hazards to the system defining variables. Especially, the time constants of these variables play an important role.

4. Stochastic automata concept of ecosystems

Following the above mentioned statements from general system theory, ecosystems may be considered as stochastic Mealy-automata (Gnauck, 1976; Gnauck and Straškraba, 1980). The whole ecosystem and its subsystems will be described as a stochastic automata $A = [U, Y, X, H]$, called S-automata, whose changes of states are given by discrete measures of probability. U (input alphabet), Y (output alphabet) and X (state alphabet) are non-empty sets with $u \in U$ - input signal, $y \in Y$ - output signal and $x \in X$ - state signal. H is a function defined on $(X \times U)$ with values $H(x, u)$ which are discrete measures of probability. The transfer function of A is defined on $(X \times U)$, their values are discrete measures of probability: $F(x, u)(X^*) = H(x, u)(Y \times X^*)\ \forall X^* \subseteq X$. The output function of A is defined on $(X \times Y)$, their values are discrete measures of probability: $G(x, u)(Y^*) = H(x, u)(Y^* \times X)\ \forall Y^* \subseteq Y$. A works on a discrete time scale with infinite innumerable time strokes $t_1, t_2, \ldots$. For each time stroke: A gets an input signal, releases an output signal and has exactly one well-defined state. If $x \in X$ is the state of A at time t and $u(t) \in U$ an input signal, then for each subset $S \subseteq (Y \times X)$ the value $H(x, u)(S)$ is a discrete measure of probability for the set S. For the output signal $y(t)$ and the state $x(t+1) = x^*$ the relation $(y, x^*) \in S$ with probability $H(x^*, u)(y, x^*)$ holds. Because ecosystems are observable systems A is considered as an observable stochastic automata. Its state at time $t+1$ will be described by the situation $(x(t), u(t))$ at time t. The output signal $y(t)$ is then given by the equation $y(t+1) = \delta(x, u, y)$. Otherwise, if A is observable then there is exactly one function F with $x(t+1) = F(x, u, y)$.

The functioning of a stochastic automata is given by the set of input variables $\{U\}$ which act on the set of state variables $\{X\}$ by means of transfer functions F_i The state variables will be changed during these actions. The feedback of this changing system behaviour or the ecosystems response to its environment will be realised by an output function G_i on the set of output variables $\{Y\}$. A works stepwise in a time scale which consists of a series of different time intervals produced by internal ecological processes and by external forcing functions. Therefore the ecosystem behaviour cannot be determined by observations uniquely or by deterministic automata with a unique function H.

In automata language time series of values of defining variables may be called words. $w(u) \in W(U)$, $w(y) \in W(Y)$ and $w(x) \in W(X)$ denote sets of words produced by the defining variables. To explain the macroscopic behaviour of an ecosystem the model of a non-deterministic or stochastic automata must be used. Summarising the statements of automata theory, the following definitions are useful for general ecosystem descriptions:

1. *Stochastic automaton* A:

$$H[x,w(u)](Y' \times X') = G[x, w(u)](Y') \cdot F[x, w(u)](X') \text{ with } X' \subseteq X \text{ and } Y' \subseteq Y.$$

2. *Transfer function of* A:

$$F[x, w(u)](X') = H[x, w(u)](Y' \times X') \text{ with } X' \subseteq X.$$

3. *Output function of* A:

$$G[x, w(u)](Y') = H[x, w(u)](Y' \times X) \text{ with } Y' \subseteq Y.$$

4. *Global internal and external behaviour of* A:

$$H[x, w(u)](W^*) = \sum_{(y,w(x)) \in W^*} x(X) H[x, w(u)](W^*) \quad \textit{with} \quad W^* \subseteq W(Y) \times W(X)$$

5. *Input-output-behaviour of* A:

$$B[x, w(u)](Q) = H[x, w(u)](Q \times W(X)).$$

6. *Internal change of state of* A:

$$F[x, w(u)](X') = H[x, w(u)](W(Y) \times W(X)X').$$

7. *Imbedding of an automata* A_i *into an automata* A:

$$B[x_i, w(u)] (W(Y)) = B [x_i, w(u)_i] (W(Y)).$$

8. *Loop operation L_i within A*:
   $(Y \times U) \rightarrow X$ defined on $(Y \times U)$ with $L(y, u_i) = F(x, u_r) = x$ for $u_i \neq u_r$.

For ecological purposes it is necessary to combine stochastic automata (e.g. combination of subsystems within a trophic level, combination of different trophic levels and others). This operation can be done by the direct summation of automata.
If $A = [U, Y, X, H]$, $A' = [U, Y, X' ,H']$ stochastic automata with disjunct sets of states. Then one gets for the direct sum
$A + A' = [U, Y, X \cup X', H^*]$ with $u \in U, \forall x^*, x^{**} \in X \cup X', \forall y \in Y$ and

$$H^* [x^*, u] (y, x^{**}) = \begin{cases} H [x^*, u] (y, x^{**}), & \text{if } x^*, x^{**} \in X \\ H' [x^*, u] \ (y, x^{**}), & \text{if } x^*, x^{**} \in X' \\ 0 & \text{else} \end{cases}$$

As a result from this approach of general system theory to ecology a simple example for an aquatic ecosystem is presented. The whole ecosystem is given by a stochastic automata $C = [U, Y, X, H]$. The subsystems (e.g. populations, other nonliving compartments) are also represented by stochastic automata C_{ij} which are equivalently imbedded into automata C_i, the trophic levels. Internal resources are given by stochastic automata C_G, and C_{NU},. Destruents are denoted by C_D and the sediment by C_S. The trophic levels include other subsystems which are equivalent imbedded stochastic automata. These subsystems contain other equivalently imbedded stochastic automata. In consequence, the hierarchic structure of an ecosystem will be described by imbedding a subsystem from a lower level into subsystem of a higher level. Also individuals are described by stochastic automata which are equivalently imbedded into other stochastic automata, describing a population and so on. The following general descriptions developed for an aquatic ecosystem include also automata descriptions of single ecological processes (Gnauck 1987):

Ecosystem: $C = [U, Y, X, H]$ with $X = \{ X_i, X_D, X_{NU}, X_G, X_S \}$.
Trophic level: $C_i = [U_i, Y_i, X_i, H_i]$ $(i = 1, ..., n)$ with $U_i \subseteq U, Y_i \subseteq Y, X_i \subseteq X$
and H_i is a restriction of H on $(X_i \times U_i) \subseteq (X \times U)$.
Population: $C_{ij} = [U_{ij}, Y_{ij}, X_{ij}, H_{ij}]$ $(j = 1, ..., m)$ with $U_{ij} \subseteq U_i, Y_{ij} \subseteq Y_i, X_{ij} \subseteq X_i$
and H_{ij} is a restriction of H_i on $(X_{ij} \times U_{ij}) \subseteq (X_i \times U_i.)$
Individuals: $C_{ijk} = [U_{ijk}, Y_{ijk}, X_{ijk}, H_{ijk}]$ $(k = 1, ..., l)$
with $U_{ijk} \subseteq U_{ij}, Y_{ijk} \subseteq Y_{ij}, X_{ijk} \subseteq X_{ij}$
and H_{ijk} is a restriction of H_{ij} on $(X_{ijk} \times U_{ijk}) \subseteq (X_{ij} \times U_{ij})$
Destruents: $C_D = [U_D, Y_D, X_D, H_D]$
Dissolved nutrients (organic, inorganic): $C_{NU} = [U_{NU}, Y_{NU}, X_{NU}, H_{NU}]$
Dissolved gases: $C_G = [U_G, Y_G, X_G, H_G]$
Sediment: $C_S = [U_S, Y_S, X_S, H_S]$
Predator-prey-relation: $C_{2r} = [U_{2r}, Y_{2r}, X_{2r}, H_{2r}]$, $C_{1s} = [U_{1s}, Y_{1s}, X_{1s}, H_{1s}]$
and $\xi: U_{1s} \rightarrow U_{2r} = \xi(e) = e$, $\eta: Y_{1s} \rightarrow Y_{2r} = \eta(e) = e$, $\zeta: X_{1s} \rightarrow X_{2r}$,
with $H_{2r} [\zeta(x_{2r})] (x_{1s}) = \Sigma H_{1s} [x_{1s}, w(u)_{1s}] (w(y)_{1s}, x^*_{1s})$,
$\zeta(x^*_{1s}) = x_{2r}$,
$\eta(w(y)_{1s}) = w(y)_{2r}$
Concurrence relation: $C_{11} = [U_{11}, Y_{11}, X_{11}, H_{11}]$, $C_{12} = [U_{12}, Y_{12}, X_{12}, H_{12}]$
and $\xi: U_{11} \rightarrow U_{12} = \text{id}$, $\zeta: X_{11} \rightarrow X_{12}$, $\eta: Y_{11} \rightarrow Y_{12}$
with $H_{12} [\zeta(x_{11}), \xi(w(u)_{11})] (w(y)_{12}, x_{12}) = \Sigma H_{11} [x_{11}, w(u)_{11}] (w(y)_{11}, x^*_{11})$

$$\zeta(x^*{}_{11}) = x_{12}$$
$$\eta(w_{11}) = w(y)_{12} \ .$$

5. Algebraic generalisation of ecosystems

The interpretation of an abstract automaton ***A*** as an algebraic structure ***S*** leads to an algebraic automata description and, consequently, to formulations by category theory. One goal of application of algebraic methods consists of a unified and general mathematical description of their biocoenotic-structural and functional components including the interrelationships within the system and with its environment, where the hierarchic couplings are considered also. While structural, functional and informational ecological processes will be covered, the combinatorical and homological algebraic methods are valid for analysis and modelling of systems behaviour. Ecological processes are modelled in a realistic manner by stochastic functions. Therefore it is necessary to use categories with multiplication, or more general symmetric monoidal categories (Bénabou 1963, Budach and Hoehnke 1975, Semadeni and Wiweger 1979, Gécseg 1986). These generalisations are necessary because the categories of non-deterministic and stochastic functions do not possess direct products, but instead have initial and terminal objects and also morphisms of projection, exchange and diagonalisation. For ecological purposes the category ***Ens*** of sets with the cartesian product as a tensor product, the category *T* of non-deterministic functions or transductors (the morphisms are multiple valuated representations) and especially the category ***S*** of stochastic functions will be used. The objects of ***S*** are sets and the morphisms are stochastic functions with discrete measures of probability as values.

Between categories and stochastic automata exist the following relations: $\boldsymbol{M} = \langle M, \otimes \rangle$ is a multiplicative category. A (Mealy-) automata on ***M*** is given by a quintupel $\mathbf{M} = [\,U, Y, X, F, G\,]$ with $U, Y, X \in$ ob ***M***, $F, G \in$ mor ***M*** with $G: X \otimes U \rightarrow Y$ (output morphism) and $F: X \otimes U \rightarrow X$ (state transition morphism). ob ***M*** characterises the set of all objects of the category while mor ***M*** denotes a morphism belonging to the category ***M***. Because of a series of morphisms corresponding to a multiplication of morphisms all these automata form a ***M-Aut*** category. Considering the stochastic relation

$\forall x \in X, \forall u \in U$ and $X' \subseteq X, Y' \subseteq Y$: $H[x, u](Y' \times X') = G[x, u]\,(Y') \bullet F[x, u]\,(X')$,
then the category ***M(S)*** is performed with stochastic models as morphisms.

For an ecosystem the following generalisation can be formulated:
An ecosystem is represented by a category ***C-Mat*** (or shorter ***C***) which result from ***C-Aut*** categories by an inclusion functor *Inc(**C**)*: ***C-Aut*** $\rightarrow$ ***C-Mat***. ***C-Mat*** contains special morphisms:

1. *Exchange morphism:* $c_{AB} : A \otimes B \rightarrow B \otimes A$ with $A, B \in$ ob ***C-Mat***.
2. *Diagonal morphism*: $d = d_A: A \rightarrow A \otimes A$ where $A \in$ ob ***C-Mat***.
3. *Terminal morphism*: $t_A : A \rightarrow I$ where $A, I \in$ ob ***C-Mat***.
4. *Canonical projection (product)* $p_k : A_1 \otimes \quad \otimes A_n \rightarrow A_k$ where $A_i \in$ ob ***C-Mat*** and $1 \leq k \leq n$ and $n \geq 1$.

The imbedding of sub-categories into the the category ***C-Mat*** will be realised by an imbedding functor *Imb(**C**)*. In this way it is possible to describe trophic levels or single elements of an ecosystem as sub-categories which are imbedded into other categories. Single elements will be called elementary automata, their morphisms represent elementary switching circuits. The imbedding procedure of elementary stochastic automata was shown by Gnauck (1987) for an aquatic ecosystem for the category ***P*** of phytoplankton. The category ***P*** may be imbedded in the whole ecosystem by the functor *Imb(**P**)*. The objects of ***P*** are the different algal classes which are sub-categories of ***P***. They enclose the different phytoplankton species as sub-sub-categories. The single elements of these sub-categories are expressed by elementary automata which are always imbedded into higher classes of objects. The feedbacks are given by loop morphisms when the sets of objects are connectable. Combining switching operations of elementary automata with internal and external couplings the behaviour of an ecosystem may be described algebraically where parallel and serial couplings and combinations with other automata operations are included.

As an example some ecological processes may be expressed as morphisms of a category ***C*** (a category with multiplication). X denotes a set of states of an ecosystem as a part of the set of objects belonging to ***C***. l(w) denotes the length of word and N is the set of natural numbers.

1. *Growth of an element*: grow(x): $X \rightarrow X$, $X \in \text{ob}\, \boldsymbol{C}$
   where $\forall x \in X, \forall w(x) \in W(X)$: $\exists t \in N$ with $x \bullet w(x) \rightarrow \infty$ and $l(w(x)) = t > 0$.
2. *Mortality of an element*: mort(x): $X \rightarrow X$, $X \in \text{ob}\, \boldsymbol{C}$,
   where $\forall x \in X, \forall w(x) \in W(X)$: $\exists t \in N$ with $x \bullet w(x) = 0$ and $l(w(x)) = t > 0$.
3. *Accumulation process*: acc: $X^* \otimes U \rightarrow Y \otimes X \in \text{mor}\, \boldsymbol{C}$
   where $U, Y, X, X^* \subseteq X \in \text{ob}\, \boldsymbol{C}$.
4. Competition of objects $A_i \in \text{ob}\, \boldsymbol{C}$: $A_1 \otimes Nu \otimes A_2 \rightarrow A_1 \otimes A_2$
   where $X = (A_1, Nu, A_2)$ and $p = 1(A_1) \otimes t(Nu) \otimes 1(A_2)$ is an epimorphism
   and $p(1) = 1(A_1) \otimes t(A_2)$ and $p(2) = t(A_1) \otimes 1(A_2)$.

6. Conclusion

The mathematical description and modelling of ecosystems by means of general systems theory requires a conceptual mathematical framework within which the dynamic ecosystem characteristics, the structure of couplings of subsystems, the storage effects of the whole ecosystem and the couplings between the ecosystem and its environment can be analysed, explained and predicted. In contrast, state space analysis is directed to a known physical and biocoenotical structure of an ecosystem, while model parameter values are mostly unknown. They are estimated under the assumption of stationary ecological processes by observed data. For changing ecosystem behaviour the parameter values must be exchanged.

Looking on ecosystems as information systems, their transfer behaviour depends on internal and external feedback signal changes and transformations of state signals. Within such a complex system hazards of signal arise from time delays between internal sources and sinks of information. This means that ecological processes will be changed by switching processes of the input and state signals. Such time delays extremely influence the dynamics of communication within an ecosystem. They are stochastic functions with certain input probabilities. Therefore, stochastic descriptions of the time constants of transfer processes play a significant role in ecosystem analysis. A generalised (discrete type) modelling procedure of informational processes which occur in complex systems is given by Petri nets. The advantage of this modelling technique is twofold. Firstly, the hierarchic structure of an ecosystem and the couplings of subsystems are represented by the net itself. The second advantage consists of the modelling of the information structure. In contrast to the state space approach to ecosystems modelling, the cause-effect relation is changed by a condition-event relation. As an example, the growth of an individual is considered as an event caused by preconditions which are valid before the event takes place. After this the conditions are changed. The process continues if the conditions are fulfilled that allow it to again take place. In classical state space analysis the growth of an individual is regarded as a process itself dependent on different influences. At present Petri nets are useful tools for event-based and structured modelling of ecosystems by means of cellular automata.

Ecosystems may be considered as switching networks. This point of view allows not only a generalised mathematical description of ecological processes, but also an evaluation of signal transfer processes and statements on ecosystems stability. Furthermore, based on

Boolean variables a reliability analysis of synchronous (or natural) and asynchronous (or man-controlled) ecosystem behaviour will be worked out. In addition, the connections to complexity theory are very close. Using thoughts and notions of general system theory a unified approach to ecosystems will be presented in future by the theory of categories and by investigations of the mathematical characteristics of these categories. The imbedding procedure allows statements on the hierarchic couplings between different sub-categories and higher order meta-systems. Referring back to the beginning of this chapter, it can be stated that the development of a theory of ecosystems by means of general system theory will be a powerful tool to aid in proving a better understanding of what happens in the future with anthropogenic influenced ecosystems. This will then be a sustainable development.

REFERENCES

Allen, T.F.H. and T.W. Hoekstra, 1992. Toward a Unified Ecology. Columbia Univ. Press, Columbia.

Arbib, M.A., 1968. Theories of Abstract Automata. Prentice-Hall, Englewood Cliffs.

Bénabou, J., 1963. Catégories avec multiplication. *C. R. Acad. Sci. Paris* 256: 1887-1890.

Bertalanffy, L.v. 1968. General Systems Theory: Foundations, Development, Applications. Braziller, New York.

Bossel, H. 1992. Modellbildung und Simulation. Vieweg, Braunschweig Wiesbaden.

Bossermann, R.W. and R.K. Ragade, 1982. Ecosystem analysis using fuzzy set theory. *Ecol. Modelling* 16(3): 191-208.

Brauer, W., 1979. Net Theory and Applications. Springer, Berlin.

Budach, L. and H.J. Hoehnke, 1975. Automaten und Funktoren. Akademie-Verlag, Berlin.

Casti, J.L. 1979. Connectivity, Complexity, and Catastrophe in Large-Scale Systems. Wiley, Chichester.

Collier, B.D., G.W. Cox, A.W. Johnson, and P.C. Miller, 1974. Dynamic Ecology. Prentice-Hall International, London.

Conrad, M., 1983. Adaptability. Plenum, New York.

Dubois, D.M. (Ed.), 1981. Progress in Ecological Engineering and Management by Mathematical Modelling. Cebedoc, Liege.

Duckstein, L. and Y. L. Klimontovich, 1984. Selforganization and Turbulence in Liquids. Teubner, Leipzig.

Fleishman, B.S., 1982. Fundamentals of Systemology. Radio and Kommunikation, Moskau (in Russ.).

Florkin, M., 1977. Biological Information Transfer. Elsevier, Amsterdam.

Gécseg, F., 1986. Products of Automata. Akademie-Verlag, Berlin.

Gibbons, A., 1985. Algorithmic Graph Theory. Cambridge Univ. Press, Cambridge.

Gill, A., 1962. Introduction to the Theory of Finite-State Machines. McGraw-Hill. New York.

Gluschkow, W.M., 1963. Theorie der abstrakten Automaten. Deutscher Verlag der Wissenschaften, Berlin.

Gnauck, A., 1976. Grundlagen der mathematischen Modellierung limnischer Ökosysteme. In: R. Glaser, K. Unger, und M. Koch (Hrsg.). *Umweltbiophysik.* Akademie Verlag, Berlin, p. 67-75.

Gnauck, A., 1987. Kybernetische Beschreibung limnischer Ökosysteme. Habilitationsschrift, Techn. Univers. Dresden.

Gnauck, A., 1995. Systemtheorie, Ökosystemvergleiche und Umweltinformatik. In: A. Gnauck, A. Frischmuth, und A. Kraft, (Hrsg.). *Ökosysteme: Modellierung und Simulation.* Blottner, Taunusstein, p. 11-27.

Gnauck, A. and M. Straškraba, 1980. Theoretical Concepts of Ecosystem Models. *ISEM J.* 2: 71-80.

Göldner, K. and S. Kubik, 1978. Nichtlineare Systeme der Regelungstechnik. Verlag Technik, Berlin.

Halfon, E. (Ed.), 1979. Theoretical Systems Ecology. Academic Press, New York.

Innis, G.S. and R.V. O'Neill (Eds.), 1979. Systems Analysis of Ecosystems. Internat. Coop. Publ. House, Fairland.

Jeffers, J.N.R., 1978. An Introduction to Systems Analysis: With Ecological Applications. Arnold, London.

Jones, R.W., 1973. Principles of Biological Regulation: An Introduction to Feedback Systems. Academic Press, New York.

Jørgensen, S.E., 1986. Fundamentals of Ecological Modeling. Elsevier, Amsterdam.

Jørgensen, S.E., 1992. Integration of Ecosystem Theories: A Pattern. Kluwer, Dordrecht.

Kalaman, R.E., P.L. Falb, and M.A. Arbib, 1967. Topics in Mathematical Systems Theory. McGraw-Hill, New York.

Kant, I., 1781. Kritik der reinen Vernunft. J.F. Hartknoch, Riga.

Klir, G., 1969. An Approach to General System Theory. Van Nostrand, New York.
Klir, G., 1972. Trends in General System Theory. Wiley-Interscience. New York.
Klir, G., 1985. Architecture of Systems Problem Solving. Plenum Press, New York.
Laue, R., 1970. Elemente der Graphentheorie und ihre Anwendung in den biologischen Wissenschaften. Geest & Portig, Leipzig.
Levich, A.P., 1982. Set Theory, Language of the Theory of Categories and Their Use in Theoretical Biology. Moscow State Univ. Press, Moscow (in Russ.).
Lewis, E.R., 1977. Network Models in Population Biology. Springer, Berlin, Heidelberg, New York.
Litz, L., 1983. Modale Maße für Steuerbarkeit, Beobachtbarkeit, Regelbarkeit und Dominanz - Zusammenhänge, Schwachstellen, neue Wege. *Regelungstechnik* 31: 148-158.
Locke, M., 1984. Grundlagen einer Theorie allgermeiner dynamischer Systeme. Akademie-Verlag, Berlin.
Lückel, J. and R. Kasper, 1981. Strukturkriterien für die Steuer-, Stör- und Beobachtbarkeit linearer, zeitivarianter, dynamischer Systeme. *Regelungstechnik* 29: 357-362.
Ludyk, G., 1979. Theorie Dynamischer Systeme. Elitera Verlag
Ludyk, G., 1981. Time-Variant Discrete-Time Systems. Vieweg, Braunschweig.
Mac Arthur, R.H., 1972. Geographical Ecology. Harper and Row, New York.
Matthes, K., B. Breckling, and C. Ekschmitt, 1996. Systemtheorie in der Ökologie. ecomed, Landsberg.
May, R.M., 1974. Stability and Complexity in Modell Ecosystems. Princeton Univ. Press, Princeton.
May, R.M., 1981. Theoretical Ecology: Principles and Applications. Blackwell, Oxford.
Mesarovic, M.D. and Y. Takahara, 1975. General System Theory: Mathematical Foundations. Academic Press, New York.
Mesarovic, M.D., D. Macko, and Y. Takahara, 1970. Theory of Hierarchical Multilevel Systems. Academic Press, New York.
Menshutkin, V.V., 1974. Theoretische Grundlagen der mathematischen Modellierung aquatischer Ökosysteme. *Z. allg. Biol.* 35::32-42 (in Russ.).
Milsum, J.H., 1966. Biological Control Systems Analysis. McGraw-Hill, New York.
Morowitz, H.J., 1968. Energy Flow in Biology. Academic Press, New York.
Müller, F., 1997. State-of-the-art in ecosystem theory. *Ecol. Modelling* 100: 135-161.
Müller, P.C. and J. Lückel, 1984. Modale Maße für Steuerbarkeit, Beobachtbarkeit und Störbarkeit dynamischer Systeme. *Z. Angew. Math. Mech.* 54: 57-58.
Nicolis, G. and I. Prigogine, 1977. Self-Organisation in Non-Equilibrium Systems: From Dissipative Structure to Order Through Fluctuations. Wiley, London.
Nisbet, R.M. and W.S.C. Gurney, 1982. Modelling Fluctuating Populations.Wiley, Chichester.
Odum, H.T., 1983. Systems Ecology. Wiley, Chichester.
Pahl-Wostl, C., 1995. The Dynamic Nature of Ecosystems. Wiley, New York.
Patil, G.P., E.C. Pielou, and W.E. Waters, 1971. Statistical Ecology. Penns. State Univ. Press, Univ. Park London, Penns.
Patten, B.C. (Ed.), 1971. Systems Analysis and Simulation in Ecology. Vol. 1. Academic Press, New York.
Patten, B.C. (Ed.), 1972. Systems Asnalysis and Simulation in Ecology. Vol. 2. Academic Press, New York.
Patten, B.C. (Ed.), 1974. Systems Analysis and Simulation in Ecology. Vol. 3. Academic Press, New York.
Patten, B.C. (Ed.), 1976. Systems Analysis and Simulation in Ecology. Vol. 4. Academic Press, New York.
Patten, B.C., 1991 . Network Ecology. Indirect Determination of the Life-Environment relationship in Ecosystems. In: M. Higashi and T.P. Burns, (Eds.). *Theoretical Studies of Ecosystems - The Network Approach*. Cambridge Univ. Press, Cambridge, p. 288-351.
Patten, B.C. and S.E. Jørgensen (Eds.), 1995. Complex Ecology. The Part-Whole-Relationship in Ecosystems. Prentice Hall, Englewood Cliffs, N.J.
Peschel, M., 1965. Kybernetik und Automatisierung. Verlag Technik, Berlin.
Peschel, M., 1970. Kybernetische Systeme. 2. Aufl., Verlag Technik, Berlin
Peschel, M., 1978. Modellbildung für Signale und Systeme. Verlag Technik, Berlin.
Peschel, M. and G. Wunsch, 1972. Methoden und Prinzipien der Systemtheorie. Verlag Technik, Berlin.
Pielou, E.C., 1969. An Introduction to Mathematical Ecology. Wiley-Inter-science, New York.
Pospelov, D.A., 1973: Analyse und Synthese von Schaltsystemen. Verlag Technik, Berlin
Reinisch, K., 1974. Kybernetische Grundlagen und Beschreibung kontinuierlicher Systeme. Verlag Technik, Berlin.
Reinisch, K., 1982. Analyse und Synthese kontinuierlicher Steuerungssysteme. Verlag Technik, Berlin.
Rosen, R., 1970. Dynamical System Theory in Biology. Vol. 1. Wiley-Interscience, New York.
Rosen, R., 1971. Relationale Biologie. Thieme, Leipzig.
Semadeni, Z. and A. Wiweger, 1979. Einführung in die Theorie der Kategorien und Funktoren. Teubner, Leipzig.
Schwarz, H., 1991. Nichtlineare Regelungssysteme. Oldenbourg München Wien.
Söte, W., 1980. Strukturelle Methoden zur Dekomposition von Großsystemen. *Regelungstechnik* 28: 37-44.
Starke, P.H., 1969. Abstrakte Automaten. Deutscher Verlag der Wissenschaften, Berlin:
Straškraba, M., 1980. Cybernetic Categories of Ecosystem Dynamics. *ISEM J.* 2: 81-96.
Straškraba, M. and A. Gnauck, 1985. Freshwater Ecosystems. Elsevier, Amsterdam.

Stugren, B., 1978. Grundlagen der allgemeinen Ökologie. Fischer, Jena.
Svirezhev, Y.M. and D.O. Logofet, 1978. Stability of Biological Communities. Nauka, Moscow.
Takahara, Y., 1981. Significances of Categorial Approach to Mathematical General Systems Theory. Report Tokyo Inst. Technol., Nagatsuta, Yokohama.
Thomann, R.V., 1967. Time series analysis of water-quality data. Proc. ASCE. *J. San. Eng. Div.* 93(1): 1-23.
Wassel, M., 1978. Neue Ergebnisse zur Steuerbarkeit linearer zeitvarianter Systeme. *Regelungstechnik* 26, 60-64.
Wiegleb, G., 1996. Hierarchische Systemtheorie in der Ökologie. In: K. Mathes, B. Breckling, and C. Ekschmitt, (Eds.). *Systemtheorie in der Ökologie*. ecomed, Landsberg.
Windeknecht, T.G., 1971. General Dynamical Processes - A Mathematical Introduction. Academic Press, New York.
Zadeh, L. and C.A. Desoer, 1963. Linear system Theory. The State Space Approach. McGraw-Hill, New York.

1.6.1 Ecosystem Research

Otto Fränzle

1. Introduction

The ecosystem concept is fundamental to the understanding of the biosphere and the examination of human impacts on life on earth. It provides a way of comprehensively looking at the functional interactions between life and environment, and is useful in resource management and as a basis for modelling. Understanding the operational and support functions of ecosystems is vital to use of the ecosystem concept for predictive purposes. Ecosystems interact in a variety of ways through their biotic and abiotic components and are always liable to change through time. Their dynamic nature operates over time scales ranging from seconds to geological time. One of the most important dimensions of this interaction is competition between individual, and populations of, organisms. Changes to ecosystems may be, and in fact largely are, caused by human actions. They act at various scales and with varying severity, and one of the most difficult problems facing environmental science is diagnosing the nature of environmental change. Not only is the extent and rate of change often difficult to assess, and even harder to reliably predict, but also it may be hard to distinguish between those components of change which are a part of natural ecosystem dynamics, and those which are a result of human impacts. Yet unravelling all of the essential issues is vital if ecosystem function is to be sustained and irreparable damage to the biomes of the earth avoided.

Clearly the ecosystem has been a key concept in the development of modern ecology, whose history shows, however, that ecologists did not share a common understanding of this concept (Fränzle 1998, Golley 1993). Indeed, from the viewpoint of theory building, the history of the ecosystem concept offers examples for conjectures, refutations, revised conjectures and additional refutations or, as Popper ([4]1972: 215) had it with regard to science in general, it can be seen as 'the repeated overthrow of scientific theories and their replacement by better or more satisfactory ones'. It is the purpose of the following contribution to outline the role of ecosystem research in the sequential refinement of ecosystem theories which is particularly well reflected in model building. These range from empirical models for practical purposes to rather abstract ones, aiming at qualitative general insights. At one end of this spectrum there is a detailed and mechanistic description of specific systems such as soil horizons or precisely defined adsorbents in interaction with one or a few pure chemicals in aqueous solutions. At the opposite end of the spectrum are relatively general models which have to sacrifice numerical precision for the sake of general principles. Such conceptual models need not correspond in detail to any single 'real world' process, but aim to provide a framework for the discussion of broad classes of phenomena or simply of contentious issues. Rationally handled, these different approaches mutually reinforce each other, thus providing reciprocally new and deeper insights.

To illustrate this the first two sections of the present chapter include an exemplary summary of ecosystem research bases and activities in the framework of the International Biological Programme (IBP) and UNESCO's Man and the Biosphere Programme (MAB). It is followed by a detailed description of Ecosystem Research in the Bornhöved Lake District. In the fifth section focal aspects of current theoretical reflections on evolution, structure and stability of ecosystems are presented, which emerged out of, or in relation to, these international programmes. They permit, in turn, to more precisely define relevant problems for both future ecosystem studies and theory building.

2. Ecosystem research in the framework of IBP and MAB programmes

The history of systems ecology illustrates the emergence of favourite orientations and foci of ecological interest as the result of an increasingly intensive feed-back between theoretical reflection or modelling and theory-based interdisciplinary ecosystem research. By the mid-1960s, teams of researchers studied whole forests or lakes, while individual scientists focused on processes within systems, such as rates of primary production, material and energy transfers between trophic levels and populations, and the ways and rates of organic decomposition. Since only a largely unorganised body of theory was available to stimulate research, ecosystems were viewed from a variety of perspectives, and frequently researchers reasoned analogically from physical, chemical, or biological systems to ecosystems (Fränzle 1998, Golley 1993, Mcintosh 1985). Thus ecosystem theory was constructed from the field of natural history (cf. Trepl 1987), from thermodynamics, from evolutionary theory, from information theory, and so on. As a consequence part of the ecological community considered systems theory as the most appropriate paradigm to organise the information about ecosystems. In practice this resulted in two complementary approaches, the first of which was characterised by ecosystem modelling from the information about system components and linkages. The other, pioneered by H. Odum (1957) at Silver Springs, considered the ecosystem in a perspectivistic manner as an object of research whose input-output relationships had to be determined in order to mechanistically explain the conversion of inputs into outputs by the system.

The opportunity to test and further develop these alternatives was at the biome research area of the International Biological Programme (IBP) which had an overall focus on biological productivity as a basis for human well-being. Ultimately, it developed into a largely ecological programme, and the scientific director of the IBP, E.B. Worthington (1975: 64) states in the synthesis volume:

> 'It is this ecosystem approach which distinguishes much of the IBP research from what had dominated ecology before. Essentially, it consists of the careful selection of a number of variables – biological, chemical and physical – about which data are collected, quantitatively as well as qualitatively. Thereby, the ecosystem can be analysed in order to ascertain which factors and processes are important in causing the dynamics of the whole. In this, the application of system analysis to biological systems has been one of the major innovations developed during IBP.'

In regard to terrestrial productivity in the U.S.A., the focal programme proposition was to study 'landscapes as ecosystems', with emphasis on production and trophic structure, energy flow pathways, limiting factors, biogeochemical cycling, and species diversity. An important feature was the claim not to confine these studies to natural areas only. Another feature was the proposal to use systems analysis as a mechanism for integrating the results of the study. In a comparative evaluation of the pertinent biome projects organised in the grasslands, tundra, deserts, coniferous forests, and deciduous forests, Golley (1993: 139) came to the conclusion that they furthered ecological knowledge but failed to essentially contribute to the development of ecosystem theory. The programmes were not designed to sort out competing or contradictory ideas. Rather, they were driven, at least initially, by the idea that ecologists could construct a mechanical systems model built on the concepts of trophic levels, the food web, or the food cycle, and then represent the dynamic behaviour of the components by data from organisms or populations that are surrogates of the component. This 'bottom-up' or 'design-up' approach did not prove possible or useful. Further, the biome projects did not effectively promote landscape ecology, as Odum had hoped. The biome was the setting for site research but was not really addressed as such in an effective manner'.

In comparison to the U.S. biome studies the German IBP project, located in the Solling Mountains, was organised in a way similar to the Hubbard Brook project, located in the White Mountains of New Hampshire (Likens et al. 1977). Both approached ecosystem analyses from the components which could then be linked together systematically in a model-based theory or as a natural object (in Popper's 1959 sense) studied by means of conventional scientific methods. Like the Hubbard Brook investigations, the Solling project also took a landscape approach from the beginning, based on an a-priori ecological knowledge of the area which probably exceeded that in most other places where IBP work was undertaken. It focused in a comparative way on natural like ecosystems, nature-resembling ecosystems, transformed ecosystems and degraded systems. To this end the study sites included acidophilous beech forest and planted spruce forest stands at several different ages, along with permanent grassland and cultivated fields. Research proceeded from a description of climate and soils, the abundance and productivity of vegetation, animals, and micro-organisms, to plant physiology, nutrient fluxes, and energetics (Ellenberg et al. 1986).

Not only in comparison to the majority of national IBP programmes, which focused on a single question or a few questions at best, but also in comparison to studies undertaken in technologically advanced countries which mostly dealt with natural ecosystems, however, the Solling project was exceptionally successful. It provided a wealth of sound data on ecosystem structure and function, thus building the scientific concept of the ecosystem. Furthermore, it formed a promising basis to reason about the causes of forest dieback, considered a serious problem in Germany in the 1980s (Umweltbundesamt 1986). Unlike the biome programmes of the United States of America, however, the final summary report of the Solling project did not attempt to force the results into a single synthesis on the basis of an abstract theoretical device or a model. Instead, each part was placed with a site-specific conceptual ecosystem model and developed a theme within its own logic. Owing to this methodology the Solling project can be considered a milestone in the development of a theory-based ecosystem research.

Both the conceptual framework and the practical experience gained in interdisciplinary research exerted a considerable influence on the conception of a comprehensive ecological surveillance system for Germany (Ellenberg, Fränzle and Müller 1978). Composed of three interrelated components, namely an ecological monitoring network, comparative ecosystem research, and an environmental specimen bank it is intended and largely implemented in Germany to promote both theoretical ecology and, in a transdisciplinary context, planning and policy. The integrative ecosystem research component of this German programme is also an essential part of UNESCO's Man and the Biosphere (MAB) Programme which was developed out of the IBP experience. In the following section it is to be considered with regard to its bearing on ecosystem research, theory and modelling.

3. Man and the biosphere (MAB) programme

The general objective of the MAB programme, as launched by UNESCO in November 1971, endorsed by the UN Conference on the Human Environment (1972), and supplemented in 1986 and 1992, has been defined as: '... to develop within the natural and social sciences a basis for the rational use and conservation of the resources of the biosphere and for the improvement of the relationship between man and the environment; to predict the consequences of today's actions on tomorrow's world and thereby to increase man's ability to manage efficiently the natural resources of the biosphere' (UNESCO 1988, p. 11). The specific aims of the programme are:

inference is clearly in contradiction to the above conventional wisdom, but it indicates that there is no general, unavoidable connection between complexity or diversity, and community stability (cf. Wissel 1981) Another question is, to what extent May's result is an artefact arising out of the particular characteristics of the model and the interpretative techniques applied (Begon et al. 1990, Jørgensen 1990), or to which extent it can be corroborated by stability-oriented diversity analyses in the field. The picture also alters if, instead focusing on local stability and correspondingly minor disturbances, 'species-deletion stability' and larger perturbations, respectively, are considered. This is particularly interesting from an ecotoxicological point of view. Under the assumption of the deletion of one species of a community owing to a persistent impact, the system is said to be species-deletion stable if all of the remaining species are retained at locally stable equilibria.

Other theoretical approaches to the stability problem are attributed to McMurtrie (1975), Pimm (1982) and Ulanowicz (1986), who established criteria for the stability of ecosystems under certain restrictive conditions, such as zero-sum games. In particular, they showed that ecosystems can become unstable against weak perturbations when the complexity increases beyond a certain value, although the transition is not always sharp (Cohen and Newman 1984). Deriving sensitivity measures from community descriptors, Pahl-Wostl and Ulanowicz (1993) found 'ascendancy' to be an appropriate index that includes both system size and structure. The application of holistic system descriptors in simulation models was also advocated by others. Jørgensen (1992) used exergy as a goal function in models to account for changes in species composition. Bachas and Huberman (1987) and Ceccato and Huberman (1988) related the complexity of hierarchical structures, as measured by their diversity, to dynamic behaviour. The results suggest that there might be a quantifiable relation between the diversity of a hierarchical structure and its stability. Furthermore findings of Doreian (1986) and Briand and Cohen (1987) suggest that the effective dimensionality of the space in which the interactions take place, and the overlaps in food webs, contribute to the nature of the structural representation of the community. Hence it may be concluded that their stability will likewise be affected, which is in agreement with the results of a comparative model analysis of unstructured co-operative interactions between arbitrary species on the one hand and ecosystems with pyramidal organisation on the other.

Returning to the original problem set out by May (1972) this author derived a condition of stability for more general systems, as a function of species diversity and the strength of interactions. The stability of such systems is determined by the behaviour of the largest eigenvalue of matrices governing the response of the system to small perturbations. As a result it could be shown in the case of non-hierarchical organisations how the removal of a zero-sum game condition can lead to a further reduction of stability. Hierarchical ecologies, by contrast, are intrinsically more stable than unstructured ones (Allen and Hoekstra 1992, Hogg et al. 1989, O'Neill et al. 1986).

The conflicting results amongst the models developed so far must be considered in the light of model structure, in particular the number and composition of elements, i.e., species simulated. In comparison to real communities with hundreds or thousands of species (Urban et al. 1987) and highly variable degrees of interaction, most models are extremely 'impoverished' in species. Furthermore, the models quite often refer to randomly constructed communities, while real communities and ecosystems are far from randomly constructed and normally display a complex hierarchical structure (Fränzle 1993; Solbrig and Nicolis 1991). Unstable communities are liable to collapse when they experience environmental conditions which reveal their instability; but the range and predictability of conditions may vary markedly from place to place. Under stable and predictable site conditions, a community will only experience limited fluctuations, and thus even a dynamically fragile one may still persist. By contrast, in a variable and largely unpredictable environment, only dynamically robust communities are likely to persist.

5.2 Systems as adapting and self-organising entities

Every open system is energetically, materially and informationally linked with its environment, therefore the dynamic forces of the system constantly change. A particularly important type of change is adaptation, and the influence eliciting it can generally be considered as a stimulus and the adaptive process as a response, i.e. a stabilisation. In greater detail the behaviour of an adapting system can be classified into (i) goal-seeking, (ii) purposive and (iii) purposeful. The first adaptation mechanism consists of a one-to-one correspondence between stimulus and response, while a purposive system displays a one-to-many correspondence, meaning that each stimulus normally elicits a number of responses. The term 'purposeful' finally indicates an adaptive mechanism which brings about a change in system's functions.

Since the system achieves, during the process of adaptation, a new kind of structural or functional order or organisation it undergoes self-organisation. This process includes not only adaptation as such, but also cognition and learning with regard to constraints involving energy and matter. The exergy concept (Jørgensen 1992) is a basis of current attempts to develop thermodynamically based models of ecosystem functioning which may, for example, predict how the biota of an ecosystem might respond to specific environmental changes. Depending on the ratio of energy inflow to outflow, a nucleation, i.e. a clustering or aggregation of system elements, may occur. In these clusters, or regions in geographical terminology, a conversion of energy into signals is carried out, whereby the activity of the energy converter is decisive for the signal connectivity which in turn defines the interactions between elements. This is a principal statement connecting thermodynamics with information theory (cf., e.g., Patten and Jørgensen 1995).

Because the regions are characterised by their states, the process of self-organisation can be considered as a transition between these states undergoing new formations of the connections between the elements, provided the stimuli exceed certain threshold values. In addition to cybernetics (e.g., De Angelis 1995, Straskraba 1995), network theory (Higashi and Burns 1991, Patten 1992), catastrophe theory and bifurcation analysis have proved useful for formalised descriptions of such systems in transition. Catastrophe theory limits systems of interest to so-called gradient systems which arise from the minimisation of some objective function and associated dynamics (or maximisation of its negative). The latter kind of branching, or jump, behaviour is relevant to systems of a more general type than the first which cannot be appropriately characterised by the gradient of a potential function. Typically, the differential equations of non-gradient systems have a small number of isolated stable equilibrium points and information about system behaviour is presented as trajectories on state space (or phase) diagrams. The stable points act as attractors, and correspondingly unstable points as repellors, and these points shape the contours of the trajectories in state space accordingly. Bifurcation is reflected in the existence of critical parameter values at which the nature of the solution to the differential equation changes. However, there is no simple classification of possible 'cases' as in catastrophe theory, and empirical inquiry into systems behaviour is essential to proceed in more precisely determining the realm of applicability of bifurcation methodology to generalised ecosystem analysis and modelling.

The studies required involved comparative analyses of community and ecosystem succession and the chronosequential approach, or space-for-time substitution, is a traditional tactic applied. As a consequence of the probabilistic character of living systems this approach may be misleading, however. Consequently, long-term studies were already recognised by founders of ecology to be necessary for a reproducible understanding of succession (see, e.g., Clements 1916). In salt marsh succession, permanent plots and sediment core analyses have documented, for instance, that the expected autogenic and deterministic patterns were not found (Clark 1986, Niering 1987).

Despite the failings of vegetational or soil chronosequences, their value is clear if certain limits are kept in mind. In the framework of system evolutionary analyses of holarctic and tropical plant associations, space-for-time substitution has been employed to assess structural or compositional aspects (Fränzle 1994 c). The same applies to oldfield successional studies which documented trends in life history types, pathways of dominant species, convenient 'stages', and regional differences (Pickett 1989 a). Many of these insights could, at least in principle, also have come from long-term studies. However, the understanding that has emerged from the few long-term investigations of oldfield succession is of a different sort, since they documented the nature of transitions, the role of year-specific conditions, the problems with end points, and the role of newly invaded species in succession (Asshoff 1997 , Bobrowski 1982, Hemprich 1991).

5.3 Unifying concepts and integrative approaches

The above (and necessarily incomplete) summary of focal points of ecological studies may indicate that the present theoretical background to ecosystem research has not yet reached the level of a comprehensive unified theory. There is, however, a commendable number of unifying concepts and integrative approaches which are described in greater detail in the following chapters of this book; with regard to ecosystem research it may suffice to quote the following which not infrequently exhibits a considerable amount of convergence.

Catastrophe theory (Thom 1975, Wilson 1981) concerns both the stability and creation of forms, and each of the pertinent concepts imply also a general concern with dynamical analysis. Jørgensen (1992, 1995) introduced the exergy principle and deduced a theory of ecosystem evolution on this basis. Exergy may be associated with holistic indicator variables such as entropy, emergy and ascendancy (Jørgensen 1997 a, b). Schneider and Kay (1994 a, b) identified thermodynamic non-equilibrium, exergy gradients and flows as equally useful indicators of ecosystem functioning. Following Odum's (1969) succession concept and touching upon cybernetic system analysis Patten (1998) ascribed to certain state functions the intrinsic quality of attractors; in Bossel's (1992) terminology they constitute ecological orientors or goal functions (cf. Müller and Leupelt 1998).

Epistemologically speaking, all of these concepts are derived from, or associated with, systems theories in general and, more specifically, with theories of self-organising ecological systems. The latter, in turn, amalgamate components of the thermodynamic theory of irreversible processes (Nicolis and Prigogine 1977, Prigogine 1967, 1976, 1985) with elements of the above catastrophe theory, furthermore with deductions from the information and network theories (Margaleff 1995, Ulanowicz 1986, 1995), the theory of games (McMurtrie 1975, Pimm 1982, Ulanowicz 1986) and hierarchy theory (Allen and Hoekstra 1992, Allen and Starr 1982, O'Neill et al. 1986).

One of the most conspicuous representations of self-organisation processes in open systems is the formation of gradients whose structural and functional analysis provides for a particularly integrative aspect in ecosystem studies (Fränzle 1977, 1994, Müller 1998, Prigogine 1976, Schneider and Kay 1994 a, b). In an open system the 'competition' between internal entropy production and entropy 'export' into the environment permits the system, subject to certain boundary conditions, to adopt new states or structures. Entropy production can be expressed in terms of thermodynamic 'forces' and rates of irreversible phenomena (Prigogine 1967); the former may be gradients of temperature or concentration, the latter would then be heat flux or chemical reaction rates. Thus gradients with their specific coefficients, e.g. coefficients of thermal conductivity or diffusion, are related to, and an expression of, exergy flows in ecological systems. For evolving biotic networks primarily isolated notional characterisations like homogenisation, amplification, synergism (Patten 1992), ascendancy (Ulanowicz 1986), power or emergy (Odum 1983) can be conveniently associated with energetic or nutritional gradients. Also ecological hierarchies can, for the sake of a unifying generic characterisation, be described in terms of scale-dependent hierarchies of gradients.

Gradients are the most conspicuous reflection of a system's heterogeneity which is, in the operational sense of the term, a scale-dependent outcome of the regionalisation procedures adopted. The generic definition and analysis of ecologically relevant fluxes of energy, matter and information, including the determination of thermodynamic 'forces' and gradients is therefore intimately linked with the appropriate definition of spatial structures which frequently undergo seasonal or other temporal variations. This fact assumes a particular quality when considering ecosystems or their major compartments from the viewpoint of self-organised dissipative structures (Prigogine 1976, Fränzle 1994 c). To most effectively maintain the whole set of intra and intersystemic fluxes essential for the negentropy-related stability the systems have continuously to degrade and re-establish a whole network of gradients (Schneider and Kay 1994 a). This implies that concentration processes are necessarily coupled with entropy-exporting dissipative processes, such that the succession of gradient formation and degradation can be characterised by a temporal cyclicity as suggested by Holling (1986).

A final point to be made with regard to all concepts of ecosystem analysis and modelling is the fact that the analyst in most cases has incomplete and partial knowledge of the system which he seeks to understand and control. The nature of his knowledge is often incremental in that greater insight is gained as analysis or control continues. This is the elementary basis for claiming long-term commitments to ecological studies (Bennett and Chorley 1978, Pickett 1989 b). But even under such favourable technical boundary conditions elements in the analysis of ecosystems will normally be necessarily indeterminate, both because of the complexity of such systems and because of the intimacy of man-environment interrelations or subjective constraints on understanding. The major limitations are due to: methods or measurement, the presence of stochastic variation and the methods of system analysis adopted, and finally the manner in which systems knowledge is obtained, accumulated and applied. Each of these points is of crucial importance, and each constitutes a major challenge to both theoretical reflection and systematic empirical research.

REFERENCES

Allen, T.F.H. and T.B. Starr, 1982. Hierarchy. Perspectives for Ecological Complexity. University of Chicago Press, Chicago.

Allen, T.F.H. and T.W. Hoekstra, 1992. Toward a Unified Ecology. Columbia University Press, New York.

Asshoff, M., 1997. Die Erschließung und Modellierung ökologischen Wissens für das Management von Feuchtwiesenvegetation. Diss. Univ. Kiel.

Bachas, C.P. and B.A. Huberman, 1987. Complexity and ultradiffusion. *J. Phys. A.* 20: 4995-5014.

Begon, M., J.L. Harper, and C.R. Townsend, 1990. Ecology: Individuals, Populations and Communities. Blackwell, Boston, Melbourne.

Bennett, R.J. and R.J. Chorley, 1978. Environmental Systems. Methuen, London.

Bertalanffy, L. von, 1968. General System Theory. New York.

Bobrowski, U., 1982. Pflanzengeographische Untersuchungen der Vegetation des Bornhöveder Seengebietes auf quantitativ-soziologischer Basis. *Kieler Geogr. Schr.* 56.

Bossel, H., 1992. Real-structure process description as the basis of understanding ecosystems and their development. *Ecol. Model.* 63: 261-276.
Briand, F. and J.E. Cohen, 1987. Environmental correlates of food chain length. *Science* 238: 956-960.
Ceccato, H.A. and B.A. Huberman, 1988. The complexity of hierarchical systems. *Physica Scr.* 37: 145-150.
Chorley, R.J., 1972. Spatial Analysis in Geomorphology. London.
Cohen, J.E. and C.M. Newman, 1984. The stability of large random matrices and their products. *Ann. Probab.* 12: 283-310.
Clark, J.S., 1986. Late-holocene vegetation and coastal processes at a Long Island tidal marsh. *Journ. of Ecology* 74: 561-578.
Clements, F.E., 1916. Plant Succession: An Analysis of the Development of Vegetation. Carnegie Institution, Washington.
De Angelis, D.L.,1995. The nature and significance of feedback in ecosystems. In: B.C. Patten and S.E. Jørgensen, (Eds.). *Complex Ecology: The Part-whole Relation in Ecosystems.* Englewood Cliffs, p. 450-467.
Doreian, P., 1986. Analyzing overlaps in food webs. *J. Soc. Biol. Struct.* 9: 115-139.
Ellenberg, H., O. Fränzle, and P. Müller, 1978. Ökosystemforschung im Hinblick auf Umweltpolitik und Entwicklungsplanung. Umweltforschungsplan des Bundesministers des Innern – Ökologie-Forschungsbericht 78-101 04 005. Bonn.
Ellenberg, H., R. Mayer, and J. Schauermann (Hg.), 1986. Ökosystemforschung. Ergebnisse des Sollingprojekts 1966-1986. Ulmer, Stuttgart.
Elton, C. S. (1958): The Ecology of Invasion by Animals and Plants. Methuen, London
Fränzle, O., 1977. Biophysical aspects of species diversity in tropical rain forest ecosystems. *Biogeographica* 8: 69-83. The Hague.
Fränzle, O., 1990. Ökosystemforschung und Umweltbeobachtung als Grundlagen der Raumplanung. *MAB-Mitt.* 33: 26-39.
Fränzle, O., 1993. Contaminants in Terrestrial Environments. Springer, Berlin, Heidelberg, Budapest.
Fränzle, O., 1994a. Representative soil sampling. In: B. Markert, (Ed.). *Environmental Sampling for Trace Analysis.* VCH, Weinheim, p. 305-320.
Fränzle, O., 1994b. Modellierung des Chemikalienverhaltens in terrestrischen Ökosystemen auf unterschiedlichen Raum-Zeit-Skalen. In: E. Bayer and H. Behret, (Eds.). *Bewertung des ökologischen Gefährdungspotentials von Chemikalien.* GDCh-Monographien 1: 45-90.
Fränzle, O., 1994c. Thermodynamics aspects of species diversity in tropical and ectropical plan communities. *Ecol. Model.* 75/76, 63-70.
Fränzle, O., 1998. Sensivity of ecosystems and ecotones. In: G. Schürmann and B. Markert, (Eds.). *Ecotoxicology*, p. 75-115. Wiley, New York. Spektrum, Heidelberg.
Fränzle, O., 1998. Grundlagen und Entwicklung der Ökosystemforschung. In: O. Fränzle, F. Müller, and W. Schröder, (Hg.). *Handbuch der Umweltwissenschaften*, Kap. II-2.1 (In press).
Fränzle, O., D. Kuhnt, G. Kuhnt, and R. Zölitz, 1987. Auswahl der Hauptforschungsräume für das Ökosystemforschungsprogramm der Bundesrepublik Deutschland. Umwelforschungsprogramm der Bundesrepublik Deutschland. Umweltforschungsplan des BMU. Forschungsbericht 101 04 043/02. Kiel, Berlin.
Fränzle, O. and W. Kluge, 1997. Typology of water transport and chemical reactions in groundwater/lake ecotones. In: J. Gibert, J. Mathieu, and F. Fournier, (Eds.). *Groundwater/Surface Water Ecotones: Biological and Hydrological Interactions and Management Options*, p. 127-134. Cambridge Univ. Press, Cambridge.
Garniel, A., 1988. Geomorphologische Detailaufnahme des Blattes L 1926 Bordesholm. Staatsexamensarbeit, Univ. Kiel.
Gigon, A., 1983. Über das biologische Gleichgewicht und seine Beziehungen zur ökologischen Stabilität. *Ber. Geobot. Inst. Eidg. Techn. Hochsch., Zürich* 50: 149-177.
Golley, F.B., 1993. A History of the Ecosystem Concept in Ecology. Yale Univ. Press, New Haven, London.
Hemprich, G., 1991. Landschaftsökologische Untersuchungen im Bereich des Belauer Sees und Schmalensees. Dipl.-Arbeit Geogr. Inst., Univ. Kiel.
Higashi, M. and T.P. Burns, 1991. Theoretical Studies of Ecosystems. The Network Approach. Cambridge.
Hogg, T., B.A. Huberman, and J.M. McGlade, 1989. The stability of ecosystems. *Proc. R. Soc.* B. 237: 43-51.
Holling, C.S., 1966. The resilience of terrestrial ecosystems: Local surprise and global change. In: W.M. Clark and R.E. Munn, (Eds.). *Sustainable Development of the Biosphere.* Oxford.
Jørgensen, S.E., 1990. Modelling in Ecotoxicology. Elsevier, Amsterdam.
Jørgensen, S.E., 1992. Development of models able to account for changes in species composition. *Ecol. Model.* 62: 195-208.
Jørgensen, S.E., 1995. The growth rate of zooplankton at the edge of chaos: Ecological models. *J. Theor. Bio.* 175: 13-21.
Jørgensen, S.E., 1997a. Thermodynamik offener Systeme. In: O. Fränzle, F. Müller, and W. Schröder, (Eds.). *Handbuch der Umweltwissenschaften.* Kap. III-1.6. Ecomed, Landshut/Lech.

Jørgensen, S.E,. 1997b. Möglichkeiten zur Integation verschiedener theoretischer Ansätze. In: O. Fränzle, F. Müller, and W. Schröder, (Eds.). *Handbuch der Umweltwissenschaften*, Kap. III-1.8. Ecomed, Landshut/Lech.

Kluge, W. and O. Fränzle, 1992. Einfluß von terrestrisch-aquatischen Ökotonen auf den Wasser- und Stoffaustausch zwischen Umland und See. *Verh. Ges. Ökologie* 21: 401-407.

Kuhnt, G., 1994. Regionale Repräsentanz. Beiträge zu einer raumorientierten Meßtheorie. Habil.-Schrift, Univ. Kiel.

Likens, G.E., F.H. Bormann, R.S. Piercem, J.S. Eaton, and N.M. Johnson, 1977. Biogeochemistry of a Forested Ecosystem. Springer, New York.

Likens, G.E. (Ed.), 1989. Long-term Studies in Ecology: Approaches and Alternatives. Springer, New York.

Locker, A. (Ed.), 1973. Biogenesis, Evolution, Homeostasis. Springer, Berlin.

Margalef, R., 1995. Information theory and complex ecology. In: B.C. Patten and S.E. Jørgensen, (Eds.). *Complex Ecology: The Part-whole Relation in Ecosystems*. Englewood Cliffs, p. 40-50.

May, R.M., 1972. Will a large complex system be stable? *Nature* 238: 413-414.

MacArthur, R. H. (1955): Fluctuations of animal populations and a measure of community stability. Ecology 36: 533-536.

Mcintosh, R.P., 1985. The Background of Ecology. Cambridge Univ. Press, Cambridge.

McMurtrie, R.E., 1975. Determinants of stability of large randomly connected systems. *Journal of Theor. Biol.* 50: 1-11.

Müller, F., 1992. Hierarchical approaches to ecosystem theory. *Ecol. Model.* 63: 215-242.

Müller, F., 1997. State-of-the-art in ecosystem theory. *Ecol. Model.* 100: 135-161.

Müller, F., 1998. Gradients in ecological systems. *Ecol. Model.* 108: 3-21.

Müller, F. and W. Windhorst, 1991. Die Modellierungsstrategie des FE-Vorhabens "Ökosystemforschung im Bereich der Bornhöveder Seenkette". *Berichte des Forschungszentrums Waldökosysteme*, Reihe B, 22: 75-93.

Müller, F. and M. Leupelt (Eds.), 1998. Eco Targets, Goal Functions, and Orientors. Springer, Berlin.

Nicolis, G. and I. Prigogine, 1977. Self-organization in Non-equilibrium Systems. New York.

Niering, W.A., 1987. Vegetation dynamics (succession and climax) in relation to plant community management. *Conservation Biology* 4: 287-295.

Odum, E.P., 1969. The strategy of ecosystem development. *Science* 164: 262-270.

Odum, H.T., 1957. Trophic structure and productivity of Silver Springs, Florida. *Ecol. Monographs* 27: 55-112.

Odum, H.T., 1983 Maximum power and efficiency: A rebuttal. *Ecol. Model.* 20: 71-82.

O'Neill, R.V., D.L. De Angelis, J.B. Waide, and T.H.F. Allen, 1986. A Hierarchical Concept of Ecosystems. Monographs in Population Ecology 23. Princeton University Press, Princeton.

Pahl-Wostl, C. and R. Ulanowicz, 1993. Quantification of species as functional units within an ecological network. *Ecol. Model.* 66: 65-79.

Patten, B.C., 1992. Energy, emergy, and environs. *Ecol. Model.* 62: 29-70.

Patten, B.C., 1998. Steps towards a cosmography of ecosystems: 20 remarkable properties of life in environment. In: F. Müller and M. Leupelt, (Eds.). *Eco-targets, Goal functions, and Orientors*, p. 137-160.

Patten, B.C. and S.E. Jørgensen (Eds.), 1995. Complex Ecology: The Part-whole Relation in Ecosystems. Englewood Cliffs.

Pickett, S.T.A., 1989a. Space-for-time substitution as an alternative to long-term studies. In: G.E. Likens, (Ed.). p. 110-135.

Pickett, S.T.A., 1989b. Long-term studies: Experience from the Institute of Ecosystem Studies and Cary Conference II. *MAB-Mitteilungen* 31: 116-141.

Pimm, S.L., 1982. Food Webs. London.

Piotrowski, J.A., 1991. Quartär- und hydrogeologische Untersuchungen im Bereich der Bornhöveder Seenkette, Schleswig-Holstein. Berichte-Reports, Geol.-Paläontol. Inst., Univ. Kiel 43.

Popper, K.R., 1959. The Logic of Scientific Discovery. Harper and Row, New York.

Popper, K.R., [4]1972. Das Elend des Historizismus. Mohr, Tübingen.

Prigogine, I., 1967. Thermodynamics of Irreversible Processes. New York.

Prigogine, I., 1976. Order through fluctuation: selforganization and social system. In: E. Jantsch and C.H. Waddington, (Eds.). *Evolution and Consciousness*. Reading Mass., p. 93-126.

Prigogine, I., 1985. Vom Sein zum Werden. Zeit und Komplexität in den Naturwissenschaften. München.

Schleuss, U., 1992. Böden und Bodenschaften einer Norddeutschen Moränenlandschaft. *EcoSys - Beiträge zur Ökosystemforschung*, Suppl. Bd. 2. Kiel.

Schneider, E.D. and J. Kay, 1994a. Life as a manifestation of the second law of thermodynamics. *Math. Comput. Model.* 19, 25-48.

Schneider, E.D., and J. Kay, 1994b. Complexity and thermodynamics: Towards a new ecology. *Futures* 26: 626-647.

Scholle, D. and J. Schrautzer, 1993. Zur Grundwasserdynamik unterschiedlicher Niedermoor-Gesellschaften Schleswig-Holsteins. *Z. Ökol. u. Naturschutz* 2: 87-98.

Solbrig, O.T. and G. Nicolis (Eds.), 1991. Perspectives in Biological Complexity. IUBS, Paris.

Straskraba, M., 1995. Cybernetic theory of ecosystems. *Umweltwissenschaften* 6: 31-52.

of model predictions with selected observations from the real system as the sole criterion for evaluation. Rykiel (1996) argues that different validation criteria are appropriate for different types of models and suggest that validation should mean simply that a model is acceptable for its intended use, Rykiel also emphasises that no generally accepted standards currently exist for the validation of ecological models and provides an excellent discussion of the semantic and philosophical debate concerning validation in which ecological modlers still are involved. Straskraba and Gnauck (1985) and Jørgensen (1994) give several illustrative examples of model validations of concrete models. Although the details of the debate may seem quite confusing, we believe that the basic ideas involved in evaluating the usefulness of a model are easy to understand, Thus we prefer to refer simply to the process of 'model evaluation', and to focus our attention on examination of the various characteristic of a model that make it a potentially useful tool.

During model evaluation we should examine a broad array of qualitative as well as quantitative aspects of model structure and behaviour. We begin by assessing reasonableness of model structure and interpretability of functional relationships within the model. Reasonableness of model structure and interpretability are defined relative to the ecological, economic, or other subject-matter context of the model. Next we evaluate correspondence between model behaviour and the expected patterns of model behaviour that we describe during conceptual model formulation. We then examine more formally the correspondence between model predictions and real-system data. These comparisons may or may not involve the use of statistical tests of significance. Finally, we conduct a sensitivity analysis of the model, which usually consists of sequentially varying one model parameter (or set of parameters) at a time and monitoring the subsequent effects on model behaviour. By identifying those parameters that most affect model behaviour, sensitivity analysis provides valuable insight into the functioning of the model and also suggests that the level of confidence we should have in model predictions.

Relative importance of these steps for any given model depends on the specific objectives of the modelling project. If we are dissatisfied with the model during any of these steps we return to an earlier step in systems analysis to make appropriate modifications to the model. The step to which we return depends upon the reasons for which we are dissatisfied.

3.4 Phase IV: Model Use

The goal of the final phase of systems analysis is to meet the objectives that were identified at the beginning of the modelling project. Most often we wish to use the model to simulate system dynamics under alternative management policies or environmental situations. The general scheme for model use follows exactly the same steps involved in addressing a question through experimentation in the real world. We first develop and execute the experimental design in favour of the apparent expediency of a 'shotgun' approach made possible by the tremendous computing capabilities of modern computers. The ability to generate voluminous results does not preclude the need for a logical approach to the problem.

Next, we analyse and interpret simulation results. For stochastic models this often includes the use of statistical procedures such as analysis of variance to compare model predictions under different circumstances. Since results of our initial simulations invariably raise new questions, we almost always run additional simulations further examining selected types of management policies or environmental situations. These results will raise more questions, which will suggest more simulations, which will raise more questions etc. At some point we will need to modify the model to address these new questions, which brings us full circle to the formulation of a new conceptual model. However, before continuing development of more models, we must complete the last step of model use which is to communicate results to the appropriate audience.

3.5 Iteration of phases

The four phases of systems analysis are highly interconnected. Although theoretically we may think of the process as proceeding sequentially in the indicated order, in practice we may cycle through several phases more than once. During any phase we may find that we have overlooked or misrepresented an important system component or process and need to return to an earlier phase, often to conceptual model formulation or quantitative model specification. During model evaluation, in particular, we examine the model to detect any inadequacies that may require us to cycle back to earlier phases. Discovery of such inadequacies in the model during its development usually provides additional insight into dynamics of the system-of-interest and is an important benefit of modelling.

4. Closing comments

Throughout the world, both in developed and developing countries, we face challenging ecological problems, often related to our attempts to achieve economic growth without damaging the functionality of the complex ecological systems that ultimately form the basis of human existence. Systems analysis refers both to the functioning of complex systems. Thus, formal exposure to systems analysis should be an integral part of the training of all ecologist interested in the development of predictive theories of ecosystem dynamics.

REFERENCES

Bertalanffy, L. von. 1968. General Systems Theory: Foundations, Development, Applications. George Braziller, New York.

Forrester, J.W. 1961. Industrial Dynamics. MIT Press, Cambridge, Mass.

Gold, H.J., 1977. Mathematical Modeling of Biological Systems: An Introductory Guidebook. Wiley, New York.

Grant, W.E. 1986. Systems Analysis and Simulation in Wildlife and Fisheries Sciences. Wiley, New York.

Grant, W.E., E.K. Pedersen, and S.L. Marin. 1997. Ecology and Natural Resource Management: Systems Analsyis and Simulation. Wiley, New York.

Haefner, J.W. 1996. Modeling Biological Systems: Principles ad Applications. Chapman and Hall, New York.

Hastings, N.A.J and J.B. Peacock. 1975. Statistical Distributions. Butterworth, London.

Holling, C.S. 1978. Adaptive Environmental Assessment and Management. Wiley, new York.

Innis, G.S. 1979. A spiral approach to ecosystem simulation, I. In G.S. Innis and R.V. O'Neill (eds) *Systems Analysis of Ecosystems*. International Cooperative Publishing House, Burtonsville, Md.

Jeffers, J.N.R. 1978. An introduction to Systems Analysis: With Ecological Applications. University Park Press Baltimore, MD

Jørgensen, S.E. 1986. Fundamentals of Ecological Modelling. 1st Edition. Elsevier, Amsterdam, Oxford, New York, Tokyo.

Jørgensen, S.E. 1994. Fundamentals of Ecological Modelling. 2nd Edition. Elsevier, Amsterdam, Oxford, New York, Tokyo

Jørgensen, S.E., Nors Nielsen, S and Jørgensen, J.A. 1991. Handbook of Ecolotoxicology and Ecological Parameters. Elsiver, Amsterdam, Oxford, New York, Tokyo

Jørgensen, S.E., Hallin-Sørensen, B. and Nors Nielsen, S. 1995. Handbook of Environmental and Ecological Modelling. Lewis Publ. New York, Boca Raton, London, Tokyo.

Kitching, R.L. 1983. Systems Ecology: An Introduction to Ecological Modeling. University of Queensland Press, St Lucia, Queensland.

Patten, B.C. 1971. A primer for ecological modelling and simulation with analog and digital computers. In B.C Patten (ed), Systems Analysis and Simulation in Ecology (Vol1). Academic, New York.

Rykiel, E J. Jr. 1966. Testing ecological models: The meaning of validation. Ecological Modeling 90:229-244

Straskraba, M. and A. Gnauck. 1985. Freshwater Ecosystems, Models and Simulations. Elsevier. Amsterdam, Oxford, New York, Tokyo.

Different Approaches in Recent Ecosystem Theory

Introduction

If we take a critical look at the chapterss presented in Section I, it becomes obvious that ecosystem analysis is still a very young branch of science. Therefore, the central role of theories, as it is realised in the traditional scientific disciplines, still has to be discovered and agreed upon in systems ecology. This valuation can be explained by the fact that ecology, in general, has been confronted with such a problematic bundle of applied environmental, political and societal questions, hence leaving only a small space for the development of ecological theory. Thus, the recent state of the theory can be described as dynamic, far from a period of scientific maturity, although theories, in fact, have to be perceived as the real destinations in science. In addition, a mature theory is the prerequisite for a wide application of science to real life problems.

Theories are sets of hypotheses which aggregate the proved and reliable knowledge of a discipline. They function as an explaining, high level of abstraction within which all single phenomena should be explainable by coherent, theory-based laws. Another function of theories results in questions, hypotheses, and ideas for the empirical branches of the respective science. Ecosystems theory, consequently, has to utilise the empirical data for the integration, organisation and improvisation of the understanding of the general features, patterns, and causalities concerning the interactions in and between ecological systems.

It has been stated before that these tasks actually can be fulfilled on the basis of many different abstract concepts only; there is not one coherent system comprehension but a multitude of interacting ideas which are brought together in the second part of this volume. There are 19 chapters which illustrate different aspects of ecosystem understanding. These single attitudes will be discussed integratively in a final section of this Section II. All these different approaches are probably necessary to give a satisfying, holistic description of such complex entities as ecosystem. They are to a high extent complementary. The bundle of approaches includes different thermodynamic and energy-based concepts, theories of self-organisation and emergence, cybernetics, information theory, hierarchy theory, network theory, and utility theory. It contains static and dynamic approaches, whereby the last group is represented by analyses of ecosystem stability, resilience, by catastrophe theory and chaos. Ecosystems are looked at from different temporal perspectives (e.g. succession) as well as from different scales (e.g. landscapes). In the final chapter, it will be tested in how far these different concepts can be scientifically integrated and in how far the resulting patterns includes a potential for a modern direction for environmental management.

II.1.1 A General Outline of Thermodynamic Approaches to Ecosystem Theory

Sven E. Jørgensen

1. Development of ecosystem theories by application of thermodynamics

The science of thermodynamics deals with relations between different forms of energy. It is based on general laws of nature, the first, second and third law of thermodynamics. In Chapter II.1.3 it will be proposed to tentatively introduce another, fourth law of thermodynamics. This has arisen out of the needs to describe reactions of biological/ecological systems properly. By logical reasoning from these laws, it is possible to correlate many of the observable properties of ecosystems. The thermodynamic approach is particularly well fitted as a tool to describe ecosystems from an holistic point of view because it is based on the macroscopic flows of energy and mass. Many ecologists use for instance the energy transfer rate dE / dt as a currency unit (Brown, 1995).

Classical thermodynamics focuses on transformations from one form of energy to another, relatively close to thermodynamic equilibrium. Several state variables are introduced to enable the user of thermodynamics to make calculations of energy transformations and to determine which physical and chemical reactions can be realised under which conditions. Classical thermodynamics cannot be applied on systems far from thermodynamic equilibrium as for instance ecological systems. A development of classical thermodynamics to far-from-equilibrium thermodynamics has been initiated by Onsager (1931) and Prigogine (1947). The latter has introduced the term 'dissipative structure' to describe far-from-thermodynamic-equilibrium systems, but a wider theoretical application of thermodynamics in ecosystem theory was not initiated until the late seventies.

Isolated systems will degenerate to thermodynamic equilibrium, which means that all gradients are eliminated and all chemical substances are transferred into the compounds that have the lowest free energy under the given circumstances (when oxygen is present in a sufficient amount, all components will be in the highest oxidation state, i.e., as nitrate, sulphate, carbon dioxide and so on). The system at thermodynamic equilibrium will not be able to perform work – all possible work has been 'squeezed out' of the system. Such a system will not bear life because it is only possible in a system that is far from thermodynamic equilibrium.

Ecosystems are non-isolated systems in the thermodynamic sense, i.e., they are at least open to energy, and therefore non-isolated. Closed systems are open to flows of energy but not to the flows of matter, while open systems are open to both flows of matter and energy. *Most* ecosystems are open systems – *all* ecosystems are non-isolated systems. Export and import of matter is based upon exchange processes with adjacent ecosystems, while the entire ecosphere with good approximation comprises of a constant total amount of matter. All ecosystems and the entire ecosphere are open to energy and receive currently a flow of energy by solar radiation. Ecosystems could, according to the basic thermodynamic laws, not develop or even survive without this flow of energy. The maintenance of ecosystems far from equilibrium requires a steady flow of energy to cover the energy needed to maintain a system far from equilibrium and hence prevent it from developing towards thermodynamic equilibrium.

In this section, we will examine how many of the properties of ecosystems, and how much of a comprehensive ecosystem theory, can be derived from the three basic laws of thermodynamics applied on open (non-isolated) systems. The presentation here is partly based on a summary of three chapters dealing with the application of the conservation principles (Patten et al., 1997), the second law of thermodynamics (Straskraba et al., 1999) and of the third law of thermodynamics (Jørgensen et al., 1998) on ecosystems, characterised as open or at least non-isolated systems.

2. Mass and energy conservation

Energy and matter are conserved according to basic physical concepts which are valid for all systems, and therefore also all ecosystems. This means that energy and matter are neither created nor destroyed. The expression 'energy and matter' is used, as energy can be transformed into matter and matter into energy. The unification of the two concepts is possible by the use of Einstein's law:

$$E = m c^2 \quad (ML^2T^{-2}), \tag{1}$$

where E is energy, m is mass and c is the velocity of electromagnetic radiation in a vacuum ($= 3 * 10^8$ m sec^{-1}). The transformation from matter into energy and energy into matter is only of interest for nuclear processes and does not need to be applied to the ecosystems on Earth. We might therefore break the proposition down to two more useful propositions, when applied in ecology:

1. **Ecosystems conserve matter**
2. **Ecosystems conserve energy**

The conservation of matter may mathematically be expressed as follows:

$$dm / dt = \text{input} - \text{output} \quad (MT^{-1}), \tag{2}$$

where m is the total mass of a given system. The increase in mass is equal to the input minus the output. The practical application of the statement requires that the boundaries of the defined system are indicated.

Concentration is used instead of mass in most models of ecosystems:

$$V\, dc/ dt = \text{input} - \text{output} \quad (MT^{-1}), \tag{3}$$

where V is the volume of the system under consideration and assumed constant.

If the law of mass conservation is used for chemical compounds that can be transformed to other chemical compounds the equation (3) should be changed to:

$$V * dc / dt = \text{input} - \text{output} + \text{formation} - \text{transformation} \ (MT^{-1}) \tag{4}$$

The principle of mass conservation is widely used in the class of ecological models called biogeochemical models. The equation is set up for the relevant elements, e.g., for eutrophication models, for C, P, N and perhaps Si (see Jørgensen et al., 1978, Jørgensen, 1982a and 1994).

For terrestrial ecosystems mass per unit of area is usually applied in mass conservation equations. The mass flow through a food chain is mapped by the mass conservation principle. The food taken in by one level in the food chain is used for respiration, wasted food, undigested food, excretion, growth and reproduction. If the growth and reproduction are considered as the net production, it can be stated that

$$\text{net production} = \text{intake of food} - \text{respiration} - \text{excretion} - \text{wasted food} \tag{5}$$

The ratio of the net production to the intake of food is called the net efficiency. The net efficiency is dependent on several factors, and can often be as low as 10-20%. Any toxic matter in the food is unlikely to be lost through respiration and excretion because it is much less biodegradable than the normal components in the food. This being so, the net efficiency of toxic matter is often higher than for normal food components, and as a result some chemicals, such as chlorinated hydrocarbons including DDT and PCB, will be magnified in the food chain. This phenomenon is called biological magnification. DDT and other chlorinated hydrocarbons have an especially high biological magnification because they have a very low biodegradability and are excreted from the body very slowly, due to dissolution in fatty tissue.

The understanding of the principle of conservation of energy, called the first law of thermodynamics, was initiated in 1778 by Rumford. He observed the large quantity of heat that appeared when a hole is bored in metal. Rumford assumed that the mechanical work was converted to heat by friction. He proposed that heat was a type of energy that is transformed at the expense of other forms of energy, in this instance mechanical energy. It was left to J.P. Joule in 1843 to develop a mathematical relationship between the quantity of heat developed and the mechanical energy dissipated. Two German physicists, J.R. Mayer and H.L.F. Helmholtz, working independently, showed that when a gas expands the internal energy of the gas decreases in proportion to the amount of work performed. These observations led to the first law of thermodynamics: energy can neither be created nor destroyed. If the concept internal energy, dU, is introduced:

$$dQ = dU + dW \quad (ML^2T^{-2}), \tag{6}$$

where

dQ = thermal energy added to the system
dU = increase in internal energy of the system
dW = mechanical work done by the system on its environment.

Then the principle of energy conservation can be expressed in mathematical terms as follows: U is a state variable which means that the integral of dU from 1 to 2 is independent on the pathway 1 to 2. The internal energy, U, includes several forms of energy: mechanical, electrical, chemical, magnetic energy, etc.

The transformation of solar energy to chemical energy by plants conforms with the first law of thermodynamics. For the next level in a food chain, the herbivorous animals, the energy balance can also be set up:

$$F = A + UD = G + H + UD, \quad (ML^2T^{-2}) \tag{7}$$

where

F = the food intake converted to energy (Joule)
A = the energy assimilated by the animals
UD = undigested food or the chemical energy of faeces
G = chemical energy of animal growth
H = the heat energy of respiration.

These considerations pursue the same lines as those mentioned in context with equation (5), where the mass conservation principle was applied. The energy content per g ash-free organic material is surprisingly uniform, about 16-24 kJ / g ash-free material which implies that an equation based on mass conservation can easily be translated into an equation expressing energy conservation.

Ecological energy flows are of considerable environmental interest as calculations of biological magnifications are based on energy flows, as mentioned above.

There is a close relationship between energy flow rates and organism size and some of the most useful of these relationships are illustrated in Peters (1983), Jørgensen (1994 and 1990) and Kooijman (1993). Any self sustaining ecosystem will contain a wide spectrum of organisms ranging in size from tiny microbes to large animals and plants. The small organisms account in most cases for most of the respiration (energy turnover), whereas the larger organisms comprise most of the biomass.

Ecological processes may be described as the driving force, X, balanced by a frictional force that develops almost in proportion to the rate of flow, J, so that there is a balance of forces:

$$X = R * J \qquad (8)$$

where R is the resistance. Equation (8) is, as seen, a parallel to Ohm's law. L = 1/R may be denoted conductivity, and the equation (9) may be reformulated according to Onsager (1931), who stated that Lij = L ji, covering the conductivity of two interacting components:

$$J = L * X \qquad (9)$$

Population dynamic processes may be described in the same manner, for instance the metabolism J of a population N:

$$J = L * N \qquad (10)$$

The flux here is the flow of food through a food chain, expressed in units such as carbon per square metre of ecosystem area per unit of time. The force is a function of the concentration gradient of organic matter and biomass.

Energy is the driving factor of ecosystems. The various energy flows, however, are of different quality with different ratios of solar energy equivalents required in the world web to generate one equivalent of that specific type of energy (refer to H.T. Odum (1983) and Chapter II.6.).

The energy and mass conservation principles may be considered as constraints on ecosystem processes. They imply that 'nothing is coming from nothing', which is so basic that all ecological models should be in accordance to these principles. It is the shortcoming of many simple population dynamic models, for instance Lotka-Volterra's equation, that they don't obey these principles (Hall, 1988).

The conservation principles are, however, embodied in the principles of limiting factors which are extensively applied in ecological growth equations, for instance in Michaelis-Menten's equation:

$$\mu = \mu_{max} S /(k_m + S) \qquad (11)$$

where S is the limiting factor (for instance the concentration of nutrients for plants, μ is the growth rate, μ_{max} the maximum growth rate and k_m is the Michaelis-Menten's constant).

The mass conservation principle implies that there are limits to the amounts of biomass per unit of volume or area. When the most limiting factors have been depleted, new biomass can only be formed by changing the composition of the biomass towards a lower content of the limiting factors, which is only feasible to a very limited extent. For instance phytoplankton needs phosphorus as an indispensable nutrient. Phytoplankton contain in average about 1% of phosphorus, but concentrations from 0.4% to 2.5% can be found in nature. It means that when the lower limit of 0.4% has been reached, further growth of phytoplankton is only possible if phosphorus is added.

Developed (mature) ecosystems have only minor amounts of inorganic nutrients, as most of the nutrients have already been applied to build up the present biomass. The growth of ecosystems are therefore completely dependent on cycling of nutrients. This illustrates the interdependence of the ecosystem components, which may be referred to as the indirect effect; see also Chapter II.7.3.

Characteristic properties of ecosystems which can be derived from the conservation principles are given in Patten et al. (1997). A brief summary of ecosystem characteristics in four points is given below:

1) Four phases of matter are recognised in the application of the conservation principles: solid (the lithosphere), liquid (the hydrosphere), gaseous (the atmosphere) and living (the biosphere). We may add the technosphere, which covers human made constructions, and the semiosphere, representing the informational aspects of the materialistic ecosphere.

2) All systems owe their cohesion to bonding (interaction). System bonding may be by conservative substance exchange (transactions) or non-conservative information exchange (relations). In both cases, identity constraints of coupling mediate the bonds.

3) A universal metamodel for system change may be described under state-space determinism. Exogenous conditions (inputs) and endogenous conditions (states) are lawfully mapped into behavioural dynamics (outputs). There are two fundamentally different kinds of state-space systems: objects (non-living, which respond reactively to physical inputs) and subjects (living, which can respond proactively based on phenomenal inputs which are models of physical inputs).

4) Time is conserved (but irreversible in accordance to the second law of thermodynamics) and space is conserved locally.

3. The second law of thermodynamics and ecology

Spontaneous changes are always accompanied by a degradation into a more dispersed chaotic form of the ecosystem, as gas expands to fill the available volume and a hot body cools to the temperature of its surroundings. The first law leads to the introduction of internal energy of the system and it identifies the permissible changes, while the second law of thermodynamics leads to the concept of entropy, S, and exergy, which identifies the natural (or spontaneous) *direction* of changes.

The second law of thermodynamics is expressed mathematically by application of the entropy concept, S:

$$dS = \partial Q / T \quad (ML^2T^{-2}) \tag{12}$$

Entropy has the property for *any* process in an isolated system that:

$$dS \geq 0 \tag{13}$$

where > refers to real, i.e., non-equilibrium processes and = refers to all equilibrium processes. Notice that S is a state variable: $\int_1^2 dS$ is independent of the path from 1 to 2.

In nature we can distinguish two types of processes: spontaneous processes, which occur naturally without an input of energy from outside, and non-spontaneous processes, which require an input of energy from outside. These facts are included in the second law of thermodynamics, which states that processes involving energy transformations will not occur spontaneously unless a degradation of energy from a non-random to a random form occurs, or from a concentrated into a dispersed form. In other words, all energy transformations will involve energy of high quality being degraded to energy of lower quality, e.g., potential energy to heat. The quality of energy is measured by means of the thermodynamic state variable entropy (high quality = low entropy). The second law may also be interpreted as 'time has only one direction and the direction is toward increasing entropy', or 'all processes are irreversible in the sense that if we want to reverse a process it will require changes in the environment (for instance input of energy from the environment)'.

From a physical standpoint the environmental crisis is an entropy crisis, as entropy is increased by dispersion of the pollutants, which can be shown by a simple model consisting of two bulbs of equal volume, connected with a valve. One chamber contains one mole of a pure ideal gas (it means that $p V = R T$) and the second one is empty. If we open the valve between the two chambers, and assume that $\Delta U = 0$ and T is constant, an increase in entropy will be observed:

$$\Delta S = \int_1^2 \partial Q / T = Q / T = W = R * \ln V_2 / V_1 = R \ln_2 \qquad (14)$$

where ΔS = the increase in entropy and V_2 is the volume occupied by the model of gas after the valve was opened, while V_1 is the volume before the valve was opened.

Thus, paradoxically, the more we attempt to maintain order, the more energy is required and the greater stress (entropy) we inevitably put on the environment, as all energy transformations from one form to another imply the production of waste heat and entropy according to the second law of thermodynamics. We cannot escape the second law of thermodynamics, but we can minimise energy waste by:

1. keeping the energy chain as short as possible

2. increasing the efficiency, i.e., the ratio of useful energy output (named exergy; further detail about this concept; see below) to the total energy input, and by

3. wasting as little heat to the surroundings as possible, e.g., by insulation, and by using heat produced by energy transfer (heat produced at power stations can be used for heating purposes).

These rules are pertinent for solutions to the problems related to the greenhouse effect and several other pollution problems.

Organisms, ecosystems and the entire ecosphere possess the essential thermodynamic characteristic of being able to create and maintain a high state of internal organisation or a condition of low entropy (entropy can be said to measure disorder, lack of information on molecular details, or the amount of unavailable energy). Low entropy is achieved by a continuous dissipation of energy of high utility (e.g. light or food) to energy of low utility (e.g. heat). Order is maintained in the ecosystem by respiration that continually produces disorder (heat).

The second law of thermodynamics may also explain why and how ecosystems can maintain organisation. A system tends to move spontaneously toward increasing disorder (or randomness), and if we consider the system as consisting of an ecosystem and its surroundings, we can understand that order (negative entropy) *can* be produced in the ecosystem *if* and *only if* – according to the second law of thermodynamics – more disorder (entropy) is produced in its surroundings. Thermodynamics often solely considers dissipation of energy. However, dissipation of matter accompanies dissipation of energy, and dissipation of information is based ultimately on degradation of 'the energy matter makers', upon which information is always carried.

It is obvious from the second law of thermodynamics that the state in which matter, energy and information are spread uniformly throughout the volume of the system is more likely to occur than any other configuration, and is particularly a lot more probable than the state in which all matter, energy or information is concentrated within a smaller region. It would appear therefore as if ecosystems violate the second law of thermodynamics, but that is of course not the case. The maintenance of ecosystems far from thermodynamic equilibrium can be explained by the presence of a constant energy source, solar radiation. This is explained in more detail below.

The chemical energy released by spontaneous chemical processes is, of course, conserved and can be found somewhere else in the system. The decomposition processes of cells and organisms are called catabolic reactions. They involve decomposition of more complex chemical compounds in the food to simpler organic or inorganic compounds. These reactions are, of course, irreversible and generate heat and entropy.

The ecosystem catabolism is the sum of catabolic processes of all organisms and cells in the ecosystem. The energy released by the catabolic processes firstly is used by the organisms and cells to maintain a 'status quo' situation. If the law of mass conservation is used on the feeding of an organism (see part 2, this chapter), a part of the food will not be assimilated but lost as faeces. Another part will be used in catabolic processes to keep the organism alive; this means that the energy released is used for the life processes and for the maintenance of a temperature and chemical potential different from the environment. The energy needed corresponds to the minimum food intake. Poikilothermic animals and hibernators reduce energy needs by reduction of activity and body temperature at low environmental temperature.

The catabolic processes involve the decomposition of food components into more simple molecules, and are in this respect equivalent with combustion. They differ, however, from combustion in several other respects. Unlike combustion they proceed at normal temperatures without the sudden liberation of large amounts of heat. The total amount of energy released from a given decomposition process is the same – regardless of whether combustion or catabolism is involved in the process. The difference lies in the fact that the catabolic processes are organised, catalysed by enzymes, and consist of many integrated step reactions, whereas combustion is an uncontrolled, disordered series of reactions proceeding at high temperature.

At ordinary temperature, as well as in the presence of atmospheric oxygen most organic compounds are relatively stable. At high temperatures, the rate of collision between molecules is increased and the internal chemical bonds are weakened. As a result the compounds decompose most often by reactions with oxygen, and heat is released in the process.

One of the major functions of enzymes in cells and organisms is to eliminate the necessity for high temperature by removing the requirement for high energy activation. The combination of an enzyme with its substrate, the chemical changes that occur in the enzyme-bound substrate, and the release of the end product, are all reactions that have a low energy of activation. The entire sequence of processes involved in catabolism can therefore proceed spontaneously at ordinary temperature and catalysis is achieved.

Another essential function of enzymes is to provide for an orderly step-wise decomposition of organic nutrients. The third essential function of enzymes is to make part of the energy that is available from the catabolic processes useful for the cells and organisms. The multiplicity of the reactions and the energetic coupling between them prevent the release of very large amounts of heat in any one step and permit the accumulation of chemical bond energy in the form of ATP within the cell.

The catabolic reactions are the source of energy for maintenance of life in the ecosystem. These reactions generate heat and entropy that must be transferred, however, to the environment to prevent a constant increase in the temperature of the ecosystem. The ecosystem must therefore be non-isolated – an extremely important property.

4. Exergy

The concept of entropy and its practical implications are difficult to conceive. Exergy was therefore introduced (Evans, 1966) as an easier, conceivable, and more applicable concept. It is based upon a classification of energy: energy which is useful and can do work, that is exergy, and energy which cannot do work, as for instance heat without a temperature gradient. By measuring the energy that can do work, exergy expresses energy with a built-in measure of quality, for example the chemical energy in biomass. Exergy accounts for natural resources (Erikson et al., 1976) and can be considered as fuel for any system that converts energy and matter in a metabolic process. Ecosystems consume exergy, and an exergy flow through the system is necessary to keep the system functioning. Exergy stored in the system directly expresses the distance from the 'inorganic soup' in energy terms, as will be further explained in Chapter II.1.3. The introduction of exergy makes it also possible to avoid the artificial concept, negentropy, which has no real existence.

The exergy of a system is defined as the amount of work (entropy-free energy) a system can perform, when it is brought into thermodynamic equilibrium with its environment. Let us assume that the considered system is characterised by the extensive state variables S, U, V, N_1, N_2, N_3......., where S is the entropy, U is the energy, V is the volume and N_1, N_2, N_3......are moles of various chemical compounds, and by the intensive state variables, T, p, μ_{c1}, μ_{c2}, μ_{c3}....If the system is coupled to a reservoir, a reference state, by a shaft, the system and the reservoir are forming a closed system. The reservoir (the environment) is characterised by the intensive state variables T_o, p_o,, μ_{c1o}, μ_{c2o},, μ_{c3o}....and as the system is small compared with the reservoir, the intensive state variables of the reservoir will not be changed by interactions between the system and the reservoir. The system develops toward thermodynamic equilibrium with the reservoir and is simultaneously able to release entropy-free energy to the reservoir. During this process the volume of the system is constant as the entropy-free energy must be transferred through the shaft only. If a boundary displacement against pressure of the reference environment should take place, it would not be available as useful work. The entropy is also constant as the process is an entropy-free energy transfer from the system to the reservoir, but the intensive state variables of the system become equal to the values for the reservoir. The total transfer of entropy-free energy in this case is the exergy of the system. It is seen from this definition that exergy is dependent on the state of the total system (= system + reservoir) and not dependent only on the state of the system. Exergy is therefore not a state variable. In accordance with the first law of thermodynamics, the increase of energy in the reservoir, Δ U, is:

$$\Delta U = U - U_o \tag{15}$$

where U_o is the energy content of the system after the transfer of work to the reservoir has taken place. According to the definition of exergy, Ex, we have:

$$Ex = \Delta U = U - U_o$$

$$\text{As } U = TS - pV + \Sigma \mu_c N_i \tag{16}$$

(see any textbook in thermodynamics), and

$$U_0 = T_0 S - p_0 V + \sum_c^c \mu_{co} N_i \tag{17}$$

we get the following expression for Exergy :

$$Ex = S(T - T_o) - V(p - p_o) + \sum_c (\mu_c - \mu_{co}) N_i \tag{18}$$

As reservoir is the reference state, we can select the same (eco)system but at thermodynamic equilibrium, where all components are inorganic and at the highest oxidation state, if oxygen is present (nitrogen as nitrate, sulphur as sulphate and so on). The reference state will in this case correspond to the same ecosystem without life forms and with all chemical energy utilised or as an 'inorganic soup'. Usually it implies that we consider $T = T_o$, and $p = p_o$, which means that the exergy becomes equal to the Gibb's free energy of the system, or the chemical energy content of the system relative to the reference state (it is further discussed in Chapter II.1.3), included the content of information (see also below where the exergy content of information is discussed). Notice that – as the equation shown above also emphasises – exergy is dependent on the state of the environment (the reservoir = the reference state).

When dealing with flow processes, exergy flows are associated with mass flows as well as with flows of energy either as heat or as work across the control surface. Thus, in each application, we can identify the exergy input rate, Ex_i, to the process and the resulting (useful) exergy production, Ex_p. When we use the expression energy efficiency in general, we are actually referring to exergy efficiency, as the energy efficiency will always be 100% according to the first law of thermodynamics. The exergy efficiency, eff, is however always less than 100% for real processes according to the second law of thermodynamics:

$$eff = Ex_i, / Ex_p < 100\% \qquad (19)$$

Notice that exergy is not conserved, only if entropy-free energy is transferred, which implies that the transfer is reversible. All processes in reality are, however, irreversible, which means that exergy is lost (and entropy is produced). Loss of exergy and production of entropy are two different descriptions of the same reality, namely that all processes are irreversible. We will unfortunately always, in all real processes, have some loss of energy forms which can do work to energy forms which cannot do work. The energy is of course conserved by all processes according to the first law of thermodynamics. It is therefore wrong to discuss an energy efficiency of an energy transfer, because it will always be 100%, It is the exergy efficiency that is of interest because it will express the ratio of useful energy to total energy which is always, as expressed in equation (19), is less than 100% for real systems.

To describe the irreversibility of real processes, the application of exergy seems to be more useful than entropy; exergy has the same unit as energy and is an energy form, while the definition of entropy is more difficult to relate to concepts associated to our usual description of reality.

The second law of thermodynamics can be expressed by the use of exergy as follows:

$$\Delta Ex \text{ for any isolated system} \leq 0, \qquad (20)$$

which implies that exergy is always lost, i.e., work is lost in the form of heat that cannot do work. The two formulations of the second law of thermodynamics by entropy and exergy are of course completely consistent.

5. Thermodynamic information

In statistical mechanics, entropy is related to probability. A system can be characterised by averaging ensembles of microscopic states to yield the macrostate. If W is the number of microstates that will yield one particular macrostate, the probability (P) that this particular macrostate will occur as opposed to all other possible macrostates is proportional to W. It can further be shown that:

$$S = k * \ln W \qquad (21)$$

where k is Boltzmann's constant, $1.3803 * 10^{-23}$ J/(molecules*deg). The entropy is a logarithmic function of W and thus measures the total number of ways that a particular macrostate can be constituted microscopically.

S may be called thermodynamic information, referring to the amount of information *needed* to describe the system, and must *not* be interpreted as the information that we actually possess. The more microstates there are, and the more disordered they are, the more information is required and the more difficult it will be to describe the system.

Shannon and Weaver (1963) have introduced a measure of information which is widely used as a diversity index by ecologists under the name of Shannon's index:

$$H = - \sum_{i=1}^{n} pi \log_2 (pi) \tag{22}$$

where pi is the probability distribution of species.

Shannon's index of diversity (Shannon and Weaver, 1963) is sometimes called entropy, but should not be confused with thermodynamic information. The symbol H is used to avoid confusion. The use of Shannon's index should be limited to a measure of diversity and communication, although as mentioned above, the two concepts to a certain extent are parallel, as both S and H increase with an increasing number of possible (micro)states.

If an ecosystem is in thermodynamic equilibrium, the entropy, S_{eq}, is higher than in non-equilibrium. The excess entropy may be denoted as the thermodynamic information and is sometimes also defined as Schrödinger's negentropy NE – an expression which should be omitted after introduction of the easier understandable concept, exergy:

$$I = Seq - S = NE \tag{23}$$

In other words, a decrease in entropy will imply an increase in information and loss of information implies increase of entropy, as pointed out by Landauer (1991). Further, the principle of the second law of thermodynamics corresponds to a progressive decrease of the information content. An isolated system can evolve only by degrading its information.

I also equals Kullbach's measure of information (Brillouin, 1956):

$$I = k * \sum_{j} pj^* \ln (pj^* / pj) \tag{24}$$

where pj^* and pj are probability distributions, a posteriori and a priori to an observation of the molecular detail of the system, and k is Boltzmann's constant. It means that I expresses the amount of information that is gained as a result of the observations. If we observe a system, which consists of two connected chambers, we expect the molecules to be equally distributed in the two chambers, i.e., $p_1 = p_2$ is equal to 1/2. If we, on the other hand, observe that all the molecules are in one chamber, we have $p_1^* = 1$ and $p_2 = 0$. As seen we get the same entropy by application of equation (18) as we did by use of equation (14), since $R = k*A$, where A is Avogadro's number, and there is proportionality to the number of molecules.

It is interesting in this context to draw a parallel with the discussion of the development of entropy for the entire Universe. The classical thermodynamic interpretations of the second law of thermodynamics predict that the Universe will develop toward 'the heat death', where the entire Universe will have the same temperature, no changes will take place and a final overall thermodynamic equilibrium will be the result. This prediction is based upon the steady increase of the entropy according to the second law of thermodynamics: the thermodynamic equilibrium is the attractor. It can, however, be shown (see Frautschi, 1988, Layzer, 1976 and 1988 and Jørgensen et al., 1998) that the thermodynamic equilibrium is moving away at a high rate due to the expansion of the Universe.

Due to the incoming energy of solar radiation, an ecosystem is able to move away from the thermodynamic equilibrium – i.e., the system evolves, obtains more information and organisation. The ecosystem must produce entropy for maintenance, but the low-entropy energy flowing through the system may be able to more than cover this production of disorder, resulting in an increased order or information of the ecosystem.

Figure 1 shows the relationship, as presented by Brooks et al. (1989), Brooks and Wiley (1986), Wiley (1988) and Layzer (1976). H-max corresponds to the entropy of the ecosystem if it were in thermodynamic equilibrium, while H-obs is the actual entropy level of the system. The difference covers the information or organisation.

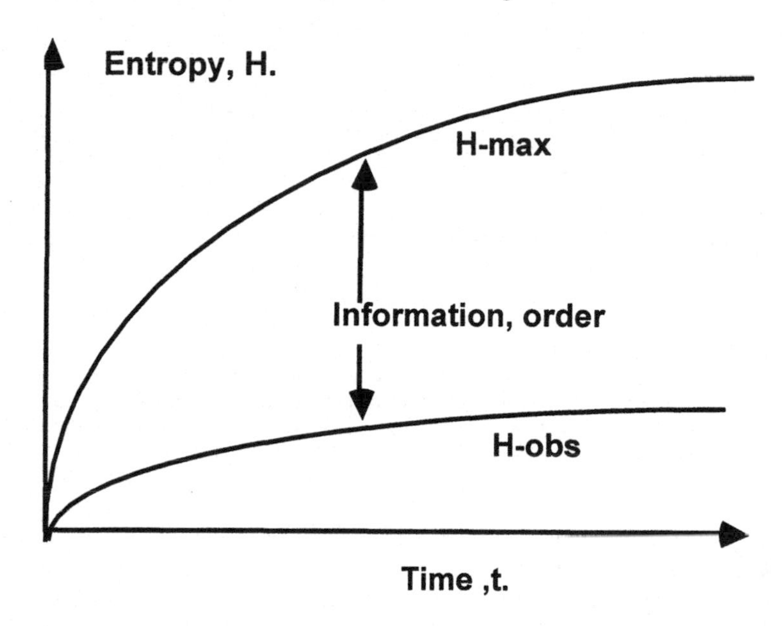

Figure: 1 H-max and H-obs are plotted versus time. The difference between H-max and H-obs represents the information, I, which increases with time, t (Brooks and Wiley ,1986).

It implies that

$$H\text{-max} = \log W, \tag{25}$$

where W is the number of microstates available to the system. H-obs is defined according to the following equation:

$$H\text{-obs} = -\sum_{i=1}^{n} pi \ln (pi) \tag{26}$$

Brooks and Wiley have interpreted this development of entropy in a variety of ways:

1. H-obs is interpreted as complexity – the higher the complexity, the more energy is needed for maintenance and therefore wasted as heat. The information in this case becomes the macroscopic information.

2. H-obs is translated to realisation, while H-max becomes the total information capacity. Notice, however, that the strict thermodynamic interpretation of H-max is H at thermodynamic equilibrium, which does not develop (change) for an ecosystem on earth.

3. H-obs represents the observed distribution of genotypes and H-max is any genotype equally likely to be found. The information becomes the organisation of organism over genotypes.

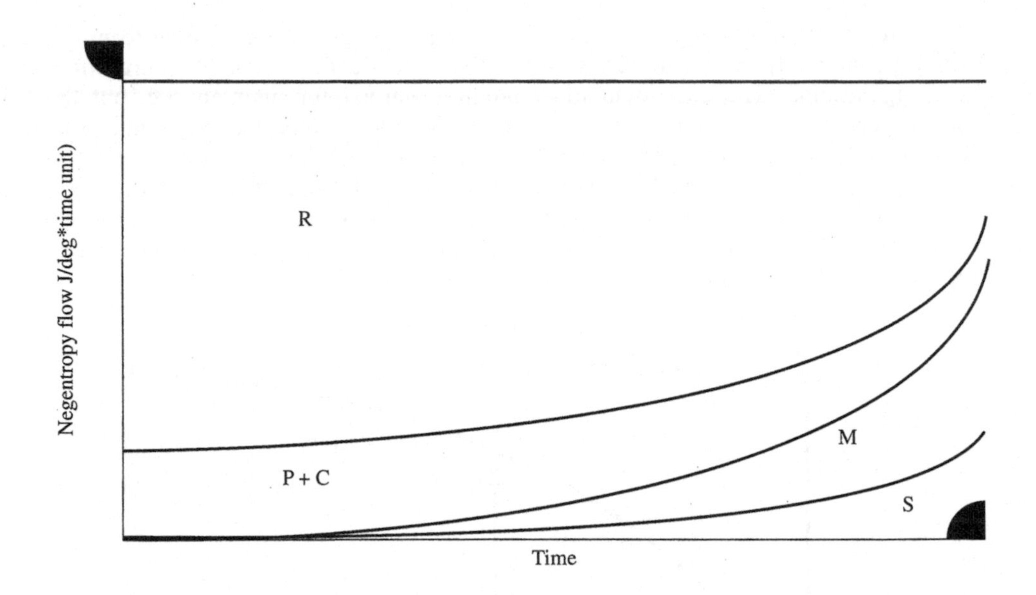

Figure 2: The figure shows a tentative development due to the biological evolution in the application of the exergy flow (or negentropy flow according to Schrödinger, 1944) (solar radiation) on earth versus time. The exergy is 1) not used, but reflected, R; 2) used by physical and chemical processes on earth, P + C, 3) used for maintenance of the biological structure, M; or 4) used to construct biological structure, S.

Brooks and Wiley's theory seems inconsistent with the general perception of ecological development: order increases – entropy therefore decreases at the cost of entropy production of the environment; see for instance Nicolis and Prigogine (1989). The misinterpretation probably lies in the translation of order to information. By increasing order, the amount of information needed to describe the system decreases. Note again, that entropy covers the amount of information which is needed.

Figure 2 attempts to give a different picture of the development in application of the exergy flow (= the solar radiation) on earth, indicated as negentropy flow according to Schrödinger (1944). The exergy flow is considered approximately constant, although the solar radiation has shown some (relatively minor) changes. Four different applications of exergy are considered: unused exergy (reflection of the radiation), exergy used for physical and chemical processes, exergy used for the maintenance of life (respiration) and exergy used for the construction of biological structures. Notice that exergy is not conserved, as most of the exergy used for chemical and physical processes, and all the exergy used for respiration is lost as heat. Notice that the exergy stored in the biological structure is increased, and hence the maintenance cost – it requires more and more exergy to maintain an increasingly developed structure.

Information contains exergy. Boltzmann (1905) showed that the free energy of the information that we actually possess (in contrast to the information we need to describe the system) is $k*T*\ln I$, where I represents the pieces of information we have about the state of the system and k is Boltzmann's constant = $1.3803* 10^{-23}$ (J / molecules*deg). It implies that one piece of information has the exergy equal to $k\ T\ \ln 2$. Transformation of information from one system to another is often almost an entropy-free energy transfer. If the two systems have different temperatures, the entropy lost by one system is not equal to the entropy gained by the other system, while the exergy lost by the first system is equal to the exergy transferred and equal to the exergy gained by the other system. Therefore, it is obviously more convenient to apply exergy than entropy.

6. Dissipative structure

As an ecosystem is non-isolated, the entropy changes during a time interval, dt can be decomposed into the entropy flux due to exchanges with the environment, d_eS, and the entropy production due to the irreversible processes inside the system such as diffusion, heat conduction and chemical reactions, d_iS. It can be expressed as follows:

$$dS/ dt = d_eS / dt + d_iS / dt \tag{27}$$

or by use of exergy:

$$Ex/ dt = d_eEx / dt - d_iEx / dt,$$

where d_eEx/dt represents the exergy input to the system and d_iEx /dt is the exergy consumed by the system for maintenance etc.

For an isolated system $d_eS = 0$, and the second law of thermodynamics yields:

$$dS = d_iS \_ 0 \tag{28}$$

In other words d_iS, the internal entropy production (increase) can never be negative, while d_eS can be negative or positive.

Equation (27) – among other things – shows that systems can only maintain a non-equilibrium steady state ($dS/dt = 0$) by compensating the internal entropy production ($d_iS /d t > 0$) with a negative entropy, or better expressed, with a positive exergy influx ($d_eS /dt <0$, or $d_eEx /dt > 0$). Such an influx induces order into the system. In ecosystems the ultimate exergy influx comes from solar radiation, and the order induced is, for example a biochemical molecular order.

A special case of non-equilibrium systems is the steady state, where the state variable does not evolve in time. This condition implies that

$$d_eS = -d_iS< 0 \tag{29}$$

Thus, to maintain a steady non-equilibrium state, it is necessary to pump a flow of exergy of the same magnitude as the internal exergy consumption into the system continuously.

If $d_eEx > - d_iEx$ (the exergy consumption in the system), the system has surplus exergy, which may be utilised to construct further order in the system, or as Prigogine (1980) calls it: dissipative structure. The system will thereby move further away from the thermodynamic equilibrium. The evolution shows that this situation has been valid for the ecosphere on a long-term basis. In spring and summer, ecosystems are in the typical situation that d_eEx exceeds $-d_iEx$. If $d_eEx < -d_iEx$, the system cannot maintain the order already achieved, but will move closer to the thermodynamic equilibrium, i.e., it will lose order. This may be the situation for ecosystems during autumn and winter or due to environmental disturbances.

An ecosystem will contain a great number of chemical compounds, which are spontaneously degrading to other compounds with lower energy levels by the production of entropy or consumption of exergy. The processes are irreversible and the amount of entropy will therefore increase and exergy will decrease. Proteins, starch and other high-energy components of ecosystems will spontaneously decompose into components with lower energy levels by the release of energy. These catabolic processes are vital for the ecosystem as they are the ecosystem's fuel.

Ecosystems have a global attractor state, the thermodynamic equilibrium, but will never reach this state as long as they are not isolated and receive exergy from outside to combat the decomposition of their compounds. As ecosystems have an energy through-flow, the attractor in nature becomes the steady state, where the formation of new biological compounds is in balance with the decomposition processes. As seen from these considerations on the second law of thermodynamics for open (non-isolated) systems, it is vital for ecosystems to be non-isolated. The consequences of this openness may be described by use of the concept of exergy as follows:

Exergy, the work potentially inherent in solar radiation, and certain geochemical compounds are built by photosynthesis and chemosynthesis into biological structure as organic compounds (charge phase). In photosynthesis, for example, if the power of solar radiation is W/t (work/time) and the average temperature of the system is T_1, then the exergy gain per unit of time from radiant energy fixation is:

$$dEx/dt = T_1 W/t\,(1/T_0 - 1/T_2), \quad [ML^2T^{-3}] \tag{30}$$

where T_0 is the temperature of the environment and T_2 is the surface temperature of the sun (Erikson et al., 1976). The spectral differences between the incoming and outgoing electromagnetic waves may also be used to express exergy gain per unit of time; see Ulanowitz, 1986 and 1997. This exergy flow generated by the photosynthetic capture of photons is used to synthesise and maintain structure (Jørgensen et al., 1998). The more structure an ecosystem has, the more exergy it can capture and utilise, but the more it also needs for maintenance. The structural compounds are progressively decomposed in catabolism (discharge phase), involving the consumption of exergy in work performed, and discharge of the residue to the environment as low exergy heat (only heat transferred from a higher to a lower temperature can produce work). The energy is conserved in accordance with the first law of thermodynamics, but it is converted from work (exergy) to heat (with near-zero exergy). In the process, due to the differences between sun and earth surface temperatures (5780 K vs. 287 K), 20 infrared photons (heat) are produced for every solar photon (Patten et al., 1997). How much work is done depends on the processes of the different pathways utilised. Hence more or less exergy may be consumed for a given quantity of radiant and chemical energy degraded to heat, more or less exergy may be stored in biomass with different levels of information – exergy is not conserved. By all real processes exergy will be lost, i.e., energy forms able to do work are converted to heat that cannot do work (except in local situations with temperatures < 287 K). The energy is conserved but not the work capacity due to the irreversibility of all real processes. The relationship can be captured with implicit function notation, writing exergy, Ex, as a function of energy, E, through work, W: Ex(W(E)). The work, W, is context-specific, hence energy, not exergy, is conserved though its basis.

It has been stated that it is *necessary* for an ecosystem to transfer the generated heat (entropy) to the environment and to receive exergy (solar radiation) from the environment for formation of dissipative structure. The next obvious question would be: will energy sources and sinks also be *sufficient* to initiate the formation of dissipative structure which can be used as a source for entropy combating processes?

The answer to this question is 'Yes.' It can be shown by the use of simple model systems and basic thermodynamics. Morowitz (1968) showed that a flow of energy from sources to sinks leads to an internal organisation of the system and to the establishment of element cycles. The type of organisation is, of course, dependent on a number of factors: the temperature, the elements present, the initial conditions of the system and the time available for the development of the organisation. It is characteristic for the system, as pointed out above, that the steady state does *not* involve chemical equilibrium.

An interesting illustration of the creation of organisation (dissipative structure), as a result of an energy flow through ecosystems, concerns the possibilities to form organic matter from the inorganic components which were present in the primeval atmosphere. Since 1897 many simulation experiments have been performed to explain how the first organic matter (from inorganic matter) on earth was formed. All of them point to the conclusion that energy interacts with a mixture of gases to form a large set of randomly synthesised organic compounds. Most interesting is perhaps the experiment performed by Stanley Miller and Harold Urey at the University of Chicago in 1953, because it showed that amino acids can be formed by sparking a mixture of CH_4, H_2O, NH_3 and H_2, which corresponds approximately to the composition of the primeval atmosphere.

Prigogine and his colleagues have shown that open systems which are exposed to an energy through-flow exhibit coherent self-organisation behaviour and are known as dissipative structures. Formations of complex organic compounds from inorganic matter as mentioned above are typical examples of self-organisation. Such systems can remain in their organised state by exporting entropy outside the system, but are dependent on outside energy fluxes to maintain their organisation, as was already mentioned and emphasised above.

Glansdorff and Prigogine (1971) have shown that the thermodynamic relationship of far from equilibrium dissipative structures is best represented by coupled non-linear relationships, i.e., auto catalytic positive feedback cycles.

Two relatively simple physical-chemical examples of self-organisations should be mentioned to illustrate the formation of dissipative structure as a consequence of a through flow of exergy:

1. Imagine a layer of water between two horizontal parallel plates, whose dimensions are much larger than the width of the layer. It is called a *Benard cell* after the French physicist Benard, who carried out the experiment in 1900. Left to itself the fluid will rapidly tend towards a homogeneous state in which all its parts will be identical. We can induce a flow of energy, say from below. The temperature of the lower plate becomes higher than the temperature of the upper plate : $\Sigma T > 0$. Suppose that the constraint is weak, i.e., ΣT is small. The system will adopt a simple and unique state in which heat is transported from the lower to the upper plate by conduction. If we move the system farther and farther from equilibrium by increasing ΣT suddenly at a value of ΣT, that may be called critical, the fluid begins to perform bulk movement and becomes structured in a series of small convection cells: see Figure 3. The fluid is now in the regime of convection. (See for instance Nicolis and Prigogine (1989)).

2. *The so-called BZ-reaction* shows a similar tendency to self-organisation. Cerium (IV) sulphate, malonic acid and potassium bromate react in sulphuric acid. The evolution of the processes can be followed visually since an excess of Ce(IV)-ions gives a pale yellow colour, whereas an excess of Ce(III)-ions leaves the solution colourless. The three chemicals are pumped into a reaction chamber, which is well mixed.

This experimental set-up allows easy control of the rates at which the chemicals are pumped into or out of the system, i.e., we can vary the residence times of these substances within the reaction vessel. Long residence times essentially result in a closed system and we expect the system to reach equilibrium-like behaviour. Conversely, by short residence times we expect the system to manifest non-equilibrium behaviour. This is according to what the experiment shows. For long residence times, the concentrations of chemicals remain constant. If we reduce the residence time, we suddenly encounter a different pattern. A pale yellow colour emerges, indicating an excess of Ce (IV)-ions. Later the solution becomes colourless, indicating now an excess of Ce (III)-ions.

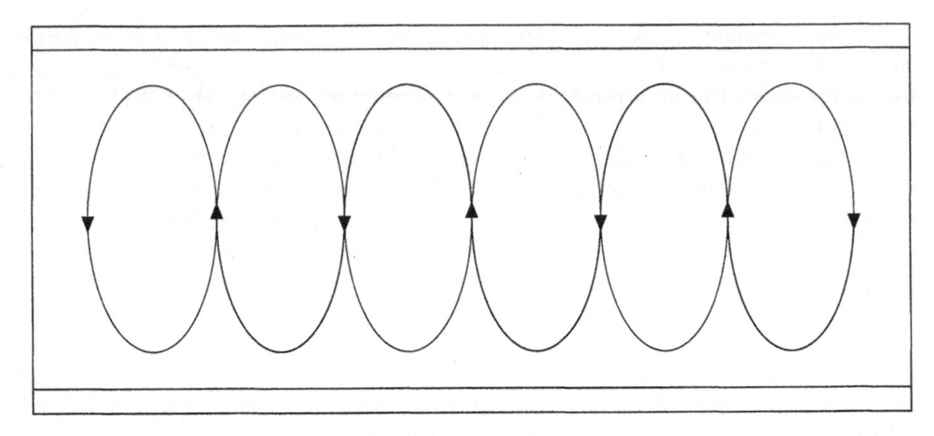

Figure 3: Formation of convection cells in a so-called Benard cell.

With enough Br^-:
$BrO_3^- + Br^- + 2H^+ = HBrO_2 + HOBr$
$HBrO_2 + Br^- + H^+ = 2\ HOBr$
With small quantities of Br^- left, Ce^{+3} is oxidised according to:
$BrO_3^- + HBrO_2 + H^+ = 2\ BrO_2^{\cdot} + H_2O$
$BrO_2^{\cdot} + Ce^{+3} + H^+ = HBrO_2 + Ce^{+4}$
$HBrO_2 = BrO_3^- + HOBr + H^+$
The first step is rate limiting, whereas HOBr disappears quickly by combining with malonic acid.

Figure 4: The individual processes of the BZ-reaction.

The process continues as a chemical clock: yellow, colourless, yellow, colourless, and so on. The amplitude depends only on the experimental parameters. The chemical reactions responsible for the observations are shown in Figure 4 and they can – as can be seen from the figure – be explained by the presence of auto catalysis.

A more comprehensive list of ecosystem properties derived from the second law of thermodynamics can be found in Straskraba et al. (1999).

7. Ecosystem perspectives from the third law of thermodynamics

The lesser-known third law of thermodynamics states that the entropies, S_0, of pure chemical compounds are zero, and that entropy production, ΣS_0, by chemical reactions between pure crystalline compounds is zero at absolute temperature, 0 K. The third law implies, since both $S_0 = 0$ (absolute order) and $\Sigma S_0 = 0$ (no disorder generation), that disorder does not exist and cannot be created at $[ML^2T^{-2}TEMP^{-1}]$ absolute zero temperature. But at temperatures > 0 K disorder can exist ($S_{system} > 0$) and be generated ($\Sigma S_{system} > 0$). The third law defines the relation between entropy production, ΣS_{system}, and the Kelvin temperature, T:

$$\Delta S_{system} = \int_0^T \Delta c_p \, d \ln T + \Delta S_0, \quad [ML^2T^{-2}TEMP^{-1}] \tag{31}$$

where Σc_p is the increase in heat capacity by the chemical reaction. Since order is absolute at absolute zero, its further creation is precluded. At higher temperatures, however, order can be created.

The consequence of the third law of thermodynamics can be more easily expressed in terms of exergy. At 0 K, from eq. (30), the exergy of a system is always 0 ($Ex = kT_0I = 0$ when $T_0 = 0$ K; see Evans, 1969); no useful work can be performed and no further order produced as it is already absolute. Entropy production implies that a degradation of energy from a state of high utility (large T) to a state of low utility (small T) occurs. Therefore, a system can only create an internal state of high exergy through energy dissipation. A system at 0 K is without any creative potential, because no dissipation of energy can take place at this temperature. A temperature greater than 2.726 ± 0.01 K (2.726 K is the temperature of deep space) is said to be required before order can be created (Turner, 1970). Given this necessary condition, simple energy flow through a system provides a sufficient condition. Creation of order is inevitable. On earth, the surface temperature difference between sun and planet guarantees this. Morowitz (1968) showed, as mentioned previously in this chapter, that a through-flow of energy is sufficient to produce cycling, a prerequisite for the ordering processes characteristic of living systems.

The input of exergy for ecosystems is in the form of solar photon flux. This comes as small portions (quanta) of exergy (= hν, where h is Planck's constant and ν is the frequency) implying that the exergy at first can only be utilised at molecular (lowest) levels in the hierarchy. The appropriate atoms or molecules must be transported to the place where order is created. Diffusion processes through a solid are extremely slow, even at room temperature, while the diffusion of molecules through a liquid is about three orders of magnitude faster than in a solid at the same temperature. Diffusion coefficients for gases are ordinarily four orders of magnitude greater than for liquids. This implies that the creation of order (and also the inverse process, disordering) is substantially more rapid in liquid and gaseous phases than in solids. The temperature required for a sufficiently rapid creation of order is considerably above the lower limit of 2,726K as mentioned above. As far as diffusion in solids, liquids and gases is concerned, gaseous diffusion allows the most rapid mass transport. However, many molecules on earth that are necessary for ordinary carbon-based life do not occur in a gaseous phase, and liquid diffusion, even though it occurs at a much slower rate, is of particular importance for biological ordering processes.

The diffusion coefficient increases significantly with temperature. The diffusion coefficient of gases varies with temperature at approximately $T^{3/2}$ (Hirschfelder et al., 1954), where T is the absolute temperature. Thus, we should look for systems with the high order characteristic of life, at temperatures considerably higher than 2.726 K. The reaction rates for biochemical anabolic processes on the molecular level are highly temperature dependent (see Straskraba et al., 1999). The influence of temperature may be reduced by the presence of reaction-specific enzymes, which are proteins formed by anabolic processes. The relationship between the absolute temperature, T, and the reaction rate coefficient, k, for a number of biochemical processes can be expressed by the following equation (Brønsted, 1943):

$$\ln k = b - A/R^*T, \quad (ML^{-3}T^{-1}) \tag{32}$$

where A is the so-called activation energy, b is a constant and R is the gas constant. Enzymes are able to reduce the activation energy (the energy that the molecules require to perform the biochemical reaction). Similar dependence of the temperature is known for a wide spectrum of biological processes, for instance growth and respiration. Biochemical and biological kinetics point, therefore, towards ecosystem temperatures considerably higher than 2.726 K.

The high efficiency in the use of low-entropy energy at the present 'room temperature' on earth works together with the chemical stability of the chemical species characteristic of life on earth. Macromolecules are subject to thermal denaturation. Among the macromolecules, proteins are most sensitive to thermal effects, and the constant breakdown of proteins leads to a substantial turnover of amino acids in organisms. According to biochemistry, an adult man synthesises and degrades approximately one gram of protein nitrogen per kilogram of body weight per day. This corresponds to a protein turnover of about 7.7% per day for a man with a normal body temperature. A temperature in the ecosystem that is too high (more than about 340 K) will therefore enhance the breakdown processes too much. A temperature range between 260 and 340 K seems the most appropriate to create the carbon-based life that we know on earth. An enzymatic reduction of the activation energy makes it possible to realise basic biochemical reactions in this temperature range, without a decomposition rate that is too high, which would be the case at a higher temperature. In this temperature range anabolic and catabolic processes can, in other words, be in a proper balance.

The conditions for creation of order out of disorder (or more specifically chemical order by formation of complex organic molecules and organisms from inorganic matter) can now be deduced from the first, second and third laws of thermodynamics:

1. It is necessary that the system be open (or at least non-isolated) to enable exchange of energy (as well as mass) with its environment;

2. An influx of exergy is both necessary and sufficient;

3. An outflow of high-entropy energy (heat originated from exergy consumption) is necessary (this means that the temperature of the system must be greater than 2.726 K) ;

4. Entropy production accompanying the transformation of energy (work) to heat in the system is a necessary cost of maintaining the order; and

5. Mass transport processes at rates that are not low rate are necessary (a prerequisite). This implies that the liquid or gaseous phase must be anticipated. A higher temperature will imply a better mass transfer, but also a higher reaction rate. An increased temperature also means a faster breakdown of macromolecules, and therefore a shift towards catabolism. A temperature approximately in the range of 260-340 K must therefore be anticipated for carbon based life.

The rates of biochemical reactions on the molecular level are determined by the temperature of the system and the exergy supply to the system. Hierarchical organisation ensures that the reactions and the exergy available on the molecular level can be utilised on the next level, the cell level, and so on throughout the entire hierarchy: molecules -> cells -> organs -> organisms -> populations -> ecosystems. The maintenance of each level is dependent on its openness to the exchange of energy and matter. The rates in the higher levels are dependent on the sum of many processes at the molecular level. They are furthermore dependent on the slowest processes in the chain: supply of energy and matter to the unit -> the metabolic processes -> excretion of waste heat and waste material. The first and last of these three steps limit the rates and are determined by the extent of openness, measured by the area available for exchange between the unit and its environment relative to the volume. These considerations are based on allometric principles (Peters, 1983 and Straskraba et al., 1999).

8. Exergy as a limiting factor

When the first life emerged from the inorganic soup on earth about 3.5 billion years ago, the major challenge to primeval life was to maintain what was already achieved. This problem was eventually solved in the present form by the introduction of DNA or DNA-like molecules which were able to store the already obtained information on how to convert inorganic matter to life-bearing organic matter. It requires, however, both matter and energy to continue the development, i.e., to continuously build more ordered structure of life-bearing organic compounds on the shoulders of the previous development. The amounts of elements on earth are almost constant, i.e., the earth is today a non-isolated (closed) system (energy is exchanged with space, but (almost) no exchange of matter takes place, if we leave out the minor input from meteorites and the minor output of hydrogen), while energy is currently supplied by the solar radiation. The concept of limiting elements (factors), see

2.0, is embodied in the conservation laws; see also Patten et al. (1997). Further development (evolution), when one or more elements are limiting, is consequently not possible by formation of more organic life-bearing matter just by uptake of more inorganic matter, but only by a better use of the elements, i.e., by a reallocation of the elements through a better organisation of the structure. This requires that more information (exergy) be embodied in the structure. More ecological niches are thereby utilised, and more life forms will be able cope better with the variability of the forcing functions. Exergy is not limiting in this phase of development, because each new unit of time brings more exergy to combat the catabolic processes and to continue the energy consuming and exergy building anabolic processes. Exergy stored in the biomass is steadily increased under these circumstances due to an increase in structures able to capture exergy. The in flowing exergy, the solar radiation, is either captured by the structure of ecosystems or reflected (unused). Kay and Schneider (1990) have shown that mature ecosystems (in the sense of E.P. Odum, 1968) capture about 80% of the incoming exergy, close to what is physically feasible (see Kay and Schneider, 1990). Most of the captured exergy is used for evapotranspiration and only a minor part is used for gross production (2% on average), which again has to cover respiration and exergy built into new biological structure.

The captured exergy will therefore also, sooner or later, when a sufficiently large structure has been formed, become limiting. Under these circumstances, the additional constraints on the processes of the ecosystem are a better utilisation of the incoming exergy, meaning that less of the exergy captured is used for maintenance and more is therefore available for further growth of the ordered structure, i.e., to increase the exergy stored in the system. Mauersberger (1983 and 1995) has used these considerations to determine the nonlinear relations between rates and chemical affinities of biological processes in an ecosystem. Mauersberger postulates that the development of bioceonosis is controlled during the finite time interval of length Σt by the chemical affinities, X, so that the time integral:

$$I = \int_{t+\Delta t}^{t} E(B(t'), X(t'))dt' \tag{33}$$

over the generalised excess entropy production, E, within $(t - \Delta t) \leq t' \leq t$ becomes an extreme value (minimum) subject to the initial values $B_o = B\ (t - \Delta t)$ and to the mass equations:

$$dBk(t) / dt = fk(B(t), Y(X(t))), \tag{34}$$

which connect B and X through the rates of the biological processes, Y, for the kth species. This optimisation principle postulates that locally, and within a finite time interval, the deviation of the bioprocesses from a stable stationary state tends to a minimum. This is similar to, and a more formal version of, Lionel Johnson's least dissipation principle (Johnson, 1995) which goes back to Lord Rayleigh. Mauersberger has used this optimisation principle on several process rates as functions of state variables. As a result, he finds well known and well accepted relations for uptake of nutrients of phytoplankton, primary production, temperature dependence of primary production and respiration. It supports the use of the optimisation principle locally and within a finite time interval.

This leads inevitably to the need for a fourth law of thermodynamics which is presented in Chapter II.1.3: how can we describe the results of the opposition between anabolism and catabolism or between the above mentioned global and local optimisation principles?

REFERENCES

Boltzmann, L., 1905. The Second Law of Thermodynamics. (Popular Schriften, essay no. 3(address to Imperial Academy of Science in 1886)). Reprinted in English in: Theoretical Physics and Philosophical Problems, Selected Writings of L. Boltzmann. D.Reidel, Dordrecht.

Brønsted, J.N., 1943. Physical Chemistry (in Danish), 498pp. Munksgard, Copenhagen.

Brooks, D.R. and E.O. Wiley, 1986. Evolution as Entropy. University Press, Chicago, IL.

Brooks, D.R., Collier, J., Maurer, B.A., Smith, J.D.H. and E.O. Wiley, 1989. Entropy and information in evolving biological systems. Biol. Philos. 4: 407-432.

Brown, J.H., 1995. Macroecology. The University of Chicago Press, Chicago, IL.

Eriksson, B., K.E. Eriksson, and G. Wall, 1976. Basic Thermodynamics of Energy Conversions and Energy Use. Institute of Theoretical Physics, Göteborg, Sweden.

Evans, R.B., 1966. A Proof that Essergy is the Only Consistent Measure of Potential Work [Thesis]. Dartmouth college, Hannover, NH.

Frautschi, S., 1988. Entropy in an expanding universe. In: B.H. Weber, D.J. Depew, and J.D. Smith (Editors). *Entropy, Information, and Evolution: New Perspectives on Physical and Biological Evolution.* MIT Press, Cambridge, MA, 11-22.

Glansdorff, P. and I. Prigogine, 1971. Thermodynamic Theory of Structure, Stability, and Fluctations. Wiley-Interscience, New York.

Hall, C.A.S. (Editor), 1995. Maximum Power: The Ideas and Applications of H.T. Odum. University Press of Colorado, Niwot, CO.

Hirschfelder, J.O., Curtiss, C.F. and R.B. Bird, 1954. *Molecular Theory of Gases and Liquids*. John Wiley and Sons, New York. 631 p.

Johnson, L., 1995. The Far-from-Equilibrium Ecological Hinterlands. In: B.C. Patten., S.E. Jørgensen, and S.I. Auerbach (Editors). *Complex Ecology. The Part-Whole Relation in Ecosystems.* Prentice Hall PTR, Englewood Cliffs, New Jersey. p 51-104.

Jørgensen, S.E., 1976. A eutrophication model for a lake. *Ecol. Modelling* 2: 147-165.

Jørgensen, S.E., Mejer, H.F. and M. Friis, 1978. Examination of a Lake Model. Ecological Modelling 4:253-279.

Jørgensen, S.E., 1982a. A holistic approach to ecological modelling by application of thermodynamics. In: W. Mitsch et. al. (Editors). *Systems and Energy*. Ann Arbor.

Jørgensen, S.E., 1982b. Modelling the eutrophication of shallow lakes. In: D.O. Logofet and N.K. Luchyanov (Editors). *Ecosystem Dynamics in Freshwater Wetlands and Shallow Water Bodies*, Vol. 2. UNEP7SCOPE (United Nations' Environmental Program/Scientific Committee on Pollution of the Environment). Academy of Sciences, Moscow, 125-155.

Jørgensen, S.E., 1990. Modelling in Ecotoxicology. Elsevier, Amsterdam.

Jørgensen, S.E., 1994. *Fundamentals of Ecological Modelling* (second edition) (Developments in Environmental Modelling, 19). Elsevier, Amsterdam, 628 p.

Jørgensen, S.E., 1995. The growth rate of zooplankton at the edge of chaos: ecological models. *J. Theor. Biol.* 175: 13-21.

Jørgensen, S.E., B.C. Patten, and M. Straskraba, 1999. Ecosystem emerging: 4. Growth. Ecol. Modelling (In press).

Kay, J.J. and E.D. Schneider, 1990. On the applicability of non-equilibrium thermodynamics to living systems. Waterloo University, Waterloo,Canada.

Kooijman, S.A.L.M., 1993. *Dynamic Energy Budgets in Biological Systems. Theory and Applications in Ecotoxicology.* 328 p.

Landauer, R., 1991. Information is Physical. Phys. *Today May*: 23-29.

Layzer, D., 1976. The arrow of time. Scientific American December. *Astrophysical J.* 206: 559-565.

Layzer, D., 1988. Growth of order in the universe: In: B.H. Weber., D.J. Depew, and J.D. Smith (Editors). *Entropy, Information, and Evolution: New Perspectives on Physical and Biological Evolution.* MIT Press, Cambridge, MA, 23-40.

Mauersberger, P., 1983. General principles in deterministic water quality modeling. In: G.T. Orlob (Editor). *Mathematical Modeling of Water Quality: Streams, Lakes and Reservoirs* (International Series on Applied Systems Analysis, 12). Wiley, New York, 42-115.

Mauersberger, P., 1995. Entropy control of complex ecological processes. In: B.C. Patten and S.E. Jørgensen (Editors). *Complex Ecology: The Part-Whole Relation in Ecosystems.* Prentice-Hall Englewood Clifs, NJ, 130-165.

Morowitz, H.J., 1968. Energy Flow in Biology. Academic Press, New York.

Nicolis, G. and I. Prigogine, 1989. Exploring Complexity: an Introduction. Freeman, New York.

Odum, E.P., 1968. Energy flow in ecosystems: a historical review. Am. Zool. 8:11-18.

Odum, E.P., 1969. The strategy of ecosystem development. Science 164: 262-270.

Odum, H.T., 1983. *System Ecology*. Wiley Interscience, New York. 510 p.

Onsager, L., 1931. Reciprocal relations in irreversible processes, I. *Phys. Rev.* 37: 405-426.

Patten, B.C., M. Straskraba, and S.E. Jørgensen, 1997. Ecosystem Emerging: 1. Conservation. *Ecol. Modelling* 96: 221-284.

Peters, R.H., 1983. The Ecological Implications of Body Size. Cambridge University Press, Cambridge.

Prigogine, I., 1947. Etude Thermodynamique des Processus Irréversibles. Desoer, Liège.

Prigogine, I., 1980. From Being to becoming: Time and Complexity in the Physical Sciences. Freeman, San Fransisco, CA. 260pp.

Schrödinger, E., 1944. What is Life? Cambridge University Press, Cambridge.

Shannon, C.E. and W. Weaver, 1963. The Mathematical Theory of Communication. (First published 1949). University of Illinois Press, Champaign, IL.

Straskraba, M., S.E. Jørgensen, and B.C. Patten, 1999. Ecosystem emerging: 3. Dissipation. Ecol. Modelling, in print.

Turner, F.B., 1970. The ecological efficiency of consumer populations. Ecology 51: 741-742

Ulanowicz, R.E., 1986. Growth and Development. Ecosystems Phenomenology. Springer-Verlag, New York, Berlin, Heidelberg, Tokyo. 204 pp.

Ulanowicz, R.E., 1997. *Ecology, the Ascendent Perspective*. Columbia University Press, New York, 201 p.

Wiley, E.O., 1988. Entropy and evolution. In: B.H. Weber., D.J. Depew, and J.D. Smith (Editors). *Entropy, Information, and Evolution: New Perspectives on Physical and Biological Evolution.* MIT Press, Cambridge, MA, 173.

II.1.2 Ecosystems as Self-organising Holarchic Open Systems: Narratives and the Second Law of Thermodynamics

James J. Kay

1. Introduction

A new understanding of complex systems, and in particular ecosystems, is emerging. (Kay (1984), Holling (1986), Kay and Schneider (1994), Kay (1997), Kay, Regier, Boyle and Francis (1999)). The hierarchical nature of these systems requires that they be studied from different types of perspectives and at different scales of examination. There is no correct perspective. Rather a diversity of perspectives is required for understanding. Ecosystems are self-organising. This means that their dynamics are largely a function of positive and negative feedback loops. This precludes linear causal mechanical explanations of ecosystem dynamics. In addition emergence and surprise are normal phenomena in systems dominated by feedback loops. Inherent uncertainty and limited predictability are inescapable consequences of these system phenomena. Such systems organise about attractors. Even when the environmental situation changes, the system's feedback loops tend to maintain its current state. However, when ecosystem change does occur, it tends to be very rapid and even catastrophic. When precisely the change will occur, and what state the system will change to, are often not predictable. Often, in a given situation, there are several possible ecological states (attractors), that are equivalent. Which state the ecosystem currently occupies is a function of its history. There is not a 'correct' preferred state for the ecosystem.

Given this understanding of the self-organising phenomena exhibited by ecological systems, it has been argued that conventional science approaches of modelling and forecasting are often inappropriate, as are prevailing explanations in terms of linear causality and stochastic properties (Holling (1986), Kay and Schneider (1994),.Schneider and Kay, 1994a, Kay (1997)). Elsewhere (Kay, Regier, Boyle and Francis (1999)) we have discussed an approach for dealing with these realities of ecosystems in the context of informing resource and land use decision makers and planners. This approach is different from the 'traditional' ecosystem approaches which are interdisciplinary in nature but focus on forecasting and a single type of entity such as a watershed or forest community. Rather this approach is in the mode of post normal science and is grounded in complex systems theory. At its heart is the portrayal of ecological systems as Self-organising Holarchic Open systems (SOHO systems).

In this approach, scientists take on the role of narrators. The task of the narrator is to scope out the array of attractors available to the SOHO system, the potential flips between them, and the underlying morphogenetic causal structure of the organisation in the domain of the attractors. This is reported as a narrative of possible futures for the SOHO. The role of the scientists is then to inform the decision makers, through the narratives, about the ecological options, the tradeoffs and uncertainties involved, and various strategies for influencing what happens on a landscape.

At the core of these narrative descriptions of ecosystems as self-organising holarchic open systems is the conceptualisation of these systems as dissipative systems. Dissipative system descriptions are in terms of how the system makes use of available energy (and other resources) to self-organise. Such a description is inherently thermodynamic in nature and as such the second law of thermodynamics plays a central role. This chapter explores the role of the second law of thermodynamics in building narratives of ecosystem self-organisation. This chapter begins with an exploration of the second law and its implications for self-organisation. This serves as the basis for a discussion of the characterisation of ecosystem attractors in terms of the their sources of exergy. This, in turn, provides the basis for formulating narratives and in particular a conceptual model of ecosystems as self-organising holarchic open systems.

2. Thermodynamics and self-organisation.

2.1 The dilemma of the random heat death of the universe

At the turn of the century, thermodynamics, as portrayed by Boltzmann, viewed nature as decaying toward a certain death of random disorder in accordance with the second law of thermodynamics. This equilibrium seeking, pessimistic view of the evolution of natural systems is in contrast with the phenomenology of many natural systems. Much of the world is inhabited by coherent systems whose time dependent behaviour is a progression away from disorder and equilibrium, into highly organised structures that exist some distance from equilibrium. Boltzmann (1886) recognised the apparent contradiction between the heat death of the universe, and the existence of life in which systems grow, complexify, and evolve. He turned to thermodynamics in an attempt to resolve this issue and suggested that the sun's energy gradient drives the living process and suggested a Darwinian like competition for entropy in living systems.

Erwin Schrödinger addressed this same dilemma in his seminal book *What is Life?* (1944). In *What is Life?* Schrödinger noted that life was comprised of two fundamental processes; one 'order from order' and the other 'order from disorder'. He observed that the gene, with its soon to be discovered DNA, controlled a process that generated 'order from order' in a species, that is the progeny inherited the traits of the parent. This observation, combined with the work of Watson and Crick, provided biology with a research framework for some of the most important findings of the last fifty years.

However, Schrödinger's 'order from disorder' observation is equally important but largely ignored (Schneider and Kay, 1995). 'Order from disorder' is about the relationship between biology and the fundamental theorems of thermodynamics. He noted that at first glance living systems seem to defy the second law of thermodynamics, as the second law insists that, within closed systems, entropy should be maximised and disorder should reign. Living systems, however, are the antithesis of such disorder. They continually evolve new order from disorder. For example, plants are highly ordered structures, which are synthesised from disordered atoms and molecules found in atmospheric gases and soils. Schrödinger solved this dilemma by turning to non-equilibrium thermodynamics, that is, he recognised that living systems exist in a world of energy and material fluxes. An organism stays alive in its highly organised state by taking energy from outside itself, from a larger encompassing system, and processing it to produce, within itself, a lower entropy, more organised state. At the same time it exports entropy to the larger encompassing system, thus contributing to the larger system's disorder. Schrödinger recognised that living systems are far from equilibrium systems that maintains their local organisation, through entropy export, at the expense of the larger system they are part of. He proposed that the investigation of living systems from a non-equilibrium perspective would help reconcile biological self-organisation and thermodynamics.

Schneider and Kay (1994a,b) have taken on the research task proposed by Schrödinger and expanded on his thermodynamic view of life. The second law of thermodynamics is not an impediment to the understanding of life but rather is necessary for a complete description of living processes. The second law underlies and determines the direction of many of the processes observed in the development of living systems. The second law mandates behaviour in systems that are a necessary (but not sufficient basis) for life itself. This re-examination of thermodynamics shows that biology is not an exception to physics, we have simply misunderstood the rules of physics.

2.2 A fresh look at thermodynamics

Comparatively speaking, thermodynamics is a young science and has been shown to apply to all work and energy systems including the classic temperature-volume-pressure systems, chemical kinetic systems, electromagnetic and quantum systems. The first law of thermodynamics arose from efforts, in the early 1800s, to understand the relation between heat and work. The first law states that energy cannot be created or destroyed and that the total energy within a closed system remains unchanged. This laws deals strictly with the quantity of energy. The second law of thermodynamics requires that if there are any processes underway in a system, the quality of the energy in that system will degrade. The second law is usually stated, as it was in 1860, in terms of the quantitative measure of irreversibility, entropy. For any process the change in entropy is greater than zero, or any real process can only proceed in a direction which results in an entropy increase. This statement of the second law is in terms of what systems cannot do *vis-à-vis* entropy change.

In 1908, Carathéodory moved thermodynamics a step forward. He developed a proof that showed that the law of 'entropy increase' is not the general statement of the second law Rather the more fundamental observation of the second law is the that all natural phenomena are irreversible. The more encompassing and fundamental statement of the second law of thermodynamics, made by Carathéodory, is that 'In the neighbourhood of any given state of a closed system, there exists states which are inaccessible along any reversible or irreversible adiabatic paths'. This implies that irreversible processes constrain systems so they cannot attain certain states previously accessible. It tells us that certain system states cannot happen spontaneously.

More recently Hatsopoulos and Keenan (1965) and Kestin (1966) have put forward a principle which subsumes Carathéodory's work as well as the 0th, 1st and 2nd Laws: 'When an isolated system performs a process after the removal of a series of internal constraints, it will reach a unique state of equilibrium: this state of equilibrium is independent of the order in which the constraints are removed' (Kestin, 1966). This is called the Law of Stable Equilibrium by Hatsopoulos and Keenan and the Unified Principle of Thermodynamics by Kestin.

The importance of the work of Hatsopoulos and Keenan and Kestin is that their statement dictates a direction and an end state for all real processes, equilibrium. All previous formulations of the second law tells us what systems cannot do. This statement tells us what systems will do. They will spontaneously move to equilibrium. Furthermore it is a statement which is in terms of directly measurable quantities, gradients in the system, such as temperature and pressure. At equilibrium differences in these quantities, over space, that is their spatial gradients, are zero. An example of this phenomena are two flasks, connected with a closed stopcock. One flask holds 10,000 molecules of a gas, the other virtually none. Upon removing the constraint (opening the stopcock) the system will spontaneously move to it's equilibrium state of 5,000 molecules in each flask, with no gradient between the flasks. This principle hold for a broad class of thermodynamic systems from chemical kinetic reactions to a hot cup of tea cooling to room temperature.

There are two points to this summary, thermodynamics is a discipline which should be thought of as part of modern physics, contemporary with quantum and relativity theorem, albeit a poor cousin in terms of attention and development. Most of thermodynamic thinking has been about equilibrium or near equilibrium situations and an integrated theory of these phenomena is less than thirty-five years old, very young indeed in terms of scientific maturity.

2.3 Non-equilibrium open systems

The same cannot be said for the phenomena of systems that are open to energy and or material flows and reside at quasi stable states some distance from equilibrium. An integrated theory for these situations does not exist. Yet it is understanding precisely these situations that Schrödinger identified as pivotal to the reconciliation of biological self-organisation and thermodynamics.

These non-equilibrium situations have been investigated by Prigogine (1955) and his collaborators (Nicolis and Prigogine, 1977, 1989). These systems are open and are moved away from equilibrium by the fluxes of material and energy across their boundary and maintain their form or structure by continuous dissipation of energy and, thus, are known as dissipative structures. This research group, along with many others, showed that non-equilibrium systems, through their exchange of matter and/or energy with the outside world, can maintain themselves for a period of time away from thermodynamic equilibrium in locally produced stable steady-states. This is done at the cost of increasing the entropy of the larger 'global' system in which the dissipative structure is imbedded; thus following the second law, that overall entropy, in the global sense, must increase. Non living organised systems (like convection cells, tornadoes and lasers) and living systems (from cells to ecosystems) are dependent on outside energy fluxes to maintain their organisation in a locally reduced entropy state. The entropy relationships in dissipative systems were put forward by Denbeigh (1951) and Prigogine (1955).

Unfortunately, Prigogineian descriptions of dissipative structures are formally limited to the neighbourhood of equilibrium. This is because this form of analysis depends on a linear expansion of the entropy function about equilibrium. This is a severe restriction on the application of Prigogineian theory and in particular precludes its formal application to living systems.

To deal with the thermodynamics of non-equilibrium systems, we propose the following corollary that follows from the proof by Kestin of the Unified Principle of Thermodynamics. His proof shows that a system's equilibrium state is a unique stable attractor in the Lyapunov sense. A consequence of Lyapunov stability theory is that there will be a domain of any attractor within which a system will resist being moved from the attractor. Bearing in mind that the equilibrium attractor is unique (in state space), a thermodynamic system will necessarily resist being removed from the equilibrium state (the attractor). The degree to which a system has been moved from equilibrium is measured by the gradients imposed on the system. The thermodynamic principle which governs the behaviour of systems as they are moved away from equilibrium is:

As systems are moved away from equilibrium, they will utilise all available avenues to counter the applied gradients. As systems are moved further from equilibrium, attractors, for (thermodynamic) non-equilibrium organisational steady states, will emerge that allow the system to be organised in a way that reduces or degrades the applied gradients. Hence, as the applied gradients increase, so does the system's ability to oppose further movement from equilibrium. If environmental conditions permit, self organisation processes are to be expected. The building of organisational structure and associated processes is such that it degrades the imposed gradient more effectively than if the dynamic and kinetic pathways for those structures were not available.

When moved away from their local (spatially) equilibrium state, systems shift their state in a way which opposes the applied gradients and moves the system back towards its local equilibrium attractor. The stronger the applied gradient, the greater the effect of the equilibrium attractor on the system. In simple terms, systems have the propensity to resist being moved from equilibrium and a propensity to return to the equilibrium state when moved from it. We refer to this principle as the 'restated second law of thermodynamics'.

Le Chatelier's principle is an example of this equilibrium seeking principle. LeChatelier's principle is about the effect of a change in external conditions on the equilibrium of a chemical reaction. 'If the external conditions of a thermodynamic system are altered, the equilibrium of the system will tend to move in such a direction as to oppose the change in the external conditions', (Fermi, 1956). Fermi noted that if a chemical reaction were exothermal, i.e. (A+B=C+D+heat) an increase in temperature will shift the chemical equilibrium to the left hand side. Since the reaction from left to right is exothermal, the displacement of the equilibrium towards the left results in the absorption of heat and opposes the rise in temperature. Similarly a change in pressure (at a constant temperature) results in a shift in the chemical equilibrium of reactions which opposes the pressure change. This equilibrium seeking nature in chemical systems is a shared aspect of all dissipative systems.

2.4 Exergy degradation

Further discussion requires the introduction of the notion of quality of energy, exergy. Over the past thirty years and particularly in the last decade, *exergy* has emerged as a central concept in the discussion of thermodynamics (Bejan (1997), Szargut, Morris, Steward, (1988), Wall, G. (1986)). Energy varies in its quality or capacity to do useful work. During any chemical or physical process the quality or capacity of energy to perform work is irretrievably lost. Exergy is a measure of the maximum capacity of the energy content of a system to perform useful work as it proceeds to equilibrium with its surroundings and reflects all the free energies associated with the system. (Brzustowski and Golem, 1978). For example, water at the top of a high cliff is a high quality energy source (high exergy) because its potential energy can be used to perform work. We can use the falling water to turn a turbine and produce high quality energy in the form of electricity. But if the high quality energy in the falling water is not run through a turbine and falls freely to the rocks below, it turns into low quality dispersed heat energy. The exergy content of the water at the top is high, but the same water, at the bottom, with the same energy content, has much less exergy. Exergy is a measure of the quality of energy.

In terms of exergy, the classical second law of thermodynamics can be stated as: during any macroscopic thermodynamic process, the quality or capacity of energy to perform work is irretrievably lost. Energy loses exergy during any real process. A traditional first law energy analysis does not account for differences in energy qualities. A first law efficiency only compares the total amount of energy put into a system to the total amount received out of the system. This realisation has, in the discipline of energy system analysis and engineering thermodynamics, as can be seen from examining any introductory text book in the field that postdates 1993 (Moran and Shapiro, 1993), resulted in the recognition that both first law (energy, quantity) and second law (exergy, quality) analysis are necessary for the understanding and development of efficient and effective energy utilisation systems. (Bejan (1997), Gaggiolo,. (1980, 1983), Hevert, H.; Hevert, S. (1980), Moran, M.J. (1982), Szargut, Morris, Steward, (1988)).

Exergy is a function of the gradients between a system and its environment. It is a summation of the free energies in the situation. It measures the distance between a system and its environment. In effect it measures how far a system is from thermodynamic equilibrium with its environment. Exergy is not a useful concept for discussing equilibrium situations, the domain of classical thermodynamics, as, by definition, it value is zero in such situations. However it is a very powerful tool for non-equilibrium situations. The larger the value of the exergy, the more out of equilibrium the situation is.

The restated second law can be formulated in terms of exergy: A system exposed to a flow of exergy from outside will be displaced from equilibrium. The response of the system will be to organise itself so as to degrade the exergy as thoroughly as circumstances permit, thus limiting the degree to which the system is moved from thermodynamic equilibrium. Furthermore, the further the system is moved from equilibrium, the larger the number of organisational (i.e. dissipative) opportunities which will become accessible to it and consequently, the more effective it will become at exergy degradation. This is the exergy degradation principle for non-equilibrium thermodynamic situations.

2.5 Dissipative structures as gradient dissipators and exergy degraders

This section examines the behaviour of dissipative structures in light of the exergy degradation principle. Prigogine and others have shown that dissipative structures self-organise through fluctuations and instabilities which lead to bifurcations and new stable system states. Glansdorff and Prigogine (1971) have shown that these thermodynamic systems can be represented by coupled non-linear relationship i.e. autocatalytic positive feedback cycles, many of which lead to stable macroscopic structures which exist away from the equilibrium state. Convection cells, hurricanes, autocatalytic chemical reactions, and living systems are all examples of non-equilibrium dissipative structures which exhibit coherent behaviour.

The formation of Bénard cells is the classic example of emergent coherent organisation in response to an external energy input. They occur when a heated fluid makes the transition from conduction to convection as the primary form of heat transfer. In Bénard cell experiments, the lower surface of an experimental fluid filled chamber is heated, the upper surface is kept at a cooler temperature, and thus a temperature gradient is induced across the fluid. The initial heat flow through the fluid is by conduction. Energy transfer is by molecule to molecule interaction. When the heat flux reaches a critical value of the temperature gradient, the system becomes unstable and the molecular action of the fluid becomes coherent and connective overturning emerges. This coherent convective overturn results in hexagonal patterns in the fluids This coherent kinetic structuring increases the rate of heat transfer and exergy degradation by the fluid.

This transition between non-coherent, molecule to molecule, heat transfer, to coherent convection and structure results in excess of 10^{22} molecules acting together in an organised manner. This seemingly improbable occurrence is the direct result of the applied temperature gradient, the dynamics of the system at hand, and is the system's response to attempts to move it away from equilibrium. At higher temperature gradients, there appear to be a number of further transitions at which the hexagonal cells re-organise themselves so that the cost of increasing the temperature gradient escalates even more quickly. Ultimately the system becomes chaotic and dissipation is maximised in this regime. Organisation in this system resides in a window between linear near equilibrium processes and the chaos of turbulent flow.

Detailed analysis of Bénard cell experiments in terms of exergy degradation and entropy production have been reported on earlier (Schneider and Kay, 1994b). Figure 1 shows a graph of exergy degradation rate versus Rayleigh number. Rayleigh number is linearly proportional to the temperature gradient across the fluid. As the system is moved further from equilibrium by the increased temperature gradient, the amount of exergy degraded by the system increases. The rate of increase of exergy degradation goes up sharply with the emergence of the Bénard cells. Furthermore for every incremental increase in temperature gradient there is an increase in the incremental increase in the exergy degradation rate (i.e. the exergy destruction rate increases non-linearly with temperature difference across the fluid.).

The point of the Bénard cell example is that in simple physical systems, new structures and processes can spontaneously emerge which better resist the application of an external gradient, in the sense that it gets harder and harder to move the system from equilibrium because the system gets better and better at degrading the external input of exergy. The more a system is moved from equilibrium, the more sophisticated its processes for resisting being moved from equilibrium. This behaviour is not sensible from a classical second law perspective, but is what is expected given the exergy degradation principle. No longer is the emergence of coherent self-organising structures a surprise, but rather it is an expected response of a system as it attempts to resist and dissipate externally applied gradients which would move the system away from equilibrium. The term dissipative structure takes on new meaning. No longer does it mean just increasing dissipation of matter and energy, but dissipation of gradients as well.

This graph shows exergy destruction during heat transfer across a fluid. In the Bénard cell apparatus the working fluid is heated from below and the top of the apparatus acts as a cold sink. Initially all dissipation through the fluid occurs via conduction and molecule to molecule interaction. When the gradient reaches a critical level (Rayleigh number 1760) the transition to organised convection (Bénard Cells) occurs.

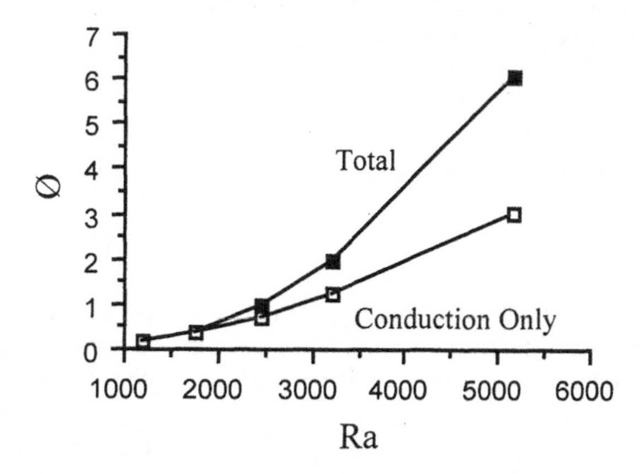

Figure 1: Exergy degradation in Bénard cells (Exergy degradation rate (ø) vs Gradient (Ra)). The graph shows the exergy (Kcal/hour) which must be provided from an external source in order to maintain the gradient versus Rayleigh number (the gradient). The bottom curve is the exergy destruction rate without Bénard Cells and the top curve is the exergy destruction rate with the formation of Bénard Cells. The point is that the emergence of the ordered structure (Bénard Cells) degrades more exergy. As the gradient increases a greater amount of work must be done to incrementally increase the gradient. The further the system is moved away from its equilibrium state the more exergy is destroyed, the system produces more entropy, and more work is required to maintain it in it's non-equilibrium state. It becomes more difficult to maintain the gradient as the system becomes more organised.

The literature is replete with similar phenomena in dynamical chemical systems. Chemical gradients result in dissipative autocatalytic reactions, examples of which are found in simple inorganic chemical systems, in protein synthesis reactions, or phosphorylation, polymerisation and hydrolysis autocatalytic reactions. Autocatalytic reaction systems are a form of positive feedback where the activity of the system or reaction augments itself in the form of self-reinforcing reactions. In autocatalysis, the activity of any element in the cycle engenders greater activity in all the other elements, thus stimulating the aggregate activity of the whole cycle. Such self-reinforcing catalytic activity is in itself self-organising, is an important way of increasing the dissipative capacity of the system and can act as an active selection process between competing elements in the cycle (Ulanowicz, 1996). Cycling and or autocatalysis are fundamental aspects of dissipative systems and represent not only the building of structure but is the source of complexity in non-equilibrium systems. Table 1 outlines some of the common behaviour or properties of dissipative systems.

2.6 Living systems as non-equilibrium dissipative systems

As discussed earlier, Schrödinger and Boltzmann, recognised the apparent contradiction between the thermodynamically predicted randomised cold death of the universe and the existence of processes (i.e. life) in nature by which systems grow, complexify, and evolve. Life does not decay into its composite parts but grows and complexifies. This apparent contradiction is resolved, as Schrödinger suggested, by considering living systems as non-equilibrium dissipative systems. Given the exergy degradation principle and the right conditions, the emergence of living systems should be expected as a means of furthering the mandate of exergy degradation.

If we view the earth as an open thermodynamic system with a large exergy flow impressed on it by the sun, physical and chemical processes will emerge to degrade the incoming exergy. Energy shifts, (conversion of short wave radiation to longer wave infrared), absorption, meteorological and oceanographic circulation will degrade much of the incoming solar exergy. However, there will still be exergy available for degradation. It has been argued elsewhere that life is another means of degrading this exergy (Kay, 1984, Kay and Schneider, 1992, Schneider and Kay, 1994ab).

Table 1: Properties of Dissipative Systems

- **Open** to material and energy flows.
- **Non-equilibrium**: Exist in quasi-steady states some distance from equilibrium.
- Maintained by energy **gradients** (exergy) across their boundaries. The gradients are **irreversibly** degraded in order to build and maintain organisation. These systems maintain their organised state by exporting entropy to other hierarchical levels.
- Exhibit material or energy **cycling**: Cycling and especially autocatalytic cycling is intrinsic to the nature of dissipative systems. The very process of cycling leads to organisation. **Autocatalysis** (positive feedback) is a powerful organisational and selective process.
- Exhibit **chaotic** and **catastrophic** behaviour. Will undergo dramatic and sudden changes in **discontinuous** and **unpredictable** ways.
- As dissipative systems are moved away from equilibrium they become organised:
  they use more exergy
  they build more structure
  this happens in spurts as new attractors become accessible.
  it becomes harder to move them further away from equilibrium

The origin of prebiotic life is the development of yet another route for the degradation of exergy. Most theories on the origin of life start with a chemical homogeneous soup. External sources of exergy (sun light, thermal or chemical gradients (e.g. deep sea vents)) can drive the system from equilibrium. Chemical or hydrophobic potentials probably developed phase transitions favouring self assembly of various molecules. A stepwise progression of stages can be recognised in the emergence of prebiotic organised structures; formation of simple molecules, the formation of biomonomers (amino acids, sugars), the formation of biopolymers (polypeptides, nucleic acids), the aggregation of bipolymers onto microspheres and the emergence of protocells as functional relationships develop among microspheres (Wicken, 1987). Life should be viewed as the sophisticated end in the continuum of development of natural dissipative structures from physical, to chemical autocatalytic, to living systems.

Life with its requisite ability to reproduce, insures that these dissipative pathways continue and has evolved strategies to maintain these dissipative structures in the face of a fluctuating physical environment. Living systems are essentially dynamic dissipative processes with encoded memories. The gene with its DNA, allows the dissipative process to continue without having to restart new dissipative pathways via stochastic events. Wicken (1987) noted that living systems are a unique example of dissipative structures, because they are self creating, rather than a product of only impressed forces. Life is a self-replicating system, operating through informed pathways of autocatalytic thermodynamic dissipation. The origin of life should not be seen as an isolated event but as a holistic process that represents the emergence of yet another class of processes whose goal is the dissipation of thermodynamic gradients, the degradation of exergy.

The dilemma which faced Schrödinger and Boltzmann is resolved in that aspects of growth, development, and evolution are the response to the thermodynamic imperative of exergy degradation. Biologic growth occurs when the system adds more of the same types of pathways for degrading exergy. Biologic development occurs when new types of pathways for degrading exergy emerge in the system. The larger the living system, i.e., the larger the system flow activity, the more reactions and pathways (both in number and type) are available for exergy destruction. This observation, derived from considering the principle of exergy degradation, provides a criteria for evaluating growth and development in living systems. All else being equal, the more effective exergy degradation pathway is preferred.

2.7 Ecosystems as exergy degraders

Following this line of logic ecosystems can be viewed as the biotic, physical, and chemical components of nature acting together as non-equilibrium dissipative processes. As ecosystems develop or mature they should develop more complex structures and processes with greater diversity, more cycling and more hierarchical levels all to abet exergy degradation. Species which survive in ecosystems are those that funnel energy into their own production and reproduction and contribute to autocatalytic processes which increase the total exergy degradation of the ecosystem. In short, ecosystems develop in a way which systematically increases their ability to degrade the incoming solar exergy (Kay, 1984, Kay and Schneider, 1992, Schneider and Kay, 1994ab).

Keeping in mind that the more processes or reactions of material and energy that there are within a system, (i.e. metabolism, cycling, building higher trophic levels) the more the possibility for exergy degradation, Schneider (1988) and Kay and Schneider (1992) showed that most, if not all, of Odum's (1969) phenomenological attributes of maturing ecosystems can be explained by ecosystems behaving in such a manner as to degrade as much of the incoming exergy as possible.

The energetics of terrestrial ecosystems provides an excellent example of the thesis that ecosystems will develop so as to degrade exergy more effectively. The exergy drop (i.e. degradation) across an ecosystem is a function of the difference in black body temperature between the captured solar energy and the energy reradiated by the ecosystem. This is discussed in detail in Fraser and Kay, 2000. Thus if a group of ecosystems are bathed by the same amount of incoming energy, the most mature ecosystem should reradiate its energy at the lowest exergy level, that is the ecosystem would have the coldest black body temperature. The black body temperature is determined by the surface temperature of the canopy of the ecosystem.

Consider the fate of solar energy impinging on five different surfaces, a mirror, a flat black surface, a piece of false grass carpet (i.e. Astroturf), a natural grass lawn and a rain forest. The perfect mirror would reflect all the incoming energy back toward space with the same exergy content as the incoming radiation. The black surface will reradiate the energy outward at a lower quality than the incoming energy, because much of the high quality ultraviolet exergy is converted to lower quality infra-red sensible heat. The green carpet will reradiate it's energy similar to the black surface but will differ because of it's surface quality and different emissivity. The natural grass surface will degrade the incoming radiation more completely than the green carpet surface, because processes associated with life, (i.e. growth, metabolism and transpiration) degrade exergy (Ulanowicz, and Hannon, 1987). The rain forest should degrade the incoming exergy most effectively because of the many pathways (i.e. more species, canopy construction) available for degradation.

In previous chapters (Schneider and Kay 1994, a, b) have discussed the work of Luvall and Holbo (1989, 1991) and Luvall et al. (1990) who conducted experiments in which they overflew terrestrial ecosystems and measured surface temperatures using a Thermal Infrared Multispectral Scanner (TIMS). Their technique allows assessments of energy budgets of terrestrial landscapes, integrating attributes of the overflown ecosystems, including vegetation, leaf and canopy morphology, biomass, species composition and canopy water status. Luvall and his co-workers have documented ecosystem energy budgets, including tropical forests, mid-latitude varied ecosystems, and semiarid ecosystems. Their data shows one unmistakable trend, that when other variables are constant the more developed the ecosystem, the colder its surface temperature and the more degraded its reradiated energy.

Work by Akbari, Murphy, Swanton and Kay on agricultural plots showed a similar trend. A lawn (single species of grass) had the warmest surface temperature, an undisturbed hay field was cooler, and a field which has been naturally regenerating for 20 years was coldest. These trends were confirmed over three years of observation. Also another field, which was regenerating for 20 years was disturbed by mowing. Its surface temperature immediately rose significantly, but very quickly returned to its cooler pre-disturbance value. Very recently, Allen and Norman have performed a set of experiments to explore the relationship between development and surface temperature in plant communities. So far their experimental results demonstrate that the surface temperature of plant communities tend to warm when they are removed from their normal conditions. That is plant communities are coldest (degrading the most exergy) when they are in the normal conditions which they are adapted to.

Clearly, there is much to be gained from examining ecosystems through the lens of exergy degradation. A number of ecosystem phenomena can be explained and hypotheses concerning ecosystem development can be generated and tested. But there is more to the story. Most ecosystems will have many different options for exergy degradation available to them. Some will have different sources of exergy available. Different combinations of exergy sources and degradation possibilities may be equivalent from a exergy degradation perspective. So the number of possible variations on ecosystem organisation, which are thermodynamically equivalent, may be significant. This quickly leads to a complicated set of possible organisational pathways and which is actually manifested may very well be a reflection of a collection of accidents of history.

The imperative of thermodynamics and exergy degradation is not the only one acting on living systems. Of equal importance is survival, an imperative which may not be consistent with maximum exergy degradation. Inevitably tradeoffs will have to be made and ecosystems as they exist on the ground will reflect these tradeoffs (Kay, 1984). There will not be single best solutions to the imperatives of exergy degradation and survival. Just solutions that work longer than others. Furthermore, to add to the complexity and uncertainty, Dempster (1998) and Kay have shown that such systems must, by necessity, be recursively nested autopoietic and synpoietic systems. Together these factors and others, summarised in Table 2, mean that the consideration of ecosystems as self-organising systems must confront the issues of complexity and uncertainty head on. Conventional scientific approaches simply are not adequate for this task (Kay and Schneider, 1994, Kay, Regier, Boyle and Francis (1999)). In the next section an alternate way to describe ecosystems, which is rooted in their nature as adaptive self-organising complex dissipative systems, is presented.

3. The self-organising holarchic open systems portrayal of ecological systems

3.1 Complex systems thinking and SOHO system descriptions

The issue of complexity has attracted much attention in the past decade. This issue emerged in the wake of the new sciences which became prominent in the 1970's; catastrophe theory, chaos theory, non-equilibrium thermodynamics and self-organisation theory, Jaynesian information theory, complexity theory etc. A number of authors have focused specifically on self-organising systems (di Castri, 1987, Casti,1994, Jantsch, 1980, Kay 1984, Nicolis and Prigogine, 1977, 1989, Peacocke, 1983, Wicken, 1987). The term *complex systems thinking* is being used to refer to the body of knowledge that deals with complexity. Complex systems thinking has its origins in von Bertalanffy's general systems theory.

Table 2: Properties of *complex systems* to bear in mind when thinking about ecosystems.

•**NON-LINEAR:** Behave as a whole, *a system*. Cannot be understood by simply decomposing into pieces which are added or multiplied together.

•**HIERARCHICAL:** Are *holarchically nested.* The system is nested within a system and is made up of systems. The 'control' exercised by a holon of a specific level always involves a balance of internal or self-control and external, shared, reciprocating controls involving other holons in a mutual causal way that transcends the old selfish-altruistic polarising designations. Such nestings cannot be understood by focusing on one hierarchical level (holon) alone. Understanding comes from the multiple perspectives of different *types* and *scale*.

•**INTERNAL CAUSALITY:** non-Newtonian, not a mechanism, but rather is *self-organising*. Characterised by: goals, positive and negative feedback, autocatalysis, emergent properties and surprise.

•**WINDOW OF VITALITY:** Must have enough complexity but not too much. There is a range within which self-organisation can occur. Complex systems strive for *optimum*, not minimum or maximum.

•**DYNAMICALLY STABLE?:** There may not exist equilibrium points for the system.

•**MULTIPLE STEADY STATES:** There is **not** necessarily a unique preferred system state in a given situation. *Multiple attractors* can be possible in a given situation and the current system state may be as much a function of historical accidents as anything else.

•**CATASTROPHIC BEHAVIOUR:** The norm

Bifurcations: moments of unpredictable behaviour
Flips: sudden discontinuities, rapid change
Holling four box cycle Shifting steady state mosaic

•**CHAOTIC BEHAVIOUR:** our ability to forecast and predict is always limited, for example to between five and ten days for weather forecasts, regardless of how sophisticated our computers are and how much information we have.

Maruyama (M.T. Caley and D. Sawada. 1994) was one of the first to examine the issues of complexity. In his 'second cybernetics' of the mid 1960s, he identified a class of systems which require explanation in terms of morphogenetic causal models, that is explanations that involve both positive and negative feedback loops and autocatalysis, mutual causality. He demonstrated how probabilistic or deterministic loops of mutual causality can increase a system's pattern of heterogeneity towards higher levels of organised complexity. He showed that traditional explanations in terms of linear causality, that is in terms of a clear cause and effect relationship, were not possible for the phenomena exhibited by this class of systems. The problem is that when feedback loops dominate a system, the effect becomes part of the cause. So the cause is not independent of the effect as is required by linear cause and effect explanations. The new fields of (first) cybernetics and general systems theory were also incapable of providing an explanation of these phenomena as they focused on systems where negative feedback leads to homeostasis. His conclusion was that a new mode of explanation, quite different from traditional scientific approaches, was needed for this class of systems.

Koestler (Koestler and Smythies 1969, Koestler 1978) focused on self-organising, holarchic and open (SOHO) attributes of systemic phenomena of the kind identified by Maruyama. A holarchy is a generalised version of a traditional hierarchy (not to be confused with Allen's notion of hierarchy, Allen and Starr, 1982, Allen and Hoekstra, 1992), with reciprocal power relationships between levels rather than a preponderance of power exerted from the top downwards. A particular system of this type Koestler termed a 'holon' because it occurs in a contextually nested or holarchic reality with mutual causality guiding reciprocal interactions between a holon and proximate contiguous holons of different scales – inside, outside and lateral to the holon of interest. The term holon or SOHO system is used herein to refer to the self-organising entity that is the subject of our inquiry.

Ulanowicz (1996, 1997) developed further some proposals by K. R. Popper (Popper, 1990) to extend a perception of indeterminacy in the quantum realm to other scales of phenomena by generalising the usual Newtonian concept of force to obtain a notion of systemic dynamical cohesion which Popper called a 'propensity'. A propensity is always contextual (as are Maruyama's morphogenetic system and Koestler's SOHO holon) rather than universalistic as in the Newtonian sense of 'force'. Ulanowicz (1995) proposed that a mutual-causal kind of autocatalysis plays a self-organising role in Popper's propensity, perhaps in generating dynamical cohesion through forces that act asymmetrically, and not symmetrically as in the Newtonian sense. Ulanowicz also implied that 'dynamical cohesion' tends to be attenuated in a step-wise manner at the interfaces of interacting holons, both with respect to nested and non-nested kinds of relationships and these attenuations may be perceived as boundary like.

It has been argued (Kay, Regier, Boyle and Francis (1999)) that ecosystems fall into the class of systems that Maruyama identified and which Koestler called SOHO systems. The dynamics of such systems are described by narratives. A central question to be addressed by the narrative description of a SOHO system is an elaboration of its propensities. The elaboration delineates the mutual causality of the feedback loops and autocatalytic process which give the system its coherence as an entity. This set of propensities, which define a holon, is referred to as its 'canon'.

The remainder of this section sketches the elements of a Self-organising Holarchic Open Systems description of an ecological system with particular emphasis on the application of the exergy degradation principle as part of a complex systems thinking approach to understanding ecosystems.

3.2 Hierarchy

The consideration of a SOHO system must begin with a hierarchical description. The first step is to define the holons, that is self-organising entities of interest. Careful attention must be paid to the issues of scale and type (Allen and Starr, 1982, Allen and Hoekstra, 1992, Allen, Bandurski, and King 1993, King, 1993, Günther and Folke, 1993). The narrator must decide if the focus is a watershed, or perhaps a community, or the home range of a species, etc. A delineation of the important processes which make up the holon and their interconnection is required. Are we talking about reproduction?, energetics?, spatial interconnection..., and of what? Key relationships between holons must be established. Most importantly, the context for the holon must be explored. This requires investigating the constraints on the holon dictated by the upper level holon of which it is a part (i.e. the implications for a lake of its watershed). The hierarchical description of a SOHO system will encompass several levels of holons (different scales: watersheds, drainage basins, sub watersheds, ...) and several holarchies (different types: a description as a watershed, as a landscape, as habitat etc.). Prior experience helps to inform how many levels and types of descriptions are sufficient to understand the SOHO system, as the array of such descriptions is only limited by the narrator's imagination.

3.3 Attractors

Having defined the SOHO system of interest, the next aspect to explore is its self-organising behaviour. A SOHO system exhibits a set of behaviours which are coherent and organised, within limits. The nexus of this organisation at any given time is referred to as an attractor. The term 'attractor' comes from the state space description of the behaviour. The system has a propensity to remain in a limited domain of state space (for example a gravity well). It behaves as if it were 'attracted' toward this domain and hence the term 'attractor'. As SOHO systems evolve they shift between attractors within the SOHO system's overall state space. The re-organisation that these shifts entail is not smooth and continuous but rather is step-wise. The system flips its organisational state in often dramatic ways. (Recall the emergence of Bénard Cells discussed earlier.).

Ecosystems have multiple possible operating states or attractors, and may shift or diverge suddenly from any one of them (See Figure 2). The notion of alternate stable states (attractors) in ecosystems is not well known in the ecological community, but it is also not new. Kay (1984), Holling (1986), Kay (1991), and Kay and Schneider (1994), Kay (1997), Ludwig et al. (1997) examine, in general, the notion of alternate stable states in ecosystems and their implications. Yet the importance of this notion for explaining ecosystem phenomena remains largely unexplored.

For example, a portion of a natural area[1] in Southern Canada is a **closed soft maple swamp** in a wetland community. However, the amount and duration of the flows of water can radically alter this operating state. Drying events, such as an extended drought, could change the operating state to an **upland forest community or grassland** with associated vegetation structure. If there are extended periods of flooding causing high water levels, the operating state would be that of a **marsh ecosystem**. This is because red and silver maple are tolerant to flooded conditions within 30% to 40% of the growing season. If flooding events are greater than this threshold, the forest trees will die, giving way to more water tolerant herbaceous marsh vegetation. The feedback mechanism which maintains the swamp state is evapotranspiration (i.e. water pumping) by the trees. Too much water overwhelms the pumping capability of the trees and not enough shuts it down. The point of this example is that the current ecosystem state is a function of its physical environment and the accidents of its history. A single dry or wet season can change what is on the landscape for decades.

The task of the narrator is to scope out the array of attractors available to the SOHO system, the potential flips between them, and the underlying morphogenetic causal structure of the organisation in the domain of the attractors.

3.4 The thermodynamics of self-organisation

A key question about attractors is what characterises them. This is equivalent to asking what gives rise to the propensities which animate the particular canon associated with a given attractor. As discussed earlier, Prigogine showed that, when dealing with open systems with an enduring (not necessarily constant) flow of exergy, spontaneous emergence of coherent behaviour and organisation can occur (Nicolis and Prigogine, 1977, 1989). Prigogine showed that this occurs because the system reaches a catastrophe threshold and flips into a new coherent behavioural state. (This is evident for example in the vortex which spontaneously appears when draining water from a bathtub).

[1] More detail about this case study, The Huron Natural Area can be found on its WWW site: www.fes.uwaterloo.ca/u/jjkay/HNA/

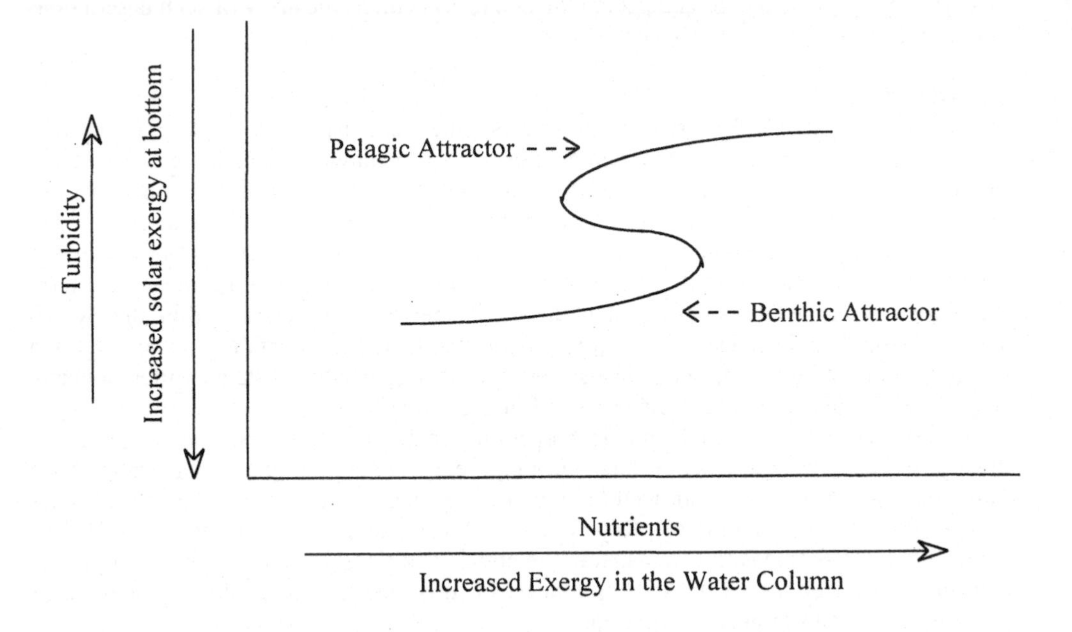

Figure 2: Benthic and Pelagic Attractors in shallow lakes
Two different attractors for shallow lakes have been identified. In the benthic state, a high water clarity bottom vegetation ecosystem exists. As nutrient loading increases the turbidity in the water, the ecosystem hits a catastrophe threshold and flips into a hypertrophic, turbid, phytoplankton pelagic ecosystem. The relationship of these two attractors, from a thermodynamic perspective, is as follows:
Let us assume that the benthic attractor is dominant and that the rate at which phosphorus is being added to the water is increasing. The benthic system has means of deactivating phosphorus. However the amount of active phosphorus will increase, albeit slowly, effectively increasing the exergy in the water column. As this exergy increases a critical threshold is passed which allows the pelagic system to self-organise to coherence. Once this occurs the exergy at bottom decreases rapidly due to shading (turbidity) thus catastrophically de-energising the benthic system. This results in the eventual re-activation of the phosphorus in the bottom muds which the benthic system had previously deactivated, thus strengthening the pelagic attractor even more.
Assuming the pelagic attractor is dominant and if the level of active phosphorus in the water column decreases, a critical threshold is again reached below which it is no longer possible to capture enough solar energy to energise the pelagic system. In effect, the exergy in the water column decreases below the minimum level for the window of vitality of the pelagic system. As this occurs the exergy at the bottom increases thus re-energising the benthic system. And so the aquatic system flips back and forth between the pelagic and the benthic regime depending on where in the water column the sunlight's exergy is available to energise the system.

In examining the energetics of open systems, Kay and Schneider (Schneider and Kay, 1994a,b) noted that as an open system with exergy pumped into it is moved away from equilibrium. But nature resists movement away from equilibrium. So in such a situation a system will organise itself so as to degrade the exergy as thoroughly as circumstances permit, thus limiting the degree to which the system is moved from thermodynamic equilibrium. When the input of the exergy and material pushes the system beyond a critical distance from equilibrium, the open system responds with the spontaneous emergence of new organised behaviour that uses the exergy to build, organise and maintain its structure. This dissipates the ability of the exergy to move the system further away from equilibrium. As more exergy is pumped into a system, more organisation emerges, in a step-wise way, to degrade the exergy. Furthermore these systems tend to get better and better at 'grabbing' resources and utilising them to build more structure, thus enhancing their dissipating capability. There is however, in principle, an upper limit to this organisational response.

Beyond a critical distance from equilibrium, the organisational capacity of the system is overwhelmed and the system's behaviour leaves the domain of self-organisation and becomes chaotic. As noted by Ulanowicz there is a 'window of vitality', that is a minimum and maximum level in between which self-organisation can occur.

The theory of non-equilibrium thermodynamics suggests that the self-organisation process in ecosystems proceeds in a way that: a) captures more resources (exergy and material); b) makes more effective use of the resources; c) builds more structure; d) enhances survivability (Kay, 1984, Kay and Schneider, 1992, Schneider and Kay, 1994a,b). These seem to be the kernel of the propensities of self-organisation in ecosystems. How these propensities manifest themselves as morphogenetic causal loops is a function of the given environment (context) in which the ecosystem finds itself imbedded as well as the available materials, exergy and information.

3.5 SOHO systems as dissipative systems: A conceptual model

The conceptualisation of self-organisation, as a dissipative system, is presented in Figure 3. Self-organising dissipative processes emerge whenever sufficient exergy is available to support them. This expectation is a consequence of the exergy degradation principle. Dissipative processes restructure the available raw materials in order to degrade the exergy. Through catalyse, the information present enables and promotes some processes to the disadvantage of others. The physical environment will favour certain processes. Therefore which specific processes emerge depends on the raw materials and exergy available to operate them, the information present to catalyse the processes, and the physical environment. The interplay of these factors defines the context for (i.e. constrains) the set of processes which may emerge. Generally speaking, which specific processes emerge from the available set are uncertain. Once a dissipative process emerges and becomes established it manifests itself as a structure.

In the case of a vortex in the bathtub water, the exergy is the potential energy of the water, the raw material is the water and there is no information, the dissipative process is water draining, the dissipative structure is the vortex. The vortex will not form until enough height of water is in the bathtub, and if too much height of water is present, laminar flow occurs instead of a vortex.

These structures provide a new context, nested within which new processes can emerge, which in turn beget new structures, nested within which... Thus emerges a SOHO system, a nested constellation of self-organising dissipative process/structures organised about a particular set of sources of exergy, materials, and information, embedded in a physical environment. The canon of the SOHO system is the complex nested interplay and relationships of the processes and structures, and their propensities, that give rise to coherent self-perpetuating behaviours that define the attractor.

In an ecological setting, examples of the structures are the individuals of species, breeding populations, forests etc. The processes are reproduction, metabolism, evapotranspiration etc. The context is the available set of nutrients and exergy sources in a physical environment. The information includes the biodiversity (See Kay and Schneider 1994). The general propensities of the ecological systems were stated above.

4. Narratives I: exergy, canons, and attractors

The task of characterising the attractors and canon of ecosystems then becomes one of describing how the local context of exergy, materials and information and biophysical environment, and the global propensities of capturing more resources (exergy and material), making more effective use of the resources; building more structure, and enhancing survivability, give rise to the emergence of the nested structures and processes which constitute a self-organising holarchic open system. This characterisation takes the form of a narrative[2], literally a story of how the system might unfold over time. The narrative is qualitative, multi threaded (multiple threads of explanation) and multi pathed (that is portrays a number of possible pathways for development or storylines). Following are two overly brief examples of how narratives can be used to discuss ecosystem organisation

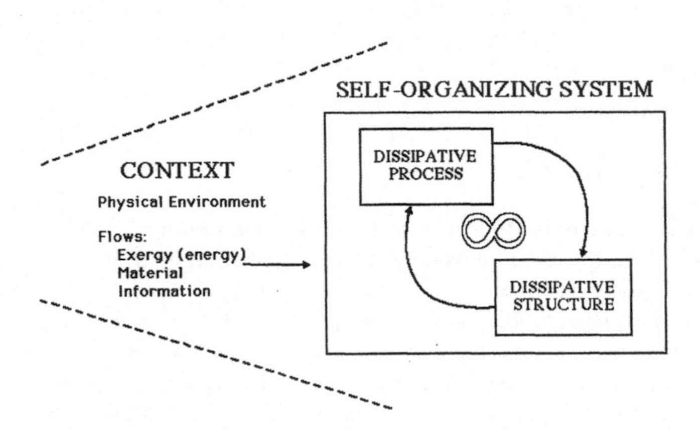

Figure 3: A conceptual model for self-organising systems as dissipative structures. Self-organising dissipative processes emerge whenever sufficient exergy is available to support them. Dissipative processes restructure the available raw materials in order to dissipate the exergy. Through catalyse, the information present enables and promotes some processes to the disadvantage of others . The physical environment will favour certain processes. The interplay of these factors defines the context for (i.e. constrains) the set of processes which may emerge. Once a dissipative process emerges and becomes established it manifests itself as a structure. These structures provide a new context, nested within which new processes can emerge, which in turn beget new structures, nested within which. Thus emerges a SOHO system, a nested constellation of self-organising dissipative process/structures organised about a particular set of sources of exergy, materials, and information, embedded in a physical environment. The canon of the SOHO system is the complex nested interplay and relationships of the processes and structures, and their propensities, that give rise to coherent self-perpetuating behaviours, that define the attractor.

4.1 A narrative of Lake Erie

Consider the case of shallow lakes, such as Lake Erie in North America. Regier and Kay (1996) interrelated empirical generalisations from aquatic ecology by R. Margalef, R. A. Vollenweider and others, together with notions from complex system theory to propose a two-attractor catastrophe cusp model (Figure 4) as a way of integrating much empirical information of how aquatic systems might transform under powerful, careless human interventions. Two different attractors for shallow lakes have been identified. (Scheffer(1990), Blindow et al. (1993), Scheffer et al. (1993), Regier and Kay (1996), and Carpenter and Cottingham (1997) Scheffer (1998), Kay and Regier (1999)). In the oligotrophic/pelagic state, a high water clarity bottom vegetation ecosystem exists. As nutrient loading increases the turbidity in the water, the ecosystem hits a catastrophe threshold and flips into a hypertrophic, turbid, phytoplankton oligotrophic/benthic ecosystem. Each has its own canon. Different locations in a lake will be organised about one of these attractors depending on the holarchic context for the specific portion of the lake. Lakes which flip between these attractors on a regular basis have been found. Lake Erie appears to be currently in the midst of such a flip, from pelagic to benthic. At least three quite different descriptions of such a lake will be needed, one for the pelagic state,

[2] Thanks to David Waltner-Toews for suggesting this term.

one for the benthic, and one for the intermediate stage as the system flips between attractors.

The essence of the canon of the benthic system is that it depends on solar energy reaching the bottom for the exergy necessary to energize the system. The solar exergy is captured by the green matter on the bottom and is transformed into forms appropriate to power the benthic processes. These include predation and grazing of the pelagic system, thus suppressing it. Various means emerge to maintain the ecosystem at the benthic attractor. Notable of these are means for keeping the water clear so solar energy will reach the bottom and means for keeping the water column free of sufficient exergy which would empower the pelagic attractor.

The pelagic system, on the other hand, depends on exergy in the water column to energise it. Solar energy may be in the water column. However, unless the materials necessary for the existence of dissipative processes, which can utilise the solar energy, are present in the water column, nothing can be done with the solar energy, so it has no exergy. For example, in many lakes, available phosphorus in the water column limits the level of photosynthesis by phytoplankton. Beyond a critical level of available phosphorus in the water column, there is enough availability of solar energy (i.e. sunlight exergy) to support the phytoplankton bloom necessary for the activation of the pelagic attractor. Once this occurs, the solar energy capture happens near the surface water instead of at the bottom and means emerge for promoting and maintaining the pelagic attractor. Of course, by its very presence the pelagic system shades the benthic from irradiation by the sun, thus decreasing the exergy at the bottom.

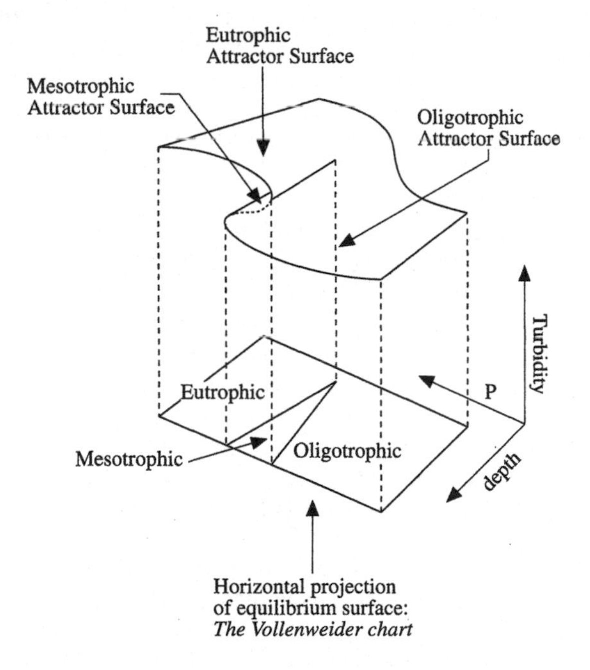

Figure 4: A catastrophe model for aquatic ecosystems: One attractor, associated with high loading of phosphates into relatively shallow waters with high residence time, relates to an eutrophic state. The other attractor associated with low loadings into deep waters of low residence time, is associated with an oligotrophic state. The narrow zone of mesotrophic conditions may be related to the unstable overlapping part of the catastrophe fold. The horizontal projection of the surface (control space) is the so called 'Vollenweider chart'. The vertical projection gives a fold catastrophe surface similar to that presented in Scheffer et al. (1993). The axis labelled P, the phosphorous loading in the lake (or other nutrient loading), corresponds to the Y axis is the Vollenweider chart and Scheffer fold catastrophe diagram. Depth is the mean depth of the lake as in the X axis in the Vollenweider chart. Turbidity is the z-axis and corresponds to the x-axis in the Scheffer fold catastrophe diagram.

The relationship of these two attractors, from a thermodynamic perspective, is shown in Figure 2. This thermodynamic relationship is based on the nutrient-turbidity relationship reported by Scheffer et al (1993) for shallow lakes and generalised to deeper systems (Regier and Kay, 1996) and in particular for Lake Erie (Kay and Regier, 1999). Let us assume that the benthic attractor is dominant and that the rate at which phosphorus is being added to the water is increasing. The benthic system has means of deactivating phosphorus, as discussed in Kay and Regier (1999). However, the amount of active phosphorus will increase, albeit slowly, effectively increasing the exergy in the water column. As this exergy increases a critical threshold is passed which allows the pelagic system to self-organise to coherence. Once this occurs the exergy at bottom decreases rapidly due to shading (turbidity) thus catastrophically de-energising the benthic system. This results in the eventual re-activation of the phosphorus in the bottom muds which the benthic system had previously deactivated, thus strengthening the pelagic attractor even more.

Assuming the pelagic attractor is dominant and if the level of active phosphorus in the water column decreases, a critical threshold is again reached below which it is no longer possible to capture enough solar energy to energise the pelagic system. In effect, the exergy in the water column decreases below the minimum level for the window of vitality of the pelagic system. As this occurs the exergy at the bottom increases[3] thus re-energising the benthic system. And so the aquatic system flips back and forth between the pelagic and the benthic regime depending on where in the water column the sunlight's exergy is available to energize the system.

In the numerous shallow parts of the Great Lakes, it appears that the phosphate rich runoff from human activities, usually in combination with other cultural stresses, empowered the pelagic attractor decades ago. Sewage treatment plants, agricultural management practices and other remediative measures, now seem to have reduced the phosphate content of the runoff sufficiently that the benthic attractor is reasserting itself.

4.2 A narrative of the Holling four box

Another example (see Figure 5) of the notion of canon and attractor and the ability to characterise them in terms of the form of exergy utilised is Holling's four-box model of the dynamics of terrestrial ecosystems. (See Holling 1986, 1992). All living systems exhibit a birth-renewal-growth-death cycle and it is this characteristic which makes life more sophisticated than non living dissipative systems (Wicken, 1987). We are all familiar with the death and reproduction at the cellular level and the birth-growth and death of individuals, but it is only recently that Holling has made us aware that this cycle occurs at many temporal and spatial scales, including ecosystems. (Holling, 1992). The idea of a forever stable climax stage of succession has been abandoned. Fire, pests, new species and other disturbances produce a shifting mosaic pattern of ecosystem succession on the landscape.

Using the restated second law of thermodynamics and in particular the exergy degradation principle as a basis, the dynamics of the Holling four box model can be described to proceed as follows:

Referring to Figure 5, the horizontal axis can be taken as 'stored exergy'. This is the amount of exergy stored in biomass and is related to the amount of nutrients bound in the biomass. The vertical axis is the 'exergy consumption'. This is the rate at which exergy is utilised by the system, that is the rate at which the incoming exergy is degraded.

Starting at **exploitation**, if there is sufficient materials and biological information available, then dissipative processes will emerge which utilise the exergy in the solar energy. In other words, some organisms will take advantage of the available resources. The thermodynamic direction of all self-organising processes is to increase the rate of exergy

[3] As the pelagic system unravels the shading of the bottom decreases.

degradation. Thus the exergy consumption rate will increase as the ecosystem proceeds to develop towards **conservation.** In this case (as noted by Jørgensen and Meyer, 1979; Jørgensen, 1992) this developmental pathway also involves increasing biomass and hence stored exergy. The more exergy stored, the bigger the structure, the better able it is to utilise exergy, the bigger it gets, ... This is the direction of the **first thermodynamic branch**[4]. The exergy source is solar energy.

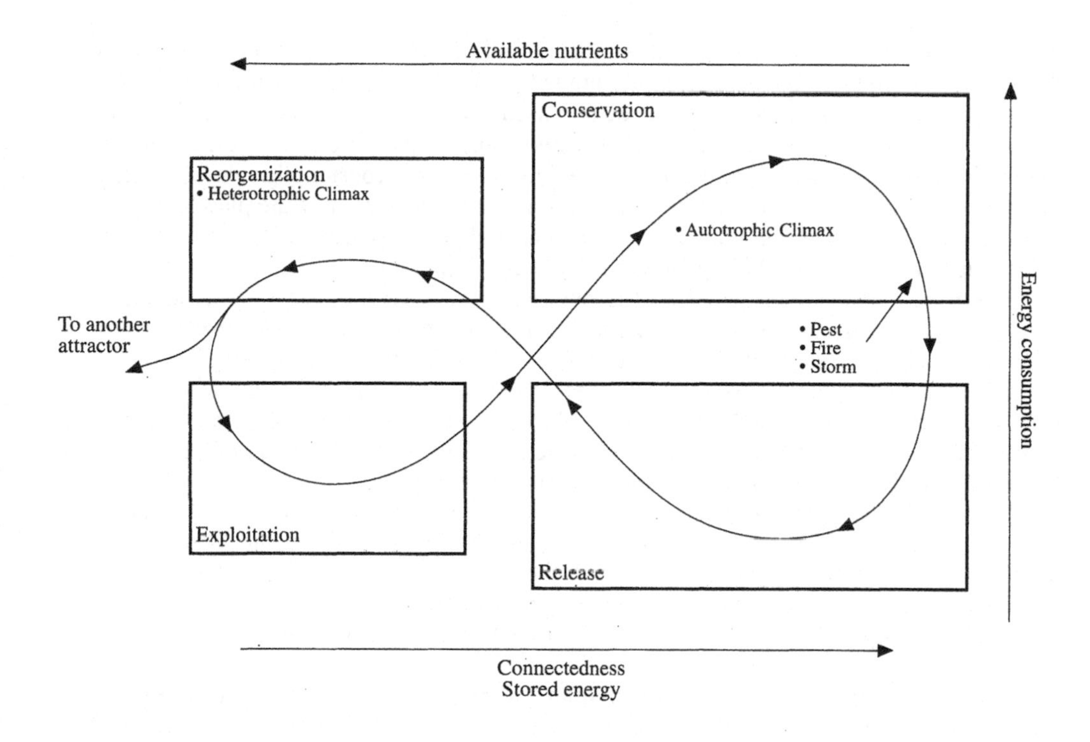

Figure 5: Holling's four box model as a dual thermodynamic branch system.
The first trajectory is the 'exploitation' to 'conservation' thermodynamic branch which culminates in the 'climax' community. The biological attractor is the autotrophic system (i.e. a forest). The canon is expressed, for example, as the growth of a forest to maturity and this is energised by solar energy. However, in the process of increasing the utilisation of solar energy and hence building more structure, much exergy is stored in the biomass. This has the effect of moving the system further and further from thermodynamic equilibrium as it develops.
When, as Holling puts it, the inevitable accident (fire, windstorm, or pest outbreaks) happens, suddenly much exergy is available in the form of dead biomass. This exergy energizes a new biological attractor, the heterotrophic or decomposer system. This is the thermodynamic branch which runs from 'release' to 're-organisation'. As the system progresses along this path, it releases the stored nutrients while using the stored exergy. Eventually the stored exergy runs out and the heterotrophic system collapses. However, in the process it has released the nutrients necessary for the re-emergence of the solar energy-powered system. This interplay between two biological attractors, which are organised around different forms of exergy, materials, and information is played out giving rise to the landscape we see.

[4] Thermodynamic branch: the developmental path taken by a self-organising system as it develops

However there is a fundamental contradiction inherent in this developmental pathway. The more exergy that is stored in the system, the more likely (according to the restated second law of thermodynamics) that some dissipative process will emerge to take advantage of it. So fire, pest outbreaks etc., occur that take advantage of all the exergy stored in the biomass. The paradox is that the more effective the ecosystem is at consuming exergy, that is the more organised it is, the more exergy it contains and hence the more likely it is to be consumed by another self-organising process (i.e. fire, pest outbreaks etc.). So **conservation** represents a point of maximum thermodynamic organisation in the sense that the system is utilising the available exergy as fully as possible. But it also represents a point of maximum thermodynamic risk as it as far out of equilibrium as is possible. (Recall that distance from thermodynamic equilibrium is measured by exergy content of a system).

In the language of attractors, there are two attractors, the attractor of maximum exergy consumption and the attractor of local thermodynamic equilibrium. For this particular thermodynamic branch the attractor of maximum exergy consumption is moving in opposition to the local equilibrium attractor. The **conservation** point is the place where the two attractors are in balance. For some systems this balance is precarious for others less so, but in the end the local equilibrium attractor is always dominant

Once the inescapable happens, that is **release,** a new source of exergy is available for use, that is the exergy in the stored biomass. Again it inevitable that this new exergy source will be utilised. As always, the self-organising process unfolds in a direction of increasing exergy utilisation, except that the processes involved are fundamentally different and instead of storing biomass and hence exergy, they release the exergy in the stored biomass and at the same time release the stored nutrients. This is the direction of the **second thermodynamic branch**. The exergy source is stored energy.

Eventually the **reorganisation** point is reached, that is point where the stored exergy runs out. But now the raw materials are available to start along the first type of thermodynamic branch again. Which specific branch is followed is a function of the biological information, nutrients and current environmental conditions. And this is the point where biodiversity is so crucial, as this is the point where resiliency matters.

To summarise:

The **exploitation point** is one of minimum exergy use and storage. The **conservation point** is one of maximum exergy use and storage. The **release point** is one of minimum exergy use and maximum storage. The **reorganisation point** is one of maximum exergy use and minimum storage.

There are two thermodynamic branches, that is self-organising pathways that are followed. One (from **exploitation** to **conservation**) is driven by the exergy in solar energy and involves increasing biomass and hence stored exergy. The other (from **release** to **reorganisation**) is driven by stored exergy and involves the release of the stored exergy and hence biomass. The direction of both is increased exergy consumption. The ecosystem alternates between these two sources of exergy and hence follows two qualitatively different pathways of self-organisation. The specifics are determined by the environmental conditions, available resources and biological information, the latter usually being the determining factor.

The first branch has been traditionally referred to as succession, or growth and development and culminates in the 'climax' community of Clementsian succession theory. Biologically it is the attractor for the autotrophic system (i.e. a forest). The canon is expressed, for example, as the growth of a forest to maturity and this is energised by solar energy. The second branch is about creative destruction, that is decomposition. Biologically it is the attractor for the heterotrophic system, a system whose canon is the decomposition of organic material. The interplay between these two biological attractors, which are organised around different forms of exergy, materials, and information is played out giving rise to the landscape we see.

These two examples demonstrate the role of thermodynamics, particularly the dissipative system conceptual model, and the restated second law in generating a narrative description of ecological systems as SOHO systems.

5. Narratives II: morphogenetic causal loops

Having discussed the hierarchical structure of the SOHO system, it's attractors, flips, propensities and canons in terms of the dissipative system conceptual model, as above, there is still the matter of describing the internal causal schemes which maintain the attractor and which make up the canon of the ecosystem. The description of these causal schemes explains the local propensities of the ecosystem and are the kernel of the narrative. Their description is in terms of morphogenetic causal loops which are made up of positive and negative feedback loops, some of which generate autocatalysis. Ulanowicz (1997) discusses the importance of these morphogenetic causal loops to the understanding of ecosystems. DeAngelis (DeAngelis et al, 1986, DeAngelis, 1995) has written extensively about these feedback loops and has given many ecological examples. Two simple examples, taken from DeAngelis (1986), are presented here to illustrate narration in terms of morphogenetic causal loops.

Consider forests in dry mountainous areas of the world. Often, as moisture laden clouds pass over bare mountains, they will not drop rain because of the heat reflected from the bare rocks. However as a forest develops on a mountain the re-radiated heat decreases. (Schneider and Kay, 1994b). As the re-radiated heat decreases more rain falls, which promotes more forest growth, which promotes more rain fall....

In southeastern Australia the dominant trees are sclerophyllous eucalypti, but the undergrowth consists of lush mesophytic vegetation. Normally these circumstances would give rise to a temperate rain forest. However these systems are subject to frequent fire, which would not occur if the mesophytic vegetation dominated. Fire increases soil leaching and sclerophylls are better adapted to poorer soils than mesophylls. Thus the dominance by sclerophyllous forest depends on fire and the occurrence of fire depends on the dominance by sclerophyllous forest. The morphogenetic causal loop of sclerophyllous dominant forest, fire, and soil infertility obstructs the development of temperate rain forests.

Another important aspect of SOHO systems to be scrutinised is the role of morphogenetic causal loops in maintaining the canon of a system in spite of a changing context (Rapport and Regier, 1985, 1995). Consider for example the acidification of lakes. The acidity in the precipitation did not suddenly change, but rather incrementally changed over the years. In our terms the context (pH of precipitation) of the SOHO system changed substantially over time. However the pH of the lake water did not change substantially, relatively speaking, over the same period (Stigliani, 1988). The lake maintained the canon through a series of feedback loops that largely buffered the lake from the environmental change. Eventually the runoff reached a level of acidity which exceeded the compensatory capacity of these loops. Once this happened, the effectiveness of the SOHO system decreased, which in turn decreased the capacity of the loops to compensate, which decreased the effectiveness of the SOHO system.... and then quickly the canon unraveled and the SOHO system flipped to another attractor, in this case a 'dead' lake. The narrative description of a SOHO system must not only delineate the morphogenetic causal loops, but also the contextual circumstances in which the loops can and cannot operate. Doing this in effect defines the domains of the attractors, the resiliency of the canon, its window of vitality.

Elsewhere (Kay and Regier, 1999) a more detailed partial narrative sketch of Lake Erie as a SOHO system is presented. It suggests that some issues are not as important as is currently thought and that others of importance have not been examined at all. This narrative of Lake Erie weaves together the themes of organism, species, ecosystem, landscape and biome in the context of physical environment, climate and human habitation and the changes therein. Some of the crucial morphogenetic causal loops, particularly those involving phosphorus, and their relationships to the canon of the pelagic and benthic attractors are outlined. The narrative takes the form of a multilayered account of the ecosystem's operation from different perspectives and scales. While some individual elements of the narrative consist of traditional scientific models and descriptions, the synthesise of these elements together into a narrative transcends normal scientific descriptions.

In this narrative of Lake Erie, the feedback loops, which buffer the system from changes in external influences, are of particular importance. The benthic attractor has elaborate feedback schemes, operating at different spatial and temporal scales, for limiting the phosphorous in the water column[5]. The pelagic attractor has elaborate schemes to accomplish just the opposite. The way in which changes in context enable and disable these feedback loops, and their associated canons, thus reinforcing attractors or triggering flips between them, has received little attention from the scientific community whether it be for this example or for ecosystems in general.

Yet our work would suggest that it is precisely these issues (that is describing the 'flip' from one attractor to another through accounting for how environmental influences (context), acting at different spatial and temporal scales, disable one feedback system while enabling another) that we must understand, if we are to comprehend the relationships between human activities and changes in the ecology of our planet. Understanding about attractors and flips is a necessary prerequisite for sustainability. It is the capability to address these issues which is at the core of the utility of the SOHO ecosystem description. This chapter has argued that thermodynamics and in particular, the revised second law, plays a central role in discussing ecosystem self-organisation and building SOHO descriptions. The concept of exergy is the focal point for the construction of narratives of the possible futures of ecosystems on this planet.

Acknowledgements

This chapter represents a synthesis of much of what I have thought about over the past twenty years. As such, it is important to acknowledge that this paper reports on the results of extensive collaborations with several individuals whose ideas are so intertwined with my own that it is not possible to tease them apart. The section on Thermodynamics and Self-organisation draws heavily on my collaboration with Eric Schneider. The section on SOHO models owes much to my collaboration with Henry Regier and Michelle Boyle. I have also depended heavily on Roydon Fraser for discussions about exergy as a thermodynamic concept. Many conversations over the past twenty years with Bob Ulanowicz and Tim Allen have helped to shape my thinking about complex systems. More recently, members of the Dirk Gently gang have also contributed much to my thinking. (Silvio Funtowicz, Jerry Ravetz, Mario Giampietro, Martin O. Connor, David Waltner-Toews, Tamsyn Murray and Gilberto Gallopin). I thank my grad students, Nina-Marie Lister, Kate Oxley, Richard Martell, and Beth Dempster (UW) and Charlotte Sunde (Massey University, N.Z.) for keeping me honest. Finally, on this day of his retirement (1 August, 1999) I wish to acknowledge the inspiration throughout my career of the work of Buzz Holling and I dedicate this chapter to him.

[5] Kay and Regier (1999)

REFERENCES

Allen, T.F.H.; Bandurski, B.L.; King, A.W.1993. The Ecosystem Approach: theory and ecosystem integrity: International Joint Commission. (Report to the Great Lakes Science Advisory Board).

Allen, T. F. H.; Hoekstra, T.W. 1992. Toward a Unified Ecology. New York: Columbia University Press.

Allen, T. F. H.; Starr, T. B. 1982 Hierarchy: Perspectives for Ecological Complexity: University of Chicago Press.

Akbari, M., Murphy, S., Kay, J.J., Swanton, C., 1999. 'Energy-Bases Indicators of (Agro)Ecosystem Health'. In: D. Quattrochi and J. Luvall (eds) *Thermal Remote sensing in Land Surface Processes.* Ann Arbor Press

Bejan, A. 1997. Advanced Engineering Thermodynamics. New York: John Wiley and sons.

Blindow, I.; Andersson, G.; Hargeby, A., and Johansson, S.1993. Long-term pattern of alternative stable states in two shallow eutrophic lakes. Freshwater Biology. 30:159-167.

Boltzmann, L. 1886. The Second Law of Thermodynamics. In: McGinness, B. (1974). Ed. *Ludwig Boltzmann, Theoretical Physics and Philosophical Problems.* New York: D. Reidel.

Brzustowski, T.A., and P.J. Golem 1978. Second Law Analysis of Energy Processes Part 1: Exergy-An Introduction. *Transactions of the Canadian Society of Mechanical Engineers.* 4:209-218.

Caley M.T. and D. Sawada (eds.). 1994. Mindscapes: the Epistemology of Magoroh Maruyama. Gordon and Breach Science Publishers, Langhorne, Pennsylvania. xxiii + 206 pp.

Carathéodory, C. 1976. Investigations into the foundations of thermodynamics. In: Kestin, J. (Ed.) *The Second Law of Thermodynamics:* (Benchmark Papers on Energy; v. 5). New York: Dowden, Hutchinson, and Ross. 229-256.

Carpenter, S. R., and K. L. Cottingham. 1997. Resilience and restoration of lakes. *Conservation Ecology* 1 (1): Article 2.

Casti, J. L. 1994 Complexification: Explaining a Paradoxical World Through the Science of Surprise. NY: Harper Collins.

DeAngelis, D. L.; Post, W. M.; Travis, C. C. 1986. Positive Feedback in Natural Systems. Berlin: Springer-Verlag.

Dempster, M.L. 1998. A Self-Organising Systems Perspective on Planning for Sustainability, Masters Thesis, School of Planning, University of Waterloo.

Denbeigh, K. G. 1951. The Thermodynamics of the Steady State. London: Methuen LTD.

di Castri, F. 1987. The Evolution of Terrestrial Ecosystems. In: Ravera, O., (Ed.). *Ecological Assessment of Environmental Degradation, Pollution and Recovery.* Elsevier Science. pp.1-30.

Fraser, R., Kay, J.J. (in press) Surface temperature and Solar exergy. In: D. Quattrochi and J. Luvall (eds) *Thermal Remote sensing in Land Surface Processes.* Ann Arbor Press

Fermi, E. 1956; c1933. Thermodynamics. London. Dover Publications.

Gaggiolo, Richard A. 1980. Thermodynamics: Second Law Analysis. American Chemical Society.

Gaggioli, Richard A., (Ed.) 1983. Efficiency and Costing; Second Law Analysis of Processes. Washington, D.C.: American Chemical Society.

Glansdorff, P.and I. Prigogine 1971. Thermodynamic Theory of Structure, Stability, and Fluctuations. New York:Wiley-Interscience.

Günther, F., Folke, C. 1993. Characteristics of Nested Living Systems. *Biological Systems.* 1(3): 257-274.

Hatsopoulos, G. and J. Keenan. 1965. Principles of General Thermodynamics. New York:John Wiley.

Hevert, H.; Hevert, S. 1980. Second Law analysis: An alternative indicator of system efficiency. *Energy-The International Journal* 5(8-9): 865-873.

Holling, C.S. 1986. The Resilience of Terrestrial Ecosystems: Local Surprise and Global Change. In: W.M. Clark and R.E.. Munn (eds.) *Sustainable Development in the Biosphere.* Cambridge: Cambridge University Press. pp. 292-320.

Holling, C.S. 1992. Cross-scale Morphology, Geometry, and Dynamics of Ecosystems. *Ecological Monographs.* 62:4 , pp. 447-502.

Jantsch, Erich. 1980. The self-organising universe : scientific and human implications of the emerging paradigm of evolution. Pergamon Press.

Jørgensen, S.E. and H. Mejer 1979. A holistic approach to ecological modeling. *Ecol. Modeling.* 7:169:189.

Jørgensen, S. 1992. Integration of Ecosystem Theories: a Pattern. Elsever Publishing, London, U.K. 383 pp.

Kay, J.J. 1984. Self-organisation in living systems. Ph.D. thesis. Systems Design Engineering, University of Waterloo, Waterloo, Ontario. 458 pp.

Kay, J.J. 1991. A non-equilibrium thermodynamic framework for discussing ecosystem integrity. *Environmental Management* 15(4): 483-495.

Kay, J.J., Schneider, E.D. 1992. Thermodynamics and Measures of Ecosystem Integrity. In: *Ecological Indicators*, Volume 1, D.H. McKenzie, D.E. Hyatt, V.J. Mc Donald (eds.), *Proceedings of the International Symposium on Ecological Indicators*, Fort Lauderdale, Florida, Elsevier, pp.159-182.

Kay, J.J. and E. Schneider. 1994. Embracing complexity, the challenge of the ecosystem approach. *Alternatives* 20(3): 32-39.

Kay, J.J. 1997. Some notes on: The Ecosystem Approach, Ecosystems as Complex Systems. In: T. Murray, and

G. Gallopin (eds.) *Integrated Conceptual Framework for Tropical Agroecosystem Research Based on Complex Systems Theories* pp. 69-98. Cali, Colombia. Centro Internacional de Agricultura Tropical, Working Document No. 167.

Kay. J., Regier, H., Boyle, M. and Francis, G. 1999. An Ecosystem Approach for Sustainability: Addressing the Challenge of Complexity' *Futures* Vol 31, #7, Sept. 1999, pp.721-742.

Kay. J., Regier, H. 1999. An Ecosystem Approach to Erie's Ecology. In: M. Munawar, T.Edsall, S.Nepszy, G. Sprules and B. Shute (eds), *International Symposium. The State of Lake Erie (SOLE) - Past, Present and Future. A tribute to Drs. Joe Leach and HenryRegier* Backhuys Academic Publishers, Netherlands, pp.

Kestin, J. 1968. A Course in Thermodynamics. New York: Hemisphere Press.

King, A.W. 1993. Considerations of Scale and Hierarchy. In: Woodley, S.; Kay, J. ; Francis, G., (eds). *Ecological Integrity and the Management of Ecosystems*. St. Lucie Press; 1993: 19-46.

Koestler, A. and Smythies, J. R. (eds.) 1969. Beyond Reductionism. London: Hutchinson.

Koestler, A. 1978. Janus: a Summing Up. Hutchinson, London, U.K. vii + 354 pp.

Ludwig, D., B. Walker, B and Holling, C. S. 1997. Sustainability, stability, and resilience. Conservation Ecology. 1(1): Article 7.

J. C. Luvall, and H. R. Holbo, 1989. Measurements of short term thermal responses of coniferous forest canopies using thermal scanner data. *Remote Sens. Environ.*, 27, pages 1-10.

J. C Luvall, et.al. 1990 Estimation of tropical forest canopy temperatures, thermal response numbers, and evapotranspiration using an aircraft-based thermal sensor. *Photogrammetric Engineering and Remote Sensing*, 56: 10 (1393-1401).

J. C. Luvall, and H. R. Holbo 1991. Thermal Remote Sensing Methods in Landscape Ecology. In: M. Turner, and R.H. Gardner (eds.) *Quantitative Methods in Landscape Ecology*, Ch.6. New York: Springer-Verlag,

Moran, M.J. 1982. Availabilty Analysis: A Guide to Effecient Energy Use. Prentice-Hall.

Moran, M. and Shapiro, H. 1993. Fundamentals of Engineering Thermodynamics. New York: John Wiley and sons.

Nicolis, G., Prigogine, I. 1977. Self-Organisation in Non-Equilibrium Systems. Wiley-Interscience.

Nicolis, G., **Prigogine, I.**1989. Exploring Complexity. Freeman.

Odum, E.P. 1969. The strategy of ecosystem development. *Science* 164: 262-270.

Peacocke, A. R. 1983. The Physical Chemistry of Biological Processes. Oxford University Press.

Prigogine, I. 1955. Thermodynamics of Irreversible Processes. New York: John Wiley.

Popper, K.R. 1990. A World of Propensities. Thoemmes, Brussels.

Rapport, D.J., H.A. Regier and T.C. Hutchinson. 1985. Ecosystem behaviour under stress. *American Naturalist* 125: 617-640.

Rapport, D.J. and H.A. Regier. 1995. Disturbance and stress effects on ecological systems. In: S.E. Jorgensen and B.C. Patten (eds.) *Complex Ecology: the Organisation, Feedback and Stability.* Prentice-Hall, New York. Pp. 397-414.

Regier, H.A. and J.J. Kay. 1996. An heuristic model of transformations of the aquatic ecosystems of the Great Lakes - St. Lawrence River Basin. *Journal of Aquatic Ecosystem Health* 5: 3-21. Kluwer Academic Publishers.

Scheffer, M. 1990. Multiplicity Of Stable States In Freshwater Systems. *Hydrobiologia* 200/201: 475-486.

Scheffer, M. 1998. Ecology of Shallow Lakes. London: Chapman and Hall.

Scheffer, M. S.H. Hosper, M.L. Meijer, B. Moss and E. Jeppesen. 1993. Alternative equilibria in shallow lakes. *TREE* 8(8): 275-279.

Schneider, E. D. 1988. Thermodynamics, information, and evolution: new perspectives on physical and biological evolution. In Weber, B.H.; Depew, D.J.; and J.D. Smith, J. D., (ed.) *Entropy, Information, and Evolution: New Perspectives on Physical and Biological Evolution.* Boston: MIT Press, 108-138.

Schneider, E.D, Kay, J.J. 1994a. Complexity and Thermodynamics:Towards a New Ecology. *Futures* 24 (6) pp.626-647, August 1994.

Schneider, E.D, Kay, J.J. 1994b. Life as a Manifestation of the Second Law of Thermodynamics'. *Mathematical and Computer Modelling*, Vol 19, No. 6-8, pp.25-48.

Schneider, E.D, Kay, J.J. 1995. Order from Disorder: The Thermodynamics of Complexity in Biology. In: Michael P. Murphy, Luke A.J. O. Neill (ed) *What is Life: The Next Fifty Years. Reflections on the Future of Biology* Cambridge *University Press,pp. 161-172.*

Schrödinger, E. 1944. *What is Life?:* Cambridge University Press.

Stigliani, W.M. 1988. Changes in valued 'capacities' of soils and sediments as indicators of non-linear and time-delayed environmental effects. *Int. J. Env. Mon. and Assess* 10: 95-103.

Szargut, J.; Morris, D. R.; Steward, F. R. 1998. Exergy Analysis of Thermal, Chemical, and Metallurgical Processes: Springer-Verlag; 1988.

Ulanowicz, R.E. 1995. Ecosystem integrity: a causal necessity. In: L. Westra and J. Lemons (eds.) *Perspectives on Ecological Integrity*. pp. 77-87 Kluwer Academic Publishers, Dordrecht, The Netherlands, vii + 279 pp.

Ulanowicz, R.E. 1996. The propensities of evolving systems. Pp. 217- 233 In: E.L. Khalil and K.E. Boulding (Eds.) *Evolution, Order and Complexity*. Routledge, London. 276p.

Ulanowicz, R.E. 1997. Ecology: The Ascendent Perspective. Columbia University Press, New York.

Ulanowicz, R. E. 1997. Limitations on the Connectivity of Ecosystem Flow Networks. In: A. Rinaldo, and A. A. Marani (eds), *Biological Models: Proceedings of the 1992 Summer School on Environmental Dynamics.*, pp.125- 143. Venice, Istituto Veneto di Scienze, Lettere ed Arti, Venice.

Wall, G. 1986. Exergy- A Useful Concept. Goteborg, Sweden: Physical Resource Theory Group, Chalmers

University of Technology.
Wicken, J. S. 1987. Evolution, Thermodynamics, and Information: Extending the Dawinian Program. Oxford University Press.

II.1.3 The Tentative Fourth Law of Thermodynamics

Sven E. Jørgensen

1. How can we describe the development of ecosystems as a consequence of their energy through flow?

Ecosystems are steadily changing due to reactions on the steadily changing, forcing functions. They are very dynamic systems that are maintained far from thermodynamic equilibrium and may even attempt to move further away from thermodynamic equilibrium. The only way to move systems away from equilibrium is to perform work on them. The available work of the system, i.e., the exergy stored in the system, is furthermore a measure of the distance from thermodynamic equilibrium.

As we know that the ecosystems, due to the through-flow of energy, have a tendency to move away from thermodynamic equilibrium losing entropy or gaining negentropy and information, it is possible to formulate the following proposition: **Ecosystems attempt to develop toward a higher level of exergy.**

Ecosystems are soft and adaptable systems as they are able to meet changes in external factors or impacts with many varying regulation processes on different levels. The results are often that only minor changes are observed in the *function* of the ecosystem, despite the relatively major changes in environmental conditions. It means that the state variables – but not necessarily the *species* – are maintained almost unchanged, in spite of changes in external factors.

It has been widely discussed during the last years (H.T. Odum, 1983, Straskraba, 1980, Jørgensen, 1982, Straskraba and Gnauck, 1983 and Jørgensen, 1992a, b and 1997) how it is possible to describe these regulation processes, particularly those on the ecosystem level – i.e., the changes in ecological structure and the species composition.

The modernised Neo-Darwinian theory expanded to include all the recent developments in genetics and in evolutionary ecology is able to describe the very complex competition among species. Darwin's theory states that the species that are best fitted to the prevailing conditions in the ecosystem will survive. This formulation may be interpreted as a tautology. We should therefore prefer the following formulation: life is a matter of survival and growth. Given the conditions, determined by the external and internal functions, the question is: which of the available organisms and species (and there are more available species than needed, some of them waiting in the wings) have the combinations of properties that give the highest probability for survival and growth? Those species, or rather this combination of species, may be denoted the fittest and will be selected due to their properties. Darwin's theory may, in other words, be used to describe the changes in ecological structure and species composition, but cannot be directly applied quantitatively with the present formulation, for instance in ecological modelling. Thermodynamics may, however, offer a concept which can be used to give a quantitative description of ecosystem developmental tendencies.

2. Application of exergy in ecosystem theory

Exergy is an obvious candidate to be applied as a quantitative description of ecosystem development. It expresses the distance from thermodynamic equilibrium and covers therefore both the size of the organised structure and its content of thermodynamic information (see Subsections II.1.1.4 and II.1.1.5).

Growth may be defined as a quantitative expression of the increase of organised structure. In thermodynamic terms, growth means that the system is moving away from thermodynamic equilibrium by increasing its organised structure. At thermodynamic equilibrium, the system cannot do any work, the components are inorganic and have the lowest possible free energy, and all gradients are eliminated (as already stated in Chapter II.1.1). We use the expression growth of a crystal, growth of a society and growth of an economy to indicate that the structure in one way or another is getting larger. Biological systems in particular have many possibilities to grow or to move away from thermodynamic equilibrium. It is therefore crucial in ecology to know which pathways among the possible ones an ecosystem will select for growth.

Development of the content of information of an organised structure, as opposed to it's size, is covered by Boltzmann's equation for the free energy of information (see Subsection II.1.1.5).

The ecosystems must be open, or at least non-isolated. This is absolutely necessary for their existence. A flow of energy through the system is sufficient to form an ordered structure (also named a dissipative structure (Prigogine and Nicolis, 1988)). Morowitz (1992) calls this latter formulation the fourth law of thermodynamics, but it would be more appropriate to expand this law to encompass a statement about *which* ordered structure among those available the system will select, or *which* factors determine how an ecosystem will develop. This expanded version was formulated as a tentative fourth law of thermodynamics in Jørgensen (1992), but was already expressed with a slightly different formulation in Jørgensen and Mejer (1977), Mejer and Jørgensen (1979) and in Jørgensen (1982). It is presented in more detail below as a candidate for the description of ecosystem development.

3. Exergy and Darwin's theory

Darwin and neo-Darwinism provide the answer to the question raised above, when one species is considered: the best fitted species, i.e., with the properties best co-ordinated to the prevailing conditions, will survive. Survival means that the biomass of the species will be maintained or maybe even increased. An organism or a population is exposed to many constraints, determined by the forcing functions on the ecosystems and the other organisms living in it. The question is: who is winning the resource competition? The winner's award is survival and even growth. In thermodynamic terms it means that the organisms that have their properties better co-ordinated to the prevailing conditions will be able to contribute most to the work content or exergy of the system due to their biomass with embodied information (which also in accordance with Boltzmann, 1905 represents work (exergy); see Subsection II.1.1.5 and Jørgensen, 1992a and 1997 and Jørgensen et al., 1998). The exergy or chemical energy which can be used to do work in mineral oil is about 42 kJ/g. The chemical energy (free energy, exergy) of biomass with an average composition of proteins, carbohydrates and fat can be calculated to be approximately 18.7 kJ/g (the details of this calculation are given below and in Jørgensen et al., 1995).

Brown (1995) defines fitness as the rate at which resources in excess of those required for maintenance can be utilised for reproduction. He uses dW/dt, called reproductive power, to find the optimal body mass, W. So, he is asking the question: which size is best fitted? The question is answered by determining the size with the highest growth potential. i.e., the size yielding the biggest increase of the biomass corresponding to the biggest increase of the exergy in the system.

An ecosystem encompasses, however, many species. They cannot all obtain the biggest biomass independently of the other species – the species are interdependent. Darwin considered this complication and expressed it as 'prevailing conditions', and is anticipated to include all the abiological and biological constraints imposed on the species, i.e., including constraints originating from other species. The evolution and co-evolution over a very long period have, however, implied that the species have adapted to each other. They were able to find ways to move further away from thermodynamic equilibrium (get more growth) if they co-operated by adjusting their properties to each other and the prevailing external factors. The effect of this co-operation is consistent with Patten (1991). He shows that the indirect effect often exceeds the direct one. For instance, a predator- prey relationship may also be beneficial for the prey due to a number of factors including faster circulation of the nutrients; see also Chapter II.7.3.

Darwin's Theory presumes, that populations consist of individuals, which:

1) on average produce more offspring than needed to replace them upon death – these are the properties of high reproduction. Translated into thermodynamics: more possible pathways for utilisation of the energy flow are developed than the system and its energy flow can sustain. It implies that a competition even among pathways, that are only slightly different, will be established.

2) have offspring which resemble their parents more than they resemble randomly chosen individuals in the same population – this is the property of inheritance. Thermodynamically, it means that the properties that showed a better ability to utilise the energy flow to move as far away from thermodynamic equilibrium as possible, by construction of more biomass, to a large extent will be preserved. Genetics can explain how this is possible.

3) vary in heritable traits influencing reproduction and survival due to differences in fitness to the prevailing conditions – this is the property of variation. Modernised neo-Darwinism is able to give a long list of mechanisms that can create new pathways. It implies that new possibilities are steadily created to meet the challenge of utilising the energy flow to move away from thermodynamic equilibrium. These possibilities are tested under the prevailing conditions, and those which are successful are preserved according to 2.

Evolution can therefore continue on the shoulders of the already found successful solutions and therefore steadily find new and better solutions, i.e., select the best genes, among all the present genes including those that are continuously emerging by mutations and sexual recombinations.

It implies that the properties are being changed by selection processes to facilitate 'best possible survival' under the prevailing conditions. For plants this includes grazing and for the grazer, the availability of food. The species cannot change the properties of other species directly, but all species must consider all other species in their effort to find a feasible combination of properties that is able to offer a higher probability of survival. This explains how species adapt to each other (coevolve), and how they can co-operate on the joint goal to move, as much as possible, away from thermodynamic equilibrium. In principle, each of the species is striving toward its own goal: to get the highest possible growth for its own species. As these goals cannot be reached if the species don't adapt to other species (as they are also part of the life conditions), the result will be that the species together move away from thermodynamic equilibrium, i.e., give the system the highest possible exergy. It will often coincide with the highest, or close to the highest, biomass for most species, at least on a long term basis.

The conclusion from these considerations is that exergy, as it measures the distance from thermodynamic equilibrium of the entire ecosystem, seems to be a good candidate to quantify survival and growth in the Darwinian sense for the entire ecosystem. Calculations of the exergy of ecosystems make it possible to unite the chemical energy of the organic matter and the information, in the sense of Boltzmann, embodied in the species.

4. Estimation of the exergy of an ecosystem

If we assume a reference environment that represents the system (ecosystem) at thermodynamic equilibrium, which means that all the components are inorganic at the highest possible oxidation state (as much free energy as possible is utilised to do work) and homogeneously distributed in the system (no gradients), the situation illustrated in Figure 1 is valid. As the chemical energy embodied in the organic components and the biological structure contributes by far the most to the exergy content of the system, there seems to be no reason to assume a (minor) temperature and pressure difference between the system and the reference environment. Under these circumstances we can calculate the exergy content of the system as coming entirely from the chemical energy: _c (μ_c - μ_{co}) Ni. We find by these calculations the exergy of the system compared with the same system at the same temperature and pressure but in the form of an inorganic soup without any life, biological structure, information or organic molecules. As (μ_c - μ_{co}) can be found from the definition of the chemical potential replacing activities by concentrations, we get the following expressions for exergy:

$$Ex = RT \underset{i=0}{\overset{i=n}{\_}} C_i \ln C_i / C_{i,0} \tag{1}$$

where R is the gas constant, T is the temperature of the environment (and the system; see Figure 1), while C_i is the concentration of the i'th component expressed in a suitable unit, e.g. for phytoplankton in a lake C_i could be expressed as mg /l or as mg /l

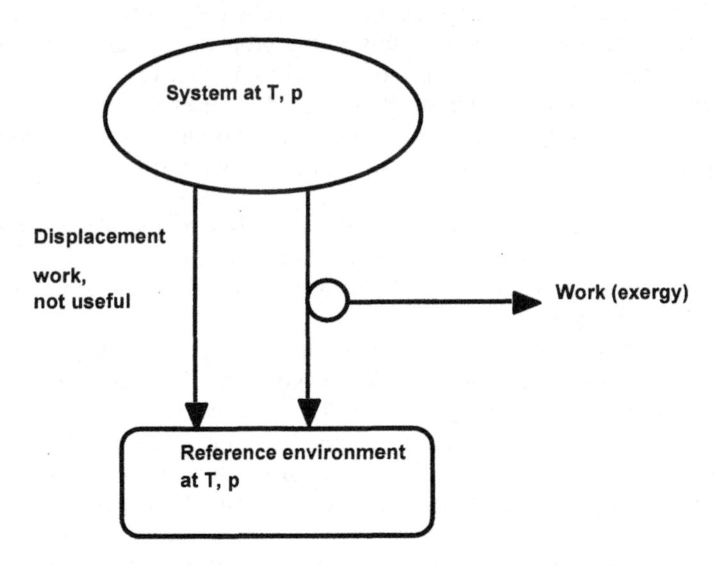

Figure 1: The exergy content of the system is calculated in the text for the system relatively to a reference environment of the same system at the same temperature and pressure, but as an inorganic soup with no life, biological structure, information or organic molecules.

of a focal nutrient. $C_{i,o}$ is the concentration of the i'th component at thermodynamic equilibrium and n is the number of components. $C_{i,o}$ is of course a very small concentration (except for i = 0, which is considered to cover all the inorganic compounds), corresponding to a very low probability of forming complex organic compounds or even biological components spontaneously in an inorganic soup at thermodynamic equilibrium. Notice the accordance between equation 1(above) and 24 (from Chapter II.1.1). Exergy can be interpreted as the sum of the chemical exergy of the components and Kullbach's measure of information (see also Mejer and Jørgensen, 1979 and Svirezhev, 1998).

The problem by application of equation (1) is related to the size of C_{io}. The problem related to the assessment of C_{io} has been discussed and a possible solution has been proposed in Jørgensen et al. (1995). For dead organic matter, detritus, which is given the index 1, it can be found from classical thermodynamics that:

$$\mu_1 = \mu_{1o} + RT \ln C_1 / C_{1o}, \tag{2}$$

where μ indicates the chemical potential. The difference $\mu_l = \mu_{lo}$ is known for organic matter, e.g., detritus, which is a mixture of carbohydrates, fats and proteins. If we use these figures, we get approximately 18.7 kJ/g detritus.

It may be shown (Jørgensen et al., 1995 and 1998 and Jørgensen 1997) that exergy of all the components in the ecosystem including biological components with approximations can be computed as:

$$Ex = \beta_i C_i \tag{3}$$

where C_i is the biomass concentration of species i and β_i is the weighting factor expressing the information that the ith species is carrying. Exergy is obtained in kJ / unit of volume or area if ß- values are based on 18.7 kJ/g for detritus and correspondingly for other components. It may, however, be convenient to express exergy in the unit detritus exergy equivalent (ß for detritus = 1 and ß for inorganic components at the highest oxidation state = 0). The biological components then have ß-values > 1 corresponding to the information they carry. Table 1 gives ß- values for various biological components, calculated from their non repetitive genes, i.e., the information they carry. If these ß-values are used, the exergy in the detritus equivalent is obtained but can easily be converted to kJ by multiplication by 18.7.

The equation can be applied to calculate exergy for important (known from measurements or models) ecosystem components. As can be seen from the equation and the ß-values in Table 1, exergy is dominated by the contributions coming from information, originated from the genes of the organisms. The total exergy of an ecosystem can of course not be calculated exactly, as we cannot model or measure the concentrations of all components of an ecosystem, but we can calculate the contributions from the dominant components, that are of interest for a particular problem.

The calculation of exergy accounts for the chemical energy in the organic matter and for the information that originates in the extremely small probability to form living components, for instance algae, zooplankton, fish, mammals and so on, spontaneously from inorganic matter. The weighting factors could also be considered quality factors reflecting how developed the various groups are due to their embodied information. In principle, we should include all the components in the calculations. All inorganic compounds should also be included but their contribution to the exergy of the system is minor and may therefore be excluded from these approximate calculations.

The calculations are also consistent with the classical application of thermodynamics on chemical equilibria. If we consider the chemical reaction zooplankton + oxygen <--> carbon dioxide + water + nutrients, the mass constant, K, can in principle be defined in the usual way. The very low zooplankton concentration, due to the low probability of the presence of zooplankton at thermodynamic equilibrium, will be reflected in a huge K-value, which can be translated to a high free energy equal to exergy in this case. In this context, information contributes considerably to the exergy of the system, as already mentioned in Chapter II.1.1.

Table 1: Approximate number of non repetitive genes

Organisms	**Number of Information genes**	**Conversion Factor**
Detritus	0	1
Mimimal Cell (Morowitz, 1992)	470	2.7
Bacteria	600	3.0
Algae	850	3.9

Sources: T.Cavalier-Smith (1985), Li and Grauer (1991) and Lewin (1994).

*) based on number of information genes and the exergy content of the organic matter in the various organisms, compared with the exergy contained in detritus. 1 g detritus has about 18.7 kJ exergy (=energy which can do work).

The presented calculations do not include the information embodied in the structure of the ecosystem network, i.e., the relationships between the various components, which is represented by the network. The information of the network encompasses the information of the components and the relationships of the components. The latter is calculated by Ulanowicz (1991) as a contribution to ascendancy.

It is important to emphasise that *all* computations of exergy will have the following shortcomings:

1) The computations will be based upon either a model or a limited number of measurements. The results of the computations are therefore more appropriate for finding a *relative* difference in exergy by a comparison of an ecosystem under different conditions.

2) The calculations – as with all calculations in thermodynamics – are based upon approximations and assumptions. But as we draw conclusions on the basis of the differences in exergy rather than on the basis of absolute values, the results may still be applicable in ecosystem theoretical contexts.

In addition, the application of equation (1) for the computations of exergy assumes the reference state shown in Figure 1. It implies that the computed exergy will be entirely related to the chemical composition and the biological structure with its information. This is the major contribution under all circumstances, as only minor temperature and pressure differences are realistic between an ecosystem and its environment. Due to the above mentioned shortcomings and approximations, the exergy calculated in accordance with equation (3) should be denoted an exergy index.

5. The tentative fourth law of thermodynamics

These considerations, applying Darwin's theory simultaneously on all components of an ecosystem by translating survival and growth into the concept of exergy, makes it possible to formulate a tentative fourth law of thermodynamics: **If a system receives a through-flow of exergy, the system will utilise this exergy to move away from thermodynamic equilibrium. If the system is offered more than one pathway to move away from thermodynamic equilibrium, the one yielding most stored exergy, i.e. with the most ordered structure or the longest distance to thermodynamic equilibrium by the prevailing conditions, will have a propensity to be selected.**

As it is not possible to prove the first, second and third laws of thermodynamics by deductive methods, the tentative fourth law can only at the best be proved by inductive methods. It implies that the fourth law should be investigated in as many cases as possible. Several modelling cases have been examined and they have all approved the tentative law; see Jørgensen (1986, 1988, 1990 and 1992a and b). The law has particularly been used successfully to develop models with dynamic structure, see Jørgensen (1986), Nielsen (1992), Jørgensen (1992a and b), Jørgensen and Padisak (1996), Jørgensen and de Bernardi (1997) and Caffaro et al. (1997). The term 'have a propensity to be selected' is used in the formulation to indicate that it is hardly possible to make deterministic statements about the development of an ecosystem (see also Ulanowicz, 1997), due to the random character of many of the factors and to the above mentioned shortcomings in our calculations. Ulanowicz (/1997) uses the term that an ecosystem is ontic open.

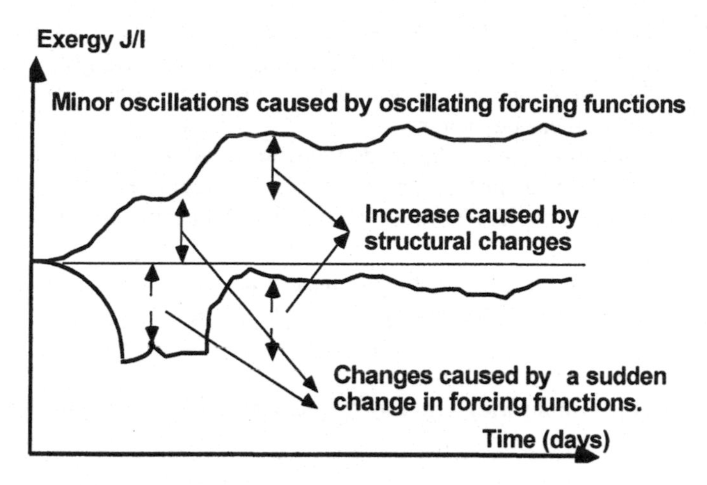

Figure 2: Exergy response to increased and decreased nutrient concentration.

Several observations have supported the tentative fourth law of thermodynamics. A few are mentioned below.

It may be considered strong support that the order of microbial processes at pH = 7.0 and 25°C is oxidation by oxygen (aerobic respiration), by nitrate, by Mn(IV), by iron (III), by sulphate and by carbon dioxide in accordance with a declining number of produced ATPs which measure the exergy stored in the cells (see for instance Schlesinger, 1997). Table 2 gives the exergy available for storage in ATPs and released by these oxidation processes. Note that the potential energy in aerobic heterotrophs is only slightly greater than that from denitrifiers, which often coexist in upland soils (Carter et al., 1995).

Numerous experiments have been performed to imitate the formation of organic matter in the primeval atmosphere on earth $4 * 10^9$ years ago (see for instance Jørgensen, 1997) . Various sources of energy have been sent through a gas mixture of carbon dioxide, ammonia and methane. Analyses have shown that a wide spectrum of various compounds, including several amino acids, is formed under these circumstances, but generally only compounds with rather large free energy (i.e., high exergy storage) will form an appreciable part of the mixture (Morowitz, 1968).

The entire process of evolution has worked towards organisms with an increasing number of information genes (genes actually utilised) and more types of cells, i.e., towards storage of more exergy due to the increased information content.

Table 2: kJ /equiv available to build ATP for various oxidation processes of organic matter at pH = 7.0 and 25°C

Reaction	Available Kj/equiv
$CH_2O + O_2 \rightarrow CO_2 + H_2O$	125
$CH_2O + 0.8\ NO_3^- + 0.8H^+ \rightarrow CO_2 + 0.4\ N_2 + 1.4\ H_2O$	119
$CH_2O + 2MnO_2 + H^+ \rightarrow CO_2 + 2\ Mn^{2+} + 3H_2O$	85
$CH_2O + 4FeOOH + 8H^+ \rightarrow CO_2 + 7H_2O + Fe^{2+}$	27
$CH_2O + 0.5SO_4^{2-} + 0.5H^+ \rightarrow CO_2 + 0.5HS^- + H_2O$	26
$CH_2O + 0.5CO_2 \rightarrow CO_2 + 0.5CH_4$	23

At the biochemical level, we find that different plants operate three different biochemical pathways for the process of photosynthesis: a) the C3 or Calvin Benson cycle b) the C4 pathway and c) the crassulacean acid metabolism (CAM) pathway. The latter pathway is less efficient than the two other possible pathways, measured as grams (g) of plant biomass formed per unit of energy received. Plants using the CAM pathway can, however, survive in harsh, arid environments, which plants following C3 and C4 pathways cannot. Photosynthesis will, however, switch to C3 as soon as the availability of water is sufficient; see Shugart (1998). The CAM pathways give, in other words, the highest exergy storage under harsh, arid conditions, while the two other pathways give the highest exergy storage under other conditions. These observations are completely in accordance with the tentative fourth law of thermodynamics.

Many more studies are obviously needed to offer a full acceptance of the hypothesis or even better; experiments for instance, in microcosms should be carried out to attempt to violate the hypothesis. It could, however, be considered a useful working hypothesis at this stage.

The tentative fourth law of thermodynamics may also be considered as an extended version of 'Le Chatelier's Principle'. Formation of biomass may be described as:

Energy + nutrients = molecules with more free energy (exergy) and organisation (4)

If we pump energy into a system the equilibrium will, according to Le Chatelier's Principle, shift towards a utilisation of the energy. It means that molecules with more free energy and organisation are formed. If more pathways are offered, the pathways, that give most relief, i.e., use most energy and thereby form molecules with most embodied free energy (exergy) will win according to the proposed tentative law of thermodynamics.

Notice that the change in exergy is not necessarily > 0, as it depends on the changes in the resources of the ecosystem. The proposition claims, however, that the ecosystem attempts to reach the highest possible exergy level under the prevailing condition and with the available gene pool ready to challenge the currently changed conditions imposed by the currently changed forcing functions. Compare this with Figure 2, where the reactions of exergy to an increase and an decrease of the nutrient concentration in a lake are shown.

The different pathways available in an ecosystem compete in a very complex way, because the processes are many, complex and dependent on many factors:

1) **at least somewhere in the order of 20 elements, of which some may be limiting,**
2) **competition from the other possible pathways,**
3) **temperature,**
4) **light (for photosynthetic pathways only),**
5) **ability to utilise the combined resources,**
6) **initial conditions,**
7) **variations of conditions in time and space.**

This implies also, that the history of the system plays an important role in selection of the organisation that gives the highest exergy. Two systems with the same prevailing conditions will consequently not necessarily select the same species and food web, because the two systems most probably have a different history and therefore different initial conditions.

Decomposition rates are of great importance, as they determine not only the ability of the different products (organisms) to maintain their concentrations, but also at which rate the inorganic compounds are recycled and can be reused. Measurements in ecosystems (see for instance Vollenweider, 1975) show that the most abundant resource (nutrient), relative to its need, recycles fastest. Model studies have shown that this is completely consistent with the tentative fourth law of thermodynamics. Ecosystems that recycle the most abundant nutrient (resource) fastest, will obtain the highest storage of exergy.

The two hypotheses, 'to optimise exergy dissipation or exergy captured' (see Chapter II.1.2) and 'to optimise the exergy of the system under the prevailing conditions' (this section) are largely just two sides of the same coin, when we describe the development of an ecosystem from the early stage to the mature stage, see also the next subsection. Increased exergy stored in the structure of the system under development will also enable the system to capture more exergy, as already mentioned, but the discrepancy between the two theories occurs when the system has attained the maximum rate of capturing exergy (as mentioned above 85-90% of the incoming solar radiation). The next subsection will demonstrate and conclude that the tentative fourth law of thermodynamics can be used as an orientor to describe the development both for a developmental system where the exergy captured and the energy dissipation may also be used, and for a mature system, where minimum entropy production and minimum energy dissipation can also be used as descriptor.

6. Growth and development of ecosystems

We may describe the results of a flow of energy through an ecosystem by the following equation

Ex + Ex(t) -> Ex(t+1) + energy utilised (lost by dissipation) for maintenance + reflected exergy (5)

The successional development of ecosystems from an early to a mature stage (see for instance E.P. Odum, 1969), illustrates that the two concepts, exergy storage and exergy utilisation are parallel. An ecosystem at an early stage of development, for instance an agricultural field has only a small exergy storage and exergy utilisation. The biomass per square meter is small (compared with the mature system), i.e., the exergy storage is small. The structure is simple and only little energy (exergy) is needed for respiration or growth as they are both to some degree proportional to the biomass. The total surface area of the plants is furthermore small, which implies that they catch and utilise only a fraction of the incoming solar radiation.

As the system develops, the structure becomes more complicated, animals with more information per unit of biomass, i.e., with more genes, populate the ecosystem and the total biomass per square metre increases. It implies that exergy storage as well as the exergy needed for maintenance both increase. A very mature ecosystem, for instance a natural forest, has a very complex, organised structure and very well organised food webs. It contains a very high concentration of biomass per square meter and contains a lot of information in a wide variety of organisms. The entire structure tries to utilise solar radiation either directly or indirectly, resulting in a high utilisation of the solar exergy flux. The catabolic energy demand is related to the total biomass and the overall organisation. It represents the exergy needed for maintaining the ecosystem far from thermodynamic equilibrium, in spite of the second-law tendency of the system to develop towards thermodynamic equilibrium. This is a parallel to what is experienced by man-made systems: A large town with many buildings of different types (skyscrapers, cathedrals, museums, scientific institutes, etc.) needs obviously much more maintenance than a small village consisting of a few almost identical farmhouses.

The development of ecosystems may also be described (Kay and Schneider, 1990) as a steady growth of a gradient between the ecosystem and thermodynamic equilibrium. The force to break down the gradient will increase with increasing gradient. This tendency to break down the gradient is represented by respiration and evapotranspiration, that spend exergy and produce entropy. As long as the exergy received from solar radiation can compensate for this need of exergy to maintain the gradient, it is possible for the system to stay far from thermodynamic equilibrium. If even more exergy can be captured than needed for maintenance of the gradient, the surplus exergy increases the stored exergy, which means that the system moves further away from thermodynamic equilibrium and thereby increases the gradient even more.

When an ecosystem by this development has reached the state of maturity where the system captures almost all the solar energy that is physically possible, it is no longer possible to increase the amount of energy captured. The system can therefore in this situation not increase further the dissipation and it cannot exceed the energy captured on a long term basis. Moreover, the amount of inorganic matter available for further growth must in this stage be very limited as most inorganic matter has been utilised to build up the very developed structure. The amount of information stored in biomass may, however, still increase in a mature ecosystem due, for example, to:

1) immigration of (slightly) better fitted species, and

2) emergence of new genes or genetic combinations. This latter possibility is covered by the concept of 'evolution'.

The system stops the growth of biomass when the most limiting inorganic component has been fully utilised for biomass construction. Nutrients and water are often the limiting factors in growth of plants. These resources cycle which gives possibilities for formation of new biomass with perhaps more information, but the total biomass cannot be changed by this reallocation of resources. This constraint by the laws of conservation is essential for the development of more and more complex living structures. As living organisms compete for limited food supplies, they invent and develop thousands of new and ingenious strategies (Reeves, 1991). Some species invest in movement; speed can be a valuable asset both for capturing prey and for avoiding predators. Others use protective armour or chemical poisons. Each species defines the terms under which it engages in the harsh business of life. Better feedbacks to assure maintenance of a high biomass level to changed circumstances, better buffer capacities, better specialisation to populate all possible ecological niches and better adapted organisms to meet the variability in forcing functions are all developed. Thus, biomass is maintained at the highest level over a longer time and the information level will increase due to the steady development of better feedbacks and more self-organisation. Both contributions are reflected in a higher exergy. The exergy of the mature system can therefore still grow further, namely by increase of the information. In other words, the system better utilises its resources and becomes more fitted to the prevailing conditions. Adaptation and specialisation require information, which implies that a better fitness to prevailing conditions is more probable by a system with more stored information in the genes.

Neither exergy nor information is conserved. Exergy is lost by all transfers of energy; but exergy and information are also lost by the death of an organism as ß in equation (19) decreases from a value >> 1 to 1 (detritus exergy equivalent is applied as unit). Exergy and information may, however, be gained when phytoplankton is converted to zooplankton by grazing. **Exergy and information are therefore not cycling in the same manner as we know for mass and energy. Their distribution in ecosystems will as, energy and matter, anyhow follow a complicated pattern, determined by the life processes.**

The two hypotheses of maximisation of exergy storage and of exergy capture are, as already pointed out, not consistent when we have to describe the further development of an ecosystem which has reached maturity. In this domain Mauersberger's minimum principle may be valid, (see Mauersberger, 1983 and 1995): Ecosystems locally decrease entropy production by transporting energy-matter from more probable to less probable spatial locations. Thereby less exergy is lost and more exergy can be stored.

A possible hypothetical formulation of the tentative fourth law of thermodynamics trying to unite the tentative fourth law of thermodynamics with hypotheses of Kay and Schneider (1990 and 1992) valid for a system under development and with the hypotheses of Mauersberger for mature ecosystems may be:

If a system is moved away from thermodynamic equilibrium by application of a flow of exergy, it will utilise all avenues available, to build up as much dissipative structure (store as much exergy) as possible. An ecosystem at an early stage will try to store more exergy by increasing the amount captured – decrease the reflection, while a mature ecosystem will gain more exergy by decreasing the exergy lost by the maintenance processes – mainly achieved by increase of the information content of the system.

The overall description of the development of ecosystems is illustrated in Figure 3. In the first phase the structure and its biomass increases rapidly mainly due to the rapid growth of r-strategists. The gradients, the amount of exergy captured by the system and the exergy required for maintenance are all increasing. A transition phase is shown between the first and third phase. In the third phase, the limiting elements are practically used up, which implies that a further increase in the physical structure measured by the biomass is not possible. A better use of the resources and more storage of information can however continue and will work hand in hand. More exergy cannot be captured or dissipated but the resources can be reallocated to give more exergy stored and a development towards relatively less dissipation of exergy (Mauersberger's principle).

A development towards K-strategists is therefore favoured as they are bigger in size and therefore have less specific exergy needs. Micro-organisms also shift from r-strategists with quick exploitation of the resources to K-strategists with a slow exploitation of resources (Gerson and Chet, 1981). Late successional plants with K-strategy produce litter which is poor in nutrients and simple sugars but high in lignin (Heal and Dighton, 1986), which change the physico-chemical environment of the top soils. These processes cause modifications in the composition as well as in the activity of microbial communities, favouring the K-selected organisms. The shift from r-strategists to K-strategists may therefore be considered a process with synergistic effects.

Figures from satellite measurements support the description of ecosystem development. A forest captures a lot more exergy than a desert or a grassland, but a 50 year old forest or a 200 year old forest or even an old rain forest capture approximately the same amount, almost 90% of the incoming solar radiation (Kay and Schneider, 1992). It can however been shown (Jørgensen, 1997), that the stored exergy increases when a forest gets older, and a rain forest has more exergy stored than a temperate forest.

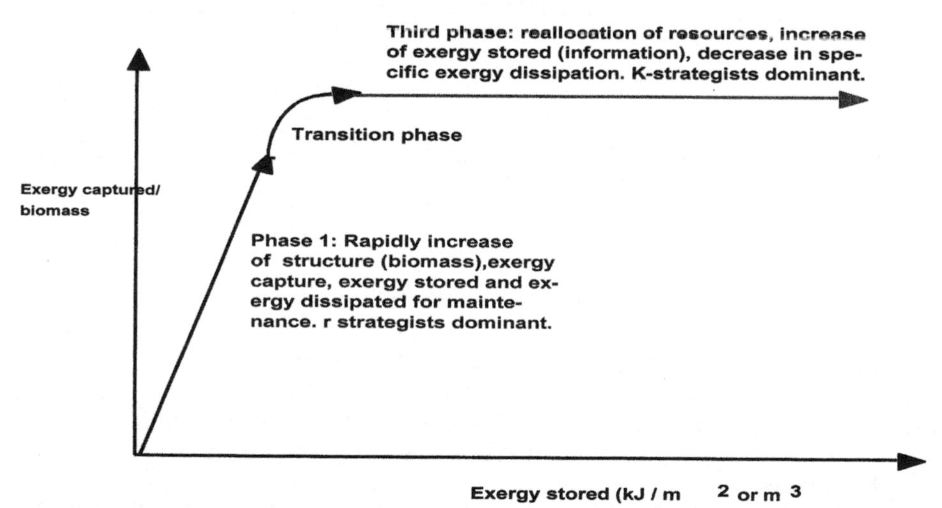

Figure 3: The plot shows the exergy captured (or biomass) versus the exergy stored as biomass and information in the ecological structure. When the exergy captured has reached the physically possible level, further development of exergy storage is still possible by more cycling and by increase of the information embodied in the biomass.

Table 3 shows the exergy utilisation for different types of systems (Kay and Schneider, 1992). In the same table the exergy storage is shown for some typical 'average' systems. The results of Table 3 are completely consistent with the illustration of the development of ecosystem shown in Figure 3. **As seen, there is a steep linear relationship between exergy storage and exergy capture, when the system is under development from the early to the mature stage, but a mature system may still develop its exergy storage, although the exergy captured has attained the practical maximum of about 70 - 80 % of the total solar energy received by radiation.** This points toward Ex_{bio} (exergy storage) as a general optimiser during the entire development from the early stage to the mature stage.

Johnson (1990) concludes that 'ecosystem structure is a function of two antagonistic trends: one toward a symmetrical state resulting in least dissipation and the other toward a state of maximum attainable dissipation'. The latter is obtained by a rapid growth of biomass and structure, implying a more effective capture of the exergy contained in the solar radiation. The first trend is obtained by a reallocation of the biochemical elements, which due to mass conservation limit the amount of biomass and structure. A development towards more effective organisms (less dissipation relative to the biomass which for instance is the case for bigger organisms; see also Straskraba et al., 1999), requiring less exergy for maintenance, will imply that although (almost) the same amount of exergy is captured, the stored exergy can still increase. Inevitably, the configurations of the interactions become more mutualistic, self-reinforcing and self-entailing.

It is also of importance to notice the basic difference between the two descriptors, exergy storage and exergy degradation. The latter is a differential quantity and is measured in a unit which includes time, for instance $kJ/(m^2 24h)$. As the available exergy per unit of time in the form of solar radiation has an upper limit, exergy degradation per unit of time seems not an appropriate descriptor of long-term development of ecosystems, although it may be used to describe the development from an early to a mature stage. Exergy storage is an integrated quantity, measured in a unit not including time, for instance kJ/m^2. It has in principle no upper limit, as the storage of exergy can steadily increase. Each new day brings more exergy which may be added to the already obtained storage of exergy.

Table 3: Exergy utilisation and exergy storage

Ecosystem	% Exergy utilisation	Exergy storage KJ/m^2
Quarry	0	6
Desert	73	2
Clear Cut	594	49
Grassland	940	59
Fir Plantation	12700	70
Natural Forest	26000	71
Old deciduous forest	38000	72
Tropical rainforest	64000	70

Patten and Fath (1999) have shown that increased cycling implies increased exergy storage at steady state conditions. It was shown as a general mathematical consequence of steady state network theory. Ecosystems are dynamic. It is therefore a simplification to assume steady state. On the other side, steady state may be interpreted as a freezing of the system which gives a realistic comparison of two situations: with and without cycling. It seems therefore appropriate (according to Patten and Fath, 1999) to set up the following pertinent hypothesis:

Increased cycling implies that the exergy storage and the exergy through flow increase simultaneously with a decrease in specific exergy dissipation. It is in accordance with the interpretation of the tentative fourth law of thermodynamics and Mauersberger's principle applied for mature ecosystems.

The role of natural disturbances such as fire or storms should be discussed in this context. When a forest is burned (details see Botkin and Keller, 1995), complex organic compounds are converted into inorganic compounds. Some of the inorganic compounds from the wood are lost as particles of ash that are blown away or as vapours that escape into the atmosphere and are distributed widely. Other compounds are deposited on the soil surface These are highly soluble in water and readily available for vegetation uptake. Therefore, immediately after a fire there is an increase in the availability of chemical elements, which are taken up rapidly, especially if there is a moderate amount of rainfall.

The pulse of inorganic nutrients can then lead to a pulse in growth of vegetation. This in turn provides an increase in nutritious food for herbivores. The pulse in chemical inorganic elements can therefore have effects that extend through the food chain. Challenges to find new opportunities to move even further away from thermodynamic equilibrium are therefore created which may explain why natural disturbances may have a long-term positive effect on the growth of ecosystems in the broadest sense of this concept.

This description is according to Holling's cycle (Holling, 1986). Figure 4 is a modified version of this cycle presented by Ulanowicz (1997), where the Holling cycle is modified in almost the same way as Figure 4, but the x-axis is not specific exergy = exergy / total biomass in Ulanowicz presentation of Holling's cycle, but instead mutual information of flow structure. The basic idea is, however, the same. The renewal phase corresponds to rapidly increased biomass exploitation phase to a rapid increase in the level of information and conservation to a very slow increase is both biomass and information. The destruction phase will, due to an external impact (forcing function), reduce both the amount of biomass and the information stored in this biomass, but thereby new possibilities are created for the utilisation of emergent mutations and sexual recombinations. After each round in Holling's cycle the biomass can probably hardly be higher as it is limited by the presence of essential elements, but due to the current test of new mutations and sexual recombination, new and perhaps better combinations of properties will emerge. Consequently, there is a *propensity* that the exergy and the specific exergy may have increased.

It can be concluded that the tentative fourth law of thermodynamics can explain the development of ecosystems in all phases. Several case studies including many structural dynamic models and case studies presented in this section support the tentative law of thermodynamics, but due to its general character it is absolutely necessary to test the tentative law with many more case studies before it can be recommended to use it more generally.

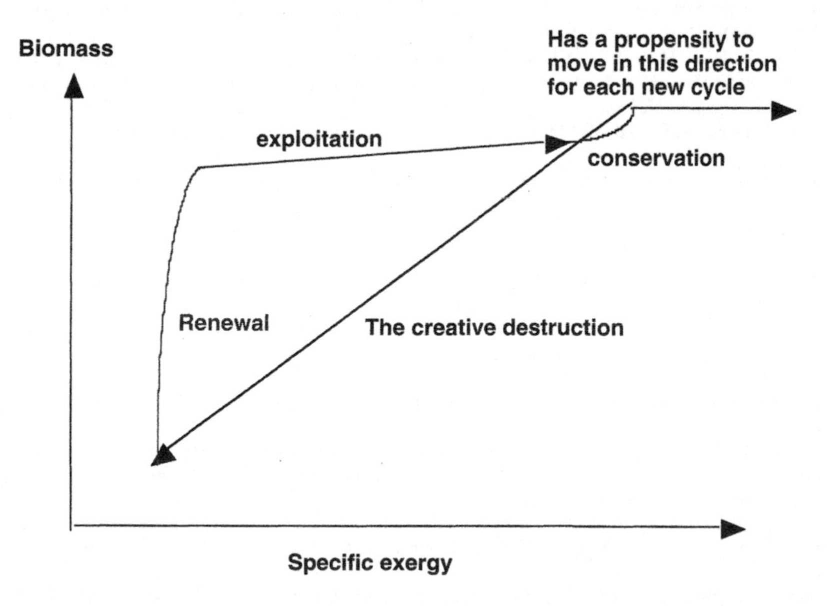

Figure 4: The so called Holling's cycle is interpreted by use of the tentative fourth law of thermodynamics. A mature system has only very scarce available inorganic matter to build new biomass and therefore to test new possible (and better) solution to moving away from thermodynamic equilibrium. By introduction of a destruction phase is created possibilities in the long run to move further away from thermodynamic equilibrium.

REFERENCES

Boltzmann, L. 1905. The Second Law of Thermodynamics. (Populare Schriften, Essay no. 3 (address to Imperial Academy of Science in 1886). Reprinted in English in: Theoretical Physics and Philosophical Problems, Selected Writings of L. Boltzmann. D. Reidel, Dordrecht.

Botkin and Keller. 1995. Environmental Science, Earth as a living Planet. John Wiley and Sons, NY, 630 pp.

Brown, J.H. 1995. Macroecology. The University of Chicago Press, Chicago, IL.

Carter, J.P., Hsiao, Y.H., Spiro, S. and Richardson, D.J. 1995. Soil and sediment bacteria capable of aerobic nitrate respiration. Applied and Environmental Microbiology 61: 2852-2858.

Cavalier-Smith, T. 1985. The Evolution of Genome Size. Wiley, Chichester. 480 pp.

Coffaro, G. and Bocci, M. 1997a. How *Ulva rigida* and *Zostera marina* compete for resources: a quantitative approach with application to the Lagoon of Venice. *Ecol. Modelling* 102:81-96.

Gerson, U. and Chet, I. 1981. Are allochthonous and autochtonous soil microorganisms r- and K-selected? Rev. *Écol. Biol. Sol.* 18: 285-289.

Heal, O.W. and Dighton, J. 1986. Nutrient cycling and decomposition in natural terrestrial ecosystems. In: M.J. Mitchell, J.P. Nakas (eds.): Microfloral and faunal interactions in natural and agro ecosystems. Nijhoff & Junk Dordrecht, pp. 14-73.

Holling, C.S. 1986. The resilience of terrestrial ecosystems: Local surprise and global change. In: W.C. Clark and R.E. Munn (eds.). Sustainable Development of the Biosphere. Cambridge University Press, Cambridge, pp. 292-317.

Johnson, L. 1990. The thermodynamics of ecosystem. In: O. Hutzinger, (ed.). The Handbook of Environmental Chemistry, vol. 1. The Natural Environmental and The Biogeochemical Cycles. Springer-Verlag, Heidelberg. pp: 2-46.

Jørgensen, S.E. 1982. A holistic approach to ecological modelling by application of thermodynamics. In: W. Mitsch et. al. (eds.). Systems and Energy. Ann Arbor Press, Ann Arbor, MI.

Jørgensen, S.E. 1986. Structural dynamic model. *Ecol. Modelling* 31: 1-9.

Jørgensen, S.E. 1988. Use of models as an experimental tool to show that structural changes are accompanied by increased exergy. *Ecol. Modelling* 41: 117-126.

Jørgensen, S.E. 1990. Modelling in Ecotoxicology. Elsevier, Amsterdam.

Jørgensen, S.E. 1992a. Development of models able to account for changes in species composition. *Ecol. Modelling* 62: 195-208.

Jørgensen, S.E. 1992b. Parameters, ecological constraints and exergy. *Ecol. Modelling* 62: 163-170.

Jørgensen, S.E. and de Bernardi, R. 1997. The application of a model with dynamic structure to simulate the effect of mass fish mortality on zooplankton structure in Lago di Annone. *Hydrobiologia* 356: 87-96.

Jørgensen, S.E. and Mejer, H.F. 1977. Ecological buffer capacity. *Ecol. Modelling* 3: 39-61.

Jørgensen, S.E. and Mejer, H.F. 1979. A holostic approach to ecological modelling. *Ecol. Modelling* 7: 169-189.

Jørgensen, S.E. and Padisák, J. 1996. Does the intermediate disturbance hypothesis comply with thermodynamics? *Hydrobiologia* 323: 9-21.

Jørgensen, S.E., Nielsen, S.N. and Mejer, H. 1995. Emergy, environ, exergy and ecological modelling. *Ecol. Modelling* 77: 99-109.

Jørgensen, S.E., Patten, B.C. and Straskraba, M. 1998. Ecosystem emerging: 3. Openness. *Ecol. Modelling*, in press.

Jørgensen, S.E. 1997. Integration of Ecosystem Theories: A Pattern. 2. edition. Kluwer Academic Publ. Dordrecht. Boston. London. 400 pp. (1. edition, 1992).

Kay, J. and Schneider, E.D. 1990. On the applicability of non-equilibrium thermodynamics to living systems [internal paper]. Waterloo University, Ontario, Canada.

Kay, J. and Schneider, E.D. 1992. Thermodynamics and measures of ecological integrity. In: Proc. Ecological Indicators, Elsevier, Amsterdam. pp. 159-182.

Lewin, B. 1994. Genes V. Oxford University Press, Oxford. 620 pp.

Li, W.-H. and Grauer, D. 1991. Fundamentals of Molecular Evolution. Sinauer, Sunderland, Massachusetts. 430 pp.

Mauersberger, P. 1983. General principles in deterministic water quality modeling. In: G.T. Orlob (ed.). Mathematical Modeling of Water Quality: Streams, Lakes and Reservoirs (International Series on Applied Systems Analysis, 12). Wiley, New York, 42-115.

Mauersberger, P. 1995. Entropy control of complex ecological processes. In: B.C. Patten and S.E. Jørgensen (editors). Complex Ecology: The Part-Whole Relation in Ecosystems. Prentice-Hall Englewood Cliffs, NJ, 130-165.

Mejer, H.F. and Jørgensen, S.E. 1979. Energy and ecological buffer capacity. In: S.E. Jørgensen (ed.). State-of-the-Art of Ecological Modelling. (Environmental Sciences and Applications, 7.). Proceedings of a Conference on Ecological Modelling, 28. August - 2. September 1978, Copenhagen. International Society for Ecological Modelling, Copenhagen, 829-846.

Morowitz, H.J. 1968. Energy Flow in Biology. Academic Press, New York.

Morowitz, H.J. 1992. Beginnings of Cellular Life. Yale University Press, New Haven and London.

Nicolis, G. and Prigogine, I. 1989. Exploring Complexity: an Introduction. Freeman, New York.

Nielsen, S.N. 1992. Application of maximum exergy in structural dynamic models. Ph.D. Thesis, DFH, Institute A, Section of Environmental Chemistry, Copenhagen, Denmark. 51 pp.

Odum, E.P. 1969. The strategy of ecosystem development. *Science* 164: 262-270.

Odum, H.T. 1983. System Ecology. Wiley Interscience, New York. 510 pp.
Patten, B.C. 1991. Network ecology: indirect determination of the life-environment relatinship in ecosystems. In: M. Higashi and T.P. Burns (eds.). Theoretical Studies of Ecosystems: The Network Perspective. Cambridge University Press, 288-351.
Patten, B.C. and Fath, B.C. 1999. Environ Theory and Analysis. Submitted.
Reeves, H. 1991. The Hour of our Delight. Cosmic, Evolution, Order and Complexity. Freeman, New York. 246 pp.
Schlesinger, W.H. 1997. Biogeochemistry. An Analysis of Global Change, 2nd. edition. Academic Press, San Diego, London, Boston, New York, Sydney, Tokyo, Toronto. pp 680.
Schneider, E.D., Kay, J.J. 1994. Life as a Manifestation of the Second Law of Thermodynamics. *Mathl. Comput. Modelling* Vol. 19, No. 6-8, pp. 25-48.
Shugart, H.H. 1998. Terrestrial ecosystems in changing environments. Cambridge University Press, 534 pp.
Straskraba, M. 1980. The effects of physical variables on freshwater production: analyses based on models. In: E.D. Le Cren and R.H. McConnell (eds.), The Functioning of Freshwater Ecosystems (International Biological Programme 22). Cambridge University Press, 13-31, Cambridge.
Straskraba, M. and Gnauck, A. 1983. Aquatische Ökosysteme - Modellierung und Simulation. WEB Gustav Fischer Verlag, Jena. English translation: Freshwater Ecosystems - Modelling and Simulation (Developments in Environmental Modelling, 8). Elsevier, Amsterdam.
Straskraba, M., S.E. Jørgensen, and B.C. Patten, 1999. Ecosystem emerging: 3. Dissipation. Ecol. Modelling, in press.
Svirezhev, Y. 1998. Thermodynamic Orientors: How to Use Thermodynamic Concepts in Ecology. In: F. Müller and M. Leupelt (eds.). Eco Targets, Goal Functions, and Orientors. Springer. pp 102-122.
Ulanowicz, R.E. 1991. Formal agency in ecosystem development. In: M. Higashi and T.P. Burns (eds). Theoretical Studies of Ecosystems: The Network Perspective. Cambridge University Press, 58, Cambridge
Ulanowicz, R.E. 1997. Ecology, the Ascendent Perspective. Columbia University Press, New York, 201 pp.
Vollenweider, R.A. 1975. Input-output models with special reference to the phosphorus loading concept in limnology. Schweiz: Z. Hydrol. 37: 53-84.

II.2.1 Ecosystems as Subjects of Self-Organising Processes

Felix Müller and Søren Nors Nielsen

Self-organisation is a consequence of emergence.
Emergence is a consequence of self-organisation.

1. Introduction

Since the beginning of philosophical and scientific thinking, the question of the origin of the various forms of order in nature and society has been a focal problem. Many of concepts have been suggested in order to find a convincing answer, a lot of these ideas have developed into religions or myths, and a lot of them have failed. A modern attempt to discuss creation, maintenance, and dynamics of ordered organisations in very distinct entities originates in a combination of physics, chemistry, biology, synergetics, thermodynamics, computer science and systems analysis. This interdisciplinary enterprise (see Table 1) has been following the tracks of the philosopher Kant who suggested in 1790 that nature apparently is able to create order, expediently. Kant nominated the corresponding processes as self-creation, self-regulation, and self-organisation, elements, that in his opinion are characteristics of all living systems as they interact co-operatively for the mutual benefit of the whole entity (Fischer, 1990; Krohn and Kuppers, 1990). From our recent point-of-view these are very modern ideas that are still progressive. Due to the lack of respective scientific methods, Kant and his contemporaries could not explore the self-organised phenomena in a detailed, empirical mode. These opportunities have not been available until the beginning of this century, but in that period important steps were taken towards an understanding of self-organising systems. Some of them are documented in Table 1.
Discussing the concepts of self-organisation and ecosystems in the present situation very often leads to the following questions: 'Are the ecosystems in our present situation really self-organised systems?', 'What are the features a system has to fulfil to be called a self-organising system?', 'Do ecological systems really exhibit these features?', 'Is not the influence of man in our highly used and economically managed landscapes dominating so extremely that 'strange'-organisation would be the right term?', and 'If we really call the human used ecosystems 'self-organised', doesn't that concept become too broad to be fixed to any concrete case?' We will discuss these questions, report on the corresponding theoretical ideas, describe the ecological applications of the theory, and we will derive some ideas for a sustainable environmental management. The text will start with an attempt to define 'self-organisation', it is continued by a description of general features of self-organised systems and a thermodynamic interpretation. Afterwards the development of the self-organisation concept is summarised, and some examples are given referring to biological and ecological entities and ecosystems. Then the concept will be discussed in relation with other ecosystem theories, and finally some remarks will be made about consequences for environmental management.

Table 1: Some milestones for the development of the concept of self-organisation, after Muller et al. (1997a).

1900	Bénard: Convection cells
1923	Taylor: Instability of a liquid layer between 2 rotating cylinders
1925	Lotka: Physical biology and simple population models
1931	Volterra: Self-organisation in predator-prey models
1934	Gause and Vitt: Population oscillations
1938	Langmuir: Roll cells on water
1940	Bertalanffy: Organisms as open systems in steady state
1944	Schrödinger: Thermodynamic interpretation of life processes
1952	Turing: Theory of morphogenisis
1960	Von Förster: Cybernetic approach in systems theory
1964	Zhabotinskii: Chemical waves
1967	Prigogivne and Nicolis: Theory of dissaptive structures
1971	Zhabotinskii and Zaikin: Chemical spiral waves
1971	Haken: Synergetics, laser as a prototype of self-organisation
1971	Eigen: Theory of hypercycles
1972	Gierer and Meinhard: Theory of biologcial formation
1973	Holling: Concept of elastic ecosystems
1975	Maturana and Varela: Concept of autopoesis
1963......1977	Lorenz, Feigenbaum, Mandelbrot, May, a.o.: Theories if deterministic chaos and fractal geometry

2. What is self-organisation?

Let us start the search for a definition of self-organisation with the example of BÉNARD-cells, as illustrated in Figure1: When a fluid with a suitable viscosity is filled into a container and heated from below – like a pan on a stove – a heat gradient will rise which affects heat conduction from molecule to molecule, if the heat difference is small. When the gradient reaches a critical level (Rayleigh number), a transition from the unordered conduction to a highly organised convection occurs spontaneously producing many small convection cells that reduce the gradient in the inner container while the heat difference is carried by thin boundary layers at the hot and cold sinks. With an increasing degree of self-organisation, the efficiency of the heat flux increases as well as the energy dissipation. Schneider and Kay (1994) conclude that the higher the gradient, the more energy is necessary to degrade it and to maintain an ordered structure. If the energy input exceeds a further critical level, the system will behave chaotically; the observed self-organisation is only possible within a certain energetic window.

Generalising the patterns observed by the BÉNARD experiment, we can define self-organisation as the spontaneous formation of ordered spatio-temporal and functional structures (e.g. convection cells) from microscopic disorder (e.g. non-structured liquid in the case of conduction). Ebeling (1989) states that self-organisation is an irreversible processes in non-linear dynamic systems within which complex structures of the whole arise on the basis of co-operative interactions of the parts. The structures are results of internal irreversible processes. They are not forced upon the systems from the outside. Besides many other definitions (some of them can be found in Table 2), Müller (1997) has proposed to understand self-organisation as the spontaneous formation of a system of internal gradients. The gradients are supported by external gradients from higher hierarchical levels. This definition is illustrated by the examples in Figure 2.

Table 2: Some definitions of self-organisation

An der Heiden (1992, 72)	Those features and structures of a system that are induced by the dynamical interrelationships of the components are self-organised features.
Ebeling (1989, 118)	Self-organised processes are irreversible processes in non-linear dynamical systems which lead to complex structures of the whole entity on the basis of co-operative effects of the subsystems. The resulting structures are not forced from the outside but arise due to internal interactions.
Krohn and Küppers (1992 a , 165)	All systems in which the causes for changes are induced internally and in which external influences are not significant are self-organised systems. Such systems exhibit boundaries which delimit the system from its environment.
Krohn and Küppers (1992 b , 395)	A system is self-organised if its spatial and temporal structures are created by their internal dynamics, only.
Roth (1990, 169)	Self-organised processes are such physical and chemical processes which lead to an ordered state or an ordered sequence of states from a broad range of potential initial conditions.
Salthe (1993, 323)	Self-organisation is a dynamic process in natural systems which leads to an increase of complexity, size, influence and throughput.
Von Förster (1990, 88)	Self-organised processes are processes which cause an increase of the degree of complexity within a limited area due to self-referential interactions.

3. Characteristics of self-organising processes

Figure 2 represents a simple scheme of gas molecules that are initially included in 5 of 10 distinct chambers (state A in Figure 2). When the boundaries between the cells are opened, the incoming energy is transformed into diffusive processes, the molecules are distributed homogeneously, and the most probable state – thermodynamic equilibrium – is reached, disorder and entropy being high (state B in Figure 2). We can compare this state with the situation of heat conduction case of the BÈNARD experiment.

In contrast to the equilibrium state, open systems, that import and export energy, matter, and information, are able to attain states of high improbability: They can develop and persist far from equilibrium without receiving directive impulses from the outside. In this case, the energy is transformed into the emergence of structure, order, heterogeneity, and information (state C in Figure 2). The corresponding process is called 'dissipative self-organisation'[6] (Nicolis and Prigogine, 1996 and 1989). During this process, the system exhibits specific emergent properties which arise on certain developmental steps as a consequence of the subsystems' interrelations (Müller et al., 1997b). Therefore, emergent properties are consequences of self-organising processes (Nielsen and Müller, this volume).

[6]There is also the process of 'conservative self-organisation'. Ordered structures are formed while energy is exported, when a heated system cools down slowly (e.g. formation of crystalline patterns).

Such interactions can be characterised by the properties listed in Table 3. Many of these features also function as prerequisites for self-organisation. For example, without the import of high quality energy, which can be transformed into mechanical work (exergy-input)[7], without internal energy transformations and without the export of non-usable energy (entropy-output), dissipative self-organisation would not appear. Furthermore, the parts have to be co-operative: there must be a network of interactions which develops on the basis of exergy degradation and simultaneous entropy production. In the course of the self-organisation process the qualities of the characteristics listed in Table 3 emerge, and their quantities in many cases increase with the developmental state of the system. As these rules also come true in biological systems, many ecological properties can be derived from these general principles of systems analysis.

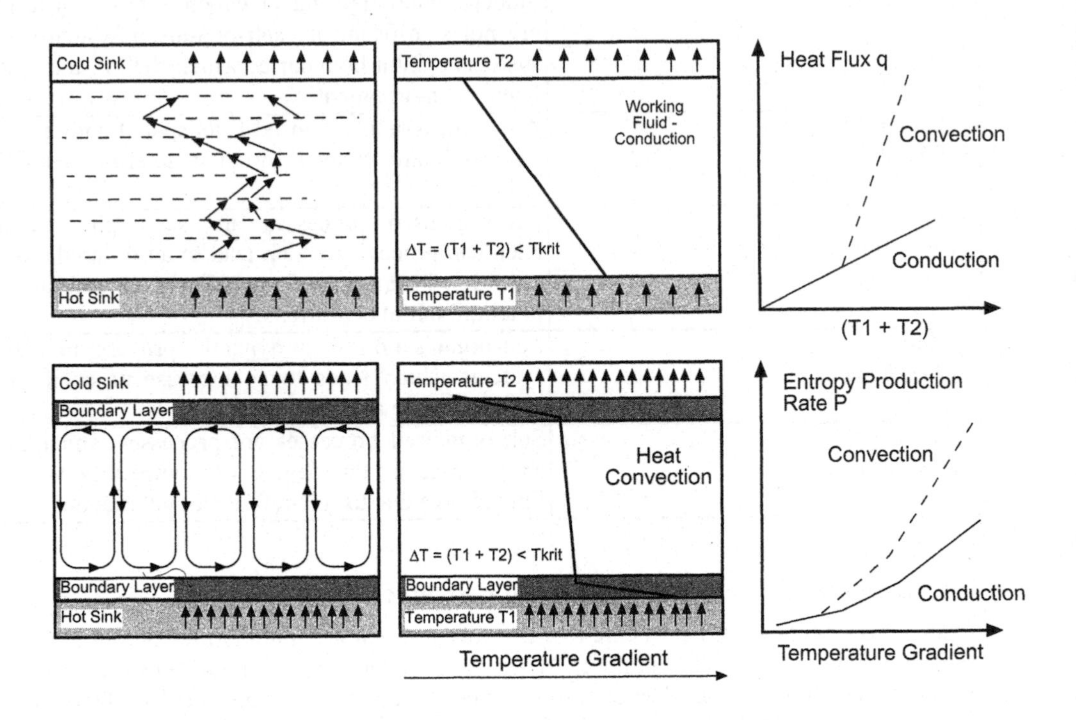

Figure 1: Self-organised creation of convection cells (BÉNARD cells) in a liquid that is heated from below and some thermodynamic features of the structural formation (after Schneider and Kay 1994). The upper part of the scheme describes heat conduction, the corresponding heat gradients and the heat flux for both processes, convection and conduction. The lower part illustrates the conditions while heat convection appears. The right figure shows that the efficiency of the gradient degradation is much higher in the self-organised case.

[7]Exergy is a measure for the quality of energy (See Jørgensen in chapter 2.1 of this volume). It represents the difference of free energy of a system in contrast to its environment. This is the energy fraction which can be transformed into mechanical work. In other words, exergy is that part of a system's energy which can be utilised. In ecosystems, the solar radiation has a very high exergy status. That energy is transformed in the food chains and webs of the ecosystem. During these processes the exergy is degraded and the system produces an energy fraction which cannot be internally utilised any longer (e.g. by respiration or transpiration). That energy form is called entropy.

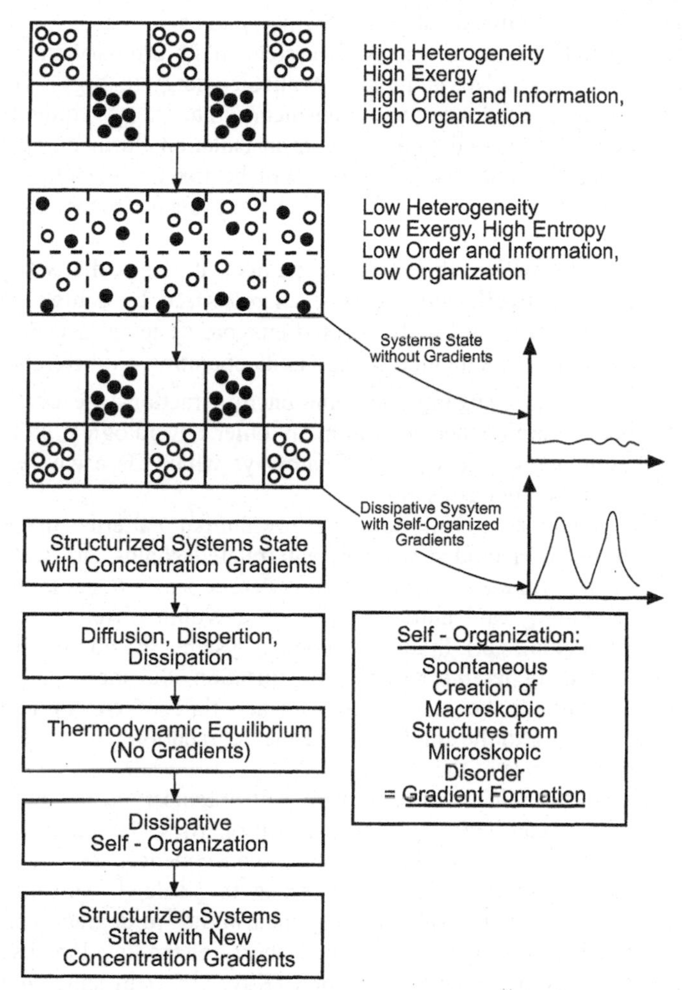

Figure 2: A simplified scheme of self-organised processes and their features referring to the dynamics of ecosystemic gradients. The three boxes illustrate an experiment with different gas molecules which are separated (state A), dissipated (state B) and re-organised (state C). The surrounding elements of the figure illustrate the characteristics of the states and combine the process description with an interpretation from the viewpoint of the gradient concept (see Müller 1998).

4. A thermodynamic interpretation of self-organisation

As a result of the experiments mentioned in Table 1 and other thermodynamic investigations, Schneider and Kay (1994a) and Jørgensen, Kay, Nielsen and Schneider (unpublished) have postulated the thermodynamic non-equilibrium principle: *"If an (eco)system is moved away from thermodynamic equilibrium by the application of a flow of exergy, it will utilise all avenues available, that is build up as much dissipative structure as possible, to reduce the effects of the applied gradient."* We can also formulate that living systems degrade and simultaneously utilise gradients by the self-organised formation of a hierarchy of small, internal, interrelated gradients. These gradients can be characterised by biomass patterns, by different concentration profiles, and by a complex structurisation. These features form the basis for the storage of the imported exergy, that increases with a growing degree of self-organisation. The corresponding thermodynamic correlations have been formulated in the 'ecological law of thermodynamics' by Jørgensen (Jørgensen, 1992; Jørgensen in chapter 1.1 of this volume). These principles provide an additional level of system properties, that integrate the features mentioned in Table 3 into a general hypothesis about the structural and functional patterns of living systems.

In order to illustrate this idea, we take a system in its ordinary state, for example a chlorophyll molecule that receives an input of available energy (exergy). The molecule will be transformed into an excited state, as it can incorporate exergy equivalents for a short time. If these photons are applied to a molecule in isolation, further transformations will rapidly take place and the initial solar exergy will be transferred into phosphorescence, fluorescence and heat; the energy will be lost for the system because the gradient is too high to be processed.

In contrast to this situation, the energetic gradient in plant cells is degraded within a functional biochemical network which can be characterised by a high number of small internal gradients, performed by highly adapted enzyme systems (see Figure 4). Besides this degradation chain, the cells include storage units operating on different scales (enzyme systems, ATP, $NADPH+H^+$, sugars) and some energy fractions are cycled inside of the system. These processes are connected with many other physiological units that provide a highly organised structure with a variety of pathways which, as a whole, provide a high efficiency of energy and matter utilisation.

The carbon fluxes at the ecosystem scale portray similar patterns: the energy gradient provided by the plants is degraded in many interrelated chains and webs (Higashi and Burns, 1991), there are storage components on different scales (Chapin et al., 1990), and within this network energy and nutrients are cycled within different functional entities (Patten, 1992; Kappen et al., i.p.). In these ecological systems the non-equilibrium principle is realised as well: the higher the system's organisation, the more pathways of exergy degradation are constituted; the longer the residence time of energy and nutrients, the smaller the losses referring to the single steps of the degradation web, and the more energy is needed for the maintenance of the structure.

Resuming the thermodynamic interpretation of self-organisation and including the non-equilibrium concept, we can derive two groups of properties, the general features of self-organised systems (Table 3) and the functional consequences of the non-equilibrium principle. These are summarised in the properties in Table 4. Self-organised systems develop in a way that is strongly influenced by their ability to degrade energy and matter gradients. This degradation is necessary for the stability of the developed structure, i.e. for the resistance near an optimum operating point (Kay, 1993) in spite of the pertubating effects of the high applied energy gradients (see e.g. the temperature optimisation model of Lovelock, 1988). Thus, during the development the resilience of the system is increased, if the gradient degradation scheme is optimised. This is reached by an improvement of the system's exergy budget. Systems are successful, if they incorporate the applied exergy (exergy capture) as long as possible and if the losses of high quality energy are reduced. In ecosystems such states are attained by the emergence of an increasing diversity of exergy degrading structures that are organised hierarchically in food chains (diversity of gradients and pathways, network formation) and storages (exergy storage).

Table 3: Prerequisites and features of self-organisation, working principles after Ebeling (1989) and Müller et al. (1997a). The characteristics will be interpreted ecologically in chapter 4.3.

Openness	Self-organising systems exchange energy, matter and information with their environments.
Import of convertable energy	Self-organising systems need energy imports that are transferable into mechanical work (exergy).
Suitable quantities of energy input	Self-organising systems can develop and maintain only within a certain energetic window.
Internal energy transformations	Self-organising systems transform the imported energy in physical and chemical, anabolic and catabolic processes.
Export of non-usable energy	Self-organising systems produce non-usable energy forms as a result of the irreversible processes. That energy fraction is exported into the environmment (entropy).
Distance from equilibrium	Self-organising systems build up internal structures and hereby increase their distances from thermodynamic equilibrium.
Non-linearity	Self-organising systems can only develop if the changes of one state variable lead to non-linear changes of other state variables
Amplification	Self-organising systems exhibit high fluctuations when their states approach phase transitions due to the non-linear interactions.
Self-referentality and co-operation	Self-organising systems are not regulated by external forces. Their dynamics arise as a result of the internal interactions.
Hierarchies and constraints	Self-organising systems are internally co-ordinated by processes on different scales, whereby there is a structural and functional hierarchy of constraints.
Stability	Self-organising systems are able to buffer minor inputs when their dynamics are situated near steady state.
Historicity and irreversibility	Self-organising systems develop on the basis of irreversible processes. Therefore, their actual state can only be understood in its historical context.

As a consequence, the exergy demand for the maintenance of the structure, which becomes more and more complex, increases as well. Therefore, entropy is produced with a growing intensity. Highly developed ecosystems have to transfer an increasing amount of degraded energy (entropy) to their environment. As a result, the total entropy production (e.g. respiration and transpiration) increases with the developmental stage, while the specific entropy production, which is related to the single steps of the energy transformation chain, decreases as a result of the system's organisation and the mutual adaptation of the parts (see Jørgensen, 1992; Müller and Fath, 1998; Müller and Jørgensen in chapter 3.2 of this volume).

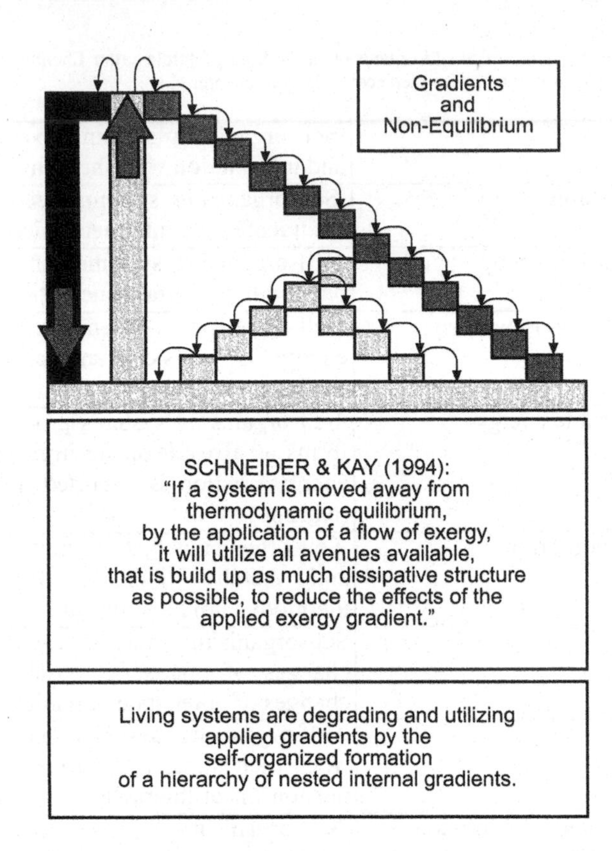

Figure 3: An illustration of the thermodynamic non-equilibrium principle. The stairs represent the eco-physiological consequences of the 'reformulated second law of thermodynamics': A high gradient is applied to the system. If its structure is not suitable for the capture of the energetic potential the exergy is lost. If the system builds up co-operative structural units – that is what happens in self-organised systems – the gradients can be transformed.

Table 4: System properties resulting from the implications of the non-equilibrium principle after Schneider and Kay (1994a) and Müller et al. (1997a): With an increasing intensity and duration of self-organisation processes (eco)systems tend to increase the following properties:

Property	Potential ecosystem indicator	References
Exergy capture	Radiation used for production Surface temperatures	Luvall and Hobro 1991 Schneider and Kay 1994
Total entropy production	Respiration Evapotranspiration Nutrient loss	Ripland Hildmann i.p. Schimming et al. i.p Schneide and Kay 1994
Diversity of gradients and pathways	Number of elements and flows Community diversity	Ulanowicz 1986 Margalef 1968
Storage capacity	Gause and Vitt: Population Oscillations	Kappen et al. i.p. Chapin et al. 1990 Reiche et al. i.p.
Cycling of energy and materials	Cycling index Residence times	Finn 1976 Herendeen 1989
Network formation	Network articulation network homogenisation	Patten 1991 Patten 1992

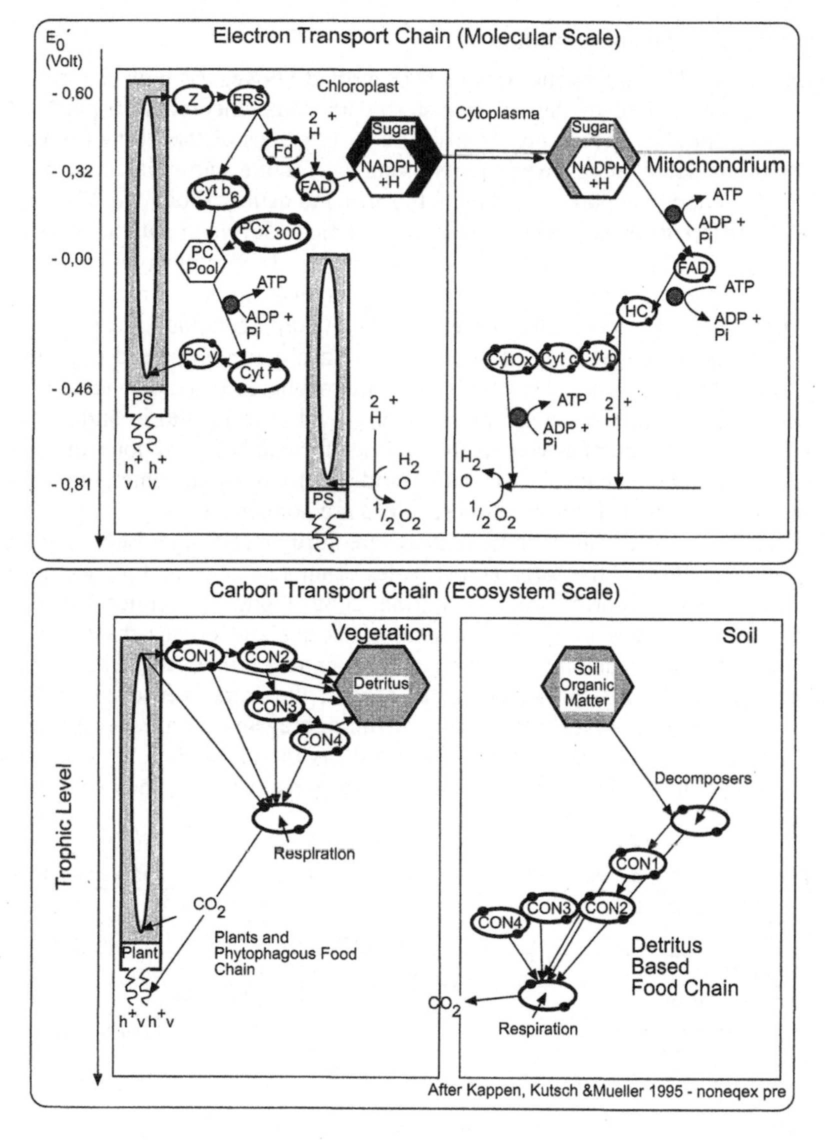

Figure 4: An illustration of the gradient concept and the non-equilibrium principle on different scales (plant cells and ecosystems), according to Kappen et al. (i.p.), after Müller (1996); explanation in the text.

5. Self-organisation in biological systems

With regard to biological systems, numerous applications of the self-organisation concept have been carried out. They are based on the initiating ideas of Bertalanffy (1940), Schrödinger (1944) and Turing (1952) who introduced biomorphogenesis as an important case study to observe self-organising processes. Further applications take into account physiological processes, such as glycolysis (e.g. Hess and Markus, 1987), neurophysiology (e.g. Roth, 1990), physiological rhythms (e.g. Haken and Haken-Krell, 1989), morphogenesis (e.g. Gierer and Meinhardt, 1972) or activity regulation of animals (e.g. Bünning, 1977).

5.1 Self-organisation and evolution

To illustrate the thermodynamic approach described above, we will choose an aspect of long-term self-organised evolution. We start with some theoretical aspects of chemical evolutionary processes, which occurred in the early stages of the Earth's development. In the 'primordial soup' atoms were functioning as mobile sub-units in a non-structured solution with energy imports, similar to the gas molecules shown in Figure 2. We can distinguish different levels of the historical chemical development that provide specific properties, leading from unstable molecular compounds to collaborating chemical units. Consequently, the systems exhibit an increasing complexity and structure, and already in the first unorganised reaction chains the stability of some compounds increases, as well as the spatio-temporal organisation of the sub-units in more complicated organic molecules. Higher stages of this chemical evolution are microspheres or coazervates (Oparin, 1938; Fox, 1969). They comprise chemical boundaries, internal equilibria and the abilities for selective imports or exports as well as the abilities for catalytic reactions and growth. After all, these 'primitive' chemical systems provide self-organised, emergent characteristics which are not predictable if the reaction schemes are not considered.
Introducing early biological evolution, there are further self-organised properties arising from the interrelations of the parts. Hypercycles of nucleic acids and proteins, for example, enabled their carriers to profit from regulation, control, and self-replication (Eigen, 1971). And with the emergence of cells, cell fusions, tissues and organs the biological entities were able to set up metabolic equilibria and to enhance their efficiencies by continued compartmentalisation and specialisation. In organs, the parts even abandon their isolated stability by building up co-operative, mutualistic aggregates. Consequently, the properties of life came into existence as a result of the self-organised, evolutionary processes (Table 5), and – following the thermodynamical definition given above – they comprise emergent properties.

Table 5: Self-organised (emergent) properties of biological systems, a compilation from various authors

Identical self-replication	Metabolic efficiency
Genetic variation	Stabilisation of structure and function
Self-regulation	Reactivity
Ability for evolution	Functional processing of information
Hierarchical organisation	Internal chemical steady states
Growth and development	Variability

A central level of development of organic evolution is the organism. Wicken (1987) has applied the non-equilibrium for biological self-organisation when he defined organisms as informed auto-catalytic systems possessing – in virtue of information stored in macromolecules – an internal organisation of kinetic relationships. Organisms thus are able to maintain themselves by pulling environmental resources into their own production and faithful reproduction by dissipating unusable energy to appropriate sinks. We take this definition as a statement of self-organised properties at the organism level, including the demands that organisms are highly organised entities of the whole system's exergy degradation chain (Weber et al. 1989; Jørgensen, 1992).

5.2 Self-organisation in ecological systems

On the trip through the relevant scales of ecological systems, the next level of development comprises groups of organisms like swarms or schools that gain advantages as a whole by practising a co-ordinated mutualistic coexistence. They can be regarded as diminished prototypes for the evolutionary criteria postulated by Weber et al. (1989), who interpret autocatalytic cycles (for example the nitrogen cycle) as self-organised holistic units of selection. The authors show that within specific developmental phases new elements (for example bacteria which are able to transform nitrogen compounds) will be integrated into ecological flow nets if the probability of nutrient and energy loss of the whole system decreases (for example by nitrogen cycling instead of nitrogen leaching), internal cycling optimised (for example by root uptake of the mineralised nitrogen compounds), and the signal dampening capacity (for example by buffering high nitrogen applications) improved. This specification of the non-equilibrium principle mainly concerns the levels of populations, communities and ecosystems. As a consequence, we can add to the list of self-organised features developmental attributes such as the minimisation of exergy loss, the optimisation of buffer capacities, and the optimisation of the total system efficiency.

5.3 Self-organisation in ecosystems

In this section, the general features of self-organisation which have been summarised in Table 3, will be resumed and applied to the ecosystem level to describe the basic characteristics of self-organisation at this scale.

Openness of the system: Dissipative self-organisation can only take place in open systems which exchange energy, matter, and information with their environments. In ecosystems, these prerequisites are fulfilled because of the energetic input-output dynamics of exergy (for example. energy inputs with solar radiation) and entropy (for example energy outputs by respiration and transpiration), by water, matter and nutrient flows (for example rain vs. drainage, fertilisation vs. harvesting, or deposition vs. leaching), and by biocenotical exchange processes between corresponding ecosystems (see Jørgensen in chapter II.1.1 or Odum et al. in chapter II.6 in this volume).

Import of exergy: Self-organisation processes are 'pumped courses of events' which consume usable energy. The imported energy must be transferable into mechanical or biochemical work, thus the ecosystem compartments convert it as part of their physiological biochemical metabolisms. The 'high qualitative energy' which can be utilised by the respective system is called exergy (see Jørgensen in chapter II.1.3 in this volume). In ecosystems the highest portion of the exergy input is received from the solar radiation which is converted into usable chemical energy by photosynthesis. Thus, the capacity of ecosystems to perform self-organised processes is a function of the plant-ecophysiological potentials.

Suitable quantities of exergy inputs: The imported exergy can be converted into self-organising processes only within a certain energetic window. If the import is too small, the self-organising potential cannot be activated, and if the exergy input is too high, the system will not be able to convert the exergy into an ordered structure. Schneider and Kay (1994) have demonstrated this 'energetic window of self-organisation' on the basis of Bénard cells (Figure 1): If the exergy gradient is small, the energy is conducted from molecule to molecule, and no self-organised patterns can occur. If the gradient is too high, turbulent mixtures take the stage, and the self-organised cells can neither be created nor maintained. Parallel to that, ecosystems have to be able to capture the gradients that are offered by the environment. This potential varies and is influenced at many different scales. For example, as an interpretation of the Gaia Hypothesis, the climatic system of the Earth's ecosphere guarantees a general, evolutionary optimised, meta-stable and self-organised environment which enables the subsystems to operate optimally on their specific spatio-temporal scales. These subsystems can be easily distinguished by their potential to degrade the incoming exergy as they have adapted (self-referentially) to their specific geographical situation, all of them optimising their ability of energy-uptake. The respective potentials differ due to site specific constraints and limitations which are existing because of the multitude of living necessities and physiological needs.

Therefore, desert ecosystems are utilising the solar exergy (gradient) optimally, as well as the tropical rain forests do; the difference arises from the distinctive sizes of the 'exergy capture window' which defines the site-specific potential of an ecosystem to transform environmental (external) gradients into internal structure (gradients).

Internal energy transformations: The imported exergy is converted into the catabolic and anabolic processes of the ecosystem. It is degraded in many interrelated reaction chains and cycles, whereby numerous energetic transformation can take place (see Jørgensen, 1992; Odum, 1983). These reactions are executed on different spatial and temporal scales (in organells of cells as well as in the matter of long-term soil carbon balance), and as a result, the incoming exergy (for example solar radiation) is stored in the living biomass (for example energy content of organisms) and in the abiotic compartments (for example energy content of soil organic carbon), it is transformed into information and structure (for example genetic variety), can be exported (for example by transpiration) or converted into entropy (for example respiration). All ecosystem processes are elements in these transformation networks that contribute to the breakdown of the incorporated exergy as well as to the successional complexifications.

Export of entropy: As a result of the ecosystemic processes some components of the incoming energy are reflected or cannot be stored by the system. Other energy portions are utilised for life-supporting energetic processes in the organisms. They affect a qualitative degradation of the imported exergy. These fractions of energy balance, the produced entropy, are re-transported to the environment (see Steinborn and Svirezhev i.p.). Important entropy sources originate in the heat which is lost with evapotranspiration and which is produced by respiration processes. But also the loss of nutrients by evaporation and leaching, or the output of soluted, weathering mineral substances can be interpreted as parts of entropy export (see Jørgensen in chapter II.1.1 in this volume).

Distance from thermodynamic equilibrium: The developing system which utilises the imported exergy fractions for the creation of an increasing internal structure accumulates complexity and information (see Nielsen, this volume). With these processes, the system enhances its distance from thermodynamic equilibrium which represents a homogeneous distribution of the parts. Thus, the distance can be indicated by the system's heterogeneity, which is the amount of gradients that are created and maintained. The more complex the system's structure is, the more interrelationships operate, the higher is the distance from equilibrium (see Jørgensen in chapter 1.1, this volume).

Non-linearity: The dynamics of dissipative structures are dominated by non-linear processes: The changes of one state variable can lead to very different dynamics of other variables or parameters. As these variables are standing in complex relationships steadily, non-linear systems can affect complicated chains of reactions, indirect effects and exhibit numerous steady states. Therefore, their historical origins take high influences on the actual situations. In non-linear systems, positive feedbacks, chaotic dynamics or catastrophes can rapidly provoke significant changes. Consequently, non-linearity is an important precondition for the self-organised changes of system structures. The extreme consequences of indirect effects (Patten, 1991 and 1992) are ecosystemic examples for the evidence of non-linearity. Other examples can be taken from population dynamics where oscillating dynamics of different species are regularly interrelating. In steady state periods they are following the principles of Lotka (1925) and Volterra (1931) which include long-term stability. But during periods with high external fluctuations, also chaotic behaviour can be found (May, 1976). Further details concerning this feature can be found in the chapters of Svirezhev, Gundersson et al., Bendoricchio, and Jørgensen in the chapter II.8 of this volume.

Amplification: Dissipative structures can show extreme fluctuations when phase transitions and structural changes are approached. Ecotones are exemplary systems where the spatial dynamics of neighbouring systems lead to high amplifications of structural instabilities. Of course, these high variability's are consequences of the non-linear coupling of functional subsystems. In order to fulfil the basic demands of life and as a consequence of the observable optimisation trends, the consequences of phase transitions are systems which are, in whole, more able to cope with environmental situations. Such adaptations can guide the system to structures which often exhibit optimised degrees of ecological orientors (see Müller and Jørgensen, this volume).

Self-referentially and co-operation: The self-organised processes operate on a basis of a non-regulated, internal network of interrelations. They are not externally influenced, but the degrees of freedom of their self-referential dynamics are constrained by the systems' environments. For example, natural ecosystems develop uninfluenced from human land-use activities. Their dynamics are the results of complex ecological interactions. Nevertheless their developmental potential is restricted by natural constraints or can be changed by human activities in their surroundings. In spite of these potential external inputs or interruptions, the basic structuring processes are conducted and regulated internally. Thus, also agricultural ecosystems exhibit self-organised interactions; but in contrast to nature-near systems, their limitations are very high (see chapters of Dierssen, Rapport, or Barkmann and Windhorst in chapter III of this volume).

Hierarchies and constraints: The self-referential co-ordination of self-organised processes is based on the interactions of processes at very different scales. Within those functional hierarchies, the potential state spaces of rapidly changing processes are constrained by slow processes which operate on broader spatial extents. Examples for these asymmetric interactions can be read in the chapter on hierarchy theory by Hári and Müller in this volume.

Stability: Self-organised systems are able to buffer and filter small disturbances when their dynamic trajectories are a steady state (optimum operating point). Thus ecosystems can react invariantly after minor pulses from their environment for extremely long temporal periods. The corresponding theories have been dominant ideas in ecology. For a further inclusion of these arguments about the stability principle we can refer to Svirezhev and to Gundersson et al. both in this volume, or to the summarising chapters of Gigon and Grimm (1996) or Grimm and Wissel (1992).

Historicity and irreversibility: Self-organisation is based on irreversible sequences of processes. Their development can only be understood on the basis of the system's history. Furthermore, a specific state can never be reached again. The temporal dimension excludes reversible reactions in all living systems. This aspect has gained more and more attention since the paper of Holling (1986) which introduced destruction and change as new paradigmatic elements in a world which was filled up with stability concepts (see Gundersson et al, this volume).

Self-organised features and ecosystem development: The features proposed up to this point can also be understood as very general properties of ecological systems. They represent the ecosystems' potentials for self-organisation and they can be used to indicate the degree of self-organisation an ecosystem has attained. Consequently, their respective quantities can be taken as references for the state of an ecosystem. In this context, since Odum's presentation of the maturity concept in 1969, a lot of authors have discussed the corresponding hypotheses about the change of ecological properties during the self-organised ecosystem development (e.g. Baird et al., 1991; Christensen, 1992; Herendeen,

1989; Jørgensen, 1992; Schneider, 1988; Schneider and Kay, 1994; Ulanowicz, 1986). The concept has become a central part of our basic knowledge in ecology, although conceptual problems arise when the maturity model, which has been elaborated to show orderly and directed processes on the (abstract) ecosystem scale, is used for accurate predictions on a species or population level (Breckling, 1993). Due to the large effects of the initial conditions, small variations at the starting point of a succession can lead to very different communities, when we observe the species composition. Therefore, the model cannot be used to predict the exact spatio-temporal sequence of specific plant or animal populations. Furthermore, the uniqueness of a specific spatial situation, the high dynamics of the forcing functions, and the lack of comprehensive long term studies make it very difficult to test Odum's hypotheses. But the model offers a general, unifying scheme of change during ecosystem development, and as it has been developed for this scale, it is still broadly accepted in the scientific community (Christensen, 1992). Younger branches of the maturity ideas that incorporate Holling's cycles as well as thermodynamic principles and that do not base on the idea of a climax community are fixed in the idea of orientor approach which tries to construct an abstract description of ecosystem development as an example of a self-organised process sequence (see Müller and Jørgensen in the last chapter of this volume).

6. Self-organisation and ecosystem theories

As self-organisation can be comprehended as a meta concept which is capable of integrating many different theoretical approaches and many different theories (see also Jørgensen, 1992 or Müller, 1997), a short explanation of the linkages to those theories (which are all presented in this volume) will be given in the following chapter.

Self-organisation and thermodynamics: This theory-linkage obviously is an essential relation, because the principles of open system development are constraints for the dynamics of self-organised ecological systems. Ecosystem development thus is one example for the dynamics of dissipative structures far from thermodynamic equilibrium.

Self-organisation and information theory: The information accumulated in a self-organised system increases with the duration of the dissipative processes. In ecosystems we consequently find high levels of heterogeneity and diversity, high processual interrelations and high flow densities as results of self-organised processes.

Self-organisation and cybernetics: Self-organised systems realise the principles of self-regulation and self-referential control. This co-ordination is based on positive and negative feedbacks which can be reflected as cybernetic systems. The dynamics of ecosystems can be modelled on the basis of cybernetic approaches.

Self-organisation and hierarchy theory: The concepts illustrated in chapter 4 show that self-organisation can be understood as a process sequence which continuously creates interacting, nested hierarchical levels. Their relationships play an important role for the stability of the system. They also form the basis for emergent ecosystem properties.

Self-organisation and network theory: One result of the ecosystemic self-organisation is a very complicated network of energy, material and information pathways. Their complexity increases as a function of a rising degree of self-organisation on the one hand, and on the other, the web of interactions which is analysed by network theory forms the processual basis for the self-organising process sequences.

Self-organisation and ecosystem dynamics: As we have seen in chapter 4.3, in the run of a self-organised ecosystemic development, there is a tendency for an increase of ecosystem properties such as diversity, complexity, efficiency, information, and hierarchical structure. But besides these and similar orientors, we can find no regular correlations with attributes such as stability, buffer capacity or resilience. These problems can be solved by introducing a multiscale aspect including death and destruction to the successional development. Holling (1986) has shown that living systems generally proceed in sequences of four different functions (Figure 5). In a juvenile stage (function 1) the exploitation of the existing resources is the dominating feature, and – following the property sequence in table 3 – afterwards there is a slow, consolidating development towards a conserving adult stage (function 2) which is characterised by a high degree of internal connectedness. There are a lot of well known cases where such high internal organisation turns into Overconnectedness (Holling, 1986; O' Neill et al., 1986). Thus, the adaptability of the system is decreased, and there is a high sensitivity against external perturbations. Furthermore, the system itself can provoke a change from this mature stage to a sudden breakdown. On a small scale, this phenomenon is found in the transition from an algal bloom to the clear water stage of lakes, or in incorporated disturbances which have frequently been reported in fire ecology (Shugart and Urban, 1988). Therefore, the passage of the 3rd Holling function (creative destruction) is a normal state of systems' self-organised development which has been neglected in ecology for a long time, although we know it from the short life spans of bacteria, the changing dominance of populations, the strong (and partly regular) changes and extinctions in the evolutionary history of life (e.g. Raup, 1992), or the fate we have to face ourselves. Finally, mineralisation (function 4) opens the door to a new sequence of the system's life cycle.

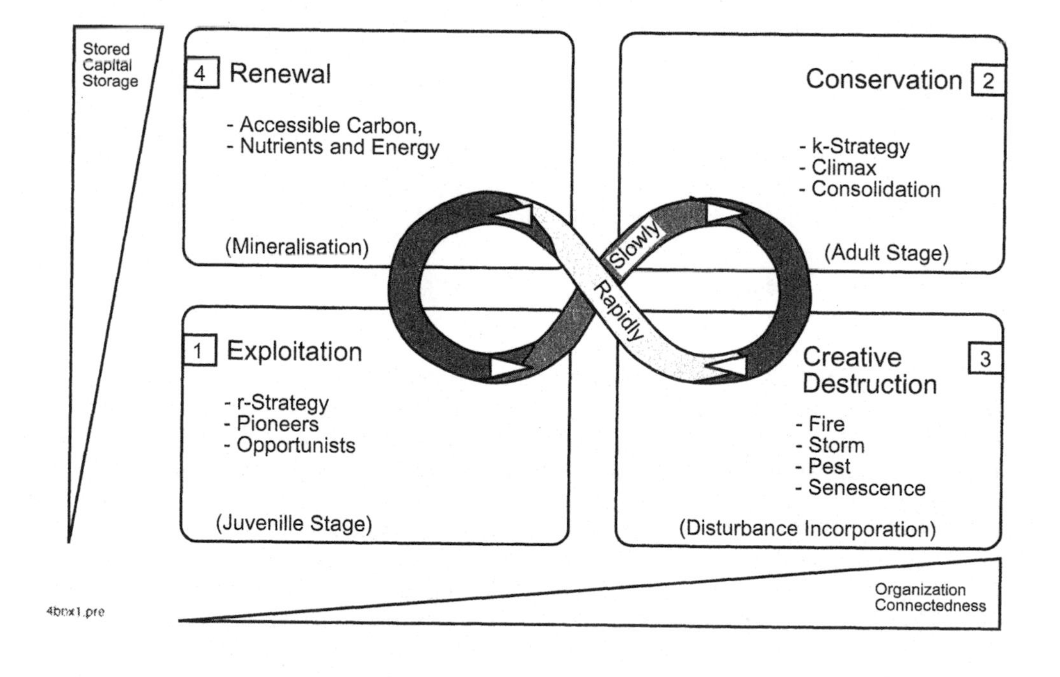

Figure 5: Four ecosystem functions after Holling (1986)

Self-organisation and the gradient concept: Self-organised processes are responsible for the creation, maintenance and dynamics of ecosystemic gradients (Müller 1998). They arise as potentials for energetic, nutritional, functional and structural processes. Therefore, gradients form a handable level of integration for all activities in self-organised ecological systems.

Self-organisation and emergence: Self-organisation produces new structures, new networks, new gradients, and new interrelationships, and therefore it forms new hierarchical levels with new hierarchical interactions. As defined in Nielsen and Müller (this volume) the specific features of such levels which can only be observed and assigned to these specific subsystems and which are coming into existence on the basis of self-organisation, are called emergent properties. *We can conclude that self-organisation is a consequence of emergence and – on the other side – emergence is a consequence of self-organisation.*

7. Self-organisation and ecosystem management

One focal point to learn from the text ahead, is that ecological systems in fact, are self-organised entities. Under natural conditions they follow natural constraints, they therefore behave thermodynamically and continuously optimise the multitude of emergent properties which can be used to indicate the state of ecosystem development. In these states, ecosystems realise the legal demands from nature conservation, for example the preservation and development of the capacity of the natural balance, the ensuring of the utility of the natural resources, goods and services, and the assurance of plant and animal species, their diversity, the landscape beauty and its individuality. All these targets of environmental management are very strictly connected with important criteria for self-organisation. Similar consequences arise by looking at the targets of sustainability: If a sustainable landscape management has to be carried out, it must develop strategies which are

- long-term oriented,
- integrating spatial and temporal scales,
- interdisciplinary,
- holistic,
- realistic, including information about uncertainties,
- nature-oriented,
- theoretically based, and
- goal - oriented.

These criteria implicitly carry the demand for a high adaptability of the natural systems and resources and for the maintenance of an optimal potential for self-organisation. Corresponding concepts have been developed in different locations, for example ecosystem health (see Rapport or Costanza in this volume, Costanza et al. 1993), ecological integrity (see Barkmann and Windhorst in this volume, Woodley et al. 1993) or process protection (e.g. Plachter 1995). In the centre of all these target systems and attempts, there is the ability of ecosystems to maintain their organisations after small impulses, to adapt to the dynamics of the constraints and to continue their long term development with a high degree and efficiency of self-organising processes. Thus, the capacity of self-organisation is one of the central ecological objectives which should find applications in environmental policy and management if this refers to a holistic concept, based on the idea of ecological systems.

Finally, returning to the questions raised in the beginning of this chapter we are now able to find the respective answers: Ecosystems are self-organised systems. They fulfil all prerequisites of self-organisation, and they exhibit all necessary features of self-organised systems. Human resource utilisation often puts heavy constraints on the self-organising processes. An intelligent resource management will not fight against these forces but will use them and take benefit from their performance. In the centre of the correlating ideas for ecosystem management will be the target to maintain and develop the ability of ecosystems to continue their development through effective self-organised process sequences.

Acknowledgements

We want to thank Paulette Clowes, Gerti Rosenfeld, and Regina Hoffmann-Kroll for their strong support during the (not always self-organised) procedure of writing this text. We are also grateful to our colleagues Broder Breckling, Mike Bredemeier, Volker Grimm, Horst Malchow and Ernst-Walter Reiche who co-authored a German version.

REFERENCES

An der Heiden, U. 1992. Selbstorganisation in dynamischen Systemen. In: Krohn, W. and G Küppers (eds.) *Emergenz: Die Entstehung von Ordnung, Organisation und Bedeutung.* Suhrkamp, Frankfurt/Main.

Baird D., J.M. McGlade, R.E .Ulanowicz 1991. The comparative ecology of six marine ecosystems. *Phil Trans R Soc London* 333:15-29.

Bertalanffy, L. v. 1940. Der Organismus als physikalisches System betrachtet. *Naturwissenschaften,* 28, 521.

Breckling, B. 1993. Naturkonzepte und Paradigmen in der Ökologie - einige Entwicklungen. Veröff.-R. Abt Normbildung und Umwelt, Wissenschaftszentrum Berlin, FS II:93-304.

Bünning, E. 1977. Die physiologische Uhr. Springer. Berlin, Heidelberg, New York.

Chapin, F.S. III, E.D. Schulze and H.A. Mooney 1990. The ecology and economics of storage in plants. *Annu. Rev. Ecol. Systems* 21: 423-447.

Christensen, V.1992. Network analysis of trophic interactions in aquatic ecosystems. PhD Thesis, Royal Danish School for Pharmacy, Copenhagen.

Costanza, R., L. Wainger, C. Folke, and K-G Mäler. 1993. Modeling complex ecological economic systems: toward an evolutionary, dynamic understanding of people and nature. *BioScience* 43: 545-555.

Ebeling, W. 1989. Chaos - Ordnung - Information - Verlag Harri Deutsch, Frankfurt/M, 118 pp.

Eigen, M. 1971. Selforganisation of matter and the evolution of biological macromolecules. *Naturwissenschaften* 58: 465-523.

Finn, J.T. 1976. Measures of ecosystem structure and function derived from analysis of flows. *Journal of Theoretical Biology* 56: 363-380.

Fischer, H.R. 1990. Selbstorganisation. Kritische Bemerkungen zur Begriffslogik eines neuen Paradigmas. In: Kratky, K.W. and F. Wallner (eds.) *Grundprinzipien der Selbstorganisation.* Wiss. Buchges. Darmstadt.

Foerster, H. von 1990. Kausalität, Unordnung, Selbstorganisation. In: Kratky, K.W. and F. F. Wallner (eds.) *Grundprinzipien der Selbstorganisation.* Wiss. Buchges. Darmstadt.

Fox, S.W. 1969. Self-ordered polymers and propagative cell-like systems. *Naturwissenschaften* 56.

Gierer, A. and H. Meinhardt 1972 A theory of biological pattern pattern formation. *Kybernetik* 12: 30-39.

Gigon, F. and V. Grimm 1996. Stabilitätskonzepte in der Ökologie. Typologie und Checkliste für ihre Anwendung. In: Fränzle, O., F. Müller, and W. Schröder (eds.) Handbuch der Handbuch der Ökosystemforschung. Ecomed. Landsberg.

Herendeen, R.A. 1989. Energy intensity, residence time, exergy, and ascendency in dynamic ecosystems. *Ecological Modelling* 41: 117-126.

Haken, H. and Haken-Krell, M. 1989. Entstehung von biologischer Information und Ordnung. Wiss. Buchges., Darmstadt.

Hess, B. and M. Markus 1987. Ordnung und Chaos in chemischen Uhren. In: Küppers, B.O. (ed.) *Ordnung aus dem Chaos.* Piper, München, Zürich. 157-176.

Higashi, M. and T. Burns 1991. Theoretical studies of ecosystems: The network perspective. Cambridge University Press. Cambridge.

Holling, C.S. 1986. The resilience of terrestrial ecosystems: local surprise and global change. In: Clark, W.M. and R.E. Munn (eds.) *Sustainable development of the bioshpere.* Oxford University Press pp 292-320.

Jørgensen, S.E. 1992 Integration of ecosystem theories: A pattern. Dortrecht, Boston, London.

Kay, J.J. 1993. On the nature of ecological integrity: Some closing comments. In: Woodley,S., Kay, J. and Francis, G. (eds.) *Ecological integrity and the management of ecosystems.* University of Waterloo and Canadian Park Service, Ottawa.

Krohn, W. and Küppers, G. 1992. Emergenz: Die Entstehung von Ordnung, Organisation und Bedeutung. Suhrkamp-taschenbuch. 984. 414 pages.

Krohn, W. and G. Küppers 1992a. Zur Emergenz systemspezifischer Leistungen. In: Krohn, W. and G. Küppers (eds.): *Emergenz: Die Entstehung von Ordnung, Organisation und Bedeutung.* Frankfurt

Krohn, W. and G. Küppers 1992b. Glossar. In: Krohn, W. and G. Küppers (eds.) *Emergenz: Die Entstehung von Ordnung, Organisation und Bedeutung.* Frankfurt

Lotka, A.J. 1925. Elements of physical biology. Baltimore

Lovelock, J. 1988. The ages of gaia: A biography of our living earth. Norton, New York.

Luvall, J.C. and H.R. Hobro 1991. Thermal remote sensing methods in landscape ecology. In: Turner, M. and R.H. Gardner (eds.) *Quantitative methods in landscape ecology*. Berlin, Heidelberg, New York, 127-152

Magalef, R. 1968. Perspectives in Ecological Theory. Chicago. University Press. Chicago, IL.

May, R.M. 1976 Simple mathematical models with very complicated dynamics. *Nature* 261: 459-467.

Müller, F. 1996. Emergent properties of ecosystems - consequences of self-organising processes? *Senckenbergiana Maritima* 27 (3/6), S. 151-168.

Müller, F. 1997. State-of-the-art in ecosystem theory. *Ecological Modelling* 100, 135-161.

Müller, F. 1998. Gradients in ecological systems. Proceedings of the Eco-Summit 1996 in Copenhagen. *Ecological Modelling* 108: 3-21.

Müller, F. and B. Fath 1998. The physical basis of ecological goal functions. In: Müller, F. and M. Leupelt(eds.) *Eco targets, goal functions and orienters.* Springer. Berlin, Heidelberg, New York, 269-285.

Müller, F., B. Breckling, M. Bredemeier, V. Grimm, H. Malchow, S.N. Nielsen and E.W. Reiche 1997a. Ökosystemare Selbstorganisation. In: Fränzle, O., F. Müller and W. Schröder (Hrsg.) *Handbuch der Ökosystemforschung*. Landsberg.

Müller, F., B. Breckling, M. Bredemeier, V. Grimm, H. Malchow, S.N. Nielsen and E.W. Reiche 1997b. Emergente Ökosystemeigenschaften. In: Fränzle, O., F. Müller and W. Schröder(Hrsg.) *Handbuch der Ökosystemforschung*. Landsberg.

Nicolis, G. and I. Prigogine 1967. Self-organization in non-equilibrium systems: from storage in plants. *Annu. Rev. Ecol. Systems* 21: 423-447.

Nicolis, G. and I. Prigogine 1989. Exploring complexity: An introduction. W.H. Freeman and co. New York.

Odum, E.P. 1969. The strategy of ecosystem development. *Science* 164:262-270.

Odum, E.P. 1983. Grundlagen der Ökologie. Thieme, Stuttgart.

O'Neill, R.V., D.L. DeAngelis, T.F.H. Allen, and J.B. Waide, 1986. A hierachical concept of ecosystems. *Monographs in Population Biology* 23. Princeton, NJ Princeton University Press.

Oparin, A.L. 1938. Origin of Life. Macmillan, New York.

Patten, B.C. 1982. Environs: relativistic elementary particles for ecology. *Americal Naturalist* 119:179-219.

Patten, B. C. 1991. Network ecology: indirect determination if the life-environment relationship in ecosystems In: Higashi, M. and Burns, T.P. (eds.) *Theoretical studies of ecosystems.* Cambridge University Press, Cambridge.

Patten, B.C. 1992. Energy, emergy, and environs. *Ecological Modelling* 62: 29-70.

Plachter, H. 1995. Naturschutz. Stuttgart.

Raup, D.N. 1992. Ausgestorben - Zufall oder Vorsehung? - vgs Verlagsgesellschaft, Köln, 240 pp.

Roth, G. 1990. Gehirn und Selbstorganisation. In: Krohn, W. and G. Küppers (eds.) *Emergenz: Die Entstehung von Ordnung, Organisation und Bedeutung.* Frankfurt.

Salthe, S. N. (1993): Development and evolution. Complexity and change in biology. MIT Press, Cambridge.

Schneider, ED 1988. Thermodynamics, information, and evolution: new perspectives on physical and biological evolution. In: Weber BH, Depew DJ, Smith JD (eds) *Entropy, Information, and Evolution: New Perspectives on Physical and Biological Evolution.* Cambridge.

Schneider, E.D. and J. Kay 1994. Life as a manifestation of the second law of thermodynamics. *Math. Comput. Modelling* 19: 25-48.

Schrödinger 1944. What is Life? Cambridge University Press. Cambridge.

Shugart, H.H. and D.L. Urban. 1988. Scale, synthesis and ecosystem dynamics. In: Pomeroy, L.R. and J.J. Albers (eds.) Concepts of ecosystem ecology. *Ecological Studies* 67: 279-290.

Turing, A.M. 1952. On the chemical basis of morphogenesis. *Phil. Trans. R. Soc. Lond.* B237, 37-72.

Ulanowicz, R.E. 1986. Growth and development: Ecosystems phenomenology. Berlin, Heidelberg, New York.

Volterra, V. 1931a. Théorie mathématique de la lutte pour la vie. Gauthier. Villars, Paris.

Volterra, V. 1931b. Leìons sur la théorie mathématique de la lutte pour la vie. Gauthier.

Weber, B.H., D.J. Depew, C. Dyke, S.N. Salthe, E.D. Schneider, R.E. Ulanowicz and J. S. Wicken 1989. Evolution in thermodynamic perspective: An ecological approach. *Biology and Philosophy* 4: 373-405.

Wicken, J.S. 1987. Evolution, thermodynamics and information. Extending the Darwinian program. Oxford University Press.

Woodley, S., J. Kay, and G. Francis (Eds.), 1993. Ecological Integrity and the Management of Ecosystems. Ottawa: St. Lucie Press.

II.2.2 Emergent Properties of Ecosystems

Søren Nors Nielsen and Felix Müller

"Emergence is a consequence of self-organisation"

1. Introduction

The concept of emergent properties has found widespread use in ecology, particularly after it was proposed as a concept for the set up of research strategies to be used in ecosystem research by E.P. Odum. Since then, it has occurred in many papers which present ecosystem studies. In addition, it has also been described and discussed for several types of biological systems in the scientific literature during the last decades.

The concepts themselves, emergent properties and emergence, have their origin in the 19th century or possibly even earlier, finding the earliest roots back in Kantian philosophy. Throughout this century, several scientists have addressed the concept from a more philosophical point of view (Garnett, 1942, Henle, 1942), resulting in the appearance of different explanations of the concept. The definitions given often have a heuristic character. They have, in general, been referring to subjective arguments, such as surprise, unexpectancy, thus being clearly observer dependent in their definition. This has strongly influenced the present use and approaches taken to analyse the concepts within different scientific areas.

Thus, the use of the terms in the current biological and ecological literature reflects the unclear and diverse definitions mentioned in the various treatments used (Emmeche et al., 1993). This causes severe problems in clarification and revelation due to the diverse use of this term and the related concept of emergence. The problems leave us with some principal questions that need to be clarified before determining whether emergent properties in biological or ecological systems actually. According to various definitions, various phenomena observed can be identified as emergent properties and in other cases not. At the one extreme, any convinced reductionist will claim that they do not exist at all. At the other end, other researchers, e.g. Green and Bossomaier (1993) state that holistic strategies are the only way of understanding complex, biological systems such as ecosystems.

Another important issue to be raised is the need for the concept to be loosened from its dependency on the subjective elements within the definition, like surprise, unpredictability, unexpectancy, etc. or for that matter any tendencies toward creativity within the definition. This work has taken a great effort to avoid and get rid of these elements. Meanwhile, those parts of the definitions, which refer to emergent properties as something unexpected, form an important connection to a possible quantification of the concept. A suggestion for the first steps in such a quantification will be presented.

As a starting point, when discussing emergence, many biologists will recognise the statement that *"the whole is more than the sum of the parts"* as a very commonly used phrase when referring to emergence. This formulation in turn refers to *"the idea that there are systems which possess additional qualities or quantities, beyond easily measurable or predictable physical parameters"* (Müller, 1996).

Emergence may occur on many levels of the ordinary, well known biological hierarchy, going from pure physical systems, through to simple organic molecules being organised in increasingly complex patterns, to cells, organisms and constellations hereof, ending at the ecosystem or biosphere level, see Table 1.

Table 1: Examples of biological hierarchies very often found in literature. The biological hierarchy is commonly used throughout literature, going from the physio-chemical to the biosphere level. The ecological hierarchy is widespread in works on ecosystem dynamics, and discussions on, for instance, top-down or bottom up control of ecosystems.

Biological hierarchy	Ecological hierarchy
Biosphere	top carnivore
Ecosystem	Carnivore
Societies	Herbivore
Populations	primary producers
Organisms	Bacteria
Organs	Nutrients
Cells	
cell organelles	
proteins, enzymes	
amino acids	
organic molecules	
inorganic molecules	

Emergence has been stated on many levels of biological hierarchy, starting from examples of organisation of physical systems, like laser beams, to the organisation of the whole biosphere through the Gaia concept of Lovelock. The emergent property at one level is in general finding its causality at the subsystems components and the interaction with them (see statements about causality later). An organised form of cell function, stemming from self-organised transformation of cellular compounds, known as hyper-cycling may be considered as an emergent property. At the physiological levels we are dealing with mating behaviour of organisms as results of hormone interactions. Likewise, motion, feelings or intelligent behaviour occur as a consequence of special couplings of neurones. Patterns in the development of ecosystem societies or the ecosystem itself may not be easily predicted from knowledge of organisms alone. The patterns might be stable and have a special balance with the surroundings, repetitive as through Holling cycles, or destructive as it is the case with chaos and catastrophes. Those are just a few examples of the cases we will be dealing with in this chapter.

The history, and some philosophical aspects of the concepts of emergence have been presented. Together, this leads to a framework of an operational definition, which may be used to analyse reports on emergent properties found in the literature. A review of papers giving reports on emergent properties at several levels of hierarchy, spanning from pre- and protobiological systems to the global system, is presented, as it is difficult to exclude any factors as being totally unimportant at the ecological level. This overview allows one to establish arguments for prerequisites and other conditions important to emergence. At the end, a framework making a distinction between collective properties, artefacts, and various types of emergent properties is also established.

2. The concept and its history

As previously mentioned the concept of emergence mainly found its way into ecology through the proposal of E.P. Odum that the study of emergence in ecology should be a 'new integrative discipline' (Odum, 1977). This was probably due to the fact that studies of complex systems had shown that the investigations of the details alone were not adequate or effective in predicting ecosystem function and behaviour. Neither were they sufficient to explain a more advanced pattern like behaviour and performance of ecosystems, such as they for instance were presented in his own '24 principles' observed during the succession of systems from *young, immature* systems towards an *older*, more *mature* state. The ultimate state is often referred to as the *climax society*. Thus, according to his statements *"science should not only be reductionistic"... "but also synthetic and holistic"*.

The concepts of emergent properties and emergence themselves date back to the last century or early this century (Morgan, 1923) and probably relate to views represented by Kantian philosophy. Authors however vary in their historical citations of the concept. Thorough treatments which have a more philosophical point of view are few, like in Emmeche et al. (1993) but several authors have been dedicating papers (Garret, 1942, Lowry, 1974, Morgan, Müller, 1996, Müller et al., 1997, Wicken, 1986, Wieglieb and Bröring, 1996) or sections of books to this topic (Allen and Hoekstra, 1992, Küppers, 1990).

2.1 History

The use of the terms emergence or emergent properties has found widespread use in the biological sciences during the last decades. Especially, the terms that are clearly connected with the growing implementation of the system approach in the ecological sciences and is thus very much related to the works of Eugene P. Odum's brother, Howard T. Odum, whose efforts in systems ecology are widely known (Odum, 1983).

The systems approach resulted from the claim, and the necessity, for a new holistic approach to ecological science in order to deal with the high complexity of these types of systems. The need for a holistic approach was due to the failure of the traditional reductionistic research strategies to explain the properties of ecosystems by the knowledge of the behaviour and the properties of the ecosystem constituents alone. Such viewpoints are often reflected in current literature, e.g. (Green and Bossomaier, 1993, Lowry, 1974). It has become clear that ecosystems are highly complex, medium numbered systems (O'Neill et al., 1986) dominated by non-linear relationships between their constituents. In such systems, things are bound to happen that are not easy to predict from the basic knowledge of the system, no matter how extensive this knowledge is. The traditionally quoted sentence : *"that the whole is more than the sum of the parts"* has been quoted into triviality but nevertheless cannot be avoided in this treatment. One should also keep in mind one important message of the new science of chaos: we do not need complex systems at all in order to observe complex behaviour.

The concept of emergent properties and its corollary, the substantive emergence, have, in accordance with vague definitions, found a widespread and diffuse use in the current biological and ecological literature. According to some scientists the use has been even too diffuse, as seen from the debates that have taken place in literature (Salt, 1979, Edson, et al., 1981). The widespread, but undefined, use has led to criticisms towards the concepts such as being a 'pseudocognathe', having the future risk to be 'lost in the semantic miasma', sharing the destiny of other ecological concepts such as the niche concept (Salt, 1979). So the concepts to a high extent suffer from problems of definition, which shall be seen in the following.

The concept of emergence seems to have far older roots than one may judge from the above, some authors stating the origin to Kantian philosophy from the 18th century, namely Kritik der Reinen Vernunft. Meanwhile, according to Emmeche et al. (1993) quoting C. Lloyd Morgan the actual term was coined by G.H. Lewis as far back as in 1875. A common definition can be found from different authors around this century, namely that *"emergence is the denomination of something new which could not be predicted from the elements constituting the preceding condition".*

The term, as seen, has often been connected with a scent or flavour of mysticism, which may have been caused by the rather vague definition quoted above, and with a use of words connected to subjective observations and emotions, such as sudden appearance, surprise or inexplicability. The clarification and unified formulation of the concept, which makes it applicable in all scientific areas, together with an elimination of the subjective elements, calls upon a more philosophical treatment of the issue which shall be briefly included here.

2.2 Definitory problems

The definitions given in current literature often have an heuristic character when referring to subjective arguments and, being clearly observer dependent, (see Table 2.), are strongly influencing the present use within different scientific areas. Thus, the use of the terms in the current biological and ecological literature reflects the unclear and diverse definitions given in the mentioned treatments.

Table 2: Definitions of the concept of emergence and/or emergent properties found in literature

Definition statement (quotation)	**Reference**
groups of several non-linear interacting elements evolve in space and time displaying new higher order phenomena	Solé et al., 1993
an emergent property of a whole which is produced by properties of its parts but is not qualitatively similar	Harré, quoted in Salt, 1979
an emergent property of an ecological unit is one which is wholly unpredictable from that observation of that unit	Salt, 1979
simple mechanisms begin to 'emerge' in the sense that the composite system can behave in new ways, none of which are characteristics of its constituent elements	Mikuleckey, 1988

The two core problems of the concept are thus to find an objective definition of the concept, which if possible should be valid for all biological systems, and to define a clear materialistic basis to the observed phenomena in order to remove any mystic elements from the observations and interpretations. Such a basis may be found within non-linear thermodynamics.

Few reviews of the concept that concentrate on definitory problems, ontological aspects and related epistemilogical consequences have been found in the literature. Some Danish researchers, Emmeche et al. (1993), have done this quite recently. In Germany, Krohn and Küppers (1992) have made an extensive attempt to relate the concept of emergence to order and organisation. Very few definitions seem to form the core use which the concept has found in the current biological and especially, the ecological literature. At the core, we have Harre's 1972 definition (quoting Salt, 1979) "*Many groups or aggregates have properties that are not properties of the individuals of which they are a collection.*

Such properties are called 'emergent' properties. ...Emergence: the property of the whole is produced by properties of the parts but is not qualitatively similar..."

To this, Salt, (1979), makes the following remark: *"For ecologists, I suggest the following operational definition: 'An emergent property of an ecological unit is one which is wholly unpredictable from observation of the components of that unit' "*. The corollary is: *"An emergent property of an ecological unit is only discernible by observation of that unit itself"*. If the property or properties can be deduced from the components, such properties should be referred to as *collective* rather than emergent.

From this, it directly follows that emergent properties are observable at the systems level alone, and will not, and cannot, be found by following the reductionistic strategies of research in accordance with the normal positivistic ways of performing science. The 'secrets' of emergent properties are to be revealed by holistic research strategies only.

Presenting the concept by the two terms, emergence and emergent properties, it is already clear that a treatment must address two aspects. Firstly, the phenomenon itself – emergence – that something appears, what appears and from what does it appear? What are the components of the new structure? This part of the problem is clearly ontologically dependent.

Secondly, a phenomenological description of how the emergence is expressed, in structure and especially in terms of its new functionality, how is this observed and measured, i.e. the more qualitative aspects of the concept of emergence, which are tightly connected to epistemological questions.

A philosophically comprehensive presentation, dealing with these perspectives, is given in Emmeche et al., 1993. The line in their work follows the definition of Lloyd Morgan's emergence, focusing on one single sentence: the 'creation of new properties'. Thus, the treatment takes its starting point in a discussion of the three words constituting the definition given above: *property, creation and new.* According to the authors, it is possible thereby 'to grasp the primary topics in the concept of emergence'.

Following this scheme, *properties* in general refer to the phenomena observed and being identified as an emergent property, raising the question: '*how particular res general'* the observed phenomenon has to be in order to be identified as being truly emergent. The authors themselves state that *"every new property whenever it is created and every time it is created is emergent"*, but also make the remark that this definition may be broader than most users of the word have intended. This wide and tolerant definition is important in order not to be prejudiced when evaluating emergent properties.

In the treatment it is also pointed out that the concept is generally used in three different ways 'to designate primary levels, sub-levels and aspects of single entities'. The concept of emergence can be used to describe the creation of any of the three types, which are not necessarily identical and leaves us with different implications.

The second issue is the *creation*, where the authors omit a religious, i.e. vitalistic, idealistic, or creationist, interpretation of this word, but restrict themselves to its scientific use. This implies at least two different ways of viewing the concept of emergence – either 1) that the conditions creating the emergent properties exist side by side with the created, emergent properties, a situation to which they refer as the *historical* explanation, or 2) the creation at the same time creates both conditions and product, i.e. the emergent properties. This is referred to as the *structural* explanation. The discussions at this point clearly illustrate an important connection between the emergent properties and a hierarchical view of the biological systems, which we shall come to later.

Highly relevant to biology and ecology seems to be a discussion regarding when, i.e. for what time, the emergent property appears. This leads to the concepts of *primary* and *secondary* emergence. Primary emergence being the first time an emergent property appears. To be conserved the property can be reproduced again and again but in this case as a secondary emergence.

The last word in the definition, – new, is considered to lead to even worse philosophical and terminological problems simply due to the difficulties in answering the question – what is new? Is something new because you haven't heard it before, because you haven't seen it before, because you have not been able to measure it up to now, etc. This point refers to the gain of knowledge. How this knowledge is gained will be discussed later together with the possibilities for measuring emergent properties (see section 6).

Discussing the recent approaches to emergence the authors refer to the works of Cariani and Baas. Cariani (1992) has come up with three notions of emergence: computational emergence, thermodynamic emergence, and emergence relative to a model. The computational emergence deals with the not always predictable patterns produced by different computer programs, e.g. cellular automata systems developing certain patterns out of simple rules from game theory. Thermodynamic emergence covers the establishment of highly complex, self-organised structures and their relations to the non-linear, far from equilibrium thermodynamics of Prigogine and co-workers (Prigogine and Wiame, 1946, Nicolis and Prigoine, 1971, 1977). Emergence relative to a model defines emergence as the deviation of the actual behaviour of a physical system to an observers' model of it.

From the above, we have come out with the following definition to be operational from a viewpoint of ecological systems:

- As emergent properties we will consider properties of a system which are not possessed by component subsystems alone.
- The properties emerge as a consequence of interactions within the system (for the character of these interactions, see section 4.)
- Two fundamental types of interactions are found that may be characterised as intra- and inter-connectedness, i.e. connections within and between levels including controls. This point does not consider the direction of the intra-level interactions.
- The emerged properties a considered 'new' (even occurring repeatedly).
- These new properties appear at one level of a system and are not immediately deducible from observation of the levels or units of which the system consists.
- The emergence or emergent property may be qualitative or quantitative in character and expression, and is a result of quantitative and/or qualitative properties of compartments at any, but usually a lower, level.

3. Emergence and hierarchy

Emergence has been described at many levels of the biological hierarchy. As argued above the reason for emergence is to be found in the structural, often hierarchical organisation, of the system and the quantitative and qualitative character of the 'linkages' within the structure. As biological structures are often complex, this makes it hard to determine the actual cause of emergence (see Discussion): which interconnection(s) exactly, and what level(s) of the hierarchy, are the real determinators of the emergent properties?

These facts also make it hard to tell exactly where ecology does start and when one has to begin the discussion of emergence and emergent properties within context of ecology. Accepting the examples of hypothetical, emergent properties found in literature (for summary see Table 2), it is hard to find an example which will be totally neutral in an ecological context, i.e. which would have no ecological implications at all.

3.1 Prebiological emergence

Several examples of emergent properties can be found in the physical and chemical sciences (Stephens, 1997). The chemical examples, especially those from organic chemistry and biochemistry form the interface of biological examples, such as the self-regulating mechanisms of photosynthesis and related processes which have been argued as emergent properties (Geiger and Servantes, 1994). They form an important border and transition to protobiology and thus evolutionary processes.

Within the area of physics some examples are nearly classical in character. The sense of 'wetness' of water, and many other properties of water, which in many ways is a molecule with complex behaviour, that is unpredictable from knowledge about oxygen and hydrogen alone, have often been mentioned as examples of emergent properties. Likewise, the sense of colours by the eyes is not predictable by knowing a certain wavelength of light. What makes the leaves appear green to us?

Two other famous examples related to self-organised behaviour of systems may also be mentioned, the Bénard cells and the Bhelusov-Zabotinsky- (BZ) reaction. In the case of Bénard cells, during certain specific conditions, hexagonal, convective cells form a fluid when a thermal gradient is imposed on the experimental set-up containment. In the BZ-reaction a special ratio of chemicals causes a chemical mixture to perform a pulsing pattern in colours with a period of about one minute. Both examples have posed a rather fascinating problem to scientists for years and several models have been established in order to understand the reactions.

The structure of these physico-chemical processes, gradients resulting in convective cells, pulsing patterns, together with other observations like the occurrence of Turing structures in chemical fluids, spontaneous formation of lipid coacervates, are important to keep in mind. They might be crucial to our understanding of the emergence of life and the function of earliest, primitive, life forms and therefore of larger importance than first thought.

3.2 Protobiologcial emergence

The appearance of the earliest life forms has often been referred to as emergence in the sense of primary emergence mentioned above. Although many of the properties occurring during this phase of evolution have since then been repeated over and over again they are still to be considered as emergent properties. As examples, one can mention the emergence of life (Nussinov et al., 1997), emergence of animals (McMenamin, 1987) or the emergence of bird feathers from reptile scales (Brush, 1996) as situations of primary emergence.

Many of the examples found in the literature are dealing with the formation of the earliest cells. Biochemical cycles, the organisation and exchanges of information of the genetic material DNA or RNA and the compartmentalisation of the material within membranes are but a few examples. Molecular complementarity, complementarity defined as 'non-random, reversible coupling of the components of a system', has been argued to be a widespread mechanism in biological systems and important for the understanding of the processes lying behind emergent properties (Root-Bernstein and Dillon, 1997). The seemingly (self)organisation of molecules observed in prebiotic systems, such as Turing structures and hypercycles can be seen as emergent properties already at a very low level. Autocatalysis form a basis of the hypercycle type of organisation, e.g. Hogeweg, (1993), see also Müller and Nielsen (this volume).

3.3 Emergence at higher levels of biological systems

Emergent properties really come into play when biological systems are reaching higher levels. This becomes evident already at the cellular level, where cells or groups of cells are communicating with each other as in the case of hormones and natural neural networks. Organs are composed of cells, their individual functions are important only to the organism as a whole. A heart, kidneys or lungs, are vital but worth nothing on their own. Organisms get together and form populations and societies, have properties none of which can be explained by properties of the individual organisms alone. They all go together in what we consider as ecosystems and thus are a part of the biosphere of our planet earth.

3.3.1 Cellular level

Emergence at the sub-cellular level was dealt with in one of the previous sections. Here we will regard what happens when more cells interact in various ways. Not surprisingly many of the studies here have been concentrating on the organisation of neuronal systems (Gilbert and Jorgenson, 1998), which result in unexpected properties like the ability to move, to sense, to be intelligent and to feel. In the biological sense, they will belong to the area of ethology.

The sensory systems, being connected to visual, auditory, or other communicative processes are all playing a major role in how successful we, or other living organisms, are in performing specific life strategies, throughout our life cycles. Reliable senses, and responding the right way to the received stimuli are crucial to the existence of many life forms, in processes like finding food, knowing when and where to escape, or creating bonds to other members of the species, for example, during reproduction.

Senses are complex responses to complex stimuli and emergent properties are bound to exist. In the olfactory bulb of the Salamander, the combined response of the cells belonging to this sensory organ is creating complex patterns as a response to odour stimuli (White et al., 1992). The responses are so far characterised as emergent properties. The interactions meanwhile seem to emerge from one level of interactions, inter-level interactions, – self-organisation, only, not depending on any super-system.

Neural networks, like in our brains, consisting of a huge number of interconnected neurons, are so complex that unforeseen patterns in responses are bound to occur and have also been reported to exist (Fujii et al., 1996). During the evolution of the brain, emergent properties, together with new cell types, local and large circuits have added up to the increasing complexity of brain function (Bullock, 1993). Motor control, the control and co-ordination of motor activity are taken care of by our brain passing on signals to the limbs or organs involved (Higgins, 1985, Mitniski, 1997). Model studies have been used in order to understand how this co-ordination comes about, like in the case of six legged spider robot (Cruse, et al., 1998).

3.3.2 Organ level

More cells, often during morphogenesis differentiated in certain, specialised directions can go together, forming organs, the performance of which is often that of taking up a particular task of the organism, like for instance liver cells secreting enzymes, kidney cells (in the broadest sense) filtering an cleaning the coelom. Although, the formal 'layout' for this is existent in the genetic material of all cells, the eventual determination is occurring during the development of the organism and the actual function of the organs may be viewed as emergent. The brain as an organ may serve as an example of this emergence. Here cells differentiated, with highly specialised physiological properties, go together and create activity patterns that are far more complex than expected from knowing the physiology of neural cells alone. The whole becomes more than the sum of the parts.

3.3.3 Organism level

Complex behaviour occurs among the individual organisms that cannot be determined exclusively by internal factors. The sending, reception and interpretation of signals from interagent organisms, the relationship(s) to the outside, and thus semiotics (Emmeche et al., 1993) are playing an important, if not a major, role in this play, creating patterns impossible to foresee.

In trees, the formation of branches and leaf mosaics have often been studied with modelling approaches. The allocation of resources between above and below ground biomass and the related physiological mechanisms has always been studied this way. A modelling study of this problem indicates a 'complex integrated growth pattern' which may only be understood as an emergent property as it is claimed to have no direct or indirect mechanistic basis and related to sub-cellular activities, e.g. Reggia et al., (1993). In a similar manner Colasanti and Hunt (1997) showed that whole-plant behaviour is an emergent property arising from a rule-based model of the system.

The factors leading to the reproductive success of the European Robin *Erithacus rubecula* were investigated in a modelling study using an individual oriented programming (IOP) approach (Breckling et al., 1997). The implementation of IOP was caused by earlier findings which indicated that an aggregation of the properties was preventing the understanding of individual reaction to the environment (Breckling and Reuter, 1996). The life history of the Robins were shown to emerge from 4 factors: behaviour, selection activity, temperature and availability of food (Reuter and Breckling, 1996).

Communication between individuals, i.e. their social interactions within a population are important to the function of the organism as a whole and at this level indistinguishable from the emergence at the ethological level. As an example, according to Tegeder (1994), emergence of co-operation and communication was correlated to fluctuations in resources and spatial patterns of distribution. Stressing the importance of communication, may lead to an interpretation of the communicative process between individuals as an interpretation of signs, which is described within the area of semiotics. Recently, studies of ecosystem semiotics, where ecosystem flows and feedbacks are interpreted as communicative processes, have been proposed as a fruitful strategy for future ecosystem studies (Nielsen, in print).

3.3.4 Population level

Populations are composed of individual organisms, interacting in a large number of ways, differing in quantity and quality, throughout the biological system. The interaction may vary in character according to the complexity. At the one end of the spectrum, we find the single cell organisms interacting in the ways described above where interactions are mostly material (matter fluxes) in character. At the other end, there are colonial organisms forming complex societies, where brains, senses, memory, (Godsmark and Brown, 1997) and thus informational interaction becomes dominating, not to mention humans, where mind and feelings also become important (Voneida, 1998, Porges, 1998).

Emergent properties as a result of individual level behaviour and interactions in populations of social insects has been argued (Karsai and Wenzel, 1998, Page and Mitchell, 1998). Several studies report on the population level, (Crossey and La Point, 1988, Starfish (Van der Laan and Hogeweg, 1992) and the society level (Solé et al, 1993). Whereas, Clarck, (1986) found that the development and dynamics of a tree population was explained by the life histories of the species in combination with changing environmental parameters, and thus that emergent properties, from this point of view, seemed to be non apparent.

Social insects are extremely obvious objects where emergent properties can be expected with great likelihood. The distribution of food to larvae of the Fire Ant has been argued as emerging from interactions between individuals, workers and larvae (Cassill et al., 1998). Cellular automata models were used to study the short time oscillations in ant colonies (Solé et al., 1993). The non-linear dependencies describing the relationships between, and the movement of, individuals was explaining the behaviour. The oscillations were found to be emergent properties of the colony.

In the same way, patterns of Crown-of thorns starfish outbreaks at the Great Barrier Reef were studied (Van der Laan and Hogeweg, 1992). Investigations were carried out to illustrate whether the waves were caused by the physical, environment, i.e. the current patterns. The results showed that the waves were independent of the current pattern leading to the conclusion that the outbreaks were taking place in a wavelike pattern characterised as an emergent property of the system.

3.3.5 Community level

The productive and respiratory parameters of periphyton communities were studied by Crossey and LaPoint (1988). They concluded that their observations contributed to information about emergent properties that would not be derived from the studies on individual population levels alone. The studies were carried out at polluted sites exposed to pollutants comparing non polluted to exposed sites. While parameters frequently measured, like biomass increased and other parameters did not change significantly compared to the controls. A decrease in the P/R-ratio was the only parameter indicating a change in the functionality of the community affected by pollutants. In another study of the responses of an inter-tidal, algal community were investigated and it was concluded that variation in responses could be explained by the 'life history characteristics of component species rather than emergent properties'.

3.4 Emergent properties of ecosystems

Ecosystems are inherently complex as they are composed of an embedded hierarchy of all the previously mentioned subsystems. Emergence is to be expected, but surprisingly few reports exist at this level, in spite of the recommendations of E.P. Odum many years ago.

Emergent properties have been discussed at levels that can be characterised as ecological by only a few authors: Microcosms (Wilson and Botkin, 1990, Drake, 1991), Forest ecosystems (Prentice and Leemans, 1990), predator-prey relationships (Fauth and Resetarits Jr., W.J., 1991) and Foodwebs (Winemiller, 1990). Drake (1991) observed the development in the organisation of aquatic communities by manipulation of the invasion of various assemblages of the system, thereby giving different possibilities of the outcome. The spatio-temporal differences in outcomes may be considered as emergent properties of the system.

A coupling of mathematical and experimental models, i.e. microcosm studies of planktonic societies was used for putting new lengths on the existence of emergent properties (Wilson and Botkin, 1990). The authors using the definition of Salt (1979), simply put up two questions to be answered by their experiments. The questions were if patterns of species' abundance could be predicted from knowledge of their requirements, and whether the nutrient content of the biomass could be predicted from abundance? Their result was that community and ecosystem level properties could not be predicted from knowledge of single species alone and therefore should be characterised as emergent. The same conclusions were made by Sousa (1980), studying an inter-tidal algal community, since 'community responses an non-linearities in responses were attributable to life history characteristics' as in the previous example. The dynamics of a forest ecosystem in Sweden was studied by Prentice and Lemans (1990). The requirement of disturbance in this type of system resulting in multiple tree-gaps to be produced was seen as an emergent property of that very vegetation. Predator-prey relationships have been analysed in a study by Fauth and Restetarits, 1991, asking the question if the property of being a keystone predator was also an emergent property of that species.

Trophic chains and networks are even more complex systems, including the possible interactions of all mechanisms mentioned above. Trophic networks have been claimed to posses unique emergent properties (Winemiller, 1990), which has a spin-off of what is named potentially informative macrodescriptors.

As complex systems, ecosystems behaviour is often analysed through modelling studies. The relation to emergent properties becomes clear when looking at recent efforts of structural dynamic modelling, where the changes in ecosystem composition and structure over time are analysed. Other examples may be the works of B.C. Patten on the propagation of energies through the ecosystem network, leading to the discovery of the importance of *indirect effect*, *quantitative* and *qualitiative utility* of the system, results that are highly surprising and unexpected, and as such are candidates of emergent properties (see Patten and Jørgensen, 1995). Both examples link to the higher level information expressions such as ascendency (Ulanowicz, 1986), different kinds of entropy or information derived descriptors such as exergy (see Jørgensen, this volume).

The ability of the ecosystem to perform with systematic and directional, and predictable changes in some macroscopic characters, not predictable from knowledge about the ecosystem members alone has been known since 1967 from the 24 principles of ecosystem development during succession in the second edition of E.P Odums 'Fundamentals of Ecology'. Many other factors, known as indicators, orientors or goal function have been presented since then. A more elaborate list is given in Table 3.

The story might not stop at this level, but continue to even higher levels of hierarchy when considering the regulating mechanism as proposed in Lovelock's Gaia Theory (Lovelock, 1989).

4. How emergence emerges

The concept of emergent properties refers very clearly to, and must be seen in tight connection with, at least two other concepts often occurring in literature on modern ecosystem theory. Those are the concepts of hierarchy and self-organisation (Müller and Nielsen, this volume). In connection with hierarchy, the emergent properties are seen as outcomes of a certain kind of organisation where super-systems are formed with subsystems as constituents and where the properties are observable at the super-system level only. Here the emergent property is an outcome of a certain way of organisation. On the other hand the ability of biological systems to arrange themselves in a special manner, e.g. in a hierarchical way, is in itself a property which emerges as a consequence of the properties of its constituents, but the organisation and the function for sure cannot always be foreseen. Thus, the capability of self-organisation can be seen as an emergent property itself. According to Emmeche et al. (1993) the notion of self-organisation was needed to fill a gap left when it was recognised that neo-Darwinism alone was not able to explain the complex organisational patterns shown by biological systems.

4.1 Some pre-requisites for emergence

Looking at the above descriptions of properties argued to be emergent in character allows us to make a list of prerequisites that are, if not compulsory then, adding up to the increasing probability of the occurrence of emergence.

- Structure of systems - Emergent properties result from a system's structure and the relations to the external and internal environment. The systems are performing with a certain order, that is often self-organised as a result of the thermodynamic relationships to the immediate environment, i.e. the material, energetic and informational fluxes.

- Organisation - The structures or systems are organised/ordered from the material and energetic fluxes. This deals with inter level organisations, as well as the following intra-level relations. The systems are often showing a hierarchical form of organisation where systems are themselves subsystems of a super-system, that in itself is a subsystem and so on. Hierarchical organisation in definitely an important if not crucial condition of emergent properties to emerge, e.g. Valentine and May, (1996).

Table 3: The emergent properties of ecosystems

IMMATURE STATE	MATURE STATE
Properties of the dominating species	
rapid growth	slow growth
r-selection	K-selection
quantitative growth	qualitative development
small size	large size
short life spans	long life spans
broad niches	narrow niches
Properties of production	
small biomass	large biomass
high P/B ration	low P/B ratio
low respiration	high respiration
small gross production	medium gross production
Properties of nutrient flows and cycles	
simple, rapid and leaky	complex, slow, and closed cycles
small storage	large storage
extrabiotic	intrabiotic nutrient distribution scheme
small amounts of detritus	large amounts of detritus
rapid nutrient exchange	slow nutrient exchange
short residence times	long residence times
minor chemical heterogeneity	high chemical heterogeneity
loose network articulation	high network articulation
low diversity	high diversity of flows
undeveloped symbiosis	developed symbiosis
Properties of the community	
low diversity	high diversity
poor feedback control	developed feedback control
poor spatial patterns	developed spatial patterns
Thermodynamic and integrative system properties	
poor hierarchical structure	developed hierarchical structure
close to equilibrium	far from equilibrium
low exergy storage	high exergy storage
small total entropy production	high total entropy production
high specific entropy production	small specific entropy production
small level of information	high level of information
small internal redundancy	high internal redundancy
small.path lengths	high path lengths
low ascendancy	high ascendancy
poor indirect effects	developed indirect effects
small respiration and evapo-transpiration	high respiration and evapo-transpiration
small energy demand for maintenance	high energy demand for maintenance

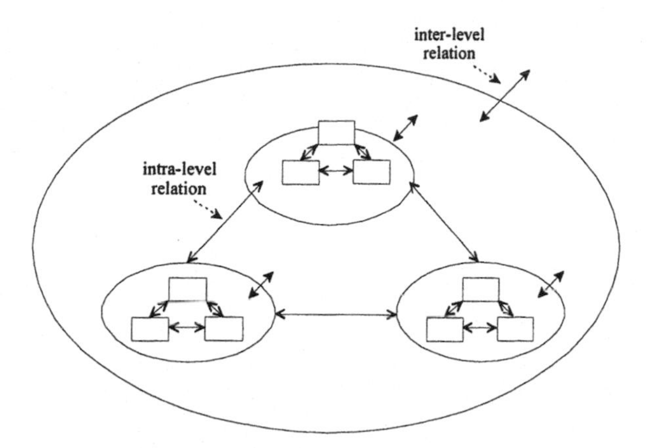

Figure 1: Biological entities are often organised in a hierarchical manner; in this case the subsystems are shown in internal relations and even some self-similarity between levels exists.

- Interaction(s) - The system, in addition to its component elements, exhibits a variety of interactions, varying in quantity as well as quality. The interaction may be inter-level in character as with molecular complementarity, autocatalysis, or the hyper-cycling of Eigen (1971), where the emergent structure only serves a functional purpose and not in itself forms a new higher structure (see Figure 1). As a result of the hierarchical form of organisation, interactions between hierarchical levels of organisation also occur, referred to as intra-level interaction (see Figure 2). In general, the more interactions the higher the probability of emergence
- Communication - Interactions may vary. In general, the more sophisticated the communication the higher chance of emergence. Communication is hereby understood very widely as a phenomenon that may be carried out by chemical compounds as well as signal processing, auditory, visual, etc., including the corresponding interpretation of the processes.

The above statements are far from being complete as one has to remember that even simple structures may perform with unexpected complexity.

More conditions might add up to the chances that emergent properties will appear. Among other things, instabilities seem to be an important factor, especially to evolutionary emergence. The numbers of such instabilities are many as they may occur in the form of 'stable' oscillations, as limit cycles or Holling cycles, but also more unpredictable situations like chaos and catastrophes. The unpredictability of these situations partly makes them emergent properties themselves, which will bring us close to a tautology and thus into definitory problems.

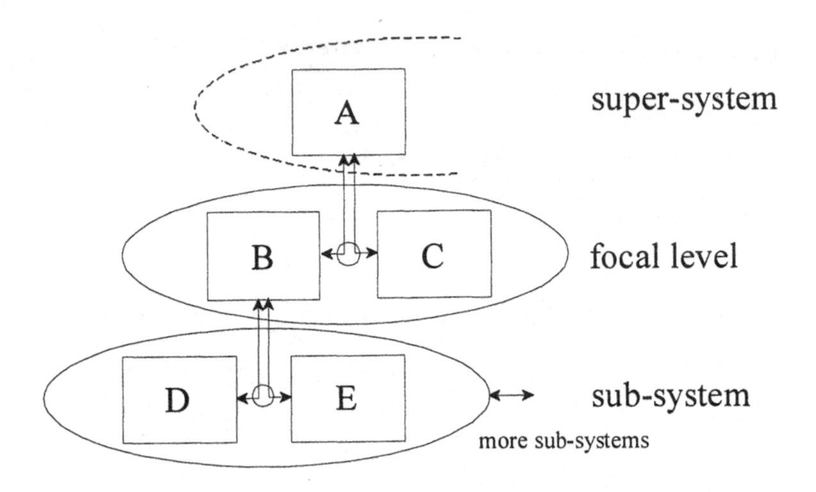

Figure 2: Interactions at one level of hierarchy, intralevel, propagates to higher levels through interlevel interactions. Coarse graining is taking place as one moves up the hierarchy and lower level phenomena are levelled out.

Even stable periods may lead to emergence of new structures through bifurcations following the theories of Prigogine and Wiame (1946) and Nicolis and Prigogine (1977). As systems move toward the state of minimum dissipation they are, at the same time, moving towards bifurcation points with possibilities of further evolution to occur. Likewise broken symmetries, complementarity has been proposed as a global mechanism (Stephens, 1997).

The actual causes of emergence or emergent properties may be hard to give as the causality is hardly ever unambiguous. Meanwhile, the cause, in the Aristotelian sense, (see e.g. Ulanowicz, 1997) are mainly *material* and *efficient* in character. The material cause is found in the building blocks of life and the efficient cause lies in the agents and the interacting network. When it comes to the *formal* cause, it becomes difficult to accept that any plan or blueprint for the emerged properties exists. Rather, emergence is a consequence of self-organisational processes where 'chance and necessity' come into play together. There seems to be no 'goal' and thus no *final* cause, which does not mean no direction, for whatever property is actual emerging, other than a demand for the emerging structure to fit the conditions under which it emerged, and we do not see emergent properties as teleological entities. Causes are always to be found in the interagent compartments, not considering level and their interactions. Emergent properties without any causality do not exist! Whether the causality is identifiable to us does not affect this argument. There may be many reasons for the cause not to be identifiable.

This will probably mean that holistic approaches, like modelling, will be the only strategy to study emergent properties if also pursuing causality, e.g. Lloyd and Gould, (1993), Ball and Gimblett, (1992). This is partly also indicated by the many modelling studies dealing with this matter.

4.2 Emergent properties and methods

We have briefly mentioned the complexity of systems, which is not a mandatory pre-requisite to emergence but adds up to the likelihood of its occurrence. At the same time, complexity reduces our possibilities to investigate whether a property is really emergent or not. As a consequence, we have argued the chances of reductionism to be very small, if not non-existent, in order to solve this debate.

As a consequence many authors have taken up modelling in various forms to analyse complex behaviour at more or less all the levels mentioned above, e.g. the individual oriented population models of the European Robin, *Erithacus rubecula*, (Reuter and Breckling, 1996) or the seemingly 'co-ordinated' movement of fish schools (Breckling and Reuter, 1996). Many other examples will be found throughout this text.

Meanwhile, there seems to be no clear strategy for studying emergence or emergent properties. Therefore, evaluations of properties that seem quite similar, may come out with different results. That is, the result of the evaluation(s) become dependent on methodology. A more objective methodology therefore needs to be developed.

5. Emergent properties and management

Many of the problems we face in management today are related to emergent properties as they are dealing with the unpredictable behaviour of nature and its unexpected responses to the way we interact with it through exploitation, pollution, etc. Our chances of understanding these 'unpleasant', emergent, events better are small if we do not turn to untraditional ways of studying nature, as illustrated by the network analysis introduced by Patten and described in brief above. For a relatively recent introduction to the area see for instance Higashi and Burns (1991), It also follows that our chances to deal with management problems and eventually improve our future interaction with nature, in terms of eliminating or minimising, unwanted effects of our actions, totally relies on an improved understanding of the mechanism underlying this emergence.

Agriculture, for instance, represents a highly specialised way to manage nature and forms fundamental units within most human societal structures today. The ultimate goal of the sector is the food for the population which makes it indispensable to any nation. The specialised and balanced way of production, called the 'syndrome of production' (Vandermeer, 1997), found in farming countries, is more or less unique and may serve as emergent property. This in turn means that attempts to shift to other production equilibria is bound to pose problems, with regards to stability, in terms of 'dramatic and unpredictable' shift from one syndrome to the other (Vandermeer, 1997).

The possible existence of emergent properties, taken as maybe the simplest definition of all – properties we do not understand – contain a severe warning to our politicians and managers. In spite of the severity of the warnings, it is our opinion that the area has been vastly overlooked. In the future the warnings should be looked upon more seriously than they have been hitherto.

6. Discussion

From the presentation of the concepts above it can be seen that emergence and emergent properties will not easily find a clear, consistent and unifying definition for covering all the cases described. The widespread and 'loose' use of the concepts over a vast range of areas at a first glimpse simply seems too confusing to achieve this. One may perhaps ask the question if emergent properties exist at all or whether they simply emerge as a result of the complexity of the problem and the way it is addressed.

Meanwhile, it seems possible to establish some typology of the areas where, and the ways in which, the concepts have been used, following, Müller (1996), partly building on Allen and Star, (1982).

- Emergent properties are observer dependent and thus related to the classical debates about subject-object relations. On one hand, statements of surprise or unexpectancy are highly subjective in character. On the other, the use of proper methodology, linking and relating emergence to the process of observation might clear the way for objectifying the use of the concepts.

- This brings in the scale problem which relates to problems of scale not only in space but also in time, or in short the graining of the system. Emergent properties may appear or vanish as time and space scales are changed, as one moves from one resolution to the other, up or down in the spatio-temporal hierarchy.
- The properties are not to be derived from the subsystems alone, i.e. the properties are not possessed by the subsystems alone and deducible from *a priori* observation. This again brings in the role of time. In this case, time seems to be involved in two different manners.
- First, emergent properties might appear through evolution of the systems, primary emergence, hereafter only being repeated. This may be called evolutionary emergence. This involves time, short term or long term, depending in time units of the systems behaviour.
- As structures go together, new organisational forms, as previously mentioned often hierarchical, are occurring (hierarchical emergence). Time may enter into another form as emergence becomes dependent to observe a spatial structure and to determine the difference in function, i.e. the emergent property itself.

All four points have been modified, the first three taken from Allen and Star (1982) the fourth from Müller (1996).

Taking the view that emergent properties do exist and that the reductionistic approach to science will not (dis)solve the problem so it eventually disappears may allow us to establish a schematic relationship between the various categories of use (see Figure 4).

One major line follows a direction of research problems, the search for the unexplained and not understood. This lies close to using the study of emergent properties as research strategy and at the extreme leads to the reductionist approach. This is more or less the situation at the second line, where properties are collective, i.e. that the properties **are** the sum of the whole, and may be explained at subsystem level, provided sufficient knowledge exists. At the other end, the attitude that only holistic studies will lead to increased understanding might be taken.

Along the third line, we find the core of emergence, and following the above four points they may be divided in an evolutionary line and in a hierarchy of organisation. That is, emergence represented as basically a function of time and space. The evolutionary process was described above and deals with primary and secondary emergence. The organisational, hierarchical line includes four areas described in the previous sections: *global emergent properties* as a function of local rules and local interactions, *thermodynamic emergent properties* dealing with the emergence as a consequence of mainly the second thermodynamical laws, the emergence of (dissipative) structures as a result of thermodynamical gradients. *Computational emergence* are global patterns emerging form local rules. As mentioned above, emergent properties are also appearing as a result of models being used to analyse the problem, which is called *emergent property relative to a model* (for more discussions of these typologies, see. Müller et al, 1997, Müller, 1996).

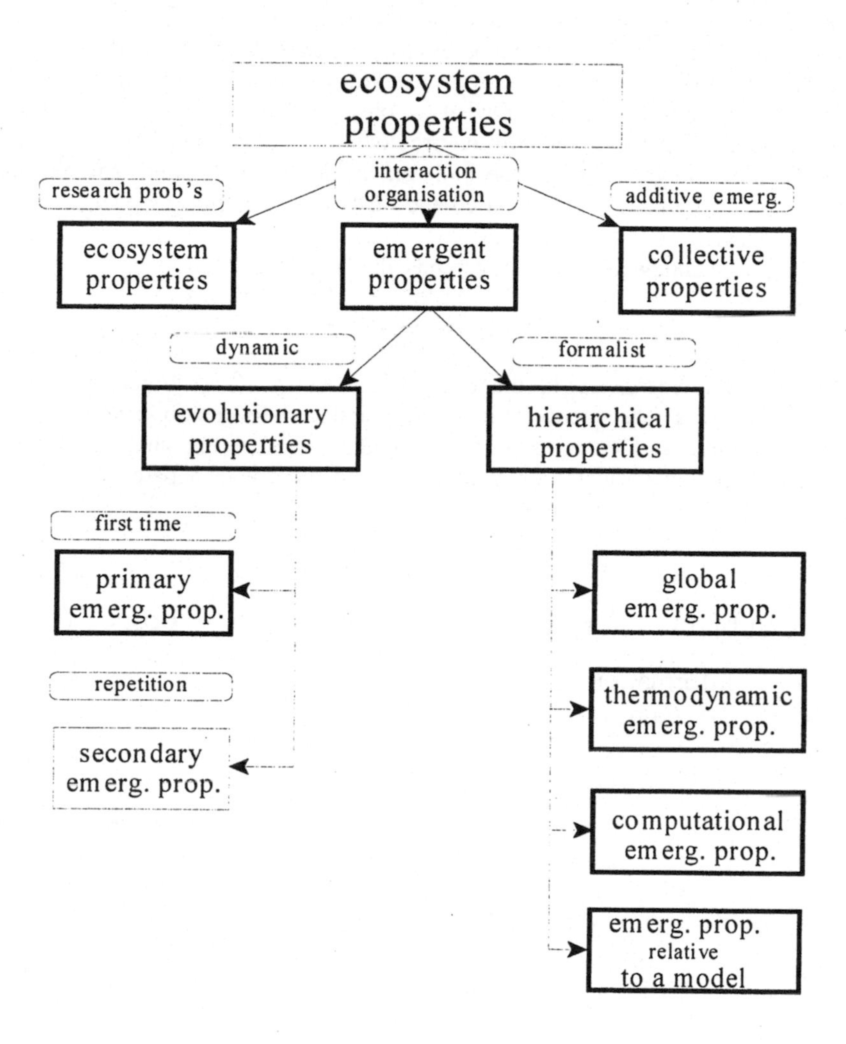

Figure 3: An attempt to group and form a typology of the various definitions and descriptions of emergence and emergent properties found in literature (after Müller et al., 1997, Müller, based on Cariani, 1992, Emmeche et al, 1993, Salthe, 1985, 1993).

To round out the discussion, it is tempting to raise the question, if it were possible to reach a full objectification of the concepts, i.e. to quantify emergent properties. Several authors argue that any attempts to formalise the concepts (Gunji et al., 1997), or discuss that true emergent properties should be observer independent. This does not necessarily mean that emergent properties should be observation independent. Observations undertaken by different methods result in differences in acquired knowledge. This means that emergent properties can be defined as the difference in knowledge gained by the observation of a system by two different methods. This is partly reflected by the computational emergence described by several authors in Emmeche et al., 1993.

It is this observer dependency that leaves a way open for the quantification of emergent properties. Emergent properties could then be expressed in a semi-quantitative way by the use of an 'index' derived of Kullbacks measure of information. This involves moving the normal reference frame in information theory assuming the *a priori* knowledge of the system to be zero, which is not necessarily the case (Nielsen, in print).

Rather in ecology, we do possess some knowledge about the system and what we usually refer to are the deviations in what we observe in the systems or models of systems compared with our expectations built on previous knowledge (e.g. Drake, 1991). The way of quantifying emergence has to be built on the use of computers and models. If our knowledge gained hitherto is synthesised and treated in a computer model (from traditional ecological science) is p^*, and the outcome of an experiment or observations of a system differs by p^{**} the emergent properties can be calculated by the following

$$emergence = \sum p^{**} \ln \frac{p^{**}}{p^*}$$

which correlates emergence to the concept of exergy, (see Jørgensen, this volume). Emergence is now a consequence of information gained between observations.

The question is if emergence in this manner will, at the end, dissolve itself and disappear as knowledge increases, which refers to the above debate of reductionism vs. holism.

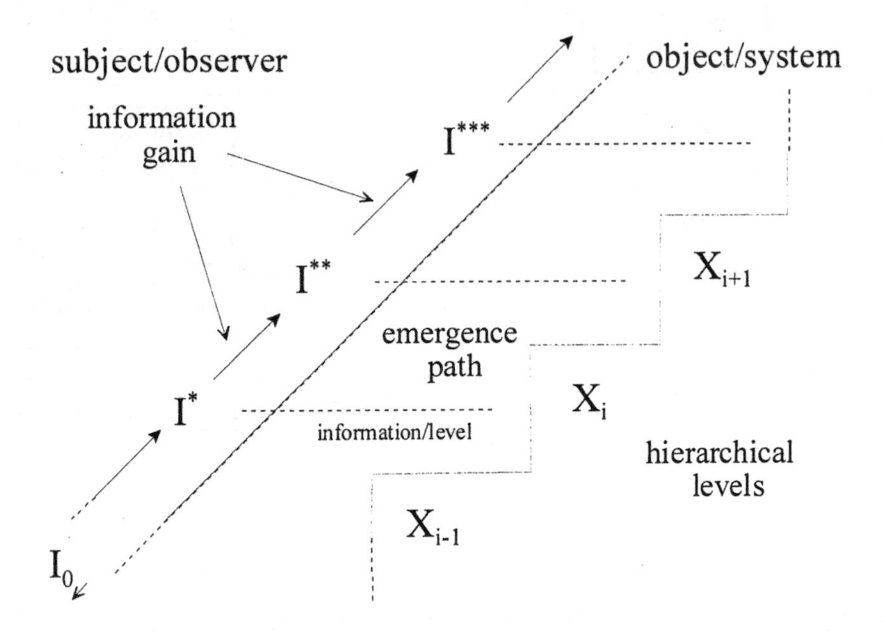

Figure 4: Quantification of emergence, based on Kullbacks measure of information, might be carried out from quantifying the difference between actual observed, a posteriori, behaviour or composition of a system and what may be predicted from a priori knowledge about subsystems. The analysis may be carried out at various levels of hierarchy, differing in emergence value.

As pointed out by Salt (1979), many of the concepts used to characterise ecosystems are based on various numerical treatments of data observed on the ecosystem. Since the concepts are immediately deducible (calculable) from certain knowledge about the components of the ecosystem, e.g. numbers, species, biomass etc. such concepts cannot be coined as emergent property but rather as a *collective* property of the system. An interesting corresponding analogue in this context is the macroscopic properties from thermodynamics such as entropy and parallels in formulation of formulas. Reductionism cannot win the debate since it will be impossible to achieve enough knowledge. If not for anything else, then for thermodynamical reasons, since the achievement of more and more detailed knowledge becomes more and more expensive in terms of energy but also dissipation.

Meanwhile, what strikes is that such a traditional, vertical organisation of systems is not mandatory in order to produce emergent behaviour. Vertical, here, refers to levels being either higher or lower in the hierarchy. Rather only parts are needed, of which none have actual regulatory functions and therefore should be evaluated or ranked higher than the other(s). Emergent properties can occur also in horizontally organised systems, emergence appearing alone as a consequence of interactions at the same level. The study of these intra-level relationships and their consequences to the higher levels in the hierarchy may be important to investigate in the future.

7. Conclusions

In the above, we have used a presentation of some definitory problems together with examples of what has been presented as emergent properties of a series of systems spanning from prebiological to ecosystem examples. In spite of the definitory problems, it has been possible to distinguish classes of emergent properties from mere artefacts and collective properties. The properties may be divided as having emerged as either evolutionary emergence or hierarchical emergence.

Emergent properties emerge as a consequence of a set of prerequisites meeting a set of the right conditions, much like hazard and necessity. Hazard being the spontaneous, self-organisation of compounds and connections, a set of internalities interacting with the adequate, sufficient and necessary environment. The environment is necessary in terms of delivering the thermodynamical gradients, sources as well as sinks. As a consequence, self-organisation is emerging.

Acknowledgements

We wish to thank Simone Bastianoni and S.E. Jørgensen for careful reading and criticism of the manuscript. We are also grateful to our colleagues Broder Breckling, Mike Bredemeier, Volker Grimm, Horst Malchow and Ernst-Walter Reiche with whom a co-authored paper emerged after many inspiring discussions on this topic.

REFERENCES

Aguda, B.D., 1996. Emergent properties of coupled enzyme reaction systems. 1. Switching and clustering behaviour. *Biophysical Chem.* 61: 1-7.

Allen, T.F.H. and Hoekstra, T.W., 1992. Toward a unified ecology. Columbia University Press, 384 pages.

Allen, T.F.H. and T. B. Starr, 1982. Hierarchy: Perspectives for Ecological Complexity. University of Chicago Press, Chicago, IL.

Ball, G.L. and Gimblett, R., 1992. Spatial dynamic emergent hierarchies simulation and assessment system. *Ecol. Modelling.* 62: 107-121.

Baretta-Bekker, J.G., Baretta, J.W. and Ebenhöh, W., 1997. Microbial dynamics in the marine ecosystem model ERSEM II with decoupled carbon assimilation and nutrient uptake. *J. Sea. Res.* 38, 195-211.

Beckner, M., 1968. The Biological Way of Thought. University of California Press.

Bernstein, H., Byerly, H.C., Hopf, F.A. and Michod, R.E., 1985. Sex and Emergence of Species. *J. Theor. Biol,* 117, 665-690.

Bhalla, U.S. and Iyengar, R., 1999. Emergent Properties of Networks of Biological Signalling Pathways. *Science*

283, 381-387.

Breckling, B. And Reuter, H., 1996. The Use of Individual Based Models to Study the Interaction of Different Levels of Organization in Ecological Systems. *Senckenbergiana maritima* 27(3/6): 195-205.

Breckling, B., Reuter, H. and Middelhoff, U., 1997. An object oriented modelling strategy to depict activity pattern of organisms in heterogeneous environment. *Env. Modelling and Assessment* 2: 95-104.

Brush, A.H., 1996. On the origin of feathers. *J. Evol. Biol.* 9, 131-142.

Bullock, T.H., 1993. How are more complex brains different. *Brain Behav. Evol.* 41: 88-96.

Buzsáki, G., 1991. The thalamic clock: emergent network properties. *Neuroscience* 41(2/3): 351-364.

Calderone, N.W., 1998. Proximate mechanisms of age polyethism in the honey bee, Apis mellifera L. *Apidologie* 29, 127-158.

Cariani, P. 1992. Emergence and artificial life - In: Langton, G., Taylor, C., Farmer, J.D. and Rasmussen, S. (eds.) *Artificial life II.* Addison-Wesley, Redwood City, 775-797

Cassil, D.L., Stuy, A. and Buck, R.G., 1998. Emergent Properties of Food Distribution Among Fire Ant Larva. *J. theor. Biol.* 195, 371-381.

Cheeseman, J.M., 1993. Plant growth modelling without integrating mechanisms. *Plant, Cell and Environment,* 16: 137-147.

Clark, J.S., 1986. Coastal Forest tree populations in a changing environment, Southeastern Long Island, New York. *Ecol. Monogr.* 56(3): 259-277.

Colasanti, R.L. and Hunt, R., 1997. Resource dynamics and plant growth: a self-assembling model for individuals, populations and communities. *Funct. Ecol.* 11: 133-145.

Conrad, M. And Hastings, H.M., 1985. Scale Change and the Emergence of Information Processing Primitives. *J. Theor. Biol.* 112, 741-755.

Conrad, M. And Rizki, M.M., 1989. *The artificial worlds approach to emergent evolution. BioSystems* 23: 247-260.

Crossey, M.J. and La Point, T.W., 1988. A Comparison of periphyton community structural and functional responses to heavy metals. *Hydrobiologia* 162: 109-121.

Cruse, H., Kindermann, T., Schumm, M. Dean, J. and Schmitz, J., 1998. Walknet - a biologically inspired network to control six-legged walking. *Neural Networks* 11: 1435-1447.

DeAngelis, D.I., 1995. The Nature and Significance of Feedback in Ecosystems. In: Patten, B.C. and S.E. Joergensen (eds.). *Complex ecology: The part-whole relation in ecosystems.* Englewood Cliffs, 450 - 467

DeBoer, R.J. and Perelson, A.S., 1991. Size and Connectivity as Emergent Properties of a Developing Immune Network. *J.theor. Biol.* 149: 381-424.

DiMichele, W.A. and Phillips, T.L., 1996. Clades, ecological amplitudes, and ecomorphs: phylogenetic effects and persistence of primitive plant communities in the Pennsylvanian-age tropical wetlands. Palaeography, Palaeoclimatology. *Palaoecology,* 127, 83-105.

Drake, J.A., 1991. Community-assembly mechanics and the structure on an experimental species ensemble. *Am.Nat.* 137(1): 1-26.

Ebenhoeh, W., 1996. Stability in Models Versus Stability in Real Ecosystems. *Senckenbergiana mraitima,* 27(3/6), 251-254.

Edson, M.M., Foin, T.C. and Knapp, C.M., 1981. Emergent properties and ecological Research. *Am.Nat.* 118: 593-596.

Eigen, M. 1971. Self-organization of matter and the evolution of biological macromolecules. *Naturwiss.* 58:465-523.

Emerson, S.B., 1996. Phylogenies and Physiological Processes – The evolution of sexual dimorphism in southeast Asian frogs. *Syst. Biol.,* 45(3), 278-289.

Emmeche, C., Køppe, S. and Stjernfelt, F., 1993. Emergence and the ontology of levels. In search of the unexplainable. Arbejdspapir. Afdeling for litteraturvidenskab. University of Copenhagen.

Fauth, J.E. and Resetarits Jr., W.J., 1991. Interactions between the salamander Siren intermedia and the keystone predator notophtalmus viridescens. *Ecology,* 72(3): 827-838.

Fujii, H., Ito, H., Aihara, K., Ichinose, N. and Tsukuda, M., 1996. Dynamic Cell Assembly Hypothesis - Theoretical Possibility of Spatio-temporal Coding in the Cortex. *Neural Networks,* 9(8), 1303-1350.

Funtowicz, S. and Ravetz, J., (in print). Emergent Complex Systems. Submitted to Futures.

Fussy, S., Grössing, G. and Schwabl, H., 1997. A simple model for the evolution of evolution, *J. Biol. Syst.,* 5(5), 341-357.

Gaedke, U., Barthelmess, T. And Straile, D., 1996. Temporal Variability of Standing Stocks of Individual species, Communities, and the Entire Plankton in Two Lakes of Different Trophic State: Empirical Evidence for Hierarchy Theory and Emergent Properties? *Senckenbergiana maritima,* 27(3/6): 169-177.

Garnett, A.C., 1942. Scientific method and the concept of emergence. *J. Philosophy,* 39(18): 477-486.

Geiger, D.R. and Servaites, J.C., 1994. Dynamics of self-regulation of photosynthetic carbon metabolism. *Plant Physiol. Biochem.,* 32(2), 173-183.

Germana, J., 1996. A Transactional Analysis of Biobehavioral Systems, Integr. *Physiol. Behav. Sci.,* 31(3), 210-218.

Gilbert, S.F. and Jorgensen, E.M., 1998. Wormholes: A Commentary on K.F. Schaffner's 'Genes, Behavior, and Developmental Emergentism'. *Philosophy of Science,* 65, 259-266.

Godsmark, D. And Brown, G.J., 1997. A computational model of auditory organization II: Grouping by emergent properties. *Brit.J. Audiol.,* 31, 117.

Green, D.G. and Bossomaier, T. (eds.), 1993. Complex Systems: From Biology to Computation. IOS Press.

Gunji, U.-P., Ito, K. And Kusunoki, Y, 1997. Formal model of internal measurement: Alternate changing between recursive definition and domain equation.

Halloy, S.R.P. and Mark, A.F., 1996. Comparative leaf morphology spectra of plant communities in New Zealand, the Andes and the European Alps. *J. Royal Soc. New Zealand*, 26(1): 41-78.

Harding, S.P. and Lovelock, J.E., 1996. Exploiter-mediated Coexistence and Frequency dependent Selection in a Numerical Model of Biodiversity. *J. Theor. Biol.*, 182, 109-116.

Harris, G.P., 1998. Predictive models in spatially and temporally variable freshwater systems. *Austr. J. Ecol.*, 23, 80-94.

Henle, P., 1942. The status of emergence. *J. Philosophy*, 39(18): 486-493.

Herz, A.V.M., 1994. Collective Phenomena in Spatially Extended Evolutionary Games. *J. Theor. Biol.*,169: 65-87.

Higashi, M. and T. Burns 1991. Theoretical studies of ecosystems: The network perspective. Cambridge University Press. Cambridge

Higgins, S., 1985. Movement as an emergent form: Its structural limits. *Human Mov. Sci.*, 4, 119-148.

Hogeweg, P., 1993. As Large as Life and Twice as Natural: Bioinformatics and the Artificial Life Paradigm. In: Green, D.G. and Bossomaier, T. (eds.), 1993. *Complex Systems: From Biology to Computation.* IOS Press, pages 2-11.

Johnson, L., 1995. The Far-from-Equilibrium Ecological Hinterlands. In: Patten, B.C. and S.E. Jørgensen (eds.) *Complex Ecology. The part-whole relation in Ecosystems.* Prentice Hall. pp 51-103

Karsai, I. and Wenzel, J.W., 1998. Productivity, individual-level and colony-level flexibility, and organisation of work as consequences of colony size. *Proc. Natl. Acad. Sci.*, 95, 8665-8669.

Karsai, I., Pénzes, Z. And Wenzel, J.W., 1996. Dynamics of colony development in Polistes dominulus: a modelling approach. *Behav. Ecol. Sociobiol.*, 39: 97-105.

Krink, T. and Vollrath, F., 1998. Emergent properties in the behaviour of a virtual spider robot. *Proc. R. Soc. Lond. B.*, 265, 2051-2055.

Krohn, W. and Küppers, G., 1992. Emergenz: Die Entstehung von Ordnung, Organisation und Bedeutung. Suhrkamp-taschenbuch. 984. 414 pages.

Küppers, B.O., 1990. Information and the Origin of Life. MIT Press, 215 pages.

Lauder, G.V., 1981. Form and Function: structural analysis in evolutionary morphology. *Paleobiology*, 7(4), 430-442.

Lauder, G.V., 1982. Historical Biology and the Problem of Design. *J. Theor. Biol.*, 97, 57-67.

Lee, D.H., Severin, K. And Ghadiri, M.R., 1997. Autocatalytic networks: the transition from molecular self-replication to molecular ecosystems. *Curr. Opinion in Chem. Biol.*, 1, 491-496.

Lee, M.S.Y., 1996. On the emergent properties of species and ecosystems (A response to Johnson). *J. Nat. Hist* 30(4): 629-631.

Lennon, J.J., Turner, J.R.G. and Connel, D., 1997. A metapopulation model of species boundaries. *OIKOS*, 78: 486-502.

Lloyd, E.A. and Gould, S.J., 1993. Species selection on variability. *Proc. Natl. Acad. Sci.* USA. 90, 595-599.

Lomnitz, C., 1997. Mexico, San Francisco, Los Angeles and Kobe: What Next? *Natural Hazard*, 16(2-3), 287-296.

Loucks, O.L., 1985. Looking for Surprise in Managing Stressed Ecosystems. *BioScience*, 35(7): 428-432.

Lovelock, J.E., 1989. Geophysiology. Trans. Royal Soc. Edinburgh. *Earth Sciences*, 80: 169-175.

Lowry, A., 1974. A note on emergence. *Mind* 83:276-277.

McMenamin, M.A.S., 1987. The Emergence of Animals. *Sci. Am.*, 256(4): 84-92.

Marzluff, J.M. and Balda, R.P., 1988. The advantages of, and constraints forcing, mate fidelity in Pinyon Jays. *The Auk*, 105: 286-295.

Matsuno, K., 1997. Biodynamics for the emergence of energy consumers. *BioSystems*, 42: 119-127.

Mitnitski, A.B., 1997. Kinematic models cannot provide insight into motor control. *Behavioral and Brain Sciences*, 20(2), 318-319.

Morgan, C.L., 1923. Emergent Evolution. Williams and Norgate.

Moreno, A., Etxeberria, A. And Umerez, J., 1993. Semiotic and interlevel causality in biology. *Riv. Biol.*, 86(2): 197-209.

Müller, F., 1996. Emergent properties of ecosystems - consequences of Self-Organising processes? *Senckenbergiana maritima*, 27(3/6): 151-168.

Müller, F., 1997. State-of-the-art in ecosystem theory. *Ecol. Modelling*, 100: 135-161.

Müller, F., 1998. Gradients in ecological systems. *Ecol. Modelling*, 108: 3-21.

Müller, F., Breckling, B., Bredemeier, M., Grimm, V., Malchov, H., Nielsen, S.N. and Reiche, E.W., 1997 Emergente Ökosystemeigenschaften. *Chap. III-2.5 in Handbuch der Umweltwissenschaften.* Ecomed

Nagel, E., 1950. Mechanistic explanation and organismic biology. *Philos. Phenom. Res.* 11: 327-38.

Nicol, C.J., 1996. Farm animal cognition. *Animal Science*, 62, 375-391.

Nicolis, G. and I. Prigogine. 1977. Self-Organization in Non-Equilibrium Systems: from dissipative structures to order through fluctuations. Wiley Interscience, New York, N.Y.

Nussinov, M.D., Otroshchenko, V.A. and Santoli, S., 1997. The emergence of the non-cellular phase of life on the fine-grained clayish particles of the early Earth's regolith. *BioSystems*, 42: 111-118.

Odum, E.P., 1977. The emergence of ecology as a new integrative discipline. *Science*, 195: 1289-1293.

O'Grady, R.T., 1984. Evolutionary Theory and Teleology, 107: 563-578

O'Neill, R.V., D.L. DeAngelis, J.B. Waide, and T.F.H. Allen, 1986. A Hiearchical Concept of Ecosystems.

Princeton University Press, Princeton, NJ.

Page, Jr., R.E. and Mitchell, S.D., 1998. Self-organzation and the evolution of division of labor. *Apidologie* 29, 171-190.

Patten, B.C. and Jørgensen, S.E. (eds.), 1995. Complex Ecology. The part-whole relation in Ecosystems. Prentice Hall. 705 pages.

Phipps, M., 1981. Entropy and Community Pattern Analysis. *J. Theor. Biol.* 93: 253-273.

Porges, S.W., 1998. Love: An emergent property of the mammalian autonomic nervous system. *Psychoneuroendocrinology*, 23(8), 837-861.

Prentice, I.C. and Leemans, R., 1990. Pattern and process and the dynamics of forest structure: A simulation approach. *J. Ecol.*, 78: 340-355.

Prigogine, I. and Wiame, J.M., 1946. Biologie et thermodynamique des phénomènes irréversibles. *Experientia* 2(11): 451-53.

Reggia, J.A., Armentrout, S.L., Chou, H.-H. Peng, Y., 1993. Simple systems that exhibit Self-directed replication.

Root-Bernstein, R.S. and Dillon, P.F., 1997. Molecular Complementarity I: the Complementarity Theory of the Origin and Evolution of Life. *J. Theor. Biol* 188, 447-479.

Reuter, H. And Breckling, B., 1996. Emerging Properties on the Individual Level: Modelling the reproduction Phase of the European Robin Erithacus rubecula.

Salt, G.W., 1979. *A comment on the use of the term emergent properties*. Am. Nat. 113(1): 145-148.

Salthe, S. N. 1993. Development and evolution. Complexity and change in biology. MIT Press, Cambridge

Salthe, S.N. 1985. Evolving hierarchical systems - their structure and representation - Columbia University Press, New York, 343 pp

Solé, R.V., Miramontes, O. and Goodwin, B.C., 1993. Oscillations and Chaos in Ant Societies. *J. Theor. Biol.* 161: 343-357.

Sousa, W.P., 1980. The responses of a community to disturbance: The importance of successional Age and Species' life Histories. *Oecologia* (Berl.) 45: 72-81.

Stephan, A., 1998. Varieties of Emergence in Artificial and Natural Systems. *Z. Naturforsch.*, 53c, 639-656.

Stephens, G., 1997. Superconductiong current enhancement as an emergent property of a driven broken symmetry system. *Il nuovo cimento,* 19(1), 87-94.

Tegeder, R.W., 1994. On the emergence of primitive cooperation in an environment of fluctuating resource supply. *J. Math. Biol.*, 32: 645-662.

Testa, B., Kier, L.B. and Carrupt, P.-A., 1997. A Systems Approach to Molecular Structure, Intermolecular Recognition, and Emergence-Dissolvence in Medicinal Research. *Med. Res. Rev.*, 17(4), 303-326.

Thierry, B., Theraulaz, G., Gautier, J.Y. and Stiegler, B., 1996. Joint memory. *Behavioural Processes,* 35, 127-140.

Ulanowicz, R.E., 1986. Growth and development. Ecosystem phenomenology. Berlin: Springer.

Ulanowicz, R.E., 1997. Ecology, the Ascendent Perspective. Columbia University Press, New York, 201.

Valentine, J.W. and May, C.L., 1996. Hierarchies in biology and paleontology. *Paleobiology,* 22(1), 23-33.

Van der Laan, J.D. and Hogeweg, P., 1992. Waves of crown-of-thorns starfish outbreaks- where do they come from? *Coral Reefs,* 11: 207-213.

Vandermeer, J., 1997. Syndromes of Production: an Emergent Property of Simple Agroecosystem Dynamics. *J. Env. Manag.*, 51: 59-72.

Voneida, T.J., 1998. Sperry's concept of mind as an emergent property of brain function and its implications for the future of humankind. *Neuropshchologia,* 36(10), 1077-1082.

Watts, J.M., 1998. Animats: Computer-Simulated Animals in Behavioral Research. *J. Anim. Sci.*, 76, 2596-2604.

Westerhoff, H.V., Jensen, P.R., Snoep, J.L. and Kholodenko, B.N., 1998. Thermodynamcis of complexity. The live cell. *Thermochimica Acta,* 309, 111-120.

White, J., Hamilton, K.A., Neff, S.R. and Kauer, J.S., 1992. Emergent Properties of Odor Information Coding in a Representational Model of the Salamander Olfactory Bulb. *J. Neurosc.*, 12(5): 1772-1780.

Wicken, J.S., 1986. Evolution and Emergence. A structuralist perspective. *Riv. Biol. Biology Forum 79*(1): 51-73.

Wieglieb, G. and Bröring, U., 1996. The Position of Epistemological Emergentism in Ecology. *Senckenbergiana maritima* 27(3/6), 179-193.

Wilson, M.W. and Botkin, D.B., 1990. Models of simple microcosms: emergent properties and the effect of complexity on stability. *Am. Nat.* 135(3): 414-434.

Winemiller, K.O., 1990. Spatial and temporal variation in tropical fish trophic networks. *Ecol. Monogr.* 60(3): 331-367.

Wu, J., and Loucks, O.L., 1995. From balance of nature to hierarchical patch dynamics: A paradigm shift in ecology. *Quart. Rev. Biol.* 70(4), 439-466.

Yeakley, J.A. and Cale, W.G., 1991. Organizational Levles of Analysis: A Key to Understanding Processes in Natural Systems. *J. Theor. Biol.*, 149: 203-216.

II.3 Ecosystems as Information Systems

Søren Nors Nielsen

1. Introduction

The science of information theory, or communication theory as it is often referred to, in its present form is a relatively young science, although it undoubtedly finds its roots in the works of Boltzmann and Gibbs. An early platform was laid out by Hartley at the beginning of this century (Hartley, 1928), and Cherry (1951) states the modern formulation to be a 'logical extension' of this work. The science, as we know it today, goes back no earlier than the late 1940's, and is based especially on the works of Claude Shannon (Shannon, 1948, Shannon and Weaver, 1949) and his advocate Warren Weaver (1949). This was the time where the theory found its application, the period after the end of the Second World War, under which undoubtedly much of the basic work and foundation of the theory was laid out. For other tutorial treatments see also Renyi, (1970).

Few other scientific disciplines have led to so many controversies during its short life time. Information theory has been presented as a universal theory with a general validity and possibilities of implementation. And indeed it has succeeded in penetrating into many biological areas, including ecology.

Meanwhile, the way the word 'information' itself is used in information theory differs so much from its use in everyday life. This poses one major obstacle tó the understanding and the working of the theory. First of all, we are in many ways used to associating information with meaning, or actual knowledge, and to use the two words as more or less synonymous. This, to some extent, makes sense in the way that information has no use without a meaning, i.e. can be interpreted in some way. This point of view should not be retained while reading this text as those perspectives have been left totally out of the science of information theory. In fact, pertaining this view often will be directly misleading in the understanding of the concepts. As an example, consider a sequence of 4 letters, known to be either p, r, o, and c. Out of this knowledge, 256 possible combinations exist, out of which only a few, like crop or porc, will have any immediate meaning. Other possibilities, like proc or corp, will make some sense as abbreviation, either to a computer programmer or to a corporate employee. Most of the rest will be considered as nonsense or misspellings. The important thing is that to information theory they will all be alike, containing the same amount of information. This is the first, and maybe a major constraint to the comprehension of this theory. Unfortunately, when digging further into this area, one will find that other fundamental problems to the perception of this theory also exist.

The most controversial statement has probably been the statements about the concept of Shannon Information (section 2.1) being analogous with or even equivalent to the concept of entropy used in thermodynamics. This statement has lead to much confusion caused by, as we shall see, the unfortunate mixing of the two concepts. The confusion is reflected in much of the current literature from where it cannot be removed. Many authors have tried to sort out the conflict and clarify the differences between the two concepts and eventually combine them (see later). This development will probably continue for years. The present chapter will not aim at solving this debate, but concentrate on the use of concepts derived from information theory, in its widest sense, on biological systems.

One may expect that information theory would have had only little time and chance to have an impact on other sciences seemingly so remote as biology. But, the truth is that information theory has been introduced in a wide variety of areas of biology. Work carried out in the 1950's, by the brothers E.P. and H.T. Odum, introducing the Systems theory of von Bertalanffy into ecology, made it possible and valid to formulate a topology of ecological networks and relate these maps or digraphs to studies based on cybernetics.

Meanwhile, the introduction of the concept of information seems to have taken place in a rather heuristic manner, where a combination of several, more or less casual circumstances have played a role. First of all, the immediate adaptability and applicability of the theory to the actual branch of biology where information should be introduced seems to have played a major role. One, obvious example is the wide range of application of Shannon-information derived approaches to the analysis of genetic material. The nucleotides, arranged on strings, as a sequence of symbols or letters (A,T,G,C) made the approach easy to apply. Thus, the application was easier to understand intuitively, and hence to introduce.

Likewise, the concept of *thermodynamical information* has a more obvious connection to the biologically important elements that constitute organisms, like C, H, N, O, P and S (Morowitz, 1979). This concept has thus found its way into our understanding of the biogeochemical state and the cycles of the ecosystems through the concepts of thermodynamical information and average mutal information (e.g. Jørgensen and Mejer, 1977, Mejer and Jørgensen, 1979, Jørgensen and Mejer, 1981, Ulanowicz, 1986, 1997, Nielsen and Ulanowicz, in print).

The two approaches, Shannon Information and Thermodynamical information, in this author's opinion, represent two, more or less, opposite poles in the application of the information concept *sensu lato* to biological systems. The above mentioned unfortunate mixing of the concept of Shannon-information (or Shannon-entropy) and the thermodynamical entropy have caused a lot of intermediate varieties of the concepts to be found in the literature, e.g. the Brooks-Wiley proposal of 'Evolution as Entropy' (Brooks and Wiley, 1986, 1988), with the possibility of confusion to follow. This chapter will attempt to be consistent in its distinction between the concept of Shannon-information, or Shannon-entropy, on the one side and the concepts of thermodynamic information and (thermodynamic) entropy on the other. When the words are used alone they are always used in their widest sense (*sensu lato*).

The classical information theory in this chapter will mainly be represented by the papers of Shannon and Weaver (Shannon, 1948, Shannon and Weaver, 1948, Weaver, 1949) and the later refinements and specification by Brillouin (1962).

One major issue and basic question derived from these papers, that needs to be addressed here, is the question whether or not information has to be conveyed in order to exist. That is, if it exists also even without being conveyed, observed or interpreted. Does a system possess an informational level or content compared to its surroundings? It is clear that the concept of Shannon information with its starting point in communication theory has focused its attention and concentrated on the process of transmission. In his works, Brillouin (e.g. 1949, 1962) opens up for other approaches and possibilities. Information is connected to the process of observation, i.e. that information is what is *a posteriori* gained of knowledge after observation of a system, assuming no *a priori* (or a certain level of) knowledge. The question of the role of this a priori knowledge will be discussed later.

As just mentioned, the Shannon approach seems to have enjoyed a wide acceptance within the area of genetics. The application areas range from being of a pure genetic kind analysing from a level of amino acids, to being applied to aspects of the whole genetic apparatus.

In the mid sixties, the approach was taken a step further to form the platform for a more general understanding of the evolutionary process. The most pronounced advocates of such a possibility are to be found among, Brooks and Wiley (see section 3.1), Wicken (section 3.2) and Yockey (section 3.3) although the approaches may definitely not be seen as equivalent. These concepts must be seen as representing various attempts of drawing the consequences of not only Shannon's communication theory, but also Schrödingers *negentropy* (Schrödinger, 1944) and his order from order and disorder principles, and the *far from equilibrium* principles of systems introduced by Prigogine and co-workers (e.g. Prigogine and Wiame, 1946, Prigogine and Nicolis, 1971). In fact, in many cases information is set equal to negentropy (Popper, 1979). The above mixing led to the obvious idea to join the concepts of information and entropy into a 'grand unifying theory' of evolution. But, as an inevitable consequence of the fact that the two concepts are **not** the same, the result has been that the outcomes of the attempts have been varying in success. Especially the Brooks-Wiley attempt reflects the above mentioned possibilities of confusion all too well. This happens when the concepts of information and entropy are used in a manner not sufficiently stringent. Their proposal received much criticism, such as the accusation of abuse of the connection to entropy, from several authors (e.g. Bookstein, 1983, Løvtrup, 1983, Wicken, 1987b). This is in spite of the fact that these authors were working with seemingly the same kind of perspectives.

When it comes to ecology, many biologists will be familiar with the use of the so-called Shannon-Wiener index as a measure of (bio)diversity. Meanwhile, in addition two recent applications of information derived indexes, *ascendency* and *exergy* have been applied to ecosystems. The ascendency, initially applied to ecosystems by Ulanowicz, is based on calculation of the *average mutual information* of the ecosystem network, thereby stressing the importance of the flows of the ecosystem. The exergy approach, in its classical formulations, builds on the deviation of the elementary composition of the system compared to thermodynamical equilibrium (universal composition of elements). This deviation in entropy state is often referred to as *thermodynamical information*. This approach offers some major methodological problems. Therefore, another exergy index has quite recently been proposed, in which the calculation is based on biomasses weighted with the measure of the size of the genome of respective organisms or compartments.

When talking about information, quite often a very objective attitude is taken. Subjective values like meaning and interpretation are left out of the discussion. Nevertheless, knowing biology, and disciplines such as animal physiology and ethology, it would be hard to say that signalling, right interpretation and response to signals should be of no importance in nature. Recently, through the application of perspectives and methods of analysis from the science of semiotics the importance of transmission of signals and correct interpretation of signals together with existence of such processes in biological systems has been recognised. Thus, the existence of communicative processes between ecosystem parts must be acknowledged and the understanding of these processes has to be stressed in the future, in order to understand ecosystem behaviour, emergent properties, etc. (see Nielsen and Müller, this volume).

At the end of this introduction, I will take yet another opportunity to stress the attempts of various authors to join the two directions of information concepts, and if treated in a manner stringent enough it may eventually, according to the opinion of this author, be possible that they turn out to be consistent, complementary, or two sides of the same coin.

Because of this possible connection between information and entropy, many works have been dedicated to clarify the direct relations, such as Tribus (1961), Tribus et al. (1966), Jaynes (1957a, 1957b, 1962), Ebeling and Volkenstein (1990), Ebeling and Feistel (1992), Kubát and Seeman, (1975), and Wicken (1983). This author prefers a more indirect path indicating a possible connection between, one side, knowledge (or information) gained by picking one, single microstate out of many possible and, on the other, the deviation in its entropy state from the maximum possible, i.e. thermodynamical equilibrium.

Table 1: Various indices met in literature which are used for calculation of information of systems.

index	expression
Boltzmann	$S = k \ln W$
Boltzmann/Gibbs equation	$S = -k \Sigma p_i \ln p_i$
Shannon information	$H = -k \sum_{i=1}^{nhorz} 20 p_i \log_2 p_i$
Kullback	$I = \sum_j p_j \ln \frac{p_j}{p_j^*}$
Exergy, classic	$Ex = RT \sum_{c=0}^{n} (\xi_c \ln \frac{\xi_c}{P_{eq,C}} - (\xi_c - P_{eq,c}))$
Exergy, internal	$RT C_0 \sum_{i=1}^{n} x_i \ln(\frac{x_i}{x_{i,ref}})]$
Exergy, light	$Ex = RT \sum_{c=0}^{n} \beta_i C_i$
Ascendancy	$A = TST \sum_i \sum_j (\frac{T_{ij}}{T}) \log \frac{(T_{ij} T_{..})}{(T_{.j} T_{i.})}$

The intuitive coupling also often made between entropy and disorder (Morowitz, 1955), leads one to the conclusion that information and order (and complexity) are somehow correlated. These thoughts have been pursued in the works of Papentin (1980, 1982), Hinegardner and Engelberg (1983) and Stonier (1990b). Thus, a connection between information and complexity, may be, in turn, another way of describing the state of a biological system, that has been established (Hinegardner and Engelberg, 1983).

Several works may be found that review the whole spectrum of concepts, information, entropy, order, complexity and meaning exist. Some of them try to make a consistent pattern of the whole sequence, like Stonier (1990a, 1992, 1997), or just deal with the different perspectives without major efforts of unification (Weber et al., 1988). Others are linking information to the evolutionary process, both in an extended Darwinian sense (Brooks & Wiley, 1988, Wicken, 1987b) or in the sense of evolution of life on earth (Küppers, 1990), physical aspects of chaos and order (Ebeling, 1991) or universal structure (Frautshci, 1988, Layzer, 1988). For an early review of the history of information theory refer to Cherry (1951).

The first part of this chapter (section 2) considers information theory from the classical point of view, taking its entrance point in the concept of Shannon and Weaver and introducing later extensions. The area of connecting the evolutionary process with changes, usually increasing information at a macroscopic level, is dealt with in section 3. The implementation to biotechnology, at microscopic level, is basically omitted. Biological systems may be seen as structures showing a large deviation in distribution of elements from thermodynamical equilibrium and thus having a high thermodynamical information. The correlation to the thermodynamical efficiency measure, exergy, is introduced (section 4). Ecosystem flows may be seen as communicative processes, and as an indicator of the state of ecosystem development based on the average mutual information of the flows in the system that has been introduced in ecosystem theory through the concept of ascendancy (section 5). The importance of the understanding of these processes through transmission and interpretation is leading to a proposal for a new concept of ecosystem semiotics to be pursued in the future efforts within this area (section 6).

2. Information theory - the basis and beginning

As referred in the introduction, most of the scientific discipline, today known to us as information or communication theory, was laid out in the late 1940's through the works of Claude Shannon. Although, Shannon as stated in the introductory comment by Weaver (Shannon and Weaver, 1949) builds on works going back to Boltzmann (1905), Hartley (1928) and other authors through the first part of the 20th century. The work was at first presented in a paper in Bell techn Journal (Shannon, 1948). But just one year later, together with Weaver that immediately seems to have captured the wide relevance of Shannon's work, the whole theory was put forward in a book 'The mathematical theory of communication (Shannon and Weaver, 1949). The book contains an introductory chapter by Weaver, which is an extended version of his paper in Scientific American in July the same year (Weaver, 1949) as well as a revised and corrected version of Shannons original paper.

As a very unfortunate side effect, the word entropy was introduced in this discussion. Several authors have attributed this to be due to von Naumann, who convinced Shannon to call his new concept entropy, since the formulas looked alike. The argument was the following. Since nobody knows what entropy is, he would have an advantage and probably win any discussion.

Instead of focusing entirely on transmission, Brillouin connected the information concept to the process of observation. Thus information is basically a measure of the knowledge gained by observing a system.

Information has also been coupled to the concept of complexity through the works of especially Kolmogoroff and Chaitin (e.g. Chaitin, 1992). Information, or to be exact, algorithmic information complexity (AIC) is hereby understood as the minimum algorithm needed to describe a binary sequence, thereby taking its entrance point in area of computer science.

As a last introductory remark, through this work, the concept of information has been objectified, and loosened, maybe even freed totally, from any connotations to meaning and thus semantic aspects. So determination of the meaning and the semantic form of a sequence of letters has to take place at another level (e.g. Stonier, 1990, 1992, 1997).

2.1 Communication theory

The foundation of Information theory as a science started with the pioneering work of Claude Shannon in 1949 as described in his book: 'The mathematical theory of Communication' with Warren Weaver. Perhaps it would be more correct to consider this as the foundation of a special sub-discipline of Information theory, often referred to as communication theory. This would be justified by the fact that the work as described in the above book clearly deals with the transmission of signals with the scope of doing this as safe and as efficient as possible.

Weaver (1949) states communication as having problems on three levels dealing with separate problems of the communication process:

A. Technical problems, - accuracy of transfer

B. Semantic problems, -interpretation of meaning

C. Influential problems, - success of conduct of receiver

of which Shannon's theory mainly deals with the first level, A. Shannon (1948) views the communication systems as consisting of five parts: 1 An information *source*, that via 2 a *transmitter* is able to send a signal through 3 a *channel*. The transmitted signal may be registered by 4 a *receiver* and eventually find its way to 5, its *destination*. During transmission the signal will be subjected to the introduction of noise from a *noise source*, and thus possibly differs from the signal sent.

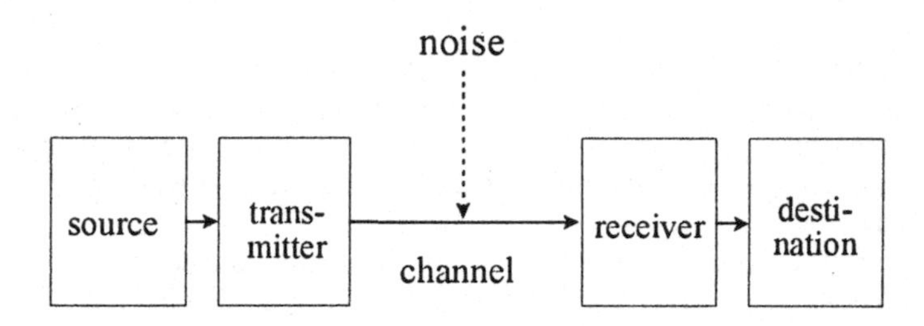

Figure 1: Diagram showing the communication system of Shannon and its 5 parts: sender, transmitter, channel, receiver and destination. Noise may be introduced disturbing the correct reception of a given message.

The signals transmitted are in general viewed as being composed of a sequence of symbols, e.g. letters, binary digits, etc. The symbols are taken out of a certain number of possible symbols, which may be illustrated by the following quotations from Weaver (1949):

"The information source selects a desired message out of a set of possible messages." ....

"In fact, in this new theory the word information relates not so much to what you do say as to what you could say."

The symbols may vary from being equiprobable and independent in their frequencies, the sum of probabilities equals 1, to having being non-equiprobable and have a successively higher levels of interdependency (see for instance Shannon, 1948). Various algorithmic approaches have been used to get closer and closer to the construction of sentences increasingly similar to the English language, but not necessarily carrying any meaning.

In his analysis of discrete information sources as Markovian processes Shannon eventually comes out with the result that the function H

$$H = -K\sum_{i=1}^{n} p_i \log p_i \tag{1}$$

is the only function satisfactory for the fulfilment of the proof, such as demands for continuity and monotony. The constant, K, is positive and refers to choice of units. The similarity to the Boltzmann-Gibbs equation is indeed striking.

In the scientific literature several problems crucial to the future of information theory are also touched upon. As a minor stand, for instance, the choice of the scale to some logarithmic expression. More fundamental and major stands the unclarity in the formulation of the relation between information and entropy (as indicated by Weaver to be caused by von Neumann). This confusion has contributed to misunderstandings in the debate about the relationship ever since. How different the two concepts are and how much they differ from the concept of thermodynamical information will be clear by reading the following:

'The situation is highly organised; it is not characterised by a large degree of randomness or of choice-that is to say, the information, or the entropy, is low' ...
'Information is used here with a special meaning that measures freedom of choice and hence uncertainty as to what choice has been made'.

The previous statements connect Shannon information with uncertainty, and a widespread attitude among researchers today seems to be that Shannons information should really be used as a measure of uncertainty or indeterminacy of the array of signs. Weaver (1949) seemed to come close an agreement with this attitude in stating:

"Thus greater freedom of choice, greater uncertainty and greater information all go hand in hand." ...
"But if the uncertainty is increased, the information is increased, and this sounds as though the noise was beneficial"

With these last statements, Weaver touches exactly on one of the dilemmas, but refers to this as a 'semantic trap' if one does not remember the new definition of information. The argumentation comes close to a tautology needed to save the new concept. At least, these quotations shows how far the concept of Shannon information is from our normal understanding of information.

The choice of scale to be logarithmic is to quote Shannon (1949) for three reasons. Firstly, it is more useful in practice, secondly, it lies close to ones intuition and, thirdly, it is mathematically convenient. Nevertheless, this choice is crucial in creating the resemblance to previous formulations and mathematical expressions of the thermodynamcial concept of entropy as stated in the formulations by Boltzmann and Gibb's, see section 2.1. The scale might be important in the situation where one has to compare systems and results from various authors.

2.2 Information by observation

Another milestone is to be found in the works of Brillouin and presented in his book 'Science and Information theory' (Brillouin, 1962). The scope of his work formed a rigid science of information theory, based on probability theory, combining both the mathematical and practical aspects of the problem. At the same time, it is claimed that this approach is solving both the problem of Maxwells Demon (see later) and demonstrating a clear connection between information and entropy, thus solving the problems introduced by Shannon and von Neumann described above.

Brillouins definition of information is based on 'a situation in which P_0 different possible things might happen'[1] and information $I_0 = 0$. Furthermore, in this *initial situation* the outcomes are equally probable[2]. In the final situation an outcome is selected $P_1 = 1$, and $I \neq 0$ (probably also >0) any longer, but defined as

$$I_1 = K \ln P_0 \tag{2}$$

where K is a constant and ln the natural logarithm. The logarithmic conversion is chosen so as to give the information an additive property.

The previous equation is derived from the more general form

$$I_1 = K \ln \frac{P_0}{P_1} = K \ln P_0 - K \ln P_1 \tag{3}$$

which reduces to equation 2 since $\ln P_1 = 0$. The equation is peculiar since equation 3 is never directly related to the gained information, i.e. the difference between the *a priori* and the *a posteriori* situation. This case would lead to troubles with the positive or negative signs of the equation. In many ways Brillouin's approach seems to be the opposite to that of Shannon. Nevertheless, Brillouin claims his derivation to be consistent with the Shannon Information.

At a later point, Brillouin uses his findings to state the 'negentropy principle of information'. The conversion between information and thermodynamical entropy is related to units of the constant K in the above equation. The conversion ratio, k/K, is approx. 10^{-16} between entropy in ergs per °C and bits.

Brillouins contribution to solving the problem of Maxwell's demon (Brillouin, 1949), a small creature hypothesised to be able to make order by distinguishing the speed of molecules and sorting them, thus seemingly breaking the second law of thermodynamics. This controversy will not be dealt with here, although the debate around it is interesting in its connections to entropy and order. For further reading in this area please refer to a collection of papers edited by Leff and Rex (1990), that contains many of the major references of the debate, among others also the works of Szilard (1929).

2.3 Information and biological systems

Although, Harold Morowitz, may be better known as a thermodynamicist connecting thermodynamics to the biological area as reflected in 'Thermodynamics and Biology' (1979), several of his works contain references in the area of combining his approach or world view with information theory, probably inspired by Brillouin but again based on sending and receiving (e.g. Morowitz, 1978).

Morowitz (1970) defines *the information content of a symbol* as

$$I = \ln_2 \frac{p_r}{p_s} \tag{4}$$

where p_s is the probability of the symbol being sent and p_r the probability of the symbol being **correctly** received[3].

1 Note that (capital) P is now used to designate microstates or number of complexions (instead of W normally used in Boltzmann's equation

2 Note that this assumes some knowledge already about the possible number of outcomes, i.e. reference point has been removed from tabula rasa situation.

3This expression reverses the understanding from Brillouin, without changing sign.

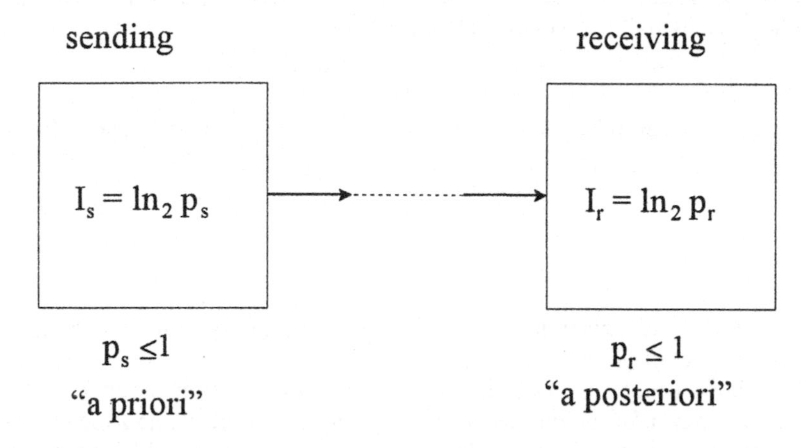

Figure 2: Diagram showing the relationship between the probablities in the sending 'a priori' situation, p_s, and the probability in the receiving, 'a posteriori' situation, p_r, of transmission

The previous equation, equation 4 may be rearranged in the following manner

$$I = -\ln_2 p_s + \ln_2 p_r \tag{5}$$

Since the probabilities are less than or equal to 1, it is seen that the first part of equation 5 in general is a positive contribution to information or zero, and the second part in general is negative or zero (see later). Provided that the channel of transmission is perfect the probability of being received if sent, p_r equals 1, and the formula reduces to

$$I = -\ln_2 p_s \tag{6}$$

If the number of possible symbols is n, then

$$\sum_{i=1}^{n} p_i = 1 \tag{7}$$

For a sequence of symbols of length of N and considering a number of i elements each occurring a number of n_i times having the probability of p_i the average information, I_{av}, may be calculated as

$$I_{av} = \sum_i \frac{n_i I_i}{N} = \sum_i \frac{n_i}{N}(-\ln_2 p_i) \tag{8}$$

which may be reduced to

$$I_{av} = -\Sigma\, p_i \ln_2 p_i \tag{9}$$

since

$$\frac{N_i}{N} \rightarrow p_i \tag{10}$$

when the string of symbols becomes sufficiently long. Equation 9 shows a striking similarity to Boltzmann-Gibbs equation and Shannon-information except for the constants and units.

The approach is extended into thermodynamics and statistical physics. The information content of a quantum state is given by

$$I = -\Sigma f_j \ln f_j \quad (11)$$

which is pointed out to be derived directly from equation 9 above. The entropy function is later derived to be

$$S = -k \Sigma f_j \ln f_j \quad (12)$$

giving the relation

$$S = 0.6932\, k\, I \quad (13)$$

through which a connection between entropy and information is established at the microscopic level of physics.

The above equations (4-6) show all too well the difference between the concept of information from information theory and an everyday, intuitive and logical understanding of the word information. If a symbol is sent with the highest possible probability, $p_s = 1$, or it is perfectly received, $p_r = 0$, it does not contribute to information according to eq. 5. Meanwhile uncertainty in sending, $p_s < 1$, and imperfect receiving of a symbol, $p_r < 1$, does lead to a positive respective negative contribution to information. This illustrates some of the reason for, not only the difficulties in explaining information theory to the layman, but, also the many debates within the scientific communities in this area.

2.4 Information and Complexity

At about the same time, a new concept, alogarithmic information theory, was introduced during the sixties. The concept seems to have emerged more or less independently (Chaitin, 1977) by several authors (Kolmogoroff, 1968, Martin-Löf, 1966, Chaitin, 1966, 1974a,b, 1975, 1977. The concept connects clearly the development of binary computers illustrated by the fact that most of the analyses deal with binary strings, i.e. sequences where the symbols are only 0 or 1's.

The concept leads to a new definition of randomness founded in information theory. The principle can be explained by the following example from Chaitin (1975). Consider two strings of binary digits:

01010101010101010101

and

01101100110111100010

The two strings differ considerably, the first being simply a sequence of 10 times of repetition of 01, whereas the second is much more difficult to describe. Probably, the latter sequence cannot be described by anything else than the sequence itself.

The fundamental idea is, that we may create a program which is able to construct either the first or the second string for us. The string will be 'created' from some instructions given to the computer. Such instructions are commonly known as algorithms. Thus, constructing an algorithm describing the first sequence is much easier and simpler than describing the second one. Now, we can define a random sequence as a set of digits that may not be compressed. The more we can compress a string the less random it is.

This is now used to define complexity by means of the size of the algorithmic information needed to describe a sequence. This measure is often referred to as the *algorithmic information complexity* or AIC. The smaller the program that may be constructed in order to describing a certain sequence the less complex it is. As an example, several other opportunities exist in describing the first sequence, for instance as 5 times 0101. This algorithm, though, will have a slightly higher AIC than the one first proposed above.

At the outer ends of the spectrum we find two minima of complexity. One string of 0's or 1's only, respectively. Maximum complexity is reached in strings which are composed of approximately 50% of each of the symbols following each other in random sequence. None of the two (three) situations are interesting, neither for computation nor for life. What would have happened with 'life' if coding of the genome would have happened with one nucleoside only? Or, what if the sequence of amino acids was really randomly composed as it has been sometimes argued? (see for instance Yockey, section 3.3 for arguments that this is not the case).

The importance of Chaitin's works also has a strong philosophical implication with its relation Gödel's theorem of incompleteness. For further discussion of this and other more philosophical matters, refer to Chaitin (1974b, 1975, 1982, 1992).

3. Information and evolution

When looking at the evolution of biological systems we traditionally choose to consider them as structures evolving in either time or space. Thus, we may for instance view them in either a historical context, as in the realms of palaeontology and/or in terms of classical Darwinian approach, or we may view the biological systems evolving as structures composed of chemical elements, organised in a scalar hierarchy. In either of the cases we tend, intuitively and very subjectively, to come out with the conclusion that the evolution of biological structures are moving in a certain direction - in the direction of higher order, more information and increasing complexity.

Information theory has attracted several scientists and tempted them to formulate a more general theory (Grand Unifying Theory) of evolution in terms of information (Küppers, 1990). This is in fact quite logical since the traditional approach introduced by Shannon primarily deals with the transfer or processing of symbols arranged in a sequential manner in a linear, one dimensional array. This view and therefore the underlying understanding of information was of course easy to transfer to the genome or the whole genetic apparatus. Here the DNA (or RNA for that matter) can be seen as a linear, sequential arrangement of nucleic acids (e.g. Rao et al., 1997), coding for a linearly arranged sequence of amino acids leading to enzymes, peptides, etc., which are some of the major compounds important to the function at the cellular level.

As the genome, or at least the part of it that is expressed during the life cycle, seems to grow with what in general is agreed upon increasing hierarchical level of the organisms, it has likewise been tempting to conclude that those organisms situated at higher levels in the hierarchy possess more information compared to organisms at lower levels.

Two things have mainly made the task of pursuing this idea and verifying that it is indeed the case, i.e. this principle is practised in nature, is very difficult and adds to the confusion. The first factor is the above mentioned possibility of confusing Shannon information and thermodynamical entropy, which in many cases seems to be a case of mixing microscopic and macroscopic entities. Second, our normal view of biological systems as being hierarchically organised really seems to encourage the probability of the first mistake, and stimulate erroneous use of the various informational concepts. In the following, a few of the basic approaches will be presented.

3.1 Evolution as Entropy

During the 80's, starting around 1982, Brooks and Wiley published a row of papers where the process of genetic evolution was to be understood as a combination of thermodynamics and information theory. The approach is partly presented as a fundamental cut with the theories of evolution as an equilibrium process, or series of punctuated equilibria, as 'evolution is best understood as an irreversible non-equilibrium phenomenon' (Brooks et al., 1984, Brooks and Wiley, 1986, 1988, Wiley and Brooks, 1982, 1983). The authors claim, on one hand, to build on the shoulders of non equilibrium thermodynamics taken from Prigogine and co-workers. On the other, when dealing with information theoretical perspective they claim to build on the 'conceptual developments' of among others Gatlin (1972), Wicken, (1979) and Ho and Saunders (1979).

They argue from a cladistic point of view, that speciation, 'the evolution of diversity in a clade', is a consequence of the individual species acting as open-ended, closed, thermodynamic systems. The systems are open in terms of energy and closed (or partly closed) when it comes to information and cohesion. The theory is seen as comprehending existing theories, as for instance neo-Darwinism and punctuated equilibria, as they 'will come to rest comfortably within' the theory presented.

Information in this context is defined as 'anything transmitted from a source, through a channel, to a receiver'. Information thus may be viewed as the capacity to execute an ontogenetic program. From this it is seen that the concept clearly lies within the framework of classical information theory (section 2) and close to the concept of Shannon. The information of the genes may be divided into a part that is *canalized*, i.e. belongs to the ontogenesis or later functionality of the organism, and a *noncanalized* part that is programming for structural products. Both the canalized and the noncanalized part may again be divided into a *stored information* part that is expressed during the life history of the organism, or a potential part, *potential information*, that is 'present but not expressed'. It is important to keep this in mind for the following presentation of their explanation of evolutionary behaviour of biological systems in terms of entropy.

At this point, it is argued that information may be added and that 'the addition of information to any system increases the entropy of that system' (quoting Denbigh, 1975). The connection to and understanding of Shannon Information as potential information, or the *information capacity* of a sequence of symbols, becomes clear as it is the same time stated: 'If this new potential information is converted into stored information, the species will exhibit a decrease in its entropy level'. This is expressed in the following manner:

$$S_p \leq S_a < S_{a+p} \tag{14}$$

where S_p is the 'entropy state of the ancestral system', S_a is the resulting 'entropy state of the descendant system' and S_{a+p} is the 'ancestral system after the addition of new potential information'. Again, it is noticed that the use of terminology and symbols is consistent with an understanding of the system in terms of Shannon's information entropy.

In an attempt to link the informational approach to thermodynamics the classical potential of free energy is introduced as:

$$F = E - TS \tag{15}$$

that also may be rearranged into the following

$$S = \frac{E - F}{T} \tag{16}$$

which according to the authors relates to the loss of order in the system.
Establishing an information theoretical analogue to eq. 15, it may be stated that

$$I_f = I_i - CS \tag{17}$$

where I_f is the *free information*, I_i is the *intrinsic information* and C is the *cohesion*. This equation may in turn be rearranged to

$$CS = I_i - I_f \tag{18}$$

where C times S, called the *entropy of cohesion*, is now renamed dS_c, and the difference on the right side of the equation is rewritten dS_i, it is argued that

$$d\,S_c = d\,S_i \tag{19}$$

at 'equilibrium'. Although at this point, these are vague similarities to Prigogines approach of understanding biological systems as dissipative structures the biological interpretation of this remains obscure.

In the further elaboration of the theory and its relations to thermodynamics through Prigogines equation for the entropy production

$$dS = d_e S + d_i S \tag{20}$$

it is argued that evolution, E, many be described by

$$E = \frac{d S_i + d S_c + d S_e}{dt} \tag{21}$$

where dS_i, change in entropy levels of information, dS_c change in entropy level of cohesion and dS_e represents changes in entropy levels of energy[4]. This equation 'represents a summary for the origin of diversity and life' whereas

$$E = \frac{d S_i + d S_c}{dt} \tag{22}$$

contains the 'relevant parameters for investigating the origin of diversity'.

Another way of understanding the evolutionary process of the genetic material and its possible connection to the Boltzmann equation is presented by the following. The dimension of the possible phase space is determined by

$$H_{\max} = \log_2 A \tag{23}$$

where A is the number of available microstates and thus corresponds to a maximum capacity of coding (compare Ulanowicz, section 5.2). Using $\log_2$ based transformation cause the system to be represented in units of bits.

From this phase space a given system will find an observed distribution H_{obs} according to

$$H_{obs} = -\Sigma p_i \log_2 p_i \tag{24}$$

where p_i is now understood as 'the probability of a randomly chosen organism occupying a particular genotype'.

During evolution not all of the capacity of the genome has been utilised, since there was no time to realise all the possibilities. Thus H_{obs} is usually lower than H_{max}. Thus, as the number of possible microstates is growing as the genome is expanding, a relatively few number of the possibilities are actually realised. Meanwhile, this is not only a question of time, the necessary niches should also be existing and vacant in order for, at least some of, the new microstates to be realised.

4The difference in notation is thought as being of no importance in this case.

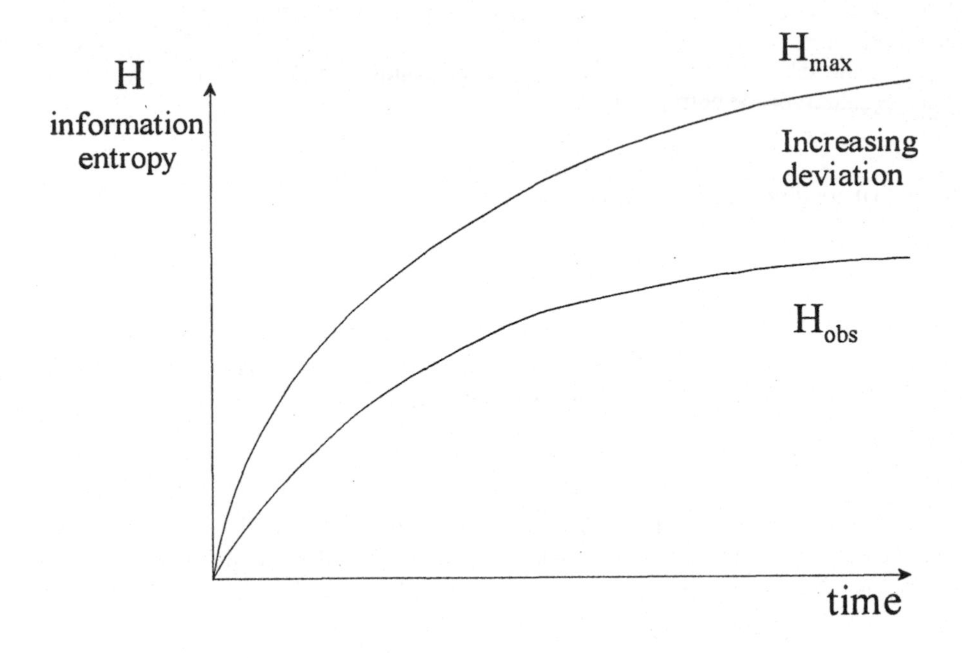

Figure 3: The coding or information capacity of the genome, Hobs, has been increasing through the period of evolution due to additions and expansion of the genetic material. Not all the possible microstates have been realised. Therefore the observed number, Hobs, is less than Hmax. Furthermore, the deviation between the two is continuously increasing.

There seems to be no doubt that the contributions by Brooks and Wiley form important pioneering work opening the debate about the importance of entropy and thermodynamics to the evolutionary process which ought to be received with the greatest appreciation. Meanwhile, the ambitions of the authors, insisting on combining the theories into one whole, seem to make them move around between many different levels in both the information theoretical, thermodynamical and biological areas. These areas are indeed difficult to combine, which is clearly seen from the above presentations. Doing this for the first time one will unavoidably receive tough criticism alone for the fact of being multidisciplinary. The approach, later presented in the book 'Evolution as Entropy' (Brooks and Wiley 1988, first ed. 1986) naturally received some, sometimes harsh, criticism with discussions to follow as reflected in papers of Bookstein, Wicken and Løvtrup (Bookstein, 1983, Løvtrup, 1983, Wiley and Brooks, 1983). For other reviews and extensive presentations of the theory see Collier (1986), Brooks et al., (1989), and Weber et al., (1989).

3.2 Evolution, thermodynamics and Information

Although, quite open to the preceding approach and acknowledging the importance and necessity of the work of the authors, Wicken has come out with an approach that seems to be more rigid and consistent in its formulation. The approach was formulated through a series of papers throughout the eighties. For more of an overall introduction to his views refer to Wicken (1987, 1988).

In this approach (Wicken, 1988) the biosphere, is to be understood as a closed system in which elements are cycling, the material fluxes being driven by the energy source of solar radiation. As 'closed systems are governed by Gibbs canonical ensemble', the probability of a macrostate can be calculated as

$$P_i = W_i e^{(A-E_i)/kT} \tag{25}$$

where E_i is the internal energy, W_i number of microstates, and A is the average Helmholz free energy of the ensemble.

As the thermodynamic information content, I_M, is - k ln P_i (compare previous expressions) one gets

$$I_M = -k \ln W_i + (E_i - A)/T \tag{26}$$

in which the second term is argued to represent the potential energetic mode of thermodynamic information I_e, refer to equation 16 The first term is negative entropy.

The number of microstates, W_i, are argued to be 'factorable' into a *configurational* and *thermal* part, W_c and W_{th}, respectively. Thus, W_i can be split in to parts

$$W_i = W_c W_{th} \tag{27}$$

The corresponding information can now be defined as

$$I_c = -S_c = -k \ln W_c \tag{28}$$

and

$$I_{th} = -S_{th} = -k \ln W_{th} \tag{29}$$

Taking this expression back to Prigogines equation equivalating the internal entropy production by the dissipation of thermodynamic information, the second law of thermodynamics may be rewritten as

$$dI_c + d I_{th} + I_e < 0 \tag{30}$$

in which the different information is interchangeable as 'increases in one of these parameters must be tied to decreases in others' (Wicken, 1988). "*The flow of thermodynamic information from its energetic and configurational modes to its thermal mode constitutes the tie between evolutionary and thermodynamical arrows*" (Wickens own italics) and thus a possible link between the change in information during evolution and thermodynamics.

3.3 Order from information - genes and proteins

Information theory has found a widespread use in the analysis of macromolecules, like proteins, and in genetics through DNA analysis as in no other biological area (e.g. Chan et al., 1992, Matsuno, 1983, Monet, 1993, Pattee, 1979, Strait and Dewey, 1996, Tsukamoto, 1979). Logically, the applications are quite recent. This is not only because information theory had to be formulated, but also the ontological platform had to be delivered, i.e. the primary structure of the genome, being composed of a sequence of nucleosides, had to be proposed and documented. Even if Schrödinger (Schrödinger, 1944) had already formulated that information about the organisation of the biological system was laid down in the system itself, in his *order from order principle*, it was too early and insufficient to find its way to application.

With the findings of Watson and Crick a platform was provided that fits nicely to information theory to be applied in the analysis of its structure. The fundamental structure of the genes was found to be a one dimensional sequence of symbols, the four nucleosides, adenine, thymine, guanine and cytosine. The nucleosides together form triplet code, and eventually the sequence if translated in to proteins. Proteins are generally composed of amino acids of up to 20 different kinds. The number of possible microstates in a sequence of nucleosides or amino acids has often been used as a measure of information. Using Boltzman's equation as an example will serve to illustrate this. A genome or gene consisting of 300 nucleotides will have a possible number of microstates (w_{nucl}) of

$$w_{nucl} = 4^{300} = 4.110^{180} \tag{31}$$

or translated into microstates of amino acids (w_{amino})

$$w_{amino} = 20^{100} = 1.3\,10^{130} \tag{32}$$

which illustrates a decrease in coding capacity between hierarchical levels stemming from the translation from gene-code to amino acid, i.e. caused by the existence of the triplet code alone! Meanwhile, although this tells something about information capacity and changes herein, this does not tell us much about the thermodynamical information of the system.

Some important papers in this area have been written by Yockey, (Yockey, 1977a, 1977b, 1977c, 1979, 1981) who participated in the criticism against the Brooks-Wiley approach to evolution. Yockey uses information theory for various purposes, but in general takes his entrance point in the analysis of protein sequences of amino acids, in particular Cytochrome c.

In a sequence of papers a formalism is laid out, that determines the information content of the genome connected to coding of proteins. The technique was developed (Yockey, 1974) to analyse functionally equivalent residues in protein sequences (Yockey, 1977a,b,c). Much of the work is concentrated around cytochrome C, a protein considered to have appeared quite early in the history of life and therefore suitable to analyse homologies between different evolutionary lines.

When applied to sequences cytochrome it is shown that the information content of the genetic message (101 sites) is 2.953 bits per residue, when including residues which are predicted to be synonymous. This is considerably lower than the 4 bits used for a duplet code **or** the information content determined on basis of using only residues that are **known** to be synonymous which has an information content of 3.701 bits per residue. The consequence of this work seems to be that Shannon information has been reduced during evolution.

The work carries a lot of implications for the discussion about possibilities of life to have occured by accidental processes. In a paper it is shown that the 1 billion years earth has existed is not enough time for a coding sequence to emerge from a racemic mixture of aminoacids by accidental or spontaneous processes. The maximum sequence to be achieved during the time of evolution is a string of 49 amino acids.

Furthermore, Yockey supports the previous existence of a duplet code in early life-forms and that a triplet code may have arised from this. If this is the case, seen in this light, the above the evolution of the triplet code might have been a consequence of a necessity to reach a further reduction of the Shannon information of the coding sequences.

4. Thermodynamical information

Another measure found in literature bringing the information concept even closer to thermodynamics is that of thermodynamical information. Thermodynamic information is explained in brief by being a measure of the deviation of the entropy state of a system from thermodynamical equilibrium. Thermodynamical information enters as an important part of the formulation of the thermodynamical efficiency function known as exergy. Exergy was proposed to be an important measure for ecosystems in the late 1970's and has since then found several applications in the evaluation of ecosystems. During the last years, other related concepts, including the exergy index and inclusive Kullback indices have been proposed. All of which are based on calculations finding a close correlation to information theory.

4.1 Exergy and thermodynamical information

The introduction of another informational concept termed thermodynamical information was introduced by Evans in his works on exergy or essergy (Evans, 1969, Evans et al. 1966). Exergy, Ex, is often defined as

$$Ex = T(S_{eq} - S_{st}) = T\,I \tag{33}$$

where T is the absolute temperature, S_{eq} the entropy of a reference state, maximal at thermodynamical equilibrium, S_{st} the entropy state of the system, and I is called the *thermodynamical information* of the system.

The thermodynamical information has been argued to be equal to the Kullback measure of information (Kullback, 1968).

$$I = k \sum_j p_j^* \ln \frac{p_j^*}{p_j} \tag{34}$$

If the entropies in equation 33 are replaced by the Boltzmann entropies (and the temperature is omitted), the thermodynamical information may be rewritten:

$$I = k \ln W_{eq} - k \ln W_{st} = k \ln \frac{W_{eq}}{W_{st}} \tag{35}$$

which gets very close to the definition of information of Brillouin.

4.2 Exergy of Ecosystems

In the early 1980's it was extended to ecosystem level (Mejer and Jørgensen, 1979, Jørgensen and Mejer, 1981), arguing that exergy of an ecosystem may be based on the chemical potentials of the most essential element(s) of the system. Quoting Evans

$$I = \frac{U + PV - TS - \Sigma x_j n_j}{T} \tag{36}$$

and Exergy, Ex

$$Ex = U + PV - TS - \sum_j x_j n_j \tag{37}$$

which is the form derived by Jørgensen & Mejer (Mejer and Jørgensen, 1979, Jørgensen and Mejer, 1981).

$$Ex = RT \sum_{c=0}^{n} (\xi_c \ln \frac{\xi_c}{P_{eq,C}} - (\xi_c - P_{eq,c})) \tag{38}$$

Thus the exergy is independent of the ecosystem network, as opposed to the concept of ascendency presented in the following (section 5.2), but the calculation is based on the values of compartments alone. The relative or absolute size of the compartments, though, are considered to be a consequence of the flows in the network.

Another index measuring the structure of the ecosystem, internal exergy, based on fraction instead of biomass was proposed by Herendeen (1989). According to Herendeen equation 38 may be rewritten

$$Ex = RT\,C_0[\ln(\frac{C_0}{C_{0,ref}}) - (1 - \frac{C_{0,ref}}{C_0}) + \sum_{i=1}^{n} x_i \ln(\frac{x_i}{x_{i,ref}})] \qquad (39)$$

where $xi = C_i/C_0$ and $xi,ref = C_{i,ref}/C_{0,ref}$.

The first part is considered to be due to external influences and the second part

$$R\,T\,C_0 \sum_{i=1}^{n} x_i \ln(\frac{x_i}{x_{i,ref}})] \qquad (40)$$

is argued to reflect the contribution to exergy of the internal structure (Herendeen, 1989).

Using either of the two expressions as goal function in ecological modelling has been shown not to have a great effect as long as the ecosystems are in a continuous state evolution as the two indices only differ when the ecosystems are undergoing abrupt changes or transitions (Nielsen, 1992).

Some fundamental problems in the previous formulations exist, and some changes bringing exergy and thus thermodynamical information even closer to information theory has recently been introduced in an attempt to answer this criticism. At the beginning, the reference levels were set at thermodynamical equilibrium. This means that one would have to find values for the concentrations of biological components at this reference level, which is difficult to estimate. Another criticism addressed the point that all compartments in the ecosystem counted equal, which does not fit with the normal intuition that for instance a fish is more complex, (and therefore should count more) than a mono-cellular algae.

4.3 An Information based exergy index

Therefore, a new *exergy index* has been proposed where these were attempted to be solved (Jørgensen et al., 1995, Bendoricchio and Jørgensen, 1997, Jørgensen, 1997). In answer to the first criticism, the reference state was moved so that the first level of the system was detritus, i.e. dead organic matter, containing an energy of approx. 18.5 kJ g^{-1}. The zero-level indicates the level of inorganic nutrients which is considered, for biological systems, to contain no useful energy or exergy. The energy content in the other compartments are taken as the same, but is now multiplied by a weighting factor, β_i, (i is the index for each compartment), see Table 2. Thus, we have

$$Ex \approx -RT \sum_{i=1}^{n} c_i \ln \beta_i \qquad (41)$$

where R is the gas constant, T the absolute temperature, c_i is the concentration of compartment and P_i is a probability value based on the amount of genes expressed during the life cycle of the organism. In this way the thermodynamic state has now been combined with the informational content of the genes (Jørgensen et al., 1995, Bendoricchio and Jørgensen, 1997). The value of the weighting factors corresponds very much to what is meant by the concept of stored information used by Brooks and Wiley above. Some weighting factors found in current literature are found in Table 2.

Table 2: Weighting factors based on the information content on the part of the genome expressed during the life history of the respective type of organisms used for calculating the recently introduced exergy index of ecosystems (after Jørgensen et al., 1995, Bendoricchio and Jørgensen, 1997

organism	number of information genes	weighting factor
Detritus	0	1
Minimal cell	470	2.3
Bacteria	600	2.7
Algae	850	3.3
Yeast	2000	6
Fungi	3000	10
Sponges	9000	26
Plants, tress	10000-30000	30-90
Worms1000	10000-100000	30-300
Insects	10000-15000	30-45
Zooplankton	10000-50000	30-150
Crustaceans	100000	300
Fish	100000-120000	300-350
Birds	120000	350
Amphibians	120000	350
Reptiles	120000	350
Mammals	140000	400
Humans	250000	700

For reviews on the application of these new indices refer to Nielsen et al., (1997).

4.4 The inclusive Kullback index

To complete this presentation of information, derived indices used in ecosystem theory another measure based on Kullback (Kullback, 1968) mentioned above is introduced. In a recent paper, Aoki, known for his approach of entropy analysis of organisms and ecosystems, proposed to modify the Jørgensen/Mejer derived exergy of ecosystems (Aoki, 1993).

Aoki states that the classical exergy expression consists of two parts, the Kullback Information, K, and Wiener entropy, L, (sic!). Thus, this exergy may be written as

$$Ex \propto (K + L) \tag{42}$$

He suggested to replace the scale-dependent, L, with the scale-independent index, M, which is considered a 'more adequate quantity' to get the index, I:

$$I = (K + M) \tag{43}$$

referred to as the *inclusive Kullback index*. He uses the new index for comparison between Lake Mendota and Lake Suwa, using the oligotrophic part and a more eutrophic part of Lake Biwa as reference system, respectively.

5. Information in networks

As seen, information seems to have had a harder time finding its way into ecology as opposed to the use in genetics presented above. This is, in particular, valid when it comes to the application of thermodynamical information. This is peculiar since many of the indices commonly used to characterise ecosystems, within the area of what we could call empirical ecology, most often find their roots in the classical information theory and probalistics (e.g. Goodman, 1975). Information theory has also been applied in the analysis of networks and their topology (Karreman, (1955), Rashevsky, (1954a,b), Wagensberg et al., (1988, 1990)). Thus the extension of this type of work to ecosystem networks seem to follow a logical development.

5.1 Diversity indexes

Ecologists have been using indices of biodiversity and similarity between societies very similar to the mathematical expressions derived in information theory for some decades now. In fact, a quite recent book in numerical ecology (Legendre and Legendre 1998) claims that the three most commonly used indices have their origin in information theory and dedicates a whole section to this point, including the almost unavoidable discussion between information and entropy.

These indices have been found to be especially valuable in the various indices of diversity, like e.g. the Shannon-Wiener index, Simpons-index, MacArthur's index, used to characterise biological societies and ecosystems. The importance of these works has been widely recognised in the efforts to improve management of nature (e.g. Burton et al., 1992) in order to preserve biodiversity of nature which is, at present, threatened by many of the activities undertaken by our society.

At a recent conference, it was proposed to base the calculation of biodiversity on biomass of the species and not on species abundance as individual numbers. This would indeed bring the calculations of diversity close to exergy calculations presented above. The outcome of the calculations of the two would at least be proportional, if not equal to each other. This is provided that biomass is expressed as energy, and exergy is calculated in its classical form.

5.2 Ascendancy and average mutual information

Another information derived indicator of ecosystem behaviour, function and distribution is the concept of *Ascendancy*, which was launched in the early eighties by R.E. Ulanowicz at the Chesapeake Bay Laboratory, University of Maryland (see Ulanowicz, 1986, 1997).

In his recent reformulation of the concept (Ulanowicz, 1997) Ulanowicz builds on the world-view of Propensities that was used by the scientific philosopher Karl Popper in one of his last works (Popper, 1990).

In this world-view Popper, makes up with a deterministic understanding of events occurring. In our normal understanding, things happen with a certain fixed probability. But in this new world view, events will happen with a non-fixed probability, i.e. the observation of things happening are connected to some uncertainty. Popper thus states that events in the real world are results of non-equiprobable events inherent in the situation combined with the history of organisms. In accepting the non-equiprobability and the combination of present and history, and further assigning an uncertainty to the non equiprobability we face a world that is dominated by **indeterminism**. The world becomes open as Popper formulates it, like a game of loaded dice.

The difference between the two situations may become clear if one compares the outcomes of a game of dice using for instance two dice, as in a game of 'black Jack', in the case of using a pair of dice that can be distinguished from each other, e.g. by colour (or using the same die for two sequential throws). In the case of making a 1000 throws with a pair of 'honest' dice, i.e. with equal outcomes of the events 1 to 6, and ideal presentation of the results of the throws will be like Figure 4.

Honest Dice b_j "a priori"

a_i "a posteriori"	1	2	3	4	5	6	s a_i
1	27	29	28	28	27	28	167
2	28	29	28	28	27	28	168
3	28	27	28	28	27	28	166
4	28	27	29	28	28	27	167
5	27	28	27	28	28	28	166
6	28	27	28	27	29	27	166
s b_j	166	167	168	167	166	166	1000

Figure 4: The outcome of 1000 throws with an unmanipulated, honest dice. The situation is 'ideal' in the sense that the distribution will probably not be that regular even with 1000 throws

Lets us now assume that somebody has manipulated the dice making them lighter or heavier on the one side or the other in order to interfere with the outcomes, i.e. the events are no longer equally probable, in which case the dice are said to be 'loaded'. In this case the outcome of a thousand throws may be represented by Figure 5.

Loaded Dice b_j "a priori"

a_i "a posteriori"	1	2	3	4	5	6	s a_i
1	150	0	0	0	0	0	150
2	102	43	0	65	8	8	226
3	0	48	5	0	0	0	53
4	0	10	87	12	0	0	109
5	0	14	34	27	20	0	95
6	226	0	0	0	17	124	367
s b_j	478	115	126	104	45	132	1000

Figure 5: Hypothetical outcome of 1000 throws with an manipulated, 'loaded' dice.
The situation is probably unrealistic since all possibilities will probably occur even with a very low frequency.

This outcome clearly deviates significantly from what should be expected, that is the honest situation. Prediction has certainly become more difficult and we will most likely have to conduct more experiments (throws) to determine the probabilities, for instance if a zero in the table is really a zero or that this outcome just has a very small probability.

In both cases it is possible to construct three types of probabilities:

1) *the marginal probability*, that is - expressing the events of all throws having either a_i or all having b_j as a result, the sequences of throws not being significant in this case. These probabilities are usually designated

$$p(a_i) \text{ or } p(b_j) \tag{44}$$

respectively. In the loaded case, the probability of having 4 in the first throw is 109/1000 = 0.109 (or 10.9%). Likewise having 3 as result of the second throw has the probability of 126/1000 = 0.126 (or 12.6%).
2) the joint probability, the probability that two events are coincident or occur together, e.g. $a_i = 4$ and $b_j = 3$. This is designated by the symbol

$$p(a_i, b_j) \tag{45}$$

which in the above situation may be calculated to be 87/1000 = 0.087 (8.7%)
3) the conditional probability, the outcome that the outcome of the second throw is a_i provided that we know that the result of the first throw was b_j, which is

$$p(a_i | b_j) \tag{46}$$

which is to be read as: the probability of a_i - provided b_j already happened.
Again calculating the probability in the above example for $a_i = 4$ provided that we knew the result of the first throw was 3, we get $p(a_i \mid b_j)$ to be 87/109 = 0.798 (or 79.8%).
This was the background of the concept in terms of probability theory. The question is now how to bring this into the ecosystem context.

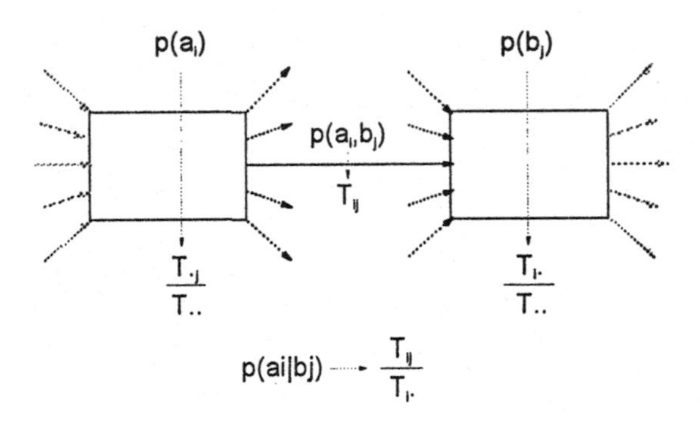

Figure 6: Diagram showing the translation between probabilities and flows which are basic to the calculation of the average mutual information of the ecosystem network.

Lets consider a subsystem now where some substance, energy or matter is transported between two compartments, a_i and b_j, for instance the transfer of energy (or matter) from phytoplankton to zooplankton by grazing. This system is visualised in Figure 6.

The system is driven by all the energy going through the system, also called the total system Throughput (TST) which is equivalent to the sum of all flows in the system, i.e.

$$TST = T_{..} = \sum_i \sum_j T_{ij} \tag{47}$$

where T_{ij} is read as the flow from i to j.

We may now construct the above probabilities of a given quantum of energy or amount of matter leaving a compartment i and entering another compartment j of a system in terms of flows. The probability of a flow leaving i, out of all flows in the system, will be a correspondent of the marginal probability. Thus,

$$p(b_j) \sim \frac{T_j}{T} \tag{48}$$

and

$$p(a_i) \sim \frac{T_{i.}}{T_{..}} \tag{49}$$

where a dot means summation over the index replaced.

The probability of a particular flow, T_{ij}, leaving compartment i and then entering compartment j of the system - an equivalent of the joint probabilities - will be

$$p(a_i, b_j) = p(b_j, a_i) \sim \frac{Tij}{T_{..}}. \tag{50}$$

Correspondingly, the marginal probability will be translated into the conditional probability

$$p(b_j|a_i) = \frac{p(b_j, a_i)}{p(a_i)} \sim \frac{T_{ij}\, T_{..}}{T_{..}\, T_{i.}} = \frac{T_{ij}}{T_{i.}}. \tag{51}$$

As the Average Mutual Information of a network, AMI, is defined as:

$$AMI \equiv \sum_i \sum_j p(b_j, a_i) \log \frac{p(b_j|a_i)}{p(b_j)} \tag{52}$$

we can express the AMI of an ecosystem network as the flows above. In this case we get

$$AMI = (\frac{T_{ij}}{T}) \log \frac{(T_{ij} T_{..})}{(T_{.j} T_{i.})} \tag{53}$$

which forms the development part of the ascendancy orientor. For more description of this derivation and the importance of this orientor to ecosystem networks and its connection to thermodynamics, please refer to (Ulanowicz and Wolff, 1991, Ulanowicz, 1998, Nielsen & Ulanowicz, in print).

An information theoretical approach has been used for analysing the effect of aggreting networks (Hirata and Ulanowicz, 1985). This is important as aggregation is necessary during the working process of modelling, where the complexity of ecosystem networks nearly always has to be reduced. It is therefore of importance to find a way to do this in an objective manner. 'This translates directly into a preference for those aggregation schemes which minimise the decrease in mutual information of the associated network'. The approach is illustrated by looking at the effect when aggregating a 17 compartment model reducing the number of compartments to seven. The approach seems obvious and the results are interesting, but more studies will be needed in order to evaluate the approach.

6. Ecosystem semiotics

In the above it has been demonstrated how the concept of information, through sciences such as cybernetics, network and graph theory, and also thermodynamics, have been applied in the area of biology and ecosystem science (Dale, 1971, Elsasser, 1983, Juhász-Nagy and Podani, 1983, Phipps, 1981). Ecosystems have been analysed as networks and average mutual information of the network have been used by Ulanowicz in a measure of ecosystem developmental state. Thermodynamical information is at the core of the exergy concept as demonstrated by (Evans, 1969 Jorgensen and Mejer, 1981). The information of the genome has been used in a recent reformulation of the exergy, with the purpose of graduating the energies diverted between the various compartments in the ecosystem (Jørgensen et al., 1995, Bendoricchio and Jorgensen, 1997).

By the introduction of these views, all based on the calculation of information derived indices, either from the compartments or the connection of the ecosystem, the road has been paved to introduce a view of the ecosystems a 'communicative systems'. The signals of the system are the flows composed of either energy or matter. The symbols being sent are the various energy quanta, biochemical or biogeochemical compounds. The information may be contained in the state variables or the flows of the system. Following these thoughts, the ecosystem may be viewed as a *semiotic* system in which the transmission and interpretation of signals is the platform on which actions and reactions of the systems take place.

The idea may be more evident at the level of population ecology, where individual organisms communicate together. The communication acts to improve and structure the society as a consequence the 'instructions' sent or received, i.e. information exchanged, between the organisms. As an example, we might take the communications taking place in beehives or in ants nests. The individuals clearly communicate with each other and, sometimes even collective actions are taken as a response to an interpretation of the messages transmitted. This communication has become an important, essential an necessary part of the life cycle of the organisms (ethology).

The idea of ecosystems as communicative systems is relatively new and has not had a widespread use, but similar thoughts can be found in Bateson (1991), Emmeche, Hoffmeyer etc. For some of the first treatments of importance of information processing, sign transmission, interpretation in nature as semiotic processes and the importance to biology, refer to Hoffmeyer (1993, 1997) or Emmeche (1988, 1990, 1991). From these works, one will easily get ideas like the ones stated above.

One important observation and comment has to be made. The communication in the ecosystem occurs at various levels of hierarchy consisting of holons (Koestler, 1969) also called a holarchy. Information exchanged at any level may interact with other levels or within the same level, and that sending of a signal, interpretation and effects may occur other places, i.e. levels of hierarchy, of the whole system.

In a recent paper, Emmeche (1998) introduces an understanding of life within a new framework of bio-semiotics and life as consisting of 'intrinsically' semiotic phenomena. The idea of interpreting the ecosystem processes as transmission and interpretation of signals fits nicely within this new and promising paradigm. Some of the steps in this direction have been taken already but for sure the next years will bring us closer to an understanding of ecosystem behaviour and communication within this framework.

Together, the above points give us a framework for analysing ecosystem communication understood as ongoing intra-level or inter-level system semiotic processes. The final result or outcome of these interactions, the ultimate interpretation and following effect might be quite difficult if not impossible to foresee. The explanation of emergent properties of ecosystems (Küppers, 1990, Müller et al., 1996, Nielsen & Müller, this volume) biological systems or for that matter of any kind of system might lie within the realm of a full understanding of information transfer, i.e. the semiotics of the system.

7. Discussion

Considering the amount of papers written on these matters, combining information and entropy to biology via application, to the genome or to the understanding of evolutionary processes, some patterns repeat themselves and thus seem to block any real solution to the problem. We must remember also, that what has been written is written and that we cannot undo history. Misunderstandings will occur and be repeated for eternity, which does not make them all right. Some authors may have been very close to elegant solutions or solutions to be agreed upon, but it is no longer possible to resolve the earlier discrepancies or points of disagreement. The scope of this paper has not been to pretend to solve the problems but it has been the intention to point out the opinion of the authors, remaining as loyal as possible, and at the same time identifying the major pitfalls that exist when comparing the various uses of the information concept.

The points that continuously repeat themselves as being cardinal points of not being formulated 'clearly' enough are things like the following. Does information exist without being transmitted? How to deal with the determination of microstates, and, what are the reference points? How do we scale the system (e.g. determine the exact exchange rate between information and entropy)? These topics will be shortly dealt with in the following discussion and some major target points are indicated.

a. Transmission or not

Due to the works of Shannon and Weaver, the concept of information seems to have been tightly connected to the transmission process or the analysis of symbols arranged in one-dimensional arrays, like strings of letters from the alphabet or binary digits. The analysis made by Brillouin and Kullback has taken this a little further by coupling the information to the process of observation of a system. This approach seems to comprehend all the others as it shall be argued here. Information is a subset of statistical mechanics or thermodynamics rather than the opposite (Tribus, 1961, Tribus et al., 1966).

Whether a string is being transmitted or not, it needs to be observed at both the sending and receiving site in order to determine the amount transmitted, noise introduced, etc. The process of transmission is taking place in space and following time. In case of the information concept nothing is necessarily separated in space, but, most certainly, in time, e.g. a certain amount of time is needed to do observations in order to determine the exact configuration of the system out of more possibilities.

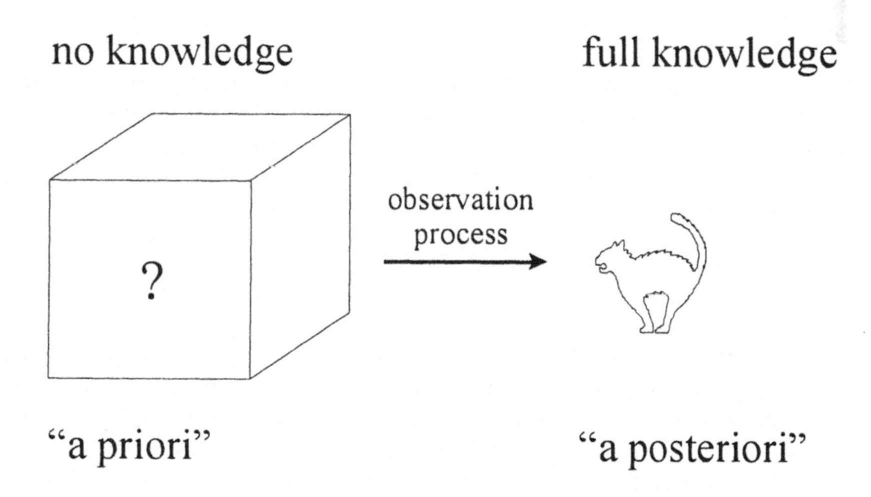

Figure 7: Figure illustrating the difference between the information concept in a transmission and observational system, compare Figure 2. An observer, e.g. a researcher, observes an unknown system and finds out that it 'hides' Schrödingers cat which is apparently still alive.

This determination has to be done in the transmitted system as well. The difference is minimal as to whether one is using Boltzmann's formula, calculations of algorithmic information content, or others of the expressions presented previously.

The amount of information gained, meanwhile, relates to the degrees of freedom we put on our system, i.e. our framework of interpreation, and on our choice of reference state. These perspectives will be described in the following section.

b. Determining microstates

The determination of microstates is partially connected to the problems of arrangement mentioned above. Briefly, the major part of the contributors to communication theory works mainly with a one-dimensional array, whereas thermodynamic information has to work with at least three dimensions and often a hierarchical organised structure has to be taken into account. The approach of Shannon, as documented above, fits nicely to the analysis of a biological system organised according to a one-dimensional view. For other arrangements the connection to communication theory is not clear and has to come via other approaches like thermodynamical information.

Determining the information content, or information gain, it seems to be important how one eventually determines the number of possible microstates of the system. This, in turn, might be a consequence of existing *a priori* knowledge of the system. This can be illustrated for various situations using the example from the introduction, a string of p, o, r, c. We determined this as having 256 possible microstates. Adding more knowledge to the system, that any of the letters will only occur once, still using Boltzmann equation, we will determine to 4! or only 24 possibilities and as a consequence get a considerable reduction. Adding another type of knowledge, that the letters are members of a sequence of not 4 but 26 possibilities, like in the English alphabet, and that they may not even have an equal probability, will get us into new formulations, like the Boltzmann-Gibbs equation. The results of the different calculations will not be the same, although the object under study has not changed. Thus, our 'a priori' knowledge affects the outcome, the information or knowledge we gain, or what we may learn about the system. Our ontological knowledge eventually in this case affects strongly our chances of perceiving the system 'right' and thus carries a huge impact on epistemology.

c. choice of reference levels

The next major point when evaluating the exact value of information is the choice of reference point, the 0-level of system, or the level to which other information values are compared. (no matter how information is calculated, it is a measure of deviation from a certain reference value (sometimes maximum - thermodynamic information vs. Shannon information. Sometimes the reference level is fixed, sometime moving as is the case of Brooks and Wiley's growing genetic material (Brooks and Wiley, 1988). In other cases, we assume no knowledge of the system as *a priori* condition vs. full knowledge *a posteriori* gained during the process of observation. Every scientist will know how little the last described situation will apply to the experimental observation process.

The understanding of this fundamental problem is of particular interest as the knowledge gain seems to differ between observers because their *a priori* knowledge may differ. Information in this context becomes a strictly subjective matter and leaves us little hope of making a real objective definition of the concept of information. The theoretical existence of an *a priori* observer representing absolutely no knowledge (*tabula rasa* poses a fundamental problem: 'How will this observer with absolutely no knowledge be able to observe and tell us something about the system?' At least some ontological knowledge **must** exist in order to observe phenomena! Likewise, the knowledge of the observer may influence the information gained so that it may vary between observers during the same observation. These matters may sound very philosophical but examples of these phenomena interfering with the observation process are found in the current literature.

d. scaling

The matter of 'scaling' the system seems in general to be neglected as almost irrelevant. Many authors reduce the choice of the constant, k, and the base of logarithm simply to a matter of convenience, e.g. taking k=1 and using $\log_2$ when working when with binary sequences. Meanwhile, the actual choice made seems to infer some complications, especially when results are compared between authors.

The *choice of constant* directly affects the units that the expression used will take at the end. It therefore bears a heavy impact on whether the results of various analyses made from and information theoretical viewpoint are comparable or not. In order to be compared between results they have to be in the same units. This refers for instance directly to Løvstrups criticism against Brooks and Wiley (Lovtrup, 1983).

The *choice of logarithm* affects directly the scaling of the system. The inconsistent, and semantically obscure, use of the concepts throughout the literature makes it difficult, if not impossible, to compare results. The logarithm taken to the base two, $\log_2$, is popular probably due to its relation to binary programming problems, whereas most calculations in general are using the natural logarithm, ln. To illustrate the problem just by calculating possible microstates at string of 10 letters will have $10^{28} = 6.3\ 10^{15}$ possibilities, but if translated into binary ASCII code (assuming 8 bits for each letter) there will be $2^{80} = 1.2\ 10^{24}$ possibilities which is increasing the Shannon information by a factor of a billion of the same string just by changing programming method. This is an effect of the wish to work with different logarithmic expression. It is true that the conversion between the various expressions may easily be done by following ordinary algebraic rules, but it needs to be done. This variation in practice, although, is also found throughout the literature.

8. Conclusions

As demonstrated above the concept of information has found widespread use in the current scientific literature. Meanwhile, the applications in various areas have not led to the clarification or unified approach that would have been expected from a theoretical framework with the powers that the concept of information is considered, by some scientists, to have.

First of all, the use of the concept differs too much from our everyday perception and use of the word information. This puts a major obstacle to the receptance of layman or for that matter even to the scientists that wish to stay outside the debate above. In other words, our everyday diffuse use of the word information blocks the way for a more precise use of the concept presented in the scientific literature. The problem of inconsistency in scientific use is dealt with in the following, but is seen as blocking effectively for communication within the scientific community.

Second, the way of use - the approach taken to implementation - within the different areas differs so much that it may seem difficult to look through the fact that the concept(s) used are actually derived from the same platform. The amount of literature in the area is enormous and probably even growing, faster and faster. The major consequence is that it is difficult to overlook the area, as well as to find consistencies, or for that matter even a pattern in the application. The discrepancies are in many cases, as seen from the discussion, more or less to be considered as artefacts introduced by inaccurate or inconsistent use of the concept.

As it appears, that a major effort in the clarification of these differences, and maybe more optimistic, to find a consistency is needed. The work should preferably result in a consensus of more scientific disciplines working together, forming a common platform for the future use of this valuable concept.

Acknowledgements

During the presentation of the above works, erroneous interpretations and (new) misunderstandings may have been introduced. Only this author is of course to blame for this. For the readers' own interpretations, please refer to the original literature. I am very grateful to, R.E. Ulanowicz for helpful comments on the introduction to his theory, to Simone Bastianoni for careful reading and constructive comments on the manuscript, to S. E. Jørgensen and to Henning Mejer, Yuri Svirezhev for the many discussions on this and related topics throughout the years.

REFERENCES

Aoki, I., 1993. Inclusive Kullback index - a macroscopic measure of ecological systems. *Ecol. Modelling,* 66: 289-299.

Bateson, G., 1991. Ånd og natur, en nødvendig enhed. Rosinante, 225 pages. (original title: Mind and Nature)

Bendoricchio, G. and Jørgensen, S.E., 1997. Exergy as a goal function of ecosystems dynamic. *Ecol. Modelling,* 102: 5-15.

Boltzmann, L., 1905. The second law of thermodynamics, Populare Schriften, Essay no. 3, address to the Imperial Academy of Sciences 1886. Reprinted in: Theoretical Physics and Philosophical Probles, Selected Writings of L. Boltzmann. D. Reidel.

Bookstein, F.L., 1983. Comment on a 'Nonequilibrium' Approach to Evolution. *Syst. Zool.,* 32(3): 291-300.

Brillouin, L. 1949. Life, Thermodynamics and Cybernetics. *Am. Sci.,* 37: 554-68. In: Leff, H.S. and Rex, A.F. (eds.), 1990. *Entropy, Information, Computing. Princeton Series in Physics,* 349 pages.

Brooks, D.R. and Wiley, E.O., 1986. Evolution as Entropy. (1st edition) University of Chicago Press, 335 pages.

Brooks, D.R. and Leblond, P.H. and Cumming, D.D., 1984. Information and Entropy in a Simple Evolution Model. *J. theor. Biol.,* 109: 77-93.

Brooks, D.R., Collier, J., Maurer, B.A., Smith, J.D.H. and Wiley, E.O., 1989. Entropy and Information in Evolving Biological Systems. *Biology and Philosophy,* 4: 407-432.

Burton, P.J., Balisky, A.C., Coward, L.P., Cumming, S.G. and Kneeshaw, D.D., 1992. The value of managing for biodiversity. *The Forestry Chronicle,* 68(2), 225-237.

Chaitin, G.J., 1966. On the length of programs for computing Finite Binary Sequences. *J. ACM* 13: 547-569.

Chaitin, G.J., 1974a. Information-Theoretic Computational Complexity. IEEE transactions on information theory, 20: 10-15.

Chaitin, G.J., 1974b. Information-Theoretic Limitations of Formal Systems . *J. ACM* 13: 547-569.

Chaitin, G.J., 1975. Randomness and mathematical Proof. *Sci. Am.,* 225(5): 47-52.

Chaitin, G.J., 1977. Algorithmic Information Theory. *IBM J. Res. Develop.,* 21: 350-359.

Chaitin, G.J., 1982. Gödels Theorem and Information. *Int. J. Theor. Physics.,* 21(12): 941-954.

Chaitin, G.J., 1992. Information-Theoretic Incompleteness. *Appl. Math. Comp.* 52, 83-101.

Chan, S.C., Wong, K.C. and Chiu, D.K.Y., 1992. A Survey of Multiple Sequence Comparison Methods. *Bull. Math. Biol.,* 54(4): 563-598.

Cherry, E.C., 1951. A History of the theory of Information. Inst. Electric. *Engin. Proceedings,* 98(3): 383-393.

Collier, J., 1986. Entropy in Evolution. *Biology and Philosophy,* 1: 5-24.

Dale, M.B., 1971. Validity and utility of information theory in ecological research. Chap 2 in Nix, H.A., (ed.). *Quantifying Ecology. Proceedings of the Ecological Society of Australia,* vol. 6.

Denbigh, K., 1975. An Inventive Universe. Hutchinson.

Ebeling, W., 1991. Chaos - Ordnung Information. Verlag Harri Deutsch, 118.pages.

Ebeling, W. and Volkenstein, M.V., 1990. Entropy and the Evolution of Biological Information. *Physica A.,* 163: 398-402.

Ebeling, W. and Feistel, R., 1992. Theory of Selforganization and Evolution: The Role of Entropy, Value and Information. *J. Non-Equilib. Thermodyn.,* 17: 303-332.

Elsasser, W.M., 1983. Biological Application of the Statistical Concepts Used in the Second Law. *J. Theor. Biol.,* 105: 103-116.

Emmeche, C., 1988. Information i naturen. Nyt Nordisk Forlag, 161 pages.

Emmeche, C., 1990. Det Biologiske Informationsbegreb. Kimære, 319 pages.

Emmeche, C., 1991. A Semiotical Reflection on Biology, Living Signs and Artificial Life. *Biol. Phil.,* 6, 325-340.

Emmeche, C., 1998. Defining life as a semiotic phenomenon. *Cybernetics And Human Knowing,* 5(1): 3-17.

Evans, R.B., 1969. A proof that essergy is the only consistent measure of potential work for chemical systems. (note: title changed during archiving to - work systems!)

Evans, R.B., Crellin, G.L. and Tribus, M., 1966. Thermoeconomic Considerations of Sea Water Demineraliszation. Chapter 2, pages 21-76 In: Spiegler, K.S. (ed.) *Principles of Desalination.* Academic Press, 566 pages.

Frautschi, S., 1988. Entropy in an expanding Universe. Chapter 1 In: Weber, B.H., Depew, D.J. and Smith, J.D. (eds.) *Entropy, Information, and Evolution.* Bradford, MIT, 376 pages.

Gatlin, L., 1972. Information Theory and the Living System. Columbia University Press, CO.

Goodmann, D., 1975. The theory of diversity-stability relationships in ecology. *Quart. Rev. Biol.,* 50(3), 237-266.

Hartley, R.V.L., 1928. Transmission of Information. *The Bell System Tech. J.*, 7:535-563

Herendeen, R., 1989. Energy intensity, residence time, exergy, and ascendency in dynamic ecosystems. *Ecol. Modelling*, 48: 19-48.

Hinegardner, R. and Engelberg, J., 1983. Biological Complexity. *J. Theor. Biol.*, 104: 7-20.

Hirata, H. and Ulanowicz, R.E. 1985. Informational Theoretical Analysis of the Aggregation and Hierarchical Structure of Ecological Networks. *J. theor. Biol.* 116: 321-341.

Ho, M.W. and Saunders, P.T., 1979. Beyond neo-Darwinism - an epigenetic approach to evolution. *J. theor. Biol.*, 78, 573-591.

Hoffmeyer, J., 1993. *En snegl på Vejen. Betydningens naturhistorie.* Omverden, 227 pages. (Translated to English as: Signs of Meaning in the Universe, Indiana University Press).

Hoffmeyer, J., 1997. SHORTS. 40 artikler om Natur Videnskab og Liv. Munksgaard, Rosinante, 96 pages.

Hons, P., 1991. Study of the Dynamic of Nutrient Uptake by Plants Using the Method of Relative Information Entropy. *Scientia agriculturae Bohemoslovaca*, 23(1), 17-26.

Jaynes, E.T., 1957a. Information Theory and Statistical Mechanics. *Phys. Rev.*, 106(4): 620-630.

Jaynes, E.T., 1957b. Information Theory and Statistical Mechanics. II. *Phys. Rev.*, 108(2): 171-190.

Jaynes, E.T., 1962. New Engineering Applications of Information Theory. Pages 163-203 In: Bogdanoff, J.L. and Kozin, F., (eds.): *Proceedings of the First Symposium on Engineering Applications of random function theory and probability.*

Juhász-Nagy, P. and Podani, J., 1983. Information theory methods for the study of spatial processes and succession. *Vegetatio*, 51, 129-140.

Jørgensen, S.E. and Mejer, H., 1977. Ecological Buffer Capacity. *J. Ecol Modelling*, 4: 253-279.

Jørgensen, S.E., 1997. Integration of Ecosystem Theories: A Pattern (2nd rev. ed). Luwer, 388 pages.

Jørgensen, S.E. and Mejer, H.F., 1981. Exergy as key function in ecological models. Pages 587-590 In: Mitsch, W.J., Bosserman, R.W., and Klopatek, J.M., (eds.*) Energy and Ecological Modelling. Developments in Environmental modelling, 1.* Elsevier, 839 pages.

Jørgensen, S.E., Nielsen, S.N., Mejer, H., 1995. Emergy, environ, exergy and ecological modelling. *Ecol. Modelling*, 77: 99-109.

Karreman, G., 1955. Topological Information Content and Chemical Reactions. *Bull. Math. Biophys.*, 17: 279-285.

Koestler, A., 1969. Beyond atomism and holism - the concept of the holon. In: Koestler, A. and Smythies, J.R., (eds.) *Beyond reductionism. New perspectives in the life sciences.* 192-232 pages.

Kolmogorov, A.N., 1968. Logical Basis for Information Theory and Probability Theory. *IEEE transactions on information theory.* IT-14(5) 662-664.

Kubát, L. and Seeman, J. (eds.), 1975. Entropy and Information in science and Philosophy. Elsevier, 260 pages.

Kullback, S., 1968 (1959). Information Theory and Statistics. Dover Publications, 399 pages.

Küppers, B.O., 1990. Information and the Origin of Life. MIT Press, 215 pages.

Laxton, R.R., 1978. The measure of Diversity. *J. theor. Biol.*, 70, 51-67.

Layzer, D., 1988. Growth of Order in the Universe. Chapter 2 In: Weber, B.H., Depew, D.J. and Smith, J.D. (eds.): *Entropy, Information, and Evolution.* Bradford, MIT, 376 pages.

Leff, H.S. and Rex, A.F. (eds.), 1990. Entropy, Information, Computing. Princeton Series in Physics, 349 pages.

Legendre, P. and Legendre, L., 1998. Numerical Ecology, 2nd English edition. Development in Environmental Modelling 20. Elsevier, 853 pages.

Løvtrup, S., 1983. Victims of Ambition: Comments on the Wiley and Brooks Approach to Evolution. *Syst. Zool.*, 32(1): 90-96.

Martin-Löf, P., 1966. The Definition of Random Sequences. *Info. Control*, 9, 602-619.

Matsuno, K., 1983. Evolutionary Changes in the Information Content of Polypeptides. *J. Theor. Biol.*, 105: 185-199.

Mejer, H. and Jørgensen, S.E., 1979. Exergy and Ecological Buffer Capacity. In: Jørgensen S.E. (ed*) State-of-the-art in Ecological Modelling* vol 7, pp 829-846

Monet, R., 1993. Quantification of a Genetic Message in a Selection. *Acta. Biotheor.*, 41: 199-203.

Morowitz, J.H., 1955. Some Order-disorder Considerations in Living Systems. *Bull. Math. Biophys.*, 17: 81-86.

Morowitz, H.J., 1970. Entropy for Biologists - An introduction to thermodynamics. Academic Press, 195 pages. (3rd printing, 1972).

Morowitz, H.J., 1978. Foundations of Bioenergetics. Academic Press, 344 pages.

Morowitz, H.J., 1979. Energy flow in biology. Ox Bow Press, 179 pages.

Müller, F., Breckling, B., Bredemeier, M., Grimm, V, Malchow, H, Nielsen, S.N. and Reiche, E.W., 1996. Emergente Ökosystemeigenschaften. In: Fränzle, O., Müller, F. and Schröder, W. (eds.): *Handbuch der Ökosystemforschung*, Chap. III-2.5

Nielsen, S.N., 1992a. Application of maximum exergy in structural dynamic models. Ph.D. thesis. National Environmental Research Institute. 52 pages

Nielsen, S.N. and Ulanowicz, R.E., in press. In the consistency between thermodynamical and network approaches to ecosystems. *Ecol Modelling.*

Papentin, F., 1980. On Order and Complexity. I. General Considerations. *J. Theor. Biol.*, 87: 421-456.

Papentin, F., 1982. On Order and Complexity. II. Application to Chemical and Biological Structures. *J. Theor. Biol.*, 95: 225-245.

Pattee, H.H., 1979. The Complementarity Principle and the Origin of Macromolecular Information. *BioSystems*, 11: 217-226.

Phipps, M., 1981. Entropy and Community Pattern Analysis. *J. Theor. Biol.* 93: 253-273.

Popper, K.R., 1979. Die Subjektivistiche Theorie der Entropie. Chap 36 In:: Ausgangspunkte. *Meine Intellektuelle Entwicklung*. Hoffman und Campe.

Prigogine, I. and Nicolis, G., 1971. Biological order, structure and instabilities. *Quart. Rev. Biophys.*, 4(2&3), 107-148.

Prigogine, I. and Wiame, J.M., 1946. Biologie et thermodynamique des phénomènes irréversibles. *Experientia* *2*(11): 451-53.

Rao, G.S., Hamid, Z and Rao, J.S., 1979. The Information Content of DNA and Evolution. *J. Theor. Biol.*, 81: 803-807.

Rashevsky, N., 1955a. Some Remarks on Topological Biology. *Bull. Math. Biophys.*, 17: 207-218.

Rashevsky, N., 1955b. Life, Information Theory, and Topology. *Bull. Math. Biophys.*, 17: 229-235.

Rényi, A., 1970. Chapter 9. Appendix. Introduction to Information theory. In: Renyi, A. (ed) *Probability theory.* North-Holland.

Ruch, E. and Lesche, B., 1978. Information extent and information distance. *J. Chem. Phys.*, 69(1): 393-401.

Schrödinger, E. 1944. What is life? Cambridge

Shannon, C.E., 1948. A Mathematical Theory of Communication. *Bell System Techn. J.*, 27(3):379-423.

Shannon, C. and Weaver, W., 1949. The mathematical Theory of Communication. University of Illinois Press, 125 pages.

Stonier, T., 1990a. Information and the Internal Structure of the Universe. Springer, London, New York. 155 pages.

Stonier, T., 1990b. Towards a new theory of Information. Telecommunications Policy, 10: 278-281.

Stonier, T., 1992. Beyond Information. The natural history of Intelligence. Springer, London, New York. 221 pages.

Stonier, T., 1997. Information and Meaning. An Evolutionary Perspective. Springer, London, New York. 255 pages.

Strait, B.J. and Dewey, T.G., 1996. The Shannon Information Entropy of Protein Sequences. *Biophys. J.*, 71: 148-155.

Szilard, L., 1929. On the decrease of entropy in a thermodynamic system by the intervention of intelligent beings. In: Leff, H.S. and Rex, A.F. (eds.), 1990. *Entropy, Information, Computing. Princeton Series in Physics*, 349 pages.

Tribus, M., 1961. Information Theory as the Basis for Thermostatics and Thermodynamics. *J. Appl. Mech.*, 28: 1-8.

Tribus, M., Shannon, P.T., and Evans, R.B., 1966. Why Thermodynamics Is a Logical Consequence of Information Theory. *A.I.Ch.E. Journal*, 12(2):, 244-248.

Tsukamoto, Y., 1979. An Information Theory of the Genetic Code. *J. Theor. Biol.*, 78: 451-498.

Ulanowicz, R.E., 1986. Growth and Development. Springer 203 pages.

Ulanowicz, R.E., (1998). Network Orientors: Theoretical and Philosophical Considerations why Ecosystems may Exhibit a Propensitiy to Increase in Ascendency. Chap. 2.10 In: Müller, F. and Leupelt, M., (eds.) 1998. *Eco Targets, Goal Functions, and Orientors.* Springer, 619 pages.

Ulanowicz, R.E. and Wolff, W.F., 1991. Ecosystem Flow Networks: Loaded Dice. *Math. Biosci.* 103: 45- 68.

Vincent, L.-M., 1993. Theory of data Transferral. Principles of a New Approach to the information Concept. *Acta. Biotheo.*, 41: 139-45.

Wagensberg, J., Valls, J. and Bermudez, J., 1988. Biological adaptation and the mathematical Theory of Information. *Bull. Math. Biol.* 50(5): 445-464.

Wagensberg, J., Garcia, A., and Sole, R.V., 1990. Connectivity and Information transfer in Flow Networks: Two Magic Numbers in Ecology., *Bull. Math. Biol.*, 52(6): 733-744.

Walker, I., 1993. Competition and Information. *Acta. Biotheor.*, 41: 249-266.

Weaver, W., 1949. The mathematics of communication. *Sci. Am.* 181(1): 11-15.

Weber, B.H., Depew, D.J. and Smith, J.D., 1988. Entropy, Information, and Evolution. Bradford, MIT, 376 pages.

Weber, B.H., Depew, D.J., Dyke, C., Salthe, S.N., Schneider, E.D., Ulanowicz, R.E. and Wicken, J.S., 1989. Evolution in Thermodynamic Perspective: An Ecological Approach. *Biology and Philosophy*, 4: 373-405.

Weizsäcker, E. von, and Weizsäcker, C. von, 1972. Wiederaufnahme der begrifflichen Frage: Was ist Information? Nova Acta Leopoldina, 37(206): 535-555.

Whittaker, R.H., 1972. Evolution and measurement of species diversity. *TAXON*, *21*(2/3), 213 -251.

Wicken, J.S., 1978. Information Transformations in Molecular Evolution. *J. Theor. Biol.*, 72: 191-204.

Wicken, J.S., 1983. Entropy, Information, and Nonequilibrium Evolution. *Syst. Zool.*, 32(4): 438-443.

Wicken, J.S., 1987a. Entropy and Information: Suggestions for a Common Language. *Philosophy of Science*, 54: 176-93.

Wicken, J., 1987b. Evolution, Thermodynamics and Information. Extending the Darwinian Program. Oxford, 243 pages.

Wicken, J.S., 1988. Thermodynamics Evolution, and Emergence: Ingredients for a New Synthesis. Chapter 7 in Weber, B.H., Depew, D.J. and Smith, J.D. (eds.): *Entropy, Information, and Evolution.* Bradford, MIT, 376 pages.

Wiley, E.O. and Brooks, D.R., 1982. Victims of history - A nonequilibrium approach to evolution. *Syst. Zool.*, 31(1): 1-24.

Wiley, E.O. and Brooks, D.R., 1983. Nonequilibrium Thermodynamics and Evolution: A Response to Løvtrup.

Syst. Zool., 32(2): 209-219.
Yockey, H.P., 1977a. A Calculation of the Probability of Spontaneous Biogenesis by Information Theory. *J. Theor. Biol.*, 67: 377-398.
Yockey, H.P., 1977b. A prescription which Predicts Functionally Equivalent Residues at Given Sites in Protein Sequences. *J. Theor. Biol.*, 67: 337-343.
Yockey, H.P., 1977c. On the Information Content of Cytochrome c. *J. Theor. Biol.*, 67: 345-376.
Yockey, H.P., 1979. Do Overlapping Genes Violate Molecular Biology and the Theory of Evolutionand *J. Theor. Biol.*, 80: 21-26.
Yockey, H.P., 1981. Self Organisation Origin of Life Scenarios and Information Theory. *J. T heor. Biol.*, 91: 13-31.

II.4. Ecosystems as Cybernetic Systems

Sven E. Jørgensen and Milan Straskraba

1. Introduction of cybernetics and its terminology

Norbert Wiener called cybernetics 'control and communication in the animal and machine' but this control and information transfer function of systems is rather neglected in descriptions and simulation models of ecosystems. Ecosystem models are mostly based on mechanistic approaches, assuming fixed system function and structure. This is limited not only with respect to theory, but also to practical applications. Fleishman (1976) recognised that ecosystems belonged to the highest category of cybernetic, self transforming systems, not connected by steady material carriers. According to Fleishman (1976), the mechanistic increase of the number of system parameters leads to the problem of multidimensionality which cannot be solved by even the greatest of computers.

Cybernetic, particularly control theory, terminology pays tribute to the original economic application of cybernetics and is therefore rather strange to the physically oriented biologist or ecologist (Kozlowski et al., 1980). Terms like 'goal function', 'control action' or 'optimiser' have an anthropomorphic flavour and cause one to look for a thinking creature able to follow goals, to decide and to act. However, these terms, in contemporary cybernetics have much broader meanings, showing only some similarity to economic systems. The above names are used for historical reasons, although they cover rather than elucidate the contemporary meanings of the corresponding cybernetic terms (Straskraba, 1982). Some ecologists dislike the idea of optimisation, because, as stated by Innis and Clark (1977); all properties that require optimisation are to be rejected as the basic organisational principles of ecosystems, as they require some kind of objective or goal for the ecosystem. Optimality in mathematics originated on a purely physical basis, intended as description of physical phenomena, completely independent of man. The fear of biologists of anthropomorphism when using the term goal or goal function is therefore without any real basis. The classical example of a ball reaching its optimum position in the deepest hole accessible demonstrates this point very clearly. Since the 17th century mathematicians have formulated the reaching of the hole by a ball using a goal function, and no mathematician was ever worried about where the ball gets a brain to 'find' the hole and 'solve' the optimality problem. Certainly, there may be applications in which humankind is involved, and there may be scientists using optimality notions for ecosystems with an improper anthropomorphic justification. However, such examples have nothing in common with the mathematical meaning of optimality. An optimum is simply a maximum or a minimum, subject to constraints that are inevitable components of the optimality formulation. Without constraints no optimality is possible. Discussion such as whether it is the history of organisms or optimality that drives organisms are based on complete ignorance of the mathematical optimality notions. History of organisms determines their characteristics and abilities given by their evolutionary history, and these are, among others, the constraints in which a certain minimum or maximum is searched. Another problem may be that we are unable to specify the constraints with sufficient accuracy or state the goal function appropriately. But this is our fault, not the fault of the goal function principle. It would be non-scientific of ecologists to reject the notion of optimality on the basis of ignorance about its meaning. Cybernetics, technology and other branches of science start in an increasing extent to use biological and ecological notions as demonstrated by the existence of 'adaptive systems', 'learning systems', and 'viability theory'. It would be negligence if we ecologists *a priori* reject the methodological offers of cybernetics, particularly its major branch, control theory. Cybernetics is today a science which takes many directions and is divided into a number of branches, which extend very rapidly into new ramifications and thus losing a unified framework. However, the rapid development into new directions shows great potential of what was once a well defined science. What is particularly appealing from the ecosystem point of view is that generalities rather than specifics, and whole systems rather than simply separate parts are the focus. This is exactly what system ecology is aiming for, but has not yet reached.

At the recent workshop 'Eco Targets, Goal Functions and Orientors' (Müller and Leupelt, 1998), it was agreed among the participants that the term *goal function* should be used solely in a modelling context, while the term *orientors* covers the propensity that characterises the development of ecosystems. A third term, *ecological indicators*, is applied to indicate the health or integrity of ecosystems. For further details on this discussion see Müller and Leupels, 1998, Jørgensen, 1997 and Ulanowicz, 1997.

Mesarovic (1968) had already demonstrated that the 'goal seeking' notion does not imply any 'spirit' and that it can be deduced from a clear physical basis. Tribus and McIrwine (1971), when discussing the two independent notions of entropy in thermodynamics and in information theory, demonstrated the identity and transferability of both definitions. They suggested that the differences between thermodynamics and cybernetics, however strange they may seem, originated from historical rather than subject reasons.

A contemporary mechanistic approach is suitable for treating ecological systems as systems with fixed structures and behaviours (which they are not), but has not developed ways of helping us to understand the changes in the structure and functions that are characteristic of ecosystems. Cybernetics, in contrast, provides us with a developed methodology for describing and understanding the dynamics of such systems. The methodology is based on generalising the common threads of fairly divergent classes of systems independently of their physical nature (Straskraba, 1982).

It appears difficult to see how the many factors of importance for the development of ecosystems could be maximised, minimised or optimised at two or more levels simultaneously. However, contemporary cybernetics elaborates directions suggesting that in society and in technical systems unified goals do not always exist. Moreover, very often the opposite is true. Branches dealing with the mathematical formulation of such systems are called 'theory of hierarchical system' (Mesarovic et al., 1970), 'multi-objective optimisation' (Peschel and Riedel, 1976) and 'theory of games' (Germeyer, 1976). In an economic system, multidimensional optimisation is often relatively easy because the system uses a common currency, money.

Modern control theory recognises a number of different approaches which take into account the uncertainties present in the determination of constraints and goals. Different formulations of sub-optimality do not pose criteria as strict as those in classical optimality which seems not to be reached by any biological assembly, with the exception of the most complex one, an ecosystem. Viability theory is a branch of science with a name indicating its biological background. However, with evolution in mind its author Jean Pierre Aubin was targeting the theory primarily for use in economics. Only later the approach was applied to ecological problems.

We may be able to use the thermodynamic concept of exergy as a common currency for the organisms in the ecosystem (for presentation of this concept see Chapter II.1), Exergy represents the organisational level of the entire ecosystem by including biomass, the information carried in the organisms (the genes) and the information carried in the network, thus allowing it use as a common currency. Exergy also represents the distance from thermodynamic equilibrium, i.e. from the state when the system is dead and without any work potential (see Chapter II.1).

The need for an introduction of cybernetics to system ecology is due to the presence of a high number of feedback mechanisms (see Chapter I.1.1.) which are hierarchically organised like the ecosystem itself. The problem in short is: how can we describe the development of an ecosystem when we have to account for all these feedback processes which give it flexibility, adaptability and self-organisation. An ecosystem is not a rigid physical system, instead it is steadily changing in a very changeable world.

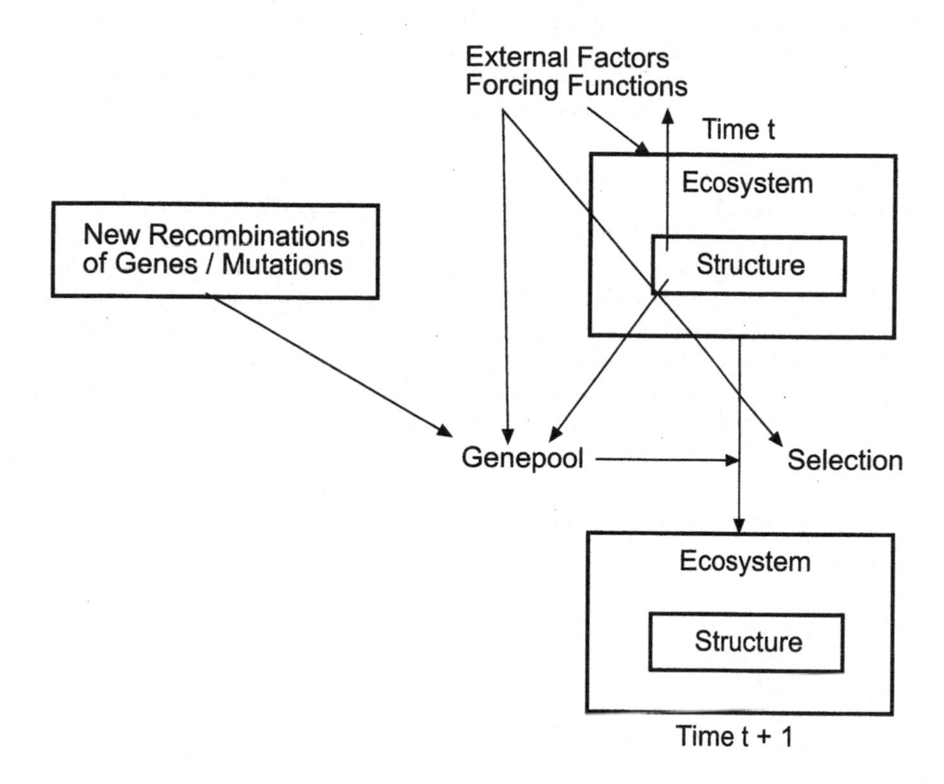

Figure 1: Conceptualisation of how the external factors steadily provoke phenotypic adaptation and change of the species composition. The steadily changed conditions are in the first hand met by species adaptation, but adaptation of the species present in the ecosystem is only possible to a certain extent. If adaptation is not sufficient to cope with the external factors, possible shifts in species composition occur and mutual relations between species change (= changes of the ecosystem structure). The properties of the species are determined by phenotypic adaptations and by the gene pool, which is steadily changed due to mutations and new sexual recombinations of genes. The development is, however, more complex. This is indicated by 1) arrows from 'structure' to 'external factors' and then to 'selection' to account for the possibility that the species are able to modify their own environment and thereby their own selection pressure; 2) an arrow from 'structure' to 'gene pool' to account for the possibilities that the species can to a certain extent change their own gene pool which occur within life-cycle of a species. The same sequence of changes may be provoked by internal factors like species interactions.

All species in an ecosystem are struggling for survival and trying to increase their population numbers under the prevailing conditions. The prevailing conditions are considered as *all* factors influencing the species, i.e., all external and internal factors including those originating from other species. This explains coevolution as: any change in the properties of one species will influence the evolution of the other species.

All natural external and internal factors of ecosystems are dynamic, i.e., are also steadily changing. The present species attempt to adapt to the changing conditions. When their adaptation is not sufficient to meet the current changes, there are always many species waiting in the wings, who are better fitted to the emerging conditions and are ready to 'take over' from the species dominating under the present conditions. They are present in the system and its surroundings in a low number (sometimes below the detection limit). There is a wide spectrum of species representing different combinations of properties available for the ecosystem. The question is, which of these species are best able to survive and grow under the present conditions, and which species are best able to survive and grow under the conditions one time step further, two time steps further and so on? The necessity in Monod's sense is given by the prevailing conditions – the species must have genes, or perhaps rather phenotypes (meaning properties), which match these conditions in order to survive. But the natural external factors and the genetic pool available for the 'test' may change randomly or by 'chance'.

Steadily, new mutations (misprints are produced accidentally) and sexual recombinations (the genes are mixed and shuffled) emerge and thus provide new material for selection.

These ideas are illustrated in Figure 1. It is shown, that the external factors are steadily changing (some even relatively fast) – partly at random e.g. the meteorological or climatic factors. The species dominating in the system are selected among the species available and represented by the genetic pool, which is slowly but surely changed at random or by 'chance'. What is defined as ecological development is the changes over time in nature caused by the dynamics of the external factors.

In contrast, evolution is related to the genetic pool. It is the result of the relation between the dynamics of the external factors and the dynamics of the genetic pool. The external factors steadily change the conditions for survival and the genetic pool steadily changes, resulting in new solutions to the problem of survival.

2. Control mechanisms in ecosystems

It is possible to distinguish six types of control in natural systems (Straskraba 1980 and 1995):

1) Direct control
2) Feedback control
3) Systems self adaptation
4) Systems self organisation
5) Systems evolution
6) Hierarchical systems self control

Like any classification, this one is somewhat arbitrary, as there will always be cases which will not fit exactly in any of these categories.

Direct control corresponds to direct relations between system elements and the system with its broader environment. Examples include; the effect of temperature on different biological processes like respiration, photosynthesis, excretion etc.

Feedback control corresponds to feedback relations between system elements, both negative and positive. The basic type of feedback control is represented in the ecosystems by the predator-prey relations; see Figure 2.

However, trophic control is far from the only type of feedback control we can distinguish. Andzejewski (1977) distinguished an intrapopulation feedback control by internal population changes in density dependence of organisms, for example, control of the environment and paratrophic control by which he denominated the change in prey selection by competition among predators. The complexity of environmental control by organisms can be demonstrated by the example of self shading by algae; the self shading causes complex hydrophysical changes leading to a temperature increase and augmentation of algal growth (Straskraba, 1995).

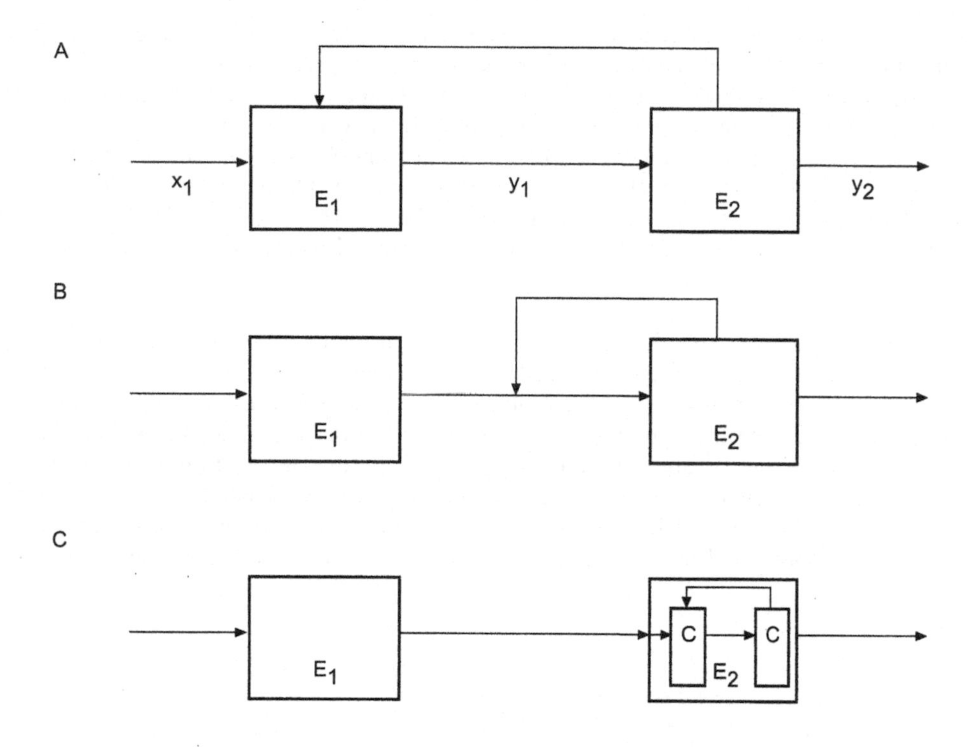

Figure 2: Trophic feedback control in aquatic ecosystems. A shows the feeding control by feeding of E2 on the lower element in the food chain, E1. B indicates the control of food uptake and feeding utilisation efficiency. C illustrates the adaptive trophic feedback control. C and P are the controller and the process, respectively. Reproduced from Straskraba (1995)

Other types of feedback control can be distinguished in addition to those named by Andrzejewski, for example feedback control by learning (food selection by some visual predators during changing food availability is known to recognise an 'image' which is followed even when the prey density decreases and only then the predator learns to recognise the now more dense food items switching of feeding to this new food occurs) and the most complicated hierarchical feedback control.

An example of hierarchical control is the feeding of *Daphnia.* We can distinguish three levels of control. The explanation is based on the angular shape of dependence of grazing rate on food concentration. The first level of control acts when food concentrations are below the critical level, labelled incipient concentration level, ICL. At this level the filtration rate is constantly independent on food concentration and reaches the maximum possible rate for the given *Daphnia* species and clone; temperature and food composition being the primary controls. In this range, the grazing rate, which expresses the amount of food eaten per unit of time, therefore rises linearly with food concentration. There is no feedback between *Daphnia* and phytoplankton food concentration. The second level of control is activated when ICL is reached. The organism tunes the filtration rate exactly, so as to keep the grazing rate almost constantly at saturation level. This is achieved by the hyperbolic decrease of the filtration rate. The control at that time can be schematised by the negative feedback between *Daphnia* and filtration rate. The third level of control is reached only in nature and perhaps in long term experiments. We can assume that it starts to act when the food concentration is such that the energetic gain from feeding exceeds the needs of the organism for respiration and movement. At this, moment the negative feedback between *Daphnia* and filtration efficiency starts to operate, leading to an adaptation causing the filtration efficiency to decrease. We assume that the filtration rate increases slightly, to enable the maintenance of constant gain if concentration increases further.

Selfadaptive systems are characterised by the change of internal parameters, without changing the organisms, some of the environmental variables and reactions of the organisms. The inclusion of adaptation into ecosystem models can be achieved by the use of adaptable parameters in several ways. One way is to develop empirical relations of the adaptable parameter to the variable(s) causing the change. The causing variable may be an external forcing function or an internal abiotic or biotic variable. As an example we give the dependence of one parameter, optimum temperature of the environment which they inhabit (Straskraba, 1976).

Another example is the dependence of the slope of photosynthesis – light relationship to the light history such as found by Straskraba and Gnauck (1985). However, general use of such empirical relations has to be considered with caution (similarly to other empirical relationships) because the material which serves for the derivation of the relationship is usually restricted, and because of the multi-variable nature of most of the relationships in nature. Under different conditions, with different species composition and other differences, the influence of another variable may completely change the otherwise well established empirical relationship. The different adaptation mechanisms of algal species is a good example.

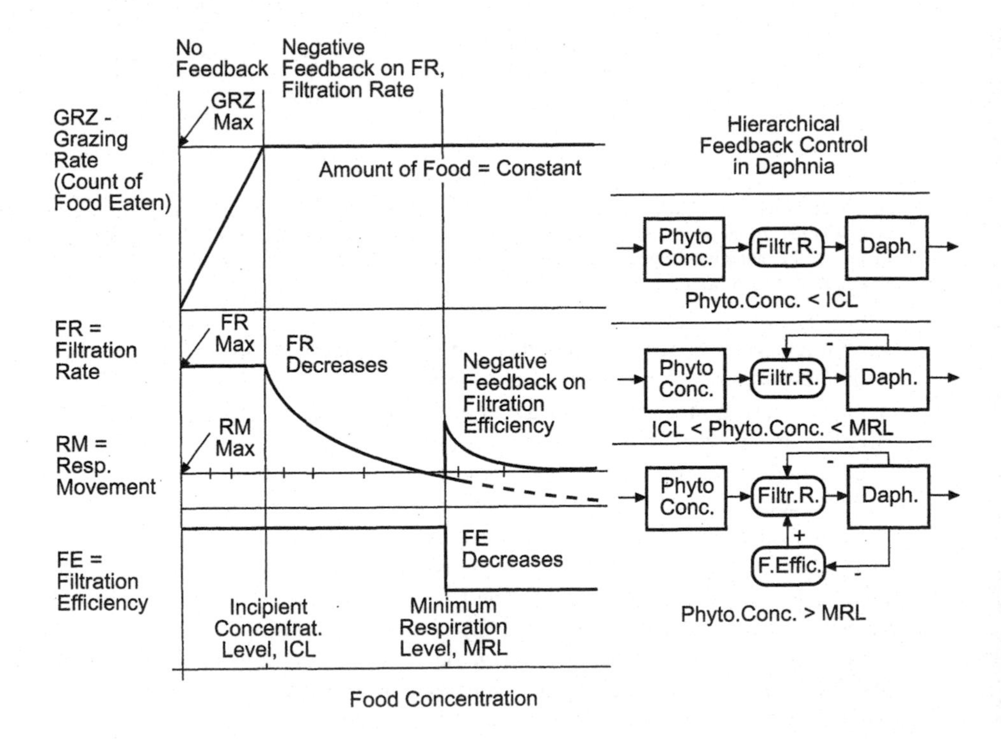

Figure 3: The left part of the figure shows the dependence of the grazing rate, filtration rate and filtration efficiency on food concentration. The right part shows the control mechanisms. In the left part three regions of food concentrations are distinguished, below the incipient concentration level (ICL), below the concentration saturating the energetic needs of the organisms, the minimum respiration level (MRL) and above this level. For more detailed explanation refer to text.

In spite of many detailed processes participating in the temperature adaptation process, the resistance of organisms, and in the photosynthesis light adaptation, there is a fairly uniform summary reaction for such relationships. This provides strong support for the hypothesis that some kind of optimum performance of the organisms is the goal for which various ways are utilised. The consequence is that we may ecologically grasp the temperature adaptation of organisms without going into the tedious job of discovering the functioning of all the participating processes and their enzymes. Using cybernetics and control theory methods, we are able to correctly formulate the goals and constraints, and to holistically estimate the result under a broad set of conditions.

Self adaptation can also be included by going deeper and describing the generalised mechanism of adaptation by means of an optimisation sub-model. As an example, Kmet et al., 1993, presented an optimisation model of the adaptation of phytoplankton photosynthesis. However, computations with the present-day optimisation methods are usually time-consuming and it will make the running of ecosystem models impractical. Therefore, the best way will be to develop, via the optimisation model, appropriate empirical relations and use them for the ecosystem model.

Self organisation of systems is understood as an improvement of the relations between system elements. For ecosystems this means first of all changes to the structure of the systems, including the presence and absence of different organisms and their relations. A typical ecosystem characteristic is to switch from one prey to another. The selection process illustrated in Figure 1 represents a typical process of self organisation in ecosystems.

The term *self-organisation of systems* was given at the time when the notion of self-organisation in the sense of Nicolis and Prigogine (1977) was not widely in use. The term as used here is not in full accord with the use of that used by Prigogine and followers, although the symptoms may be similar: more specifically, change in the structure of the system when external conditions change. The change can be relatively 'sudden' and is documented best in a study by Scheffer, 1990 discussing the multiple steady states in the interactions between nutirent-turbidity-macrophytes-fish for shallow lakes.

Evolutionary systems (Kotek et al., 1973) are systems which improve their behaviour by improvement of the optimality criterion. Evolution of biological system is a common notion, covering evolution of new species and higher taxonomic groups. Characteristic is the longer time scale of evolution. This control mechanism is represented by the change in the gene pool in Figure 1.

Chapter II.5 will present ecosystems as hierarchical systems. **Hierarchical systems self control** will deal with the problems in co-ordination of multi-layer control and information transfer between layers. No layer has the possibility to reach the global goal, reached only when every layer functions properly. In a two-layer system with n elements at the lower layer, every element has its own goal depending on the co-ordination parameter obtained from the higher level. The goal of the higher level is different from the global goal This layer selects the value of the co-ordination parameter in a way to realise its own goal.

Using co-ordination the system reaches integrity. The understanding of the mechanism of integration is connected with the determination of the goal of the co-ordinator, understood as a co-ordinator only with respect to its function; the existence of a structurally separated co-ordinator is not necessary.

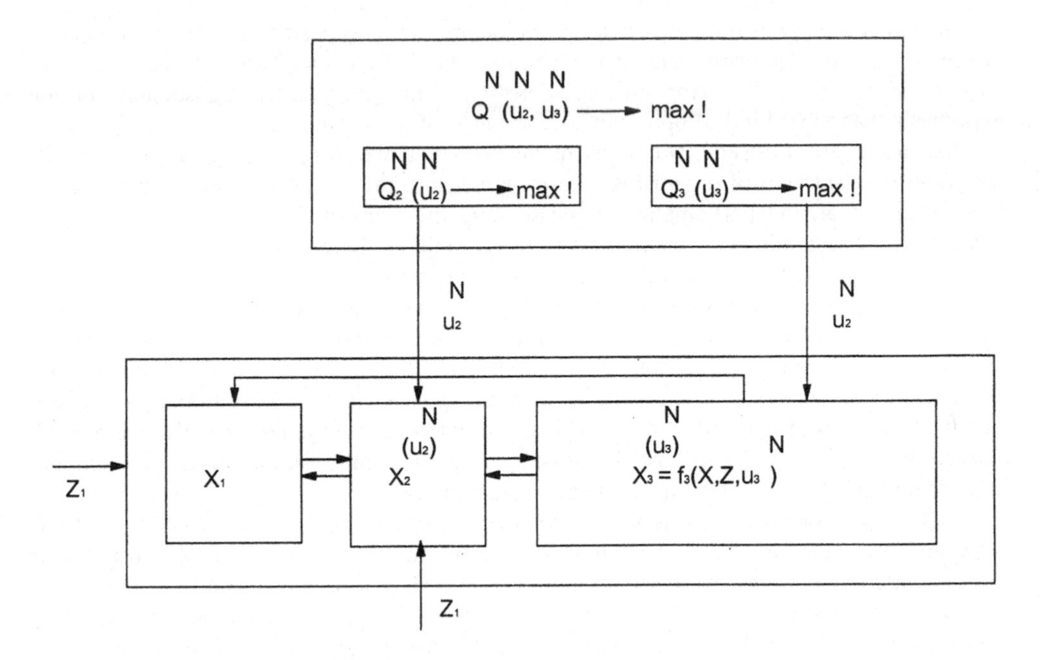

Figure 4: A decision hierarchy of an ecosystem is shown. Two subsystems, x_2 and x_3 pursue independent objectives functions, Q_2 [u_2] and Q_3 [u_3] with each of these being optimised. Relations between both objective functions are materialised by the dynamics of the system. Descriptions of whole systems can be formulated as a control problem with accomplishment of both sub-objectives being ensured by statement of an overall objective. Common objectives stand for concomitant accomplishment of both objectives.

3. Cybernetic types of ecosystem models

For covering changes in ecosystems, usually also associated with different time scales, the notion of different orders of ecosystem dynamics was introduced into ecology and ecosystem modelling (Straskraba, 1980 and 1995). By system dynamics we usually understand a change of the system state in time and space. In a complex system, however, not only the state is changed, but also the systems function, structure and tactical goals. Ruberti and Mohler (1975) distinguish for technical and economical systems:

1) systems with fixed parameters and fixed structure,
2) systems with variable parameters,
3) systems with variable structure.

Ecosystems belong to the highest category and this should be reflected in development of ecological models. Straskraba (1980) defined, with reference to ecological models, four orders of dynamics. The first order covers matter and energy circulation and flows. It encompasses direct and feedback control. The second order considers ecosystem function, while the third order accounts also for the dynamics of ecosystem structure. The last order is related to the dynamics of systems evolution. Characteristics of the four orders are summarised in Table 1.

The first order represents the only dynamics usually covered in ecosystem models. It means that each, usually aggregated, variable does not change its characteristics (properties, represented by the parameters in the model) – there is no adaptation.

The second order system changes the function according to the parameter changes, resulting from adaptation processes, while the entire structure, represented in ecosystems by the dominant species and their interactions, remains constant.

The third order of dynamics varies the structure with time, with the change of the systems state or of external inputs and impacts. Species with other parameters (properties) replace present species.

Table 1: **Orders of system dynamics**

Order	Formulation*)	Strategy	Highest control
First	f(x,z,p)	Fixed parameters and structure	feedback
Second	f(x,z,u,p(u))	Variable parameters, fixed structure	self adaptation
Third	f(x(u),z,u,p(u))	Variable structure	self organisation
Fourth	f(x(u),z,u(x),p(u))	change of tactical goals	self evolution

*) x = state variables, z = forcing function, u = control variables and p = parameters.

Because real ecosystems behave as if to follow these orders of dynamics within the time frames usually covered in our ecological models, it is important to develop ecological models considering dynamics of order two and three. The need for inclusion of the structural changes (higher order dynamics) is not only theoretical, but has deep practical implications. Recently, van Straten (1998) challenged, in respect to water quality models, the common belief that a calibrated and validated model is sufficient to warrant a technically sound application to guide water quality planning measures. He stresses that the goal of water quality planning is to remedy an existing unsatisfactory situation, or to create a new system. Both tasks are characterised by the fact that the model is not adequate or satisfactory for the present situation due to the change of the operation range of the model, for example less pollution or increased flushing. The following possibilities to overcome these difficulties are proposed:

1) Oversimplification plus cross-comparison which go into details based on a collection of broad information under different situations (not feasible for the reductionistic model approach).
2) Cybernetic modelling.
3) Educated speculation on the basis of qualitative knowledge about the likely direction of parameter change. This is close to the empirical methods outlined above.
4) Flexibility by feedbacks and model post-audit. It is suggested to confront the earlier models with recent data to reveal weaknesses of the available models.

The systems evolution is considered in the fourth order system. The evolutionary time scale in the Darwinian sense is included in the dynamic description. For ecosystem models this time scale will often be out of interest. Our models are usually applied to make predictions only a couple of years up to maybe one hundred years ahead, while evolution is a slow process requiring a very long time to show resulting changes.

4. Modelling structural dynamics

If we follow the modelling procedure proposed in most modelling literature, we will attain a model that describes the processes in the focal ecosystem, but the *parameters* will represent the properties of the state variables as they are in the ecosystem *during* the examination period. This description of processes is not necessarily valid for another period of time, because we know that an ecosystem is able to regulate, modify, and change them if needed as a response to the altered situation. Our present models have rigid structures and a fixed set of parameters, reflecting that no changes or replacements of the components are possible. They cover only the first order of dynamics. We need to introduce parameters (properties) that can change according to changing, forcing functions. Furthermore, we need to introduce general conditions for the state variables (components) that are able to optimise continuously the ability of the system.

These type of models can be developed when we can test if a change of the most crucial (= most sensitive) parameters is able to produce a higher value of the control or goal function. If that is the case, these set of crucial parameters can be used. Thereby, we obtain a better description of the regulation mechanisms in our model.

These models that are able to account for the adaptation processes and the change in species composition are sometimes called structural dynamic models, indicating that they are able to capture structural changes. They may also be called the next generation of ecological models to emphasise that they are radically different from previous modelling approaches and can do more, namely describe changes in species composition. What we develop in this section are systems where adaptation phase and self-organisation phase (the change of species composition) are both represented by parameter changes. Smaller parameter changes are assumed to characterise the same species. Different species appearing only as characterised by larger changes.

It could be argued that the ability of ecosystems to replace present species with other, better fitted, species can be modelled by the construction of models that encompass all actual species for the entire period that the model attempts to cover. This approach has, however, two essential disadvantages. The model becomes first of all very complex, as it will contain many state variables for each trophic level. It implies that the model will contain many more parameters that have to be calibrated and validated. This will introduce a high uncertainty to the model and will render the application of the model very case specific (Nielsen 1992a and 1992b, and van Straten, 1998). In addition, the model will still be rigid and not include the property of the ecosystems to adapt to the parameters in ways other than changing the species composition (Fontaine, 1981).

This section will present a few results of the proposed application of the exergy optimisation principle for a continuous change of the parameters, i.e. exergy is applied as the goal function. For the presentation and definition of exergy see Chapter II.1. Other goal functions have been proposed as shown in Table 2 but only very few models that account for a change in species composition or for the ability of the species to change their properties within some limits have been developed.

Bossel (1992) uses what he calls six basic orientors or requirements to develop a system model that is able to describe the system performance properly. The six orientors are:

1) Existence. The system's environment must not exhibit any conditions which may move the state variables out of its safe range.
2) Efficiency. The exergy gained from the environment should exceed over time the exergy expenditure.
3) Freedom of action. The system is able to react to the inputs (forcing functions) with a certain variability.
4) Security. The system has to cope with different threats to its security requirement with appropriate, but different, measures. These measures either aim at internal changes in the system itself or at particular changes in the forcing functions (external environment).
5) Adaptability. If a system cannot escape the threatening influences of its environment, the one remaining possibility is a change in the system itself in order to cope better with the environmental impacts.

6) Consideration of other systems. A system will have to respond to the behaviour of other systems. The fact that these other systems may be of importance to a particular system may have to be considered with this requirement.

Bossel (1992) applies maximisation of a benefit or satisfaction index based upon balancing weighted surplus orientor satisfactions on a common satisfaction scale. The approach is used to select the model structure of continuous dynamic systems and, although it seems very promising, has only been applied to ecological systems in one case.

The application of exergy as goal function for models corresponds to the application of the orientors 2, 4 and 5, while the ecological model, on which this constraint is imposed, should be able to cover orientors 1 and 3. Orientor 6 can only be accounted for by expansion of the model to include at least the feedbacks to the environment from the focal system.

Radtke and Straskraba (1980) use a maximisation of phytoplankton biomass as the governing principle. The model computes the biomass and adjusts one or more selected parameters of algae to achieve the maximum biomass at every instance. The combination that gives the maximum biomass over an annual cycle is selected by an optimisation. The use of optimisation approaches is very widespread for the study of individual organisms, their strategy and interrelations. One investigation with relevance for ecosystems is mentioned in Straskraba, 1995). He found that cybernetic models can best describe the available experimental observations of microbial growth in mixed cultures with quantitative accuracy in both batch and continuous cultures. An ecosystem approach toward self-organisation of the aquatic community based on simulation was recently presented by Kompare, 1996.

The idea of the new generation of models mentioned here is to find continuously a new set of parameters (limited for practical reasons to the most crucial (=sensitive) parameters) which are better fitted for the prevailing conditions of the ecosystem. Table 2.

Table 2: Goal functions proposed.

Proposed for	Objective function	Reference
Several systems	Maximum useful power or energy flow	Lotka (1922), Odum and Pinkerton (1955)
Several systems	Minimum entropy	Glansdorff and Prigogine (1971)
Networks	Maximum ascendancy	Ulanowicz (1980)
Several systems	Maximum exergy	Mejer and Jørgensen (1979)
Ecological systems	Maximum persistent organic matter	Whittaker and Woodwell (1971) and O'Neill et al.(1975)
Ecological systems	Maximum biomass	Margalef (1968)
Economic systems	Maximum profit	Various authors

The use of exergy calculations to vary the parameters has now been used successfully in ten cases; see Jørgensen (1986, 1992, 1995 and 1997) and Jørgensen et al. (1994, 1996 and 1997). In one such case a structural dynamic model was developed by use of exergy as goal function in a lake study. The results from Søbygaard Lake (Jeppesen et al., 1989) are particularly fitted to test the applicability of the described approach to structural dynamic models. As an illustration to structural dynamics of ecosystems and the possibilities to capture the flexibility of ecosystems, the case study of Søbygaard Lake will be presented.

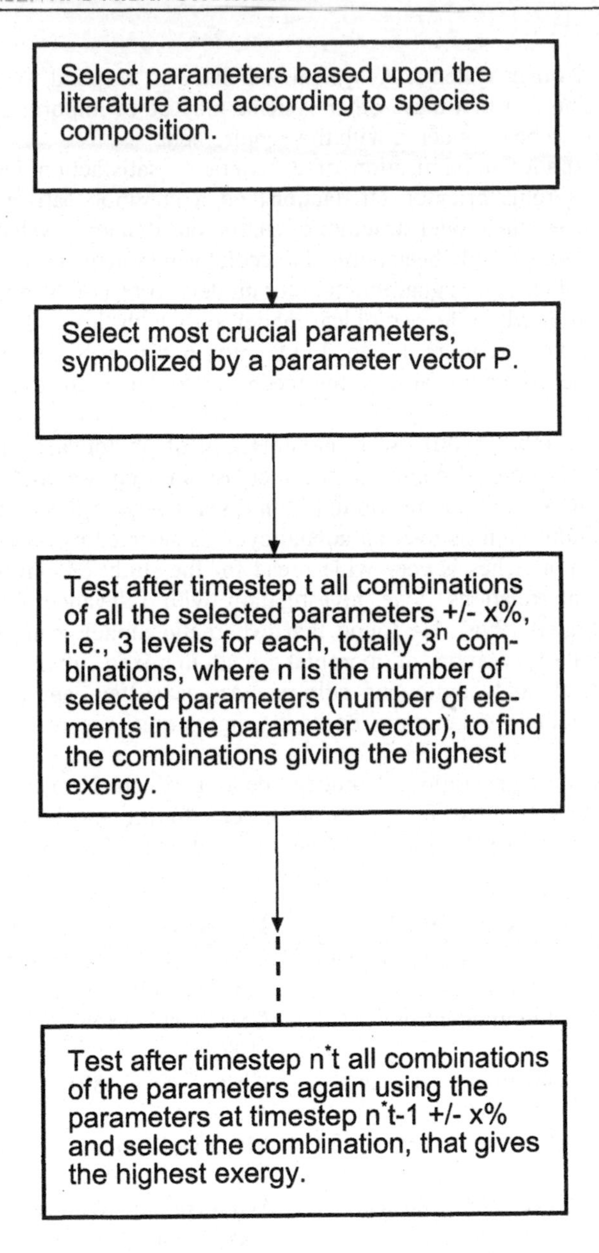

Figure 5: The procedure used for the development of structural dynamic models.

Søbygaard Lake is a shallow lake (depth 1 m) with a short retention time (15-20 days).The nutrient loading was significantly reduced after 1982, particularly for phosphorus from 30 gP/ m^2y to 5 gP/ m^2y. The reduced load did, however, not cause reduced nutrients and chlorophyll concentrations in the period 1982-1985 due to an internal loading caused by the storage of nutrients in the sediment (Søndergaard, 1989 and Jeppesen et al., 1989).

However, radical changes were observed in the period of 1985-1988. The recruitment of planctivorous fish was significantly reduced in the period of 1984-1988 due to a very high pH caused by the eutrophication. As a result zooplankton increased and phytoplankton decreased in concentration (the summer average of chlorophyll A was reduced from 700μ g/l in 1985 to 150 μ g/l in 1988). The phytoplankton population even collapsed in shorter periods due to extremely high zooplankton concentrations. Simultaneously the phytoplankton species increased in size. The growth rate decreased and a higher settling rate was observed (Kristensen and Jensen, 1987). The case study shows pronounced structural changes. The primary production was, however, not higher in 1985 than in 1988 due to a pronounced self-shading by the smaller algae in 1985. It was therefore very important to include the self-shading effect in the model, which was not the case in the first model version, hence giving incorrect figures for the primary production.

Simultaneously intensive feeding of the zooplankton was observed, caused by a change in zooplankton from *Bosmina* to *Daphnia.*

The model applied has 6 state variables: N in fish, N in zooplankton, N in phytoplankton, N in detritus, N as soluble phosphorus and N in sediment. As seen, only the nitrogen cycle is included in the model, but as nitrogen in this case is the nutrient controlling the eutrophication, it may be sufficient to include only this nutrient.

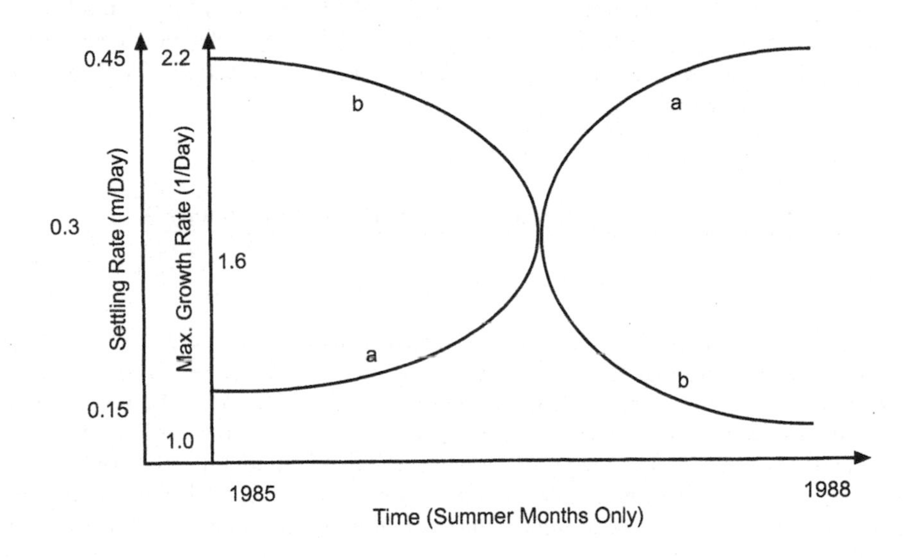

Figure 6: The continually changing parameters obtained from the application of a structural dynamic modelling approach on Søbygaard Lake are shown. a covers the settling rate of phytoplankton and b the maximum growth rate of phytoplankton.

The aim of the study was to describe the continuous changes in the most essential parameters using the procedure shown in Figure 5. The data from 1984-1985 was used to calibrate the model and the two parameters that it is intended to change from 1985 to 1988 received the following values from this calibration:

Maximum growth rate of phytoplankton: 2.2 day $^{-1}$

Settling rate of phytoplankton: 0.15 day $^{-1}$

The state variable fish-N was kept as constant = 6.0 during the calibration period, but an increased fish mortality was introduced during the period 1985-1988 to reflect the increased pH. The fish stock was thereby reduced to 0.6 mgN/l.

A time-step of t = 5 days and x (see Figure 5) = 10% was applied. This means that 9 runs were needed for each time step to select the parameter combination that gives the highest exergy.

The results are shown schematically in Figure 6 and the switch in parameters from 1985 to 1988 (summer situation) are summarised in Table 3.

The proposed procedure is able to simulate approximately the observed change in structure. The maximum growth rate of phytoplankton is reduced by 50% from 2.2 day^{-1} to 1.1 day^{-1}, which approximately corresponds to the increase in the individual size. It was observed that the average size was increased from a few 100 - 300μm^3 to 500 - 1000 μm^3 of the dominant phytoplankton species, which is a factor of 2-3 (Jeppesen et al. 1989). It would correspond to a specific growth reduction by a factor $f = 2^{2/3} - 3^{2/3}$ (see Jørgensen and Johnsen, 1989).

This means that:

growth rate in 1988 = growth rate in 1985 / f,

where f is between 1.58 and 2.08, while 2.0 is found by use of the structural dynamic modelling approach.

Kristensen and Jensen (1987) observed that the settling was 0.2 m day^{-1} (range 0.02-0.4) in 1985, while it was 0.6 m day^{-1} (range 0.1-1.0) in 1988. By the structural dynamic modelling approach an increase from 0.15 day^{-1} to 0.45 day^{-1} was found, the factor being the same – three – but with slightly lower values. The phytoplankton concentration as chlorophyll-A was simultaneously reduced from 590 µg/l to 206 µg/l, which is (only) 7% than the observed reduction.

Table 3: Parameter Combinations giving the highest Exergy.

	Maximum growth rate (day^{-1})	Settling Rate ($m*day^{-1}$)
1985	2.0	0.15
1988	1.2	0.45

All in all, it may be concluded that the structural dynamic modelling approach gave an acceptable result and that the validation of the model and the procedure in relation to structural changes was positive. It is, however, necessary to expand the model to account for all the observed structural changes, including zooplankton, to be able to demonstrate a completely convincing case study. The structural dynamic modelling approach is of course dependent on the underlying model applied. The presented model may be criticised for being too simple and not accounting for the structural dynamic changes of zooplankton.

For further elucidation of the importance to introduce a parameter shift, the model was run for the 1985 situation with the parameter combination found to fit the 1988 situation and vice versa. These results are shown in Table 4. The results demonstrate that it is of great importance to apply the right parameter set to given conditions. If the parameters from 1985 are used for the 1988 conditions, a lower exergy is obtained and the model, to a certain extent, behaves chaotically, while the 1988 parameters used on the 1985 conditions give a significantly lower exergy. As discussed in Chapter II.8.4, the change in parameters may play a role in avoidance of chaotic conditions of the system.

The results show that it is important for ecological and environmental models to contain the property of flexibility, which we know ecosystems possess. If we account for this property in the models, we obtain models that are more capable of producing reliable predictions, particularly when the forcing functions on the ecosystems change and thereby provoke changes in the properties of the important biological components of the ecosystem. In some cases we get completely different results when a current change of parameter is applied, where before a fixed parameter was used. In the first case, we get results that are more in accordance with our observations in nature. As we know that the parameters do actually change in the natural ecosystems, we can only recommend the application of this approach as far as possible in ecological modelling.

Table 4: Exergy and Stability by different Combinations of Parameters and Conditions

Parameter	Conditions	
	1985	**1988**
1985	75.0 Stable	39.8 (average) Violent fluctuations . Chaos
1988	38.7 Stable	61.4 (average) Only minor fluctuations.

The property of dynamic structure and adaptable parameters (the cybernetics of ecosystems) is crucial in our description of ecosystems and should therefore be included in all descriptions of the system properties of ecosystems.

REFERENCES

Andzejewski, R. 1977. Populacja jako system ekologiczny. *Wiadom. Ecol.*, 23: 3-33 (in polish).

Bossel, H. 1992. Real structure process description as the basis of understanding ecosystems. In: Workshop 'Ecosystem Theory', 14-17 October 1991, Kiel. (Special Issue). *Ecol. Modelling* 63: 261-276.

Fleishman, B.S. 1976. Philosophy of Systemology. Cybernetica, 4: 261-272.

Fontaine, T.D. 1981. A self-designing model for testing hypotheses of ecosystem development. In: D. Dubois (editor). *Progress in Ecological Engineering and Management by Mathematical Modelling. Proceedings of the 2nd International Conference on State-of-the-Art Ecological Modelling,* 18-24 April 1980, Liége, Belgium, 281-291.

Germeyer, Ju.B. 1976. Games with nonantagonistic interests. Nauka, Moscow (in Russian).

Glansdorff, P., and I. Prigogine 1971. Thermodynamic Theory of Structure, Stability, and Fluctuations. Wiley-Interscience, New York.

Innis, G.S., and W.R. Clark 1977. A self organizing approach to ecosystem modeling. In: Innis, G.S. (Ed.), *Grassland Simulation Model.* Springer Verlag, Berlin: 179-187.

Jeppesen, E., E. Mortensen, O. Sortkjær, P. Kristensen, J. Bidstrup, M. Timmermann, *et.al.* 1989. Restaurering af søer ved indgreb i fiskebestanden. Status for igangværende undersøgelser. Del 2: Undersøgelser i Frederiksborg slotssø, Væng sø og Søbygård sø. Danmarks Miljøundersøgelser, Silkeborg. Denmark.

Jørgensen, S.E. 1986. Structural dynamic model. *Ecol. Modelling* 31: 1-9.

Jørgensen, S.E. 1992. Development of models able to account for changes in species composition. *Ecol. Modelling* 62: 195-208.

Jørgensen, S.E. 1997. Integration of Ecosystem Theories: A Pattern. Second Edition. Kluwer Academic Publishers, Dordrecht, Boston, London.

Jørgensen, S.E. 1995. The growth rate of zooplankton at the edge of chaos: ecological models. *J. Theor. Biol.* 175: 13-21.

Jørgensen, S.E., and R. De Bernardi 1997. The application of a model with dynamic structure to simulate the effect of mass fish mortality on zooplankton structure in Lago di Annone. Hydrobiologia, in press.

Jørgensen, S.E., and I. Johnsen 1989. Principles of Environmental Science and Technology (Studies in Environmental Science 33). Elsevier, Amsterdam.

Jørgensen, S.E., and S.N. Nielsen 1994. Models of the structural dynamics in lakes and reservoirs. *Ecol. Modelling* 74: 39-46.

Jørgensen, S.E., and J. Padisák 1996. Does the intermediate disturbance hypothesis comply with thermodynamics? *Hydrobiologia* 323: 9-21.

Kompare, B., 1996. Use of artificial intelligence in Ecological Modelling. Thesis, DFH, Copenhagen.

Kmet, T., M. Straskraba, and P. Mauersberger 1993. A mechanistic model of the adaptation of phytoplankton photosynthesis. *Bulletin of Mathematical Biology,* Vol 55, No. 2: 259-275.

Korinek, V., and J. Machacek 1980. Filtering structures of Cladocera and their ecological significance. I. Daphnia pulicaria. *Vest. Csl. Zool. Spol.* 44: 213-218.

Kotek, K., S. Kubik, and M. Razim 1973. Nelinearni dynamické systémy. Statní Nakladatelství Technické Literatury, Praha, (in czech).

Kozlowski, J., A. Lomnicki, H. Warkowska-Drahual, and J. Weiner 1980. Kilka ewag o cybernetyczaym sterowanin ekosystemow. *Wiadom. Ekol.*, 26 (3): 307-309, (in polish).

Kristensen, P., and P. Jensen 1987. Sedimentation og resuspension i Søbygård sø. (Univ. Specialerapport). Miljøstyrelsens Ferskvandslaboratorium & Botanisk Institut, Univ. Århus.

Lotka, A.J. 1922. Contribution to the energetics of evolution. *Proc. Natl. Acad. Sci.* USA 8: 147-150.

Magalef, R. 1963. On certain unifying principles in ecology. *Am. Nat.* 97: 357-374.
Magalef, R. 1968. Perspectives in Ecoligical Theory. Chicago. University Press. Chicago, IL.
Mejer, H.F., and S.E. Jørgensen 1979. Energy and ecological buffer capacity. In: S.E. Jørgensen (editor). *State-of-the-Art of Ecological Modelling.* (Environmental Sciences and Applications, 7). Proceedings of a Conference on Ecological Modelling, 28. August - 2. September 1978, Copenhagen. International Society for Ecological Modelling, Copenhagen, 829-846.
Mesarovic, M.D. Systems Theory and Biology. 1968. Proceedings of the 3rd Systems Symposium at Case Institute of Technology, Springer-Verlag, Berlin.
Mesarovic, M., D. Macko, and Y. Takahara 1970. Theory of hierarchical multi-level systems. Academic Press, New York-London.
Müller, F., and M. Leupels (Eds.). 1998. Eco Targets, Goal Functions, and Orientors. Springer, 618.
Nicolis, G. and I. Prigogine. 1977. Self-Organization in Non-Equilibrium Systems: from dissipative structures to order through fluctuations. Wiley Interscience, New York, N.Y.
Nielsen, S.N. 1992a. Application of Maximum Exergy in Structural-Dynamic Models [Thesis]. National Environmental Research Institute, Denmark.
Nielsen, S.N. 1992b. Strategies for structural-dynamical modelling. *Ecol. Modelling* 63: 91-102.
Odum, H.T., and R.C. Pinkerton 1955. Time's speed regulator: the optimum efficiency for maximum power output in physical and biological systems. *Am. Sci.* 43: 331-343.
O'Neill, R.V., W.F. Hanes, B.S. Ausmus, and D.E. Reichle 1975. A theoretical basis for ecosystem analysis with particular reference to element cycling. In: F.G. Howell, J.B. Gentry and M.H. Smith (editors). *Mineral Cycling in Southeastern Ecosystems.* NTIS pub. CONF-740513.
Peschel, M., and C. Riedel 1976. Polyoptimierung, eine Entscheidungshilfe für ingenieur-technische Kompromisslösungen. Verlag Technik, Berlin.
Radtke, E., and M. Straskraba 1980. Self-optimization in a phytoplankton model. *Ecol. Modelling,* 9: 247-268.
Rothaupt, K.O., and W. Lampert 1992. Growth-rate dependent feeding rates in Daphnia pulicaria and Brachionus rubens: adaptation to intermediate time scale variations in food abundance. *J. Plankt. Res.* 14: 737-751.
Ruberti, A., and R. Mohler 1975. Variable Structure Systems with Applications to Economics and Biology. Lecture Notes in Economics and Mathematical Systems, Springer, Berlin, Vol. 111.
Scheffer, M. 1990. Simple Models as Useful Tools for Ecologists. Elsevier, Amsterdam.
Straskraba, M. 1976. Development of an Analytical Phytoplankton Model with Parameters Empirically Related to Dominant Controlling Variables. In: R. Glaser., K. Unger und M. Koch (Hrsg.): Umweltbiophysik, Arbeitstagung 19.10 bis 1.11.1973 Kühlungsborn, Akademie Verlang, Berlin, pp. 33-65.
Straskraba, M. 1979. Natural control mechanisms in models of aquatic ecosystems. Ecological Modelling, 6: 305-322
Straskraba, M. 1980. Cybernetic categories of ecosystem dynamics, ISEM Journal, 2, 81-96.
Straskraba, M. 1982. Cybernetic Formulation of Control in Ecosystems. Ecological Modelling, 18: 85-98.
Straskraba, M., and A. Gnauck 1985. Freshwater Ecosystems, Modelling and Simulation, Elsevier, Amsterdam.
Straskraba, M. 1995. Cybernetic Theory of Ecosystems. In: Gnauck, A., Frischmuth, A., Kraft, A. (Eds.). *Ökosysteme: Modellierung und Simulation.* Eberhard Blottner Verlag, Taunusstein, Germany, 253.
van Straaten, G., 1998. Personal Communication
Søndergård, M. 1989. Phosphorus release from a hyperthropic lake sediment; experiments with intact sediment cores in a continous flow system. *Arch. Hydrobiol.* 116: 45-59.
Tribus, M., and E.S. McIrvine 1971. Energy and information. *Sci. Amer.* 225 (3): 171-188.
Ulanowicz, R.E. 1980. A hypothesis on the development of natural communities. *Ecol. Modelling* 85: 223-245.
Ulanowicz, R.E. 1997. Ecology, the Ascendent Perspective. Columbia University Press, New York, 201.
Whittaker, R.H., and G.M. Woodwell 1971. Evolution of natural communities. In: J.A. Weins (editor). *Ecosystem Structure and Function.* Oregon State University Press, Corvallis, OR. 137-159.

II.5 Ecosystems as Hierarchical Systems

Stefani Hári and Felix Müller

1. Introduction

One of the most important purposes of theories is abstraction, a scientific condensation of many single cases into one nucleic, focal system of hypotheses, which may function as aggregated guidelines for the empirical branches of science. Theories also enable the scientist to draw general conclusions and to transfer knowledge between different cases, locations and points of time. Therefore theories, locations and points of time can also be regarded as basic elements of a science's application and even as a potential step toward practice.

Adapting these ideas to ecology and taking a system's approach, and thus taking an holistic view to environmental questions, a new problem arises. The investigated systems are so complex that the temptation to return to reductionistic fundamentals becomes nearly overwhelming. Consequently, there is a greater need to reduce the complexity we have to face when confronted with ecosystemic interactions as far as possible. Although this task is one focal purpose of theories, not all theoretical focal approaches are able to cover it. An extraordinary suitable concept to fulfil these demands of complexity reduction is hierarchy theory. It is not only designed in an aggregating manner, suggesting rules for integration, it is also a most suitable approach to close the conceptual gaps between top-down-techniques and bottom-up-conceptions. Furthermore, the hierarchical construction of the theory is obliging our way of thinking and classifying.

In the following text, this promising approach will be briefly introduced. The presentation starts with an outline of the theory, including the short history of the concept and its central hypotheses. These ideas will be applied to spatial, temporal and functional contexts, using different examples, concepts and empirical results for illustration. Taking this route, some methods will be described, that can be used to prove the theory. A brief discussion of potential applications completes the paper.

2. An outline of hierarchy theory

The concept of hierarchy theory was developed as a describing tool by general system's theory in the sixties and seventies. Having been known for a much longer time in social communities and social sciences, hierarchical structures and their applications can be found in ancient societies as well as modern ones. Taking these structures and a multitude of further applications of hierarchical interrelations into account, the concept represents such a common idea that it seemed to be useful to adopt it to ecology.

Ecological systems are very complex, non-linear and self-organised entities. They are middle-number systems according to O'Neill et al. (1986) and as such, they can neither be described by simple models nor by statistical approaches that demand high quantities of individual cases. On the other hand, the need of managing ecosystems is increasing with growing environmental problems. The complexity of those systems makes it very hard to put adequate environmental methods into practice. In this problem area, the concept of hierarchy theory may provide scientific support for applied environmental management and adjacent disciplines.

3. Historical development of the concept

Hierarchies are typical traditional classification methods. The hierarchical approach seems to be founded in the general patterns of man's structured of thinking. Thus it is extremely tempting and very natural to use these conceptions in theoretical approaches also. Referring to research on complex systems, general system theory has adopted the hierarchical viewpoint and invented hierarchy theory in the early sixties and seventies.

The term 'hierarchy' originally referred to a vertical authority structure in human organisations (Simon, 1973). Applied in general system theory, it means that one system contains another, smaller one, which contains an even smaller system and so on. One can easily depict the fundamental ideas of a hierarchy by using the image of a 'Matrujschka', a Russian doll. When one doll is opened, another doll appears, which can be opened again, revealing a smaller doll. This can be continued until the last doll is too small to get another one in. The difference between a 'Matrujschka' and a hierarchy is that the hierarchical order is not a straight line but branched like a tree (Koestler, 1970; Simon, 1973). One system is divided into smaller subsystems, that are split up into smaller subsystems and so on (see figure). During the development of general system, theory hierarchies were used to structure systems in order to make them more understandable. In 1970 Koestler presented the theory in its entirety. He combined structural features with functional aspects so that a more general concept was developed (see below). The concept has been utilised in the discussion of evolution and functioning of systems by many authors (i.e. Mesarovic et al., 1970; Pattee, 1973; Simon, 1973).

The first application in ecology was carried out by Allen and Starr in 1982. They pointed out that one of the biggest problems in ecology is the problem of scale. In the field of biology the question of size and generation/lifetime was already discussed in the 18th century and has been integrated into evolutionary concepts (Allen and Starr, 1982). In the case of the development of biological science and ecology, it became more and more obvious that important processes in natural systems take place at different scales, a fact that gives an immense rise to the complexity of the system. Allen and Starr (1982) among others suggested hierarchy theory to be a basic approach to deal with scales and to analyse the connected complexity. Many scientists followed their suggestions so that hierarchy theory was completely integrated in theoretical ecology in general and also particularly in ecosystem theory (Salthe, 1985; O'Neill et al., 1986; Allen and Hoekstra, 1992; Müller, 1992; Klijn, 1997).

4. System analytical hypotheses of hierarchy theory

As stated above, a hierarchy defines a vertical structure of authority, size, regulation or, more general, a relationship between superior and inferior organisational levels. On each level there are subsystems, that are hierarchically and vertically connected with each other and that also comprise horizontal interactions between subunits at the same level. Each subsystem consists of subsystems itself, that are getting smaller and smaller.

Each of the subsystems is at the same time part of a superior level and an organisational unit for the inferior level, which represents a kind of 'janusfaced' duality of the single levels (Koestler, 1970). High levels are formed and created by the (horizontal) processes that relate to the structural units of the lower levels. This means that all properties of the lower level can also be found on the next higher one where their sum is the feature of the whole. These additive characteristics are called 'collective properties'. New properties also emerge on the higher level, which can only be found at that very position in a hierarchy. It becomes clear that apart from collective features, hierarchies also produce 'emergent properties' which can be assigned to the results of self-organising processes (see Nielsen et al., this volume, Müller, 1996, Müller et al., 1997a,b).

Subsystems are interacting. There are two kinds of interactions, a vertical case on the same level and a horizontal one between different levels of hierarchy. Interactions on the same level usually have minor significance for the system's behaviour and development. More interesting and more significant are the relationships between different levels. The interactions can be viewed as signal transfers, that are controlled by certain signal filtering rules. Signals in the ecological context are fluxes of information, energy or matter.

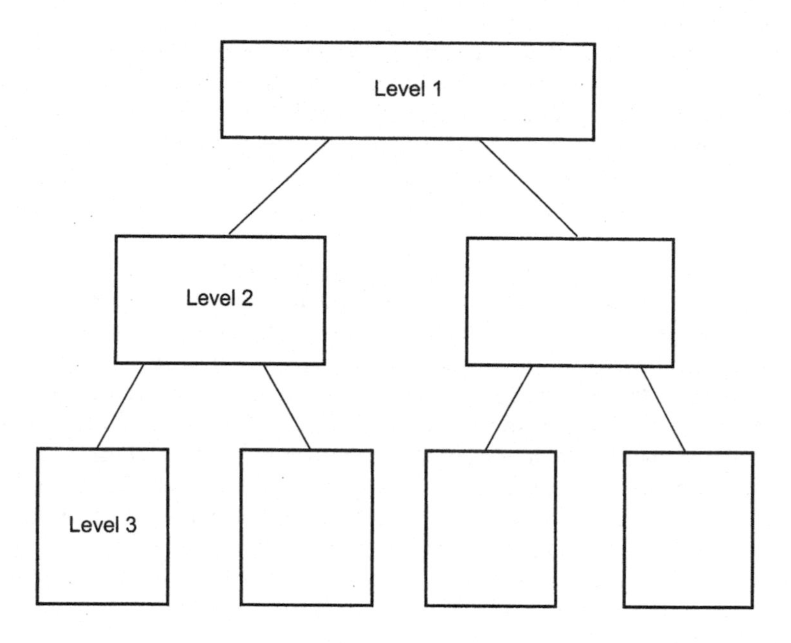

Figure 1: Simple form of a hierarchical structured system

When a signal from a subsystem on an inferior level is passed up the hierarchy, the superior subsystem is able to filter, smooth and buffer the signal before it is either translated into a message, integrated or passed on to the next higher level. For example, the rapid temperature changes of the atmosphere are buffered by our skins as well as smoothed by the compounds of the upper soil layer. Both, organisms and soils are able to regulate the environmental energy inputs and reduce the high velocity of external changes to a viable dynamism for the living systems.

When a signal is transferred down the hierarchy, inferior levels are not able to adjust it to their properties. They have to cope with the signal and its dynamics in its original way. An example for that can be taken from the theory of nutrient cycles. If there are not enough nutrients in the soil solution, plants are only able to grow in a reduced manner and at a decreased rate. The plants have to adapt to the restricted availability. If they don't manage to do so, they die. Returning to the example of temperature, this type of regulation may be exemplified by poikilotherm organisms or badly insulated soil combinations.

The specific rules for signal transfer and filtering lead to another important feature of hierarchy theory, the relations of space, time, and the scale of subsystems. There are two significant properties that characterise the different levels – the extension of the subsystems in space and the connected temporal behaviour of those subsystems. Superior levels have a broad spatial extent and they act or react very slowly. On the other hand, inferior levels have a small spatial extent and their behavioural characteristics in time are much faster. This is of great significance for the understanding of the system's behaviour. The first important relation is that the large and slow acting levels provide constraints for the small and fast levels, while the small and fast operating niveaus together constitute the biological potential of the system. The second important concept is that time and space are irrevocably connected with each other in terms of scales. The scale of a subsystem is the temporal and spatial period that is needed to integrate, smooth or dampen a signal before it is converted into a message (Allen and Starr, 1982). It is important to recognise, that time and space are interrelating units in this definition.

In addition to above described characteristics, there is another feature resulting from the system's transferring signal that is found to be typical for hierarchies. Each level of the hierarchy has individual, specific characteristic process rates and interactions. As already mentioned, the process frequencies of superior levels are long-waved, while the frequencies of inferior levels are short-waved. Going back to the image of the 'Matrujschka' we can pretend that each doll is a dynamic process. The big doll constrains the dynamics of the small doll, it is not possible for the small doll to 'move' very much. The degrees of freedom of the inferior levels are thus constrained by the superior level. On the other hand, it is not possible for the small doll to have very much influence on the large one. But it is able to produce an impulse towards the big doll. O'Neill et al. (1989) transferred this symbolic figure to ecological systems and shaped the *concept of constraints.* They state that the inferior levels form the biotic potential of the system, while superior levels are seen as environmental limits. The following two examples help to make this more clear: The biotic potential of the system limits the dynamics of the whole system through the components, e.g. atmospheric nitrogen cannot be fixed by a forest if the requisite organisms are missing (O'Neill et al., 1989). Limitations can also result as environmental limits in the features of the superior levels. Population growth for example is constrained by food supply (O'Neill et al., 1989). In other words, macroscopic structures control microscopic processes, which themselves form these structures (Müller, 1992). It turns out that everything is somehow related to the duality of the single levels stated above.

A term that is very often used in hierarchy theory is the *holon*. It is an artificial term created by Koestler (1970) originating in Greek. The term *holos* means whole and the suffix *-on* means part. At a given scale of resolution in a system a holon describes a subsystem that is at the same time part of a superior unit and contains all inferior subsystems. Holoi are semi-autonomous, self-regulating, open entities. Holoi can be symbolised as dual units, like a two-way window through which the environment influences its parts and through which the parts are communicating with the environment (Allen and Starr, 1982). We already mentioned the features of holon, but we used the term subsystem as a synonym for it. Figure 2 depicts the different features of hierarchy theory, described above.

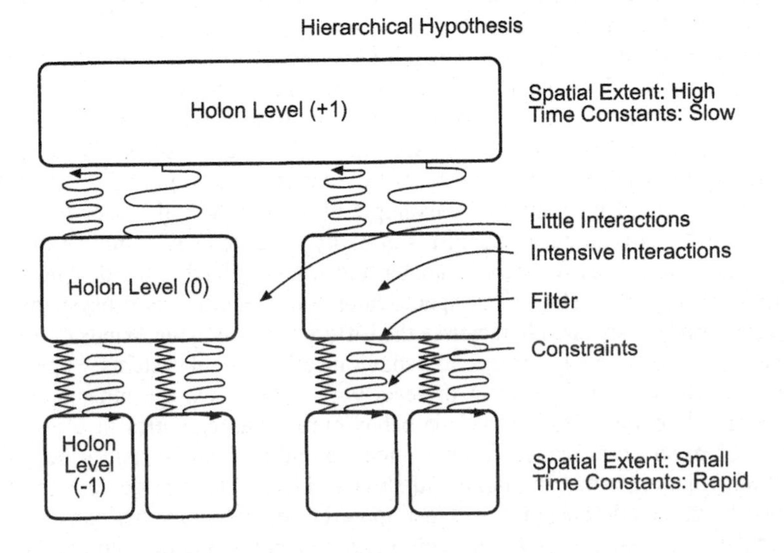

Figure 2: Hypotheses of hierarchy theory

All the properties of hierarchy mentioned are only valid if the system is operating in the neighbourhood of a steady-state, where large scale processes determine the systems development. Small perturbations deflect the system only for a short time from its main attractor, to which it quickly returns. Prolonging the distance from steady-state, bigger disturbances can exceed certain thresholds and the system reaches a bifurcation point. The hierarchy breaks and the small scale processes take over the developing direction. The mechanisms of signal transfer and filtering operate no longer efficiently and so the whole system structure changes. Residing near this state the system is easily affected by small perturbations, which now can have important consequences. The result of the hierarchy break is that the high levels lose their controlling functions over the lower levels They take over the developing direction until the system has found a new attractor (O'Neill et al., 1986; Müller, 1992).

Resuming the features of a hierarchy that is owned by an ecological system under steady-state, we can build up the following list:

- the system consists of subsystems, that are connected with each other
- subsystems are organisational units for inferior levels and parts of a superior level at the same time
- at each level collective and emergent properties occur
- the subsystems interact with each other by passing signals
- interactions between subsystems of the same level are not very intense, but there are important interactions between different levels
- superior levels constrain inferior levels by specific rules of signal transfer and filtering
- processes of superior levels operate on a large spatial area and have slow temporal fluctuations, while processes on inferior levels operate on small spatial areas and have high temporal fluctuations
- the scale of a level is a feature of its spatial and temporal characteristics
- macroscopic structures constrain the microscopic processes, that form them.

The described hypotheses can be applied to ecosystem research. Some examples will be given in the next section.

5. The application of hierarchy theory in ecology

The application of hierarchy theory in ecology has a short history. The theory was introduced to ecology in 1982 by Allen and Starr. Since that time, it has become more and more popular to use hierarchy theory for description of ecological systems, and the ideas have been widely spread in the natural sciences.

Hierarchies as tools for structuring subjects have been used, in other areas of science, for a longer period of time. Systematical hierarchies as classification methods can for example be found in botany or zoology, in geology or soil science. Here, most of the internationally used classifications (e.g. FAO, US Soil Taxonomy) are based on hierarchical distinctions of soil units ranging from orders to suborders to groups and smaller units. The biological taxonomy is a distinction in hierarchies, too. In this chapter some qualitative examples from the literature will be presented. We will also include some examples from recent empirical research in order to demonstrate quantitative applications of the theory and to show its potential applicability to environmental management.

6. Spatial hierarchies of structures and gradients

Example 1

An example for a simple qualitative hierarchy is what Allen and Hoekstra (1992) call the 'conventional hierarchy of levels of organisation'. It shows a ranking of ecological levels from cell to biosphere (Figure 1). A ranking is represented which meets the basic hypothesis of hierarchical structuring: each level is part of an integrated superior level which consists of the inferior sub-units. Some basic principles of hierarchy theory are optional in this concept. They are not necessarily adopted by the observer of the system. The series of ordering conventional levels is not very useful for the observer, because it implicates a broad range of scales throughout the system analysis is made. That is the reason why problems occur when transferring results from one scale to another (O'Neill et al., 1986).

Level 8	Biosphere	e.g. Global carbon balance incl. Marine systems
Level 7	Biome	e.g. Climatically constrained carbon dynamics
Level 6	Landscape	e.g. Carbon balance watershed
Level 5	Ecosystem	e.g. Carbon balance ecosystem
Level 4	Community	e.g. Food web
Level 3	Population	e.g. Carbon dynamics of one species
Level 2	Organism	e.g. Digestion
Level 1	Cell	e.g. Respiration

Figure 3: Conventional hierarchy of levels of organisation from cell to biosphere (after Allen and Hoekstra 1992) and an example for carbon flows on different levels. The carbon budget units make clear that throughout the passage from a lower level (e.g. carbon dynamics of one species) to a higher niveau (e.g. food web) additional qualities appear (e.g. predator-prey-relations), that cannot exclusively be described on base of a lower level. Such hierarchical system's features (e.g. the integration of the soil carbon dynamics to reach the ecosystem's level) emerge from the lower level's horizontal interactions.

Example 2

A hierarchy of biocoenotic interactions is described by Kolasa (1989). He tried to explain the interactions between the species composition and the habitat requirements by a conceptual model, which was validated by his research. The environment can be viewed as a nested hierarchy of habitat units, described by a multidimensional set of biotic and abiotic variables called habitat factors. Species are divided into 'specialists' and 'generalists', with each having resource requirements. The abundance's of species vary with the habitat factors, narrow range, intermediate range and broad range species can be distinguished. They are assigned to different levels in the hierarchy (see Figure 4). Regarding their habitat requirements, the specialists have a narrow range, so the spatial extend of their habitats are very small. On the other hand the generalists can cope with big variations in their habitat, they can exist in a broader spatial area than the specialist. The compartment size of the habitats and species abundance is a function of their position in Kolasa's hierarchy.

Example 3

An example for spatial hierarchies of abiotic and biotic gradients and heterogeneities is taken from a forest ecosystem (Müller, 1998; Reiche et al., in press; Hári et al., in press). Gradients and heterogeneities can be seen as a hierarchy in form of an ordered set which is interrelated by asymmetric interactions (Müller, 1998). The hypotheses of hierarchy theory can be applied to gradients and heterogeneous structures. Gradients are the spatial expression of an order of dynamics ranging from slow to fast changes in structure. Heterogeneous structures are delineated as self-organised entities that arise from gradient dissipation (Reiche et al., in press). Both gradients and heterogeneous structures exist at different scales. So, they can be viewed as an expression of hierarchy theory.

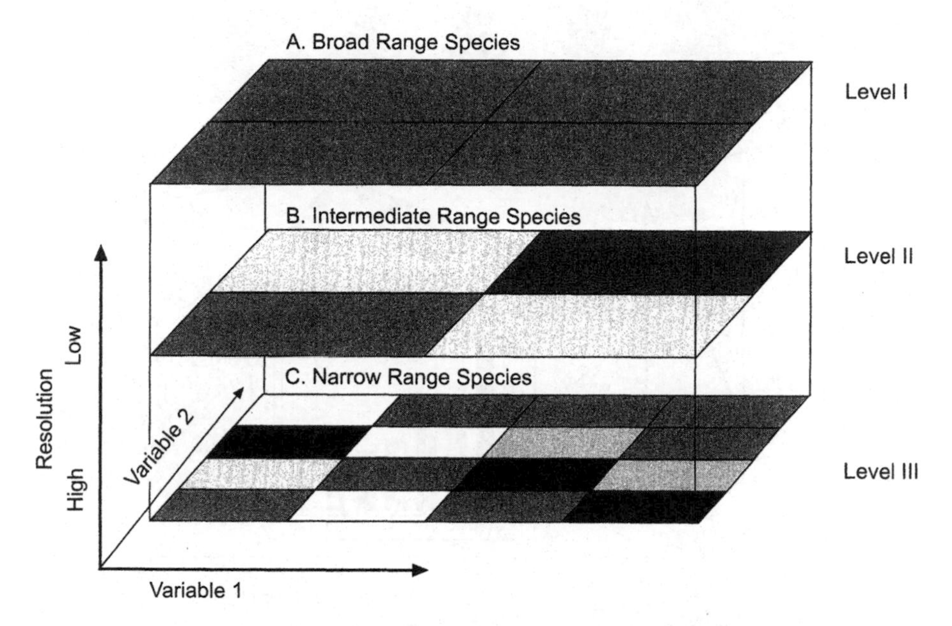

Figure 4: Hierarchical distinctions of habitat structures (after Kolasa 1989).Homogeneity or heterogeneity depend on the resolution with which species see their habitat. The space all species' use is physically the same, but they operate on different niveaus. The first level is used by only one species (generalist), the second is used by four different species and the third level is used by 16 different species (specialists).

In this context, emphasis is placed on the fact that the structural gradients at different scales are related to different processes that also operate with hierarchical distinctions. For example the gradients between tree top areas and clearings are based on the activities of one tree within the extent of its top area and the duration of its life time. The Ca concentrations in the soil come from a marled forest path which has existed for approximately 100 years, and the soil texture is a result of interglacial processes which have been influential since about 10000 years. Thus, the spatial extents of holons are in fact correlated with their temporal characteristics. Gradients and heterogeneous structures can be observed on three different scales in the examined beech forest. The smallest can be found between tree trunks and clearings, evolving from root uptake, exsudation, stemflow and litterfall. Because there is a direct connection to the growth of the tree, the gradients are built up as long as the tree is living. The degradation begins with the dying of the tree and is influenced by the new generation of dominating species.

On an intermediate scale of observation (10 m x 10 m) other types of gradients and structures exist. On this scale influences apart from tree growth like edge effects, concentration profiles between different trees or anthropogenic influences dominate. In this special case, an example easy to remember is the influence coming from the marling of a forest track. The track was built a century ago and mainly affected the calcium concentration and soil acidity. The original linear structure was degraded by the root uptake of Ca from beech trees, which transport and spread the element regularly, by litterfall.

The third type of gradient exists on a broad scale (50 m x 50 m). It can be related to influences of relief and soil texture, that are formed by processes on a higher system scale. Those processes, like climatic conditions or relief forming, show an extremely slow change and operate on large spatial areas. At this scale influences from human activity and inputs from adjacent systems are also effective and visible. The relative size of the gradients gets smaller compared to the intermediate scale but it has a much larger effect in space.

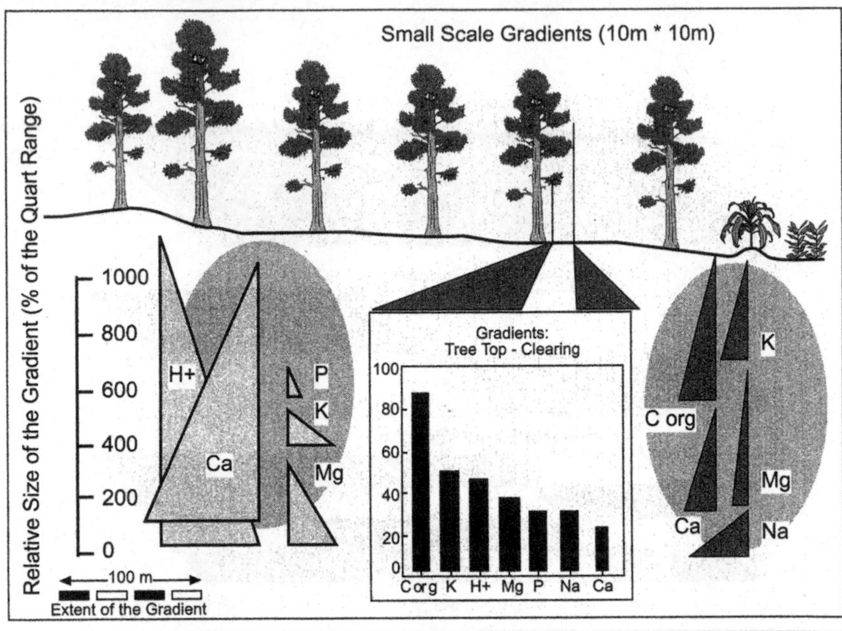

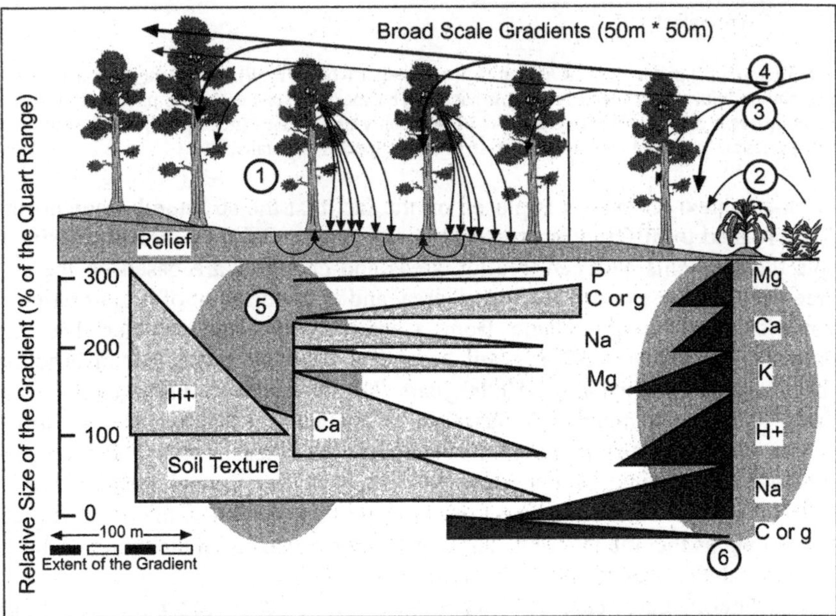

Figure 5: Spatial gradients in the upper soil layer of a beech forest in the Bornhöved Lake Region. In the upper part gradients on a 10 m grid scheme and on smaller scales are shown, while the lower part demonstrates the gradients of the 50 m sampling grid. The gradients have been calculated as relative percentile portions of the ranges between the first and the third quartile values. The numbers refer to: (1) marled forest track; (2) dust input from neighbouring field; (3) ammonia input due to slurry fertilisation; (4) atmospheric input by long distance transport; (5) gradient centre around the forest track; (6) gradient centre around the forest edge. For detailed description see Reiche et al. (in press).

Example 4

The gradients form structures in the area and because gradients represent differences in concentrations the structures are heterogeneous in space. Processes on a high scale in ecosystems form gradients which have a big extent and the resulting structures are more homogenous. The lower in hierarchy the forming process, the smaller the gradient, and the more heterogeneous the structure.

A useful example can be found in the soil calcium concentration, the soil acidity and the soil fraction in the beech forest. The existing structures were analysed with semivariograms and fractal methods (Hári et al., in press). In order to calculate the fractal dimension of the structures two different methods were used, one (Fractal Dimension A) was an aggregation of the semivariances described by Boroughs (1983), Milne (1990), and others; the other method (Fractal Dimension B) was done by a re-sampling and blocking of the original data as described by Loehle and Li (1994). Table 1 shows the results which will be described below.

Table 1 : Results of three different methods for quantifying spatial heterogeneity in a beech forest in the Bornhöved Lake Area, Germany. Fractal dimension A is gained by aggregating the corresponding semivariograms, fractal dimension B by re-sampling and blocking the original data. The middle sand fraction was chosen as an example for the soil fractions.

Method Parameter	Fractal Dimension A		Fractal Dimension B		Semivariogram Range
	10 m	50 m	10 m	50 m	
Calcium	1.74	1.65	1.95	1.68	200 m
pH	1.73	1.93	1.97	1.76	160 m
Phosphate	1.86	2.07	1.94	1.71	< 10 m
Soil fractions	1.88	1.74	1.97	1.76	360 - 380 m

As we know from above, the existing forest track was marled one century ago. This had great influence on the distribution of calcium in this forest. Therefore, the structure of calcium concentration in the soil was initialised by a great anthropogenic influence. The variograms show statistical dependencies in a range of 200 meters. Within this range the values depend upon each other, for this reason this is a homogeneous structure in itself. Looking at the forest as a whole, the structure of the calcium concentration is heterogeneous. The same result is gained by the fractal methods. In the 10 metre scale homogeneity is shown while in the 50 metre scale a highly heterogeneous structure is visible. The processes responsible for the distribution of calcium in the soil are root uptake from the beech and decomposition of the litter fall. There were no significant abiotic lateral transports measurable, that could also be influencing. The two processes operate at an intermediate scale in the hierarchy in this ecosystem (see above).

The soil calcium concentration has of course influence on the pH value, so the structure should be similar. Regarding the gradient structure, the fractal analysis B and the semivariograms, this assumption can be confirmed (see Figure and above), but looking at the fractal analysis A, *it does not show up*. This seems to be a question of method. What we can derive from the research in this forest is, that there is a broad scale acidification of the soil. This process is also visible in other places in the forest and is described in detail by Wetzel (1998) For the presented purpose we can derive that the processes that lead to soil acidification are operating on an hierarchical intermediate scale.

An example for a structure driven by processes on a low hierarchical scale is the content of phosphate in the soil. With the methods used it is impossible to detect a clear structure, because they all show contrary results. The most reliable result can be gained using variograms, that show very high nugget effects and a random distribution of the structure. Both methods for calculating fractals result in very high values of the fractal dimension, an indicator for homogeneity. From this we can derive that heterogeneities only occur at a scale smaller than 10 metre. The conclusion is that the processes responsible for the phosphate content of the soil operate at a low scale.

Quite different is the behaviour of the soil fractions. None of the three methods show a clear structure at the 10 metre scale, while at the 50 metre scale, a structure is visible. That means that at the smaller scale, the structure seems to be homogeneous and that heterogeneity only shows up on the higher scale. The semivariances at the 10 metre scale have also high values, therefore a statistical dependence is not measurable. At the 50 metre scale the semivariances are rising and an interdenpency can be shown in the variogram up to 380. The fractal dimensions at the 10 metre scale are very high (sign of homogeneity), while on the higher scale the values indicate heterogeneity. Resulting from that we can say that the processes responsible for this structure operate on a high hierarchical scale.

Summary

With the examination of gradients and structures we can show that spatial hierarchies exist in ecosystems and that they provide great influences on the ecosystem performances. The examples show three different structural levels that are formed by processes of different scales. Of course, interactions happen between those levels, so sometimes it may be quite complicated to exactly reveal the structures themselves. They may be covered by each other.

7. Temporal hierarchies of processes

In addition to spatial hierarchies of structures, temporal hierarchies are predicted by theory (see Chapter 2.2). This respective part of the chapter presents examples to investigate whether theory and application meet.

Example 5

Gaedke et al. (1996) performed empirical investigations on the temporal variability of biomass at different hierarchical scales. They examined the temporal fluctuations of three levels in two lake ecosystems: the species level, the community level and the system level. In order to quantify the variations of biomass in time, the coefficient of variation was used. The temporal variability of standing stocks is not only a function of their hierarchical level but also of external abiotic factors, so called 'external effects'. This is taken into account by choosing two ecosystems of different trophic states. Lake Constance (south-west Germany)which is mesotrophic and the oligotrophic Königssee (south-east Germany). With regard to the foodweb structure both systems are comparable. The same techniques and sampling frequencies were used to avoid any variables that can influence the results.

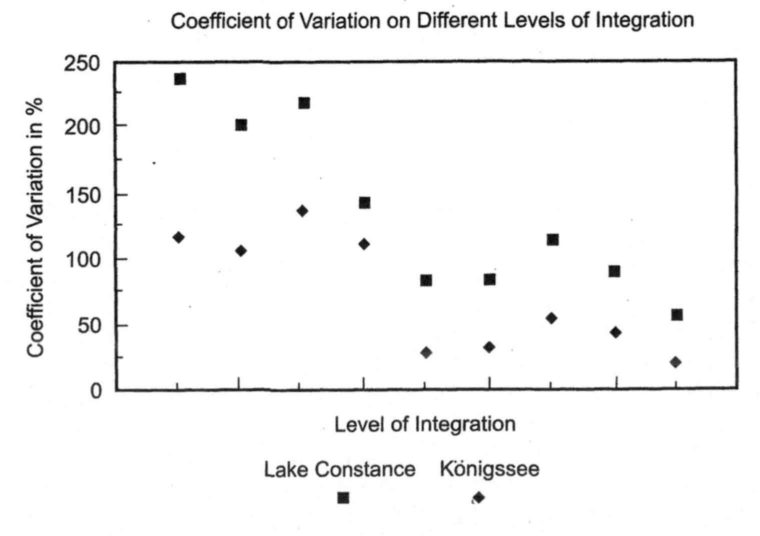

Figure 6: Coefficient of variation in two lake ecosystems (after Gaedke et al., 1996).

The authors found that in both lakes the temporal variability decreases with increasing hierarchical level. Of course, there are differences in the absolute values of the coefficient of variation resulting from the different trophic states, but a general tendency in the temporal behaviour could be detected. Although there are many influences on the biomass e.g. trophic state or generation time, it is clear that the higher you get in the hierarchy the lower the temporal variation in biomass (see Figure 3.).

Example 6

Water level dynamics at different scales can illustrate a temporal hierarchy (Müller, 1998). Time series data of the groundwater table in two distances from a lake and time series data of the lake surface's water level dynamics were analysed and then compared according to their temporal characteristics. The dynamics of the three locations are shown in the next figure.

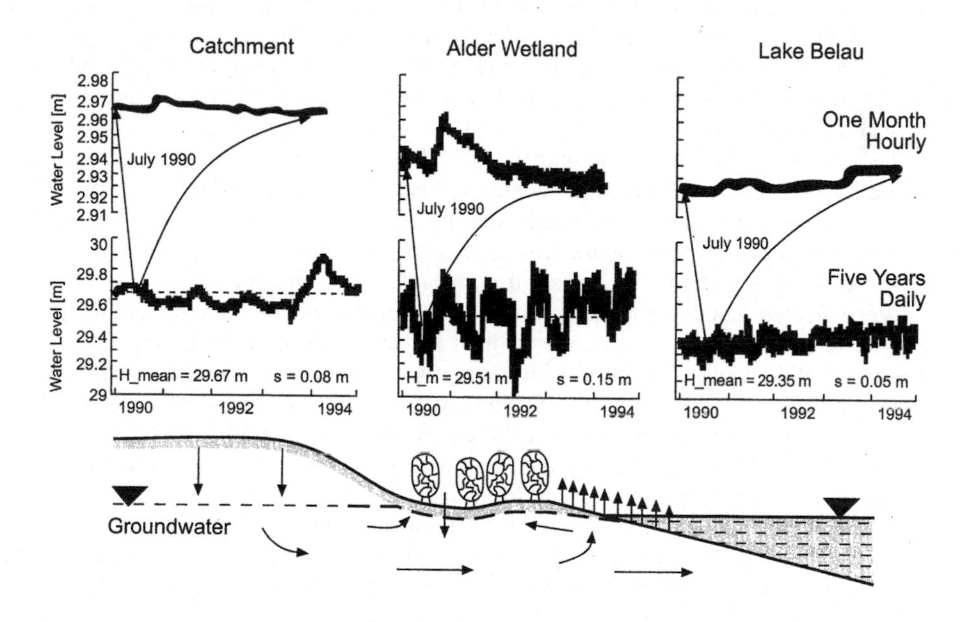

Figure 7: Water level dynamics of groundwater and lake on different scales (after Müller, 1998). The dynamics of the groundwater table and the lake surface are shown for the periods (1) from 1990 to 1994 (daily resolution) and (2) for July 1990 (hourly resolution). Data from Kluge (1993) and Kluge (unpublished data).

The hypotheses of hierarchy theory predicted that high level processes operate in areas of big size over a slow temporal frequency. The fluctuations of these processes are significantly small. The lower the level in the hierarchy the higher the fluctuations. This is depicted by the water level dynamic of the catchment, Lake Belau and the alder wetland. The alder wetland is influenced by more processes (e.g. lake water table, groundwater input, transpiration dynamics of the alders) than the lake (e.g. inflow and discharge of the river, surrounding wetlands) and the catchment (e.g. buffering capacity of the vegetation, soil and unsaturated zone), and therefore the fluctuations are higher. In other words, the lower levels are more effectively constrained than the higher levels. The latter are buffered by various factors; consequently the reaction time is much slower (Müller, 1998). The water level dynamics of the catchment represent a constraint for the dynamics of the lake, which itself constrains the dynamics of the alder wetland, and thus a dynamic hierarchy visible.

Example 7

A further example for a temporal hierarchy originates in abiotic climatic processes and deals with the different influences of soil temperature on a maize field (Hári in prep.). The basic question in this investigation was how different characteristic processes behave in time and if a hierarchy of those processes can be found. For this purpose the temporal characteristics have to be analysed. This was achieved using wavelet transformations, a time series analysis method.

Processes in ecosystems operate on certain, more or less frequent, fluctuations e.g. daily or annual courses. Usually Fourier transformations are used for detecting these frequencies. Unfortunately, this procedure has quite a disadvantage, in that important localisation information corresponding to the original signal gets lost by transforming the data. So when the Fourier transformation is used, only the major frequencies in the signal can be detected but not the exact time of occurrence. The wavelet transformation is more useful for this purpose, because characteristic frequencies and their exact occurrence in the original signal are shown (Clemen, 1997 and 1999). A detailed description of this relatively new and complicated method can be found in Burke Hubbard (1997), Torrence and Compo (1998) and Clemen (1998).

The data sets used for the wavelet analysis depicted in Figure 8 were derived from an agricultural field and the above atmosphere. As part of a methodological test, the following parameters were analysed in hourly resolution for the period of time August 1989 until July 1993: soil temperature, air temperature, air moisture, evaporation and radiation. Figure 3. shows a resulting wavelet for the period from August 1989 until July 1990.

We expected that the radiation would be a controlling factor over air temperature, humidity and evaporation, components that in return have an influence on the soil temperature. A cascade of temporal influences was assumed to show up in the different wavelets in form of a shift in the positive and negative waves' location until it reached the soil surface layer. The results of the different wavelet matrices were not as obvious as expected. The radiation only influenced the air and soil temperature when there were strong positive events occurring. It is most trivial that the air temperature rises with rising radiation. But this can only be recognised in summer, when the days are longer. In winter the radiation does not have a significant effect on the air temperature. The air temperature is then influenced by the moisture visible in the third and fourth wavelet. There is also an effect from the radiation on the air moisture, which works in contrast. When there are high values in radiation, there is very low air humidity (see August 89 and April/May 90).

This is a typical characteristic for the investigated location. Calculations showed the same structures over years. But even if it may not seem that clear, a hierarchy of processes can be found. There are processes which have a high frequent behaviour (like air temperature, soil temperature) and others which are low frequent (evaporation, radiation). The low frequent processes influence the high frequent ones as shown above, representing a higher scale in the system's hierarchy.

Summary

These few examples from biotic and abiotic field research show that there is a hierarchical structure in ecosystems. Furthermore, as with spatial hierarchies, the theory of temporal hierarchies requires more field research in order to provide more evidence as to their existence and function.

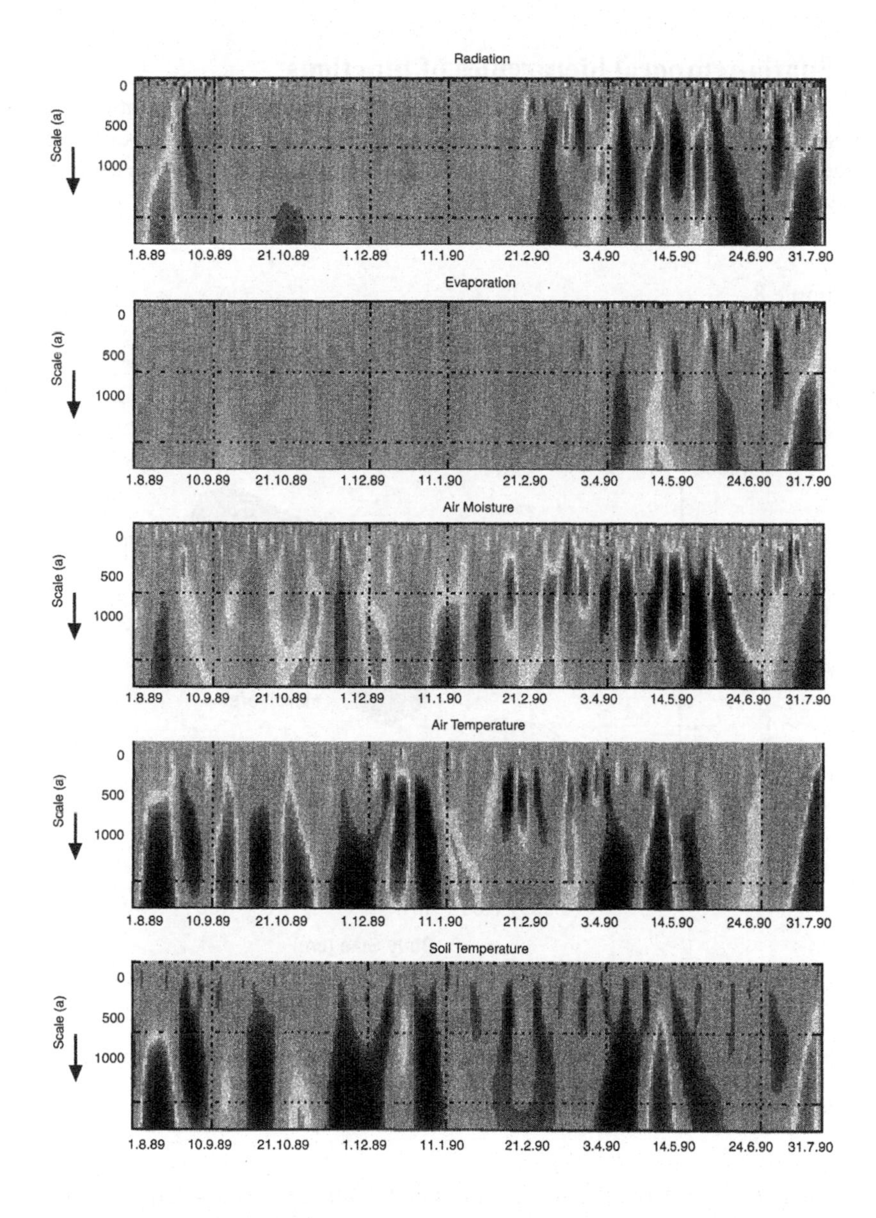

Figure 8: Temporal hierarchy of soil temperature influencing processes. The x-axis represents time in hours, the y-axis is the scale in hours. White structures mark positive waves, while grey structures mark negative waves. Positive waves represent a high correspondence of the original signal with the wavelet function, negative waves show the inversion.

8. Spatio-temporal hierarchies of functions

The greatest challenge posed by hierarchy theory is the combination of time and space in structuring the organisational units. It seems to be easy to find examples in space or in time but it is more complicated to find the combination. The reason for that is that respective data sets are extremely rare.

Example 8

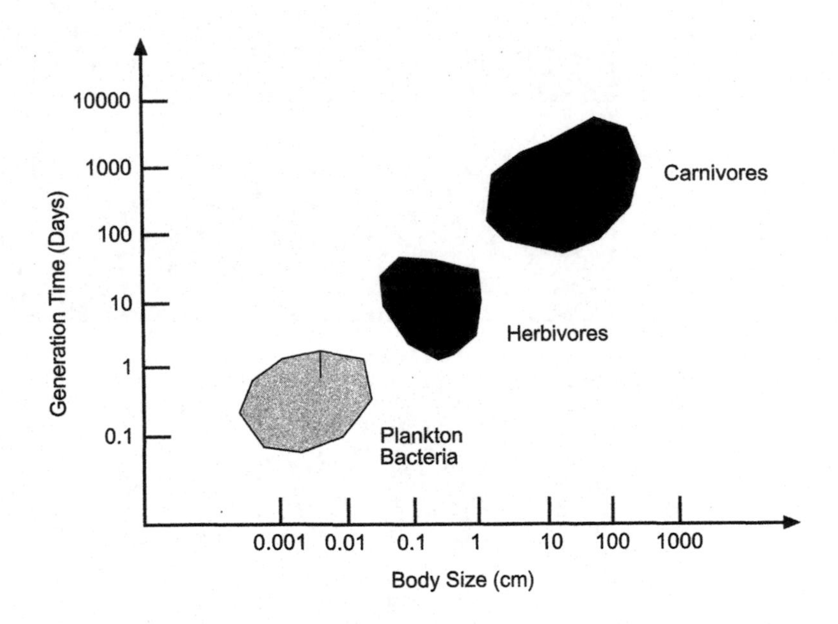

Figure 9: Relationship of body size-to-life time (after Steele 1992). For marine populations it is valid that the lowest trophic level (phytoplankton and bacteria) have the smallest body size and the shortest life time while top predators have the longest life time and the biggest body sizes.

Therefore we restrict the illustration to one case study. An example for a simple spatio-temporal hierarchy can be found in marine biocoenocis. It is often described how body size and generation time are connected (Figure 9). There is a high correlation between the body size of different species like fish or plankton and their life time (Steele 1989). Small organisms generally have a short life time while big organisms are living longer. Other characteristic features can be added to this linkage of spatial and temporal scales. The bigger the body size of an organism, the higher are the demands for the habitat. Furthermore, the spatio-temporal features are correlated with functional regulations represented by the controlling functions in the food web.

Summarising, it can be stated that spatial hierarchies can be quantified as well as temporal hierarchies. There are case studies and data sets which indicate that the combination of spatial and temporal features are in accordance with the theory. In addition, good hints make us suggest that the concept of functional constraints is valid which can be assigned to spatio-temporal scales. To prove the detailed evidence of these suggestions, further empirical tests will be carried out and as result the 'space of validity' of the hierarchical hypotheses will be defined on the basis of empirical data gathered from ecosystem research approaches.

9. Prospect for environmental applications

The preceding chapters have shown that hierarchy theory has a high potential to become an integrating branch in ecosystem theory. Although empirical tests are still necessary to adjust the hypotheses and to determine their specific validities, there is already a rather high applicability in ecological and environmental problem fields. Due to its level-conducting fundamentals, hierarchical methods are useful in all classifying activities. They include new ideas to implement spatio-temporal interactions and they reveal an interesting relationship between structural, functional and organisational system features. Especially landscape-ecological problems might find new solutions if the theory is applied. This concerns all activities of extrapolation and interpolation as well as the 'everlasting problem' to derive regional statements from punctual data.

Another potential field of application deals with decision making processes in environmental planning or in environmental compatibility tests. These procedures could be improved by the simple question for the scale of an intervention. If the spatial extent and the temporal duration of a measure is known, the prognosis of potential consequences can be carried out much easier. From the spatio-temporal features it can be derived for which ecological processes the intervention might function as a constraint, for which interactions the scale is too small to be effective and where holi of the same level are met. Also, the extents of the concerned ecological structures and functions should be reflected to define the potentially effected area of a planning measure. In a lot of cases it will turn out that the border lines of the environmental effects do not at all coincide the borders of the environmental management because they are not defined ecologically but on an administrative basis. Furthermore, a hierarchical analysis will show vested spatial structures that cannot be investigated by the present administrative structures and the present system of responsibilities. In this context, three scales should be taken into account, the focal scale of the measure and its direct implication, the next-higher level which provides the constraints of the focal niveau, and the next-lower level which qualifies the biotic potential of the level of interest.

Finally, the theory can also be useful to evaluate time series of ecological data, e.g. in environmental monitoring. On the one hand, it is obvious that a disturbance is effective especially against processes that operate on similar temporal characteristics. Thus – besides the spatial conditions – a hierarchical analysis of temporal features of an intervention and of the potential acceptors can be helpful to avoid harmful consequences. On the other hand, the theory predicts unstable temporal variability's in case systems are leaving their steady states. Thus, the height of amplitudes, the variances, and the regularity of a state variable's dynamics are good indicators for the integrity of an ecological system which is only given under hierarchical conditions. Whenever phase transitions are approached, the dominating hierarchies are broken and systems will find a new steady state, where a new regime of spatial and temporal, hierarchical interaction will be established.

REFERENCES

Allen,T.F.H. and Starr,T.B. 1982. Hierarchy - Perspectives for Ecological Complexity. The University of Chicago Press, Chicago - London, 310 pp.

Allen, T.F.H. and Hoekstra, T.W. 1992. Toward a Unified Ecology - Complexity in Ecological Systems. Columbia University Press, New York, 384 pp.

Burke Hubbard, B. 1997. Wavelets - Die Mathematik der kleinen Wellen. Birkhäuser Verlag, Basel - Boston - Berlin, 308 pp.

Burrough, P.A. 1981. Fractal dimensions of landscapes and other environmental data. *Nature* 294, pp 240-242

Burrough, P.A. 1983. Multiscale sources of spatial variation in soil I: the application of fractal concepts to nested levels of soil variation. *Journal of Soil Science* 34, pp. 577-597

Clemen, T. 1997. Integrating simulation models into environmental information systems - model analysis. In: Denzer, R., Swayne, D.A. and Schimak, D. (Eds.) *Environmental Software Systems. Volume 2*, pp 292-299

Clemen, T. 1999. Zur Wavelet-gestützen Validierung von Simulationsmodellen in der Ökologie. - Dissertation. Universität Kiel.

Gaedke, U., Barthelmess, T. and Straile, D. 1996. Temporal variability of standing stocks of individual species, communities, and the entire plankton in two lakes of different tropic state: empirical evidence for hierarchy theory and emergent properties? *Senckenbergiana Maritima* Band 27 3/6, pp. 169-177.

Hári, S. 1999. Die Anwendung der Hierarchitätstheorie in der Ökosystemforschung. - Dissertation. Universität Kiel.

Hári, S., Fränzle, O., Li, B.L., Müller, F. and Reiche, E.-W. (in press): Spatio-temporal heterogeneity and scales of forest ecosystem processes. *Ecological Studies.*

Klijn,F. 1997. A Hierarchical Approach to Ecosystems and its Implications for Ecological Land Classification. Ponsen and Looijen BV, Wageningen, 186pp.

Kluge, W. 1993. Der Einfluß von Uferfeuchtgebieten auf den unterirdischen Wasser- und Stoffaustausch zwischen Umland und See. *Mitteilungen der Deutschen Bodenkundlichen Gesellschaft* 72, S. 151-154.

Koestler, A. 1970. Jenseits von Atomismus und Holismus - der Begriff des Holons. In: Koestler, A. and Smythies, J.R. (Hrsg.) *Das neue Menschenbild.* S. 192-229.

Kolasa, J. 1989. Ecological systems in hierarchical perspective: breaks in community structure and other consequences. *Ecology* 70: 36-47.

Loehle, C. and Li, B.L. 1994. Statistical properties of ecological and geologic fractals. *Ecological Modelling* 85: 271-284.

Mesarovic, M.D., Macko, D. and Takakara, Y. 1970. Theory of Hierarchical, Multilevel Systems. Academic Press, New York, 294 pp.

Milne, B. 1990. Lessons for applying fractal models to landscape patterns. In: Turner, M.G. and Gardner, R.H. (Eds.) *Quantitative Methods in Landscape Ecology.* pp 199-235.

Müller, F. 1992. Hierarchical approaches to ecosystem theory. *Ecological Modelling* 63: 215-242.

Müller, F. 1996. Emergent properties of ecosystems - consequences of self-organising processes? - *Senckenbergiana Maritima* 27 (3/6), S. 151-168.

Müller, F. 1998. Gradients in ecological systems. *Ecological Modelling* 108: 3-21.

Müller, F., Breckling, B., Bredemeier, M., Grimm, V., Malchow, H., Nielsen, S.N. and Reiche, E.W. 1997a. Emergente Ökosystemeigenschaften. In: Fränzle, O., Müller, F. and Schröder, W. (Hrsg.). *Handbuch der Umweltwissenschaften.* Kapitel III-2.5, 21 S.

Müller, F., Breckling, B., Bredemeier, M., Grimm, V., Malchow, H., Nielsen, S.N. and Reiche, E.W. 1997b. Ökosystemare Selbstorganisation. In: Fränzle, O., Müller, F. and Schröder, W. (Hrsg.) *Handbuch der Umweltwissenschaften.* Kapitel III-2.4, 19 S.

Nielsen, S.E. et al. (this volume).

O'Neill, R.V., DeAngelis, D.L., Waide, J.B. and Allen, T.F.H. 1986. A Hierarchical Concept of Ecosystems. Princeton University Press, Princeton - New Jersey, 253 pp.

O'Neill, R.V., Johnson,A.R. and King, A.W. 1989. A hierarchical framework for the analysis of scale. *Landscape Ecology* 3: 193-205.

Pattee, H.H. 1973. Hierarchy Theory - The Challenge of Complex Systems. - Braziller, New York, 156 pp.

Reiche et al. in press. Components of spatial heterogeneity in a forest ecosystem. *Ecological Studies.*

Salthe, S.N. 1985. Evolving Hierarchical Systems - Their Structure and Representation. - Columbia University Press, New York, 343 pp.

Simon, H.A. 1973. The Organisation of Complex Systems. In: Pattee, H.H. (Ed.) *Hierarchy Theory.* pp 1-27.

Steele , J.H. 1989. The ocean landscape. *Landscape Ecology* 3: 185-192.

Torrence, C. and Compo, G.P. 1998.A practical guide to wavelet analysis. - Bulletin of the American *Metereological Society* 79: 61-78.

Wetzel, H. 1998. Prozeßorientierte Deutung der Kationendynamik von Braunerden als Glieder von Acker- und Waldcatenen einer norddeutschen Jungmoränenlandschaft - Bornhöveder Seenkette. - EcoSys Suppl.Bd 25, 132 S.

II.6 Ecosystems as Energetic Systems

H.T. Odum, M.T. Brown and S.Ulgiati

1. Introduction

This chapter is a brief synopsis of energy systems representation of ecosystems, roughly defined as an environmental system with living populations. Like a short course given with limited time, this text with limited space touches on what is most important.

For concepts to be useful to human minds, that part of the continuum of complexity of the universe which we can sense has to be simplified. Once we define the scale of our interest, we abandon this interest if we simplify by reducing scale to smaller parts and mechanisms. Instead, we have to retain the whole, larger-scale system and simplify by aggregating parts and processes into fewer units and pathways. Because many of the concepts depend on system connections, a diagrammatic network language is required to keep an holistic perspective about design, cycles, and the controls of the larger scale while also viewing the parts with analytic precision. We start by introducing concepts using energy systems diagram to make network relationships clear. In Figure 1 are the energy systems symbols. How they are used to represent ecosystems is shown in Figure 2.

1.1 Concepts and measures

The following are procedures and principles used to represent ecosystems with energy systems concepts and measures (Figure 2):

1. A window frame of attention is defined.
2. Adopting a boundary identifies all the inputs from the continuum of structure and processes outside the window frame as sources.
3. Position in the diagram reflects the scale of turnover and territory. Structures and processes are arranged from left to right in order of increasing scale.
4. Entities of scale smaller than the main window of interest are included but aggregated. For example, processes of a population of micro-organisms might be aggregated as a single pathway.
5. Since energy accompanies everything, pathways are flows of energy with or without material or information. Energy inflowing either increases storages inside or exits as outflows. Pathway flows are in power units, since useful energy per time is power.
6. Total resource available to the defined area is the sum of the inputs expressed on a common basis as emergy (spelled with an 'm'). Emergy is defined as the available energy of one kind of energy previously used up to make the product. Flow of emergy per time is empower.
7. Inflows include those of the smaller scale, items of similar scale located elsewhere and items of the larger scale such as those from the economy, human society, and the earth.
8. Energy transformations from left to right illustrate first and second energy laws, with available energy passing out through the heat sink. Energy flows decrease from left to right, but the capabilities of that energy increase.
9. As a result of self organisation for maximum empower, structure of relationships has autocatalytic loops within components and between aggregates. These 'feedback' loops pass from right to left as inputs to production-transformation processes and are drawn with counter-clockwise arcs to production (interaction) symbols.
10. Effective reinforcements of production processes require matching of high quality feedbacks (small energy flows) interacting and amplified by larger flows of lower quality energy.

11. Production symbols combine necessary flows of different kinds and usually imply multiplicative production functions between items of different scale. The reinforcement of the feedback loops at the system level tends to eliminate the tendency for isolated production processes to become limited by one or more factors.
12. Position in energy hierarchy from left to right is measured by the increasing transformity from left to right. Transformity is emergy required per unit of transformed energy.
13. To the right energy components and their flows are of a higher quality and capable of greater action per unit in feedback reinforcements to other parts of the system.
14. Materials are bound into structure of higher quality to the right, being released by consuming processes there and circulating back to lower concentrations on the left where they can be incorporated into production processes again.
15. Items to the right form into more concentrated centres, spatially; higher in energy hierarchy.
16. On each scale; production generates storages that grow followed by pulses of consumption and transformation of these stores by other units to the right. Thus pulsing pairs appear during self organisation.
17. The larger the scale of the pulse pair, the longer the interpulse period and the sharper the pulse's feedback actions.
18. The multiplicative interaction of dynamic pulsing at many scales produces patterns of variation resembling skewed statistical distributions (Log normal, Weibull).
19. Where it is desirable to show in the diagram the self organising mathematical relationships that generate pulsing, connections between pulsing pairs include 3 parallel competing pathways (a linear pathway, autocatalytic loop, and quadratic autocatalytic loop).
20. Where it is desirable to make the system quantitative, values of flows and storages are placed on the diagram in whatever units are appropriate. These may be values for a given time or averages as if the system were in steady state.
21. Where a time dimension is desirable, the numerical values are used to calibrate the coefficients of equations for computer simulation programs. The equations may be included at the foot of the diagram, thus translating the kinetic aspect of the model.
22. Information has high transformities and belongs on the right of the systems diagrams where their feedback control loops may be the most important. To maintain information requires a circle of duplications, reapplication and selection.
23. Humans, their economy, and their information processing interact as part of a larger scale (high transformity) and are shown on the right.

Energy systems diagrams are concentrated ways of showing many systems relationships.

1.2 Appropriate complexity

For a top down overview, introductory education, public policies, and simple computer simulations of the broadest features, a system of ten or less components is desirable. Diagrams with 30 or more components may serve as an inventory, but this level of complexity is hard to evaluate, simulate, or anticipate consequences. Very complex diagrams may be useful as a pre-aggregation exercise. The complex diagrams help consider more items in the aggregating process. Complex diagrams have been used as a system checklist, an impact statement, and sometimes on walls as aesthetic art of complexity.

ENERGY SYSTEMS SYMBOLS

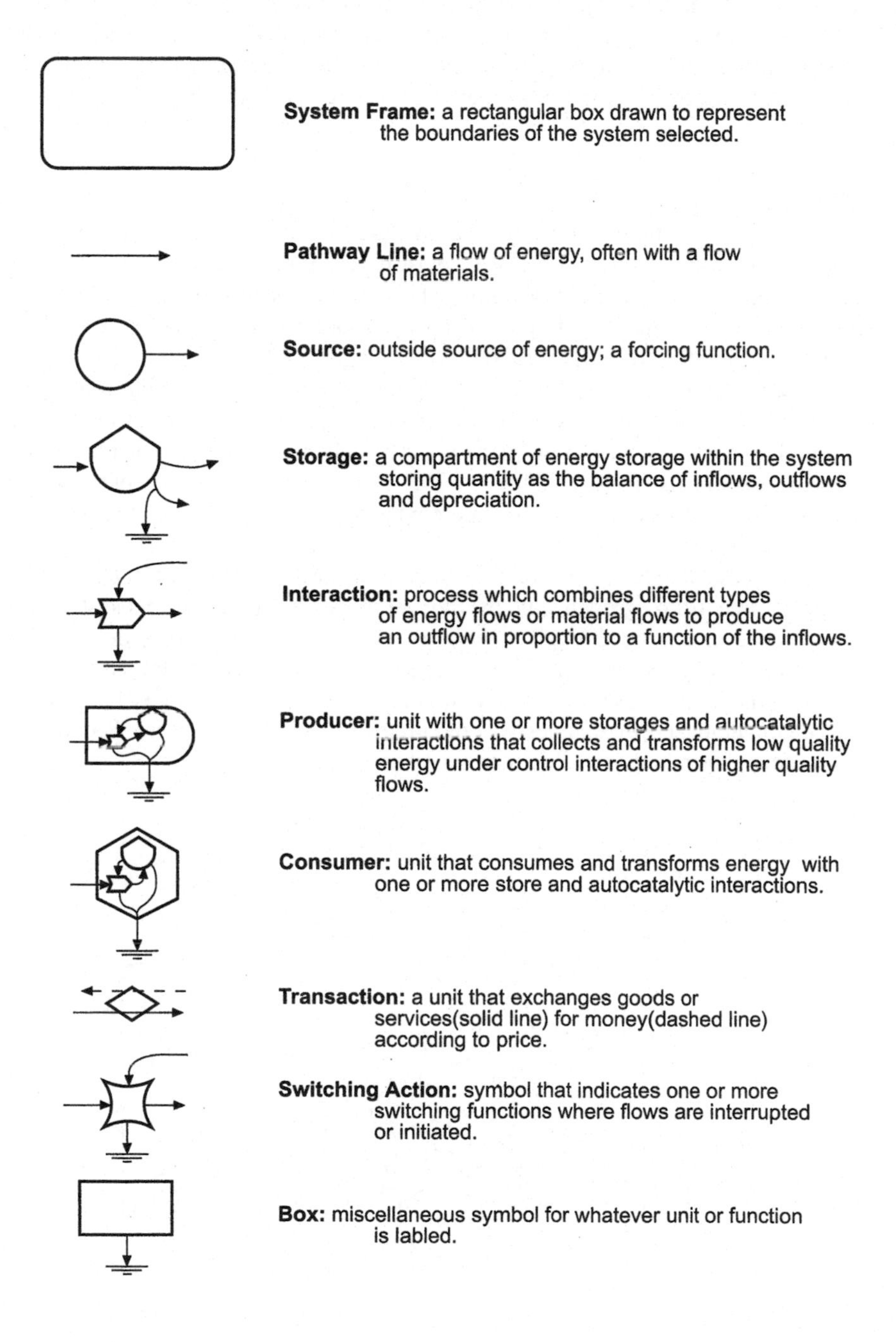

System Frame: a rectangular box drawn to represent the boundaries of the system selected.
Pathway Line: a flow of energy, often with a flow of materials.
Source: outside source of energy; a forcing function.
Storage: a compartment of energy storage within the system storing quantity as the balance of inflows, outflows and depreciation.
Interaction: process which combines different types of energy flows or material flows to produce an outflow in proportion to a function of the inflows.
Producer: unit with one or more storages and autocatalytic interactions that collects and transforms low quality energy under control interactions of higher quality flows.
Consumer: unit that consumes and transforms energy with one or more store and autocatalytic interactions.
Transaction: a unit that exchanges goods or services(solid line) for money(dashed line) according to price.
Switching Action: symbol that indicates one or more switching functions where flows are interrupted or initiated.
Box: miscellaneous symbol for whatever unit or function is labled.

Figure 1: Energy Systems Symbols

2. Energy hierarchy and transformity

Geological processes, atmospheric systems, ecosystems, and societies are interconnected, each receiving energy and materials from the other, interacting through feedback mechanisms to self-organise in space, time, and connectivity. While processes of energy transformation throughout the geobiosphere build order, cycle materials, sustain information and degrade energy in the process, they organise units in an energy hierarchy.

2.1 Energy hierarchy

When many of one type of unit are combined to form a few of another, the relationship is hierarchical. Since there is energy in everything including information, and since there are energy transformations in all processes, most if not all things form a hierarchical series. The scale of space and time increases along the series of energy transformations. Many small scale processes contribute to fewer and fewer of larger scale. In our systems diagrams scale increases from left to right (Figure 3).

Energy is converged to higher order processes where with each transformation some energy is passed along the web and much is degraded as a consequence of the 2nd Law of Thermodynamics. For each transformation step, much energy loses its availability, and only a small amount is passed along to the next step. For some purposes the energy transformation processes normally interconnected in webs can be aggregated as simpler transformation chains.

Examples include water streams converging in a watershed, the leaf cells processing energy to tree trunk growth and the convergence of energy in ecosystem food chains. Convergence of energy through a series of energy transformations yields a final product which carries less energy than invested to start the chain, due to the entropic degradation. However, the higher position of the item in the energy hierarchy makes it more valuable, as a large convergence of resources was required to support the process. We may say that the final product has a higher quality than initial products.

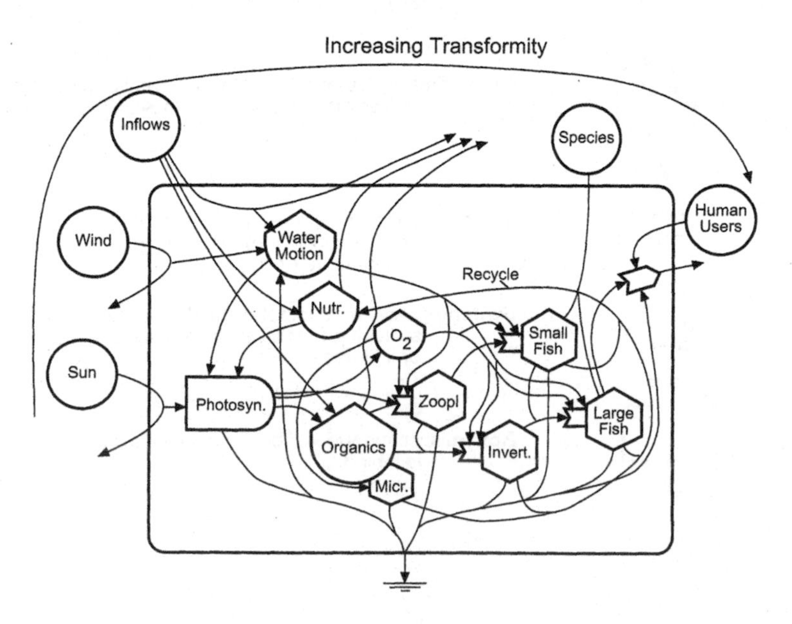

Figure 2: Systems diagram of typical ecological systems (a) Terrestrial ecosystems; Aquatic ecosystems

Hierarchical Levels

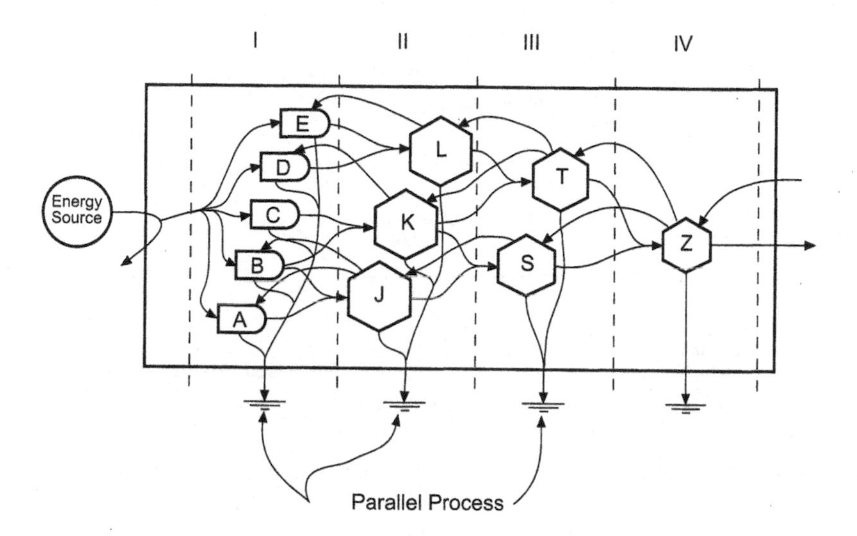

Figure 3: An energy web organised hierarchically, showing levels and parallel processes, and reinforcing feedbacks (right to left)

2.2 Emergy, empower, and transformity

Whereas different amounts and qualities of energy are found along the energy hierarchy, the concept of emergy, spelled with an 'm' (Odum, 1983, 1996) is used to express all available energy flows in a comparable way. By definition, emergy is the amount of energy of one form that is required, directly and indirectly, to provide an energy flow or storage. Emergy stands for energy memory (Scienceman, 1987). In many papers we standardise by expressing everything as solar emergy. The unit of emergy is the emjoule or emcalorie. In this paper solar emergy is used and expressed as solar emjoules (abbreviated sej).

The flow of emergy is empower in units of emjoules per time. Solar empower is in solar emjoules per time (abbreviated sej/t).

The emergy required to make one unit of available energy is called transformity. It is calculated from observed ratios of emergy to energy (either in storages or in flows). Solar transformity is expressed in solar emjoules per joule (abbreviated sej/J).

3. Production

Energy transformations generate a stream of products. Usually the production process occurs with interaction of two or more different kinds of inputs with different transformity. As shown with the interaction symbol in Figure 4, there is a matching between a high transformity, controlling feedback from the right with the abundant, low quality, low transformity energy from the left. In order to maximise its effect commensurate with the energy which went into its formation, high transformity items have to interact so as to mutually amplify a flow with larger energy quantity to produce outputs of intermediate transformity.

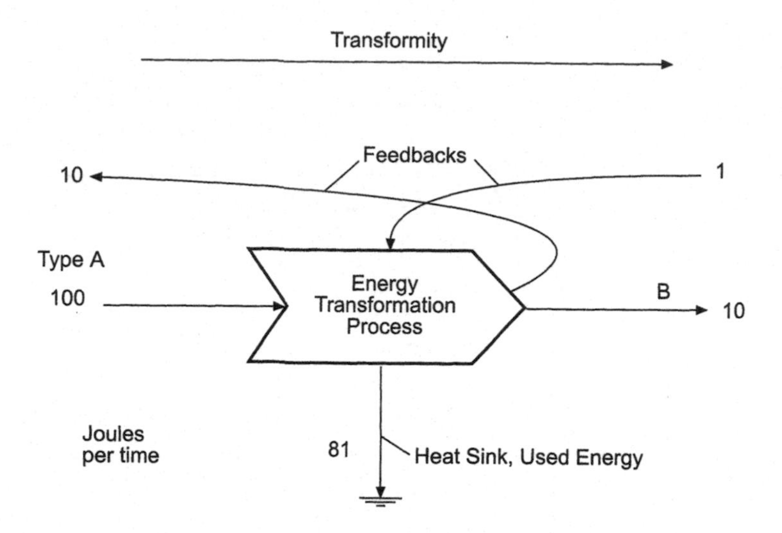

Figure 4: Energy systems diagram of a production process with values of energy on the pathways.

3.1 Autocatalytic reinforcement and maximum power

Following Boltzmann (1886), Lotka (1922) enunciated the principle of maximum power "...that in the struggle for existence, the advantage must go to those organisms whose energy-capturing devices are most efficient in directing available energy into channels favourable to the preservation of the species." Further, he states "...natural selection tends to make the energy flux through the system a maximum, so far as compatible with constraints to which the system is subjected. Therefore, power maximisation was suggested by Lotka as the measure of fit designs and offered as the 4th Law of Thermodynamics.

One of the common design mechanisms for maximising power is the autocatalytic loop from stored product back interacting and amplifying the production process. In energy systems diagrams the feedback from high quality passes back to the left (Figure 5). Living and non-living units of the environment develop autocatalytic reinforcement loops as part of production processes. Examples here include leaves capturing sunlight, animals catching their food, watershed land forms capturing runoff.

3.2 Maximum empower principle

Relative to the energy hierarchy, however, these definitions would imply that self organising processes develop the low transformity scales where energy flow is greater. We restated the principle as maximising empower, which means maximising useful work on all scales equally at the same time.

Maximum Empower Principle: On each scale, systems designs prevail that maximise empower, first by reinforcing inflows of available energy and second by increasing the efficiency of useful work.

Energy dissipation without useful contribution to intake and efficiency is displaced because such pathways are not mutually reinforcing. For example, drilling oil wells and then burning off the oil may use oil faster (in the short term) than refining and using it to run machines, but that design is replaced with a system that uses oil to develop and run machines, increase drilling capacity and ultimately the rate at which oil can be supplied.

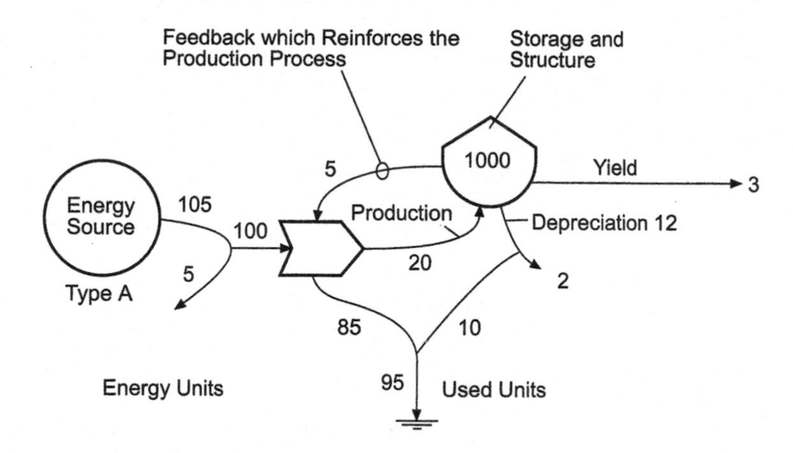

Figure 5: Configurations of productions that reinforce by feeding back higher transformity products in an autocatalytic loop

3.3 Splits and co-products

The pathways of networks depend on the human perceptions with which the complex real systems are simplified. If a product flow is divided into branches of the same kind, it is a 'split', and each branch has the same transformity and emergy per mass (Figure 6a). For example some of the fruits produced by a tree split when some wash away and some remain for local birds.

On the other hand, we may have aggregated the real world into units where there are outputs of different kind, which we call 'co-products' (Figure 6b). For example, trees produce fruits and leaf litter. Different kinds of products have different transformities. Although we may aggregate units so that two different transformity products are co-products, they may be at different levels of energy hierarchy within the unit as suggested by the dashed line details within the box (Figure 6b).

3.4 Annual emergy budget of the earth

For evaluating production within the earth's geobiosphere, the abundant general input energy is that of solar insolation, which is taken as the reference for calculating transformities. By defining average insolation at the earth's surface as reference, the transformity of solar radiation becomes one. Transformities of the main natural flows in the biosphere (wind, rain, ocean currents, geological cycles, etc.) are calculated as the ratio of total emergy driving the biosphere as a whole to the actual energy of the flow under consideration. Figure 7 shows the total emergy driving the biosphere as the sum of solar radiation, energy from the deep earth, and tidal momentum. The total of these (equal to 9.44 E24 sej/yr) is used as the base emergy because the biosphere processes that produce winds, rains, ocean currents, and geologic cycles are coupled and cannot generate one without the other.

In addition in this century, there is high emergy coming from use of fuels and mineral reserves accumulated at an earlier time (Table 1). The changes observed in global climate suggest that this emergy should also be included (Brown and Ulgiati, 1999).

Table 1: Flux of renewable and non-renewable energies driving global processes

Note Source	Energy Flux (J/yr)	Transformity* (sej/J)	Solar Emergy Flux (E24 sej/yr)	Emdollars# (E12 Em$)
Global Renewable Energies				
1. Solar insolation	3.94E+24	1	3.94	3.57
2. Deep earth heat	6.72E+20	6055	4.07	3.69
3. Tidal energy	8.52E+19	16842	1.43	1.30
	Subtotal		9.44	8.56
Society Released Energies (non-renewable)				
4. Oil	1.38E+20	5.40E+04	7.45	6.75
5. Natural gas	7.89E+19	4.80E+04	3.79	3.43
6. Coal	1.09E+20	4.00E+04	4.36	3.95
7. Nuclear energy	8.60E+18	2.00E+05	1.72	1.56
8. Wood	5.86E+19	1.10E+04	0.64	0.58
9. Soils	1.38E+19	7.40E+04	1.02	0.93
10. Phosphate	4.77E+16	7.70E+06	0.37	0.33
11 Limestone	7.33E+16	1.62E+06	0.12	0.11
12. Metals	992.9E+12g	1.0E+09sej/g	0.99	0.90
	Subtotal		20.46	18.54
	TOTAL		29.91	27.10

* Transformities from Odum (1996)
# Emdollars obtained by dividing Emergy in column 5 by 1.1E12 sej/$ (Table 4)

1. Sunlight	Solar constant, 2 cal/cm^2/min 70%absorbed Earth cross section facing the sun = 1.278 E14m^2 Energy Flux = (2 cal/cm^2/min) (1.278E18cm^2)(5.256E5min/yr)(4.186J/cal)(0.7) =3.936E24J/yr	(Von der Haar and Suomi, 1969)
2. Deep earth heat	Heat released by crustal radioactivity = 1.98 E20J/yr Heat flowing up from the man = 4.74 E20J/yr Energy Flux = 6.72 E20J/yr	(Sclater et al., 1980) (Sclater et al., 1980)
3. Tidal Energy	Energy received by the earth =2.7 E19erg/sec Energy flux = (2.7 E19erg/sec)(3.153E7sec/yr)/(1E7 erg/J) = 8.513E19J/yr	(Munk and Macdonald, 1960)
4. Oil	Total production = 3.3 E9Mt oil equivalent Energy flux = (3.3E9 t oil eq.)x (4.186E10J/t oil eq.) = 1.47 E20 J/yr oil equivalent	(British Petroleum, 1997)
5. Natural gas	Total production = 2.093 E9m^3 Energy flux = 82.093 E12m^3)x(3.77E7J/m^3) = 7.89 E19J/yr	(British Petroleum, 1997)
6. Coal	Total production (soft) = 1.224 E9 t/yr Total production (hard) = 3.297 E9 t/yr Energy Flux = (1.224 E9 t/yr) (13.9 E9J/t) + (3.297 E9 t/yr)(27.9 E9 J/t) =1.09 E20 J/ yr	(British Petroleum, 1997) (British Petroleum, 1997)
7. Nuclear energy	Total production = 2.39 E12 kwh/yr Energy Flux = (2.39 E12 kwh/yr) (3.6 E6 J/kwh) = 8.60 E18 J/yr elec. equivalent	(British Petroleum, 1997)
8. Wood	Annual net forest area loss = 11.27E6 ha/yr	(Brown et.al, 1997)

	Biomass = 40 kg/m2	(Lieth and Whittaker, 1975)
	Energy Flux = (11.27 E6 ha/yr)(1 E4 m2/ha) (40 kg/m2)(1.3 E7 J/kg) = 5.86 E 19 J/yr	
9. Soil ersosion	Total soil erosion = 6.1 E10 t/yr	
	Based on conservative soil loss estimate of 10 t/ha/yr and 6.1 E9 ha agricultural land = 6.1 E16 g/yr (assume 1.0% organic matter), 5.4 kcal/g	
	Energy Flux = (6.1 E 16g)(.01)(5.4 kcal/g)(4186 J/kcal)	
	=1.38 E19 J/yr	
10. Phosphate	Total global production = 137 E6 t/yr	
	Gibbs free energy phosphate rock = 3.48 E2 J/g	(USDI, 1996)
	Energy Flux = (137 E12 g)(3.48 E2 J/g)	(Odum, 1996 p125)
	= 4.77 E16 J/yr	
11. Limestone	Total production = 120 E6 t/yr	
	Gibbs free energy phosphate rock = 611 J/g	(USDI, 1996)
	Energy Flux = (120 E12 g)(6.11 E2 J/g)	(Odum, 1996 p47)
	= 7.33 E16 J/yr	
12. Metals	Total global production of Al, Cu, Pb, Fe, Zn (1994) = 992.9 E6 t/yr = 992.9 E12 g/yr	(World Resources, 1997)

The total emergy driving a process measures the self-organisation work that is converged to make a product, service or stored reserve. By measuring work in emergy units, work in different parts of the energy hierarchy are comparable. Details on the procedure to perform an emergy account are published in many places (Odum, 1996).

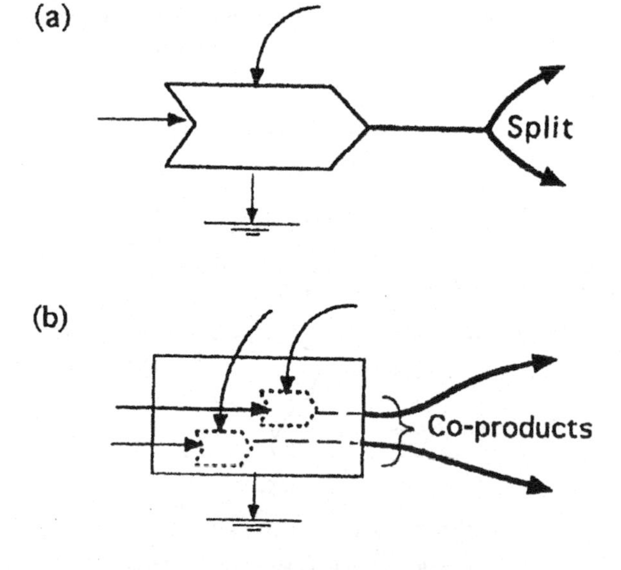

Figure 6: Configurations of production flows: (a) splits and (b) co-products

4. Material cycles

Material cycles in systems of the biosphere as part of the processing of energy transformations. There are cycles of the chemical elements, water, sediments, and waste products of society. Recycle is a part of all systems, from the scale of long term geologic cycles to the scale of ecosystems where nutrients and organic matter are recycled in relatively short term storages of plant and animal tissue. Material recycling is another way that processes are mutually reinforcing. Increasingly materials used by human economic systems are being recycled better as economic systems get better organised.

Material pathways are a regular part of an energy systems diagram, since there is always some energy in materials that are more concentrated than background. Sometimes it is useful to highlight pathways of a particular material using special shading, colour, or lines on a transparent overlay. See Figure 8.

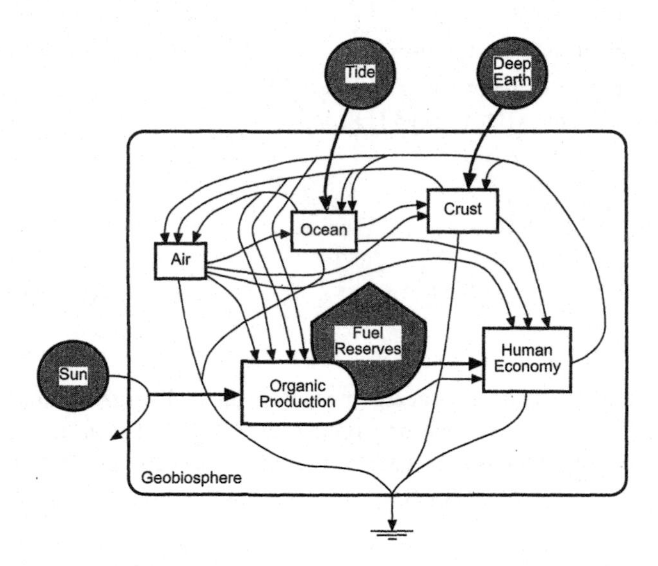

Figure 7: Systems diagram of the geobiosphere showing main emergy sources: solar insolution, tidal energy and deep earth inputs and use of mineral reserves in interconecctiong pathways of the atmospheric, oceanic and crustal systems

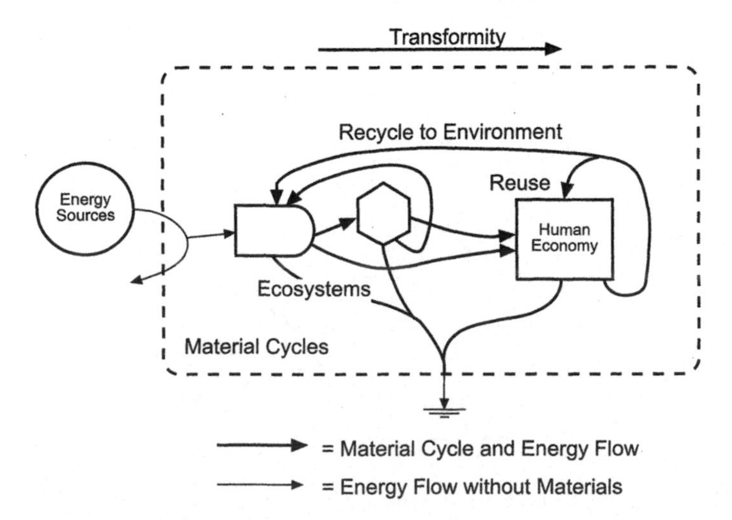

Figure 8: Systems diagram of recycling

4.1 Mass emergy

Each cycle occupies a zone in the energy hierarchy for its cycle from a dilute to more concentrated state and back again. The concentrating process requires use of available energy. The material gains emergy as it is concentrated. Emergy per unit mass (mass emergy) is a useful measure of material state which increases with concentration.

In part of each cycle the material may be incorporated into a product which has the emergy of its inputs. For example, phosphorus is incorporated into plant structures as it is concentrated. Where the material is released and dispersed into air, water, or solids at lower concentration, it loses some of its emergy. For example, the emergy per mass in phosphorus dispersing from plant decomposition decreases. The amount can be estimated from the inputs required for its reconcentrating.

Although everything on earth is more concentrated and has emergy relative to the general background of the universe, for our earth evaluations, however, we use the geobiosphere as our reference base. For many primary substances such as chemical elements, there is a general background concentration in the geobiosphere. When a recyling material disperses to this concentration, it has no remaining available energy and thus no emergy relative to the earth. For example, phosphorus at the background concentration of phosphorus in the surface layers of the ocean has no emergy.

4.2 Self regulation of cycles

Cycling allows for the continuous convergence and divergence of materials. Without continual flows of input energy that build structure, concentrations degrade away, falling into entropic disorder. It is through cycling that systems remain adaptive and vital. Materials sequestered in unreachable or unusable storages detract from function.

As Lotka (1925) showed, material cycles self regulate by developing larger storages upstream from the parts of their cycles where unit rates are slow (bottlenecks) until no part of the cycle is limiting except the driving source of available energy. Self organisation can also eliminate bottlenecks by supplying energy to accelerate the limiting step, accelerating dispersal. We might regard some products as important and the rest by-products when they appear to have no use. However, well organised systems process, cycle, and reuse all products, leaving none as waste.

4.3 Cycle management

Humans increasingly manage material cycles through the environment as part of municipal utilities and industrial ecology. Transformity of a material may be a good indicator of the appropriate part of the earth hierarchy for its use and recycle. Matching of transformities so that a material can amplify indicates an appropriate zone of the energy hierarchy for interactions. For example, it may be economic for concentrated metals and chemicals to be reused, whereas dilute organic materials and nutrients are appropriately recycled through the environment as concentrations usable by the ecosystems (an ecological engineering practice).

5. Emergy and information

The combination of parts and connections that make a system work is useful information. As used here it is something easier to copy for reuse than to remake anew. This information can be in a system's network, where its configuration makes the operation go. Or the plan for the system can be isolated as a code, message, or plan held by a 'carrier' with a very small energy content. For example, information is carried on paper, on computer disks, in human memory, in television transmissions, etc.

5.1 Information circle

Information depreciates by developing unrepaired error. Considerable to be sustainable in the long run an information storage has to be supported by a duplication and testing cycle. Emergy is required to maintain information with a cycle of repeated duplication, reusing, retesting and selecting to eliminate errors, a process that sometimes adds improvements. The circular life history diagrams taught in biology courses are examples. Very large emergy is required to generate the systems information the first time, especially genetic information.

5.2 Representing information in an energy systems diagram

Information can be represented in energy systems models in several ways. There is some information in the parts and connections of the model. Information is shown as a storage tank in Figure 9. It receives information inputs, has a copying loop, feedback control actions to the left, and depreciation pathway representing information is lost when carriers depreciate. For some purposes all this may be aggregated within a consumer symbol (Figure 9).

5.3 Emergy and complexity

Complexity of systems has long been represented by 'information theory' measures. For example, complexity in bits is the number of yes-no decisions required to define a configuration and is expressed on a logarithmic scale. In short, the information theory measure is the logarithm of the possibilities among the parts and connections. A storage may be used to represent such complexity in the aggregate. Since the possible arrangements and connections increase roughly as the square of the number of items, the emergy requirement to generate and maintain such storage may be proportional. However, information theory measures don't differentiate between useful complexity that operates a system and happenstance complexity with the same number of parts that can't do anything. The contribution of complexity to the system is best evaluated from its emergy content and empower required to sustain it.

There is great complexity at the small scales of molecules and heat where information theory measure on a logarithmic scale is molecular entropy. Information theory measures don't distinguish the same complexity on small molecular scale from that found on a large ecological scale. However, emergy does increase with scale of the units.

5.4 Scale and the hierarchy of information

Various categories of information can be placed in appropriate position in the left to right energy hierarchy by calculating their transformity. Transformities of valuable information such as the human genome and globally shared religious documents are very large because the emergy used to generate and maintain them is large and the energies of the carrier materials used to hold and carry the messages are tiny. In general, the information of higher transformity feeds back and controls items of lower transformity. For example, the human mind controls the computer. Public opinion controls individual information. To achieve the highest transformity status, information must have great generality, territory, and utility as a reinforcing control. Consider information in relation to scale starting with populations of heated molecules on a molecular scale. There the energy flows are large. The complexity on the scale of molecular chemistry has usually been viewed as degraded, disorderly statistical assemblages rather than as organised networks to be maintained. The transformities are small.

Useful information as controls emerges at the larger scale of complex molecules and living processes. The structures and information processes are facilitated by their extreme miniaturisation in living and electronic systems, but what happens on this scale is controlled by controls and selection from the larger scale of the ecosystems. Larger scale networks with long term memories control the rapid small scale information processes. For example, global changes in the self organisation of the biogeosphere control the molecular biological processing. High transformity information of life with emergy accumulated over many billion years (on the right in diagrams) controls the duplication and distribution of that information in rapid (microsecond) physiological processes (on the left in diagrams).

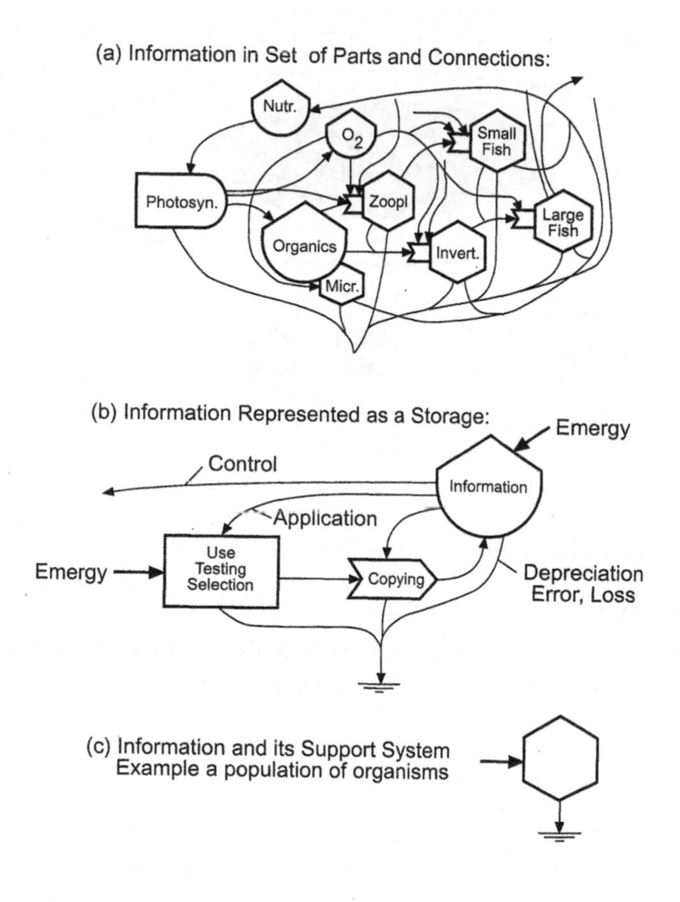

Figure 9: Ways of representing information on systems diagrams

5.5 Information as a Maxwell demon

When information developed and stored from energy flow systems feeds back to control and regulate thermal and chemical processes on a smaller scale, it fits the definition of a 'Maxwell Demon'. The traditional idea is that a Maxwell Demon cannot derive enough available energy from its own scale to select its inputs. However, real energy systems develop energy hierarchies converging resources spatially, accumulating emergy into useful information that reinforces the network maximising empower. Thus humans are the earth's information processor learning to utilise their genetic and learned information to reinforce the system which supplies their basis of life support.

6. Spatial scales and energy systems

The energy hierarchy has a spatial pattern which is exhibited within ecosystems. Many small units converge products and services to spatial centres (Figure 10). These in turn converge on even larger centres.

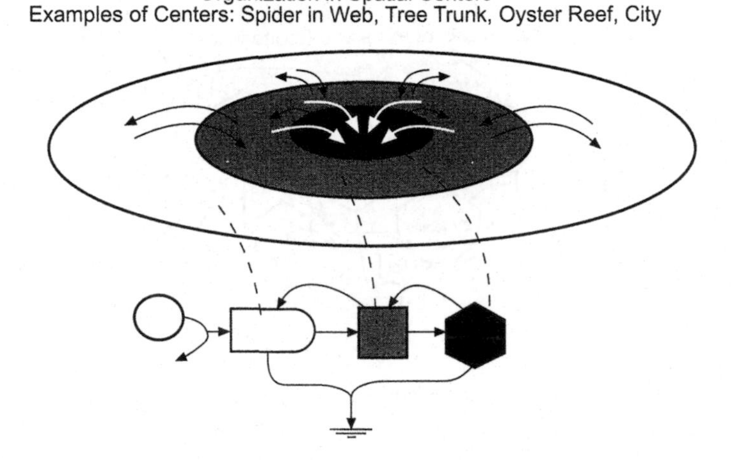

Organization in Spatial Centers
Examples of Centers: Spider in Web, Tree Trunk, Oyster Reef, City

Figure 10: Spatial centres organised according to energy hierarchy

6.1 Empower density

Empower density is a measure of the intensity of activity and is evaluated as the total emergy per unit area per time (Sej/m^2/yr). Values are higher in centres in ecosystems. Highest values occur in cities (Table 2 and 3).

Table 2: Spatial Organisation of Landscape in North Florida (after Brown, 1980)

Land Use Type	Empower Density (E12 sej/m^2 *yr-1)	Emergy of Structure (E12 sej/m^2)	Structural mass (E3 kg/m^2)	% of city area
Single family residential (med. density)	20.7	149.1	181.5	81.3
Multi-family residential (avg. 4 floors)	126.6	1135.4	1170.0	4.5
Commercial strip	46.4	517.1	720.4	3.5
Commercial mall	220.7	1248.7	1429.4	0.9
Central business district (avg. 4 floors)	294.2	2026.8	2067.0	0.8

Table 3: Characteristics of urban systems in North Central Florida, USA (after Brown, 1980)

Urban Class	Population (E3 people)	Area (km^2)	Annual Empower (E21 sej/yr)	Empower Density (E12 sej/m^2 *yr-1)	Support Region (km^2)
Class 1	504	399.9	30.6	76.4	9855.1
Class 2	99	254.5	14.9	58.6	4778.3
Class 3	37	55.5	2.8	50.4	896.6
Class 4	13	18.6	0.8	44.2	263.1
Class 5	1.8	2.8	0.1	35.8	32.2

6.2 Spatial organisation of cities

Cities are points of convergence in the landscape that represent large concentrations of people, structure, and information. Energies and materials inflow from surrounding regions producing large volumes of wastes in air, water, and solid waste dumps. It has long been demonstrated that landscapes of cities are organised hierarchically, where there are many small rural towns, fewer small cities, and fewer and fewer larger cities. It has been suggested (Christaller, 1966; Losch, 1954) that the hierarchy results from the distribution of goods having varying market regions. It was found that market regions increased with city size and the array of goods increased because of the larger market areas from which to draw demand.

Probably just as important is the convergence of energies and materials into cities. Environmental support of cities must be converged from larger and larger support regions as city size increases. The larger the city the greater the area of support required to produce necessary inputs or from which inputs are extracted. Natural ecosystems provide resources like water, wood, clean air and biodiversity, while agricultural systems provide food and fibre. Heavily managed urban green spaces provide important inputs directly to humans in the form of recreation, education, and psychological relief from stress. The wastes generated by all aspects of the urban system are recycled back to the surrounding environment, some stimulating production, others having a negative effect.

The hierarchy of cities results from the interplay of both market regions and support regions. Energy and materials are concentrated in pathways of convergence and information and goods are fedback in diverging pathways of control and amplifier actions. Table 3 lists several characteristics of classes of cities in Florida, USA. Class 1 cities are the largest urban centres in the region, serving as central places and having populations of over half a million people. Class 5 cities are the smallest incorporated towns found scattered throughout the landscape having typical populations of about 2000 people. Annual empower is the total inflow of emergy per year consumed within the city. Empower density is the flow of emergy per unit area of the city per year. Annual use of emergy varies from 30.6 E21 sej/yr to only 0.1 E21 sej/yr for the class 1 and class 5 cities respectively.

The most intensely developed cities are the class 1 central places where commercial and industrial uses make up a greater proportion of the total city area than in the smaller cities. Intensity of activity can be measured by empower density (empower per unit area, sej/m^2 * yr-1). The empower density of the Florida cities ranges from about 76 E12 sej/ m^2 * yr-1 to about 36 E12 sej/ m^2 * yr-1 as given in Table 3.

Within cities spatial organisation is hierarchical from the low intensity rural fringe to the high concentrations of information and business in the Central Business District. The central city, where buildings and populations are largest, is surrounded by rings of decreasing intensity. When structure and land use 'metabolism' (energy use) are expressed as emergy, the increasing intensity of activity is obvious. Given in Table 4 are characteristics of several typical urban land uses in Florida cities and the percent of city area that is devoted to these uses. The empower density and emergy in structure increases with increasing intensity of activity while area decreases.

7. Interface of ecological systems with economic use

With the spread of human population and economic development, landscapes and waterscapes are usually controlled by the human economy. Energy Systems view of the landscape includes the environmental systems, the interface with the economy and the circulation of money and other exchanges between people.

7.1 Interface with the economy

Figure 11 includes ecosystems on the left and the interface with the economy on the right. Human society draws materials, energies, and 'services' from the environment. The materials and energies are easily understood to be things like wood, water, fruit, animals, and so forth. The services are things like waste assimilation, flood protection, or aesthetic qualities. Money circulates through the interface in payment for products and services passing to the economy again in payment for goods, services, fuels, and materials purchased from the economy. Storages of environmental products are natural capital. Some natural capital is an active part of continuing productive processes. Other storages are part of long term pro-cesses on a geologic time scale (non-renewable).

Renewability is a relative concept, since it depends on how quickly a material or energy is used compared to the speed at which it is generated. Wood, for instance, can be a renewable resource, if the rate of harvesting is matched with the regeneration rate. In contrast fossil fuels and most mineral resources are not renewable. Even though they are being constantly regenerated their rate of use is much faster than the regeneration rate. In Figure 11 the emergy flow N is from non-renewable storage. E is the emergy from renewable processes and short term storages.

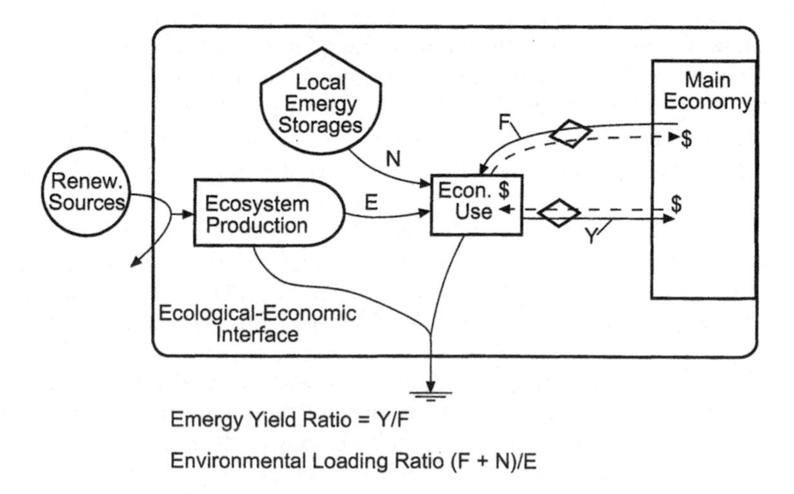

Figure 11: Systems diagram of the interface between ecosystems and economy, showing the flows of money, environmental contributions, and indices.

Agroecosystems are examples of the ecological-economic interface. Table 4 summarises emergy evaluations of thirteen crop systems in Italy. The emergy yield ratio (defined in Figure 11) measures the contribution to the economy beyond the emergy required to process it. The values between 1.1 and 2.2 are much less than those for primary energy sources such as fuels and electricity. The environmental loading ratio (defined in Figure 11) measures the economic empower impact on the local environmental processes. Fruit trees and vineyards had the greatest environmental impact. However, these values were less than average ratios within the United States.

7.2 Evaluating environmental contributions

All processes require material and energy inputs. The systems diagram of an economic use interface in Figure 11 is aggregated to show three inputs: (N) the use of non-renewal storages; (E) the use of renewable product of environment; and (F) the purchased items that 'feedback' (F) from the larger economy for the processing. The purchased items include human labour and services plus purchased non-renewable inputs brought in from other areas like minerals, and fuels, and renewable environmental inputs from other areas.

Examples are soils and water reservoirs. These have high emergy values because it can take hundreds of years to make 1 cm of topsoil (Pimentel, et al., 1995). Many environmental inputs are involved in collecting the waters used to cool power plants (Brown and Ulgiati, 1999).

Other environmental services are provided by the environment in absorbing and recycling waste products. They are often not accounted for because they are free.

Table 4: Indices for agro-ecosystems of Italy

Agro-ecosystem	Solar Transformity (E4sej/j)	Emergy-Yield ratio	Environmental loading ratio
Rice	5.58E+04	1.62	1.77
Forage	6.33E+04	2.20	0.94
Corn	7.74E+04	1.76	3.00
Sunflower	9.25E+04	2.18	1.96
Soybeans	9.33E+04	2.16	1.65
Wheat	1.00E+04	1.62	1.77
Sugar beet	1.05E+05	1.33	5.18
Rapeseed	1.05E+05	2.26	1.78
Fruits	2.16E+05	1.16	6.78
Oranges and Lemons	2.46E+05	1.15	7.26
Vineyard	3.02E+05	1.23	4.60
Olive	4.02E+05	1.35	3.10
Almonds	7.37E+05	1.42	2.59
Nation-wide crop production	9.14E+04	1.51	2.11

(Modified from Ulgiati et al., 1993)

When the environment becomes overloaded, the costs are recognised when the free service from the environment has to be replaced by technology. For production of by-products to be balanced with the environment's ability to absorb and recycle them, an adequate support area is necessary. When the environment is accounted for, performance of a production process is more time and location dependent, as it should be. In essence, by accounting for the 'load' on the environment and providing a support region for environmental recycle of by-products, a carrying capacity for economic uses of the environment can be determined.

7.3 Carrying capacity for economic investments

One theory for determining carrying capacity is that the scale or intensity of development in relation to its environment may be critical in predicting its effect and ultimately its sustainability (Brown et al. 1995, Ulgiati et al. 1995, Brown and Ulgiati 1997). Large-scale developments can be integrated into the environment, if there is sufficient regional support area to balance their effect. Much like the ecological concept of carrying capacity, where differing environments require different aerial extent of photosynthetic production for support of a given biomass of animals, environmental carrying capacity for economic investments depends on the area of 'support' over which a development's effect can be integrated. As the intensity of development increases (and therefore its consumption of resources and environmental impacts increase), the area of natural undeveloped environment required for its support must increase. All other things being equal, the more intense a development, the greater the area of environment necessary to balance it.

When environmental services are accounted for, materials and non-renewable energy inputs no longer appear to be the only important inputs. The ability of the environment to absorb impacts and recycle material by-products assumes a larger role in a process's sustainability. If the goal is to avoid changes of the environmental physio-chemical characteristics over time, the only way without increased investments in abatement technology is to expand the spatial scale of the process's supporting environment. By doing this, a strong constraint is placed on the size of the economic process allowed for a given area according to the carrying capacity of the local environment. Procedures and examples are given elsewhere (Ulgiati and Brown, 1999).

8. Time dimension for energy systems

Humans zoom up and down scales mentally when they consider policy decisions, but usually with verbal models that are not quantitative and cause semantic confusion. Energy systems concepts provide a pulsing paradigm for the time dimension of ecosystems and general systems designs for dynamic computer simulation.

8.1 Pulsing paradigm

By now enough knowledge has been accumulated in most fields to conclude that what is normally sustainable is a sequence of repeating pulses. At larger scales pulses are more widely separated but sharper. Current research seeks reasons – evidences that self-organising systems can increase their performance, maximising empower by pulsing. The reason pulsing is adaptive has to do with optimal loading of autocatalytic transformation processes. A simple steady state system of production and consumption may not transform as much as repeating surges of production and consumption. For whatever reasons the normal pulsing pattern is repetitive and oscillatory (Figure 12). Chaotic patterns may be a special pulsing design for maximising energy transformations.

Like the textbook prey-predator model, pulsing develops with a producer-consumer pair, in which a slow accumulation of one storage is followed by a frenzied consumption and momentary storage of high transformity structure and feedback actions. Self organisation reinforces those pairs that set up pulses, since more work is done by those pathway designs.

Like other autocatalytic units connected in series, information systems may also develop frenzied pulses. For example, if there is a source of species, biodiversity (information storage) may pulse as it draws emergy from accumulations of biomass (Figure 12). This is a way of looking at the increase of diversity in ecosystem succession.

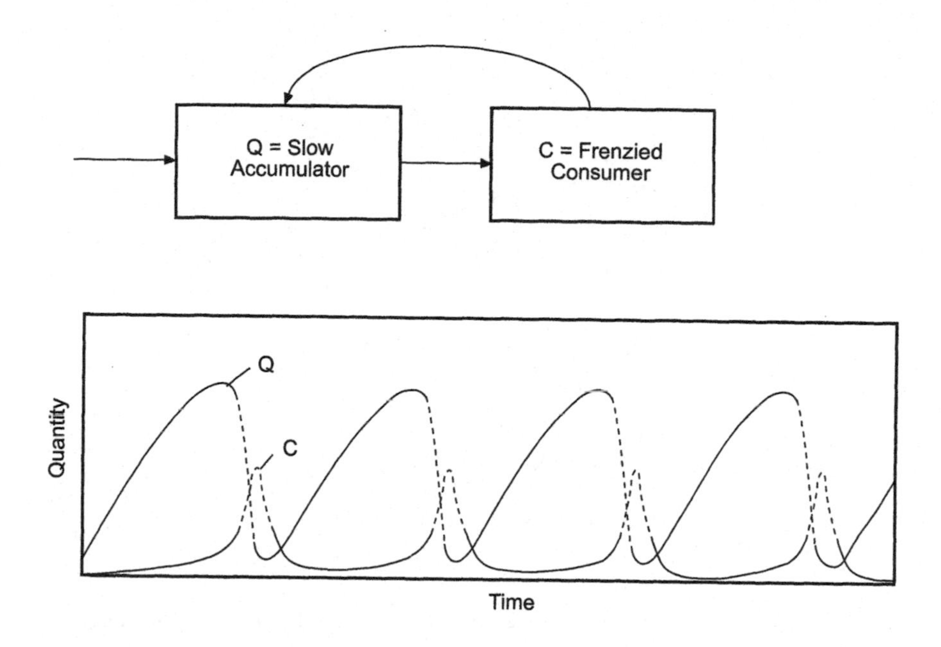

Figure 12: Pulsing pairs that dominate self organisation one each scale of size and time

8.2 Steady state aggregation of the small scale components

When a window of study is selected, scales of territory and oscillation are selected also. For example, if the view of an ecosystem selected is on the scale of hectares and years, small fast processes of photosynthesis, microbial bursts, and daily weather oscillations need to be aggregated, but not eliminated. What is relevant for the diagrams and for calculations is the running average effect of these fast processes on the larger scale of interest. As part of the aggregation process, the small, fast, and high energy inputs from the left can be included in the system diagram with a symbol (and mathematical characteristics) that aggregates the very fast, fine detail oscillations into a production function appropriate for the larger scale model. In other words, steady state of small scale inputs is a useful aggregation artefact by the beholder who is considering a larger scale.

8.3 Flow junction modelling of the small scale

While simulating a system on a time and space scale appropriate to the window of interest (days and years), it is a useless distraction to be simulating the tiny physiological parts of the ecosystem that fill and discharge in seconds. For example, simulating light-chlorophyll dynamics in a forest model wastes computer time and adds numbing complexity. In Figure 13 a flow is shown as the sum of inputs minus outputs with the remainder passing out of the system or into another storage. The limiting factor-type equation that results from this configuration is very similar to one that comes from setting the derivative of a storage equation to zero. Professor Katherine Ewel of the University of Florida found this configuration a way to input sunlight in analogue computer simulations of forests. Where small nutrient storages in oligotrophic lake models in Euler integration tend to go into errors of artificial chaos, they can be replaced with the configuration in Figure 13 which shows the nutrient flows without a storage. The computer output provides a running average of the flows that would exist with the storages in the real world. The flow junction is a way to keep models appropriate and understandable.

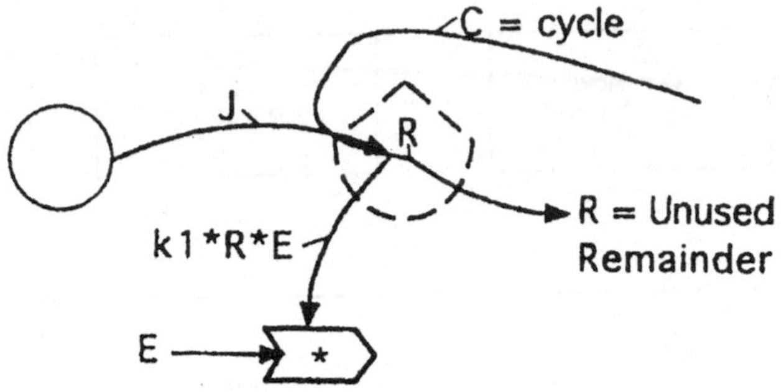

$$R = J + C - K1*R*E$$

$$R = (J + C)/(1 + K1*N)$$

Figure 13: Flow junction mechanism for contributions from a smaller scale than the primary window interest

8.4 Input pulses from larger scales

At the upper end of the systems window and its diagram are the pulses coming from the large scale. The larger scale includes the window of interest plus a much larger area in which it is embedded. In other words, pulses come from the surrounding system. These pulses are so infrequent that they appear on the ecosystem scale as catastrophe s such as hurricanes, earthquakes, and economic depressions. Ecosystems and their models appropriately respond by repairing, restoring, and repeating succession. However, these pulses from the large scale are just as regular and recurring as pulses of the smaller scale. Examinations of ecosystems in the field and their modelling need to consider repeating destruction and the repair mechanisms that make the system adaptive. The transformity of large scale pulses is great, and the best adapted systems derive useful emergy. Ecosystems strategy may be to resist small storms but to guide the destruction by large storms into patterns that add structure and favour rapid re-growth.

8.5 Pulses of global scale

The global ecosystem uses the main environmental systems of the atmosphere, ocean, and earth processes. Although it has mostly been studied with separate disciplines (meteorology, oceanography, geology), the geobiosphere is a highly integrated single system. Appropriate diagramming requires all the main processes of each as necessary to the others. When diagrammed in this way, flows are co-products and have the same empower. Increasingly, historical research in geology is finding large pulses in the earth systems. The energy systems theory suggests a means for calculating transformity and determining the scale of phenomena driving these pulses. The largest pulses may be those in evolutionary information processing or continent formation.

8.6 The pulse of the civilised human economy

One of the global pulses underway is the human economy accelerated in the last two centuries into a surge of civilisation achievement based on previous accumulation of fossil fuel and other slowly renewable resources. This global pulse is one of the large scale 'catastrophic' inputs to smaller ecosystems.

Human society adapting to the capitalistic patterns that maximise empower during rapid growth is certainly unprepared for the different type of system that will be required as the peak of the fuel-economic pulse is passed. It may be possible to rapidly change our priorities and ethics from the growth regime to one we hopefully designate 'The Prosperous Way down'. This will require reduction of population, selection of what is worth saving, increased co-operation, efficiency, and sharing of products and effort.

8.7 Selecting shared information to carry to the next pulse

To have a long existence in a global environment for the long range, humanity and its civilised information must pulse when resource accumulations permit and shift to a low emergy regime between.

In ecosystems we have two precedents to study: (a) the system that comes down in a crash as in fire and restarts, and (b) systems that have a smooth program of adapting to the winter and re-emerging in the summer. In both instances the future is based on information reserves in the larger surrounding system.

Information is increasingly shared globally through television and the internet. Global sharing makes some information of large scale and long duration. However, large scale information will have its own pulses – which we call information storms. To make civilisation sustainable in the long run through periods of pulsing and decent, perhaps we need to understand and manage information storms better. Through selection of what is worth sharing, perhaps a core of the civilisation can be placed in long term memory during periods of coming down for expanded use during the up cycles. Should we worry about these information storms, their creation and spread and the effects they have on human emotions and societies behaviour?

REFERENCES

Boltzmann, L., 1886. The second law of thermodynamics. Address in English to Imperial Academy of Science in 1886. Populare Schriften. Essay 3; Selected Writings of L. Boltzmann. D. Reidel, Dordrecht, Holland.

British Petroleum, 1997. BP statistical review of world energy, 1997. The British Petroleum Co., London. 41p.

Brown, M.T., 1980. Energy basis for hierarchies in urban and regional systems. Ph.D dissertation. Department of Environmental Engineering Sciences, University of Florida. Gainesville, FL. 357p.

Brown, M.T, H.T. Odum, R.C. Murphy, R.A. Christianson, S.J. Doherty, T.R. McClanahan, and S.E. Tennenbaum, 1995. Rediscovery of the World: Developing an Interface of Ecology and Economics. p. 216-250. In: CAS Hall, (Ed). *Maximum Power: The Ideas and Applications of H.T. Odum*. University of Co. Press. 393p.

Brown, L.R., M. Renner, and C. Flavin, 1997. Vital Signs 1997: the environmental trends that are shaping our future. W.W. Norton & Co., NY 165 p.

Brown, M.T. and S. Ulgiati, 1999. Emergy Evaluation of the Biosphere and Natural Capital. AMBIO. [In Press]

Brown, M.T. and S. Ulgiati, 1997. Emergy Based Indices and Ratios to Evaluate Sustainability: Monitoring technology and economies toward environmentally sound innovation. *Ecological Engineering* 9: 51-69.

Christaller, W., 1966. Central places in southern Germany. (Trans, by G.C.W. Baskin) Prentice Hall, Englewood Cliffs, NJ.

Lieth, H. and R.H. Whittaker, 1975. Primary productivity of the biosphere. Springer-Verlag, New York. 339p.

Losch, A., 1954. The economics of Location. (Trans by U. Waglom and W.F. Stalpor) Yale Univ. Press, New Haven CT.

Lotka, A.J., 1922. Contributions to the energetics of evolution. *Proc. National Academy of Science*, Vol 8: 147-151.

Lotka, A.J., 1925. *Physical Biology*. Williams and Wilkins, Baltimore, MD 460 p.

Munk W.H. and G.F. McDonald, 1960. The rotation of the Earth: a geophysical discussion. Cambridge Univ. Press, London. 323 p.

Odum, H.T., 1983. Systems Ecology. John Wiley and Sons, NY. 644p.

Odum, H.T., 1996. Environmental Accounting. Emergy and Environmental Decision Making. John Wiley & Sons, N.Y. 370 p.

Pimentel, et al., 1995. Environmental and economic costs of soil erosion and conservation benefits. *Science* 267: 1117-1123.

Prado-Jartar, M.A. and M.T. Brown, 1996. Interface Ecosystems with an Oil Spill in a Venezuelan Tropical Savannah. *Ecol Eng* 8: 49-78.

Scienceman, D., 1987. Energy and Emergy. p. 257-276. In: Pillet, G. and T. Murota, (Eds.). *Environmental Economics*, Roland Leimgruber, Geneva, 308 p.

Sclater J.F., G. Taupart, and I.D. Galson, 1980. The heat flow through the oceanic and continental crust and the heat loss of the earth. *Rev. of Geophysics and Space Physics* 18: 269-311.

Ulgiati S., H.T. Odum, and S. Bastianoni, 1994. Emergy use, environmental loading and sustainability. An emergy analysis of Italy. *Ecol Modelling* 73: 215-268.

Ulgiati S., M.T. Brown, S. Bastianoni, and N. Marchettini, 1995. Emergy based indices and ratios to evaluate the sustainable use of resources. *Ecol Eng* Vol.5(4): 519-31.

Ulgiati S., M.T. Brown. 1999 The role of environmental services in electricity production processes. *J. of Cleaner Production.* (in press).

USDI, 1996. Mineral Commodity Summaries, January 1997. U.S. Department of Interior, Washington, D.C.

Von der Haar, T.H. and V.E. Suomi, 1969. Satellite observations of the earth's radiation budget. *Science* 169: 657-669.

World Resources Institute, 1997. World Resources 1996-97. Oxford University Press. N.Y.

II.7.1 Ascendancy: A Measure of Ecosystem Performance

Robert E. Ulanowicz

1. Introduction

One phenomenon central to ecology is that of ecosystem succession – the more or less repeatable temporal series of configurations that an ecosystem will take on after a major disturbance or upon the appearance of new areas of the given habitat. Initially, succession was described in terms of natural history (e.g., Clements 1916), but more recently ecosystem scientists have attempted to describe succession, or ecosystem development, in more formal terms (Odum 1969). The goal in quantitative ecology eventually is to describe the process of succession in purely numerical terms.

The quantification of succession is unlikely to prove easy, for, despite prevailing temporal regularities, the process is not as deterministic as many first portrayed it. Clements' almost mechanical description of succession was challenged almost immediately by Gleason (1917), who saw community assembly to be more stochastic by nature (Simberloff 1980). Contingencies, or novel perturbations, are very much a part of any ecosystem's history, and a quantitative theory of ecosystem succession cannot assume *a priori* that such chance events will always average out. What follows is a description of one particular attempt to quantifying the process of ecosystem development. The approach falls under the rubric of ecosystem *ascendancy*, so named after the key index spawned by the theory. Ascendancy was derived to gauge the activity and organisation inherent in an ecosystem. The approach is neither purely mechanical, nor unconditionally stochastic – extremes which to date have characterised most quantitative endeavours in ecosystems science. Rather, the formulation of ascendancy resembles Popper's (1990) call to develop a 'calculus of conditional probabilities.'

Popper regarded the processes of life as almost 'lawful' in the sense that they are guided by sets of 'propensities' – generalisations of Newtonian like forces that are constantly being disrupted by contingent events. Chance does not act on individual component processes in isolation, however, as is assumed in genetic theory (Fisher 1930). Ecosystem processes, almost by definition, are coupled to one another – a situation which allows for the effects of chance events to be incorporated into the ongoing history of the system. How a chance event affects a process will depend in part on conditions elsewhere in the system. Whence the need to describe chance, not in terms of the ordinary statistics common to most of contemporary biology and physics, but in terms of Bayesian, or *conditional* probabilities.

The trick, then, in constructing a broad, quantitative description of ecosystem development is to focus first upon the agency behind the 'law-like' progression towards a developed configuration, and thereafter to quantify the actions of this agency, not in conventional, deterministic fashion, but in contingent, probabilistic terms that can incorporate historical and non-local events.

2. A Vehicle for development

There is a growing consensus that life processes are so difficult to explain because they involve highly reflexive, self- referencing and, ultimately, self-entailing behaviours (Rosen 1991). While negative feedback is the crux of most internal system regulation, theorists now acknowledge that the pressures behind the proliferation and evolution of living forms have more to do with positive feedbacks, and with autocatalytic activities in particular (e.g.,

Eigen, 1971, Haken 1988, Kauffman 1995). Before going further, it is necessary to specify more precisely how the term, 'autocatalysis' will be used here.

Autocatalysis is a special case of positive feedback (DeAngelis et al. 1986). Positive feedback can arise according to any number of scenarios, some of which involve negative interactions. (Two negative interactions taken serially can yield a positive overall effect). By 'autocatalysis' we mean 'positive feedback comprised wholly of positive component interactions.' A schematic of autocatalysis among three processes or members is presented in Figure 1. In keeping with the idea of an open or contingent universe, we do not require that A, B and C be linked together in obligatory fashion. To achieve autocatalysis, we require only that the propensities for positive influence be stronger than cumulative decremental interferences. The plus sign near the end of the arrow from A to B indicates that an increase in the rate of process A has a strong propensity to increase the rate of B. Likewise, growth in process B tends to augment that of C, which in its turn reflects positively back upon process A.

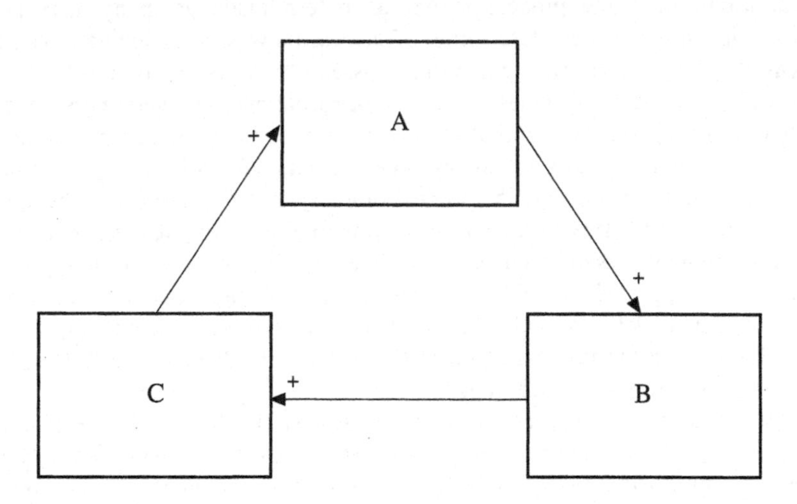

Figure 1: Schematic of a hypothetical three-component autocatalytic cycle.

Autocatalysis traditionally has been viewed in rather mechanical terms, but in the face of environmental contingenices, autocatalytic activities behave in ways that transcend mechanism (Ulanowicz 1997). For example, there is a *selection pressure* which the overall autocatalytic form exerts upon its components. If a random change should occur in the behaviour of one member that either makes it more sensitive to catalysis by the preceding element or accelerates its catalytic influence upon the next compartment, then the effects of such alteration will return to the starting compartment as a reinforcement of the new behaviour. The opposite is also true. Should a change in the behaviour of an element either make it less sensitive to catalysis by its instigator or diminish the effect it has upon the next in line, then even less stimulus will be returned via the loop.

Unlike Newtonian forces, which always act in equal and opposite directions, the selection pressure associated with autocatalysis is inherently *asymmetric*. Autocatalytic configurations impart a definite sense (direction) to the behaviours of systems in which they appear. They tend to ratchet all participants toward ever greater levels of performance.

Perhaps the most intriguing of all attributes of autocatalytic systems is the way they affect transfers of material and energy between their components and the rest of the world. Figure 1 does not portray such exchanges, which generally include the import of substances with higher exergy (available energy) and the export of degraded compounds and heat. The degradation of exergy is a spontaneous process mandated by the second law of thermodynamics. But it would be a mistake to assume that the autocatalytic loop is itself passive and merely driven by the gradient in exergy. Suppose, for example, that some arbitrary change happens to increase the rate at which materials and exergy are brought into a particular compartment. This event would enhance the ability of that compartment to catalyse the downstream component, and the change eventually would be rewarded. Conversely, any change decreasing the intake of exergy by a participant would ratchet down activity throughout the loop.

The same argument applies to every member of the loop, so that the overall effect is one of *centripetality*, to use a term coined by Sir Isaac Newton (Figure 2). The autocatalytic assemblage behaves as a focus upon which converge increasing amounts of exergy and material that the system draws unto itself (cf Jorgensen 1992). Taken as a unit, the autocatalytic cycle is not acting simply at the behest of its environment. It actively creates its own domain of influence. Such creative behaviour imparts a separate identity and ontological status to the configuration above and beyond the passive elements that surround it.

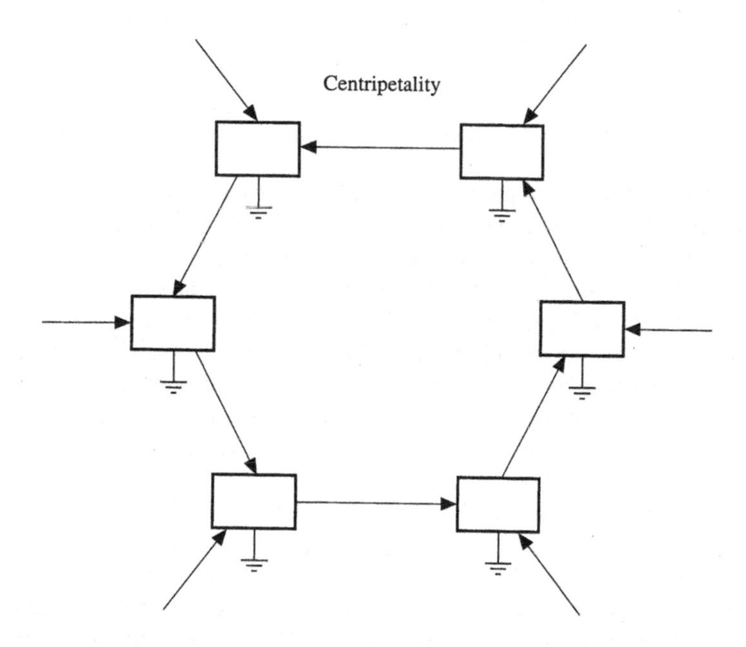

Figure 2: Autocatalytic cycle exhibiting centripetality.

To be sure, autocatalytic systems are contingent upon their material constituents and usually also depend at any given instant upon a complement of embodied mechanisms. But such contingency is not, as strict reductionists would have us believe, entirely a one-way street. By its very nature autocatalysis is prone to *induce competition*, not merely among different properties of components (as discussed above under selection pressure), but its very material and (where applicable) mechanical constituents are themselves prone to replacement by the active agency of the larger system. For example, suppose A, B, and C are three sequential elements comprising an autocatalytic loop as in Figure 3a, and that some new element D: (1) appears by happenstance, (2) is more sensitive to catalysis by A and (3) provides greater enhancement to the activity of C than does B (Figure 3b). Then D either will grow to overshadow B's role in the loop, or will displace it altogether (Figure 3c).

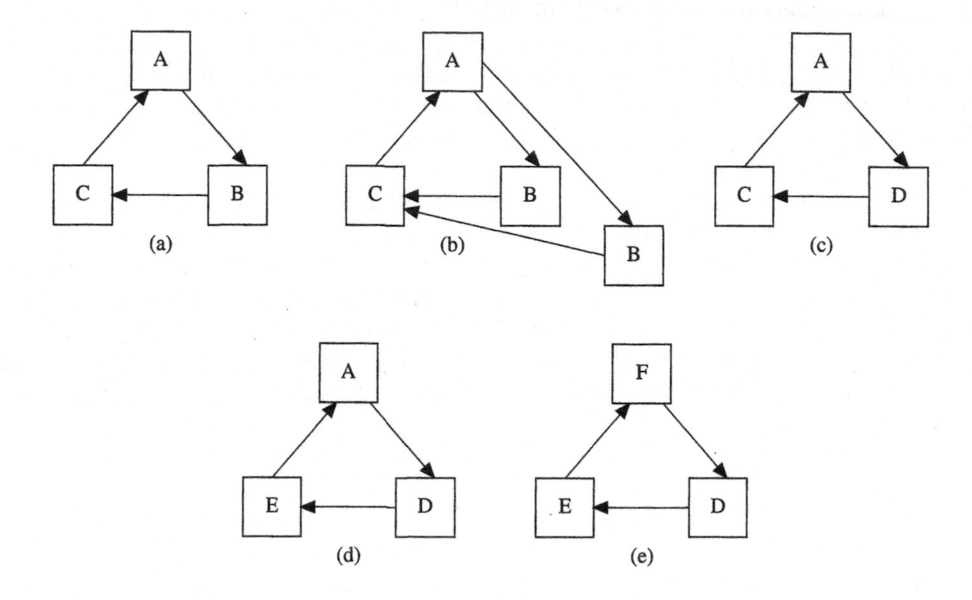

Figure 3: Successive replacement of the components in an autocatalytic loop

In like manner one can argue that C could be replaced by some other component E (Figure 3d), and A by F, so that the final configuration D-E-F contains none of the original elements (Figure 3e) (Simple induction will extend this argument to an autocatalytic loop of n members). It is important to notice in this case that the characteristic time (duration) of the larger autocatalytic form is longer than that of its constituents. Persistence of active form beyond present makeup is not an unusual phenomenon. One sees it in the survival of corporate bodies beyond the tenure of individual executives or workers; of plays, like those of Shakespeare, that endure beyond the lifetimes of individual actors. But it also is at work in organisms as well. One's own body is composed of cells that (with the exception of neurons) did not exist seven years ago.

Overall kinetic form is, as Aristotle believed, a causal factor. Its influence is exerted not only during evolutionary change, but also during the normal replacement of parts. For example, if one element of the loop should happen to disappear, for whatever reason, it is (to use Popper's own words) 'always the existing structure of the pathways that determines what new variations or accretions are possible' to replace the missing member (Popper 1990).

The appearance of centripetality and the persistence of form beyond constituents are decidedly non-Newtonian behaviours. Although a living system requires material and mechanical elements, it is evident that some behaviours, especially those on a longer time scale, are, to a degree, *autonomous* of lower level events (Allen and Starr 1982). Attempts to predict the course of an autocatalytic configuration by ontological reduction to material constituents and mechanical operation are, accordingly, doomed over the long run to failure.

It is important to note that the autonomy of a system may not be apparent at all scales. If one's field of view does not include all the members of an autocatalytic loop, the system will appear linear in nature. Under such linear circumstances, an initial cause and a final result will always seem apparent (see Figure 4). The subsystem can appear wholly mechanical in its behaviour. Once the observer expands the scale of observation enough to encompass all members of the loop, however, then autocatalytic behaviour with its attendant centripetality, persistence and autonomy *emerges* as a consequence of this wider vision.

In our consideration of autocatalytic systems, however, we have seen that agency can arise quite naturally at the very level of observation. This occurs via the relational form that processes bear to one another. That is, autocatalysis takes on the guise of a *formal cause*, sensu Aristotle. Nor should we ignore the directionality inherent in autocatalytic systems by virtue of their asymmetric nature. Such rudimentary *telos* is a very local manifestation of final cause that potentially can interact with similar agencies arising in other parts of the system.

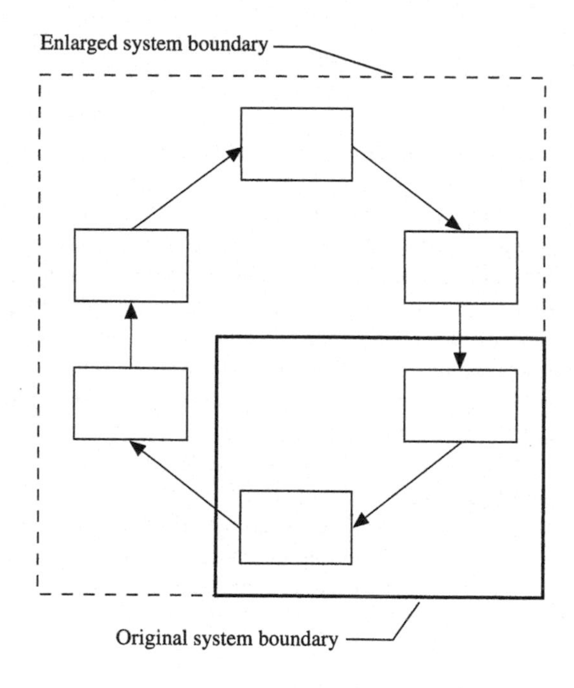

Figure 4: The emergence of non-mechanical behaviour as scope of observation is enlarged.

Finally, autocatalytic configurations, by definition, are *growth enhancing*. An increment in the activity of any member engenders greater activities in all other elements. The feedback configuration results in an increase (growth) in the aggregate activity of all members engaged in autocatalysis over what it would be if the compartments were decoupled.

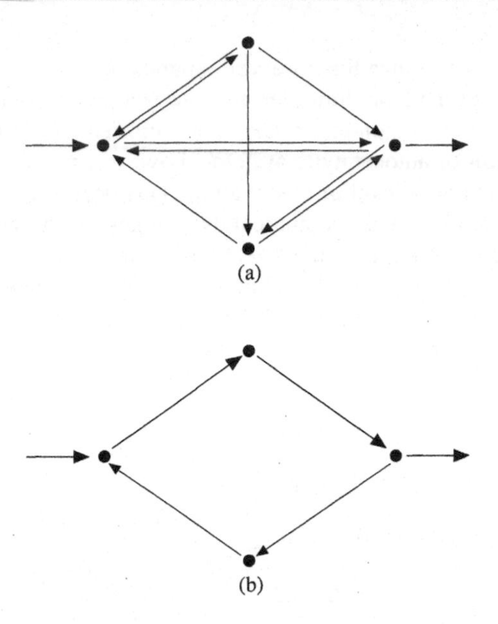

Figure 5: Schematic representation of the major effects that autocatalysis exerts upon a system.

To recapitulate, autocatalytic systems can exhibit at least eight behaviours, which, taken together, mitigate against viewing them as mechanical systems that will yield to reductionistic analysis. Autocatalysis induces (1) growth and (2) selection. It exhibits an (3) asymmetry that can give rise to the (4) centripetal amassing of material and available energy. The presence of more than a single autocatalytic pathway in a system presents the potential for (5) competition. Autocatalytic behaviour is (6) autonomous, to a degree, of its microscopic constitution. Its attributes (7) emerge whenever the scale of observation becomes large enough, usually in the guise of an Aristotelian (8) formal cause.

The overall effects of autocatalytic behaviour are exhibited both extensively (as a function of system size) and intensively (independent of size). The former is expressed as an increase in total system activity, while the latter resembles the topological 'pruning' of those processes that participate less effectively in autocatalytic activities. The combined result is depicted schematically in Figure 5. The task now at hand is to quantify both aspects of growth and development.

3. Quantifying growth and development

The extensive nature of growth is rather easy to quantify. To do so, we denote the magnitude of any transfer of material or energy from any donor (prey) i to its receptor (predator) j by T_{ij}. Then one measure of total system activity is the sum of all such exchanges, a quantity referred to in economic theory as the 'total system throughput', T.

$$T = \sum_{i,j} T_{ij} \tag{1}$$

If reckoning the 'size' of a system by its level of activity seems at first a bit strange, one should recall that such is common practice in economic theory, where the size of a country's economy is gauged by its 'gross domestic product'.

Quantifying the intensive process of development is somewhat more complicated. The object here is to quantify the transition from a very loosely coupled, highly indeterminate collection of exchangesto one in which exchanges are more constrained by autocatalysis to flow along the most efficient pathways. One begins, therefore, by invoking information theory to quantify the indeterminacy, h_j, of category j,

$$h_j = -k \log p(B_j) \tag{2}$$

where $p(B_j)$ is the marginal probability that event B_j will happen, and k is a scalar constant. Roughly speaking, h_j is correlated with how surprised the observer will be when B_j occurs. If B_j is almost certain to happen, $p(B_j)$ will be a fraction near 1, making h_j quite small. Conversely, if B_j happens only rarely, $p(B_j)$ will be a fraction very near zero, and h_j will become a large positive number. In the latter instance the observer is very surprised to encounter B_j.

Constraint abrogates indeterminacy. That is, the indeterminacy of a system with constraints should be less than what it was in unconstrained circumstances. Suppose, for example, that an *a priori* event A_i exerts some constraint upon whether or not B_j subsequently occurs. The probability that B_j will happen in the wake of A_i is defined as the conditional probability, $p(B_j|A_i)$. Hence, the (presumably smaller) indeterminacy of B_j under the influence of A_i (call it h_j^*), will be measured by the Boltzmann formula as

$$h_j^* = -k \log p(B_j|A_i). \tag{3}$$

It follows that one may use the decrease in indeterminacy, $(h_j - h_j^*)$, as one measure of the intensity of the constraint that A_i exerts upon B_j. Call this constraint h_{ij}, where

$$h_{ji} = h_j - h_j^* = [-k \log p(B_j)] - [-k \log p(B_j|A_i)] = k \log [p(B_j|A_i)/p(B_j)] \tag{4}$$

One may use this measure of constraint between any arbitrary pair of events A_i and B_j to calculate the amount of constraint inherent in the system as a whole: one simply weights the mutual constraint of each pair of events by the associated joint probability, $p(A_i,B_j)$, that the two will co-occur, and then sums over all possible pairs. This yields the expression for the average mutual constraint, A, as

$$A = k \sum_{i,j} p(A_i, B_j) \log\left[\frac{p(A_i, B_j)}{p(A_i)p(B_j)}\right] \tag{5}$$

In order to apply A to quantify constraint in ecosystems, it remains to estimate $p(A_i,B_j)$ in terms of measurable quantities. To keep matters strictly operational, we shall henceforth focus upon trophic exchanges. Then, a convenient interpretation of A_i becomes 'a quantum of medium leaves compartment i', and of B_j ,'a quantum enters compartment j'. The T_{ij} may be regarded as entries in a square events matrix, similar to Tables 1 and 2. The joint probabilities can be estimated by the quotients T_{ij}/T, and the marginal probabilities become the normalised sums of the rows and columns,

$$p(A_i) \sim \sum_j T_{ij} / T, \tag{6}$$

and

$$p(B_j) \sim \sum_i T_{ij} / T \tag{7}$$

In terms of these measurable exchanges, the estimated average mutual constraint takes the form

$$A = k\sum_{i,j}\left(T_{ij}/T\right)\log\left[\frac{T_{ij}T}{\sum_k T_{ik}\sum_l T_{lj}}\right] \tag{8}$$

That A indeed captures the extent of organisation created by autocatalysis can be seen from the example in Figure 6. In Figure 6a there is equiprobability that a quantum will find itself in the next time step in any of the four compartments. Little is constraining where medium may flow. The average mutual constraint in this kinetic configuration is appropriately zero. One infers that some constraints are operating in Figure 6b, because medium that leaves any compartment can flow to only two other locations. These constraints register as k units of A. Finally, Figure 6c is maximally constrained. Medium leaving a compartment can flow to one, and only one, other node.

4. System Ascendancy

Having quantified separately the extensive and intensive effects of autocatalysis, it remains to combine them into a single index. This amalgamation follows in a very natural way, because we have elected to retain the scalar constant 'k' in all the information measures just cited. (The conventional practice in information theory is to designate the base to be used in calculating the logarithms [usually 2, e or 10] and set the value of k=1).

The units of A would then appear as 'bits', 'napiers' or 'hartleys', respectively. The problem with this convention is that the calculated value conveys no indication as to the physical size of the system. By retaining k in the formulae, one now has a convenient way to impart physical dimensions to the measure of organisation (Tribus and McIrvine 1971, Ulanowicz 1980). That is, we set k=T, and the dimensions of A will contain the units used to measure the exchanges. For example, if the transfers in Figure 6 had been measured as $g/m^2/d$, and the base of the logarithm was 2, then the values of A would be expressed in the units g-bits/m^2/d. The ascendancy expressed in terms of trophic exchanges becomes,

$$A = \sum_{i,j} T_{ij}\log\left[\frac{T_{ij}T}{\sum_k T_{ik}\sum_l T_{lj}}\right] \tag{9}$$

To signify that the scaled measure has changed its qualitative character, we choose to rename A as the system 'ascendancy' (Ulanowicz 1980). It measures both the size and the organisational status of the network of exchanges that occur in an ecosystem. In an attempt to characterise what it means for an ecosystem to develop, Eugene Odum (1969) catalogued ecosystem attributes that were observed to change during the course of ecological succession. His list of 24 properties can be sub-grouped according to whether they pertain to speciation, specialisation, internalisation or cycling – all of which tend to increase during system development. But increases in these same four features of network configurations lead, *ceteris paribus*, to increases in ascendancy. Whence, Odum's phenomenology can be quantified and condensed into the following principle:

In the absence of major perturbations, ecosystems exhibit a propensity towards configurations of ever-greater network ascendancy.

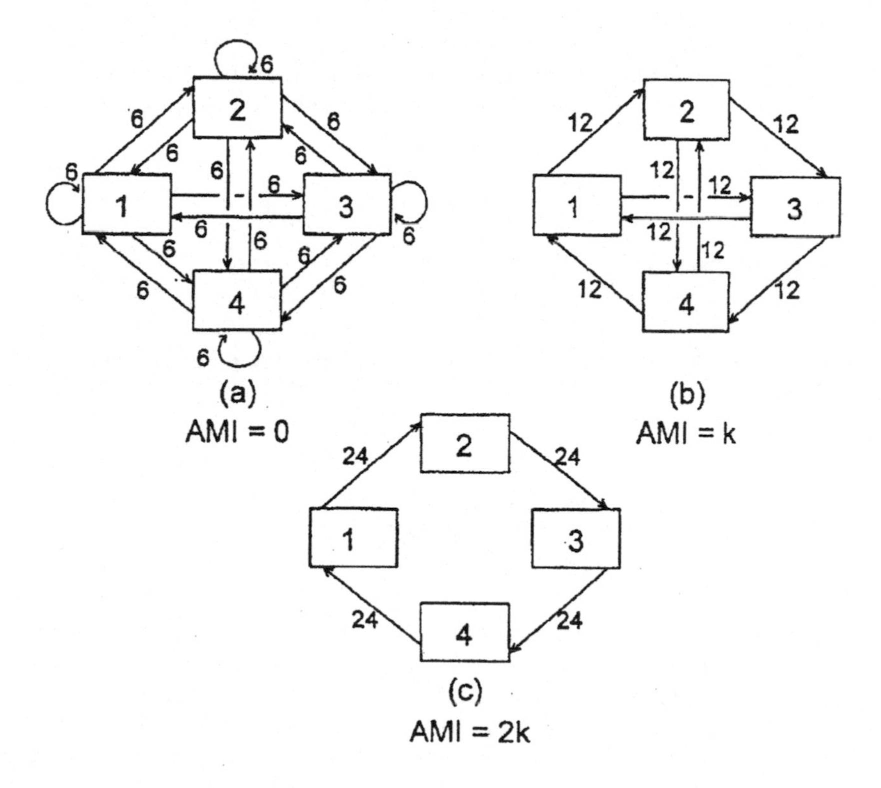

Figure 6: The increase in mutual information as flows become progressively constrained.

5. Ecological Persistence

The tendency towards increasing ascendancy, if allowed to progress unimpeded, would result in a very rigid, mechanical-like configuration. With ecosystems, matters never reach such a pass because of the contingent nature of the world in which they reside. That is, the accretion of system ascendancy is always being disrupted by chance perturbations. In time, depending upon how rigorous and stochastic the surrounding environment is, advances and setbacks will roughly balance. This is not to say the system will achieve equilibrium as regards to the list of species, which may continue to change.

The inability of the system to reach an arbitrarily high ascendancy looks at first like a glass half-empty. All does not appear quite so negative, however, once one realises that any rigid, mechanical ecosystem would be a catastrophe waiting to happen (Holling 1986). Such 'brittle' systems lack sufficient freedom to reconfigure themselves when beset by novel impacts. There is nothing left for them to do but collapse. A less-ordered configuration, by virtue of the ambiguities in its makeup can access these very inefficiencies to reconfigure itself in a way that mitigates, nullifies or incorporates the disturbance. The glass is really half-full.

It is possible to quantify the residual freedom in a system using the same informational calculus we just employed to develop the system ascendancy. One begins with the theoretical result from information theory that the mutual information is always bounded by the functional indeterminacy. This functional indeterminacy is simply the diversity of the flows that occur in the system. That is, if T_{ij}/T is the joint probability that matter both leaves i and enters j, then the formula

$$H = -\sum_{i,j}\left(\frac{T_{ij}}{T}\right)\log\left(\frac{T_{ij}}{T}\right) \tag{10}$$

quantifies the system's functional diversity. After this measure has been scaled by T in exactly the same manner as was done with A, the result is called the system 'capacity',

$$C = -\sum_{i,j} T_{ij}\log\left(\frac{T_{ij}}{T}\right) \tag{11}$$

and it can be proved that

$$C \geq A \geq 0. \tag{12}$$

The amount by which the capacity, C, exceeds the measure of constraint, A, is called the system 'overhead',

$$\Phi = -\sum_{i,j} T_{ij}\log\left(\frac{T_{ij}^2}{\sum_k T_{kj}\sum_l T_{il}}\right) \tag{13}$$

and this quantity signifies the potential for the system to recover from novel perturbation. The overhead, F, may be decomposed into four components representing the indeterminacies in the inputs, exports, dissipations and internal connections, respectively (Ulanowicz and Norden 1990). The limits that each of these terms imposes upon any increase in ascendancy can be parsed along hierarchical lines. Indeterminacy in the internal connections, or *redundancy*, represents an encumbrance upon the system for maintaining secure the internal lines of transfer. The indeterminacy in the exports has been likened to *tribute* and quantifies the 'tax' the given system must contribute to the next higher level to maintain system integrity there. Conversely, the indeterminacy among the dissipations represents the cost of maintaining kinetic order in structures at the next lower level. Finally, the indeterminacy among the inputs to the system represents the extent to which inefficient sources must be tapped in order to insure adequate sustenance for the system.

6. Using ascendancy

Ascendancy is a rather abstract concept, and much has been packed into a small set of indices. But that same richness makes the measures useful in any number of practical circumstances. To begin with, ascendancy was created to assess the developmental status of an ecosystem. If the manager of an ecosystem suspects that a particular impact has negatively affected his/her area, that hypothesis could be put to a quantitative test whenever sufficient data were available to construct the network of exchanges before and after the impact. In like manner, the developmental stages of disparate ecosystems can be compared with one another (e.g., Ulanowicz and Wulff 1991). One is now able to say quantitatively whether a system has grown or receded, developed or disintegrated. Furthermore, particular patterns of changes in the information variables can be used to identify processes that hitherto had only verbally been described. The process of eutrophication, for example, is characterised by a rise in ascendancy that is due to an overt increase in the activity of the system (T) which more than compensates for a concomitant decrease in its developmental status (average mutual information). This particular combination of changes in variables allows one to draw quantitative distinction between instances of enrichment and cases of eutrophication (Ulanowicz 1986).

The concepts of ecosystem 'health' and 'integrity' have been written into legislation in the U.S. (Costanza 1992) and Canada (Westra 1994) apparently before anyone had investigated whether those attributes can somehow be defined, quantified and measured. Because the conventional notion of health is normally associated with system vigour, performance and resilience (Costanza 1992), ascendancy and its associated indices become natural variables with which to give these metaphors real quantitative significance (Mageau et al. 1995).

If indeed ecosystems do exhibit an intrinsic direction in their development, then quantifying that direction using ascendancy might also provide a way to attach an 'intrinsic value' to the contribution that a particular process or taxon makes in that direction. Ascendancy, for example, has the same mathematical form as a 'production function' in economic theory. A production function is the sum of products of each process activity multiplied by the value added by that process. In a way yet to be specified, the logarithmic terms in the ascendancy formula are homologous to the values-added by the processes, and hence to any putative values put on the taxa themselves.

Although the ascendancy and related variables were invoked to quantify systems that are subject to contingencies, there is no reason why the same set of measures cannot be used to evaluate the performance of mechanical models (Field et al. 1989) or even networks of computational machines. The latter, for example, could be cast as a network of individual computers that exchange data at quantifiable rates. The overhead of this network, calculated according to the formula given above, should be lowest for those configurations that perform most effectively (Ulanowicz 1997).

7. Extending Ascendancy

Because ascendancy initially was formulated on the basis of steady-state snapshots of homogenous ecosystems, some might be inclined to regard the index as restricted to only equilibrium situations. Such an attitude, however, would do grave injustice to the robustness and broad relevance of contemporary information theory. The days when information theory was limited to the Shannon formula as applied to a communications channel are long gone. Information indices have been formulated to extend the basic notions to cover temporal and spatial inhomogenities as well (Pahl-Wostl 1992, Ulanowicz 1997). What mostly limits the extension of ascendancy theory into these realms is the extremely data-intensive nature of any such endeavour. One needs to know the full configurations of trophic exchanges at each time or spatial point (or both).

Failing sufficient data, one could still employ models to generate suites of data that could be used to test the capabilities of multi-dimensional information indices at identifying those times and places where system dynamics are most interesting and influential. Recently, a cellular automaton was programmed to represent the migration of animals across a landscape interspersed with barriers. The dynamic patterns were analysed using the components of the ascendancy to quantify the contribution that each spatial point makes to the overall ascendancy. The 'field' of components was plotted over the landscape, and the resulting profile indicated those points at which the critical actions were taking place (Ulanowicz, in press).

Ascendancy and its ancillary indices were originally defined (as above) entirely in terms of process rates. The full dynamics of systems, however, are known to depend also upon the biomass stocks in the taxa. Only recently has a way been found to incorporate biomass stocks into the calculation of ascendancy in a way fully consonant with the algebra of information theory (Ulanowicz and Abarca 1997): If B_i represents the amount of biomass in taxon i, and B, the total amount of biomass in the system, then B_i/B will estimate the a priori likelihood that a quantum of material is leaving taxon i. Similarly, B_j/B will estimate the a priori probability that a quantum will enter j. The a priori joint probability that material both leaves i and enters j thereby becomes B_iB_j/B^2. One may compare this estimate with the observed *a posteriori* joint probability, T_{ij}/T, in what is known as the Kullback- Leibler index,

$$I = k\sum_{i,j}\left(\frac{T_{ij}}{T}\right)\log\left(\frac{T_{ij}B^2}{B_iB_jT}\right) \tag{14}$$

As with the original ascendancy, one may scale I by T to obtain a biomass-inclusive ascendancy,

$$A_b = \sum_{i,j} T_{ij}\log\left(\frac{T_{ij}B^2}{B_iB_jT}\right) \tag{15}$$

It is possible to demonstrate that the original ascendancy is bounded from above by the biomass-inclusive ascendancy, i.e., $A_b \geq A$, and that the difference between the two is due to the departure of the biomass distribution from what it would be at chemical equilibrium. This difference, therefore, should be related to the exergy content of the system (S.E. Joergensen, personal communication).

Unfortunately, the Kullback-Leibler index possesses no upper bound, and therefore does not yield expressions homologous to either the capacity or the overhead. The advantage that the new index does afford, however, is that biomass dynamics become implicit in A_b (Ulanowicz and Baird, In press). For example, one may construct separate but parallel networks, each pertaining to one of several chemical elements (e.g., C, N, and P) circulating within the same ecosystem. With multiple elements, T_{ijk} can represent the amount of element k flowing from i to j, and B_{ik}, the amount of element k incorporated into i. One may then use the sensitivities of the resultant A_b with respect to each of the B_{ik} to determine which element is limiting the activity of each taxon. In other words, the principle of increasing ascendancy subsumes Liebig's Law of the Minimum. Such 'theory reduction' is one of the hallmarks of a robust theory, but the advantages of A_b don't end there. If one further calculates the sensitivities of A_b to the flows, T_{ijk}, one can then determine which input of the limiting element plays the pivotal role to that taxon. Liebig's principle offers no guidance on how to identify which flow might be limiting, so that this method yields a theoretical prediction to be compared with experiment.

8. Ascendancy, the New Perspective

As the concepts surrounding ascendancy evolve, it becomes ever clearer that ecosystems exhibit very non-traditional dynamics – or what Eber et al (1989) have called 'infodynamics.' As Popper discerned, new dynamics require a new calculus, and it now appears that the cluster of variables defined using information theory might be prime candidates with which to begin the development of a post-Newtonian ecology (Ulanowicz 1997).

REFERENCES

Allen, T.F.H. and T.B. Starr. 1982. Hierarchy. University of Chicago Press, Chicago. 310p.

Clements, F.E. 1916. Plant Succession: An Analysis of the Development of Vegetation. Carnegie Institution of Washington, Washington, D.C. 340p.

Costanza, R. 1992. Toward an operational definition of ecosystem health. 239-256. In: R. Costanza, B.G. Norton, and B.D. Haskell, (Eds.). *Ecosystem Health: New Goals for Environmental Management*. Island Press, Washington, DC. 269p.

DeAngelis, D.L. W.M. Post, and C.C. Travis. 1986. Positive Feedback in Natural Systems. Springer-Verlag, NY. 290p.

Eigen, M. 1971. Self-organization of matter and the evolution of biological macromolecules. *Naturwiss*. 58:465-523.

Field, J.G., C.L. Moloney, and C.G. Attwood. 1989. Network analysis of simulated succession after an upwelling event. 132-158. In: F.W. Wulff, J.G. Field, and K.H. Mann, (Eds.). *Network Analysis in Marine Ecology: Methods and Applications*. Springer-Verlag, Berlin.

Fisher, R.A. 1930. The Genetical Theory of Natural Selection. Oxford University Press, Oxford, UK. 272p.

Gleason, H.A. 1917. The structure and development of the plant association. *Bull. Torrey Botanical Club*. 44:463-481.

Haken, H. 1988. Information and Self-organization. Springer-Verlag, Berlin. 196p.

Holling, C.S. 1986. The resilience of terrestrial ecosystems: local surprise and global change. 292-317. In: W.C. Clark, and R.E. Munn, (Eds.). *Sustainable Development of the Biosphere*. Cambridge University Press, Cambridge, UK.

Jorgensen, S.E. 1992. Integration of Ecosystem Theories: A Pattern. Kluwer, Dordrecht. 383p.

Kauffman, S.A. 1995. At Home in the Universe: The Search for the Laws of Self- Organization and Complexity. Oxford University Press, Oxford. 321p.

Mageau, M.T., R. Costanza, and R.E. Ulanowicz. 1995. The development, testing and application of a quantitative assessment of ecosystem health. *Ecosystem Health*. 1(4):201-213.

Odum, E.P. 1969. The strategy of ecosystem development. *Science*. 164: 262-270.

Pahl-Wostl, C. 1992. Information theoretical analysis of functional temporal and spatial organization in flow networks. *Mathl. Comput. Modelling*. 16 (3): 35-52.

Popper, K.R. 1990. A World of Propensities. Thoemmes, Bristol. 51p.

Rosen, R. 1991. Life Itself: A Comprehensive Inquiry into the Nature, Origin and Foundation of Life. Columbia University Press, NY. 285 p.

Simberloff, D. 1980. A succession of paradigms in ecology: Essentialism to materialism and probabilism. *Synthese*. 43:3-39.

Tribus, M. and E.C. McIrvine. 1971. Energy and information. *Sci. Am*. 225: 179-188.

Ulanowicz, R.E. 1980. An hypothesis on the development of natural communities. *J. theor. Biol*. 85: 223-245.

Ulanowicz, R.E. 1986. A phenomenological perspective of ecological development. 73-81. In: T.M. Poston and R. Purdy, (Eds.). *Aquatic Toxicology and Environmental Fate: Ninth Volume*, ASTM STP 921. American Society for Testing and Materials, Philadelphia.

Ulanowicz, R.E. 1997. Ecology, the Ascendent Perspective. Columbia University Press, New York. 201p.

Ulanowicz, R.E. (In press). Quantifying constraints upon trophic and migratory transfers in spatially heterogeneous ecosystems. In: L.D. Harris and J.G. Sanderson (Eds.). *Series in Landscape Ecology I*. St. Lucie Press.

Weber, B.H., D.J. Depew, C. Dyke, S.N. Salthe, E.D. Schneider, R.E. Ulanowicz, and J.S. Wicken. 1989. Evolution in thermodynamic perspective: an ecological approach. *Biology and Philosophy* 4:373-405.

Westra, L. 1994. An Environmental Proposal for Ethics: The Principle of Integrity. Rowman & Littlefield, Lanham, MD. 237p.

II.7.2 Ecosystems as Dynamic Networks

Claudia Pahl-Wostl

1. Concepts of regulation and control

Early concepts of ecosystems were determined by the idea that ecosystems were homeostatic entities where self-regulating forces drive the systems towards a time-invariant stability point. Fluctuations were mainly perceived as being imposed by the physical environment on an otherwise time-invariant biological system. Lakes were important for the development of such a concept. Because of their relative closure they were perceived as microcosms (Forbes 1887; Thienemann 1925) where the influence of internal processes dominates external influences from the environment regarding flows of nutrient and energy. At the same time, lakes are highly influenced by environmental fluctuations. Because of the weakness of spatial structure, fixed patterns of organisation in space are usually weak. Therefore, the organisation of the biological community has mainly been perceived as being determined by the vagaries of the physical environment. To understand the importance of self-regulation in ecological networks and the link between biological organisation and energy and matter flows one has to consider dynamic patterns of organisation. This chapter introduces a conceptual framework accounting for self-regulation and structural complexity in dynamic ecological networks and discusses the implications for ecosystem function.

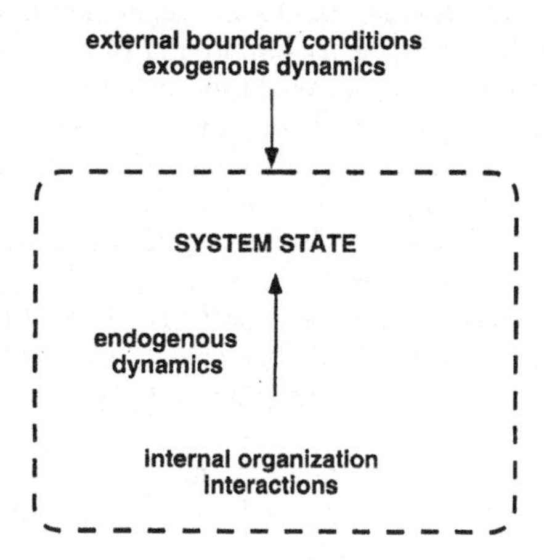

Figure 1: Essential requirements to account for self-regulation

Self-regulation can be defined as control of a system's properties by its internal feedback mechanisms. Figure 1 summarises the essential elements required to account for a concept of self-regulation. Regulation can only be explained within a system's perspective and relative to descriptors of a system's state – e.g. total biomass, productivity, stability. The state of a system can be changed by endogenous and exogenous sources of influence. Endogenous influence is derived from endogenous processes such as nutrient recycling or species succession. Exogenous influence is derived from processes which have their origin outside the system, such as the seasonal dynamics of environmental variables or nutrient inputs. Self-regulation is only meaningful in ecosystems which have, at least, a minimal degree of autonomy. What concepts of autonomy are useful in this context and what is required for an ecosystem to be autonomous?

The concepts of autonomy and, thus, self-regulation are dependent on scales in time and space. Consider the example of pelagic ecosystems. One can talk about pelagic ecosystems as units of self-regulation when the deep water body is sufficiently large for the interactions (material exchanges) within this habitat to dominate over those which operate across habitat boundaries, especially with respect to the littoral zone and the sediments. The difference between small ponds and large systems such as Lake Baikal may be greater than the difference between Lake Baikal and some marine system. Considerations of scale are of vital importance in determining the characteristics of lacustrine habitats and their relationships (e.g. volume/surface area, water retention time, epilimnion depth/total depth determine the influence of sediments and nutrient gradients).

Before discussing community structure and organisation it is useful to consider a plausible reference state of an ecosystem, maintained by regulatory processes. One of the major founding principles was the belief in 'the balance of nature' with its strong emphasis on equilibrium and stability in terms of a complete absence of variation. It is based on the assumption that the properties of systems are maintained and temporal variability is suppressed. The hypothesis underlying many stability concepts can be summarised as: ecosystems are organised in such a way that any deviations from the equilibrium state are counteracted by negative feedback forces which prevent the system from deviating from a preferred state.

Traditional concepts of ecosystem structure and regulation have emanated from such thinking. The whole field of food web theory has, to a large extent, been motivated by the hypothesis that patterns observed in nature ought to correspond to the patterns that were shown to yield stable equilibrium points in model investigations. The finding of an inverse relationship between the number of species and food web connectance has been taken as support for the theoretically-derived trade-off between food web complexity and stability. Deterministic models, mainly of the Lotka-Volterra type, were used to derive the further structural properties to be expected in systems attaining stability, as in the absence of positive omnivory or the absence of positive feedback. The lack of these properties in data-sets from natural food webs was taken as further evidence for natural systems to exhibit configurations assuring stability (Lawton and Warren 1988; Cohen, Briand et al. 1990; Pimm, Lawton et al. 1991).

However, food web theory may be seriously compromised by ignoring largely spatial and temporal scales, by the weak data basis exhibiting e.g., a highly heterogeneous level of aggregation (taxonomic and trophic species, functional groups) and by the absence of criteria to record a link or not (e.g. Hastings 1988; Paine 1988). The diversity (complexity)/ stability discussion has always suffered from the lack of coherent definitions for the two properties and the resulting heterogeneity and incompleteness of both theoretical arguments and field studies. It is therefore useful to have first a closer look at stability concepts.

Pimm 1984; Pimm, Lawton et al. 1991 identified five concepts that have been referred to as stability by theoretical and empirical ecologists (equilibrium refers to an equilibrium point where all variables are time invariant).

In the mathematical sense a system is considered to be stable if and only if the variables return to equilibrium conditions after displacement from them.

Resilience corresponds to how fast variables return to their equilibrium after having been displaced from it. The higher the resilience the faster the recovery following a perturbation. Systems with high resilience may be said to be 'stable'.

Persistence is how long a variable lasts before it is changed to a new value. Systems with high persistence may be called to be stable.

Resistance measures the degree to which a variable is changed, following a perturbation. Systems with high resistance may be described as stable.

Variability is the degree to which a variable varies over time. Systems exhibiting little variability may be described as stable.

There does not yet seem to be a general agreement on the consistent use of these various terms. In addition, depending on one's research interests, one may consider individual species abundances, species composition or trophic level abundance as variables of interest. Hence, when talking about stability and equilibrium of ecosystems one should always be explicit regarding the variables, the spatio-temporal scales, and the concept one refers to.

It is doubtful whether the reference state of time-invariant ecosystems is appropriate in a variable environment. Such ecosystems would be static, rigid entities without any potential for adaptation and change. Traditional stability concepts have been developed for ecosystems in a homogenous environment and for single levels of ecological organisation. If these assumptions are relaxed, one may come to quite different conclusions. It has, for example, been assumed that species fluctuations lead to extinction. A plausible conclusion is therefore to claim species able to maintain stable biomass levels are evolutionary superior to those which fluctuate and come close to the threshold of extinction. However, such an argument is only valid in a perfectly homogeneous equilibrium world. Simulations with meta-population models show that low densities also lead to more frequent extinction at the local level, while chaotic oscillations reduce the degree of synchrony among populations and thus reduce also the risk of their simultaneous extinction (Allen, Schaffer et al. 1993). A desynchronised, spatially distributed pattern of species may therefore be evolutionary superior to a single, stable, spatially homogenous population.

Discussions about stability have often been limited to an isolated view on one level of ecological organisation. However, regulation in ecological systems is decentralised and needs to invoke several levels of organisation. The maintenance of ecosystem function depends on the diversity and flexibility of the component populations.

The hypothesis can be stated that the trade-off between global and local variability is just the stabilising principle of ecosystem organisation (Figure 2). It is supported by empirical evidence in work reported by Tilman and co-workers (Tilman and Downing 1994; Simon Moffat 1996; Tilman 1996) which shows that species-rich grassland systems have higher resistance to disturbances with respect to functional properties at the system level, whereas the biomass of the individual species exhibited higher fluctuations than in species-poor systems.

Reynolds (1996) argued that similar patterns can be observed in pelagic systems during the course of ecosystem succession. Yet it seems quite astonishing that there is not more empirical evidence on this issue from aquatic ecosystems where the shorter generation times of the component organisms would favour such experimental approaches.

In conclusion, one can state that stabilisation does not imply time invariance. On the contrary, for ecosystems to maintain the balance between flexibility and functioning, a delicate balance is required between internal degrees of freedom and constraints, a combination of positive and negative feedback effects. Positive feedback leads to a constant tendency of a system to explore the limits of its boundary conditions and to use opportunities. Positive feedback is thus an essential ingredient for adaptive ecological and evolutionary change. Negative feedbacks keep the system within delimited bounds.

To further improve our understanding of ecosystem organisation, along the lines of reasoning, requires the investigation of the relationship between levels of ecosystem organisation and between its functional and structural properties.

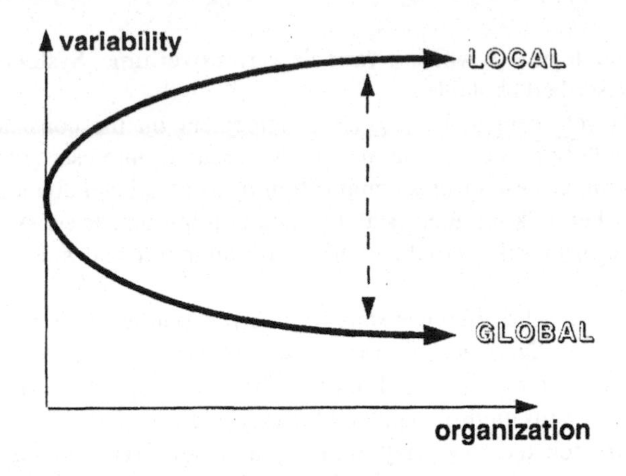

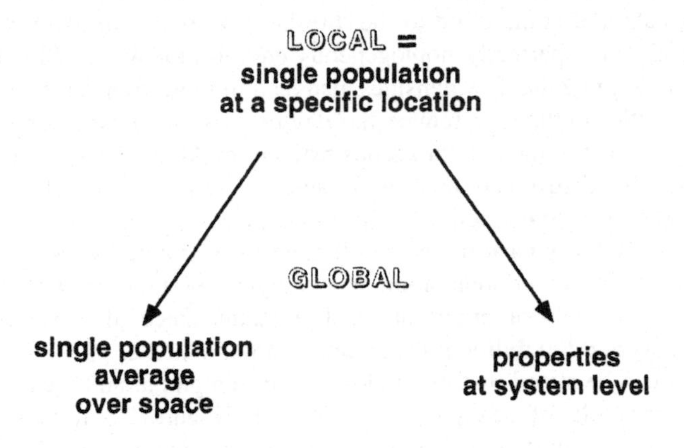

Figure 2: Tradeoff between local and global system properties as essential elements of ecosystem dynamics

Figure 3 depicts ecological systems as multi-level dynamic networks of energy and matter flows. At the macro-scale the systems-level, an ecosystem can be described by aggregate extensive variables, such as biomass and productivity, or by intensive variables such as adaptability and structure. At the meso-scale, the network level, the structure of the ecological networks may be characterised by the connectivity, the niches. At the micro-scale, the species-level, species differ in phylogeny and function. Theoretical concepts for relationships among levels of organisation require a consistent definition of functional diversity. For a start, it is useful to distinguish between the horizontal and vertical dimensions of organisation along the trophic gradient (Figure 4). Horizontal organisation refers to ensembles of species at the same trophic level or the same interval of trophic position, respectively. The vertical axis refers to the distribution of species along the trophic gradient – the simplest case would be given by discrete trophic levels of primary producers, herbivores and carnivores.

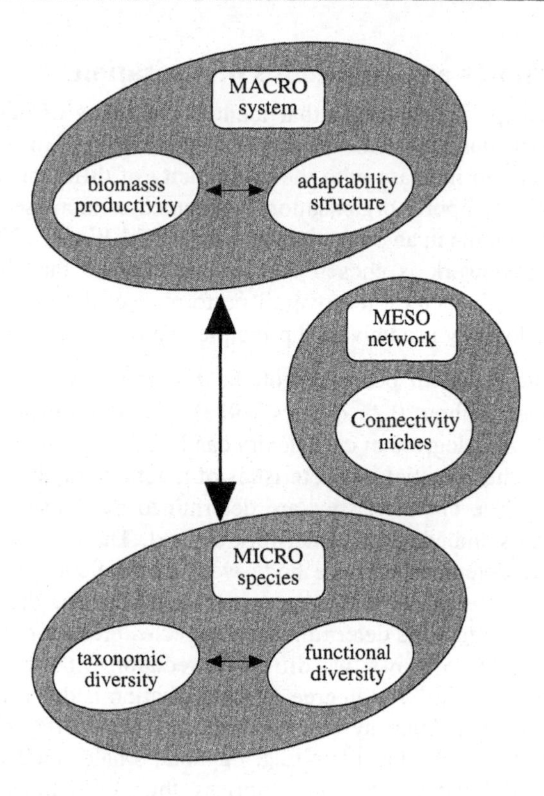

Figure 3: Different levels of ecosystem organisation and their interdependence. Functional properties at the level of the ecosystem as a whole are stabilised and maintained in a variable environment by changes in the ecological network comprised of the component species and functional groups. A high diversity in functional pathways endows thus an ecosystem with a high degree of flexibility and adaptability. At the same time such organisation renders ecosystem development difficult to predict and control.

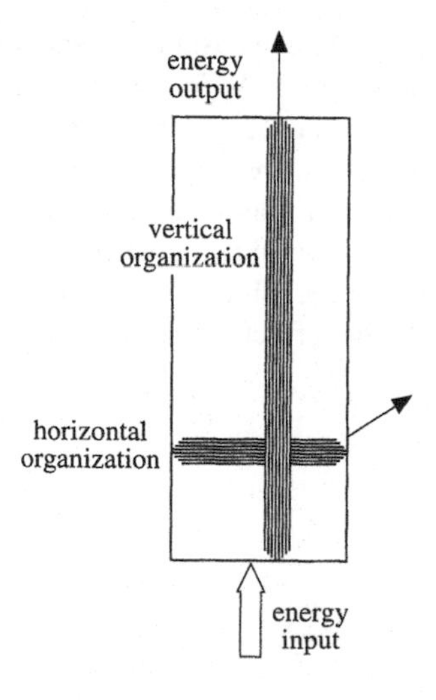

Figure 4: Horizontal (within a trophic level/interval of trophic positions) and vertical (along the trophic gradient) dimensions of ecosystem organisation defined by the direction of energy flow through an ecological network.

2. Metrics to quantify an ecosystem's organisation

In this section, a concept is introduced that identifies patterns of vertical and horizontal organisation, and that quantifies the degree of organisation in ecological networks. Organisation is assumed to increase with the multiplicity of different functional patterns in an ecological network. Temporal organisation is assumed to increase with the multiplicity of different temporal patterns in an ecological network (Pahl-Wostl, 1995).

An allometric framework is chosen that can investigate the change in patterns of organisation and their influence on the overall performance of an ecological network in a systematic fashion. Allometry is a universal principle that describes the scaling of metabolic and physiological rates, r, with body weight, w: $r = aw^{-\alpha}$ where α is the allometric exponent with a value of about 0.25 (Peters, 1983). Using allometric principles, generic network models of varying degrees of complexity can be constructed based on simple rules.

Figure 5a depicts the essential characteristics of a generic model species in the chosen allometric framework. The characteristics are determined by a species location along the body weight axis and its embedding in a network context. The metabolic rates and thus the typical time scales are determined by the body weight of the species. Function within the ecological network is determined by the species embedding in the network context of feeding relationships. The input is determined by a species predation window characterised by width and shape. Figure 5 shows two different predation windows of a generalised (1) and of a specialised (2) species. The degree of specialisation is determined by the width of the predation window. In pelagic systems this reflects two different feeding modes of raptorial (e.g. copepods, shark) and filter (e.g. daphnia, whale) feeders, respectively. The organisation of the ecological network comprises thus two dimensions: the dynamic diversity that increases with the number of time scales and the functional diversity that increases with the multiplicity of links and predation windows in the food web.

Figure 5b shows the simplest network structure one may think of ñ parallel food chains that are shifted along the body weight axis. In such a network structure the functional organisation is not high since the overall trophic structure can be described by an aggregated chain with discrete trophic levels. However, such networks may display temporal organisation.

The principle underlying temporal organisation is shown in Figure 6 for two predator-prey pairs sharing a common nutrient pool. At first glance the pairs compete for the limiting nutrient, and are therefore linked by a negative feedback interaction. However, the nutrient is again recycled which results in a positive feedback. Figure 6b shows that the interpretation of the type of interaction may change when the temporal activity pattern is accounted for. The pairs engage in temporal resource partitioning which leads to a temporal shift between nutrient utilisation and recycling. As a consequence, the interaction between the two pairs may be dominated by the effects of positive rather than negative feedback.

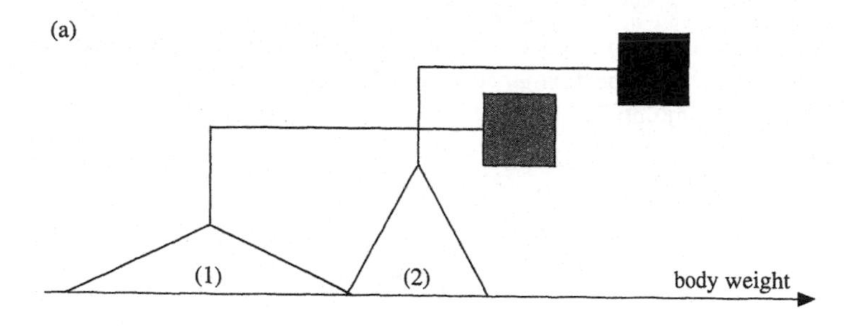

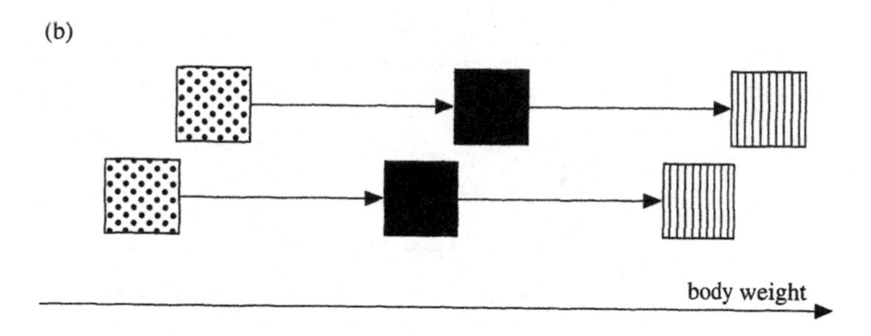

Figure 5: (A) Characteristics of a model species along the body weight axis. A species is characterised by its location along the body weight axis and by the width and shape of its predation window. (B) Parallel food chains as simplest structure of a network with a high functional redundancy (discrete trophic levels) but high dynamic diversity (different body weights and thus many temporal scales).

A measure was derived to quantify temporal organisation in ecological networks (Pahl-Wostl, 1995). The measure of temporal organisation I_t can be expressed in terms of flows as:

$$I_t = \sum_{j=0}^{n} \sum_{i=1}^{n+2} \sum_{k=1}^{r} \frac{T_{jik}}{T} \log\left(\frac{T_{jik}^2 \, T}{T_{ji.} T_{j.k} T_{.ik}} \right) \qquad (1)$$

where
a point denotes the sum over the corresponding index
n total number of compartments in network
index of a virtual external input compartment
$n+1$ index of virtual external output compartment
r total number of time intervals over time period
T_{jik} flow from compartment j to i in time interval k
T total system throughflow
and T the total system throughflow is defined as

$$T = T_{jik} \qquad (2)$$

I_t quantifies the decrease in redundancy of compartmental outputs and inputs upon resolution of the temporal flow pattern. It is thus a measure of the reduction in competitive interaction due to temporal organisation. Several species share a functional niche by temporal resource partitioning. Similarly, the measure may be extended to include the organisation along one or more dimensions of space.

It was claimed that an increase in temporal organisation is positively correlated with an increase in the network's performance (Pahl-Wostl, 1995). Support for this claim can be provided by model simulations.

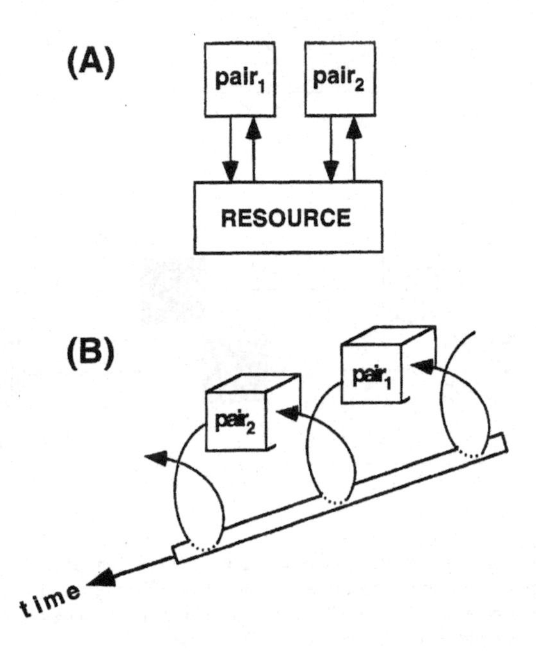

Figure 6: (A) Two predator prey pairs sharing a common nutrient pool. (B) The activity of each pair is confined to a different time interval. The rectangle along the time axis denotes the nutrient pool. What appears as two separate feedback cycles in the time averaged representation reveals to be a feedback spiral linking the pairs across time.

3. Self-regulation in simple chain-like networks

The trophic-level concept constitutes an important abstraction in the clarification and organisation of our understanding of energy transfer in ecosystems. It is often assumed that the structure of complex food webs can be reduced to a sequence of discrete trophic levels which include trophically homogeneous groups. Concepts of trophic structure have emphasised the analyses of time-invariant equilibrium states in linear food chain models. Hairston, Slobodkin and Smith viewed whole trophic levels as dynamically equivalent to single species and assumed there to be no direct effects of population density or a given trophic level on the per capita growth rate of that level (Hairston, Smith et al. 1960; Slobodkin, Smith et al. 1967). In this way increased nutrient input increases the density of the top level and levels that are an even number of levels below the top. The abundance of other levels remains unchanged. Figure 7 shows that this leads to a change in the pattern of the relative distribution of total biomass among different trophic levels. Subsequently, a number of related concepts based on food chain dynamics have been derived (Fretwell 1987; Oksanen 1991; Carpenter, Frost et al. 1993

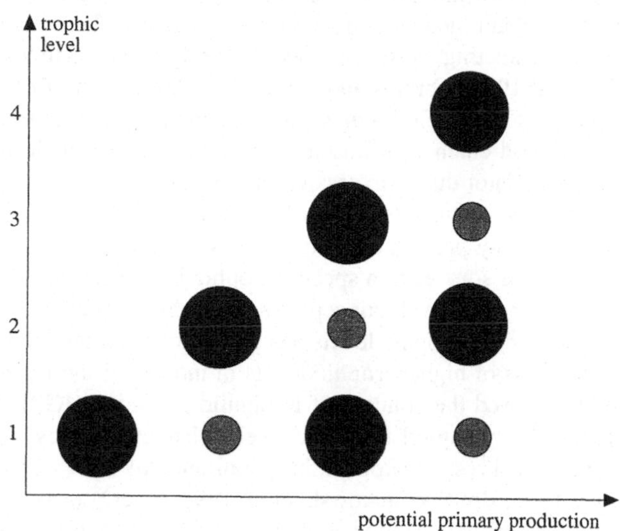

Figure 7: Change in biomass distributions with increasing nutrient input and thus increasing potential primary production. The change in the size of the different spheres indicates the shift in the relative distribution of biomass across trophic levels.

Considerations of food-chain dynamics have enhanced our understanding of the multiple causal pathways governing community dynamics. The traditional emphasis on food or habitat-limited populations has gradually been replaced by a more balanced perspective accounting for the importance of varying levels of predation. Theoretical predictions have stimulated a number of empirical investigations (e.g. Vanni and Findlay 1990; McQueen, Mills et al. 1992; Persson, Diehl et al. 1992; Diehl 1993; Brett, Wiackowski et al. 1994; Mazmuder 1994; Elser, Luecke et al. 1995; Mateev 1995). However, the empirical results are by no means unequivocal. Since indirect evidence often needs to be used to support the hypotheses made, results may be explained by different interpretations, depending on the perspective of the observer. The presence or absence of planktivorous or piscivorous fish in a lake is, for example, used to classify lakes as being two-, three- or four-link systems, respectively (e.g. Hansson 1992; Persson, Diehl et al. 1992). Due to lack of data, the classification cannot be based on quantitative analyses of the trophic structure itself.

A recent debate in the literature on the validity of biomanipulation as a management tool exemplifies some of the current controversies that hinge on trophic structure (Carpenter and Kitchell 1992; DeMelo, France et al. 1992). The concept of cascading trophic interactions has been basic to deriving predictable patterns for regulatory processes which have been employed to devise schemes of biomanipulation. By specific removals and/or additions of trophic levels, one sets out to control algal biomass in eutrophic lakes. For example, reducing the predation pressure of herbivores by the removal of planktivorous fish is assumed to increase the grazing on algae and thus reduce algal biomass. However, a recent evaluation of a set of whole lake experiments has revealed that food web complexity may lead to quite unexpected outcomes (Carpenter and Kitchell 1993).

The practical problems in testing hypotheses about trophic level dynamics are not surprising, because of the sensitivity of the predictions to the underlying assumptions. Below, important concerns about this approach are summarised A comprehensive and balanced overview of the current discussion concerning structure and dynamics of food webs can be found in Polis and Winemiller (1996).

Analyses based on equilibrium assumptions have been challenged by investigations of two- and three-level food chain models with complex dynamics. These may provide results different from, even contradicting those obtained in models with a time-invariant steady state (Hastings and Powell 1991; Abrams and Roth 1994; Abrams and Roth 1994; Hastings and Higgins 1994; Hastings 1996). Enrichment, for example, leads in general to an increase in the bottom level of a food chain with unstable dynamics. However, it may even lead to the extinction of the top predator due to excessive fluctuations.

An important argument against the use of integer trophic levels is the ubiquitous presence of omnivory. In many practical situations, it is not possible to assign consistently organisms with different food sources to a specific trophic level. Cousins (1985, 1987) even suggested replacing the trophic level concept by a trophic continuum. Abrams (1993) showed that heterogeneity within trophic levels has an influence on the relationship between productivity and the biomass of higher trophic levels in models of the Lotka-Volterra type. Polis and Holt (1992) reviewed the concept of intraguild predation (IGP). They concluded that IGP systems cannot be collapsed to shorter webs characterised by cascading trophic interactions. Recent empirical results support the importance of food web complexity (e.g. Carpenter and Kitchell 1993; Brett, Wiackowski et al. 1994; Elser, Luecke et al. 1995).

Consumers may affect primary producers by means other than direct consumption. The importance of consumer-mediated nutrient recycling and transport has been emphasised in empirical and theoretical investigations (DeAngelis, Bartell et al. 1989; Sterner 1990; Vanni and Findlay 1990; DeAngelis 1992; Vanni 1996; Vanni and Layne 1997; Vanni, Layne et al. 1997). While predation imposes a direct negative feedback on prey populations, nutrient recycling has an indirect positive effect. In plant-herbivore systems, grazing has been observed to first increase net primary production for moderate levels of grazing and then to lead to a decrease for a further increase of the grazing pressure (review in DeAngelis 1992).

Abrams 1996, emphasises that theory dealing with the response of trophic levels to nutrient input is largely based on models that assume trophic levels to be homogeneous, non-adaptive entities. He makes the point that since it is not a simple matter to predict the direction of change in a particular trophic level in response to nutrient enrichment, even when the trophic levels are homogeneous. It will require more study of the types of changes that occur in the composition and functional response of the trophic levels before anything beyond speculation is possible.

One may deduce that the interpretation of experimental results and the empirical support of hypotheses on regulation, based on the paradigm of a simple food chain, are difficult because a range of processes are of importance:

- resource limitation resulting in bottom-up effects on consuming populations,
- competition within groups of functionally similar populations,
- predation resulting in top-down effects on prey populations,
- positive feedback via nutrient recycling and mutualistic interactions.

Even when these processes all act in concert, their relative importance varies over the seasonal cycle which further complicates an unequivocal interpretation.

In addition to the varying importance of the processes, one has to consider that the simple model of a food chain may be appropriate only in exceptional cases. Real food webs are more complex in structure. The next section will discuss a systematic approach to explaining food web complexity and its consequences.

4. Self-regulation in complex web-like networks

One of the major topics in current ecological research is the investigation of the relationship between ecosystem function and diversity (Schulze and Mooney 1993). Pelagic ecosystems are good candidates for developing conceptual models where these relationships are addressed in a systematic fashion (Pahl-Wostl 1995).

The structural complexity of ecological networks can be approached systematically by making use of the fact that body weight determines to a large extent an organism's physiological and ecological characteristics (Peters 1983; Pahl-Wostl 1995). In most food webs, energy flow is directed along a gradient of increasing body weight. Energy flow is thus directed along a gradient of increasing scales in time and space due to the decrease in metabolic rates and the increase in generation time and radius of activity associated with increasing body weight. Pelagic systems constitute a prime example for organisation along a continuum in both trophic function and spatio-temporal scales. Figure 8 sketches in a stepwise fashion the transition from a food chain to a food web. The diagrams on the left of each figure show the distribution of biomass in a 2-dimensional framework of trophic level and body weight which correspond to the model of trophic transfers depicted on the right. A circle always denotes a weight class along a logarithmic body weight axis.

The simplest and most often used model structure corresponds to a food chain as depicted in Fig 8A. This model is based on the assumption that a trophic level is a dynamic entity which can be adequately represented by an averaged body weight and thus by a state variable with well-defined dynamic properties. The regulatory concepts that can be derived based on such a model structure have been discussed in the previous section. Food-web complexity can be increased in two ways - by increasing diversity along the body weight axis and thus diversity in time scales (horizontal organisation) and by increasing diversity along the axis of trophic function (vertical organisation).

The food chain depicted in Figure 8A leads to a sharply peaked biomass distribution along the body weight axis. Numerous empirical investigations have provided empirical evidence for an even rather than a peaked distribution (e.g. Sheldon, Prakash et al. 1972; Sheldon, Sutcliffe et al. 1977; Sprules and Munawar 1986; Witek and Krajewaska-Soltys 1989; Sprules 1991; Gaedke 1992). Such findings may be accounted for by the trophic structure depicted in Figure 8B: each trophic level is still well defined but now comprises a wide range in body weight. Such a pattern could be explained by an ensemble of food chains which are shifted along the body weight axis generating a continuum in body weight.

The figure indicates as well that the independent food chains may become linked by taking into account that a consumer's prey window extends over more than one body weight class. However, in the case presented one can still discern well defined trophic levels that are distinguished by the different shadings.

Figure 8C depicts schematically how one can explain a gradual transition to a continuum along the axis of trophic function. The food web fragment given on the right illustrates the gradual breakdown of defined trophic levels. The rectangular box denotes the window of a consumer in a higher weight class selecting organisms according to their size irrespective of their being autotrophs or heterotrophs. Such feeding strategies are common for planktonic predators (e.g. Brett, Wiackowski et al. 1994; Straile 1994; Gaedke, Straile et al. 1996). The trophic position of such a consumer ranges somewhere between 2 and 3. The diagram on the left depicts the continuum of trophic positions which may be generated by shifts in the prey window and variations in the relative contribution of different prey to a consumer's overall diet. The bottleneck between primary producers and herbivores (levels 1 and 2) indicates the lack of knowledge regarding the quantitative importance of mixotrophy. Even when the existence of switching between autotrophic and heterotrophic modes of production is well documented (e.g. Porter, Paerl et al. 1988; Brett, Wiackowski et al. 1994; Persson, Bengtsson et al. 1996), it would yet not be empirically well grounded to assume a continuum right away.

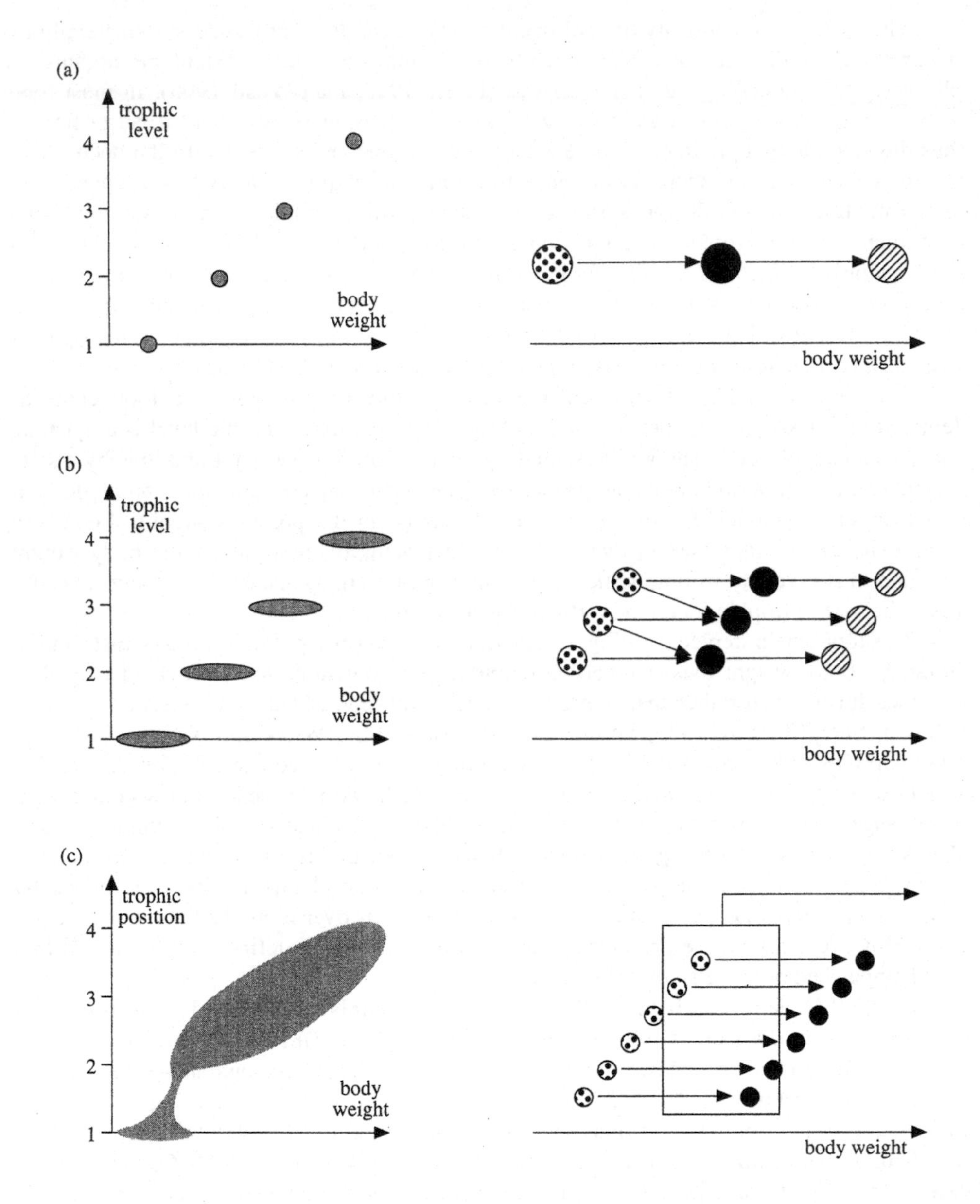

Figure 8: Left: Hypothetical distribution of biomass in a two-dimensional framework of trophic level and body weight for different models describing the energy transfer along a gradient of increasing body weight. - Right: Illustration of the transition from a food chain with discrete trophic levels to a trophic continuum. Dotted circles always refer to autotrophs, light shaded circles to herbivores, dark shaded circles to carnivores or omnivores, respectively. (A) Distribution discrete in body weight and trophic function. - Food chain with discrete trophic levels. (B) Distribution with a continuum in body weight but still discrete in trophic function. - Ensemble of food chains which differ in their location along the body weight axis. Chains may be linked due to overlapping prey windows. However, discrete trophic levels are still preserved. (C) Distribution with a continuum both in body weight and trophic function. The notion trophic level which can only take discrete integer values is replaced by trophic position which may take any value. - Omnivorous consumer whose prey window comprises both primary producers and herbivores. The consumer's trophic position may thus range between 2 and 3.

Models reflecting as a first approximation the configuration depicted in Fig 8B have been investigated previously Pahl-Wostl 1993; 1995). Figure 9 shows the structure of such a model comprising an ensemble of predator-prey pairs. The diversity in time scales increases with the number of predator(P)-prey(B) pairs differing in body weight. Figure 9A shows the network of nutrient flows. The size of the arrows representing internal exchanges indicates that most of the nutrient is recycled within the system. A predator-prey pair comprises two pathways of recycling: a fast, short one, deriving from direct losses from the prey species to the nutrient pool, and a slower, longer one deriving from the recycling via predators.

As represented in Figure 9B, the predator-prey pairs are distributed along the body weight axis which is partitioned into weight classes equally spaced on a logarithmic scale. The average body weight in the kth weight class, w_k, is expressed in fraction of w_0, the weight in class 0: $w_k = 2^k w_0$. The weight ratio between neighbouring classes, w_{k+1}/w_k, equals two. The choice of a logarithmic instead of a linear scale becomes more intelligible when one considers that the weight of a predator's prey is determined by the predator-prey weight ratio rather than by the predator-prey weight difference. A ratio is constant on a logarithmic scale. Similar considerations can be made with respect to the time scales associated with body weight. One may conceive of the body-weight axis representing the axis of a one-dimensional niche space along which species occupy niches according to their dynamic characteristics. Temporal organisation implies that species organise themselves along this axis. Model equations and parameter values are listed in Table 1.

Model simulations show that an increase in the number of different food chains and thus an increase in dynamic diversity has major effects on system dynamics with a shift from a time-invariant stable state to periodic and finally chaotic oscillations. However, an increased variability at the level of individual species resulted in a decreased variability of functional properties at the level of the system as a whole. Figure 10 shows, as an example, results obtained in model simulations with ensembles of predator-prey pairs. Such behaviour is an expression of temporal organization of the ecological network despite chaotic fluctuations at the species level. These model simulations support the hypothesis of the tradeoff between global and local variability (cf. Figure 2).

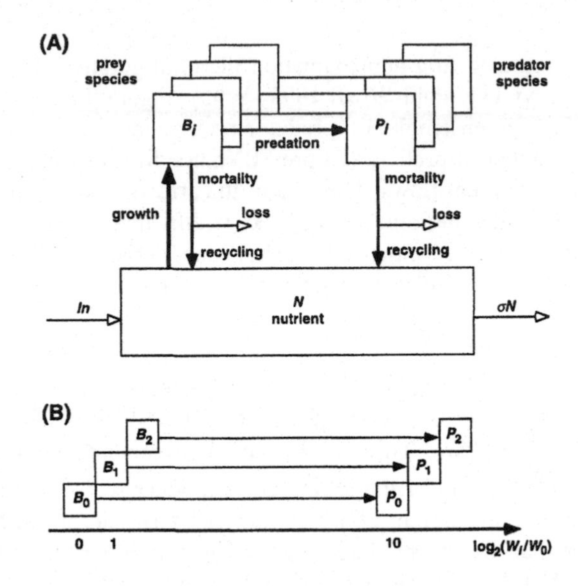

Figure 9: (A) Network representation of the basic model components. The internal exchanges are set in bold type to emphasise the quantitative dominance of recycling over the external exchanges. The exchanges with the environment are denoted by open arrow heads. (B) Arrangement of predator-prey pairs along the body weight axis. Weight is expressed in fraction of w_0, the average weight in class 0. The weight class difference between a predator and its prey was chosen to be equal to 10 corresponding to a weight ratio of 1000.

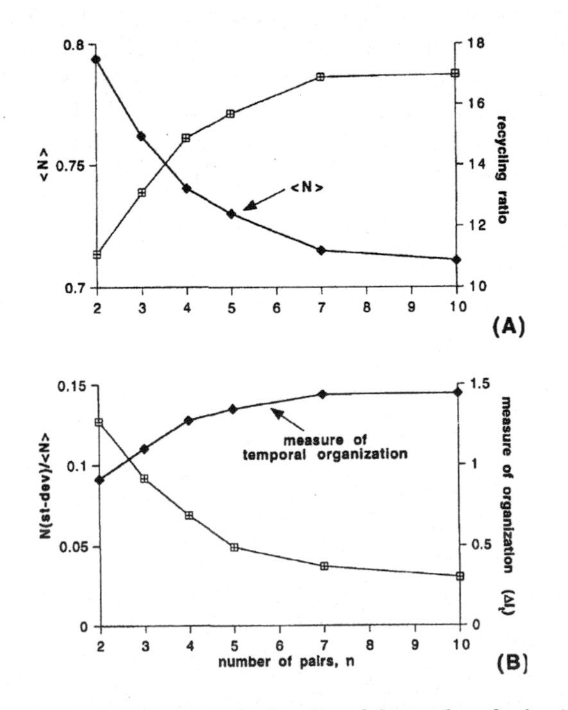

Figure 10: Global system parameters as a function of the number of pairs. (A) <N>, time average of the nutrient concentration in the pool and the recycling ratio that is expressed as the ratio of nutrient recycled within the system to the external nutrient input. (B) Ratio of the standard deviation to <*N*> and the temporal organisation quantified by the measure ΔI_t (cf. Eq. 1).

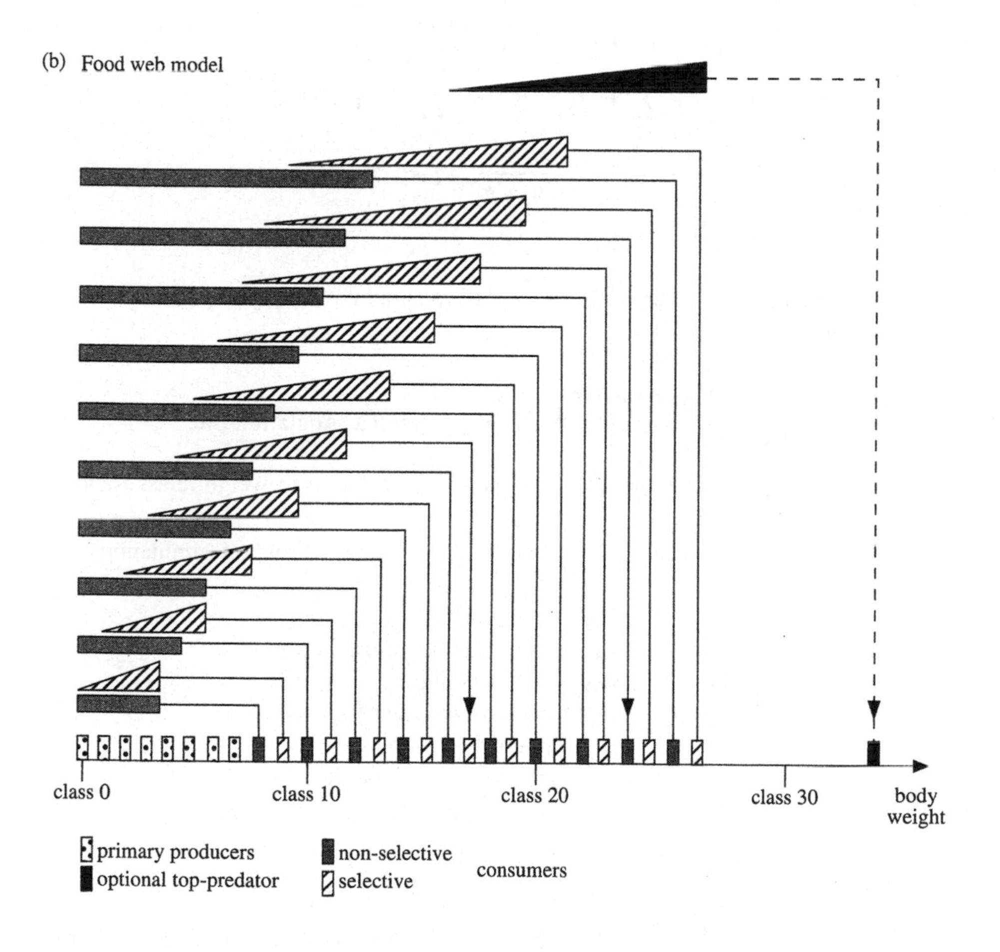

Figure 11: Trophic pathways of the two different simulation models: (A) a food chain with discrete trophic levels; (B) a food web with a high diversity in dynamic and trophic function. The body weight axis is divided into weight classes which are equally spaced on a logarithmic scale: $w_{i+1} = 2w_i$ where w_i denotes the average weight in class i. The different shadings denote different functional groups. In (B) each consumer weight class is linked to a box denoting the range of the consumer's prey window. The width of a predator's prey window increases with increasing body weight. The rectangular and triangular shape of the window denote differences in feeding modes. Non-selective filter feeders prey indiscriminately.

Table 1: Model equations and parameter values for the predator-prey pair model

(a) Mathematical equations

$$\frac{dB_i}{dt} = \Psi_i \{f(N)B_i - \rho B_i - \Omega h(B_i)P_i\} \tag{3}$$

$$\frac{dP_i}{dt} = \Psi_{i+q} \{h(B_i)P_i - (\rho + \lambda P_i)P_i\} \tag{4}$$

$$\frac{dN}{dt} = In - \sigma N - \sum_{i=0}^{n} \Psi_i f(N)B_i + rec \sum_{i=0}^{n} \Psi_i \rho (B_i + \Omega P_i) \tag{5}$$

$$where \quad f(N) = \frac{N}{N+1}, \; h(B_i) = \frac{B_i}{B_i + K}, \quad i = 0,1,\ldots,n; \; n \leq q-1.$$

The quadratic loss terms of the predators simulate density-dependent limitations of growth.

(b) Symbols and parameters

n	total number of predator-prey pairs in a simulation run
$< >$	time average
i	index of a predator-prey pair with B_i being in weight class i and P_i being in weight class $i + q$.

Model parameters and numerical values used in the model simulations presented here.

ρ	$= 0.4$	rate of respiration.
σ	$= 0.10$	rate of loss from the nutrient pool
rec	$= 0.90$	degree of recycling
K	$= 2.55$	half saturation constant for predation
In	$= 1.00$	external input of nutrient
λ	$= 0.05$	quadratic loss term of predator
ε	$= 0.25$	allometric exponent
Ψ_k	$= 2^{-\varepsilon k}$	allometric factor for the kth weight class
q	$= 10$	predator-prey weight class difference
Ω	$= 2^{-\varepsilon q}$	allometric factor for a predator relative to its prey

Allometric factors are denoted with capital Greek letters, rates are referred to with small Greek letters.

The focus is now shifted on to the effects of increasing diversity along both the horizontal and the vertical dimensions by comparing a simple food chain with a complex food web model (cf Pahl-Wostl, in press). Figure 11 depicts the pathways of trophic transfers in the food chain model (Figure 11A) with discrete trophic levels and the food web model (Fig 11B) characterized by a complex trophic network. In the following, the two models will be referred to as the chain and the web model, respectively. The models differ only in the complexity of trophic transfers. Otherwise model parameters and the chosen type of functional response are equal. Both models account for the dynamics of nutrients which are recycled within the system. Both include density dependence of the predators' mortality rates. The 'species' and thus the dynamic state variables correspond to weight classes. Model equations and parameter values are listed in Table 2.

Table 2: Equations and parameter values for the food chain and the food web model (cf. Pahl-Wostl, 1997)

Growth (γ_i) and loss (λ_i) rates for a species in weight class *i* are derived from the rates of the species in weight class 0 by allometric relationships:

$$\gamma_i = k\gamma_0 2^{-\varepsilon i} \quad and \quad \lambda_i = k\lambda_0 2^{-\varepsilon i} \tag{6}$$

where

$k = 1$ for autotrophs and $k = het$ for heterotrophs

(Moloney and Field 1989). The models are made dimensionless by expressing all rates in multiples of γ_0, the growth rate of class 0, and by expressing biomass in multiples of K_7, the half-saturation constant of the largest phytoplankton weight class. The maximum net growth rates range thus from 0.8 (class 0) to 0.0019 (class 35). In both models, the statevariables comprise the nutrient in the pool (N) and the biomasses (B_i) in the weight classes.

In the web model the dynamics of the nutrient pool is described by

$$\frac{dN}{dt} = N_{in} + rec\sum_i \lambda_i B_i - \sum_{i=0}^{7} \gamma_i \frac{N}{N+K_i} B_i - \lambda_N N \tag{7}$$

The basic web model comprises 8 classes of primary producers (classes 0-7) and20 classes of heterotrophic consumers (8-27) which feed on the prey classes within the range of their prey windows. The growth rates of primary producers depend on nutrient availability according to a Michaelis-Menten relationship. The half-saturation constant, K_i, is assumed to decrease linearly with increasing weight class. If the constant of class 7 is normalised to 1 the constant in class 0 equals 2.75. Larger phytoplankton species are thus more efficient in utilising nutrients. This pattern accounts for the combined effects of uptake dynamics and storage capacities Wirtz and Eckhardt 1996) and Gaedke, pers. comm).

The dynamics of phytoplankton species i ($0 \leq i \leq 7$) follows:

$$\frac{dB_i}{dt} = \left(\gamma_i \frac{N}{N+K_i} - \lambda_i - \gamma_j \sum_{j=8}^{26} pr_{ji} B_j \right) B_i \tag{8}$$

where

$$K_i = 1.0 + (7-i)\Delta K \tag{9}$$

and pr_{ji}, the relative contribution of prey *i* to the total predation of predator *j*, is calculated as:

$$pr_{ji} = \frac{sel_{ji}}{\sum_k sel_{jk} B_k + Kp} \tag{10}$$

The dynamics of heterotrophic species i ($8 \leq i \leq 27$) follows:

$$\frac{dB_i}{dt} = \left(\gamma_i \sum_{j<i} pr_{ij} B_j - \lambda_i - q_i B_i - \gamma_j \sum_{j>i} pr_{ji} B_j \right) B_i \tag{11}$$

where

$$\gamma_i = het\, 2^{-\varepsilon i}, \quad \lambda_i = het\, \lambda_0 2^{-\varepsilon i}, \quad q_i = het\, q_0\, 2^{-\varepsilon i}$$

The predators in the web model are characterised by weight class and by range, shape and location of their prey window along the body weight axis. A predator i feeds on all classes j that are within the range of its prey window which is defined by: $low_i = j = high_i$. Range and selectivity coefficients are a function of class and feeding mode:

filter feeders at classes $i = \{8, 10, ..., 26\}$

$$low_i = 0, \; high_i = 3 + (i-8)/2, \tag{12a}$$

all $sel_{ij} = 0.45$ for all j

raptorial feeders at classes $i = \{9, 11, ..., 27; 35\}$

$low_i = (i-9)/2, \quad high_i = 3 + (i-9)$

$$sel_{ilow_i} = 0.1 \quad and \quad sel_{ihigh_i} = 1.0 \tag{12b}$$

linear interpolation in-between the limits

An additional selective top predator ($i = 35$) may be introduced:

$$\frac{dB_i}{dt} = \left(\gamma_i \sum_j pr_{ij} B_j - \lambda_i - q_i B_i \right) B_i \tag{13}$$

The chain model:

Nutrient dynamics:

$$\frac{dN}{dt} = N_{in} + rec \sum_i \lambda_i B_i - \gamma_3 \frac{N}{N + Kn} B_3 - \lambda_N N \tag{14}$$

Dynamics of the phytoplankton species in class 3:

$$\frac{dB_3}{dt} = \left(\gamma_3 \frac{N}{N + K_n} - \lambda_3 - \gamma_{14} \frac{B_{14}}{B_3 + Kp} \right) B_3 \tag{15}$$

Dynamics of the herbivorous consumer in class 14 :

$$\frac{dB_{14}}{dt} = (\gamma_{14}\frac{B_3}{B_3 + Kp} - \lambda_{14} - 0.1 q_{14} B_{14} - \gamma_{25}\frac{B_{25}}{B_{14} + Kp})B_{14} \tag{16}$$

Dynamics of the carnivorous consumer in class 25:

$$\frac{dB_{25}}{dt} = (\gamma_{25}\frac{B_{14}}{B_{14} + Kp} - \lambda_{25} 0.1 q_{25} B_{25} - \gamma_{35}\frac{B_{35}}{B_{25} + Kp})B_{25} \tag{17}$$

A secondary carnivorous consumer in class 35 may be introduced as fourth trophic level:

$$\frac{dB_{35}}{dt} = \left(\gamma_{35}\frac{B_{25}}{B_{25} + Kp} - \lambda_{35} - q_{35}B_{35}\right)B_{35} \tag{18}$$

To be able to make quantitative comparisons, the sum of the carrying capacities of the consumers in the second and third trophic level of the chain model was set equal to the sum of the carrying capacities of the consumers in classes 8 to 27 of the web model. The aggregated carrying capacities of the total heterotrophic biomass are thus equal in the two models.

List of parameters and default numerical values

Symbol	Numerical Value	Meaning
ε	0.25	allometric exponent
λ_0	0.3	loss rate in class 0
het	1.5	ratio of heterotrophic/autotrophic rates
K_i (web)	f(class)	half-saturation nutrient concentration
ΔK	0.25	incremental change in K_i for $\Delta i = 1$
Kp	2.0	half-saturation prey concentration
Kn (chain)	1.75	half-saturation nutrient concentration
q_0	0.05	quadratic loss term of heterotrophs
s_{ji}	$0 \leq s_{ji} \leq 1$	selectivity coefficient of predator *j* for prey *i*
rec	0.8	fraction of nutrient losses which is recycled
N_{in}	0.25	external nutrient input
λ_N	0.01	rate of losses from the pool

In both the chain- and the web-model, the top predator in class 35 is optional. Comparisons of simulations with and without this additional top predator serve to investigate the influence of structural changes on the dynamics of the network as a whole. In the chain model, adding this predator corresponds to a shift from a food chain with three trophic levels to one with four. In the web model, a species does not occupy a defined trophic levels to one with four. In the web model, a species does not occupy a defined trophic level but is characterised by a trophic position which is determined as the weighted average over all feeding pathways. A species' trophic position may vary over time if the distribution of food intake over the various feeding pathways is not constant. The trophic position of the additional top predator with raptorial feeding mode may range somewhere between 3 and 4. Model versions where the predator in class 35 is absent are labelled as I, whereas model versions where this predator is present are labelled as II.

Figure 12 shows the results from model simulations for the web and the chain model with and without the additional top predator. In all simulations the nutrient input was the same and did not vary over time. A comparison of the results obtained for the chain (Figure 12A) and the web (Figure 12B) model shows that the chain model exhibits the expected effect of cascading trophic interactions. In the web model the additional top predator leads to hardly any change in the trophic structure. It is further evident that total biomass and the ratio of herbivorous to autotrophic biomass (Figure 12C) are higher in the chain-model than in the web model. Energy transfer along the trophic gradient seems to be far more efficient in a model with a web type structure than in a simple food chain.

Another major difference between the chain and the web model is the effect of fluctuations in the environment. In the model the environment is only represented by nutrient exchanges. The influence of a variable environment can be investigated by performing model simulations with a pulsed nutrient regime reflecting the annual seasonality in a temperate climate. The nutrient input was given in sine pulses over a period of 40 days at intervals of 400 days. The results obtained for a pulsed nutrient regime are shown in Figure 13. To facilitate a comparison between chain and web, the biomasses are aggregated over all primary producers and over all heterotrophic consumers for both the chain and the web model. The ratio of herbivorous to autotrophic biomass is given at the top of each figure. Figure 13A shows that the interaction of environmental variations fluctuating on a dominant frequency such as the annual cycle with a food chain oscillating on one or a few dominant frequencies has major effects. The pulsing of the nutrient input leads nearly to the extinction of the predator in level two and to a very low efficiency of transfer along the trophic gradient. The situation is different for the web model with a range of internal frequencies and redundant pathways (Figure 13B).

In conclusion, one can state that because of its dynamic and flexible network, the web model buffers, the effects of structural changes such as adding a top predator and changes in the environment such as altering nutrient input. This behaviour is in sharp contrast to the sensitivity of the chain model. A more detailed comparison can be found in (Pahl-Wostl 1997).

The framework outlined above allows the investigation of the structural complexity of ecological networks in a systematic fashion. In extension to the issues already addressed, one may for example account for differences in life history strategies. A comparison from limnetic ecosystems between daphnids and cyclopoid copepods illustrates how this can be accomplished. Daphnids are not selective and feed on a wide range of weight classes. In aggregate, the cyclopoid copepods have a similarly large prey window. However, the wide range of the prey window for copepods derives from the large ontogenetic shift in body weight during maturation. The range of individuals at a given size is much smaller in these raptorial feeders than it is for filter-feeding daphnids feeding on a large size range of food.

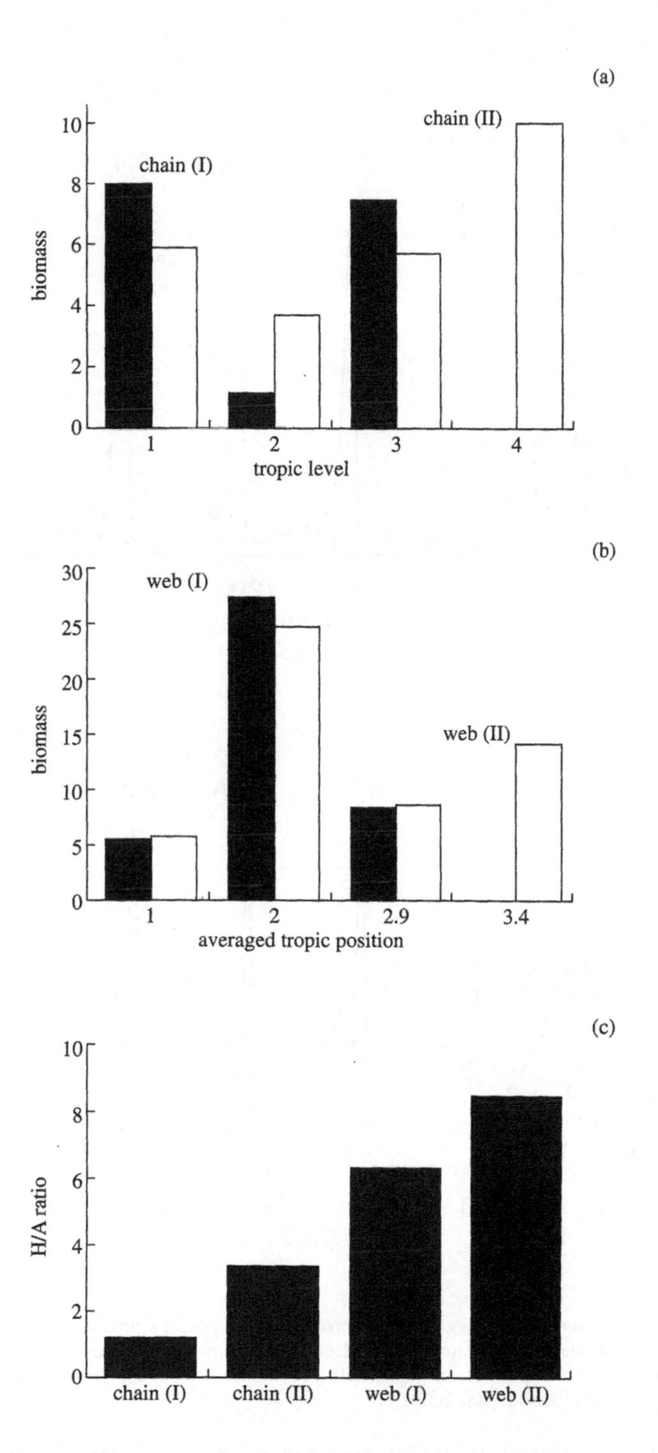

Figure 12: Distribution of biomass as a function of trophic level obtained with the food chain model (A) and as function of trophic position obtained for the food web model (B). For the web model the biomasses of the planktonic species classes were aggregated into three groups: all phytoplankton species with TP = 1, all consumers with 2 ≤ TP < 2.5 (predominantly filter feeders), all consumers with trophic positions with TP _ 2.5 (predominantly raptorial feeders). (C) The ratio of heterotrophic to autotrophic biomass obtained in the different simulations. Chain (I) refers to the three-level chain model, chain (II) refers to the four-level chain model, web (I) refers to the web model without top predator and web (I) to the web model with top predator.

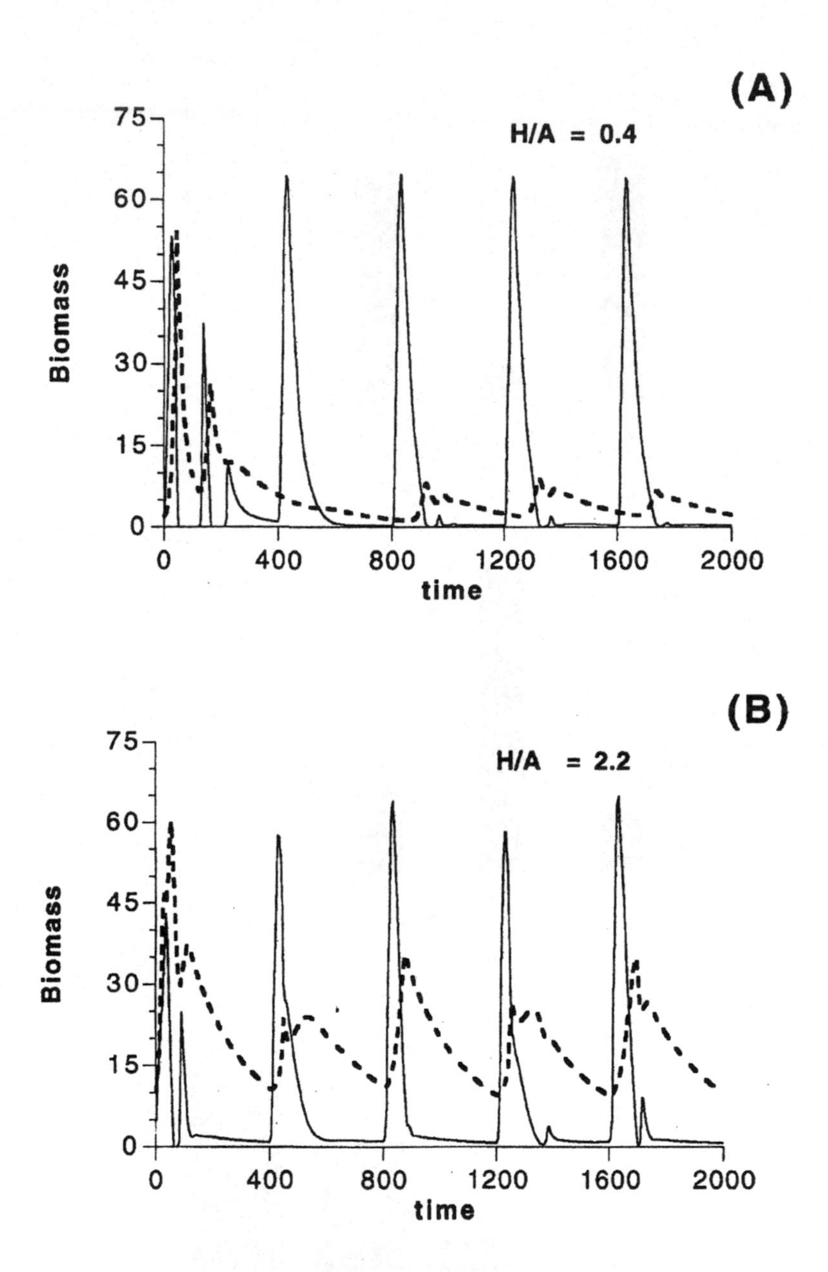

Figure 13: Temporal variations of the aggregated biomasses of all primary producers (full lines) and of all heterotrophic consumers (dashed lines) obtained for a pulsed nutrient input (sine pulses of a duration of 40 days at intervals of 400 days). (A) Three-level chain model, (B) web model without top predator. The H/A ratio was determined as average over the period 1200-2000.

For daphnids, the ontogenetic shift in size and hence the range of the prey window is of minor importance. Cyclopoid copepods move effectively one trophic level due to the more than 100 fold increase in body size during the maturation from nauplii to the adult stage. The two types of behaviour have largely different effects on the temporal organisation of an ecological network (Pahl-Wostl 1995). Daphnids couple scales, thereby reducing the variability and the dynamic diversity. This results in a stronger coupling between trophic levels. Hence, it is not surprising that the success of bio-manipulation was observed to be highly correlated with the dominance of large daphnids as major predators (Hosper and

Meijer 1993; Reynolds 1994). The simulation results support statements by (Strong 1992) that true trophic cascades are restricted to fairly low-diversity places, where great influence can issue from a single functional component.

In contrast, cyclopoid copepods may rather increase than decrease the dynamic richness. Such an increase in variability is even more pronounced for fish which cover an even wider range of the trophic gradient during ontogenetic development and where the generation times exceed the annual cycle. Carpenter and Kitchell (1993) discussed several examples of pronounced interannual effects caused by the waxing and waning of fish cohorts. Such effects add considerably to the unpredictability of ecosystem dynamics especially when one takes into account the extreme variability in reproduction. Reproductive success depends on the vagaries of the physical environment (Cushing 1990; George, Hewitt et al. 1990; Bollens, Frost et al. 1992) and a favourable timing with respect to overall community dynamics - whether the temporal niche is favourable for juveniles to thrive. Better knowledge and models of this temporal organisation will allow the derivation of potential effects that might arise from climate change.

The multi-level structure of ecosystems has implications when one goes on to consider the response of ecosystems to environmental and anthropogenic stress. Structural properties of ecosystems such as species composition seem to be more sensitive to stress than functional properties. Whole lake experiments showed functional properties such as primary production or respiration to be rather insensitive to monitor the effects of a continued exposure to stress induced by acidification (Schindler 1987; Schindler 1988; Schindler 1990). Early signs of warning could be detected at the level of species composition and morphologies, whereas functional properties showed a lagged but abrupt response. These observations can be interpreted that effects of environmental stress are first buffered by structural rearrangements to lead finally to an abrupt decline in ecosystem function. Such an effect is illustrated with simulation results from modified versions of the web and the chain model introduced in the previous section.

To investigate the influence of stress, a variable was introduced representing the effects of environmental stress which could be caused for example by a change in climate or a change in another environmental variable. Stress was assumed to reduce the growth rate of the primary producer species and thus the energy input into the network. In the web model the different primary producer species were assumed to differ in their sensitivity to stress. It was further assumed that there was a trade-off between species' growth rates in the absence of stress and their resistance to stress. The stress sensitivity of the single primary producer species in the chain model corresponds to the average of the species ensemble in the web model. Simulations of the effects of a gradually increasing stress on total biomass showed that in the web model the biomass is maintained constant over a wide stress level due to the replacement of sensitive species by less sensitive ones (Figure 14). This lead to an abrupt change in total biomass despite the gradually increasing stress level. Structural complexity may thus complicate monitoring, interpreting, and predicting the effects of stress on ecosystems, in particular, if attention is only directed at functional properties.

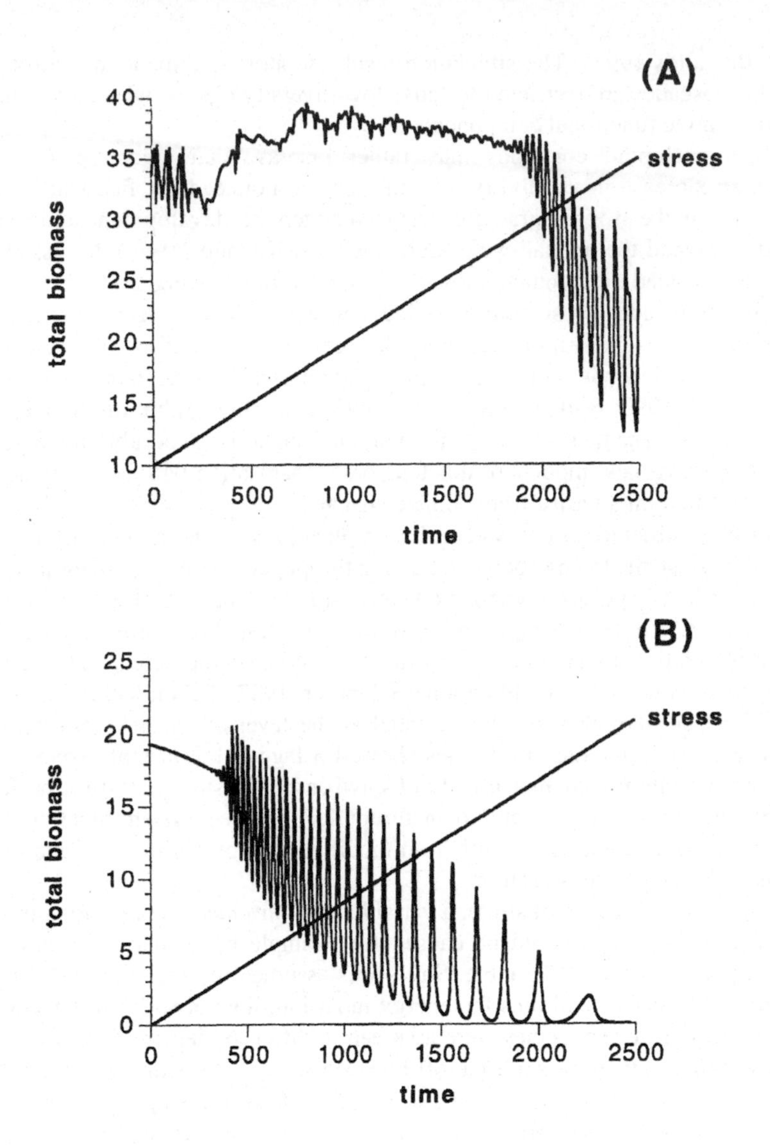

Figure 14: Results from model simulations of modified versions of the web model (A) and the chain model (B) for their response to increasing levels of stress.

5. Conclusions

Ecosystems are not static entities. Therefore, one should not seek to preserve the present state but the goals for protection must be to maintain an ecosystem's potential for self-regulation and adaptation. Understanding the relationship between the diversity of ecological networks and ecosystem properties is essential for bridging different levels of ecological organisation. Concepts for describing the complexity of ecological networks are thus a prerequisite to defining an 'ecosystem's state of health', a reference state against which to contrast the effect of anthropogenic influence. Systematic approaches for investigating the relationship between structural and functional properties as the one outlined in this chapter are required. Pelagic ecosystems may prove to be an ideal system for the investigation of general patterns which are of relevance to improve the understanding of the dynamic nature of ecosystems in general.

How to obtain reliable, empirical information on structural complexity? Taxonomic diversity is very difficult to determine empirically and may not always give the appropriate information. In the models discussed here, a functional concept for diversity based on body size was chosen. Non-taxonomic size based approaches may be of general interest in this respect. More emphasis should be given to develop concepts for functional groups that foster the understanding of structure and dynamics of ecological networks (Pahl-Wostl, 1995).

REFERENCES

Abrams, P. and J. Roth 1994. The effects of enrichment of three-species food chains with nonlinear functional response. *Ecology* 75(4): 1118-1130.

Abrams, P. and J. Roth 1994. The response of unstable food chains to enrichment. *Evolutionary Ecology* 8: 150-171.

Abrams, P. A. 1993. Effect of increased productivity on the abundances of trophic levels. *Amer. Nat.* 141(3): 351-371.

Abrams, P. A. 1996. Dynamics and Interactions in Food Webs with Adaptive Foragers. In: G. Polis and K. Winemiller (eds) *Food Webs: Integration of Pattern and Dynamics.* Pp109-112 New York, Chapman and Hall

Allen, J. C., W. M. Schaffer and Rosko, D. 1993. Chaos reduces species extinction by amplifying local population noise. *Nature* 364: 229-232.

Bollens, S., B. Frost, H. Schwaninger, C. Davis, K. Way and M. Landsteiner. 1992. Seasonal plankton cycles in a temperate fjord and comments on the match-mismatch hypothesis. *J. Plankton Res.* 14(9): 1279-1305.

Brett, M., K. Wiackowski, F Lubnow, A. Müller-Solger, J. Goldman and C. Goldman 1994. Species-dependent effects of zooplankton on planktonic ecosystem processes in Castle Lake, California. *Ecology* 75(8): 2243-2254.

Carpenter, S. and J. Kitchell 1992. Trophic cascade and biomanipulation: Interface of research and management - A reply to the comment of DeMelo et al. *Limnol. Oceanogr.* 37: 208-213.

Carpenter, S. R., T. Frost, j. Kitchell and T. Kratz 1993. Species dynamics and global environmental change: a perspective from ecosystem experiments. In: . P. Kareiva, J. Kingsolver and R. Huey (eds) *Biotic Interactions and global change*. Pp 267-279. Sinauer Assoc. Inc

Carpenter, S. R. and J. F. Kitchell, Eds. 1993. The trophic cascade in lakes. Cambridge studies in ecology. Cambridge, Cambridge University Press.

Cohen, J., F. Briand and C. Newmann 1990. Community food webs: Data and theory. New York, Springer.

Cousins, S. 1985. The trophic continuum in marine ecosystems: structure and equations for a predictive model. Ecosystem theory for biological oceanography. T. Platt and R. Ulanowicz. 213: 312-316.

Cousins, S. 1987. The decline of the trophic level concept. *Trends Ecol. Evol.* 2: 312-316.

Cushing, D. 1990. Plankton production and year-class strength in fish populations: an update of the match/mismatch hypothesis. *Advances Mar. Biol.* 26: 250-290.

DeAngelis, D. 1992. Dynamics of Nutrient Cycling and Food Webs. London, Chapman and Hall.

DeAngelis, D. L., S. M. Bartell and A.L Brenkert 1989. Effects of nutrient recycling and food-chain length on resilience. *Amer. Nat.* 134: 778-805.

DeMelo, R., R. France and D. McQueen 1992. Biomanipulation: Hit or myth? *Limnol. Oceanogr.* 37: 192-207.

Diehl, S. 1993. Relative consumer sizes and the strengths of direct and indirect interactions in omnivorous feeding relationships. *Oikos* 68(1): 151-157.

Elser, J., C. Luecke, M. Brett and C. Goldman 1995. Effects of food web compensation after manipulation of rainbow trout in an oligotrophic lake. *Ecology* 76(1): 52-69.

Forbes, S. A. 1887. The lake as a microcosm. *Illionois Nat. Hist. Surv. Bull.* 15: 537-550.

Fretwell, S. 1987Food chain dynamics: the central theory of ecology? *Oikos* 50: 291-301.

Gaedke, U. 1992. Identifying ecosystem properties: a case study using plankton biomass size distributions. *Ecol. Model.* 63: 277-298.

Gaedke, U. 1992. The size distribution of plankton biomass in a large lake and its seasonal variability. *Limnol. Oceanogr.* 37: 1202-1220.

Gaedke, U., D. Straile and C. Pahl-Wostl. 1996. Trophic Structure and Carbon Flow Dynamics in the Pelagic Community of a Large Lake. Food Webs.In: . G. Polis and K. Winemiller.(eds) *Integration of Pattern and Dynamics*. Pp109-112. New York, Chapman and Hall.

George, D. G., D. P. Hewitt , J.W. Lund and W.J. Smyly 1990. The relative effects of enrichment and climate change on the long-term dynamics of Daphnia in Eastwaite Water, Cumbria. *Freshwater Biology* 23: 55-70.

Hairston, N. F., F. Smith and L. Slobodkin 1960. Community structure, population control and competition. *Amer. Nat.* 94: 421-425.

Hansson, L.-A. 1992. The role of food chain composition and nutrient availability in shaping algal biomass development. *Ecology* 73: 241-247.

Hastings, A. 1988. Food web theory and stability. *Ecology* 69: 1665-1668.

Hastings, A. and K. Higgins. 1994. Persistence of Transients in Spatially Structured Ecological Models. *Science* 263: 1133-1136.

Hastings, A. and T. Powell 1991 Chaos in a three-species food chain. *Ecology* 72: 896-903.

Hastings, P. A. 1996. What Equilibrium Behaviour of Lotka-Volterra Models Does Not Tell Us About Food Webs. In: G. Polis and K. Winemiller (eds). *Food Webs: Integration of Pattern and Dynamics*. Pp211-217. New York, Chapman and Hall.

Hosper, H. and M.-L. Meijer 1993. Biomanipulation, will it work for your lake? A simple test for the assessment of chances for clear water, following drastic fish-stock reduction in shallow, eutrophic lakes. *Ecological Engineering* 2: 63-72.

Lawton, J. and P. Warren 1988. Static and dynamic explanations for patterns in food webs. *Trends Ecol. Evol.* 3: 242-245.

Lodge, D., J. Barko, D. Strayer, J. Melack, G. Mittelbach, R. Howarth, B. Menge and J. Titus. 1988. Spatial Heterogeneity and Habitat Interactions in Lake Communities. In: S. Carpenter (ed). *Complex interactions in lake communities.*pp229-260. New York, Springer.

Mateev, V. 1995. The dynamics and relative strengths of bottom-up vs. topdown impacts in a community of subtropical lake plankton. *Oikos* 73(1): 104-108.

Mazmuder, A. 1994. Patterns of algal biomass in dominant odd- vs. even-link lake ecosystems. *Ecology* 75(4): 1141-1149.

McQueen, D., E. Mills, J. Fourney, M. Johannes and J. Post. 1992. Trophic level relationships in pelagic food webs: comparisons from long-term data sets for Oneida Lake, New York (USA), and Lake St. George, Ontario (Canada). *Can. J. Fish. Aquat. Sci.* 49: 1588-1596.

Moloney, C. and J. Field 1989. General allometric equations for rates of nutrient uptake, ingestion, and respiration in plankton organisms. *Limnol. Oceanogr.* 34: 1290-1299.

Oksanen, L. 1991. Trophic levels and trophic dynamics: a consensus emerging. *Trends Ecol. Evol.* 6: 58-60.

Pahl-Wostl, C. 1993. Food webs and ecological networks across spatial and temporal scales. *Oikos* 66: 415-432.

Pahl-Wostl, C. 1995. The Dynamic Nature of Ecosystems: Chaos and Order Entwined. Chichester, Wiley.

Pahl-Wostl, C. 1997. Dynamic structure of a food web model: comparison with a food chain model. *Ecolog. Modelling* 100: 103-123.

Paine, R. T. 1988. Food webs: road maps of interaction or grist for theoretical development? *Ecology* 69: 1648-1654.

Persson, L., J. Bengtsson, B. Menge and M. Power. 1996. Productivity and Consumer Regulation - Concepts, Patterns, and Mechanisms. Food Webs. In: G. Polis and K. Winemiller (eds). *Integration of Pattern and Dynamics*. Pp109-112 New York, Chapman and Hall.

Persson, L., S. Diehl, L. Johansson, G. Andersson and S. Hamrin. 1992. Trophic interactions in temperate lake ecosystems: a test of food chain theory. *Amer. Nat.* 140: 59-84.

Peters, R. 1983. The implications of body size. Cambridge, England., Cambridge University Press.

Pimm, S., J. Lawton and J. Cohen- 1991. Food web patterns and their consequences. *Nature* 350: 669-674.

Pimm, S. L. 1984. The complexity and stability of ecosystems. *Nature* 307: 321-326.

Polis, G. and R. Holt 1992. Intraguild Predation: The Dynamics of Complex Trophic Interactions. *Trends Ecol. Evol.* 7(5): 151-154.

Polis, G. and K. Winemiller, Eds. 1996. Food Webs: Integration of Pattern and Dynamics. New York, Chapman and Hall.

Porter, K. G., H. Paerl, R Hodson, M. Pace, J. Priscu, B. Rieman, D. Scavia and J. Stockner. 1988. Microbial interactions in lake communities. In: S. Carpenter. (ed) *Complex interactions in lake communities*. Pp 109-227. New York, Springer.

Reynolds, C. 1994. The ecological base for the successful biomanipulation of aquatic communities. *Arch. Hydrobiol.* 130(1): 1-33.

Reynolds, C. 1996. Vegetation Processes in the Pelagic: A Model for Ecosystem Theory. Oldendorf, Ecology Institute.

Schindler, D. W. 1987. Detecting ecosystem responses to anthropogenic stress. *Can. J. Fish. Aquat. Sci.* 44(suppl): 6-25.

Schindler, D. W. 1988. Effects of acid rain on freshwater ecosystems. *Science* 239: 149-157.

Schindler, D. W. 1990. Experimental perturbations of whole lakes as tests of hypotheses concerning ecosystem structure and function. *Oikos* 55: 25-41.

Schulze, E.-D. and H. A. Mooney, Eds. 1993. Biodiversity and Ecosystem Function. Ecological Studies. Berlin, Springer.

Sheldon, R., A. Prakash and W. Sutcliffe. 1972. The size distribution of particles in the ocean. *Limnol. Oceanogr.* 17: 327-340.

Sheldon, R., W. Sutcliffe and A. Prakash 1977. Structure of the pelagic food chain and relationship between plankton and fish production. *J. Fish. Res. Board. Can.* 34: 2344-2353.

Simon Moffat, A. 1996. Biodiversity is a Boon to Ecosystems, not Species. *Science* 271: 1497.

Slobodkin, L., F. Smith and N. F. Hairston 1967. Regulation in terrestrial ecosystems, and the implied balance of nature. *Amer. Nat.* 101: 109-124.

Sprules, W. and M. Munawar 1986. Plankton Size Spectra in relation to ecosystem productivity, size, and perturbation. *Can. J. Fish. Aquat. Sci.* 43: 1789-1794.

Sprules, W., S.B. Brandt, D.J. Stewart, M. Munwar, E.H. Jin and J. Love. 1991. Biomass size spectrum of the Lake Michigan pelagic food web. *Can. J. Fish. Aquat. Sci.* 48: 105-115.

Sterner, R. W. 1990. The ratio of nitrogen to phosphorus resupplied by herbivores: Zooplankton and the algal competitive arena. *Amer. Nat.* 136: 209-229.

Straile, D. 1994. Die saisonale Entwicklung des Kohlenstoffkreislaufes im pelagischen Nahrungsnetz des Bodensees, University of Konstanz.

Strong, D. 1992. Are trophic cascades all wet? Differentiation and donor-control in speciose ecosystems. *Ecology* 73: 747-754.

Thienemann, A. 1925. Die Binnengewässer Mitteleuropas. Stuttgart, Schweizerbart'sche Verlagsbuchhandlung.

Tilman, D. 1996. Biodiversity: Population versus Ecosystem Stability. *Ecology* 77(2): 350-363.

Tilman, S. and J. Downing 1994. Biodiversity and stability in grasslands. *Nature* 367: 363-365.

Vanni, M. 1987. Effects of nutrients and zooplankton size on the structure of a phytoplankton community. *Ecology* 68(3): 624-635

Vanni, M. 1996. Nutrient Transport and Recycling by Consumers in Lake Food Webs: Implications for Algal Communities. Food Webs. In: G. Polis and K. Winemiller (eds) *Integration of Pattern and Dynamics*. Pp 109-112. New York, Chapman and Hall.

Vanni, M. and D. Findlay 1990. Trophic cascades and phytoplankton community structure. *Ecology* 71: 921-937.

Vanni, M. and C. D. Layne 1997. Nutrient recycling and herbivory as mechanisms in the 'top-down' effect of fish on algae in lakes. *Ecology* 78(1): 21-40.

Vanni, M., C. D. Layne and S.E Arnott. 1997. 'Top-Down' trophic interactions in lakes: Effects of fish on nutrient dynamics. *Ecology* 78(1): 1-20.

Wirtz, K.-W. and B. Eckhardt 1996. Effective variables in ecosystem models with an application to phytoplankton succession. *Ecol. Model.* 92(1): 33-54.

Witek, Z. and A. Krajewaska-Soltys 1989. Some examples of the epipelagic plankton size structure in high latitude oceans. *J. Plankton Res.* 11(6): 1143-1155.

II.7.3 Ecosystem Theory: Network Environ Analysis

Brian D. Fath and Bernard C. Patten

1. Environment versus Environs

Environment is used broadly to represent many different ideas depending on the situation. In network analysis, environment has two interpretations depending on scale. Within the boundaries of a predefined system, an object's local environment is made of all the other objects in the system with which it interacts. In this context, each object is seen as a partition of two mutually exclusive halves, one comprising the inflow and the other, outflow (von Uexküll, 1926). This differs from the standard organism-environment duality which separates the organism from the environment (Figure 1a). The traditional view underpins many years of reductionist science in which objects are studied as separate from their natural environment. Even ecology, which is the study of organism-environment relationships, tends to focus on the impact of environment on biota and vice versa, but less explicitly on the biotic object as an integrated part of an entire ecosystem. Kareiva (1994) reviewed 1,253 papers published in *Ecology* between 1981 and 1990 and almost one thousand of these dealt with four or fewer species. This does not represent a holistic investigation of environmental structure and function. In network environ analysis, this view is modified such that each component consists of two system-bounded *environs*, one which acts on the component, and the other which is acted upon by it (Figure 1b) (Patten, 1978). Each of these intersects the environs of the other components in a system. Therefore, the component itself is part of two environs, one received and one generated (Patten, 1978, 1982). If properly constructed, then an environ comprises a partition (exhaustive and mutually exclusively) of the surrounding environment. Therefore, the input or received environ is the set of all within-system interactions leading up to the component, and the output environ is the source or generator of new flows and future interactions. In this view, an object is inexorably linked to its surrounding world through its afferent input and efferent output environs.

Figure 1: Environment-object (=system) duality (a) versus input-output environ duality (b)

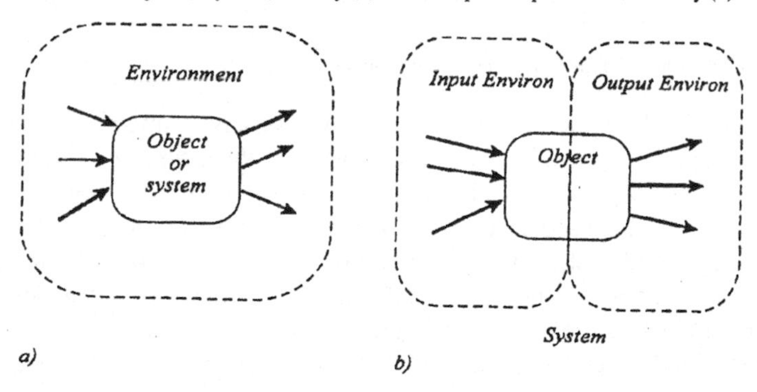

The second interpretation of environment applies to a global scale, where it more generally refers to all objects outside a predefined system's boundaries or that which is external to a model. The within-system objects and transactions are imbedded in a larger context represented by the environment. Here, an environ represents system-bounded component interactions as mediated by input and output transactions. An input environ component interactions as mediated by input and output transactions. An input environ includes transactions within the system boundaries contributing output to the environment across the boundary. Similarly, an output environ comprises the transactions to components within the system generated by inputs from the environment. A conceptual model should contain all problem – or concept – relevant entities within the ecosystem boundaries. A simple three-component model shows the dichotomy between environs and environment (Figure 2).

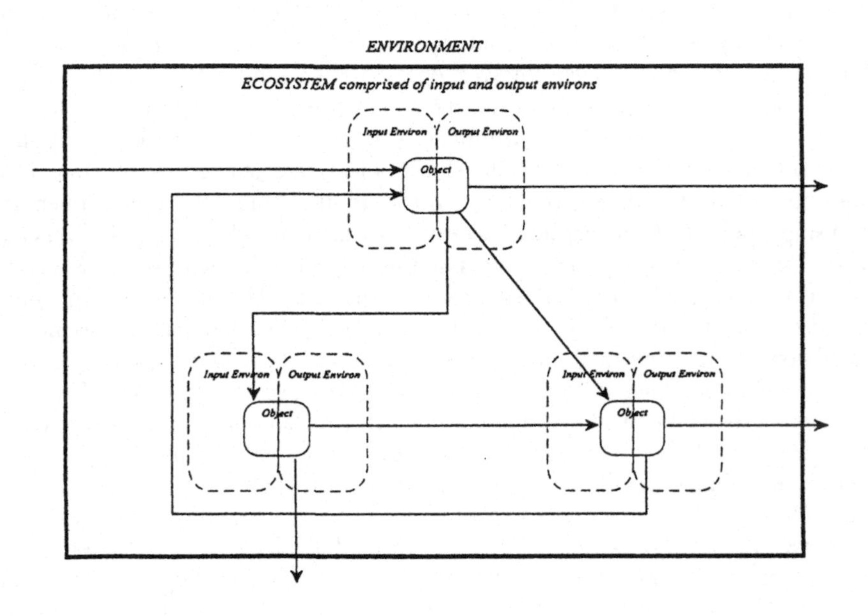

Figure 2: Component environs embedded in a larger environment

In addition to the above description of environs and environment, it is important to keep in mind, that based on physical constraints, there are three types of systems: isolated, closed and open (White *et al.*, 1992). Isolated systems have no energy or matter transfer with their environment and closed systems exchange energy but not matter. Open systems exchange both energy and matter with their environment. Every biological entity is an open system. Non-isolated systems are not at thermodynamic equilibrium because they receive, transfer, and dissipate energy. Ecological systems exist because of this energy or chemical gradient which is needed to maintain life, so at a minimum all ecological systems are non-isolated. Since most ecosystems also exchange matter, they can, for all practical purposes, be considered open systems. The energy gradient causes all open systems to exhibit behaviour, therefore, all ecological systems exhibit behaviour. This openness resurfaces later as an important contributor to the computations of network analyses and to the overall positive property of systems exhibited in *network synergism*.

2. System structure and function

Network environ analysis is predicated on a conceptual flow-storage model of any system which can be represented as a combination of conservative flows (transactions) and components. The flow diagram associated with the ecosystem contains both the structure (paths and nodes) and function (flows associated with these paths and nodes). A main advantage of network analysis is that it allows for the investigation of the relationships between components of a system without removing them from the system. A disadvantage is that it can only be applied to models. Nevertheless, using this whole-system analysis, it is possible to avoid at least in principle the reductionist shortcoming of isolating parts from an integrated whole. Direct and indirect effects are accounted for and the result is a holistic interpretation of system structure and function.

Mathematically, system structure is represented as a graph of the component interconnections. An adjacency matrix, $\mathbf{A} = (a_{ij})$ is a one-to-one mapping of the nodes and edges in the graph. It has a one in elements (i, j) and (j, i) if and only if an arc (connection) exists between vertices (objects) i and j. If the connections between the objects are directed edges are arcs, then the graph is a directed graph, or digraph. In the trophic relation i eats j, the physical flow of material is from j→i. Therefore, by convention, the adjacency matrix of a flow digraph has a one in the (i, j) position if and only if there is a flow from j→i. For example, the digraph in Figure 3 shows the flow transactions in a six-component ecological model of an oyster reef community (Dame and Patten, 1981).

This model is used in the appendix to demonstrate the four network properties. The information in the digraph is represented identically in matrix format in Equation (1). Note the orientation is from columns (j) to rows (i).

$$A = \begin{bmatrix} 0 & 0 & 0 & 0 & 0 & 0 \\ 1 & 0 & 0 & 1 & 1 & 1 \\ 0 & 1 & 0 & 0 & 0 & 0 \\ 0 & 1 & 1 & 0 & 0 & 0 \\ 0 & 1 & 1 & 1 & 0 & 0 \\ 1 & 0 & 0 & 0 & 1 & 0 \end{bmatrix} \qquad (1)$$

It is necessary to have the structural information in matrix format to perform the mathematical manipulations required for path analysis and enumeration of indirect pathways.

System function can be defined as the flow and storage of mass-energy between and at components. Function so seen is therefore, a dimensional quantity. In the case of ecological models, transactions usually have the dimensions of mass per unit area and unit time (M/L^2T), or mass per unit volume and unit time (M/L^3T). A dimensional flow matrix, $\mathbf{F} = (f_{ij})$, has a quantity of flow from j→i in the (i, j) element. There are an infinite number of possible flow matrices associated with each structure. Equation (2) is one example of a steady-state flow matrix corresponding to the adjacency matrix in Equation (1) and Figure 3.

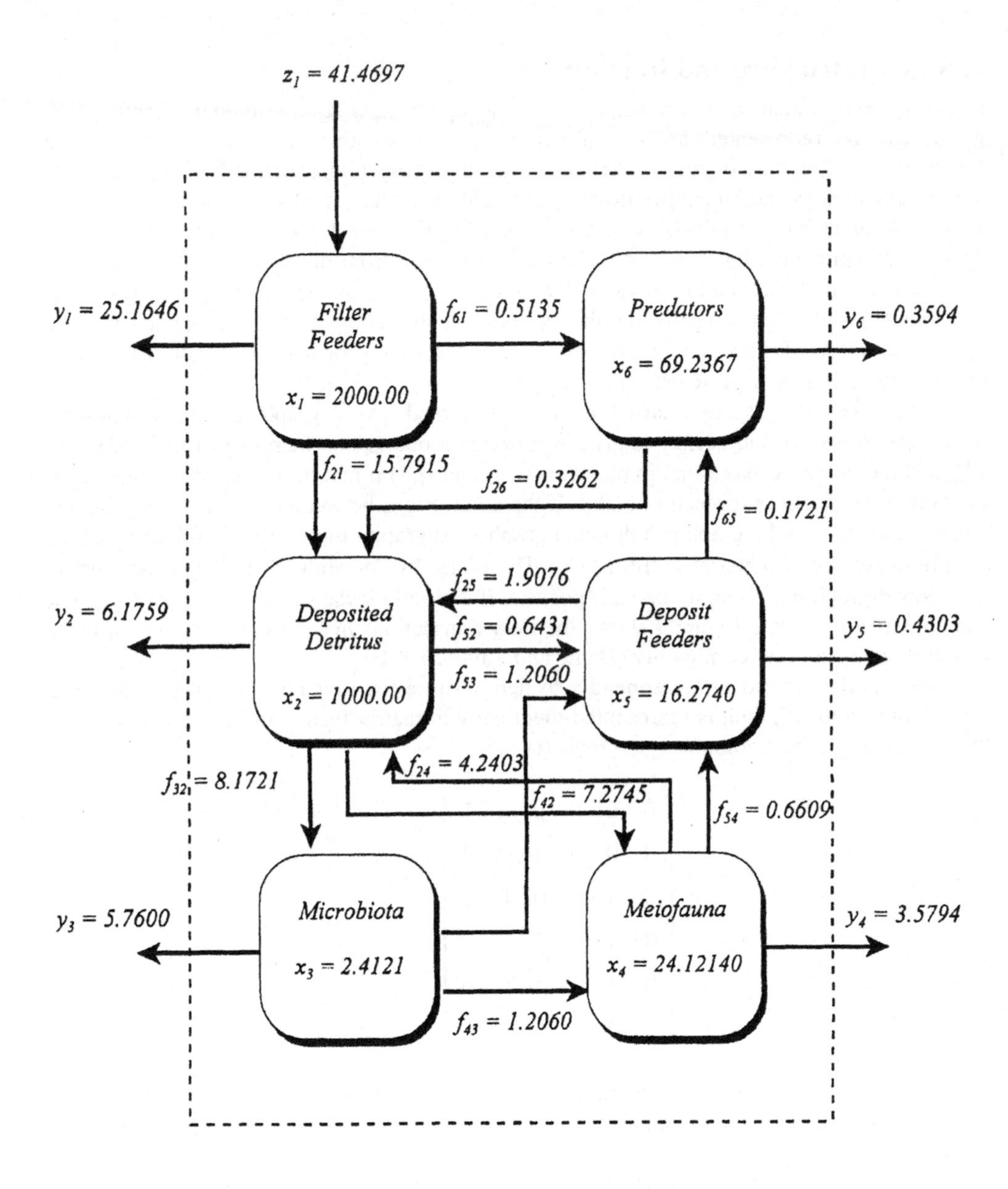

Figure 3: Six component empirical model of an oyster reef community (after Dame and Patten, 1981)

$$F=\begin{bmatrix} 0 & 0 & 0 & 0 & 0 & 0 \\ 15.7915 & 0 & 0 & 4.2403 & 1.9076 & 0.3262 \\ 0 & 8.1721 & 0 & 0 & 0 & 0 \\ 0 & 7.2745 & 1.2060 & 0 & 0 & 0 \\ 0 & 0.6431 & 1.2060 & 0.6609 & 0 & 0 \\ 0.5135 & 0 & 0 & 0 & 0.1721 & 0 \end{bmatrix} \quad (2)$$

Input is represented by a $n\times 1$ vector, **z**. Here we have **z** = $[41.45,0, 0, 0, 0, 0]^T$. When this system is at steady-state, the $1\times n$ vector, **y**, gives the outflow from each component. (**y** = [25.16, 6.18, 5.76, 3.58, 0.43, 0.36]). Steady-state storage values are represented as a $1\times n$ vector, **x** = [2000, 1000, 2.4, 24.1, 16.3, 69.2]. Together, the flow and storage values transform the structural digraph into a functional systems model. The combination of system structure and function underlies system behaviour and is sufficient to determine the values of the network properties. The contribution of indirect effects to behaviour is identified using network analysis.

In environ analysis, as stated previously, structure and function are analysed using mathematical models of flows and storages. Structural *path analysis* enumerates associated pathways (Patten *et al.*, 1982). There are presently three main lines of research in functional analysis: *throughflow-specific flow analysis* (Hannon, 1973), *storage-specific flow analysis* (Matis and Patten, 1981), and *utility analysis* (Patten, 1991) (Figure 4).

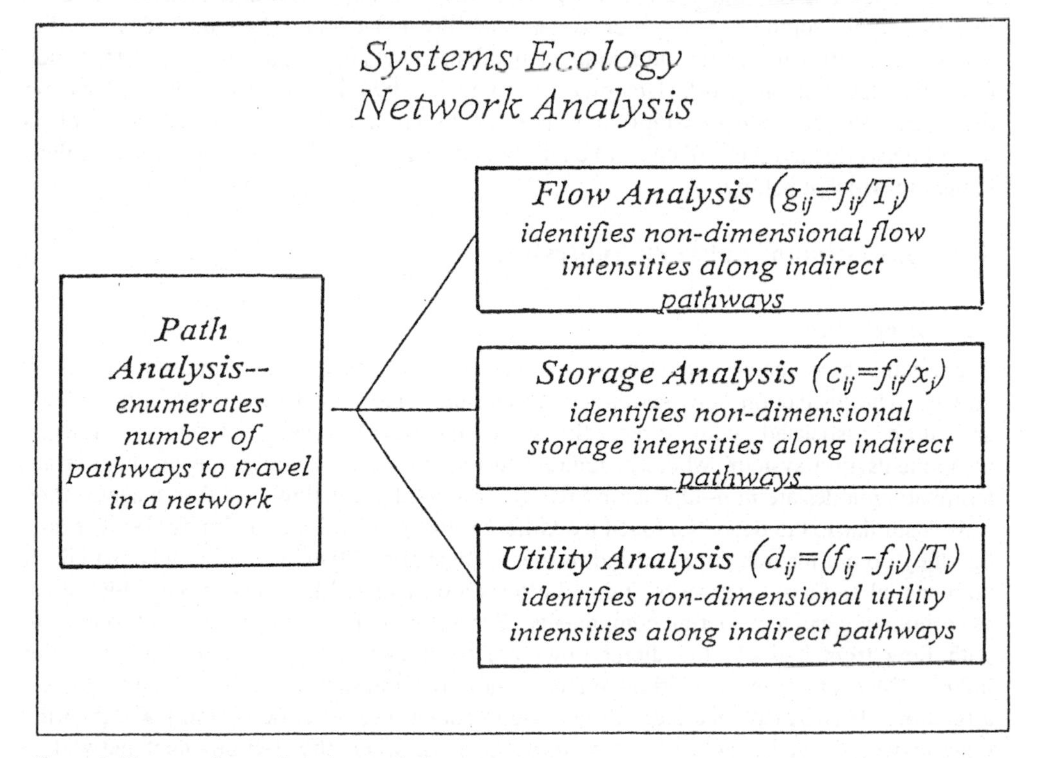

Figure 4: Diagram of network environ analysis

Throughflow is the sum of flows into or out of a component. Each of these functional analyses is based on a different non-dimensional normalisation of the flow or storage characteristics of the network. Normalisation is the process by which a set of values is divided by a constant. This makes the values relative to a standard scale, between zero and one for flow and storage analyses and between minus one and one for utility analysis. The result of normalisation is interpreted as a value intensity. Here, either the throughflows or the storages are used to normalise the flows, which thereby become throughflow-specific and storage-specific intensity values. In throughflow analysis for output environs, the flows from component j→i, f_{ij}, are normalized by the donor component throughflows, T_j, where

$$T_j = \sum_{i=0}^{n} f_{ij} = \sum_{i=0}^{n} f_{ji} \tag{3}$$

at component j, ($g_{ij} = f_{ij}/T_j$). For input environ throughflow analysis, which is the form introduced by Hannon (1973) and used by most other network analysts, the normalisation is respect to recipient component throughflows. Input and output flows are denoted by f_{0j} and f_{j0}, respectively. In output environ storage analysis, the flows are normalised by the steady-state storage at the originating component j, ($c_{ij} = f_{ij}/x_j$). For input environ analyses, recipient component normalisation is used. A time step is needed to make this quantity dimensionless and bounded between zero and one, ($p_{ij} = i_{ij} + c_{ij}\Delta t$, where i_{ij} are the elements of the identity matrix). In throughflow-specific utility analysis, the *net* flow from j to i is normalised by the steady-state throughflow at i, ($d_{ij} = (f_{ij} - f_{ji})/T_i$). A storage-specific utility analysis has thus far not been performed. Throughflow analyses identify properties of networks such as cycling rates and indirect contributions (Hannon 1973, Finn 1976, Patten *et al.*, 1982, Higashi and Patten 1989, Ulanowicz 1986). Storage analyses evaluate retention time, turnover rates, and system stability (Matis and Patten, 1981). Utility analysis identifies the direct and indirect qualitative relationships (such as competition, mutualism, etc.) in a network (Ulanowicz and Puccia, 1990, Patten 1991, 1992). Together, these four analyses, one structural and three functional, allow for the investigation of the role of networks, indirect effects, and environment on an object's behavior within a holistic system to which it belongs.

3. Formulation of network analysis

3.1 Indirect effects

There is much confusion and ambiguity among the many definitions proposed for indirect effects. The distinction between direct and indirect effects used here is that direct effects are those associated with direct flows of material between adjacently connected components in a system, whereas, indirect flows are those in which the originating and terminating nodes are non-adjacent by reason of non-adjacent flow in either space or time. This is similar to the definition used by Miller (1994) and Miller and Travis (1996) except that they do not include temporal indirectness (Higashi and Patten, 1989), only spatial. In both spatial and temporal cases, indirect effects are mediated by transactions which relate the originating and terminating components. For example, in a three component food chain with flow from i→j→k, the direct flows are from i→j and j→k, and a second-order, indirect flow is from i to k. Although there is no direct transaction between k and i, there is a relation. If analysed separately as two 2-component networks (k→j and j→i), then this relation would not be evident. It is only apparent when the 3-component network is analysed as one system. This is the meaning of holism, and understanding ecosystem behaviour and relations requires such a perspective. Path analysis counts (and to some extent) identifies the direct and indirect pathways in a network.

3.2 Structural Path Analysis

As stated above, an adjacency matrix is a matrix of zeros and ones, with a one in the ith row and j^{th} column if and only if there is a flow from j to i. The adjacency matrix gives the structural information associated with the system of interest. The complex path structure within a system is often not apparent. When the adjacency matrix is raised to a particular power, the elements of the new matrix are equal to the number of pathways of a length commensurate with the power (e.g., Patten et al., 1982, Patten 1985). For example, the number of paths to travel from node j to node i in two steps is given in the (i, j) element of $\mathbf{A}^2$. A step is a single node-arc-node sequence along a path (e.g., j→i). Similarly, the number of paths of length three from j to i is the (i, j) element of $\mathbf{A}^3$, and the number of paths of length m from j to i is the (i, j) element of $\mathbf{A}^m$. The indirect pathways are enumerated in the higher-order powers (m>1) of the adjacency matrix. Matrix multiplication gives the total number of paths between each node pair. For example, from $\mathbf{A}^4$ for the matrix in Equation (1), we see there are two ways to get from component one to component three in exactly four steps:

$$A^4 = \begin{bmatrix} 0 & 0 & 0 & 0 & 0 & 0 \\ 6 & 7 & 5 & 6 & 6 & 4 \\ 2 & 4 & 2 & 3 & 2 & 2 \\ 3 & 6 & 4 & 4 & 3 & 2 \\ 5 & 8 & 6 & 6 & 5 & 3 \\ 3 & 3 & 2 & 3 & 3 & 2 \end{bmatrix} \tag{4}$$

Notice also, that the first row is entirely zero, meaning there are no pathways of length four which terminate in the first component. This is true of all higher order pathways in this example because component one only receives input from outside the system boundary, not from within the system. In its relation to the other internal components, it would be considered a taboo state (Kemeny and Snell, 1960), meaning it is unreachable from other components within the system boundary. The opposite of a taboo state is an absorbing state (Kemeny and Snell, 1960) which is unobservable because there are no emanating pathways from an absorbing state. Simply put, a taboo state has no entrances and an absorbing state has no exits.

Another important concept is that of a cycle: a path which ends at the same node where it began. The simplest cycle has a length one and is called a loop (or self-loop). A loop can be used to represent storage (interpreted as non-flow) at that component. Loops and cycles can be embedded within other cycles. They contribute to the complexification of networks in two ways. First, they increase the total number of pathways between two components for a given path length. Second, cycles play a major role in the increase of the number pathways between two nodes as path length increases. Using network analysis, it is possible to determine if the contribution to throughflow or storage at a terminal node is from pathways which have no cycles touching that node (first passage or mode 1) or pathways with cycles at that node (recycled or mode 2). In addition, we consider dissipative or mode 3 pathways which are equal and opposite to mode 1 pathways (Higashi et al., 1993). The distinction between mode 1 and mode 2 is important because mode 2 throughflow or storage can be significant. It represents the added contribution from the system interactions and is only identified from a network perspective. Taboo and absorbing states cannot be intermediate steps in cycles and are therefore limited to first passage (taboo) or dissipative (absorbing) pathways only. In a system like the oyster reef model (Figure 3) that is well connected, the number of paths between the non-taboo and non-absorbing components increases as the path length increases. For sufficiently long path lengths, the rate of increase from path length m to m+1 is equal to the largest positive eigenvalue of **A**.

3.3 Functional Analysis

The same premise for identifying the contribution along indirect pathways is used in the functional analyses as well. A similar procedure is performed in all three functional analyses to identify system-wide interactions based on the contributions of paths of all lengths which arise from normalised direct interaction matrices. When such matrices are raised to a particular power, this gives the functional influence (expressed nondimensionally) due to all paths of lengths commensurate with the power. The network properties are based on an interpretation of the infinite power series associated with the nondimensional quantity of interest (throughflow, storage, or utility) where the higher powered terms in the series correspond to the indirect contributions of those orders. Integral interaction matrices are found by summing the infinite power series of the direct interaction matrices. The definitions of the elements of direct interaction matrices **G**, **P**, and **D**, respectively, are given earlier as $g_{ij} = f_{ij}/T_j$, $p_{ij} = i_{ij} + c_{ij}\Delta t$ (with $c_{ij} = f_{ij}/x_j$), and $d_{ij} = (f_{ij} - f_{ji})/T_i$. In Equations (5), the integral flow intensity matrix, **N**, which accounts for the contribution of all direct and indirect interactions, is found by summing all powers of **G**. **N** is interpreted to be an integral flow matrix because its elements represent the total nondimensional flow expressed across all path lengths. The integral storage, **Q**, and utility, **U**, matrices are similarly derived.

$$
\begin{array}{llcccccccccccl}
\text{Flow:} & \mathbf{N} & = & \mathbf{I} & + & \mathbf{G} & + & \mathbf{G}^2 & + & \mathbf{G}^3 & + & \mathbf{G}^4 & + \ldots & \\
\text{Storage:} & \mathbf{Q} & = & \mathbf{I} & + & \mathbf{P} & + & \mathbf{P}^2 & + & \mathbf{P}^3 & + & \mathbf{P}^4 & + \ldots & (5) \\
\text{Utility:} & \mathbf{U} & = & \mathbf{I} & + & \mathbf{D} & + & \mathbf{D}^2 & + & \mathbf{D}^3 & + & \mathbf{D}^4 & + \ldots &
\end{array}
$$

intergral = *initial input* + *direct* + *indirect*

When the power series in Equations (5) converge, Equations (6) can be used to calculate the integral matrix.

$$
\begin{array}{lll}
\text{Flow:} & N = (I-G)^{-1} & \\
\text{Storage:} & Q = (I-P)^{-1} & (6) \\
\text{Utility:} & U = (I-D)^{-1} &
\end{array}
$$

The power series methodology expressed in (5) thus allows for a quantitative description of the indirect effects as mediated by the flows, storages, and net flows in a well-connected network. Of the three analyses in Equations (5), the throughflow analysis had been used to identify three network properties (Table 1): network amplification, network homogenisation, and dominance of indirect effects. Utility analysis has been used to identify one property, network synergism. Future investigation of these analyses, in particular storage and storage specific utility analyses, may reveal additional network properties. An example worked out in detail is provided for the oyster reef model to demonstrate the development of this theory and the four network properties.

Table 1: Four emergent network hypotheses

Property	Definition	Test
Amplification	Components in a network get back more than they put in	$n_{ij} > 1$ for i_j
Homogenisation	Action of the network makes the flow distribution more uniform	$\mathrm{CV} = \dfrac{\sum_{j=1}^{n}\sum_{i=1}^{n}(\bar{m} - m_{ij})^2}{(n-1)\,\bar{m}}$
Synergism	System-wide relations in the network are inherently positive	$\dfrac{b}{c} = \dfrac{\Sigma\text{+utility}}{\lvert\Sigma\text{-utility}\rvert}$
Dominance of Indirect Effects	A system receives more influence from indirect processes than from direct processes	$\dfrac{i}{d} = \dfrac{\sum_{i=1}^{n}\sum_{j=1}^{n}(n_{ij} - i_{ij} - g_{ij})}{\sum_{i=1}^{n}\sum_{j=1}^{n} g_{ij}}$

3.4. Relation between throughflow and storage

A simple test shows that the integral matrix multiplied by the input vector returns the throughflow vector (**T** = **Nz**). This confirms that *each and every* path is unique and contributes to the overall throughflow in the network. More important, this shows that the integral matrix does not involve multiple counting of paths as is sometimes challenged (Wiegert and Kozlowski, 1984, Pilette, 1989) because each one is needed to reconstitute the throughflow. Similarly, the steady-state storage values are retrieved from mapping the inputs into the integral storage matrix (**x** = **Sz**, where **S** = -$\mathbf{C}^{-1}$ = **Q**Δt). The throughflow and storage mappings are related through turnover rates and times implicit in the following correspondences (Higashi *et al.*, 1993):

$$g_{ij} = p_{ij}/(1-p_{jj}) \tag{7}$$

$$n_{ij} = q_{ij}(1-p_{ii}) \tag{8}$$

Also, the turnover rates, τ_i^{-1}, are related to the diagonal elements of the non-dimensional storage matrix by $c_{ii} = -\tau_i^{-1}$. This turnover rate represents the quantity of flow needed to replace (1-1/e) of storage in the ith component. The inverse of the turnover rate is the turnover time, τ_i. The flows give rise to the storage as well, whereas, the storage acts as a impedance to flow similarly to a resistor in an electrical circuit. The flow and storage therefore are related to each other through the turnover time associated with the storage. Using these relations, we can rewrite the above mapping of throughflows in terms of storage coefficients:

$$T_{ij} = (n_{ij})z_j = (\tau_i^{-1} s_{ij})z_j \tag{9}$$

and also storages in terms of the throughflow coefficients:

$$X_{ij} = (s_{ij})z_j = (\tau_i n_{ij})z_j \tag{10}$$

These relations are important when considering the connection between storages and flows, and between throughflow-specific and storage-specific flow analyses.

4. Four emergent network hypotheses

Network amplification (Patten et al., 1990), homogenisation (Patten et al., 1990) and dominance of indirect effects (Higashi and Patten, 1986), from throughflow analysis and synergism (Patten, 1991) from utility analysis are the core hypotheses of network environ analysis.

The amplification hypothesis arose from the observation that terms in the integral flow matrix could be greater than one in magnitude. The values of the *direct* flow matrix **G** are in general strictly less than one because the elements are interpreted as the probability or efficiency of transfer from one component to another. Elements in the *integral* matrix **N** can be greater than one. If the system is well connected and cycling occurs, then it is possible for a particle to enter, exit, and re-enter the same component. Therefore, another interpretation of the values in the integral matrix **N** is that they represent the average number of times a unit of flow derived from a particular source reaches another particular component (Kemeny and Snell, 1960, Barber, 1978). Transactions along all the pathways are summed in the integral flow matrix, and the values can be greater than one when cycling is strong. Amplification was said to occur when any off-diagonal elements of the integral flow matrix were greater than one. This property does occur in small highly aggregated models, but Fath (1998) has recently shown it to be largely an artefact of small models.

Network homogenisation (Patten et al., 1990) also arose from examination of integral throughflow matrice, **N**. This property asserts that values within a column, which represent the leading edge output environ (or output niche; Patten and Auble, 1981), and values within a row, (input niche of the input environ) tend to be similar in magnitude. At least they appeared more similar to each other than the same comparison of the rows and columns in the direct flow matrix. The idea is that resources are well mixed by the action of cycling in the networks giving rise to a more homogeneous distribution of flow. Therefore, ecological systems are composed of material (both energy and matter) that has been highly mixed and cycled. This is a departure from the classical view of Lindeman (1942) trophic dynamics. A quantitative measure using the coefficient of variation was developed to show statistically that homogenisation is typical in well-connected systems (Fath, 1998).

Dominance of indirect effects is the third property established from the throughflow analysis (Higashi and Patten, 1986, 1989). An element in the *direct* flow matrix is the normalised amount of direct flow passing between each (i, j) pair, whereas, an element in the *integral* flow matrix is the total (direct plus indirect) normalised flow. When the sum of the elements of the direct matrix is compared to the sum of the elements of the indirect (integral minus direct and initial), there is typically a greater contribution from the indirect processes than from the direct. Higashi and Patten (1989) identified six conditions which contribute to increasing indirect effects: system size, connectivity, looping, cycling, that the feedback, and strength of direct effects. Therefore, indirect processes, as carried by the higher-order interactions, exert dominance in the system. This reinforces the notion that an object's behaviour is embedded in the network and the context in which it interacts with its surroundings. System behaviour is holistically determined by the dominance of indirect over direct effects (Patten, 1999). The property network synergism (Patten, 1991, 1992) is based on utility analysis. Here, the direct utilities are compared to the integral, not the indirect. As stated above, utility analysis is based on the normalised net flow between two components and gives the qualitative relations. Through different numerical examples, Patten (1991, 1992) noticed that the relations in the integral matrix are more positive than those in the direct utility matrix. The numerical methodology developed by Patten was given an analytical basis and it was shown that synergism occurs in all network models (Fath and Patten, 1998). Here, again, we see the action of the network altering the direct impression of system behaviour. In the integral utility, system relations are positive, providing a synergistic context in which components interact.

5. Conclusions

Network environ analysis is a systems oriented, mathematical approach to object-environment interactions with the goal of understanding the role of indirect effects and how they determine the behaviour of network properties. It is based on mapping ecosystem structure and function into a flow-storage model. The emphasis of network analysis is on model analysis, not model building. The network flows are quantified through empirical observation or mechanistic process-based models. Once system structure and function are quantified there is no need to further investigate how the flows are generated. Therefore, the system may be linear or non-linear and still be analysed at steady-state by linear matrix methods. It does not matter because the analysis takes the flows as they are, as phenomenological manifestations, not as mechanistic configurations. In order to proceed with the static numerical analyses, which these are, it is also assumed that the model processes are in steady-state (or for interpretative purposes, near). A quantified model at steady-state represents a snapshot of system behaviour. Like a photograph of a waterfall, these networks are a static description of dynamic processes, which apply in small neighbourhoods around the steady-state.

A main principle of environ analysis is that ecosystem behaviour is holistically determined by network transactions (Patten, 1999). Therefore, the best way to investigate this is by implementing a methodology which explicitly includes system-wide influences. The importance of the transactions promotes a flow oriented paradigm. The key message of this paper is that the power series analyses depicted in Equations (5), give a quantitative approach to identifying indirect effects. Matrix multiplication gives the number of pathways from $j \rightarrow i$ in which the path length is equal to the matrix power. The elements of **A** correspond to the direct paths along which flow travels. Those in $\mathbf{A}^m$, where m = 2, 3, ..., are associated with the indirect paths of mth order. In this way, the direct and indirect pathways are accounted for explicitly. Each of these paths are unique and flow along and each one contributes to the overall effect of j on i. The contribution of these indirect pathways is potentially significant, and the extent of that significance can be quantified using network analysis.

In a holistic, synthetic discipline such as ecology it is important to have tools available to determine indirect effects and techniques to investigate the system behaviour without removing the components from the network.

REFERENCES

Barber, M.C. 1978. A markovian model for ecosystem flow analysis. *Ecological Modelling* 5:193-206.

Dame, R.F., and B.C. Patten. 1981. Analysis of energy flows in an intertidal oyster reef. *Marine Ecology Progress Series* 5:115-124.

Fath, B.D. 1998. Network Analysis: Foundations, Extensions, and Applications of a systems theory of the environment. Ph.D. Thesis. University of Georgia, Athens, Georgia, USA.

Fath, B.D. and B.C. Patten. 1998. Network synergism: emergence of positive relations in ecological systems. *Ecological Modelling* 107: 127-143.

Finn, J.T. 1976. Measures of ecosystem structure and function derived from analysis of flows. *Journal of Theoretical Biology* 56: 363-380.

Hannon, B. 1973. The structure of ecosystems. *Journal of Theoretical Biology* 41: 535-546.

Higashi, M., and B.C. Patten. 1986. Further aspects of the analysis of indirect effects in ecosystems. *Ecological Modelling* 31: 69-77.

Higashi, M., and B.C. Patten. 1989. Dominance of indirect causality in ecosystems. *American Naturalist* 133: 288-302.

Higashi, M., B.C. Patten, and T.P. Burns. 1993. Network trophic dynamics: the modes of energy utilization in ecosystems. *Ecol. Modell.* 66: 1-42.

Kareiva, P. 1994. Higher order interactions as a foil to reductionist ecology. *Ecology* 75: 1527-1528.

Kemeny, J.G. and J.L. Snell. 1960. Finite Markov Chains. D. Van Nostrand Company, Inc., Princeton, New Jersey, USA.

Lindeman, R.L. 1942. The trophic dynamic aspect of ecology. Ecology 23: 399-418.

Matis J.H. and B.C. Patten. 1981. Environ analysis of linear compartmental systems: the static, time invariant case. Proceedings 42d Session International Statistics Institute, Manila, Philippines, December 4-14, 1979.

Miller, T.E. 1994. Direct and indirect species interactions in an early old-field plant community. *American Naturalist* 143: 1007-1025.

Miller, T.E., and J. Travis. 1996. The evolutionary role of indirect effects in communities. *Ecology* 77: 1329-1335.

Patten, B.C. 1978. Systems approach to the concept of environment. *Ohio Journal of Science* 78: 206-222.

Patten, B.C. 1982. Environs: relativistic elementary particles or ecology. *Am. Nat.* 119, 179-219.

Patten, B.C. 1985. Energy cycling in the ecosystem. *Ecological Modelling* 28: 1-71.

Patten, B.C. 1991. Network ecology: indirect determination of the life-environment relationship in ecosystems. In: M. Higashi and T. Burns, (eds) *Theoretical Studies of Ecosystems: the network perspective*. Pages 288-351 Cambridge University Press, New York, New York, USA.

Patten, B.C. 1992. Energy, emergy and environs. *Ecological Modelling* 62: 29-69.

Patten, B.C. 1999. Holoecology: the unification of nature by network indirect effects. Kluwer, Publishers, Inc., New York, New York, USA.

Patten, B.C., and **G.T. Auble**. 1981. System theory of the ecological niche. *American Naturalist* 117: 893-922.

Patten, B.C., M.C. Barber, and T.H. Richardson. 1982. Path analysis of a reservoir ecosystem model. *Canadian Water Resources Journal* 7: 252-282.

Patten, B.C., M. Higashi, and T.P. Burns. 1990. Trophic dynamics in ecosystem networks: significance of cycles and storages. *Ecological Modelling* 51: 1-28.

Pilette, R. 1989. Evaluating dierct and indirect effects in ecosystems. *Am. Nat.* 133, 303-307.

Ulanowicz, R.E. 1986. Growth and Development: ecosystem phenomenology. Springer-Verlag, New York, New York, USA.

Ulanowicz, R.E., and C.J. Puccia. 1990. Mixed trophic impacts in ecosystems. *Coenosis* 5: 7-16.

Von Uexküll, J. 1926. Theoretical Biology. Kegan, Paul, Trench, Tubner and Co., London, England.

Weigert, R.G. and J. Kozlowski. 1984. Indirect causality in ecosystems. *Am. Nat.* 124: 293-298.

White, I.D., D.N. Mottershead, and S.J. Harrison. 1992. Environmental Systems: an introductory text. Chapman and Hall, Inc., New York, New York, USA.

Appendix

Oyster Reef Model Example

The six component oyster reef model (Figure 3) developed by Dame and Patten (1981) has been studied extensively in the network analysis literature (e.g., Higashi and Burns, 1993). Details of the model itself are not given here. For reference, the cycling index (Finn, 1976) for this model is 0.11. Although this is a relatively low value this model exhibits all of the network properties mentioned in the text. This simple example is very useful to show how the four properties arise and can be identified. Three of the four network properties (amplification, homogenisation, and dominance of indirect effects) are identified by implementing the throughflow analysis, and one (synergism) is identified through the utility analysis. The information necessary to identify these properties is given below. In addition, we show how the throughflow and storage can be decomposed into mode 1 and mode 2 contributions.

We start with the throughflow-specific properties. The direct flow intensities are derived by normalising the flows by the throughflows ($g_{ij} = f_{ij}/T_j$). The direct flow intensity matrix is:

$$G = \begin{bmatrix} 0 & 0 & 0 & 0 & 0 & 0 \\ 0.3808 & 0 & 0 & 0.5000 & 0.7600 & 0.4758 \\ 0 & 0.3670 & 0 & 0 & 0 & 0 \\ 0 & 0.3267 & 0.1476 & 0 & 0 & 0 \\ 0 & 0.0289 & 0.1476 & 0.0779 & 0 & 0 \\ 0 & .01240 & 0 & 0 & 0.0686 & 0 \end{bmatrix} \tag{11}$$

The non-dimensional integral flow matrix is calculated from the power series associated with the direct matrix. For this example, the integral flow matrix is:

$$N = \begin{bmatrix} 1.0000 & 0 & 0 & 0 & 0 & 0 \\ 0.5369 & 1.3885 & 0.2775 & 0.7800 & 1.1005 & 0.6606 \\ 0.1971 & 0.5096 & 1.1019 & 0.2863 & 0.4039 & 0.2425 \\ 0.2045 & 0.5288 & 0.2533 & 1.2971 & 0.4192 & 0.2516 \\ 0.0605 & 0.1565 & 0.1904 & 0.1659 & 1.1241 & 0.0745 \\ 0.0165 & 0.0107 & 0.0131 & 0.0114 & 0.0771 & 1.0051 \end{bmatrix} \tag{12}$$

The easiest property to test for is network amplification. The non-dimensional integral flow matrix (Equation (14)) is used directly to identify whether or not amplification occurs. The values of the integral matrix can be interpreted as the mean number of times a unit of input will enter that particular component. A diagonal value which is greater than one, indicates that there is cycling in the system. This is expected in these systems. However, if an off-diagonal element is greater than one, then this indicates that the initial amount of unit input has travelled between those two components more than once. In other words, the initial input into the system cycles around such that on average a component receives more flow from that one component than what was initially put into the system. The flow is said to be amplified in the system by the network cycling. If any of the off-diagonal elements are greater than one, then network amplification has occurred. Looking closely at Equation (14), we see that $n_{25} > 1$, and amplification occurs.

A network homogenisation parameter can be taken as a ratio of the coefficient of variation for both the direct and integral flow matrices. This is a measure of the relative dispersion of these values about the mean. A small coefficient of variation means the elements are similar in value. The parameter is defined such that a ratio greater than one indicates that homogenisation occurs. Here, we have: CV(**G**) = 2.0195 and CV(**N**) = 1.0676 (from Table 1) giving a homogenisation parameter of CV(**G**)/CV(**N**) = 1.8916. This indicates that network homogenisation occurs in this system. Therefore, the values in the integral matrix are more tightly dispersed than those in the direct matrix. Since these values represent the distribution of flow in the network, flow is more evenly distributed in the integral case than in the direct case due to the action of feedback and cycling.

The property dominance of indirect effects states that indirect contributions are greater than those from direct transactions. The measure of direct contribution is simply the sum of the elements in **G** and the measure of indirect contribution is the sum of the elements of the integral minus the direct and initial (**N**-**G**-**I**). If the ratio of indirect to direct (Table 1) is greater than one, then the indirect contributions exceed the direct contributions. For the oyster reef example, i/d = 1.5341. This value is greater than one meaning that the contribution to system throughflow is predominantly from indirect pathways. This property can also be examined with respect to the storage-specific model (see Patten, 1985).

The fourth network property, synergism, is derived from the utility analysis. For utility analysis the net flow matrix is normalised by the throughflow ($d_{ij} = (f_{ij} - f_{ji})/T_i$). The integral utility matrix, **U**, is calculated using the same power series methodology as the flow analysis and therefore includes the direct and indirect contributions to utility. For the oyster reef model, the corresponding direct, **D**, and integral, **U**, matrices are:

$$D = \begin{bmatrix} 0 & -0.3808 & 0 & 0 & 0 & -0.0124 \\ 0.7092 & 0 & -0.3670 & -0.1363 & 0.0568 & 0.0147 \\ 0 & 1.0000 & 0 & -0.1476 & -0.1476 & 0 \\ 0 & 0.3578 & 0.1422 & 0 & -0.0779 & 0 \\ 0 & -0.5038 & 0.4805 & 0.2633 & 0 & -0.0686 \\ 0.7490 & -0.4758 & 0 & 0 & 0.2510 & 0 \end{bmatrix} \tag{13}$$

$$U = \begin{bmatrix} 0.8332 & -0.2228 & 0.0706 & 0.0128 & -0.0270 & -0.0117 \\ 0.4244 & 0.5994 & -0.1938 & -0.0359 & 0.0652 & -0.0009 \\ 0.3936 & 0.5472 & 0.7414 & -0.2000 & -0.0609 & 0.0073 \\ 0.2077 & 0.2871 & 0.0021 & 0.9457 & -0.0563 & 0.0055 \\ 0.0010 & 0.0664 & 0.4369 & 0.1663 & 0.9109 & -0.0615 \\ 0.4224 & -0.4354 & 0.2548 & 0.0684 & 0.1774 & 0.9762 \end{bmatrix} \tag{14}$$

The parameter which measures network synergism is a ratio of the total positive utility (benefit) to the total negative utility (cost) in the dimensional integral matrix. In order to calculate the absolute (dimensional) benefit-cost ratio, the non-dimensional **U** matrix is redimensionalised by multiplying by the diagonalised throughflow vector, **Y** = T ***U**, where

$$\mathrm{T} = \mathrm{diag}(\mathrm{T}) = \begin{bmatrix} T_1 & 0 & \dots & 0 \\ 0 & T_2 & \dots & 0 \\ \vdots & \vdots & \vdots & \vdots \\ 0 & 0 & \dots & T_n \end{bmatrix} \tag{15}$$

$$\mathrm{Y} = \begin{bmatrix} 34.5510 & -9.2415 & 2.9291 & 0.5320 & -1.1207 & -0.4864 \\ 9.4493 & 13.3455 & -4.3144 & -0.7997 & 1.4517 & -0.0210 \\ 3.2164 & 4.4718 & 6.0591 & -1.6346 & -0.4978 & 0.0598 \\ 1.7617 & 2.4347 & 0.0175 & 8.0205 & -0.4777 & 0.0466 \\ 0.0026 & 0.1667 & 1.0967 & 0.4174 & 2.2863 & -0.1544 \\ 0.2896 & -0.2985 & 0.1747 & 0.0469 & 0.1216 & 0.6693 \end{bmatrix} \tag{16}$$

Again, the parameter is defined such that a value greater than one indicates that the total positive utility is greater than the negative utility in the system and that synergism occurs. Here, the network synergism parameter is 4.9152, indicating that synergism has occurred.

The final section demonstrates the connection between the throughflow and storage analyses. The appropriate storage matrices are presented below. The flow matrix is normalised by the storage values, where the diagonal elements are the throughflows normalised by the storage values. The non-dimensional storage matrix for the oyster reef is:

$$\mathrm{C} = \begin{bmatrix} -0.0207 & 0 & 0 & 0 & 0 & 0 \\ 0.0079 & -0.0223 & 0 & 0.1758 & 0.1172 & 0.0047 \\ 0 & 0.0082 & -3.3879 & 0 & 0 & 0 \\ 0 & 0.0073 & 0.5000 & -0.3516 & 0 & 0 \\ 0 & 0.0006 & 0.5000 & 0.0274 & -0.1542 & 0 \\ 0.0003 & 0 & 0 & 0 & 0.0106 & -0.0099 \end{bmatrix} \tag{17}$$

In order to scale the values between zero and one, we define a new matrix **P** such that $p_{ij} = i_{ij} + c_{ij}\Delta t$. Here we select $\Delta t = 0.25$ and get:

$$\mathrm{P} = \begin{bmatrix} 0.9948 & 0 & 0 & 0 & 0 & 0 \\ 0.0020 & 0.9944 & 0 & 0.0439 & 0.0293 & 0.0012 \\ 0 & 0.0020 & 0.1530 & 0 & 0 & 0 \\ 0 & 0.0018 & 0.1250 & 0.9121 & 0 & 0 \\ 0 & 0.0002 & 0.1250 & 0.0068 & 0.9614 & 0 \\ 0.0001 & 0 & 0 & 0 & 0.0026 & 0.9975 \end{bmatrix} \tag{18}$$

Using the power series from Equation (5) gives:

$$Q = \begin{bmatrix} 192.9124 & 0 & 0 & 0 & 0 & 0 \\ 96.4560 & 249.4410 & 49.8576 & 140.1284 & 197.7126 & 118.6810 \\ 0.2327 & 0.6017 & 1.3009 & 0.3380 & 0.4769 & 0.2863 \\ 2.3266 & 6.0168 & 2.8816 & 14.7573 & 4.7691 & 2.8627 \\ 1.5697 & 4.0594 & 4.9370 & 4.3016 & 29.1522 & 1.9314 \\ 6.6783 & 4.3353 & 5.2725 & 4.5939 & 31.1332 & 406.0107 \end{bmatrix} \quad (19)$$

The throughflow is equal to the sum of all the flow in the system, including the boundary flow.

$$T = [41.4696 \quad 22.2656 \quad 8.1720 \quad 8.4806 \quad 2.5100 \quad 0.6856] \quad (20)$$

In this case there is only input into the first component so it is easy to see how the throughflow decomposes into the mode 1 and mode 2 contributions. The first passage flow is calculated as the contribution to throughflow that travels along pathways which contain no cycles touching the terminal nodes: $T_{ij}^{(1)} = (n_{ij}/n_{ii})z_j$. Some of the throughflow is carried by recycled or mode 2 pathways. This can be calculated from: $T_{ij}^{(2)} = (n_{ii}\text{-}1/n_{ii})n_{ij}z_j$.

$$T^{(1)} = [41.4697 \quad 16.0359 \quad 7.4166 \quad 6.5381 \quad 2.2330 \quad 0.6821] \quad (21)$$

$$T^{(2)} = [0 \quad 6.2297 \quad 0.7555 \quad 1.9424 \quad 0.2770 \quad 0.0035] \quad (22)$$

Using this example, we see that the sum of the mode 1 and mode 2 throughflow equals the total throughflow in the system: $\mathbf{T} = \mathbf{T}^{(1)} + \mathbf{T}^{(2)}$. All the throughflow into component one is first passage because as stated earlier this component is taboo so there cannot be any mode 2 contribution.

Similarly, we can decompose the storage into both mode 1 and mode 2 contributions using the relations $x_{ij}^{(1)} = (q_{ij}/q_{ii})z_j\Delta t$, and $x_{ij}^{(2)} = (q_{ii}\text{-}1/q_{ii})q_{ij}z_j\Delta t$.

$$X^{(1)} = [10.3674 \quad 4.0090 \quad 1.8542 \quad 1.6345 \quad 0.5582 \quad 0.1705] \quad (23)$$

$$X^{(2)} = [1989.6 \quad 996.0 \quad 0.6 \quad 22.5 \quad 15.7 \quad 69.1] \quad (24)$$

Here, we see that the storage in component three is primarily composed of mode 1 but the other components are all primarily mode 2. Note how different this storage decomposition is from the throughflow decomposition above. This is totally non- or counterintuitive, and is due to the different turnover rates (and times) in the model's components.

II.8.1 Stability Concepts in Ecology

Yuri Svirezhev

1. Introduction: Stability concepts in ecology and mathematics

It is intuitively clear that both an ecosystem and a biological community, which exists sufficiently long in a more or less invariant state (such a property is often called '*persistence*'), should possess intrinsic abilities to resist perturbations coming from the environment. This ability is usually termed 'stability'. This is some general, emergent property (so-called 'scalar invariant') of a system. Apparently, we can observe only *stable* ecosystems, since all *unstable* ecosystems have disappeared in the process of evolution. The environment destroyed them, since they could not *be adapted* to it. (You can see that we introduced the new term: 'adaptation'. Really these terms are very closed, and we can speak that only a stable system is able to be adapted to the environment, i.e. to persist sufficiently long under given environmental conditions.)

In spite of being intuitively clear, 'an ability to persist in the course of a sufficiently long time in spite of perturbations' can scarcely be defined in a unique and unambiguous way. The reason is that both the 'persistence' and the 'perturbations' (as well as the 'sufficiently long time') part of the idea needs further clarification, to say nothing of the scale of the system under concern. What is understood as 'an ability to persist' and what kind of 'perturbations' are relevant? What length of time is considered a 'sufficiently long time'? Differing in answers to these basic questions, a variety of stability concepts have been proposed and discussed in the literature on mathematical ecology (May 1973, Svirezhev 1976, Svirezhev, Logofet 1978, Svirezhev 1987, Jeffries 1988, Logofet 1993, Svirezhev, Logofet 1995), and in purely ecological literature (see, for instance, Lewontin 1969, Usher and Williamson, 1974) yet few of them have drawn proper mathematical attention (Svirezhev 1983). Though the notion seems obvious, it is quite a problem to provide it with a precise and unambiguous definition. Really, stability can be defined in quite a lot of ways, both in verbal and formal terms – either in ecology or mathematics. While neither of the 'ecological' definitions of stability can now be recognised as the most fundamental, mathematics is more lucky since it gave rise to the notion of Lyapunov stability, which appears to be inherent in or substantial for any further notion of stability – at least, within the theory of dynamic systems. However, even in mathematics, this heavily overloaded term found no established ('stable') definition. For instance, the theory of stability, which can be considered as a branch of applied mathematics and mechanics, is using about thirty different definitions of stability; we can consider stability as having some 'fuzzy' definition. Paraphrasing von Neumann we can say that '... nobody knows what stability means in reality, that is why, in the debate, you will always have an advantage.

2. Species diversity as a measure of stability

Among the different definitions, we can select two large classes differing in respect to the requirements coming under the head of 'stability'. The first group of requirements concerns preservation of the number of species in a community. A community is stable if the number of member-species remains constant over a sufficiently long time. This definition is the closest to various mathematical definitions of stability, such as those of Lagrange and Poincaré - Lyapunov.

The second group refers rather to populations than to the community, which is considered stable when numbers of component populations do not undergo sharp fluctuations. This definition is closer to the thermodynamic (or rather, statistical physics) notion of system stability. In thermodynamics (statistical physics) a system is believed to be stable when large fluctuations, which can take the system far from the equilibrium or even destroy it, are unlikely (see, for instance, Landau and Lifshitz 1964). Evidently, general thermodynamic concepts (for instance, the stability principle associated in the case of closed systems with the Second Law of Thermodynamics and in the case of open systems with Prigogine's theorem) should be applicable to biological (and, in particular, ecological) systems.

Perhaps this motivates the frequent use of specific *diversity* indices (in particular, the Shannon information entropy called also the *information diversity index*

$$D = -\sum_{i=1}^{n} p_i \ln p_i , \quad p_i = N_i / \sum_{i=1}^{n} N_i , \tag{2.1}$$

where n is the number of species (or some other groups) in a community and N_i is the population size of ith species). There is no doubt that diversity measures do carry some objective information about the properties of a system. At the beginning of the works of R. Margalef (1951) and R. MacArthur (1955) a real 'information' boom in ecology had been started. In numerous works it was postulated that diversity (or entropy) D attains a maximum in the most stable communities. It was being shown that:

- diversity increases when an ecosystem (community) evolves to a climax, i.e. to a steady-state and
- the more diverse such a community is, the more stable it is.

In accordance to this 'logic', the community is the most stable if D is maximal. But, as may readily be shown, the community structure in this case is such that specimens of any species occur with the same frequency ($\max_{p_i} D$ is attained at $p_i^* = 1/n$). In other words, the diversity of a community is maximal when the distribution of species is uniform, i.e. when there are no abundant or rare species, and no structure. However, observations in real communities, show that this is never the case, and that there is always a hierarchical structure with dominating species. What is the cause of this ecological paradox? Probably, it is the formal application of models and concepts taken from physics and information theory to systems that do not suit them. Both the Boltzmann entropy in statistical physics and the Shannon entropy in information theory make sense only for populations of weakly interacting particles. A typical example of such a system is the ideal gas, the macroscopic state of which is an additive function of the microscopic states of its molecules. The introduction of an entropy measure in such sets is well founded. Furthermore, the stability of the equilibrium, when the entropy is maximal, is associated with the Second Law. On the other hand, the stable structure of a biological community is the consequence of interactions between populations rather than the function of characteristics of individual species, etc., i.e. a biological community is a typical system of strongly interacting elements. But as soon as we become concerned with such systems the entropy measure is no longer appropriate. That is why the use of diversity indices as measures of stability can scarcely be accepted as a blameless approach.

Notice, however, that in numerous laboratory (as well as real) competitive communities in the initial stages of their successions, far from climax, an increase in diversity may be observed. The same effect is observed in aquatic communities practically in the course of the whole succession. It seems diversity is a 'good' goal function for stability. This is explained by: a) in the initial stages, far from equilibrium, the competition is still weak, its stress is low, and the community may well be regarded as a system with weak interaction; b) concerning aquatic communities (especially phytoplankton communities) we note the concentrations of biomass are really low, the intrinsic rates are very high, and therefore these communities may be also regarded as systems with weak interaction.

So, is diversity 'only the spice of life' or is it a necessity for the long life of the total ecosystem comprising man and nature? For instance, E. P. Odum (1976) has shown that there are certain relationships among the diversity level, the structure, and the functioning of energy flows in an ecosystem. Nevertheless, although the diversity measures have an obvious advantage of being determined through observable and measurable characteristics of a real ecosystem, the question of a cause-effect relationship between diversity and stability has no simple answer.

3. Model approach to definitions of stability

In contrast to intuitive understanding of stability, typical to the 'stability vs. diversity' speculations, the model approach can provide quite formal, mathematically rigorous definitions.

Let us assume that we have a 'good enough' (at the viewpoint of adequacy and descriptive completeness) mathematical model of a biological community or ecosystem, then stability properties of a real system can be deduced from investigating its model by mathematical technique of stability theory.

However, as mentioned above, there are a lot of formal definitions of stability. The task is just to decide what kind of model behaviour should correspond to a stable functioning of the real system and to choose mathematical stability definitions which are adequate both to a meaningful 'ecological', perception of stability and to the mathematics of the model. For instance, referring to ecosystem stability, E. P. Odum (1983) proposed to distinguish between two major types, namely, a resistant and a resilient stability. Resistant stability is an ability of an ecosystem to resist perturbations (or disturbances), while keeping its structure and functioning unchanged. Resilient stability is an ability of an ecosystem to restore after its structure and functioning have been disturbed. Both types are observed in nature and the evidence is growing that an ecosystem can scarcely develop both types of stability. This is probably due to the fact that a particular type of environment is likely to promote either resistant or resilient types of stability.

Stability investigations are thus dependent upon a particular mathematical model, assuming them to be adequate enough for the system under study. As far as the assumption being true, the model approach has an obvious advantage in its prognostic ability, as well as in its capability of relating stability to other systems properties such as the structure and particular mechanisms of functioning. Expressed in formal terms, stability conditions of the model promote a formulation of hypotheses concerning the functioning of the real system. In any case, the adequacy assumption itself may always be questioned and sometimes answered at least, in qualitative terms – from the outcome of stability analysis as well.

4. Lyapunov stability: formal definitions and interpretations

System dynamics can be described by the following system of an ordinary differential equation:

$$\frac{dN_i}{dt} = F_i(N_1,\ldots,N_n),\ i = 1,\ldots,n \tag{4.1}$$

where all N_i are non-negative and therefore these variables can be considered as either population sizes or their biomass for species composing some biological community, or, in the vector form as:

$$\frac{d\mathbf{N}}{dt} = \mathbf{f}(\mathbf{N}) \text{ with the initial condition } \mathbf{N}(t_0) = \mathbf{N_0} \tag{4.2}$$

While, in general, a definition of stability has to be relevant to any particular reference solution, a proper change of variables can always reduce the problem to the stability of a so-called equilibrium solution, that is, a constant vector **N*** nullifying the right-hand side.

Definition 1. Solution **N*** is said to be *locally Lyapunov stable* if for any small $\varepsilon >$ 0 there is δ > 0 such that the inequality $\left\|\mathbf{N_0} - \mathbf{N}^*\right\| < \delta$ results in $\left\|\mathbf{N}(t,\mathbf{N_0}) - \mathbf{N}^*\right\| < \varepsilon$ for any $t \geq t_0$ where $\mathbf{N}(t,\mathbf{N_0})$ designates the solution corresponding to the initial state $\mathbf{N_0}$. If, in addition, $\mathbf{N}(t) \to \mathbf{N}^*$, then it is *asymptotically Lyapunov stable*. If, moreover, this is true for any initial state of a certain domain **Ω** in the state space, then it is *asymptotically stable globally* in **Ω**, or *just globally asymptotically stable* (Barbashin 1967).

In theoretical mechanics, a locally Lyapunov stable motion is traditionally interpreted as a small 'tube' of trajectories propagating along the time direction: any disturbed trajectory remains within the tube unless the initial perturbation has shifted it out of the tube. The Lyapunov asymptotic stability then means that the 'thickness' of the tube tends eventually to zero, while the global stability turns the tube rather into a sharpened 'funnel' (see Figure1).

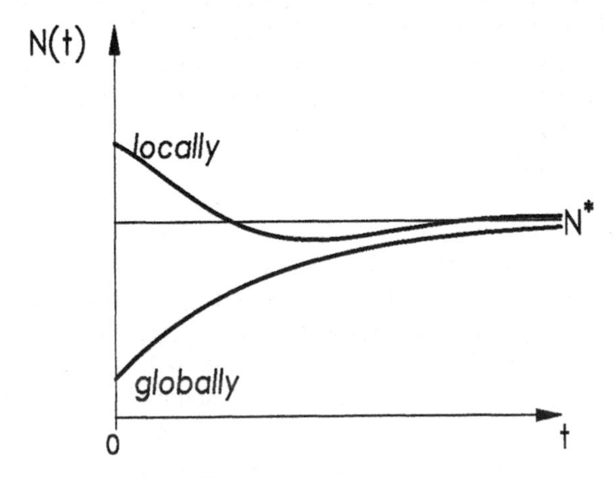

Figure 1: Example of Lyapunov stable trajectories

From an ecological point of view, Lyapunov stability means quite a special mode of 'an ability to persist in the course of a sufficiently long time in spite of perturbations': if an impact of any kind has shifted somehow the equilibrium **N** to a new state $\mathbf{N_0}$, then the state must return in the course of time, close to equilibrium again; the versions of the

Lyapunov concept modify the meaning only in what concerns the magnitude of the initial shift and the closeness of the return.

5. 'Ecological' stability (ecostability)

As a partial consequence it follows that the number of species remains unchanged in a Lyapunov stable community. This is, at the same time, the idea of an ecological meaning of stability: '...if neither species is eliminated, the community should be considered stable'. From a Lyapunov stability point of view, the latter means that the stable equilibrium must be non-trivial, i.e. all the population sizes at equilibrium are strongly positive. On the other hand, the number of species can be preserved even if the equilibrium is Lyapunov unstable but model trajectories neither go to an infinity, nor turn to zero. This weaker model property (usually it is called by *Lagrange stability*) seems to be more adequate to an ecological meaning of stability than the Lyapunov concept. In what follows this kind of stability will be referred to as *ecological stability (ecostability)*. This is illustrated in Figure 2; a formal definition is given below:

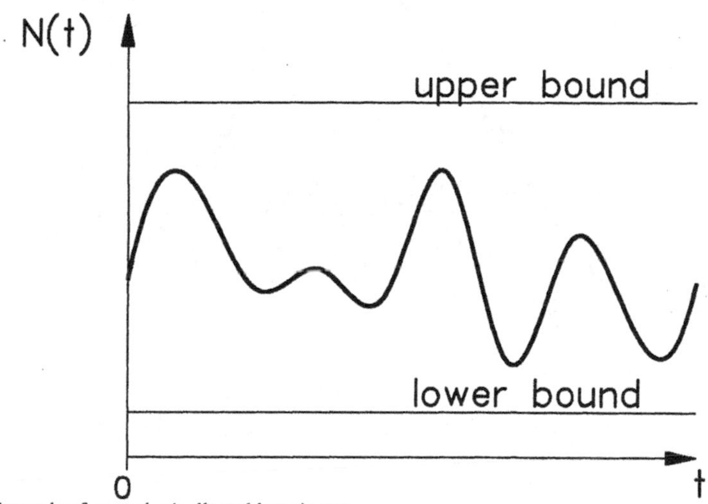

Figure 2: Example of an ecologically stable trajectory

Definition 2. Let Ω_0^n and Ω^n be closed bounded domains in the interior of the positive orthant of the state space and $\Omega_0^n \subseteq \Omega^n$. A model ecosystem (4.1) is called *ecologically stable* if any trajectory originating from Ω_0^n never goes out of Ω^n. Domain Ω_0^n is then called the *ecostability domain.*

Unfortunately, there are yet no constructive methods for ecostability analysis, although the Lyapunov stability concept is a very constructive theory. However, the Lyapunov stability is a sufficient, but not a necessary, condition of ecostability. And this is problem. However, there is a whole class of community models where both the stability notions become equivalent. These are the so-called *Volterra conservative and dissipative systems*; conservative are, in particular, the classical Lotka - Volterra pairs of 'prey - predator' species, while dissipative are the 'horizontal-structured (competitive)' communities (see Svirezhev 1983, Lögofet 1993).

We illustrate the ecostability concept by one simple example. Consider a population with three equilibria: two stable and one unstable. Its phase portrait is shown in Figure 3.

Suppose that the population dynamics is described by the equation

$$\frac{dN}{dt} = N(1 - \frac{N}{K})(N - k) \tag{5.1}$$

where N is a population size, $N_1^* = 0, N_3^* = K$ are stable equilibria and $N_2^* = k$ is a unstable one; $k<K$. Since for any $N_0>k$ $N(t) \to K$ then the domain Ω_0: $k<N<+\infty$ is the ecostability domain.

It is clear that we could solve the above ecostability problem only because of the exclusive simplicity of the problem. In more complex cases the problem becomes very serious. For this reason we shall try to reduce the ecostability problem to that of classic Lyapunov stability.

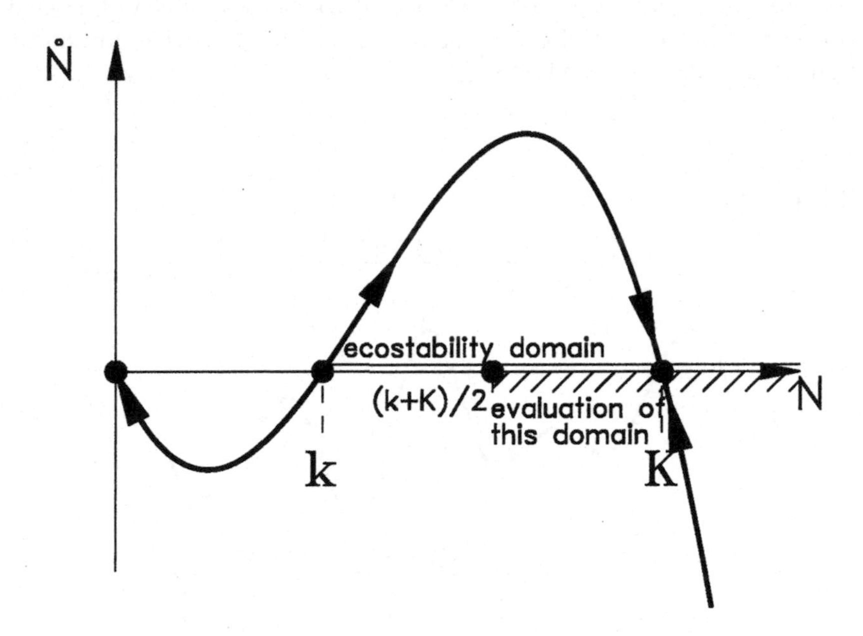

Figure 3: Phase portrait of (5.1) and a domain of ecological stability:
====== full domain, ////////// estimation with the help of the Cheteav-Malkin theorem

'Ecostability' means that any trajectory of (4.1), starting inside of the positive orthant of state space, remains there. Therefore, since all N_i and initial values N_i^0 are positive, one can always define the following change of variables:

$$u_i = N_i / N_i^0; \quad i = 1, \ldots, n \tag{5.2}$$

Substitution of u_i into (4.1) yields

$$\frac{du_i}{dt} = \phi_i(u_1, \ldots, u_n; N_1^0, \ldots, N_n^0); \ i = 1, \ldots, n \tag{5.3}$$

with the initial conditions $u_i(0) = 0$. It is obvious, if $N(t; N_0) \to +\infty$ then $u \to +\infty$, but if $N(t; N_0) \to 0$ then $u \to -\infty$. A solution to system (5.3), as a function of parameters

$\mathbf{N_0}$ is thus defined in the entire state space. Since **u*= 0** is generally not a solution to system (5.3), we transform (5.3) into

$$\frac{du_i}{dt} = \Phi_i(u_1,...,u_n;N_1^0,...,N_n^0) + B_i; i = 1,...,n \tag{5.4}$$

where

$$\Phi_i = \phi_i(u_1,...,u_n;N_1^0,...,N_n^0) - \phi_i(0,...,0;N_1^0,...,N_n^0); B_i = \phi_i(0,...,0;N_1^0,...,N_n^0).$$

It is clear that if B_i are considered as *permanently acting perturbations* (PAP)., then **u*= 0** is a solution to the disturbed system. The problem is: when would this solution be stable?

Stability makes sense only with respect to a specified class of perturbations. If, for instance, the perturbation is reduced to a single change in the initial state of a dynamic system, the well-known Lyapunov stability concept arises – either local or global, asymptotic or neutral, depending on whether the perturbations are considered sufficiently small or definitely finite and whether the perturbed trajectory is supposed to converge or just to be close enough to the disturbed one. If, in addition, sufficiently small, though permanent, perturbations are supposed to affect the right-hand sides of model equations, then this leads to the notion of *stability under permanently acting perturbations.*

If now we solve the PAP-stability problem for a particular value of parameter $\mathbf{N_0}$ and determine a region of such values in the original model state space, then certain sufficient conditions of ecological stability in system (4.1) will thereby arise. By the Chetaev-Malkin theorem (Malkin 1966), the trivial solution is stable to PAP, if it is Lyapunov asymptotically stable in the undisturbed system and the perturbations are sufficiently small. Unfortunately, there are no effective and sufficiently general means to evaluate the smallness of PAP. Nevertheless, the ecostability problem for model ecosystem (4.1) is thus reduced – at least theoretically – to the Lyapunov stability problem for a transformed system. This indicates also to the close tie between these two types of stability: ecological and Liapunov. Ecostability has a 'good' ecological interpretation and the Lyapunov – the fundamental one – has incomparably a higher developed mathematical technique to be analysed.

Let us return to equation (5.1). Applying the method of reduction we get

$$\frac{du}{dt} = (1 - \frac{N_0 e^u}{K}) \cdot (N_0 e^u - k) = \Phi(u, N_0) + B;\ u = \ln(N / N_0) \tag{5.5}$$

where $\Phi = \frac{\mathrm{N_0}}{\mathrm{K}}(\mathrm{K} + \mathrm{k} - 2\mathrm{N_0}) + \text{non-linear tems}$ and $B = (1 - \frac{N_0}{K}) \cdot (N_0 - k)$. It is obvious that the solution $\mathrm{u}^* = 0$ to the undisturbed equation $\mathrm{du/dt} = \Phi(\mathrm{u}, \mathrm{N_0})$ is asymptotically stable, if $N_0 > (K + k)/2$, then, in accordance to the Chetaev-Malkin theorem. Therefore, the population is ecologically stable if $N_0 > (K + k)/2$ (see Figure 3). In addition, B must be sufficiently small. (It takes place if, in particular, N_0 is situated in a vicinity of equilibrium K.) If we compare this solution with the exact one

($N_0>k$), we can see that they differ from each other. It is no surprise since our method gives only a *sufficient* condition for ecostability.

6. Structural stability

In contrast to physics, which uses strict laws and relationships, practitioners of ecology operate frequently with qualitative descriptions. Any attempt to 'quantify' these dependencies leads us to parametric forms. As a result, we may get different functions which must describe the same qualitative dependence. For instance, a saturation may be described both as an exponential function and as a hyperbolic Michaelis-Menten function. But in 'good' models this arbitrariness must not essentially influence the general behaviour of the system. In other words, the topology of the phase portrait of the system must not be changed at least for sufficiently small perturbations of its structure. It means that the system must be *structurally stable.*

Consider a model described by a system of ordinary differential equations. It is structurally stable, if its solution is not practically changed when we 'move about' the right sides of these equations, in particular, if we disturb their coefficients and parameters.

We consider the concept of structural stability using a 'prey-predator' model. Starting with the works of V. Volterra, this model has been a classic (canonical) subject of mathematical ecology.

Let $x(t)$ and $y(t)$ be the numbers of preys and predators respectively. Then a sufficiently general (so-called 'Kolmogoroff') model of this system takes on the form:

$$\frac{dx}{dt}=\alpha(x)x-V(x)y, \qquad \frac{dy}{dt}=[kV(x)-m]y \tag{6.1}$$

where $\alpha(x)$ is the Malthusian function for prey, $V(x)$ is the trophic function (functional response), m is the natural mortality rate of the predator, and k is the efficiency in converting the prey biomass into the reproductive biomass of the predator.

Let now $\alpha(x) = const.$ It corresponds to the case of no self-regulation mechanisms in the prey population; it is controlled only by predator. For the trophic function we use the very popular parametrisation:

$$V(x)=\frac{V_\infty x^n}{K^n+x^n},\ n=1,2,\ldots \tag{6.2}$$

If n=1, then we have the trophic function of Type I; i.e. with the constant sign of curvature; if n=2,3,... then the trophic function belongs to Type II, i.e. it is S-shaped (figure 4)

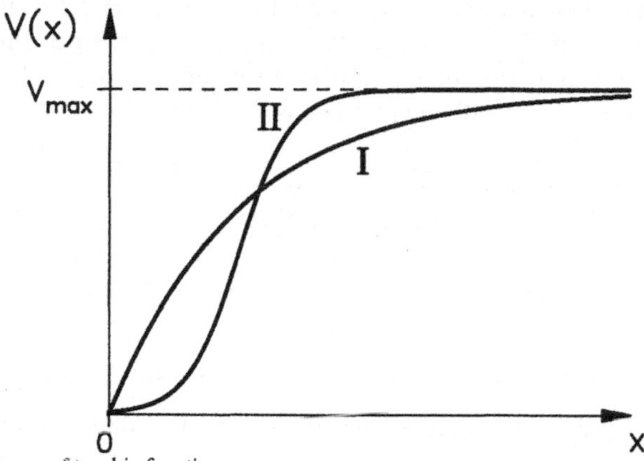

Figure 4: Two types of trophic functions

Replacing the variables:

$$\alpha t \Rightarrow t,\ x/x* \Rightarrow x,\ y/y* \Rightarrow y,\ V/V_\infty \Rightarrow V,\ m/\alpha \Rightarrow \mu,$$

we obtain from (6.1):

$$\frac{dx}{dt} = x - V(x)y, \quad \frac{dy}{dt} = \mu y[V(x) - 1]. \tag{6.3}$$

The type of equilibrium $(x*, y*)$ is determined by the value of the derivative $v = V'(1)$. This equilibrium is a topological knot; if $v < 1$ then it is unstable, and if $v > 1$, stable. When we pass through $v = 1$, we have the Andronov-Hopf bifurcation and in a 'general position' case the limit cycle is born out of this equilibrium. It seems there is no problem here.

On the other hand, if we use the description of the trophic function in the form (6.2), we can prove that the system (6.3) has no limit cycles. We consider this problem for n=2, although the final results are valid for any n>2.

For $n = 2$ in new variables we have

$$V(x) = \frac{x^2}{1 - b + bx^2},\ b = \frac{V*}{V_\infty}.$$

The bifurcation value for b is $b_c = \frac{1}{2}$. Hence the trophic function which gives the equilibrium of 'centre' type, is $V_c = 2x^2/(1 + x^2)$. In this case the system (6.3) is reduced to the Abel equation of second type and it has the integral:

$$y + Cy^{1/\mu} = (1-\mu)\frac{x}{V_c(x)},\ \mu \neq 1 \tag{6.4}$$

$$y\ln(Cy) = \frac{x}{V_c(x)}, \quad \mu = 1.$$

Therefore the limit cycle can not arise out of equilibrium (existence of the integral of Abel equation is a sufficient condition for this). We can show that the periodic regime can also not arise out of closed trajectories of centre. But if we deform (by small deformations) the trophic function, for instance, it will be presented in the form

$$V(x) = \frac{x^2}{1-b+bx^2+(b-\frac{1}{2})F(x)} \tag{6.5}$$

where $F(x)$ is a finite function and $F(1)=0, F(0)<-(1+x^2), F(\infty)<x^2$ (see Figure 5), then we can organise the birth of cycle out of closed trajectories.

Let us consider the concrete dependence $F(x)=A[(x-1)/(x+1)]^l$. It is easy to show that the cycle can be created by the appropriate choice of A. Note that the factor $(b-\frac{1}{2})$ at $F(x)$

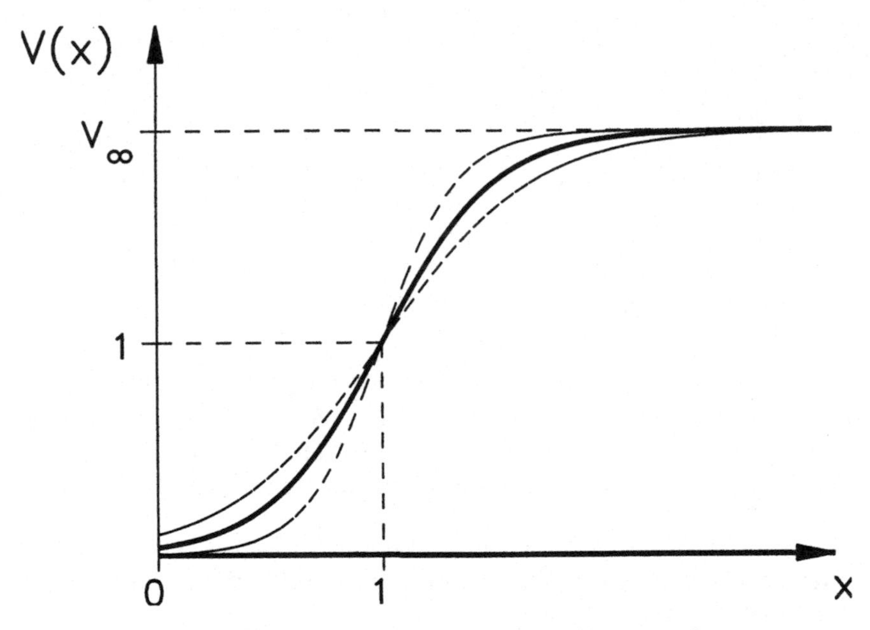

Figure 5: 'Small deformation' of trophic function: ——— before, ------- after

was introduced in order to save the Abel integral (6.4). If this factor would be changed, i.e. the Abel integral would be destroyed, we can realise the birth of (l-1) cycles out of equilibrium by means of the Andronov-Hopf bifurcation. In other words, structurally stable, while this system with the trophic function (6.2) is structurally unstable after a small deformation of trophic function (6.2) the system (6.3) becomes robust (non-robust). In robust systems the bifurcation of parameter b can provide limit cycles both out of equilibrium and out of closed trajectories, and so much as we desire.

It is very interesting that the parametric form (6.2) for trophic function generates the entire class of structurally unstable phase portraits. Really, the probability of getting into a 'non-robust' situation is very low, but we have got into it. Note when we use some asymptotic methods (for instance, Krylov-Bogolubov method) for finding the periodic solutions of (6.3), it leads to the destruction of the 'non-robust' situation, and, as a rule, we obtain more or less successful results (Svirezhev, Logofet 1978).

Thus, the system (6.1) or (6.3) with the trophic function (6.2) is structurally unstable. My question is the following: is this structural instability the principal emergent property of the prey-predator system, which ensures the high degree of adaptation and the high lability in relation to changes of environment for this system, or an artefact of parametrisation, i.e. the effect, which we can call the 'parametrisation trap'?

Let us suppose that the first answer is true and the following *Gedankenexperiment* would be carried out:

1. we imbed the structurally unstable 'prey-predator' system into a stochastic media, i.e. we consider the system behaviour under impact of random perturbations. In this case the perturbations destroy this non-robust structurally unstable situation and as a result the dynamics can be presented on the whole by a system of (stable and unstable) stochastic limit cycles. In other words, *we can predict the behaviour of the deterministic, but structurally unstable system in the principally unpredictable stochastic environment.*
2. we imbed this system in the periodically changing predictable environment and, for some values of parameters, we get a dynamic chaos, i.e. a principally unpredictable behaviour in the predictable environment.

All this allows me to formulate the following speculative hypothesis:

maybe the structural instability is a special evolutionary mechanism which allows the system to predict its own future in stochastic environment, is not it?

7. Lyapunov functions

One of the most important concepts in the theory of stability is the Lyapunov functions. Positive functions defined in a phase space of dynamic systems and possessed by the property either monotonous to increase or monotonous decrease along trajectories, they can be also considered as some special class of goal functions.

Let us remember that the system dynamics can be described by the system of ordinary differential equation (4.1). We assume also that (4.1) has the non-trivial equilibrium $(N_1^*,\ldots,N_n^*)$. Let us test the following functional class as a candidate for Lyapunov functions:

$$L=\sum_{i=1}^{n} N_i^* \varphi(N_i / N_i^*), \tag{7.1}$$

where $(\xi_i = N_i / N_i^*)$, $\varphi(1)=\frac{d\varphi}{d\xi}(1)=0; \frac{d^2\varphi}{d\xi^2} > 0$ for any $\xi \ge 0$. In other words, the function $\varphi(\xi)$ must be convex for positive ξ. It is obvious that $L(\mathbf{N}^*) = 0$. Here $\mathbf{N}^*$ is the vector $\mathbf{N}=(N_1,\ldots,N_n)$ at the equilibrium. Because the first variation of L in a vicinity of $\mathbf{N}^*$ is equal to

$$\delta L=\sum_{i=1}^{n}\frac{\partial L}{\partial L_i}\delta N_i=\sum_{i=1}^{n}\frac{\partial \varphi}{\partial \xi_i}\delta N_i, \tag{7.2}$$

then $\delta L(N^*)=0$ for any variations δN^*. Calculating the second variation we get

$$\delta^2 L=\frac{1}{2}\sum_{i=1}^{n}\sum_{j=1}^{n}\frac{\partial^2 L}{\partial N_i \partial N_j}\delta N_i \delta N_j=\frac{1}{2}\sum_{i=1}^{n}\frac{d^2\varphi}{d\xi_i^2}(\xi_i)\frac{(\delta N_i)^2}{N_i^*}>0 \tag{7.3}$$

for any non-zero variations $\delta\mathbf{N}$. Thus, L is a convex function of $\mathbf{N}$ having an isolated minimum at the point $\mathbf{N}^*$ and a monotonous increase for any $\mathbf{N} \in P^n$ where P^n is the positive orthant of state space.

The equilibrium $\mathbf{N}^*$ is stable if the derivative dL/dt is taken along the trajectory of (4.1)

$$\frac{dL}{dt} = \sum_{i=1}^{n} \frac{d\varphi}{d\xi}(\xi_i) f_i(N_1, \ldots, N_n) \leq 0. \tag{7.4}$$

for all trajectories (the Lyapunov stability theorem, see Malkin 1966). On the contrary, the equilibrium $\mathbf{N}^*$ is stable if the derivative $dL/dt \geq 0$ even for one trajectory of (4.1) (the Chetaev instability theorem).

While on the subject of stability, ie Lyapunov stability, note that the Lyapunov stability theorem and the Chetaev instability theorem give us only sufficient (not necessary) conditions.

And finally, we have the large class of functions, namely the Lyapunov functions. Let us consider different particular examples of them.

Example 1. Let

$$\varphi(\xi) = a(1-\xi)^2, a > 0, \text{ then the function } L = \sum_{i=1}^{n} a_i N_i^* (1-\xi_i)^2$$

belongs to this class. Setting $a_i = N_i^* / n$ we get

$$L = \frac{1}{n} \sum_{i=1}^{n} (N_i - N_i^*)^2, \tag{7.5}$$

i.e. the value L may be considered as a mean square measure of distance between a current state of the system and its equilibrium. If this distance decreases in time (i.e. the derivative $dL/dt < 0$ along the system trajectory), then in accordance to the Lyapunov stability theorem, we can hope that the system moves to a stable equilibrium. Note that I speak of 'hope' since we must check all the trajectories (or, a statistically 'sufficient' large number of them) in order to speak correctly about a stability of equilibrium. On the contrary, if this measure increases along some trajectory then (in accordance to the Chetaev instability theorem) the equilibrium is unstable and this trajectory moves off us from it.

All these results can be easily interpreted from a thermodynamic viewpoint. Indeed, in thermodynamics the value L is a mean square of fluctuations around an equilibrium or a *power of fluctuations*. Therefore, we can state that if a power of fluctuations decreases in time, then the system moves to a stable equilibrium. Note if the movement to a stable equilibrium can be called an *evolution of the system,* then the last statement can be reformulated in the following: a power of fluctuations decreases in the process of the system evolution.

Example 2. Let $\varphi(\xi) = \xi \ln \xi - \xi + 1$ then the corresponding Lyapunov function will be

$$L = \sum_{i=1}^{n} \{N_i \ln(N_i / N_i^*) - (N_i - N_i^*)\} \tag{7.6}$$

Looking at (7.6) we see that this is so-called Jørgensen's *exergy* (Jørgensen 1992). In Jørgensen's interpretation the equilibrium $\mathbf{N}^*$ (a reference state) is identified with a thermodynamic equilibrium when life is absent. Continuing the chain of logic we may say that the origin of life can be considered as the loss of stability for thermodynamic equilibrium and the movement of the system away from it along one trajectory. In this case, (in accordance with the Chetaev instability theorem) if the exergy increases along this trajectory, i.e. the inequality $dL/dt = d(Exergy)/dt > 0$ takes place, then the thermodynamic equilibrium is unstable. And since biological evolution can be considered as a 'moving of the system away from thermodynamic equilibrium', then we can say that the *exergy must increase in the process of evolution.* It is easy to see that this is the other formulation for Jørgensen's maximal principle.

8. Volterra equation and Lyapunov functions

Let us render concrete the right sides of (4.1) and write these equations in the Volterra form

$$\frac{dN_i}{dt} = N_i(\varepsilon_i - \sum_{j=1}^{n} \gamma_{ij} N_j);\ i = 1,\dots,n \tag{8.1}$$

with a community matrix $\Gamma = \|\gamma_{ij}\|$. Since the equilibrium **N*** is non-trivial then the vector **N*** must be the non-trivial solution of the system of algebraic linear equations $\sum_{j=1}^{n} \gamma_{ij} N_j^* = \varepsilon_i;\ i = 1,\dots,n$ and (8.1) can be rewritten in the form

$$\frac{dN_i}{dt} = -N_i \sum_{j=1}^{n} \gamma_{ij} \delta N_j);\ i = 1,\dots,n \tag{8.1'}$$

where $\delta N_i = N_i - N_i^*$.

If the function φ as $\varphi(\xi_i) = a_i(\xi_i - \ln \xi_i - 1)$ is chosen, where a_i are arbitrary positive coefficients, then the corresponding Lyapunov function will be

$$L = \sum_{i=1}^{n} (N_i - N_i^*) - N_i^* \ln(N_i / N_i^*). \tag{8.2}$$

and we get immediately

$$\frac{dL}{dt} = -\sum_{i=1}^{n} \sum_{i=1}^{n} a_i \gamma_{ij} \delta N_i \delta N_j . \tag{8.3}$$

It means that the non-trivial equilibrium **N*** is always stable if the quadratic form $\sum_{i=1}^{n} \sum_{i=1}^{n} a_i \gamma_{ij} \delta N_i \delta N_j$ is positive definite. This is the well-known result in the general theory of Volterra systems (see Svirezhev, Logofet 1978).

In this theory, if the form $F(N_1,\dots,N_n) = \sum_{i=1}^{n} \sum_{i=1}^{n} a_i \gamma_{ij} N_i N_j$ is positive definite, the system is called *dissipative*. The necessary conditions for a system to be dissipative are as follows: 1. the diagonal elements of the matrix Γ must be positive,

$\gamma_{ii} > 0;\ i = 1,\dots,n$, i.e. all species are to be self-limited; 2. det $\Gamma \neq 0$, thus implying uniqueness of the solution to system $\sum_{j=1}^{n} \gamma_{ij} N_j^* = \varepsilon_i;\ i = 1,\dots,n$. In other words, in dissipative systems the non-trivial equilibrium **N*** is always unique (but is not always positive).

Note that $dL(\mathbf{N}(t))/dt < 0$ (see (8.3)) everywhere within the positive orthant and $dL(\mathbf{N}(t))/dt = 0$ only when $\mathbf{N}(t) = \mathbf{N}^*$. Therefore, the equilibrium **N*** is globally asymptotically stable within the positive orthant, thus explaining the general pattern of trajectory behaviour in the dissipative system. It means that the existence of non-trivial stable equilibrium serves as not only a sufficient condition but also as a necessary one for stable dynamics of the system.

9. Extreme properties of Volterra systems for competing species. 'Horizontally - structured (competitive)' communities

We consider again the Volterra system (8.1) assuming the symmetry for Γ, i.e. $\gamma_{ij} = \gamma_{ji}$, and $\varepsilon_i > 0$. Such a system is a very popular model for a 'horizontal - structured (competitive)' community, i.e. for a community of competing species situated on one trophic level. The transformation $\eta_i = \pm\sqrt{N_i}$, $i = 1,\dots,n$ transfers the positive orthant P^n into the complete co-ordinate space R^n_η in which the trajectories of the system (8.1) are trajectories of the steepest ascent for the function

$$W = \frac{1}{4}\sum_{i=1}^{n} \varepsilon_i \eta_i^2 - \frac{1}{32}\sum_{i=1}^{n}\sum_{j=1}^{n} \gamma_{ij}\eta_i^2\eta_j^2 = \sum_{i=1}^{n} \varepsilon_i N_i - \frac{1}{2}\sum_{i=1}^{n}\sum_{j=1}^{n} \gamma_{ij} N_i N_j . \tag{9.1}$$

Then the system (9.1) can be rewritten in the gradient form

$$\frac{d\eta_i}{dt} = \frac{\partial W}{\partial \eta_i},\ i = 1,\dots,n; \qquad \frac{dW}{dt} = \sum_{i=1}^{n} \frac{\partial W}{\partial \eta_i}\frac{d\eta_i}{dt} = \sum_{i=1}^{n} \left(\frac{\partial W}{\partial \eta_i}\right)^2 \geq 0. \tag{9.2}$$

From (9.2) it follows that the value W increases in the process of the system evolution attaining maximum in the equilibrium, i.e. W may be considered as a goal function for the competitive community (in detail see Svirezhev, Logofet 1978). Then the function $L = W(\mathbf{N}) - W(\mathbf{N}^*) \leq 0$ is a Lyapunov function for (8.1). Note that this Lyapunov function does not belong to the introduced above class and, as opposed to those functions, the equilibrium **N*** must not necessarily be non-trivial, it may be situated on the appropriate boundary of the positive orthant. This implies that in the process of the community evolution one or several species are to be eliminated.

In that case when the equlibrium lies on the boundary and several $N_k^* = 0$, the Lyapunov function of type (7.1) cannot be used. However, we can consider for such points, e.g. for point $\breve{\mathbf{N}} = (0, \breve{N}_2,\dots,\breve{N}_n)$ the following function:

$$L(\mathbf{N}) = a_1 N_1 + \sum_{i=2}^{n} \breve{N}_i\, \varphi(N_i / \breve{N}_i) \tag{9.3}$$

as a Lyapunov function. Indeed, the minimum L is attained at $\mathbf{N} = \breve{\mathbf{N}}$ and the derivative in virtue of the system can be proved to be non-positive as before. Since the equality sign holds only at at $\mathbf{N} = \breve{\mathbf{N}}$ then such an equilibrium is asymptotically stable, globally in the positive orthant, i.e. the first species becomes extinct from the community under any initial conditions.

All these results also lead to a sensible interpretation. The value $R(\mathbf{N}) = \sum_{i=1}^{n} \varepsilon_i N_i$ in essence, accounts for the rate of biomass gain when competition and any kind of limitation of resources are absent, and the growth is only determined by the physiological fertility and natural mortality of the organisms. Therefore it is natural to define the value R as a *reproductive potential* of the community. The value $D = \frac{1}{2}\sum_{i=1}^{n}\sum_{i=1}^{n} \gamma_{ij} N_i N_j$ may be used to measure a *rate of energy dissipation* resulting from inter and intraspecific competition. Therefore we shall refer to D as the *total expenses in competition.* Hence, the increase in D in the process of evolution may be interpreted as the goal of a community to maximise the difference between its reproductive potential and the total expenses in competition. This can be done in several ways: either the reproductive potential is maximised at fixed expenses in competition (*r-strategy*) or the competition expenses are minimised for a limited reproductive potential (*K-strategy*) as well. There may be some intermediate cases.

It is interesting to clarify the role of competition in community organisation. Of fundamental importance among them are the concepts of the *ecological niche*. The ecological niche concept follows logically from a simple idea that organisms of any biological species are characterised by a certain range of physical and ecological conditions (described by a vector $\mathbf{x}$) in which they survive and successfully reproduce. Let the set of such vectors $\mathbf{x}$ form a space $\mathbf{E_x}$. Ecological niche of i-th species is therefore a domain Ω_i into $\mathbf{E_x}$, i.e. in the space of vitally important environmental factors within which the population of given species can exist and reproduce. Each species occupies its own ecological niche, while the niches of species coexisting in real communities are overlapping with each other.

Definitions of the ecological niche are fairly abundant in the literature, but from the mathematician's viewpoint we need a stricter definition.

Onto $\mathbf{E_x}$ we define the set of finite functions $p_i(\mathbf{x});\ i = 1,\ldots,n$ such that

$$0 < p_i(\mathbf{x}) \le 1 \text{ if } \mathbf{x} \in \Omega_i \text{ and } p_i(\mathbf{x}) \equiv 0 \text{ if } \mathbf{x} \notin \Omega_i . \tag{9.4}$$

In addition we set

$$\int_{\mathbf{E_x}} p_i(\mathbf{x})d\mathbf{x} = \int_{\Omega_i} p_i(\mathbf{x})d\mathbf{x} = 1 = 1;\ i = 1,\ldots,n . \tag{9.5}$$

In other words, the function f_i ($\mathbf{x}$) is the probability to detect even if one specimen of i-th species in a small vicinity of $\mathbf{x}$. The carriers Ω_i can be interpreted as *ecological niches*. In order to measure something, it is necessary to introduce some measure onto $\mathbf{E_x}$. We introduce the concept of a universal volume unit and define the function $K(\mathbf{x})$ such that the value

$$K_i = \int_{\mathbf{E_x}} K(\mathbf{x}) p_i(\mathbf{x}) d\mathbf{x}$$

is the volume of ecological niche for i-th species measured in universal units. Therefore, the value $K(\mathbf{x})$ can be considered as a volume (measured in the same units) which is accessible for occupation. We also introduce the values v_i which are elementary volumes occupied by one specimen of i-th species.

Let $N_i(t)$ be the total population size for the i-th species, then the population density is $n_i(t,\mathbf{x}) = p_i(\mathbf{x})N_i(t)$. We assume that the relative change of population size is proportional to the relative value of free (unoccupied) volume. Then

$$\frac{1}{n_i}\frac{dn_i}{dt} = \varepsilon_i \frac{K(\mathbf{x}) - \sum_{i=1}^{n} v_i n_i}{K(\mathbf{x})} \tag{9.6}$$

where ε_i is a coefficient of proportionality which can be interpreted as the coefficient of intrinsic rate. Rewriting (9.6) in the form

$$K(\mathbf{x})p_i(\mathbf{x})\frac{dN_i}{dt} = \varepsilon_i N_i \left[K(\mathbf{x})p_i(\mathbf{x}) - \sum_{j=1}^{n} v_j p_i(\mathbf{x}) p_j(\mathbf{x}) N_j \right],$$

and integrating the both parts we get

$$\frac{dN_i}{dt} = \varepsilon_i N_i (1 - \frac{1}{K_i}\sum_{j=1}^{n} \alpha_{ij} v_j N_j),\ i = 1,\dots,n \tag{9.7}$$

where the coefficients $\alpha_{ij} = \int_{\mathbf{E_x}} p_i p_j d\mathbf{x}$ can be considered as a measure of intersection of two niches. In the equilibrium $(0,\dots,N_i^*,\dots,0)$, the value $v_i N_i^* = K_i / \alpha_{ii}$. If we consider a number of specimens as the universal volume measure, then K_i will be the niche 'carrying capacity' and then $v_i = 1/\alpha_{ii} = 1/\int_{\mathbf{E_x}} p_i(\mathbf{x})^2 d\mathbf{x}$.

We came to the system of dynamic equations (9.7) which is coincident up to notations with the general-type of Volterra system. Notice that the scaling change of variables $u_i = (K_i v_i / \varepsilon_i) N_i$ makes (9.7) take the form (8.1) where $N_i \Leftrightarrow u_i$ and $\gamma_{ij} = \alpha_{ij}(\varepsilon_i \varepsilon_j / K_i K_j);\ \gamma_{ij} = \gamma_{ji}$. Since the quadratic form

$$F(\mathbf{z}) = \sum_{i=1}^{n}\sum_{j=1}^{n} \gamma_{ij} z_i z_j = \sum_{i=1}^{n}\sum_{j=1}^{n} \frac{\varepsilon_i \varepsilon_j}{K_i K_j} (\int_{\mathbf{E_x}} p_i p_j d\mathbf{x}) z_i z_j = \int_{\mathbf{E_x}} \left[\frac{\varepsilon_i}{K_i} p_i(\mathbf{x}) \right]^2 d\mathbf{x} > 0$$

for all $\mathbf{z} \neq 0$, i.e. the quadratic form is positive definite and system (9.7) is dissipative, the function

$$\begin{aligned} D &= \int_{\mathbf{E_x}} \left[K(\mathbf{x}) - \sum_{i=1}^{n} v_i n_i \right]^2 d\mathbf{x} = \int_{\mathbf{E_x}} \left[K(\mathbf{x}) - \sum_{i=1}^{n} v_i p_i(\mathbf{x}) N_i \right]^2 d\mathbf{x} = \\ &= \int_{\mathbf{E_x}} K^2(\mathbf{x}) d\mathbf{x} - 2\sum_{i=1}^{n} v_i K_i N_i + \sum_{i=1}^{n}\sum_{j=1}^{n} v_i v_j \alpha_{ij} N_i N_j \end{aligned} \tag{9.8}$$

is interpreted as the mean square difference between the really existing vital space and the space needed for a community to exist in species composition $\mathbf{N}(t)$. If every species is considered as occupying a certain volume in this space, then the function can be regarded as a measure of how densely the species are 'packed' in the given environment, the less being D, the closer being the species packing. If an equilibrium is stable, then the function attains a minimum, which is global on the positive orthant.

If we use the introduced above variables u_i then we get:

$$D = \int_{\mathbf{E}_x} K^2(\mathbf{x})d\mathbf{x} - 2\sum_{i=1}^{n} \varepsilon_i u_i + \sum_{i=1}^{n}\sum_{j=1}^{n} \gamma_{ij} u_i u_j = D_0 - 2W \text{ (see (9.1))}. \quad (9.9)$$

Since W increases everywhere along trajectories leading to the stable equilibrium and attains an isolated maximum at that point, function D decreases accordingly along the trajectories, reaching a minimum at that equilibrium. Thus, if there is a positive non-trivial equilibrium, then all the species coexist. If there is no such equilibrium but there is another one laying on the boarder of the orthant, then the one of partly stable equilibria, the minimum proves to be stable and the species vanish corresponding to the zero components of the equilibrium.

We may determine the largest number of new species that may consolidate the community after it having been invaded by a small number of individuals belonging to those species. For this we must consider an expanding sequence of state spaces with increasing dimension, i.e. the sequence of orthants $\mathrm{P}^n \subset \mathrm{P}^{n+1} \subset \mathrm{P}^{n+2} \subset \ldots$ and the associated sequence of minimal values for function $D(\mathbf{N})$:

$$\min_{\mathbf{N}\in\mathrm{P}^n} D(\mathbf{N}) \geq \min_{\mathbf{N}\in\mathrm{P}^{n+1}} D(\mathbf{N}) \geq \min_{\mathbf{N}\in\mathrm{P}^{n+2}} D(\mathbf{N}) \geq \ldots \quad (9.10)$$

The sequence (9.10) has a limit which is greater than or equal to zero. If this limit is attained at a member of the sequence with a finite number, then the dimension of the corresponding orthant is just the largest possible number of species in the community.

The results above can be formulated as the following principle of MacArthur:

A *community of species competing for the vital space (determined by function* $K(\mathbf{x})$*) evolves towards the state of the closest species packing, the density of packing always increasing in the course of community evolution and attaining, at the equilibrium, the maximal possible closeness for the given environment.*

As D is minimal at a stable equilibrium, this means that there is no free living space for introduction of a new species with characteristics similar to those of a species already present in the community. Such an introduction becomes possible only if the new species occupy the regions of the vital (ecological) space which have been left without use by the former species. However, in terms of the formal scheme, this now suggests a change in its properties i.e. the change in the form of function $K(\mathbf{x})$.

Note that the most dense packing may also be attained when one or several species are eliminated from the community.

10. Trophic chains and stability in 'vertically-structured' communities

It is already common place that both a hierarchy of trophic levels and interactions within a single level are of crucial importance to stability of the whole community structure. The vertical structure implies interactions among the levels and trophic chain, which is a structure to display typical paths for the resource flow through the ecosystem, a typical 'vertical-structured' community. This flow is realised in food energy transfers from a species (or a group of species) to another one which are both linked by a prey-predator (or resource-consumer) relation. In Figure 6 and ensuing equations, Q is the rate of the external resource inflow and R denotes the resource to be utilised by species 1 of the biomass N_i . Its specific rate of resource uptake is equal to $\alpha_0 R$. Of the total amount of the resource consumed, $\alpha_0 RN_1$, only a fraction k_1 contributes to reproducing new biomass of species 1. Concurrently, the species 1 biomass dies off at a rate m_1 and serves as a food to species 2, being consumed at a specific rate $\alpha_1 N_1$. Species 2, in turn, is consumed by species 3 at a specific rate $\alpha_2 N_2$ and so on. The chain ends up with species n, whose biomass is consumed by nobody.

This is a so-called *open* trophic chain, its dynamics is described by (10.1), where $N_0 = R$ and $N_{n+1} \equiv 0$:

$$\frac{dN_0}{dt} = Q - \alpha_0 N_0 N_1;\ \frac{dN_i}{dt} = N_i(-m_i + k_i \alpha_{i-1} N_{i-1} - \alpha_i N_{i+1});\ i = 1,\dots,n. \qquad (10.1)$$

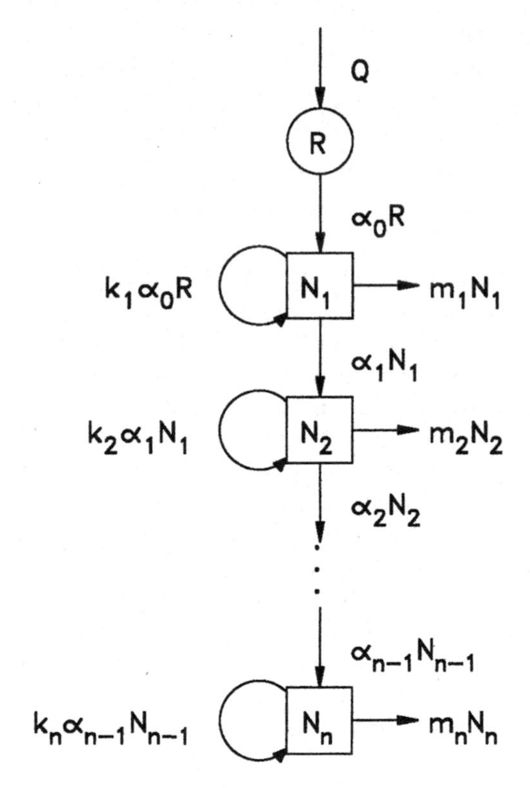

Figure 6: Schematic representaion of an open trophic chain

The system has only such type of equilibria where the first q species only are non-zero, $\mathbf{N}_q^* = (N_0^*, N_1^*, \dots, N_q^*, 0, \dots, 0)$; $1 \le q \le n$. Such a stable equilibrium will be logically called a *trophic chain of length q*. What are the conditions presented for the pertinent equilibria to exist and to be stable, i.e. the conditions for the q-length chain to exist? We can prove (see, in details, Svirezhev, Logofet 1978; Logofet 1993) that for an open trophic chain of length q to exist it is necessary and sufficient that the resource inflow rate belongs to the finite interval in the Q-axis:

$$Q^*(q) < Q < Q^*(q+1) \qquad (10.2)$$

whose bounds are defined as $Q^*(q) = \alpha_0 f_{q-1} f_q$ where

$$a) f_q = \sum_{s=1}^{q/2} \frac{\mu_{2s}}{g_2 \cdot g_4 \cdot \ldots \cdot g_{2s}} \text{ for even } q; b) f_q = \sum_{s=1}^{(q+1)/2} \frac{\mu_{2s-1}}{g_1 \cdot g_3 \cdot \ldots \cdot g_{2s-1}} \text{ for odd } q; \tag{10.3}$$

$$\mu_i = m_i / \alpha_i; g_i = k_i \alpha_{i-1} / \alpha_i; i = 1, \ldots, n.$$

The main conclusion is that the whole potential range of resource inflow rates (i.e. the Q-axis) is divided by points $Q^*(q)$; $q = 1,2,\ldots n$ into consecutive intervals within which only trophic chains of a fixed length exist (see Figure 7). In other words, the energy entering the system with the resource inflow undergoes 'quantification'; as follows from thorough investigation of sequences $\{Q^*(q)\}$, the higher being the number of the next trophic level, the greater being the energy 'quantum' that level requires for it to fix in the ecosystem.

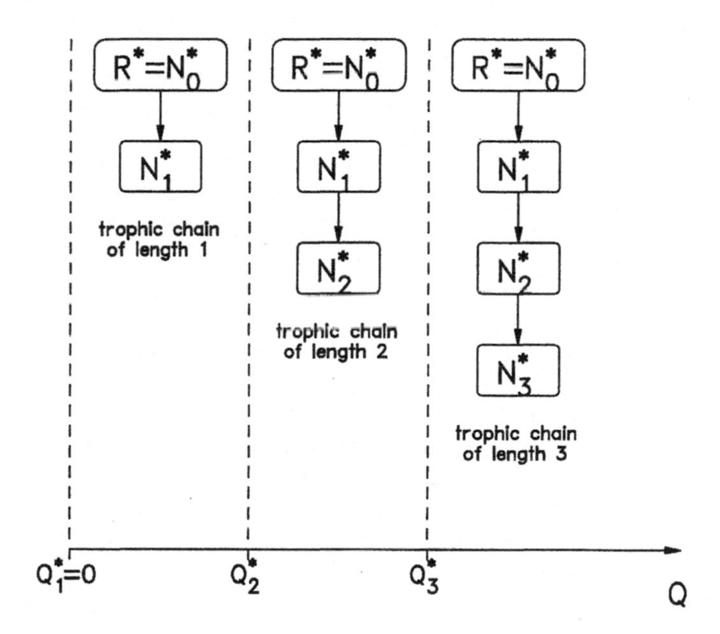

Figure 7: 'Quantification of resource inflow Q and the corresponding domains of existence for trophic chains of different lengths.

Unfortunately, our conclusions depend on the evenness or oddness of q. I think this is most likely an artefact than a 'Law of Nature'. However this effect vanishes, if we take one very realistic assumption that $k_i << 1$. Let us consider a very simple example then $k_i = k << 1$; $\alpha_i = \alpha$ and $m_i = m$. Calculating $Q^*(q)$ with the accuracy $O(k)$ we get $m^2 / \alpha k^q < Q < m^2 / \alpha k^{q+1}$, from this it follows that

$$q = E\{\ln(\alpha Q / m^2) / \ln(1/k)\} \tag{10.4}$$

where $E(x)$ is the entire function of x. It is obvious that the increase of the chain length must be accompanied by the *exponential* growth of the inflow of external energy or resource.

11. Matrix properties as a key to stability analysis

The progress in the theory and applications of the Lyapunov stability concept was predetermined by two fundamental theorems of A.M. Lyapunov: on stability by the Lyapunov function of certain properties and on stability by the linear approximation. If a Lyapunov function can judge the domain of stability in the state space then the linear approximation can establish the local stability only. Unfortunately, there is no general method to construct a Lyapunov function, whereas the second Lyapunov method is more universal, reducing the stability problem to testing of certain properties of a certain matrix, and the matrix and graph theory approaches may be used.

Revenons à nos moutons, i.e. to our general model

$$\frac{dN_i}{dt} = F_i(N_1, \ldots, N_n), \; i = 1, \ldots, n \tag{4.1}$$

linearise the system in the vicinity of some equilibrium **N*** so that ($\mathbf{x}(t) = \mathbf{N}(t) - \mathbf{N}^*$):

$$\frac{dx_i}{dt} = \sum_{j=1}^{n} \left(\frac{\partial F_i}{\partial N_j} \right)_{N^*} x_j + O(x_i); i = 1, \ldots, n; \tag{11.1}$$

or, in a vector form,

$$\frac{dx}{dt} = Ax + O(x) \text{ where } A = \left\| a_{ij} = (\partial F / \partial N)_{N^*} \right\| \tag{11.2}$$

Here $O(\mathbf{x})$ is the terms of the higher order of smallness than linear. In accordance to the Lyapunov theorem on stability in the linear approximation, if all the eigenvalues of matrix **A** (community matrix) have negative real parts, then the equilibrium **N*** in the non-linear system (4.1) is asymptotically stable(locally). If, on the contrary, there is an eigenvalue whose real part is positive, then the equilibrium **N*** is unstable. Matrix **A** is called either stable or unstable, correspondingly.

This statement still gives a powerful method to establish sufficient conditions for an equilibrium to be stable in a non-linear model. Similar to the fact of the real analysis that the linear term bears the major part of the function increment, the linear stability analysis reveals the major tendency, if any, in dynamic behaviour of the non-linear system.

12. Qualitative stability

The notion of *qualitative stability* is defined in intuitive rather than formal terms for a system of interacting components of any kind. In a general context, it means that a system holds its stability under any quantitative variations in the strength of linkage among system components which however keep unchanged the qualitative types of all interactions. In ecological terms, qualitative stability means that an ecosystem or community holds its stability under any quantitative variations in intensity of interactions among components (species) which however keep unchanged the type of intra- and interspecies relations.

Let the dynamics of an *n*-species community be governed by system (4.1). After linearisation we get the system (11.2) with a community matrix **A**. We assume also that the equilibrium **N*** is a point into state space and there is not another equilibrium manifold. This means that the system $\sum_{i=1}^{n} a_{ij}x_j^* = 0$ must have only one trivial solution **x*** = 0, i.e. det **A** ≠ 0. So, the equilibrium **N*** is *qualitatively stable* if the matrix **S** = sign**A** = $\|\mathrm{sign}(a_{ij})\|$ is stable.

We can consider **S** as a matrix which describes a qualitative structure of interaction inside a community. Note that the sign of an entry a_{ij} of the community matrix shows a qualitative nature of the effect that j-th species exerts upon *i*-th species ('+' means stimulating, '-' means suppressing, and '0' means neutral relation), while the absolute value of a_{ij} measures its quantitative effect. The pair of signs of two symmetric entries, a_{ij} and a_{ji}, gives thc basis for classification of pair-wise interactions. It is interesting that the first of such types of classification were suggested by V. Khlebnikov (1910), a very famous Russian poet. It contains the following 6 major types:

$++$ mutualism or symbiosis,

$+-$ prey-predator (or resource-consumer, or host-parasite),

$+0$ commensalism,

$--$ competition,

-0 amensalism,

00 neutralism

It is obvious that a digraph associated with matrix **S**, for instance, the digraph in Figure 8 corresponds to the trophic chain of the length 4 (see Section 10). The zero 'species' (resource) is self-regulating, since $(\partial F_0 / \partial N_0)_{\mathbf{N}^*} = -\alpha_0 N_1^* < 0$.

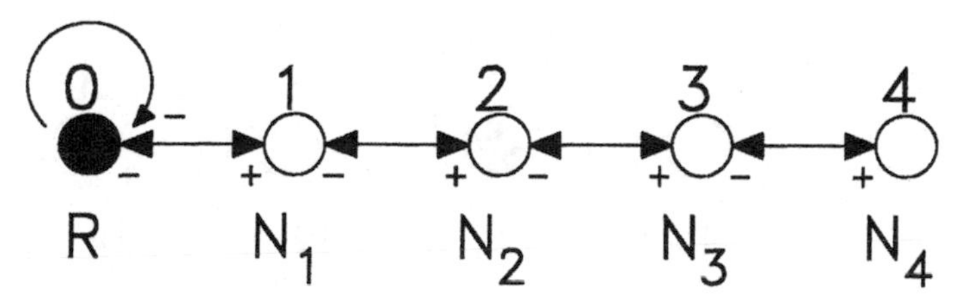

Figure 8: The digraph corresponds to the trophic chain of length 4. The zero 'species' (resource) is self regulating. The system is qualitatively unstable

The following conditions are necessary for a matrix **A** to be qualitatively stable:

1. $a_{ii} \le 0$ for all *i* and $a_{ii} < 0$ for some i_0,
2. $a_{ij} \cdot a_{ji} \le 0$ for any $i \ne j$,
3. for any sequence of 3 or more indices $i_1 \ne i_2 \ne \ldots \ne i_m$ the inequalities $a_{i_1 i_2} \ne 0,\ a_{i_2 i_3} \ne 0,\ \ldots,\ a_{i_{m-1} i_m} \ne 0$ imply $a_{i_m i_1} = 0$.

Condition 1 states that there cannot be any self-stimulating species in a qualitatively stable community and at least one species must be self-limited.

Condition 2 means that no relations of competition (--), nor mutualism (++) can be in a qualitatively stable community.

Condition 3 forbids any directed loop (or *cycle*) of length three or more in the community structure. This is probably the most severe restriction for ecological systems. In particular, it excludes all 'omnivorous' cases, where a predator feeds on two prey species one of which also is a food to another.

It is easy to be convinced that the trophic chain in Figure 8 complies with the necessary conditions of qualitative stability.

The sufficient conditions of qualitative stability can be described by the following so-called *colour test* (in details see Logofet 1993):

1. each self-regulating vertex is black,
2. there is at least one white vertex,
3. each white vertex is connected at least one other white vertex,
4. each black vertex connected to one white vertex is connected to at least one other white vertex.

Later, we introduce the notion of *predation community*. Consider a vertex (in the digraph of complete community), involved in a 2-cycle with one '+' line and one '–' line, and associate all the other vertices related to the given ones by 2-cycles of this kind; then associate with those new vertices all additional vertices related by such 2-cycles, and so on. In other words, associate into one set all the species, to form together with the given one a structure of prey - predator links. The maximal set of all such species is the *predation community*, containing the first species. For instance, the trophic chain with the structure described by digraph in Figure 8 is the predation community. A species which is not connected to any other species by prey - predator relations is a *trivial* predation community.

Finally, *for qualitative stability of a community it is necessary and sufficient that the matrix* ***A*** *satisfies the necessary conditions 1 and 2 and that the digraph of each of its predation communities fails the colour test.*

It is easy to see that the digraph of trophic chain in Figure 8 fails the test since condition 4 is failed (the zero black vertex connected to only one white vertex). On the contrary, the community with the structure described by digraph in Figure 9 passes the colour test, and it is structurally unstable.

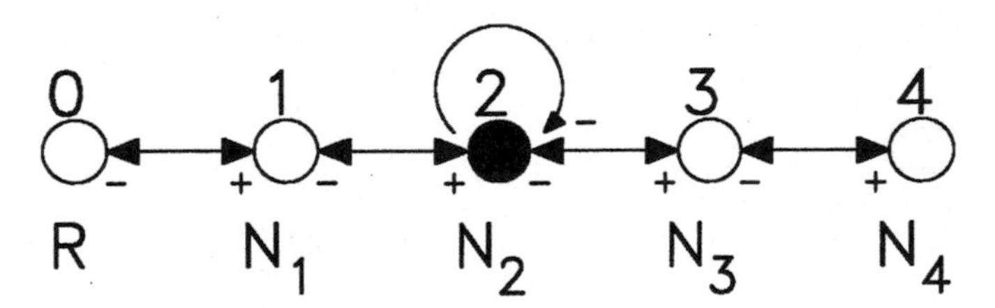

Figure 9: The digraph is very close to one of Figure 8, but the second species (not the resource) is self regulating. As a result the system is qualitatively unstable

Notice that the class of qualitatively stable communities is quite restricted. In particular, it does not embrace competitive communities. Nevertheless, analysis of qualitative stability for natural ecosystem is a fairly simple means of deriving some information about their structure. This analysis can identify, for instance, those species whose position in the community structure is of critical importance for the stability of the whole system.

13. Conclusion

In the conclusion, I would like to emphasise that the stability problem in mathematical ecology by no means may be classed with solved or nearly solved problems. Perhaps, we can say that it is only in the stage of development and formation. We are still too much restricted by the burden of ideas and concepts from the classic stability theory, and therefore any new ideas, concepts, methods may be only welcomed.

REFERENCES

Barbashin, E.A., 1967. Introduction to Stability Theory. Nauka, Moscow (in Russian).
Jeffries, C., 1988. Mathematical Modelling in Ecology: a Workbook for Students. Birkhauser, Boston.
Jørgensen S.E., 1992. Integration of ecosystem theories: a pattern. Kluwer Academic Publishers, Dordrecht.
Landau, L. and E. Lifshitz, 1964. Statistical physics. Nauka, Moscow (in Russian).
Lewontin, R.C., 1969. The Meaning of Stability. In: *Diversity and Stability in Ecological Systems, Brookhaven Symposium in Biology,* No22, National Bureau of Standards, U.S. Dept. Commerce, Springfield, VA.
Logofet, D.O., 1993. Matrices and Graphs: Stability Problems in Mathematical Ecology. CRC Press, Boca Raton.
MacArthur, R.H., 1955. Fluctuations of animal population and a measure of community stability. *Ecology* 36: 533-536.
Malkin, I.G., 1966. Theory of motion stability. Nauka, Moscow (in Russian).
Margalef, R.A., 1951. A practical proposal to stability. *Publ de Inst de Biol Apl Univ de Barselona,* 6: 5-19.
May, R.M., 1973. Stability and Complexity in Model Ecosystems. Princeton University Press, Princeton, NJ.
Odum, E.P., 1976. Diversity as a function of energy flow. In: van Dobben WH and McConnel RH (Eds.) *Unifying Concepts in Ecology.* W.Junk Publ., The Hague: 11-14.
Odum, E.P., 1983. Basic Ecology. Saunders College Publ., Philadelphia, PA.
Svirezhev, Y.M., 1976. Vito Volterra and the modern mathematical ecology. In: Volterra V *Mathematical theory of struggle for existence.* Nauka, Moscow (the postscipt to the Russian translation of this book).
Svirezhev, Y.M., 1983. Modern Problems in Mathematical Ecology. *Proc. Int. Congress of Math.,* Warsaw, Vol. II: 1677-1693.
Svirezhev, Y.M., 1987. Non-linear Waves, Dissipative Structures and Catastrophes in Ecology. Nauka, Moscow (in Russian)
Svirezhev, Y.M. and D.O. Logofet, 1978. Stability of Biological Communities. Nauka, Moscow (English version: 1983, Mir, Moscow).
Svirezhev, Y.M. and D.O. Logofet, 1995. The mathematics of Community Stability. In: Patten B.C, Jørgensen S. E. (Eds.) *Complex Ecology: the Part-Whole Relation in Ecosystems.* Prentice Hall PTR, Englewood Cliffs, NJ.
Usher, M.B. and M.H. Williamson, (Eds.) 1974. Ecological stability. Chapman & Hall, London.

II.8.2 Resilience in Ecological Systems

L.H. Gunderson, C.S. Holling and G.D. Peterson

1. Introduction

In ecological systems, resilience is the extent to which a system can withstand disruption before shifting into another state (Holling 1973). We discuss this emergent property in three sections. We begin with a review of the concept of resilience and contrast meanings used by various authors primarily around the notion of multiple stability domains. We then follow with a discussion of the relationship between resilience and system complexity or diversity. We conclude with a discussion of relationships between self-organising time dynamics in ecosystems and resilience. Each of these will be discussed, starting with resilience.

2. Resilience-Definitions and Terms

Resilience of a system has been defined in two very different ways in the ecological literature, each reflecting different aspects of stability. One definition focuses on efficiency, constancy and predictability. The other focuses on persistence, change and unpredictability. Holling (1973) first emphasised these different aspects of stability to draw attention to the tensions between efficiency and persistence, between constancy and change, and between predictability and unpredictability.

The more common definition, considers ecological systems to exist close to a stable steady-state. Resilience is the ability to return to the steady-state following a perturbation (Pimm, 1984; O'Neill et al., 1986, Tilman et al., 1994). This definition is more amenable to mathematical representation and to experimental inquiry, hence is more commonly used. We term this engineering resilience. The second definition emphasizes conditions far from any stable steady-state, where instabilities can flip a system into another regime of behaviour - i.e. to another stability domain (Holling, 1973). In this case resilience is measured by the magnitude of disturbance that can be absorbed before the system redefines its structure by changing the variables and processes that control behaviour. This we will call ecological resilience (Walker et al., 1969).The two contrasting aspects of stability - essentially one that focuses on maintaining *efficiency* of function (engineering resilience) vs. one that focuses on maintaining *existence* of function (ecological resilience) are so fundamental that they can become alternative paradigms whose devotees reflect traditions of a discipline or of an attitude more than of a reality of nature. Those using the concept of engineering resilience tend to explore system behaviour near a known stable state, while those examining ecological resilience tend to look for other stable states, and the locations of the boundaries between states.

Those who emphasise the near equilibrium definition of engineering resilience, for example, draw predominantly from traditions of deductive mathematical theory (Pimm, 1984) where simplified, untouched ecological systems are imagined, or from traditions of engineering, where the motive is to design systems with a single operating objective (Waide and Webster, 1976; De Angelis, 1980; O'Neill et al., 1986). On the one hand, that makes the mathematics more tractable, and on the other, it accommodates the engineer's goal to develop optimal designs. There is an implicit assumption that there is global stability - i.e. there is only one equilibrium steady-state, or, if other operating states exist, they should be avoided by applying safeguards.

Those who emphasise the stability domain definition of resilience (i.e. ecological resilience), on the other hand, come from traditions of applied mathematics and applied resource ecology at the scale of ecosystems - e.g. of the dynamics and management of fresh water systems (Fiering, 1982), of forests (Holling et al., 1977), of fisheries (Walters, 1986), of semi-arid grasslands (Walker et al., 1969) and of interacting populations in nature (Sinclair et al., 1990; Dublin et al., 1990). Because these studies are rooted in inductive rather than deductive theory formation and in experience with the impacts of large scale management disturbances, the reality of flips from one operating state to another cannot be avoided. Moreover, it becomes obvious that the variability of critical variables forms and maintains the stability landscape.

The heart of these two different views of resilience lies in assumptions of whether or not multi-stable states exist (Figure 1). If it is assumed that only one stable state exists or can be designed to so exist, then the only possible definition and measures for resilience are those that are near equilibrium - such as characteristic return time. And that is certainly consistent with the engineer's desires to make things work, not to intentionally make things that break down or suddenly shift their behaviour. But nature is different.

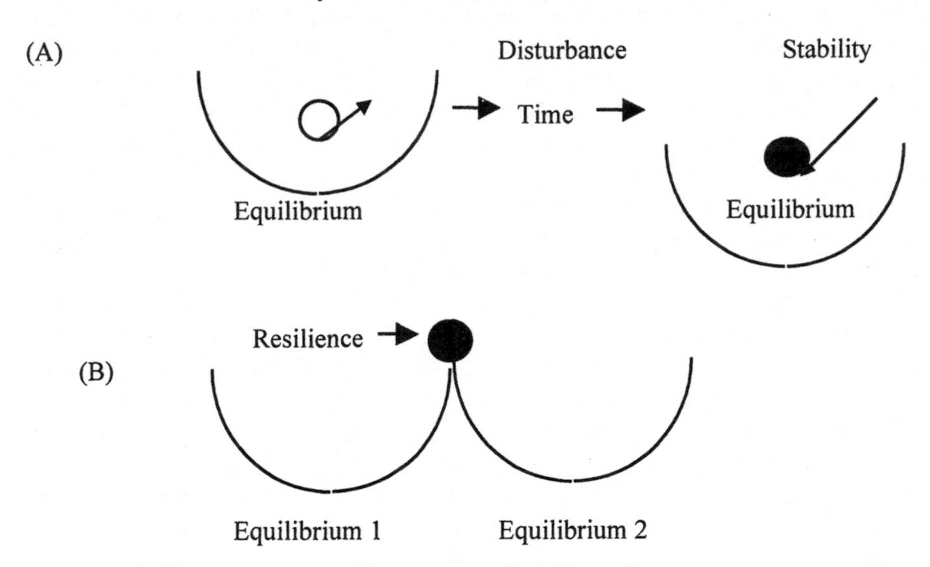

Figure 1: Graphic representation of system stability and resilience. (A) Single equilibrium with system dynamics represented by disturbances that move a ball on a surface. Stability is provided by forces that return system to global equilibrium, (B) Multiple stability domains, where resilience is defined as the amount of disturbance required to move the system into an alternative stable state.

Take the example of nutrient enrichment in the Florida Everglades. The Everglades is an oligotrophic wetland, limited primarily by phosphorus. For the past 5000 years or so, the ecosystem effectively self-organised around this low nutrient status, pulsed by annual wet/dry cycles and by decadal recycling associated with fires (Loveless, 1959, Craighead. 1971, Gunderson, 1994). The resulting landscape mosaic had small areas of enhanced nutrients or eutrophy in tree islands that were maintained by wading birds nesting or in local refugia maintained by the cycles of flooding and drydowns that first collected diffuse energy and then concentrated it locally. The remainder of the landscape (sawgrass marshes and wet prairies) adapted to low nutrient thresholds (Steward and Ornes, 1973).

In the late 1940's a plan was put into effect that divided the Everglades into three designated land uses; agriculture (in the northern third of the historic Everglades), urban (the eastern fifth) and conservation (in the southern and central remaining half of the historic system). The latent effects of these land use designations were revealed in the late

1970's and early 1980's, when large scale shifts in the vegetation were noticed in the areas immediately south of the agricultural area. After years of research, the transition from a sawgrass to cattail dominated marsh was attributed to a two part process. First, there was a slow increase in the concentration of soil phosphorus levels (Davis, 1989; 1994), followed by a fire which resulted in the replacement of sawgrass by cattail (Urban, 1994). Since the phosphorus was associated with runoff from agricultural fields, the resulting management options involved economic, human and ecological variables. At the time of this writing, plans were underway to only allow clean water to reach the areas of the Everglades set aside for conservation, but little focus had been directed to management options for the areas where resilience had been exceeded, and the vegetation community had changed from one stability domain to another.

There are different stability domains in nature and variation in critical variables test the limits of those domains. Thus a near equilibrium focus seems myopic and attention shifts to determining the constructive role of instability in maintaining diversity and persistence, and to management designs that maintain ecosystem function despite unexpected disturbances. Such designs maintain or expand ecological resilience. It is those ecosystem functions and ecological resilience that provides the ecological 'services' that invisibly provide the foundations for sustaining economic activity. And those functions and resilience that provide service are related to the complexity and diversity of a system.

3. Diversity and Resilience

The relationship between biological diversity and ecological stability has been the subject of an ongoing discussion in ecology ever since Darwin (Darwin, Elton 1958, May 1973). The central question people have explored is 'is an ecosystem that includes more species more stable than one that includes fewer species?' Recently, Tilman (1994, 1996) has demonstrated that over ecologically brief periods, an increase in species number increases the efficiency and stability of some ecosystem functions, but decreases the stability of the populations of the species. While this work is important and interesting, it focuses on how an ecosystem behaves near some steady state. As we discussed above, we feel it is important to discover the role of ecological diversity over a much broader range of variations, and that in this area the relationship between diversity and resilience has been poorly developed.

When grappling with this broader relationship between diversity and resilience two hypotheses are commonly discussed: Ehrlich's (1991) 'rivet' hypothesis and Walker's (1992) driver and passengers hypothesis. Ehrlich's hypothesis proposes that there is little change in ecosystem function as species are added or lost, until a threshold is reached. At that threshold the addition or removal a single species leads to system reorganisation. This model assumes that species have overlapping roles, and that as species are lost the ecological resilience of the system is decreased, and then overcome entirely. Walker proposes that species can be divided into 'functional groups' or 'guilds', groups of species that act in an ecologically similar way. Walker proposes that these groups can be divided into 'drivers' and 'passengers'. Drivers are 'keystone' species, that control the future of an ecosystem, while the passengers live in but do not alter significantly this ecosystem. However, as conditions change, endogenously or exogenously, species shift roles. In this model, removing passengers has little effect, while removing drivers can have a large impact. Ecological resilience resides both in the diversity of the drivers, and in the number of passengers who are potential drivers. These two hypotheses provide a start, but richer models of ecological complexity are needed that better incorporate ecological processes, dynamics and scale. Theory, models and data suggest that a small number of keystone processes create discontinuous dynamic spatial and temporal patterns in ecosystems (Holling et al. 1996 , Levin 1995) yet allow for great diversity of organisms. These keystone ecological processes produce a discontinuous distribution of structures in ecosystems, and these discontinuous structures generates discontinuous patterns in adult body masses of animals that inhabit that landscape (Holling 1992, Morton 1990). Consequently, while animals that function at the same scale are separated by functional specialisation (e.g. insectivores, herbivores, arboreal frugivores, etc.), animals that function at different scales can utilize similar resources (e.g. shrews and anteaters are both insectivores but utilize insects at different scales). This partitioning by scale and function provides many opportunities for different styles of resource use.

Local processes such as competitive relationships certainly contribute to species differences among ecosystems, but the structural differences between ecosystems from the tundra to the tropics are primarily produced by larger scale disturbance processes which are initiated locally and spread across landscapes. These contagious processes include abiotic processes, such as fire, storms, and floods, and zootic processes such as insect outbreaks, large mammal herbivory and habitat modification (Naiman, 1988, McNaughton, 1988, Pastor and Cohen 1996). These processes, interacting with topography and regional climate, form the ecosystem-specific structures that shapes the morphology and diversity of animal communities. They also generate spatial and temporal variation which increases the diversity of plant species by periodically over-riding the competitive dominance relations that occur locally (Holling, 1991). For example in the eastern boreal forest of Canada, fire and spruce budworm outbreak kill large areas of forest. In interaction, with climate, existing vegetation, and each other these processes produce a mosaic of even aged forest stands in the landscape. Since the age a stand reaches before being destroyed is primarily determined by disturbance, and what species exist within the stand is influenced by landscape pattern - these disturbance processes also strongly control what exists within stands. Consequently, these disturbance processes strongly influence the distribution and type of resources that occur in eastern Canadian boreal forest across a broad range of ecological scales.

An ecosystem that has several scales of ecological structure allows members of multi-taxa food guilds to minimize competition by having their members spread among separate body mass clumps (Figure 2). This minimizes competition for resources because different scales are exploited by the species in each body mass lump (Allen et al. 1995). In addition, because a range of food resources is exploited by the set of foragers, rapid response to sudden increases or decreases in one type of food becomes possible and introduces strong negative feed-back regulation over a wide range of densities of the food items (Holling 1986). The consequence of all that variety is that the species combine to form an overlapping set of reinforcing influences that are less like the redundancy of engineered devices and more like portfolio diversity strategies of investors (Peterson, 1994). The risks and benefits are spread widely to retain overall consistency in performance independent of wide fluctuations in the individual species.

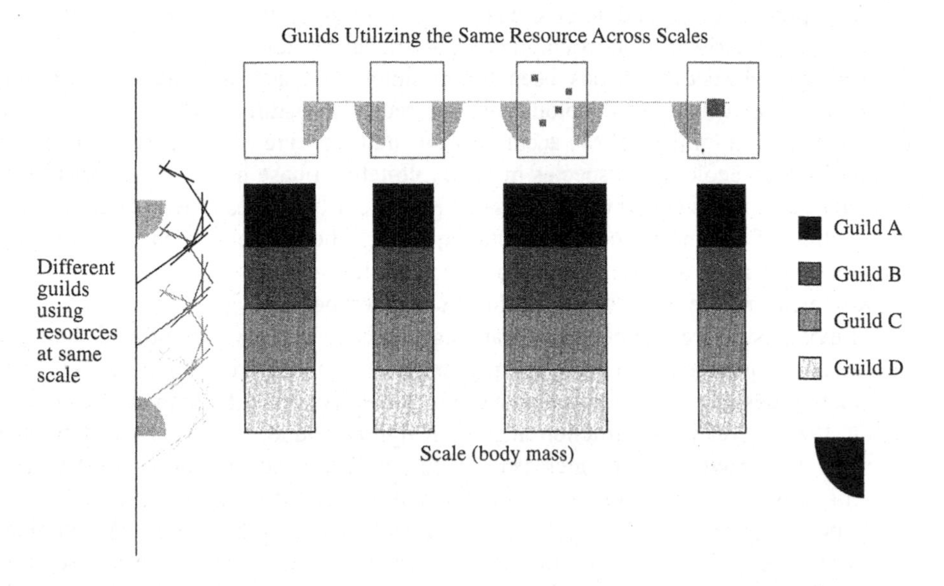

Figure 2: Animal species belonging to different ecological guilds exist at different body sizes. For example, there are both small and large insectivores. This distribution provides two forms of resilience. At the same scale, animals from different guilds can utilise the same resources with lower efficiency. Also, animals that utilise the same resources can begin to utilise resources from a lower level if they form large enough aggregations. For example, if insectivores were removed from a group, insects would become easier to catch and it would become worthwhile for animals at the same scale to switch from their normal food to insects, and it may become worthwhile for larger insectivores to eat prey items they normally would not eat.

This is at the heart of the role of the diversity of functional groups in maintaining the resilience of ecosystem structure and function (Walker 1995, Levin 1995). Such diversity provides great robustness to the functioning of the process and, as a consequence, great resilience to the system behaviour. Moreover, this seems the way many biological processes are regulated – overlapping influences by multiple processes, each one of which is inefficient in its individual effect but together operating in a robust manner. For example, those are the features of the multiple mechanisms controlling body temperature regulation in endotherms, depth perception in animals with binocular vision and direction in bird migration.

Because of the robustness of this redundancy within functional groups, and the non-linear way behaviour suddenly flips from one pattern to another and one set of controls to another, gradual loss of species involved in controlling structure initially would have little perceived effect over a wide range of loss of species. Then as loss of those species continued, suddenly, different behaviour would emerge more and more frequently in more and more places. To the observer, it would appear as if only the few remaining species were critical when in fact all add to the resilience. Although behaviour would change suddenly, resilience measured as the size of stability domains (*sensu* Holling, 1973), would gradually contract. The system, in gradually losing resilience, would become increasingly vulnerable to perturbations that earlier could be absorbed without change in function, pattern and controls.

4. Adaptive Cycle and Resilience

Over time, ecosystems demonstrate changes in structure and function, with concomitant changes in resilience. These dynamics generate a pattern of alternate periods of ordered, predictable change and periods of disordered, difficult to predict change. This alternation between predictability and unpredictability has been at the center of many ecological debates, particularly those surrounding ecosystem succession.

Ecosystem succession has been traditionally seen as controlled by two functions: *exploitation*, in which rapid colonisation of recently disturbed areas is emphasised and *conservation*, in which slow accumulation and storage of energy and material is emphasised. In ecology the species in the exploitative phase have been characterised as r-strategists and in the conservation phase as K-strategists; names drawn from the traditional designation of parameters of the logistic equation. The r-types are characterised by rapid growth in an arena of scramble competition, while the K-strategists tend to have slower growth rates and survive in an arena of exclusive competition.

Revisions in ecological understanding indicate that two additional functions are needed to adequately explain ecological change, as summarised in Figure 3. One is *release*, or 'creative destruction', a term borrowed from the economist Schumpeter (as reviewed in Elliott 1980). Creative destruction acts upon tightly bound accumulation of biomass and nutrients that have become increasingly susceptible to disturbance (overconnected in systems terms). This ecological capital suddenly released by agents such as forest fires, insect pests or intense pulses of grazing. We designate this as the omega (Ω) function.

The second functional addition is *reorganisation*, in which soil processes minimize nutrient loss and reorganize nutrients so that they become available for the next phase of exploitation. This last function is essentially equivalent to processes of innovation and restructuring in an industry or in a society - the kinds of economic processes and policies that come to practical attention at times of economic recession or social transformation. We designate these as the alpha (α) function.

During this cycle, biological time flows unevenly. The progression in the ecosystem cycle proceeds from the exploitation phase (r, Box 1, Figure 3) slowly to conservation (K, Box 2), very rapidly to release (Ω, Box 3), rapidly to reorganisation (α, Box 4) and rapidly back to exploitation. During the slow sequence from exploitation to conservation, connectedness and stability increase and ecological 'capital' such as nutrients and biomass slowly accumulate.

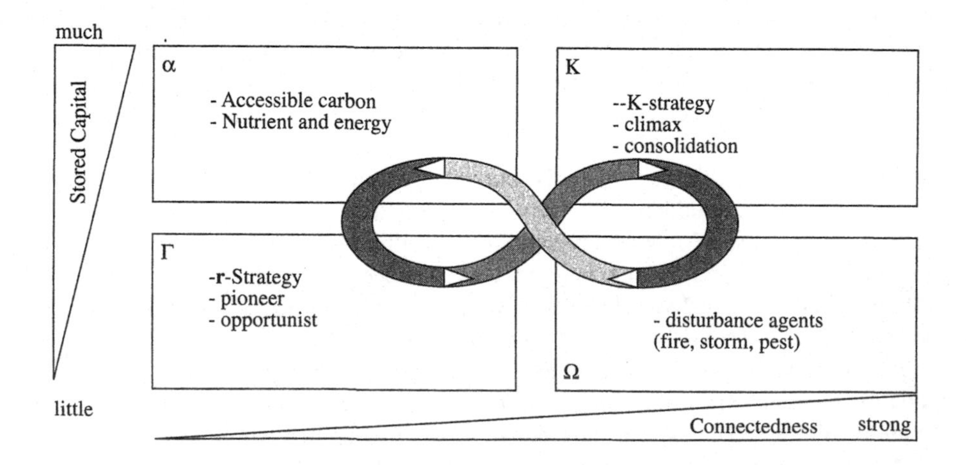

Figure 3: Holling's (1992) adaptive cycle of system dynamics, indicating four ecosystem phases (r,K, Ω,α) and the flow of events among them. The arrows indicate the speed of flow, with long arrows representing rapid change and short arrows representing slow change. The axes reflect two attributes--the horizontal (X) indicates the degree of connectedness among variables and the vertical (Y) the amount of capital stored in key variables.

The patterns produced by these four phases are discontinuous and dependent on the interaction of processes that trigger and organize the release and reorganisation functions. These interactions create multiple stable states. For example, the release role of an insect outbreak species in the spruce budworm in the eastern balsam fir forest, occurs because the maturing forest accumulates a volume of foliage that eventually dilutes the effectiveness of the search by insectivorous birds whose populations control budworm populations in younger stands. Essentially, a lower equilibrium density for budworm is set by this 'predator pit' in a stability landscape during the phase of slow regrowth of the forest (Clark et al. 1979, Holling 1988). This stability 'pit' eventually collapses as the trees mature to release an insect outbreak and reveal the existence of a higher equilibrium. A similar argument can be described for release by fire, as a consequence of the slow accumulation of fuel as a forest ages.

To summarize and generalize this example, for long periods in a regrowing forest, the slow variable (trees) controls the faster (budworm or fire) and intermediate speed ones (foliage or fuel). When budworm populations shift to the higher equilibrium, they assume control over forest dynamics. Briefly, the fast variables can assume control of behaviour and trigger a release of accumulated capital, which eliminates the higher equilibrium and the bud worm's control. Resilience and recovery are determined by this fast release and reorganisation sequence, whereas stability and productivity are determined by the slow exploitation and conservation sequence. Instabilities trigger the release or Ω phase, which then proceeds to the reorganisation or a phase where weak connections allow loosely controlled chaotic behaviour and the unpredictable consequences that can result. Stability begins to be reestablished in the r phase. In short, chaos erupts from order, and order emerges from chaos. This description is similar to Prigogine's (1980) 'order through fluctuation' model of complex system dynamics.

In nature, there is a nested set of such cycles, each occurring over its own range of scales. For example, in the boreal forest fresh needles cycle yearly, the crown of foliage cycles with a decadal period and trees, gaps and stands cycle at close to a century or longer periods. The result is an ecosystem hierarchy, in which each level has its own distinct spatial and temporal attributes. A critical feature of such hierarchies is the asymmetric interactions between levels (Allen and Starr 1982, O'Neill, et al. 1986). In particular, the larger, slower levels constrain the behaviour of faster levels. In that sense, therefore, slower levels control faster ones. If that was the only asymmetry, however, then hierarchies would be static structures and it would be impossible for organisms to exert control over slower environmental variables. However, these hierarchies are not static. They are transitory structures maintained by interaction across scales.

The birth, growth, death and renewal cycle, shown in Fig. 2, transforms hierarchies from fixed static structures to dynamic adaptive entities whose levels are sensitive to small disturbances at the transition from growth to collapse (the Ω phase), and the transition from reorganisation to rapid growth (the α phase). During other times, the processes are stable and resilient, constraining the lower levels and immune to the buzz of noise from small and faster processes. It is at the two phase transitions between gradual and rapid change that the large and slow entities become sensitive to change from the small and fast ones.

When the system is reaching the limits to its conservative growth, it becomes increasingly brittle and its accumulated capital is ready to fuel rapid structural changes. The system is very stable, but that stability is local and narrow. A small disturbance can push it out of that stable domain into catastrophe. The nature and timing of the collapse-initiating disturbance determines, within some bounds, the future trajectory of the system. Therefore this brittle state presents the opportunity for a change at a small scale to cascade rapidly through the overconnected system, bringing about its rapid transformation. This is the 'revolt of the slave variable' (Diener and Poston 1984). Collapse can be initiated by either internal conditions or external events, but typically it is internally induced brittleness (linked to over connected, and accumulated capital) that sets the conditions for collapse.

The second opportunity for small scale processes to cause system change is during the transition from reorganisation to exploitation- from α to r. During this reorganisation phase the system is in a state opposite to the conservation phase previously described. There is little local regulation and stability, so that the system can easily be moved from one state to another. Resources for growth are present, but they are disconnected from the processes that facilitate and control growth. In such a weakly connected state, a small scale change can nucleate a structure amidst the sea of chaos. This structure can then use the available resources to grow explosively and to establish the exploitative path along which the system develops and then locks into. As in Waddington's chreodic development model, there is not a stable point; rather there is a stable trajectory that progressively reinforces itself (Hodgson 1993). In Waddington's words, the system is not homeostatic (around a point), it is homeorhetic (around a path) (Waddington 1969). This transition occurs as small scale changes sow seeds of order in the larger and slower chaos within which they are embedded. The budworm example described in the previous section provides an example of these changes. The transient but critically important bottom-up asymmetry provides an opening for evolutionary change. That is, the previous system pattern may reassert itself, or the system may reorganize itself into a novel structure.

In review, the accumulating body of evidence from studies of ecosystems indicates key features of discontinuities in processes and structures, and reveals the appearance of multiple stable states. The four phase cycle of adaptive renewal captures many of these dynamics for ecological systems. Linking the adaptive cycles across scales develops a heuristic model that we dub panarchy (Gunderson, et al. 1995). Panarchy is the word we use to describe dynamic symmetries across hierarchical scales rather than an asymmetrical static relationship across scales. Two key features of panarchy are emerging that help understand resilience. These are that the events that lead to creative destruction through revolt, and those events that lead into reorganisation by remembering or carrying over elements through the period of creative destruction, are the products of cross-scale interactions.

5. Summary of Properties of Ecological Systems

In this brief review, we have attempted to highlight the theoretical foundations of resilience, and how other key properties of diversity and stability, self-organisation, and cross scale ecosystem dynamics contribute to resilience. Resilience has been defined and assessed at least two different ways in the ecologic literature. The first is more mathematically tractable and is defined as a return time to a single or global equilibrium. The second definition requires the presence of multiple equilibria and stability domains, and is the amount of disturbance that a system can withstand before it changes stability domains.

Functional diversity contributes to resilience in a redundant, overlapping manner where controls are shared among a set of components (species, taxa, numbers) across a range of scales. Self-organisation generates time dynamics that result in an adaptive cycle- where system structure and connectedness increase over time, but are subject to periods of release and reorganisation. During those periods of release and reorganisation that resilience comes into play and determines if the reorganisation phase generates development in a new stability domain.

REFERENCES

Allen, C. R., E. Forys, and C. S. Holling, 1995. Ecosystem disturbance and community transformation: gateway for invaders, exit for the endangered (In Prep.).

Allen, T. F. H. and T. B. Starr, 1982. Hierarchy: Perspectives for Ecological Complexity. The University of Chicago Press, Chicago, IL

Carpenter, S. R. and P. R. Leavitt, 1991. Temporal variation in paleolimnological record arising from a tropic cascade. *Ecology* 72: 277-285.

Clark, W. C., D. D. Jones, and C. S. Holling, 1979. Lessons for ecological policy design: a case study of ecosystem management. *Ecological Modelling* 7: 1-53.

Craighead, F. C., Sr., 1971. The Trees of South Florida. Vol. I. The Natural Environments and Their Succession. University of Miami Press, Coral Gables, Florida.

Davis, S. M., 1989. Sawgrass and Cattail Production in Relation to Nutrient Supply in the Everglades. Fresh Water Wetlands & Wildlife, 9th Annual Symposium, Savannah River Ecology Laboratory, 24-27 March, 1986, Charleston, South Carolina. U.S. Dept. of Energy.

Davis, S. M., L. H. Gunderson, W. Park, J. Richardson, and J. Mattson, 1994. Landscape dimension, composition and function in a changing Everglades ecosystem. In: S. M. Davis and J. C. Ogden, (Ed.). *Everglades: The Ecosystem and Its Restoration*. St. Lucie Press, Boca Raton, Florida.

De Angelis, D. L., 1980. Energy flow, nutrient cycling and ecosystem resilience. *Ecology* 61: 764-771.

Diener, M. and T. Poston, 1984. On the perfect delay convention or the revolt of the slaved variables. 249-268. In: H. Haken, (Editor). *Chaos and order in nature*. Springer-Verlag, Berlin.

Dublin, H. T., A. R. E. Sinclair, and J. McGlade, 1990. Elephants and fire as causes of multiple stable states in the Serengeti-mara woodlands. *Journal of Animal Ecology* 59: 1147-1164.

Ehrlich. P. R., 1991. Population diversity and the future of ecosystems. *Science* 254: 175.

Elliott, J. E., 1980. Marx and Schumpeter on capitalism's creative destruction: a comparative restatement. *Quarterly Journal of Economics* 95: 46-58.

Elton, C.S., 1958. The ecology of invasions by animals and plants. Methuen, London.

Fiering, M. B., 1982. Alternative indices of resilience. *Water Resources Research* 18: 33-39.

Gunderson, L. H., 1994. Vegetation: Determinants of Composition. In: S. M. Davis and. J. C. Ogden., (Eds.). *Everglades: The Ecosystem and Its Restoration*. St. Lucie Press, Boca Raton, Florida.

Gunderson, L. H., C. S. Holling, and S. Light, 1995. Barriers and Bridges to Renewal of Ecosystems and Institutions. New York: Columbia University Press.

Holling, C. S., 1973. Resilience and stability of ecological systems. *Annual Review of Ecology and Systematics* 4: 1-23.

Holling, C. S., 1986. Resilience of ecosystems; local surprise and global change. 292-317. In: W. C. Clark and R. E. Munn, (Eds.). *Sustainable Development of the Biosphere*. Cambridge: Cambridge University Press.

Holling, C. S., 1988. Temperate forest insect outbreaks, tropical deforestation and migratory birds. *Memoirs of the Entomological Society of Canada* 146: 21-32.

Holling, C. S., 1991. The role of forest insects in structuring the boreal landscape. In: H. H. Shugart, R. Leemans, and G. B. Bonan, (Eds.). *A Systems Analysis of the Global Boreal Forest*. Cambridge University Press, Cambridge. Chapter 6: 170-191.

Holling, C. S., 1992. Cross-scale morphology, geometry and dynamics of ecosystems. *Ecological Monographs* 62(4): 447-502.

Holling, C. S., D. D. Jones, and W. C. Clark, 1977. Ecological policy design: a case study of forest and pest management. IIASA CP-77-6:13-90. In: G. A. Norton and C. S. Holling, (Eds.). *Proceedings of a Conference on Pest Management*. October 1976,. Laxenburg, Austria.

Holling, C. S., D. W. Schindler, B. Walker, and J. Roughgarden, 1994. Biodiversity in the functioning of ecosystems: an ecological primer and synthesis. In: C. Perrings, K-G Mäler, C. Folke, C. S. Holling, and B.-O. Jansson, (Eds.). *Biodiversity Loss: Ecological and Economic Issues*. Cambridge: Cambridge University Press.

Holling, C.S., G. Peterson, P. Marples, J. Sendzimir, K. Redford, L. Gunderson, and D. Lambert, 1996. Self organization in ecosystems:lumpy geometries, periodicities and morphologies. In: B.Walker and W. Steffen, (Eds.) *Global Change in Terrestrial Ecosystems*. Cambridge University Press.

Leemans, R. and I. C. Prentice, 1989. FORSKA: a general forest succession model. *Meddelanden* 2: 1-45.

Levin, S., 1995. Biodiversity: Interfacing Populations and Ecosystems. Kyoto University Press, Kyoto.

Light, S. S., L. H. Gunderson, and C. S. Holling, 1995. The Everglades; Evolution of Management in a Turbulent Environment. In: L. H. Gunderson, C. S. Holling, and S. S. Light, (Eds.). *Barriers and Bridges to the Renewal of Ecosystems and Institutions*. Columbia University Press, New York.

Loveless, C. M., 1959. A Study of the Vegetation of the Florida Everglades. *Ecol.* 40(1): 1-9.

May, R.M., 1973. Stability and complexity in model ecosystems. Princeton University Press, Princeton, N.J.

McNaughton, S. J., R. W. Ruess, and S. W. Seagle, 1988. Large mammals and process dynamics in African ecosystems. *BioScience* 38: 794-800.

Morris, R. F., 1963. The dynamics of epidemic spruce budworm populations. *Memoirs of the Entomological Society of Canada* 21: 332.

Morton, S. R., 1990. The impact of Eurpoean settlement on the vertebrate animals of arid Australia: a conceptual model. *Proceedings of the Ecological Society* 16: 201-213.

Naiman, R. J., 1988. Animal influences on ecosystems dynamics. *BioScience* 38: 750-752.

O'Neill, R. V., D. L. DeAngelis, J. B. Waide, and T. F. H. Allen, 1986. A Hierarchical Concept of Ecosystems. Princeton: Princeton University Press.

Paine, R. T., 1974. Intertidal community structure: experimental studies on the relationship between a dominant competitor and its principal predator. *Oecologia* 15: 93-120.

Pastor, J. and W. M. Post, 1986. Influence of climate, soil moisture and succession on forest carbon and nitrogen cycles. *Biogeochemistry* 2: 3-27.

Pastor, J. and Y. Cohen, 1996. Herbivores, plant populations and the cycling of nutrients in ecosystems. American Naturalist (In Prep.).

Payette, S., 1983. The forest tundra and present tree-lines of the northern Quebec-Labrador peninsula. In: P. Morisset and S. Payette, (Eds.). *Tree Line Ecology, Proceedings of the northern Quebec Tree Line Conference*. Nordicana, Quebec, 3-23.

Peterson, G., 1994. Modelling fire dynamics in the Manitoban Boreal Forest. M.S. Thesis, University of Florida, Gainesville, Fla.

Pimm, S. L., 1984. The complexity and stability of ecosystems. *Nature* 307: 321-326.

Prigogine, I., 1980. From Being to Becoming, Time and Complexity in the Physical Sciences. W.H. Freeman, New York, New York.

Risser, P. G., 1995. Biodiversity and ecosystem function: Where shall we look first? *Conservation Biology* 9:742-746.

Shugart, H. H. and I. C. Prentice, 1992. Individual-tree-based models of forest dynamics and their application in global change research. In: H. H. Shugart, R. Leemans, and G. B. Bonan, (Eds.). *A Systems Analysis of the Global Boreal Forest*. Cambridge University Press, Cambridge. 313-333.

Sinclair, A. R. E., P. D. Olsen, and T. D. Redhead, 1990. Can predators regulate small mammal populations? Evidence from house mouse outbreaks in Australia. *Oikos* 59: 382-392.

Smith, T. M. and D. L. Urban, 1988. Scale and the resolution of forest structural pattern. *Vegetatio* 74: 143-150.

Steward, K. K. and W. H. Ornes, 1975. The Autecology of Sawgrass in the Florida Everglades. *Ecol.* 56(1): 162-171.

Tilman David andJohn A. Downing 1994. Biodiversity and stability in grasslands. *Nature* 367: 363-365.

Tilman, D., 1996. Biodiversity: Population versus ecosystem stability. *Ecology* 77: 350-363.

Waddington, C.H., 1969. The theory of evolution today. In: A. Koestler and J.R. Smythies, (Editors). *Beyond reductionism: new perspectives in the life sciences*. Hutchinson, London. 357-374.

Waide, J. B. and J. R. Webster, 1976. Engineering systems analysis: applicability to ecosystems. Volume IV, 329-371. In: B. C. Patten, (Ed.). *Systems Analysis and Simulation in Ecology*. New York: Academic Press.

Walker, B., 1995. Conserving Biological Diversity through Ecosystem Resilience. Conservation Walker, B. H., 1981. Is succession a viable concept in African savanna ecosystems? 431-447. In: D. C. West, H. H. Shugart, and D. B. Botkin, (Eds.). *Forest Succession: Concepts and Application*. New York: Springer-Verlag.

Walker, B. H., 1992. Biological diversity and ecological redundancy. *Conservation Biology* 6: 18-23.

Walker, B. H., D. Ludwig, C. S. Holling and R. M. Peterman, 1969. Stability of semi-arid savanna grazing systems. *Ecology* 69: 473-498.

Walters, C. J., 1986. Adaptive Management of Renewable Resources. New York: McGraw Hill.

II.8.3 Continuity and Discontinuity in Ecological Systems

Giuseppe Bendoricchio

1. Continuity and Discontinuity

In spite of the precise meaning of continuity (or discontinuity) in mathematics dealing with function theory, continuity of dynamic systems cannot be defined precisely because of its dependence on choice of time and space discretization.

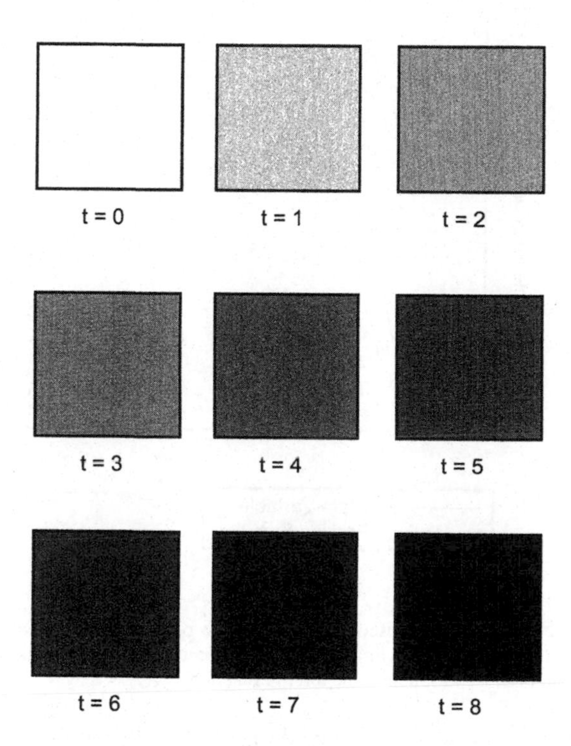

Figure 1: If $\Delta t=1$ the colour change of the square from white to grey and black in 9 time steps can be considered continuous. However, focusing the attention on the t=0 and t=8 instants the colour change seems discontinuous. Continuity and discontinuity depend on the time step used for the observation. At the space scale of a single pixel, the colour change is discontinuous even if a $\Delta t=1$ time step is considered, thus, continuity and discontinuity depend also on the considered space scale.

The system represented by the squares of Figure 1, is *discretized* in pixels. If we analyse the change of colour of the square step by step, we perceive its continuous evolution from white to black. However if we consider the entire square and the largest time step, ($\Delta t=9$), the change of colour seems discontinuous. Similarly, if we focus our attention on one specific pixel of the square, with $\Delta t=1$, the change of pixel colour is discontinuous. Continuity (and discontinuity) of the change of colour in such a dynamic system is clearly a consequence of the subjective choice of space and time steps used for the analysis.

Stressing the idea of dependence on the space scale, E. Schrödinger (1988) suggested that the evolution of a dynamic ecosystem at the subatomic space and time scale is always a sum of discontinuities. All the macro-changes are consequences of physical and chemical micro-changes, or reactions that can be regarded as rearrangements of atoms due to discontinuous jumps of electrons from one energy level to another. Hence, we might conclude that natural evolutionary changes are only effects of discontinuous steps and that, at the subatomic space at any time scale, only the discontinuity exists. Fortunately, in ecology we are interested in a less speculative approach to nature and, at the macro scale, we are forced to consider both the continuous and discontinuous changes of an ecosystem.

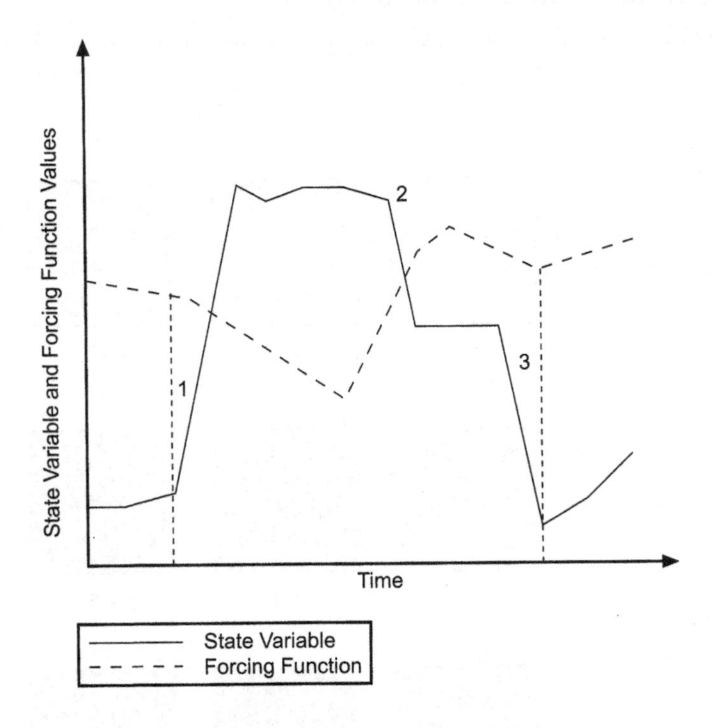

Figure 2: Change number 1 and 3 are discontinuities because of the smooth change of the forcing function compared with fast variation of the state variable. Change 2 shows a continuous and fast variation of the state variable, due to a comparably fast variation of the forcing function.

A change in a dynamic system can be defined as a 'discontinuous', 'fast', 'sudden' or 'abrupt' passage from one equilibrium state to another represented by quite different values of the state variables. If the change of the values of the state variables is a consequence of a 'continuous', 'slow', 'smooth' or 'gradual' change of the forcing function values, it can be defined as a discontinuity. A discontinuity can (but must not necessarily) be a catastrophe as defined by the catastrophe theory.

Changes number 1 and 3 of the state variable in Figure 2 are discontinuities because the variations of its values are large compared with minor changes of the values of the forcing function. Under particular conditions these discontinuities could be catastrophes . Change number 2 in Figure 2 is only a change of the values of the state variable due to a large variation of the forcing function.

The Venn diagram (Figure 3) shows that only a subset of the changes of an ecosystem contains elements of the discontinuity set and that only a well defined subset of the discontinuities has elements that can be regarded as catastrophes as defined by the catastrophe theory.

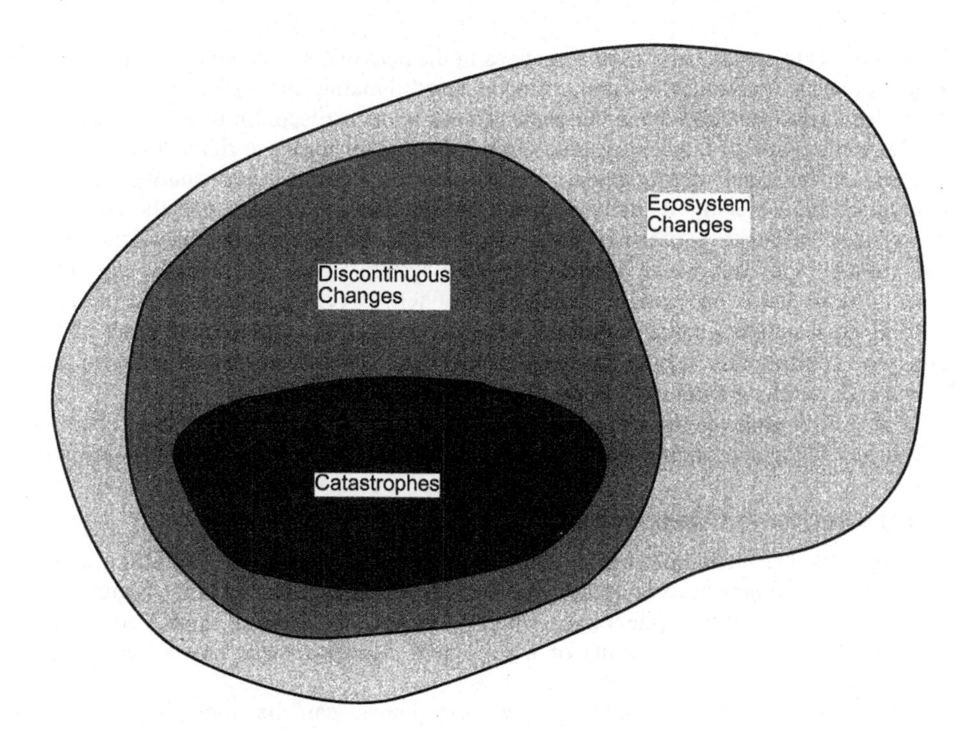

Figure 3: Venn diagram showing the sets of changes, discontinuities and catastrophes of an ecosystem.

2. Modelling Discontinuity

Ecosystems are normally complex non-linear systems and their dynamics can be mathematically described by complex models simulating the main processes involved. Nevertheless, the complexity of ecosystems cannot be completely described and simulated even if a reductionist point of view is adopted. One reason being that discontinuity is one of the main features of ecosystem behaviour.

Blooms and breakdowns of algal populations in eutrophic aquatic systems are good examples of sudden changes of the algal biomass in the system. These changes are normally a consequence of slow variations of the forcing functions such as nutrient contents in the water, light and temperature. The evolution of the state variable 'biomass' can be simulated for a long period during the year using ordinary or partial differential equations, but the bloom, and the consequent collapse, often have such a short life span that the traditional modelling approach does not fit very well in reality.

Similar discontinuities are the forest fire destroying, in a few days, a very old forest ecosystem, or the variation of the shallow lake ecosystem in a wetland as a consequence of the reduced water flow or of the increment of the solid transport. Other discontinuities in ecosystems can be recognised in population dynamics where the prey-predator succession varies sometimes very rapidly.

Ecosystem discontinuities can be described by a change in the network structure of the ecosystem. Such a change adds or cancels one or more compartments of the network and varies the fluxes within the network. Such a structural dynamic of an ecosystem network can be simulated with deterministic or stochastic models considering the complete network which includes all the compartments. The dynamic is obtained by varying the fluxes among the compartments as a consequence of the variations of the values of the forcing functions. The variations of the fluxes are usually described by variations of the parameter values in the model equations.

The dynamics of the system can be driven by an ecological goal function, the values of which the system attempts to optimise. Network analysis has also developed some dynamic tools to simulate structural changes in the network. Unfortunately, up to now, both of the approaches (structural dynamic models and dynamic networks) are not sufficiently developed to give a satisfactory description of such strong discontinuities in ecosystems.

If the network of an ecosystem does not change and a discontinuity occurs, a traditional differential mathematical model can, or cannot, reasonably describe the behaviour of the system. It usually depends on the relative velocity of the change in the state variable compared with the changes of the values of the forcing functions. While the model may fit the continuous dynamic of the system, it may fail locally in fitting the discontinuity. Usually, the present ecological models solve the problem of the local failure by adding to the model some additional state variables and equations of other processes that may be helpful in increasing the speed of the local change. Such an expedient increases the complexity of the model and stresses it to fit better the reality.

Such a discontinuous behaviour of the ecosystem could be described much better by a different modelling approach which focuses the attention only on the local discontinuity.

3. Introduction to Catastrophe Theory

For 300 years differential calculus has been the prominent method for building up models. Nevertheless, as a descriptive language, differential equations have a strong limitation: they can describe only those phenomena where changes are smooth and continuous. In mathematical terms, the solutions of a differential equation must be a function that is differentiable.

In ecosystem theory relatively few phenomena can be modelled with such a mathematical instrument, the world of ecosystems presents a lot of sudden changes and unpredictable divergence, which call for functions that are not differentiable.

A mathematical method for dealing with discontinuous and divergent phenomena has only recently been developed. It can be applied with particular effectiveness in those situations where continuous causes lead to discontinuous effects in the behaviour of a system.

This method has been named Catastrophe Theory.

Some recent and important applications of the theory are in biology and in social and environmental sciences where discontinuous and divergent phenomena are ubiquitous and where other mathematical techniques have so far proved ineffective.

Catastrophe theory was invented by René Thom. He presented his ideas in a book published in 1972 (Thom, 1972). The theory is derived from topology because the state and forcing function changes can be described by smooth surfaces of equilibrium; catastrophe occurs when the equilibrium breaks down, jumping from one to another equilibrium point on the smooth surface.

Catastrophe theory deals with the description of the shapes of all possible equilibrium surfaces. For processes controlled by no more than four factors (parameters) Thom has shown that there are exactly seven archetypal forms which he calls the elementary catastrophe.

4. The Classification Theorem: A Simplified Version

In order to keep things as understandable as possible, we begin by stating a simplified version of a part of Thom's classification theorem. The enunciate refers to the elements shown in Figure 4.

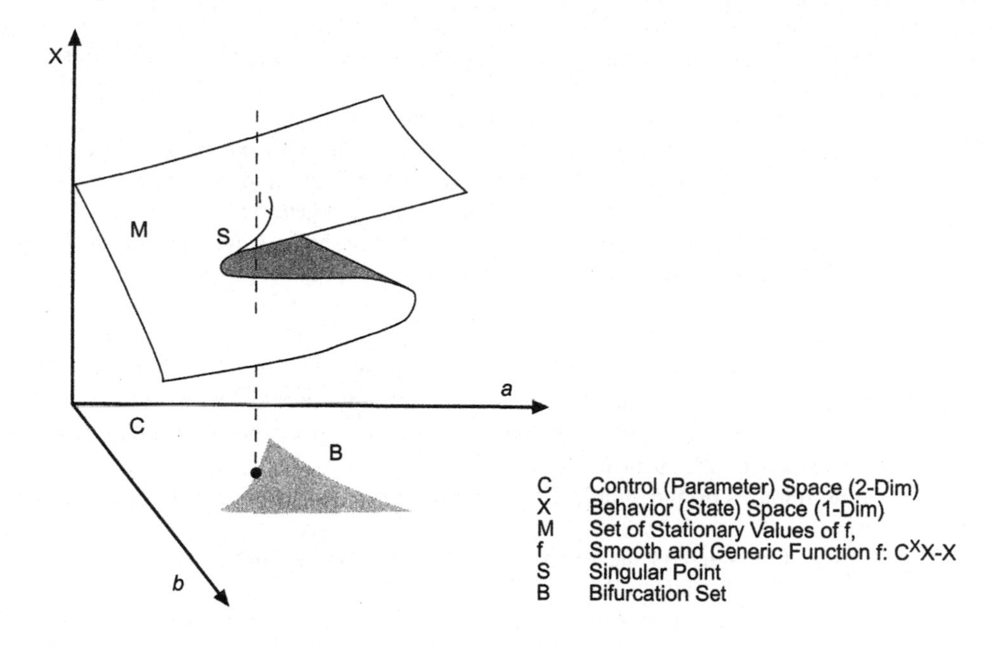

Figure 4: Three dimensional representation of the standard cusp catastrophe.

Let C be a 2-dimensional control (or parameter) space, let X be a 1-dimensional behaviour (or state) space, and let f be a smooth generic function on X parametrised by C. Let M be the set of the stationary values of f (given by$\partial f/\partial x=0$, when x is a co-ordinate for X) then M is a smooth surface in C×X and the only singularities of the projection of M on to C are fold curves and cusp-catastrophe.

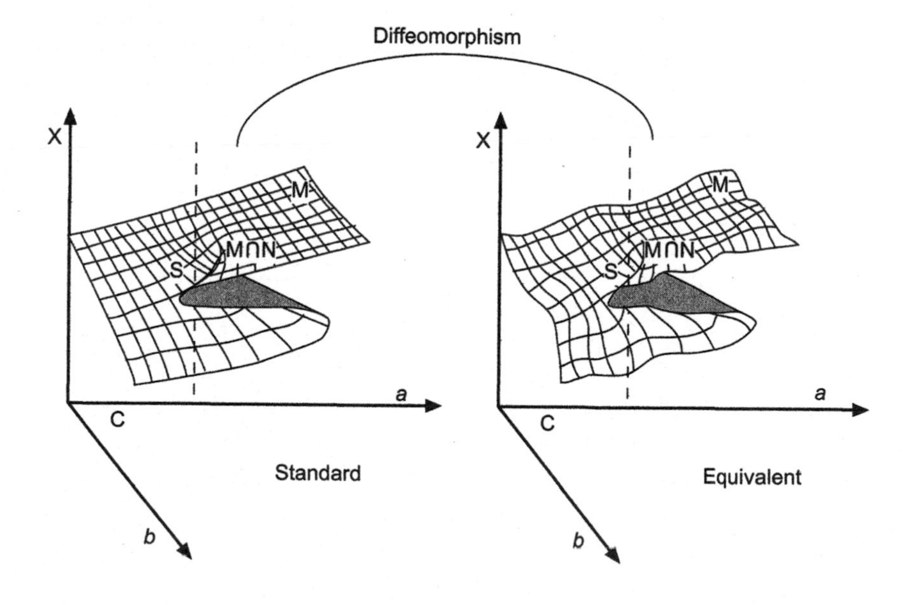

Figure 5: Just to give an idea of what equivalent means.

Before looking at the consequence of the enunciate, it is suitable to explain some of its details.

- *Smooth* means differentiable to all orders.
- G*eneric* means that the map f : CxX →X from C to the space of functions X is *transverse* to the natural stratification. Almost all smooth functions are generic. Small perturbations of a generic function remain generic.
- A point S is a singular point, and the function has a *singularity* in S, if a vertical line is tangent to the surface M in S. When we say that a singularity of M belongs to a cusp-catastrophe, we mean that near that point S, the surface M is *equivalent* to the standard surface shown in Figure 4. *Equivalent* means that there is a diffeomorphism (Figure 5) from a neighbourhood N of S in the equivalent C×X surface onto the standard C×X surface, throwing vertical lines to vertical lines and M∩N onto the standard surface. Equivalence preserves all qualitative features of the catastrophe surface.

Some remarks on the simplified version of the classification theorem can be listed as follows.

- The function f is the function whose equilibrium points are represented by the points of the surface M.
- The standard surface of Figure 4 is locally (in a neighbourhood of the singular points) the most complicated surface that can be drawn. Far from the singularities, in more stable conditions for the system, the M surface can assume different shapes that traditional differential equations are able to simulate. That is why the catastrophe theory can be used with such confidence in so many fields, and the cusp catastrophe too, whenever a process involves 2 causes and 1 effect.
- If we reduce the dimension of C from 2 to 1, then the analogous theorem, one dimension lower, says that locally, M is a smooth curve, and the only singularities are folds as shown in Figure 6.
- Thus the fold-catastrophe appears in sections of the cusp catastrophe and the latter is made up of folds together with one new singularity at the origin (critical point). Similarly, any higher dimensional catastrophe is always made up of lower dimensional ones, together with an additional singularity at the origin.

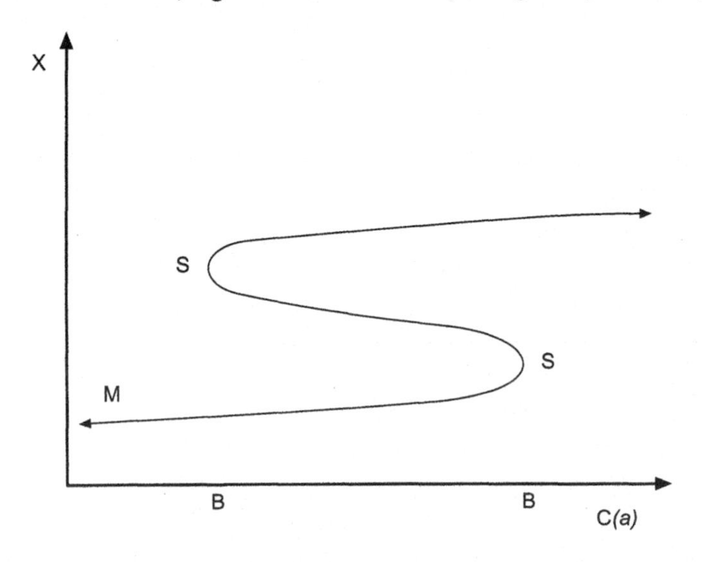

Figure 6: Representation of the standard fold catastrophe; the symbols are the same as in Figure 4.

- The cusp catastrophe has five qualitative features: *bimodality, inaccessibility, catastrophe*, *hysteresis*, and *divergence* as shown in Figure 7. These features are redundant and interrelated to define the cusp catastrophe.
- A discontinuity in a dynamic system is a cusp-catastrophe if – and only if – the system has all the qualitative feature of this surface. If the system shows a discontinuity in its behaviour and at least the features bimodality, hysteresis and divergence, the system has also the other features of the cusp surface (catastrophe and inaccessibility).
- Bimodality and hysteresis without divergence lead to a fold catastrophe curve.
- The f function of the canonical cusp catastrophe can be written with suitable values of the coefficients as: $f = \frac{1}{4}x^4 - ax - \frac{1}{2}bx^2$

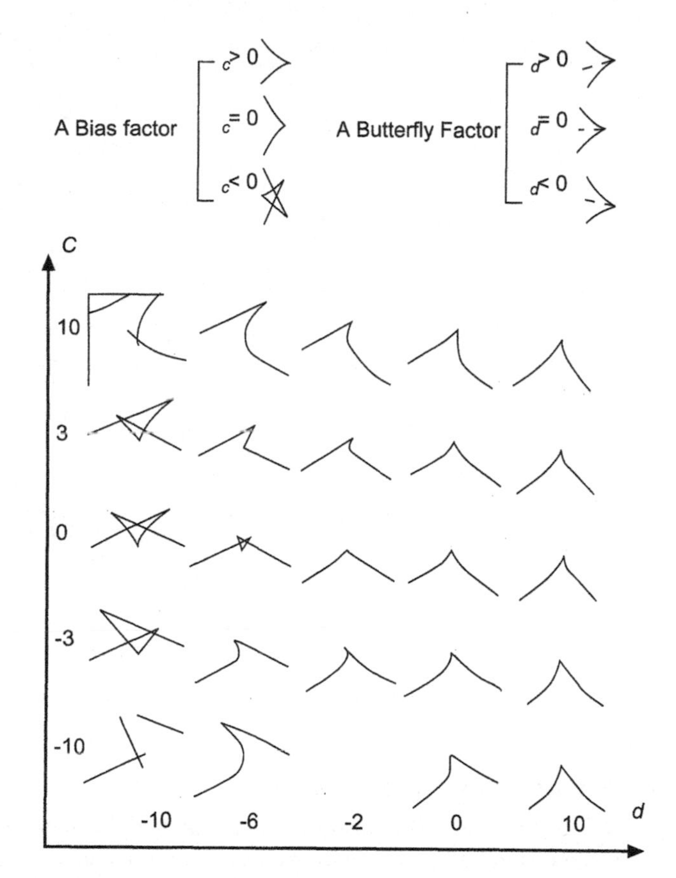

Figure 7: Qualitative features of a cusp catastrophe surface.

We can think of f as a potential energy depending on the values of the parameters *a* and *b*. The critical points (or equilibrium points) of the function f belonging to the surface M (Figure 4) are the solution of the equation:

$$\frac{df}{dx} = 0 = x^3 - a - bx\,,$$

a cubic function in x. It must have at least one and at most three real roots. The nature of the roots depends on the values of *a* and *b*: specifically on the values of the discriminate.

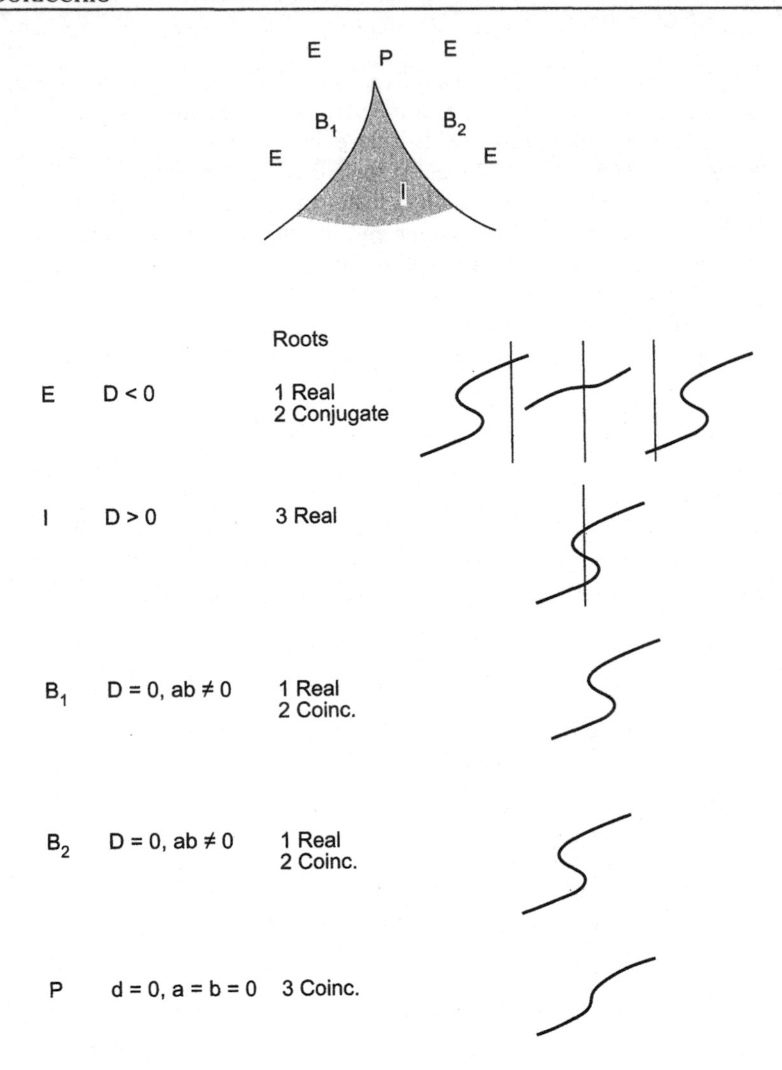

Figure 8: Solutions of the cusp catastrophe equation for different values of the discriminate depending on different values of the control parameters in the C plane.

5. The Seven Elementary Catastrophes

The dimensions of C and X of the cusp and fold catastrophe can be increased. As a consequence, the dimension of C×X could increase as well and the set of catastrophic surfaces becomes infinite. The primary interest for more relevant applications is focused on the catastrophes where the maximum dimension for C is 4 and the maximum dimension for X is 2. The equations of the surfaces (hypersurfaces) resulting from these dimensions are presented in Table 1.

Only the surfaces of fold and cusp can be drawn in a 3-dim space. The bifurcation set B is the projection of singular points S on the control space (Figures 4 and 6). The bifurcation set of the fold is formed by two points and that for the cusp by the curves shown in Figure 4.

If C×X dim^3 3, M is a hypersurface in C×X. Instead of being folded along curves (singularities), M is now folded along whole surfaces and the bifurcation sets now consist

Table 1: Equations of the f function and of the M surface for the most common catastrophes with C-dim £ 4.

CATASTROPHE		DIM C	DIM X	FUNCTION f	$\frac{\partial f}{\partial x};\frac{\partial f}{\partial y}$
CUSPOIDS	FOLD	1	1	$\frac{1}{3}x^3 - ax$	$x^2 - a$
	CUSP	2	1	$\frac{1}{4}x^4 - ax - \frac{1}{2}bx^2$	$x^3 - a - bx$
	SWALLOWTAIL	3	1	$\frac{1}{5}x^5 - ax - \frac{1}{2}bx^2 - \frac{1}{3}cx^3$	$x^4 - a - bx - cx^2$
	BUTTERFLY	4	1	$\frac{1}{6}x^6 - ax - \frac{1}{2}bx^2 - \frac{1}{3}cx^3 - \frac{1}{4}dx^4$	$x^5 - a - bx - cx^2 - dx^3$
UBILICS	HYBERBOLIC	3	2	$x^3 + y^3 + ax + by + cxy$	$3x^2 + a + cy$ $3y^2 + b + cx$
	ELLIPTIC	3	2	$x^3 - xy^2 + ax + by + cx^2 + cy^2$	$3x^2 - y^2 + a + 2cx$ $-2xy + b + 2cy$
	PARABOLIC	4	2	$x^2y + y^4 + ax + by + cx^2 + dy^2$	$x^2 + 4y^3 + b + 2dy$ $2xy + a + 2cx$

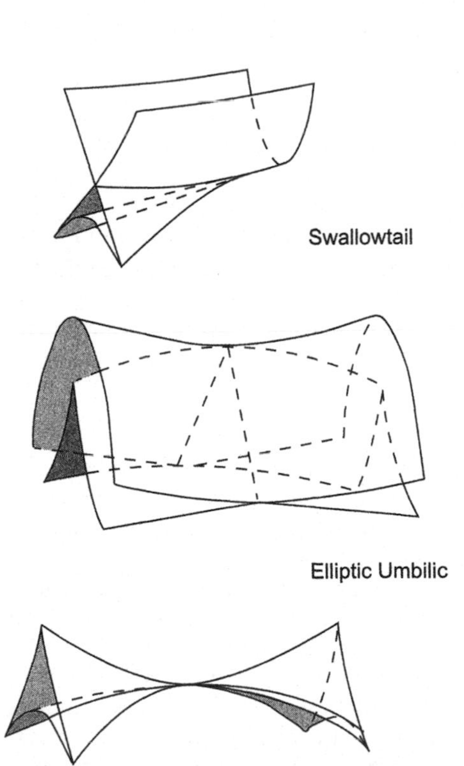

Figure 9: Surfaces of the bifurcation sets for the catastrophes with C-dim=3. In several sections of these surfaces the bifurcation set of the cusp catastrophe is easily recognisable. This is a consequence of the fact that the catastrophe with higher C-dim is a development of the catastrophe with a lower C-dim

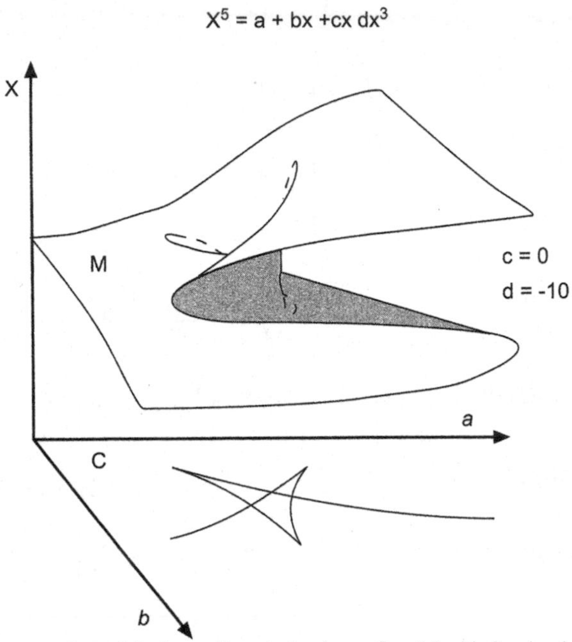

Figure 10: Graphical representation of the butterfly catastrophe surface M with fixed values of the parameter *c* and *d*.

For C-dim = 4, the butterfly and parabolic umbilic catastrophe surfaces and their bifurcation sets cannot be drawn as for the lower C-dim catastrophes.

The parameters controlling the geometry of the butterfly catastrophe (Figure 10) are: *a* (normal factor) and *b* (splitting factor) as for the cusp catastrophe, and *c* (bias factor) and *d* (butterfly factor). Increasing values of *a*, with $b < 0$, lead to smooth increasing values of x, values of $b > 0$ split M in two sheets as for the cusp. As shown in Figure 11, a variation of *c* values biases the position of the cusp; a variation of *d* values evolves the cusp in 3 other cusps which form a triangular 'pocket'; meanwhile, the equilibrium surface develops two new folds.

6. Application of Catastrophe Theory

Catastrophe theory can help in describing the discontinuity if – and only if – the features of catastrophic discontinuity are present. If we can reduce the complexity of the ecosystem adopting a holistic point of view, catastrophe theory can give a good description of the dynamics of ecosystems presenting discontinuities.

If we can apply catastrophe theory to the behaviour of an ecosystem, then we are able to transfer the power of the theory to that system and we can be sure that the ecosystem comprises all the features of the elementary catastrophe surfaces.

Up to now, there exists a sufficiently wide list of applications of the catastrophe theory to ecosystem analysis. In most cases the application deals with a qualitative approach, only a few cases within the literature attempt a quantitative simulation of the system.

An application of catastrophe theory to ecosystem analysis may follow some typical steps.

A) Can the catastrophe theory be applied? To decide, you need to ask some other questions.

1. Does the dynamic ecosystem present a discontinuity?

2. Does the behaviour of the system show at least bimodality, hysteresis and divergence ?
3. Can the behaviour of the system be summarised by at most two state variables?
4. Can the forcing functions be described by at most four parameters or at most four more complex functions in which the system parameters are included?

If you can answer positively all the previous questions, an application of the catastrophe theory may be carried out successfully.

(B) Apply the following procedure for a qualitative description of the catastrophic discontinuity:

1. Point out the state variables, the control parameters, the goal function;
2. Attempt a qualitative description of the catastrophic dynamic of the system;
3. Check if the qualitative description is satisfactory in the sense that all the feature of the catastrophic behaviour are present by selecting the appropriate variables, parameters, and goal function.

If these steps are positively completed, we can apply the force of the qualitative catastrophe theory to discover all the hidden features in the behaviour of the ecosystem. At this stage of the application we have a sufficiently deep insight into the system behaviour in the vicinity of the discontinuity. The next step is the adaptation of the traditional model to the catastrophe description of the discontinuity.

C) Apply the following procedure to obtain a quantitative model of the discontinuity.

1. Discover a mathematical transformation of the traditional differential model into a polynomium.
2. Transform the polynomium into one of the seven canonical catastrophes shown in Table 1.
3. Apply the quantitative catastrophe analysis to point out the bifurcation set points of the system.
4. Translate the values of bifurcation set points into parameter values of the traditional model.

Such a complete application of the previous procedure leads to a comprehensive description of the catastrophic behaviour of the system and returns to the scientist all benefits of the general mathematical description of a catastrophic discontinuity.

Unfortunately, the quantitative approach of the catastrophe theory is not so easily applicable to ecosystem dynamics even though it is widely used in this scientific field.

7. An Example of Application

As a consequence of pollution, algal blooms usually occur in eutrophic water bodies. For modelling purposes, a bloom can be described in terms of algal population density or chlorophyll concentration.

Contingent on the values of the forcing functions (light intensity, water temperature and external nutrient concentration): phytoplankton growth and algal blooms have two distinct equilibrium states

- the pre-bloom period with a limited and physiological productivity;
- an algal bloom period with a fast and pathological productivity.

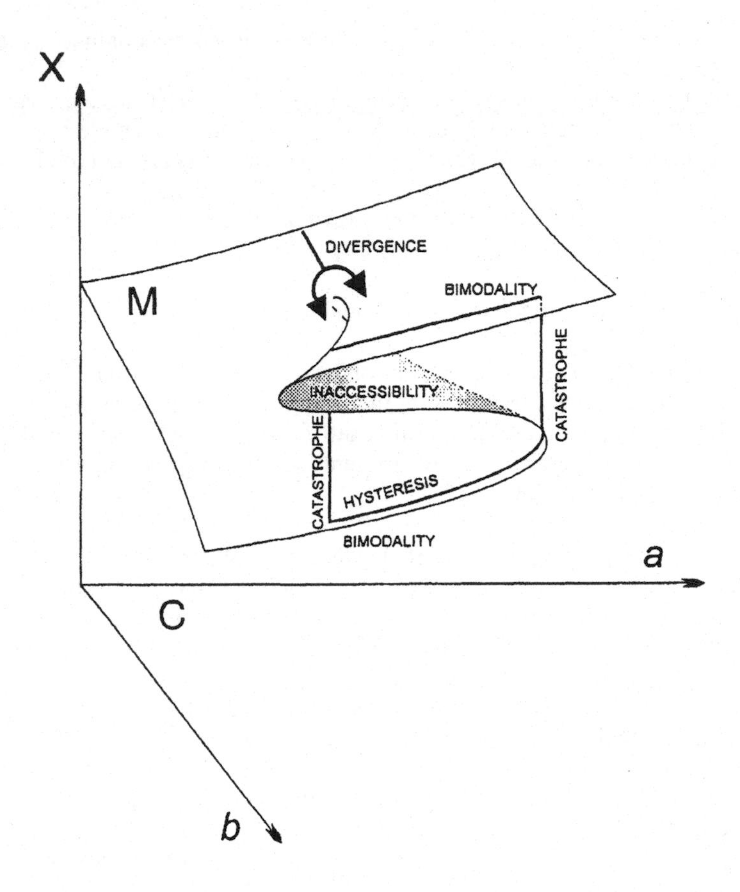

Figure 11: Bifurcation sets of the butterfly catastrophe for different values of the control parameters.

During the algal bloom, the change from one state of equilibrium to the following one occurs in a relatively short period of time compared to the longer period during which the trophic level of a water body evolves. This process reveals that a relatively small change in the boundary conditions can trigger a qualitatively large change in the algal concentration. Moreover, it indicates one of the typical characteristics of catastrophe phenomena: the discontinuity existing between two equilibrium states.

The return from the algal bloom state to the normal (lower productivity) state may be more gradual and different from the catastrophic jump to the bloom. This indicates a hysteretic behaviour typical of catastrophic phenomena. However, a collapse which is a fast change from the higher equilibrium to the lower one is also possible.

The algal bloom may develop in different ways according to the physical-chemical composition of the water body and to the entering energy. Swartzman and Bentley (1979) summarised the mathematical models and the equations that describe phytoplankton evolution, the process of eutrophication and its controlling parameters. These controlling parameters include primarily the available nutrients such as nitrogen and phosphorus, solar radiation, and water temperature (energy).

Nutrients affect phytoplankton growth in quite complex manners and according to the well known Liebig law only the nutrient which is in shortest supply limits the growth rate during the algal bloom.

In addition to the limiting nutrient, solar radiation and water temperature must be considered. The latter two variables are intercorrelated and could be expressed by one function which may be complicated due to the distribution of vertical extinction of the solar radiation (absorbance), water surface reflection, evaporation rates and other factors. In most cases, the algal bloom is triggered by increased energy inputs at higher but more or less constant nutrients levels. In general, an algal bloom is dominated by a single species.

Nevertheless, it is possible to formulate the model of a phytoplankton bloom using two parameters acting as forcing functions: *a* which is a function of solar radiation and water temperature, *b* which is the concentration of the limiting nutrient; and a state variable x which is the concentration of the phytoplankton biomass. Hence, one can then formulate a three dimensional description of an algal bloom phenomenon using the theory of catastrophe. In fact, we can say that a catastrophe (algal bloom) occurs when the stability of the first equilibrium (low growth rate) breaks down and, consequently, the system attains a new equilibrium with a jump. This rapid change happens as a result of slow and continuous variations in the forcing function values of the system.

From the large variety of models describing eutrophication, the well known biochemical diffusive model of Rabinowitch (1951) has been selected to simulate the phytoplankton evolution in a shallow water area of the Lagoon of Venice. The model assumes that:

- the net phytoplankton growth $\delta P/\delta t$ is obtained from the overall growth rate subtracting a death term;
- the overall growth rate is related to the nutrient concentration in the water body of the algae according to the Michaelis-Menten equation;
- the net growth depends on the external concentrations of nutrients and on the coefficient of the diffusive resistance time.

This model is applicable to nonstratified, completely mixed, confined and shallow water bodies in which the population is uniformly *distributed* in space and a single species dominates the population. These conditions characterise some water bodies during spring and summer when algal blooms usually occur.

Under these conditions, the Rabinowitch model can be written as follows (Van Nguyen and Wood, 1979):

$$\left(\frac{\partial P}{\partial t}\right)^2 r_d - \frac{\partial P}{\partial t}\left(\alpha I(r_x - r_d) + N_E - Rr_d\right) + \alpha I(N_E - Rr_x) - RN_E = 0 \tag{1}$$

where:

P	is the phytoplankton biomass [$mg \cdot l^{-1}$],
t	is time [day],
I	is the incident light intensity at the air water interface [$MJm^{-2} \cdot day^{-1}$],
N_E	is the limiting nutrient concentration in the water body [$mg \cdot l^{-1}$]
r_x	is the diffusive resistance time of the limiting nutrient [s],
r_d	is the diffusive resistance time [s],
α	is the photochemical efficiency [MJm^{-2}],
R	is the factor of the endogenous respiration rate (factor) [$mg \cdot (l^{-1} \cdot s^{-1})$].

The Monod equation for the algal growth rate is given as:

$$\frac{\partial P}{\partial t} = (G - D)P \tag{2}$$

where:

$G = GM\frac{N_E}{K + N_E}$ is a growth rate coefficient [s^{-1}],

$D = D_0(Q_{10})^{T_A}$ is a death rate coefficient [s^{-1}],

and:

K	is the Michaelis-Menten half saturation coefficient [$mg \cdot l^{-1}$],
GM	is the maximum uptake rate [s^{-1}],
D_0	is a constant [s^{-1}],
Q_{10}	is a factor used to model the respiration process [adim.],
T_A	is the water temperature [°C].

Integrating equation (1) from a steady state $\delta P / \delta t = 0$ to any other state and using equation (2), we obtain:

$$P^3 + fP^2 + gP + h = 0 \tag{3}$$

where:

P is the phytoplankton concentration in the water body at time t[$mg \cdot l^{-1}$];

$f(I,N_E,R) = -\beta(\alpha I(r_x+r_d)+N_E-Rr_d)$ [$mg.l^{-1}$];

$g(I,N_E,R) = -\gamma(\alpha I(N_E-Rr_x)-RN_E)$ [$mg^2.l^{-2}$];

$h(T_A) = \delta \rho T_A$ (according to the theory of temperature adaptation in thermobiology)

[$mg^3 \cdot l^{-3}$];

with:

$\beta = 3/(2r_d(G-D))$ [admin.],

$\gamma = 3/(r_d(G-D)^2)$ [s],

$\rho = 3/(r_d(G-D)^3)$ [s^2],

δ dimensional coefficient, polynomial function of T_A, P, t; $\delta = Z(T_A,P,t)$.

Applying to (3) Cardano's transformation of co-ordinates (4):

$$\pi = P + f/3 \ [mg.l^{-1}];$$

$$\xi = f^2/3 + g \ [mg^2.l^{-2}] \tag{4}$$

$$\zeta = 2\left(\frac{f}{3}\right)^3 - \left(\frac{fg}{3}\right) + h \ [mg^3.l^{-3}]$$

one can obtain the canonic equation of a cusp catastrophe

$$\pi^3 + \xi\pi + \zeta = 0 \tag{5}$$

The transformation (4) has eliminated the direct physical meaning of the control and state variables but has partially preserved their original meaning. In fact, only the normal variable ζ depends on the water temperature while the splitting variable ξ depends on the other parameter but not on the water temperature. Note also that only the transformed state variable π is a function of the phytoplankton concentration *P*.

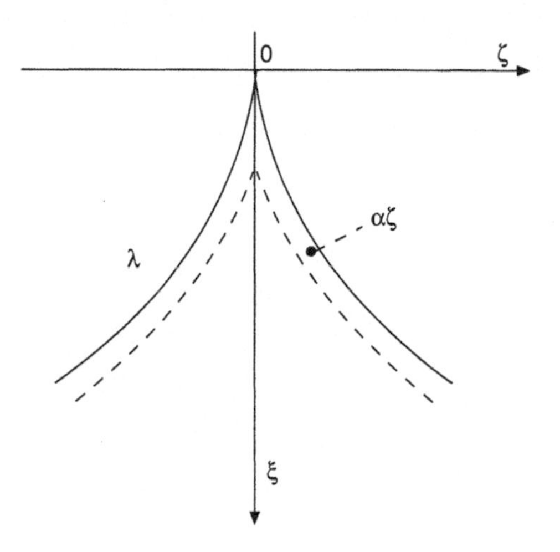

Figure 12: Bifurcation set of the cusp catastrophe. The shaded area shows the zone with a high catastrophic risk.

The bifurcation of (5) is the curve λ (Figure 12) whose equation is:

$$4\xi^3 + 27\zeta^2 = 0$$

A suitable interval $\Delta\zeta$ of the normal variable allows to highlight the shaded area with high catastrophic risk. Simultaneously, within the space of the parameters IN_E, R, T_A it singles out the hypervolume of the points P the co-ordinates of which represent an environmental situation of a water body with a high probability for an algal bloom and its successive collapse.

REFERENCES

Thom R., 1972. Stabilité structurelle et Morphogénèse: essai d'une Théorie générale des Modèles. Benjamin, New York, pp. 362.

Schrödinger E., 1988. What is Life? The Physical Aspect of the Living Cell, Cambridge University Press, pp.178

Rabinowitch E.I., (1951). Photosynthesis and related processes. Intersciences, New York, 3 Vol., pp.2088

Swartzman G. L. and R. Bentley 1979. A Review and Comparison of Plankton Simulation Models. ISEM Journal.

Van Nguyen V. and E. F. Wood 1979. On the Morphology of Summer Algae Dynamics in Non-Stratified Lakes. *Ecological Modelling* 6: 117 131

II.8.4 Ecosystems as Chaotic Systems

Sven E. Jørgensen

1. Introduction and Definitions

Chaos theory is concerned with unpredictable courses of events. The irregular and unpredictable time evolution of many non-linear and complex linear systems has been named 'chaos'. It makes strong claims about the universal behaviour of complexity. It occurs in mechanical oscillators such as pendular or vibrating objects, in rotating or heated fluids, in laser cavities, in some chemical reactions and in biological systems. The most fascinated advocates of this new science go so far as to say that this century science will be remembered for three things: relativity theory, quantum mechanics and chaos theory. Relativity has eliminated the Newtonian illusion of absolute space and time, while quantum theory has eliminated the Newtonian dream of controllable measurements. Chaos theory has eliminated the Laplacian illusion of deterministic predictability and can therefore be conceived as a ticking bomb under reductionistic science.

Chaos theory is best illustrated by Lorenz's (1963 and 1964) famous Butterfly Effect – the notion that a butterfly stirring the air in Hong Kong today can transform storm systems in New York next month. The effect was discovered accidentally by Lorentz in 1961. He was making a weather forecast and wanted to examine one sequence of greater length. He tried to make what he thought was a shortcut. Instead of starting the whole run over again, he started half way through. To give the computer its initial values, he typed the numbers from the earlier printout. The new run should therefore duplicate the old one, but it did not. Lorentz saw that his new weather forecast was diverging so rapidly from the previous run that within a few months all resemblance has disappeared. There had been no malfunction of the computer or the program. The problem lay in the number he had typed. In the computer six decimal places were stored: 0.506127, but to save time, and because he thought it was unessential, he printed a rounded-off number with just three decimals: 0.506.

The explanation is simple: Lorenz's model is very sensitive to initial conditions, and so is the weather itself. The effect today is observed in numerous relations and all ecological modellers know this problem. Therefore, the initial values of the state variables are most often included in a modeller's sensitivity analysis and he uses much effort to have the seasonal variations of the state variables repeated again and again, when the same forcing functions are imposed on the model. Figure 1 gives an ecological example of a model with high sensitivity for the initial value. As seen from the figure, a minor difference in the initial value gives two completely different curves after t = 100.

The definition of chaos implies that the difference between the two curves with slightly different initial conditions is growing exponentially:

$$d(t) = d(0) * e^{l*t}, \tag{1}$$

where d(t) is the distance at time = t, d(0) is the distance at time = 0 and l is a positive number, called the Lyapunov exponent, which is a quantitative indicator for chaos. After the time 1/l, the initial conditions are insignificant, i.e. 'forgotten'.

The Lyapunov exponent for the case study illustrated in Figure 1 can be found by plotting the logarithm to the distance between the two curves neglecting the distance at time 0 (which is 0) versus the time. The plot is approximately a straight line with a positive slope of approximately 0.012 mg / (l * day), indicating chaotic behaviour.

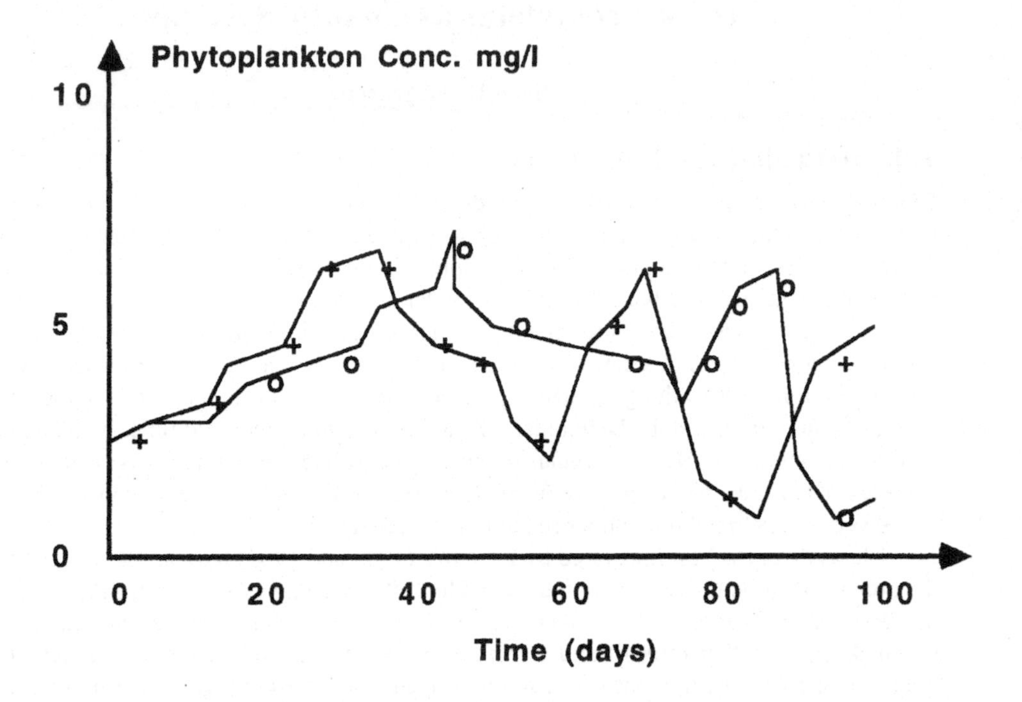

Figure 1: A six compartment lake model has been applied to illustrate that the initial value in this case of the zooplankton concentration has great influence on the final results of simulation. The plot shows the phytoplankton simulation according to the model over a period of 100 days, applying zooplankton concentrations on 0.05 (+) and 0.053 (o) as initial values. As seen the phytoplankton concentrations are completely different after 100 days, when using the two initial values of the zooplankton concentration.

A Benard cell presented in Chapter II.1.1 may expose chaotic behaviour. When the temperature difference, ΔT, between the lower and the upper plate exceeds a critical value, convection occurs as described as a series of rolls resembling rotating parallel cylinders. The rolls begin to oscillate transversely in complex ways as ΔT increases beyond a second threshold and chaotic behaviour occurs for even higher values of ΔT.

The BZ-reaction, presented in Chapter II.1.1, is an often studied example of a chemical system which exhibits both periodic and chaotic behaviour. The existence of chaos in the BZ-reaction suggests that similar behaviour might occur for other chemical oscillators such as those found in biological systems. Chaotic behaviour in these systems may indicate a pathological condition (Rapp, 1986). This may explain why chaotic behaviour does not occur more frequently in ecological systems, despite the high complexity of ecological processes. Processes and components (organisms) with properties (parameters) that will create chaos, are simply out-competed. This will be discussed further below.

2. Bifurcation

Chaos is also known in relation to bifurcation and this form of chaos can be nicely illustrated by examination of a simple model in population biology. May (1974, 1975 and 1976) has examined the behaviour of non-linear differential and difference equations, for instance:

$$N_{t+1} = N_t (1 + r(1-N_t /K)), \tag{2}$$

where N is the number of individuals in the population under consideration, r the growth rate per capita, t the time and K the carrying capacity of the environment. Notice that this equation expresses a time delay = 1 in the form the difference equation is given. As long as the non-linearity is not too severe, the time delay built into the structure of the difference equation tends to compare to the natural response time of the system and there is simply a stable equilibrium point at N# = K. However for $r \geq 2$ this point becomes unstable. It bifurcates to produce two new and locally stable fixed points of period 2, between which the population oscillates stably in a 2-point cycle. With increasing r, these two points also bifurcate to give four stable fixed points of period 4. In this way through successive bifurcations an infinite hierarchy of stable cycles of the period 2^n arises. Figure 2 illustrates the formation of bifurcations up to r = 2.75.

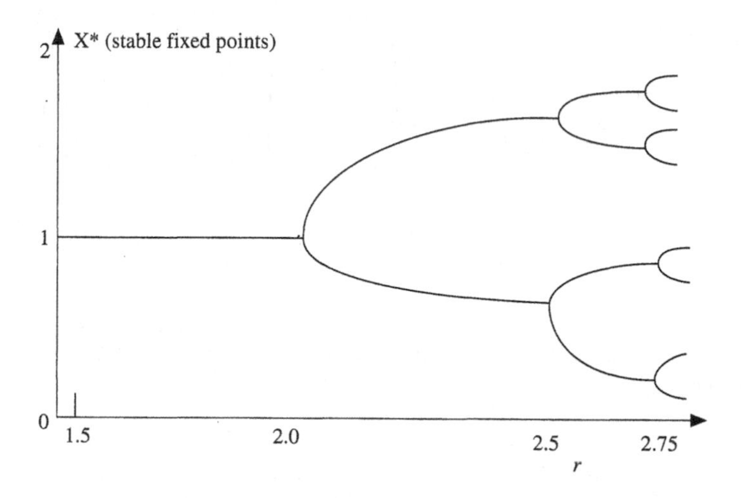

Figure 2: The hierarchy of stable fixed points of periods 1, 2, 4, 8... 2^n, which are produced from equation (10.1) as the parameter r increases. The y-axis indicates relative values.

3. Ecological Implications

When we consider the many non-linear and complex linear relationships which are valid in ecology, we may wonder why chaos is not observed more frequently in nature or even in our models.

We would expect chaos to be a rule, not an exception for ecosystems. An obvious answer could be that nature attempts to avoid chaos and, as opposed to the physical system the ecosystem has (as discussed in Chapter I.1) many possible, hierarchically organised, regulation mechanisms to avoid chaotic situations. This does not imply that chaotic or 'almost chaotic' situations are never observed in ecosystems. They are only rarer than we would expect from their complex relationships. The classical example of chaotic behaviour in ecological systems is the almost legendary lemming . According to Shellford (1943), r *T is 2.4, r being the growth rate per capita and T the time lag. From Figure 2 it can be seen that oscillations between two steady states should be expected as Shellford found; see Figure 3, which is reproduced from Shellford (1943).

Hassel et al. (1976) have culled data on 28 different populations of seasonal breeding insects. They found that the growth may be described by a difference equation as follows:

$$N_{t+1} = q \cdot N_t \, (1 - a \cdot N_t)^{-\beta} \tag{3}$$

q is here related to r as follows: r = ln q. a and ß are constants. The theoretical domains of stability behaviour for this equation are in the monotonic damping area for 26 of the 28 cases. Only one is in the chaos area (and, as indicated by Hassel et al. it is a laboratory population) and one in the stable limit cycles area.

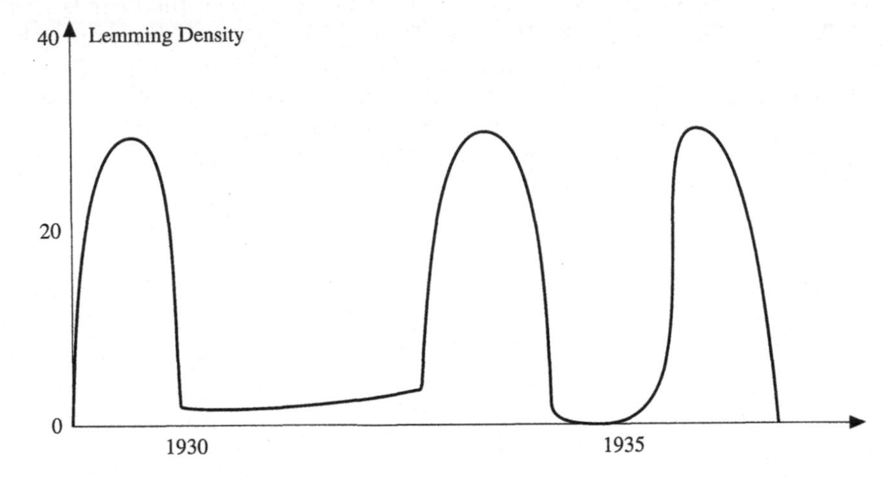

Figure 3: Data on the lemming population in the Churchill area in Canada, expressed as numbers per ha. It can be shown that a logistic equation fits nicely with the observations, assuming a relatively large growth rate of the population, (Shellford 1943).

Notice that there is a tendency for laboratory populations to exhibit cyclic and chaotic behaviour, whereas natural populations tend to have a stable equilibrium point. The laboratory populations are maintained in a homogeneous environment and are free from predators and many other natural mortality factors. It means that the indirect effect, which is found in nature (see Chapter II.7.3), is omitted. The indirect effect may very well, up to a certain level, give a stabilising effect as well.

It may be concluded that natural populations are able to avoid chaotic situations to a high extent. The long experience gained during evolution has taught the natural population to omit the properties, i.e., the parameters, that may cause chaotic situations, because it often threatens their survival. Furthermore, the natural populations have the flexibility and adaptability to select a combination of parameters that gives a better chance for survival. The relationship between the parameters (the properties) and the chaotic behaviour is the topic for the following chapter, which presents an examination of this relationship by the use of models.

4. Parameters (Properties) and Chaos

Figure 4 shows a model with only four state variables: a nutrient, phytoplankton, detritus and bacteria. The reaction of this system is extremely dependent on the parameters given to the state variables, particularly the maximum growth rates of phytoplankton and bacteria. Figure 5 shows two situations: one where the maximum growth rates of bacteria and phytoplankton are both 1 day $^{-1}$ and one where the growth rate of the bacteria is maintained on 1 day $^{-1}$, while phytoplankton was given the growth rate of 10 day $^{-1}$. If the growth rate of the bacteria is 10 day $^{-1}$, while the phytoplankton is maintained at 1 day $^{-1}$, almost the same picture is obtained as if the growth rates were inverse. Exergy is plotted versus the time in the figure to capture the biomass of the entire system. From these rather simple modelling experiments it is obvious, that if we have a very simple ecosystem with the components represented in Figure 4, a stable situation is not obtained if the two maximum growth rates for phytoplankton and bacteria are very different. Stable conditions require that almost the same maximum growth rate is used for phytoplankton and bacteria, while it is not crucial which level of growth we are using. Unstable conditions on the other hand appear when significantly different growth rates are allocated to the phytoplankton and the bacteria. The model may represent the situation on the Earth about 1.5 billion years ago (we do not know, of course, what the growth rates were then), when only phytoplankton and bacteria were present. Today it can be observed that the bacteria and phytoplankton have almost the same maximum growth rates. Therefore, it seems possible to assume, that bacteria and phytoplankton have adjusted their maximum growth rates to about 1-2 or maybe 3 day $^{-1}$. The growth rates have been forced to adjust to each-other and when they landed on the level of 1-4 day $^{-1}$, it was simply because that was biochemically feasible, considering the transfer of nutrients through the cell membrane.

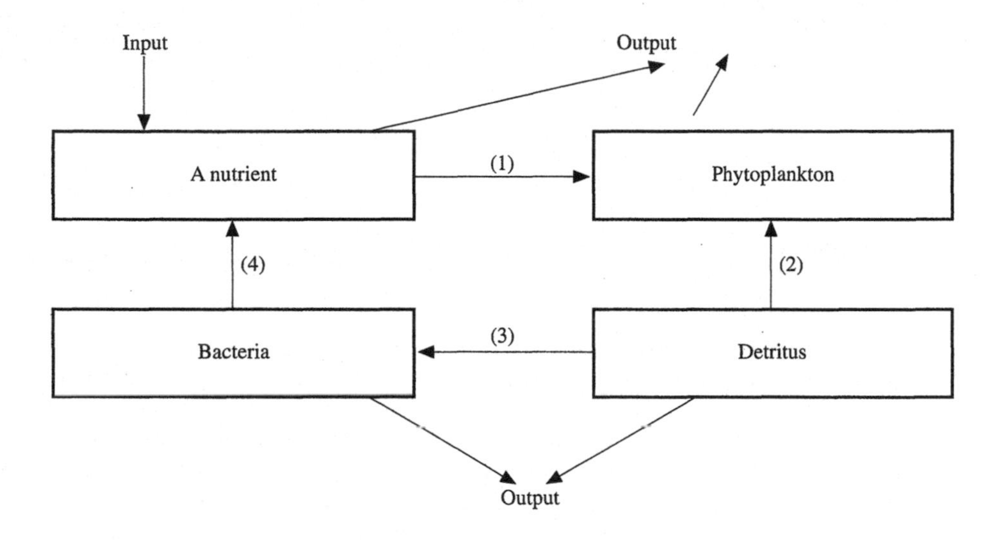

Figure 4: The model used for the parameter examination consists of four state variables: a nutrient, phytoplankton, detritus and bacteria. The processes are: growth, expressed by the use of a Michaelis Menten's equation, process number 1; mortality of phytoplankton by use of a first order expression, process 2; growth of bacteria by use of a Michaelis Menten expression, process 3; and mineralisation by use of a first-order expression, process number 4.

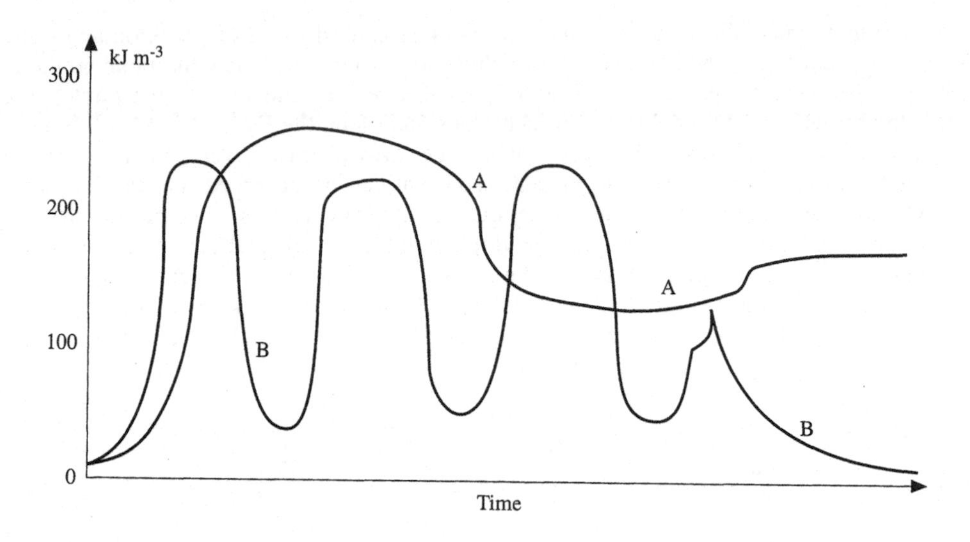

Figure 5: Exergy is plotted versus time for the model shown in Figure 4. A corresponds to an equal maximum growth rate of phytoplankton and bacteria on 1 day $^{-1}$. B corresponds to a growth rate of phytoplankton on 10 day $^{-1}$ and of bacteria on 1 day $^{-1}$.

Any attempt to increase the maximum growth rate would fail unless it was feasible for the bacteria and the phytoplankton simultaneously. Figure 6 shows another model that has been applied for similar modelling experiments. We have, however, excluded fish as state variable in the first hand, and we have given the phytoplankton and the bacteria the maximum growth rates experienced from the above-mentioned model experiments and ask now: which maximum growth rates of the two zooplankton state variables will ensure avoidance of chaotic situations? The answer, as seen in Figure 7, is that a maximum growth rate of about 0.35-0.40 day $^{-1}$ seems to give favourable conditions for the entire system, as the exergy is at maximum and stable conditions are obtained. A maximum growth rate of more than about 0.65 -0.70 1 / day seems to imply chaotic situations for the two zooplankton species.

Figure 8 shows a similar result including fish as state variable, while the two zooplankton state variables have been given maximum growth rates of 0.35 and 0.40 day $^{-1}$. A maximum growth rate of about 0.08-0.1 day $^{-1}$ seems favourable for fish, but again a too high a maximum growth rate (above 0.13-0.15 day $^{-1}$) for the state variable fish will give oscillations and chaotic situations with violent fluctuations. A more detailed examination of the relationship between the behaviour and the value of a specific parameter, in this case the maximum growth rate of zooplankton, has been made by Jørgensen (1995). Figure 9 shows the results of simulations with a model similar to Figure 6, but without fish and with only one zooplankton class. The maximum growth rate of zooplankton has been varied. The model has been run to steady state, if a steady state could be obtained. The exergy expressed as 'exergy of g organic matter' /l is plotted versus the maximum growth rate of zooplankton.

In the figure is indicated, whether a steady state can be obtained, or whether fluctuations occur. If regular oscillations occur, the average of the exergy for one oscillation is used.

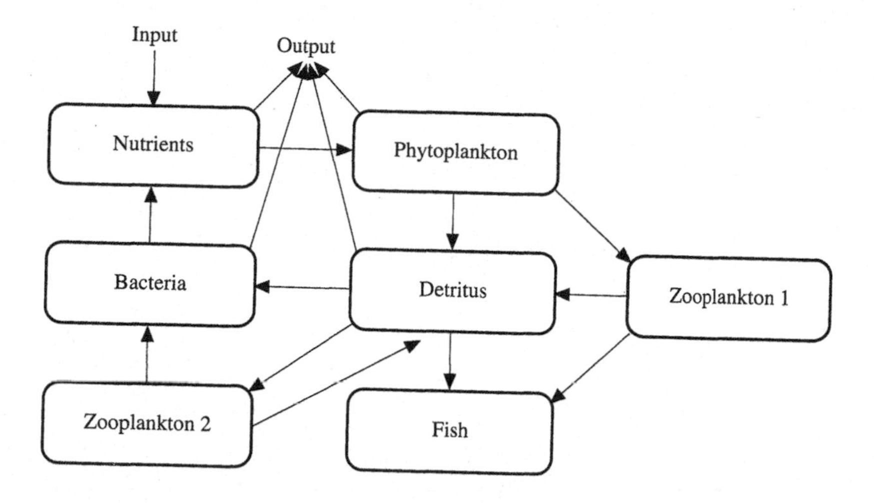

Figure 6: Model used to examine the feasible parameters. The model consists of seven state variables. Two zooplankton classes and fish are added in addition to the model in Figure 4

At a maximum growth rate of 0.5 1/day regular oscillations occur, and the average level of exergy is slightly lower than for a maximum specific growth rate of 0.425 1/day. At a maximum specific growth rate of 0.6 1/day, an even lower average exergy is obtained and the regularity is smaller. At higher growth rates, the exergy and the state variables exhibit violent and irregular changes.

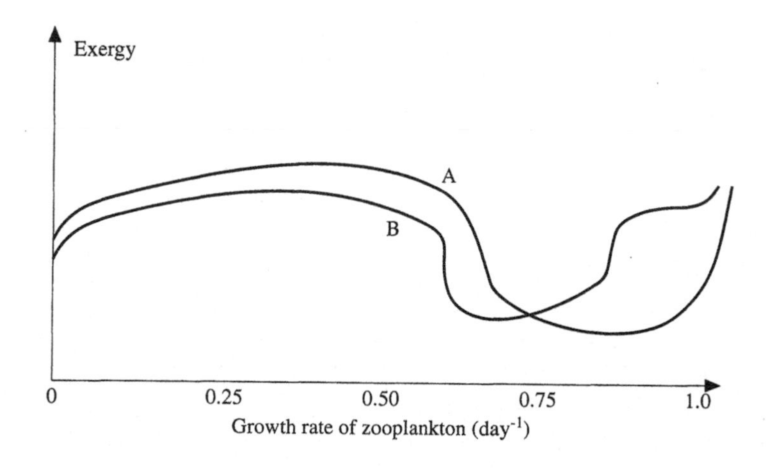

Figure 7: Exergy is plotted versus maximum growth rate for the two zooplankton classes in Figure 7. A corresponds to the state variable 'zoo' and B to the state variable 'zoo2'. The shaded lines correspond to chaotic behaviour of the model, i.e., violent fluctuations of the state variables and the exergy and a strong dependence of the initial values of the state variables. The values of the exergy fluctuate at a maximum growth rate of about 0.65-0.7 1/ day. The average values are therefore shown.

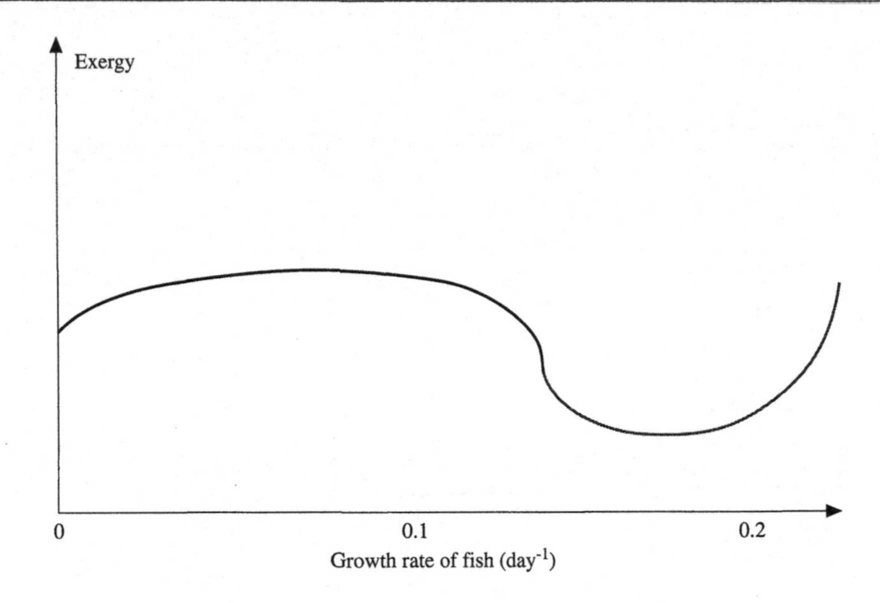

Figure 8: The exergy is plotted versus the maximum growth rate of fish . The shaded line corresponds to chaotic behaviour of the model, i.e., violent fluctuations of the state variables and the exergy. The shown values of the exergy above a maximum growth rate of about 0.13-0.15 day $^{-1}$ are therefore average values.

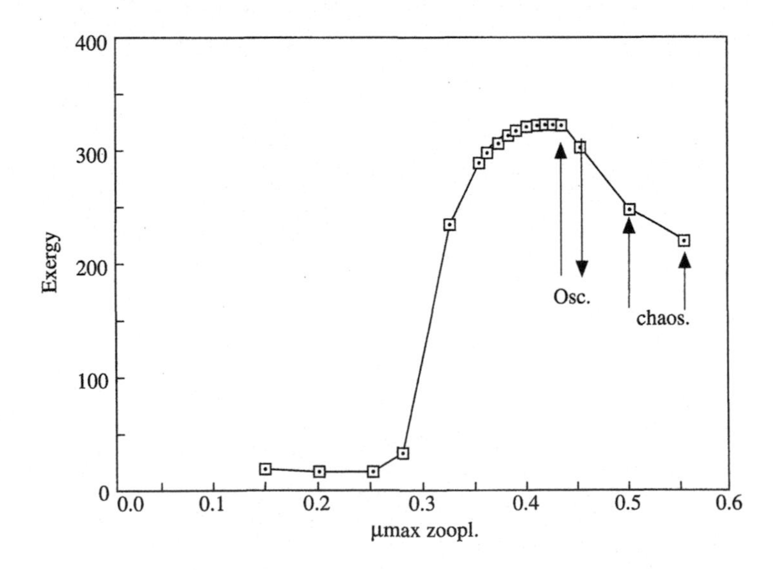

Figure 9: Exergy expressed in the unit mg detritus exergy equivalents / l is plotted versus the maximum growth rate of zooplankton for a model with nutrients, detritus, phytoplankton and zooplankton as state variables.

The highest level of exergy is obtained for maximum growth rate of zooplankton and is slightly lower than the values that exhibit chaotic behaviour; see Figure 9. The highest exergy is therefore for this particular model obtained at the 'edge of chaos' (Jørgensen, 1995). The maximum growth rate obtained at the highest level of exergy can furthermore be considered realistic, i.e., according to the range found in the literature for the maximum specific growth rate of zooplankton; see Jørgensen et al. (1991).

Figure 10 shows the same plot as Figure 9, but with introduction of fish in the model. A lower specific growth rate means that zooplankton get bigger in size following general allometric relationships; see Peters (1983). This behaviour of the model entirely follows several observations in nature: predation by fish yields zooplankton, that often have a bigger size (provided that fish doesn't have any size preference which, however, may be the case) and has slower growth rates; see Peters (1983).

The maximum specific growth rate found at maximum exergy for the model run with fish is also within the range of values found in nature: approximately 0.15 - 0.5 1/day; see Jørgensen et al. (1991), and Jørgensen (1994a and 1994b). If the fish is removed from the model again, the level of exergy decreases drastically due to the loss of the information embodied in the fish, but by increasing the maximum growth rate of zooplankton the exergy increases again. The results from Figure 9 are reproduced again with the highest level of information at a maximum growth rate of 0.425 1/day, and at the edge of chaos.

Figure11 shows the cycle of changes, plotting the maximum growth rate versus the level of exergy: for the model without fish at the edge of the chaos, for the model with fish and with the unchanged maximum specific growth rate, for the model with fish but with the maximum growth rate corresponding to the maximum level of exergy (again at the edge of the chaos for this model), for the model without fish but with the maximum growth rate from the model with fish and finally with the model without fish at the edge of the chaos again. The plot shows the hysteresis phenomena which often are observed in ecosystems, when catastrophic events take place, see for instance Chapter II.8.3, Jørgensen (1997) and Bendoricchio (1988). It is a result of adaptations to emerging conditions, and may be explained by a simultaneous self-organisation and selection.

The fractal dimension may be considered a measure of the chaotic behaviour. Fractal is a word introduced by Mandelbrot and may best be introduced by an example. The example is a typical Mandelbrot question: 'How long is the Coast of Britain?'. This question does not have an unambiguous answer, because it depends on how many details, the measurement should include. If we use a ruler of one metre, we get another result than if we use a ruler of one cm. This problem inspired Mandelbrot to introduce what he called the fractal dimension.

Suppose you want to measure the length of an irregular curve, such as for instance, the coastline of Britain. Similar considerations may apply in the measurement of an area of an irregular surface, except that the ruler is replaced by tiles. To probe the system in a manner independent of whether the system is a curve or a surface, you may divide the embedding space into cells (boxes) of side length e and count how many are intersected by the curve (or surface). For a straight line, it is possible to see that the number of boxes is proportional to e^{-1}, which indicates that the dimension is +1. For a square, the number of boxes will correspondingly be proportional to e -2 and the dimension – not surprisingly – +2. For a curve, the irregularity will prevent N (e) from ever approaching an e^{-1} behaviour and the dimension will be – dependent on the irregularity – between +1 and +2. In other words fractal dimension becomes a measure of the irregularity.

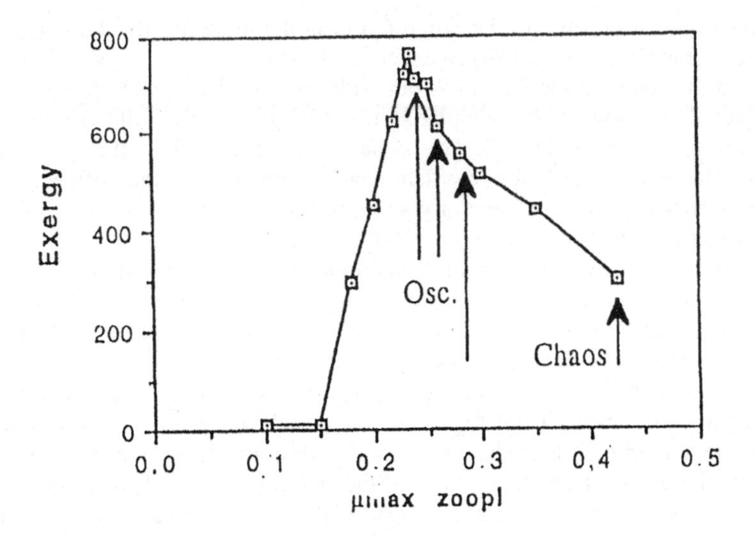

Figure 10: Exergy expressed in the unit mg detritus exergy equivalents/l is plotted versus maximum growth rate of zooplankton with nutrients, detritus, phytoplankton, zooplankton and fish as state variables. Notice that the exergy is higher than in Figure 9 due to the presence of fish, and that the maximum growth rate at maximum exergy level is lower. No size preference is assumed for the zooplankton predated by the fish.

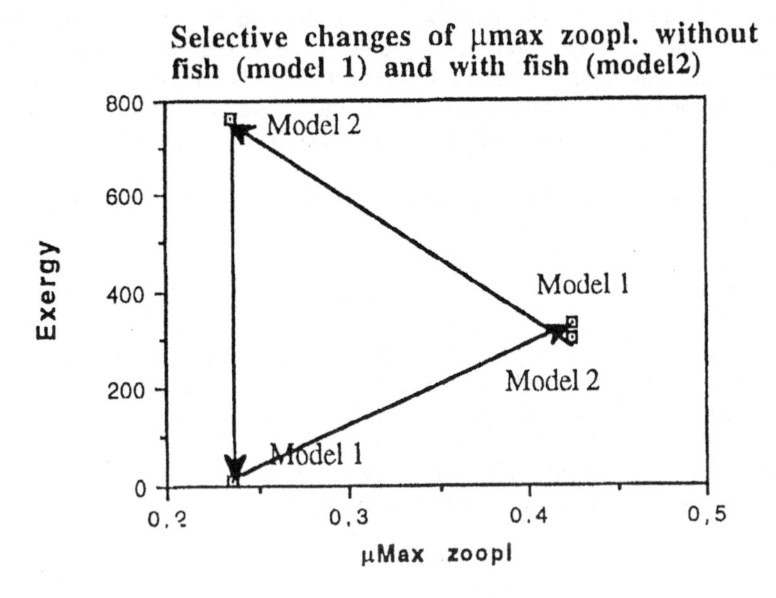

Figure 11: The maximum growth rate is plotted versus the level of exergy expressed in the unit mg detritus exergy equivalents: for the model without fish at the edge of the chaos, for the model with fish and with the unchanged maximum specific growth rate, for model with fish but with the maximum growth rate corresponding to the maximum level of exergy (again at the edge of the chaos for this model), for the model without fish but with the maximum growth rate from the model with fish and finally with the model without fish at the edge of the chaos again.

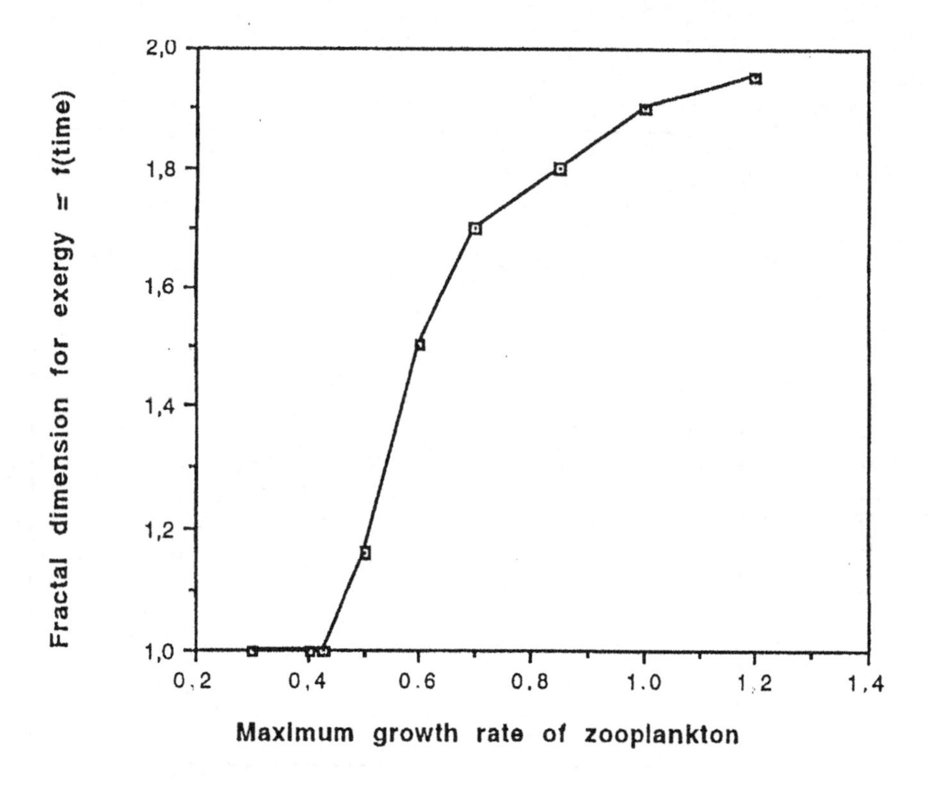

Figure 12: The fractal dimension obtained for the plots exergy = f(time) for various values of the maximum growth rate of zooplankton is shown.

The fractal dimension D is defined as D in the relationship between N (e) and e :

$$N(e) = A * e^{-D} \tag{4}$$

The fractal dimensions obtained for the plots of exergy versus the time for various levels of the maximum zooplankton growth rate for the model runs without fish are shown on the Figure 12. As seen, the fractal dimension increases with an increasing maximum growth rate of zooplankton, as expected due to the more and more violent fluctuations of the state variables and thereby of the exergy. When the maximum growth rate increases, more and more violent fluctuations result with higher and higher maximum values, smaller and smaller minimum values and increasing occurrence of the smaller values, resulting in decreasing average values of the exergy. It is illustrated in Figure 13, where the average exergy is plotted versus the fractal dimension.

In this case a fractal dimension of 1.0 for the values of the maximum growth rate of zooplankton $\leq$ 0.425 1/ day is obtained, because the model considers a steady state situation where no fluctuations of the phytoplankton due to variations in temperature and solar radiation are considered. If normal diurnal and seasonal changes are considered, these parameter values will exhibit a fractal dimension slightly more than 1, but the fractal dimension will still increase when the maximum growth rate is bigger than the maximum growth rate at maximum exergy. It is, however, interesting that the fractal dimension has a tendency to be smaller when diurnal and seasonal changes are introduced. It may be interpreted in line with the results in Chapter II.8.1.

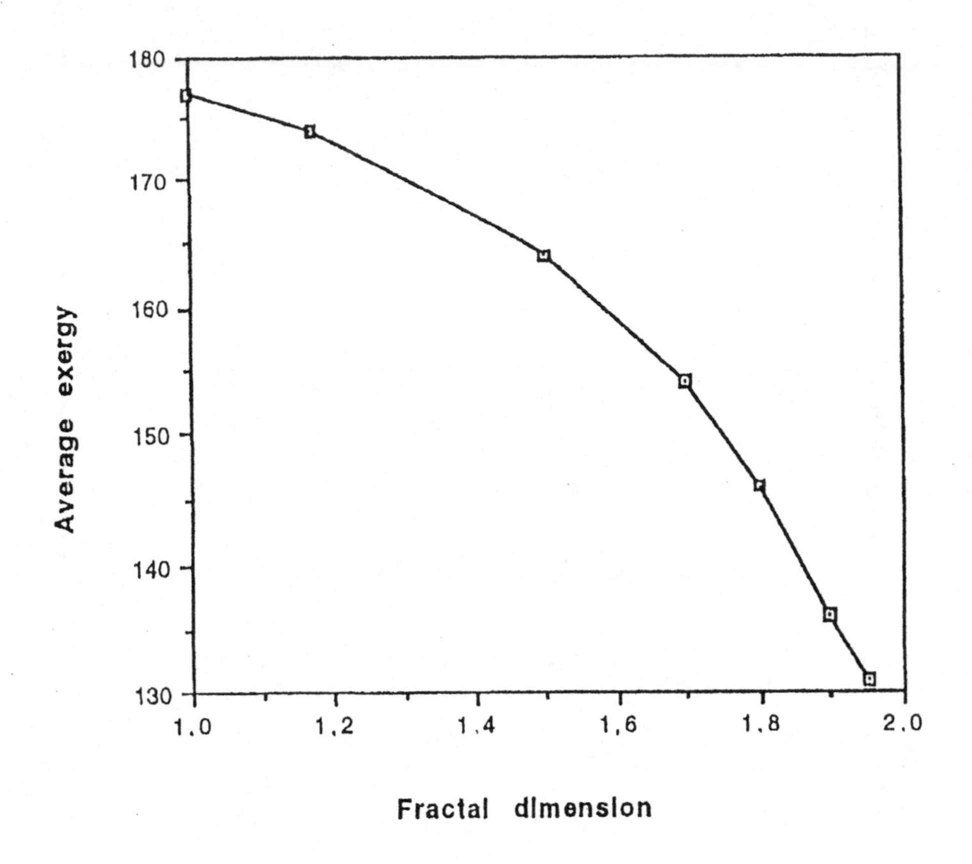

Figure 13: The average exergy is plotted versus the fractal dimension based on the examined case study .

The values of exergy and the fractal dimensions in the illustrated case study are, of course, dependent on all the selected parameter values. The shown tendency is, however, general: the highest exergy is obtained by a parameter value above which chaotic behaviour, increasing fractal dimension for the state variables and the exergy as function of time and decreasing average exergy occur.

The parameter estimation is often the weakest point for many of our ecological models due to:

1) an insufficient number of observations to enable the modeller to calibrate the number of more or less unknown parameters,
2) no or only little literature information can be found,
3) ecological parameters are generally not known with sufficient accuracy
4) the structure shows dynamic behaviour, i.e., the parameters are continuously changing to achieve a better adaptation to the ever changing conditions; see also Jørgensen (1988) and (1992a),
5) a combination of two or more of these points.

5. Ecosystem Theory and Chaos

The above-mentioned results seem to reduce the difficulties of parameter estimation by imposing the ecological facts that all the species in an ecosystem will try to obtain by selection the properties that are best fitted for survival under the prevailing conditions. The property of survival can currently be tested by use of exergy, as survival is measured by biomass and information which (see Chapter II.1.3) can be expressed by the thermodynamic concept exergy. Coevolution, i.e., when the species have adjusted their properties to each other, is considered by application of exergy for the entire system. Application of the tentative fourth law of thermodynamics (see Chapter II.1.3) as constraint on our ecological models enable us to reduce the feasible parameter range, which can be utilised to facilitate our parameter estimation significantly.

It is interesting that the ranges of growth rates actually found in nature (see for instance Jørgensen et al. (1991) are those, which give stable, i.e., non-chaotic conditions. All in all, it seems possible to conclude that the parameters that we can find in nature today, are in most cases those which assure a high probability of survival and growth in all situations; chaotic situations are thereby avoided, as they would imply that the populations from time to time would be at a very low level with the danger of extinction. The parameters that could give possibilities for chaotic situations, have therefore been excluded by selection processes. They may give high exergy in some periods, but later the exergy becomes very low due to the violent fluctuations and under such circumstances the selection process excludes the parameters (properties), that cause the chaotic behaviour.

The results are not surprising, when we consider what is behind the various processes in ecosystems. If we give the grazing rate of zooplankton a value which is too high in the sense that it will create chaotic behaviour, we will of course observe a very fast zooplankton growth, resulting in a high zooplankton biomass. It is, however, not advantageous for zooplankton, as its food source will be depleted and a high mortality due to starvation will follow the rapid growth phase. It would be more beneficial for zooplankton to adjust the growth rate to the food source to ensure that zooplankton would omit low concentrations with danger of extinction.

Markus et al. (1984, 1987 and 1988) and Markus (1990 and 1991) have examined the occurrence of chaos for populations under periodically and randomly varying growth conditions. It was found that periodically or randomly changing environmental conditions may induce a variety of unexpected dynamic behaviours (see Markus and Hess 1990a and 1990b). Systems, that are ordered under constant conditions may become chaotic. This effect lowers the threshold for chaos and may explain the fact, that the observed parameters in nature are lower than those leading to chaos in calculations which is according to the results referred to above. The reverse effect is observed, too: a system that is chaotic under constant conditions may become ordered if the conditions change periodically or randomly. Markus et al. (1987) conclude that no generalisation on the effect of temporal variation on chaotic behaviour can be made, and they assume that the statement that 'temporal variations in the environment are a destabilising influence' is much too simple in view of the diversity of coupling processes.

Systems at the edge of the chaos can co-ordinate the most complex behaviour. They can adapt gradually in typical circumstances, but mount massive changes when needed (Kauffman, 1991, 1992 and 1993). Conversely, systems in the chaotic regime are so drastically altered by even minor variations in structure that they cannot easily accumulate useful variations. At the same time, systems deep in the ordered regime are changed so slightly by minor variations that they adapt too slowly to the environment, that sometimes makes sudden and even catastrophic changes. Kauffman (1991, 1992 and 1993) has shown that this is consistent with how many cell types could appear in an organism. By the use of Boolean logic to a network, it can be shown that the number of cell types should be approximately the square root of the number of genes, if the hypothesis of anti chaos in complex systems is correct. The actual number of cell types in various organisms appears to rise accordingly as the amount of DNA increases.

Kauffman (1991 and 1992) has studied a Boolean network and finds that a network on the boundary between order and chaos may have the flexibility to adapt rapidly and successfully through the accumulation of useful variations. In such poised systems most mutations will have small consequences because of the system's homeostatic nature. Such poised systems will typically adapt to a changing environment gradually, but if necessary, they can occasionally change the properties rapidly. It explains, according to Kauffman, why Boolean networks poised between order and chaos can generally adapt most readily and therefore have been the target of natural selection.

The hypothesis is bold and interesting in relation to the results obtained by the use of exergy as an indicator in the choice of parameters. The parameters that give maximum exergy are not much below the values that would create chaos, see Figures 9 and 10. In combination with the tentative fourth law of thermodynamic, we can formulate the following proposition: Ecosystems will attempt to select parameters (properties) that would guarantee a poised system between order and chaos.

REFERENCES

Bendoricchio, G. 1988. An application of the theory of catastrophe to the eutrophication of the Venice lagoon. In: A. Marani (editor) Advances in Environmental Modelling. Elsevier, Amsterdam.

Hassell, M.P., J.H. Lawton, and R.M. May 1976. Patterns of dynamical behaviour in single species populations. J. Anim. Ecol. 45: 471-486.

Jørgensen, S.E. 1988. Use of models as an experimental tool to show that structural changes are accompanied by increased exergy. Ecol. Modelling 41: 117-126.

Jørgensen, S.E. 1994a. Fundamentals of Ecological Modelling (second edition) (Developments in Environmental Modelling, 19). Elsevier, Amsterdam, 628.

Jørgensen, S.E. 1994b. Models as instruments for combination of ecological theory and environmental practice. Ecol. Modelling 75/76: 5-20.

Jørgensen, S.E. 1995. The growth rate of zooplankton at the edge of chaos: ecological models. J. Theor. Biol. 175: 13-21.

Jørgensen, S.E. 1997. Integration of Ecosystem Theories: A Pattern. Second Edition. Kluwer Academic Publishers, Dordrecht, Boston, London. 386.

Jørgensen, S.E., S.N. Nielsen, and L.A. Jørgensen 1991. Handbook of Ecological Parameters and Ecotoxicology. Elsevier, Amsterdam.

Kauffman, S.A. 1991. Antichaos and adaptation. Sci. Am. 265 (2): 64-70.

Kauffman, S.A. 1992. The sciences of complexity and 'origins of order'. In: J.E. Mitten and A.G. Baskin (editors). Principles of Organization in Organisms. Proceedings of the Workshop on Principles of Organization in Organisms, p. 71-96. June 1990, Santa Fe. Santa Fe Institute.

Kaufmann, S.A. 1993. Origins of Order. Self Organization and Selection in Evolution. Oxford University Press, Oxford.

Lorenz, E. 1963. Chaos in meteorological forecast. J. Atmos. Sci. 20: 130-144.

Lorenz, E. 1964. The problem of deducing the climate from the governing equations. Tellus 16: 1-11.

Mandelbrot, B.B. 1983. The Fractal Geometry of Nature. Freeman, San Francisco. 420.

Markus, M. 1990. Chaos in maps with continuous and discontinuous maxima: a dramatic variety of dynamic.

Markus, M. 1991. Unvorhersagbarkeit in einer determinitischen Welt: Der Tod der Laplaceschen Dämons. (UNI Report 13). Berichte aus der Forschung der Universität Dortmund.

Markus, M., and B. Hess 1990a. Control of metabolic oscillations: unpredictability, critical slowing down, optimal stability and hysteresis. In: A. Cornish-Bowden and M.L. Cárdenas (editors). Control of Metabolic Processes. Plenum Press, New York, 303-313.

Markus, M., and B. Hess 1990b. Isotropic cellular automaton for modelling excitable media. Nature 347: 56-58.

Markus, M., D. Kuschmitz, and B. Hess 1984. FEBS Lett. 172: 235-238.

Markus, M., B. Hess, J. Roessler, and M. Kiwi 1987. In: H. Degn., A.V. Holden and L.F. Olsen (editors). Chaos in Biological Systems. Plenum Press, New York, 267-277.

Markus, M., S.C. Müller, and G. Nicolis (editors). 1988. From Chemical to Biological Organization (Springer Series in Synergetics, vol. 39). Springer-Verlag, Berlin.

May, R.M. 1974. Ecosystem patterns in radomly fluctuating environments. Progr. Theor. Biol. 3: 1-50.

May, R.M. 1975. Biolgical populations obeying difference equations: stable points, stable cycles and chaos. J. Theor. Biol. 49: 511-524.

May, R.M. 1976. Mathematical aspects of the dynamics of animal populations. In: S.A. Levin (editor). Studies in Mathematical Biology. American Mathematical Society, Providence, Rhode Island.

Peters, R.H. 1983. The Ecological Implications of Body Size. Cambridge University Press, Cambridge.

Rapp, P.E. 1986. Oscillations and chaos in cellular metabolism and physiological systems. In: A.V. Holden (editor). Chaos. Princeton University Press, 179-208.

Shellford, V.E. 1943. The relation of snowy owl migration to the abundance of the collared lemming. Auk 62: 592-594.

II.8.5 Ecosystems as States of Ecological Successions

Klaus Dierssen

1. Introduction

Dynamic processes within populations, biocoenoses, sites and ecosystems are among the most intensively discussed topics in ecology. The approaches to best understand the underlying causes and processes are among the most exciting challenges. The evaluation of the development of fallow land or nature preserves with oligo and mesohemerobic conditions needs a subtle understanding of the changes in order to adopt potentially necessary measures of conservation biology. Long term observations on sites, which represent chronosequences, that is defined successional stages (Schmidt 1981, Bakker et al., 1996; Herben, 1996), are a crucial basis for succession research. Only knowledge of the exact pattern and sequence of species in a successional sere and the relevant ecological processes allow us to identify successional pathways, to generate hypotheses on mechanisms of successions and to construct experiments which may help to understand the causes of species replacement and site development (Huisman et al., 1993; Van Andel et al., 1994).

One main conceptual problem is, that many ecologists try to solve long-term questions from short term experiments or investigations (Tilman, 1989). Consequently, an ocean of literature exists on succession, but unfortunately even prominent textbooks contribute to the confusion about the basic terminology, the knowledge of key pathways and the causes and mechanisms of succession. The same is true for understanding and modeling successional processes in order to formulate clear hypothesis, and to predict successional trends or scenarios for a given site (Connell and Slatyer, 1977; Picket et al., 1987; Peet, 1992).

The dispute at an operational level of how to describe, understand and interpret successions is as old as the study of dynamic processes in nature, especially in vegetation development. The discussion cumulated in contrasting scientific concepts of Clements (1916), who advocated the view of an organismic (deterministic) organisation of biocoenoses, and Gleason (1917, 1939), who generally denies any emergent quality of biocoenoses and ecosystems. Gleason's approach is rather stochastic and, thus, rules out the suitability of classification and typification of communities (and ecosystems) in general. Further difficulties concern the contrast between more historical and evolutional ('ultimate') explanations of dynamics in ecology, and more experimental and exact ('proximal') elucidations of present ecosystem properties underlying successions. Both attempts lead to different types of forecast: on one hand, qualitative prognoses focused on concrete objects, and on the other hand, quantitative ones focused on abstract types in various spatial and temporal scales (Harper, 1982; Valsangiacomo, 1998). In practice, ecological hypotheses cannot be constructed without invoking processes that are contingent upon past evolutionary history. Therefore, it seems feasible to combine both approaches. The evolutional preoccupations with the development of the sites and the life histories of the populations involved should be complemented by a functional approach including the dynamics of energy and nutrients (O'Neil et al., 1986; Prentice, 1998).

2. Semantics

Some commonly used terms become misleading when applied to comparisons between systems at different levels of organisation (species, populations, biocoenoses, sites, ecosystems, landscapes) or to various spatial or temporal scales. A short survey of key terms therefore seems helpful.

A comprehensive explanation of biological phenomena can only be achieved within an appropriate spatio-temporal continuum. Populations, communities and ecosystems usually range from 10^{-2} to 10^{10} m^2. The time scale for dimensions important for ecological and evolutional processes ranges from minutes to thousands of years. Succession describes the non-seasonal, directional and continuous pattern of colonisation and extinction on a site by populations of species. Successional sequences may occur over widely varying time-scales, and as a result of different underlying processes. They are 'ecoclines in time' (Whittaker, 1975). Seasonal phenological changes, annual cycles depending on climatic changes or regeneration cycles are perhaps better named fluctuations than successions (Daubenmire 1968, Müller-Dombois and Ellenberg 1974, Miles 1979, Begon et al. 1990), even though any distinction remains arbitrary to a certain degree (Bornkamm 1988).

In order to describe, explain and predict aspects of successions, it seems useful to differentiate between (i) *pathways*, which describe the temporal pattern of species change during succession; (ii) *causes,* which are the abiotical, biotical or historical agents, circumstances or actions responsible for successional patterns, and (iii) *mechanisms*, being interrelationships which contribute to successional changes (Pickett et al. 1987). A differentiation between causes and mechanisms becomes increasingly difficult from pioneer to mature stages of successions.

One may distinguish between *autogenic* and *allogenic* causes of successions. Autogenic causes occur (chiefly) as a result of intrinsic biological processes, that modify conditions and resources. With allogenic causes, the changing species composition of an investigated site is mainly driven by external influences which alter conditions, for example geophysical forces or climatic changes. External control may be exerted by seldom or singularly occurring events, by immediate causes or by long-lasting slow and indistinct environmental changes. The reaction of the biological system may differ between abrupt and gradual developments. Retarded responses are possible, because the reaction of the species composition depends on intrinsic feedbacks, and single indicating species may behave differently in various stages of their life history. Permanent plot investigations are useful in order to monitor both autogenicly and allogenicly induced changes, for instance, in impact assessment studies or to monitor the influence of management measures by application of fertilisers, different grazing regimes or herbivore exclosure (i.e. Schmidt, 1981; Beeftink, 1987; Bakker, 1989; Olff et al., 1992; Kiehl, 1997; Schreiber, 1997; Dröschmeister and Gruttke, 1998; Wolfram et al., 1998). Even relatively weak impacts, such as moderate fertilising within sensitive systems, can result in long-term vegetation changes (Egloff, 1998).

When studying the sequence of plant remnants from undisturbed profiles of lake sediments or peat cores, or by comparing subsequent vegetation maps of a landscape section, long- and short-term successional pathways will be recognised. Obviously, there is a gradient from initial (*'pioneer'*) to structurally complex *'mature'* systems (Odum, 1969; Gigon, 1975; Bazzaz, 1979, 1987; Vitousek and Walker, 1987; Armesto et al., 1991).

On newly exposed bare soil, not previously occupied by plants, *primary* successions occur. Examples are soils on lava flows following vulcanic eruptions, on sandy beaches, land slides or substrates laid bare after the retreat of a glacier or on man-made wastes (Miles and Walton, 1993; Turner et al., 1998). Their colonization is mainly driven by the accessibility of the site in question by diaspores, and by environmental changes, i.e. changes affecting soil development. Primary successions may last up to thousand years or more. Theoretically, because quaternary climate has been never stable, primary successions in the long term never end. *Secondary* successions occur when the preceding vegetation has been destroyed or the abiotic and biotic conditions have obviously changed. They are especially characteristic for, but not restricted to, cultural landscapes, where human impact increases, decreases or ceases. The further development depends on biotic processes and abiotic site conditions. In areas regenerating from fire, arable fields, meadows or pastures, the initial floristic composition usually is the predominant factor which roughly predicts the regeneration sequence (Egler, 1954).

A successional development associated with an increasing structural complexity is called *progressive*, a decrease in structural complexity is called *regressive (retrogressive)*. The succession from abandoned fields to forests is an example of progressive processes, the degradation of primary forests to shrublands and garigues by cattle grazing or burning is a regressive development.

Typically, soils undergoing primary succession have low nitrogen and organic matter contents, and lack a buried seed pool. Colonising individuals or their propagules must be dispersed to reach the site. Propagules of early colonisers have smaller seeds than late successional species (Chapin, 1993). Early and mid-successional species have higher rates of photosynthesis on an area basis and a higher potential to absorb nutrients, resulting in a high potential growth rate. Late successional species obtain higher concentrations of defensive compounds against herbivores and are therefore less palatable than early-successional species (Küppers,1984; Lambers et al., 1998). External control and selection by the physical environment, however, is strong and may override most biological processes in extreme environments or in early successional stages. The advantage of broad niches for species colonising early successional habitats is therefore obvious. Internal processes and environmental heterogeneity increase in importance in structural complex '*mature*' systems.

The results of models which simulate community assembly using Lotka-Volterra equations broadly match what one may expect to happen during succession: (i) With increasing community organisation, the webs of interaction, competition as well as co-evolutionary niche development increase. (ii) Species accumulate in the model systems with time, but at steadily decreasing rates. It gets harder for new species to invade the model communities. (iii) 'Turnover' of species slows down. Species persist in the community for longer in later model stages (Post and Pimm, 1983; Drake, 1985; Lawton, 1987). Of course, there is some uncertainty to generalize these results. It is known, however, that the niche width of late successional species as well as the rate of species loss from real communities decreases steadily as succession proceeds (Shugart and Hett, 1973; Bazzaz, 1987).

In principle, successions are open processes. Stationary species compositions – stability – may occur whenever the replacement probabilities remain 'constant' through time. Several factors may slow down successional changes. Among these are autogenous interactions in the absence of disturbance in mid-successional stages of fallow lands, the controlled management of cultivated areas, such as meadows, pastures and heathlands, or a steady state in more or less 'mature' ('climax') stages of succession. In these cases, the species involved persist, and perpetuate themselves through (vegetative) reproduction in an open equilibrium state, characterised by a more or less complete utilisation and recycling of the environmental resources. It is still a question of convention, if one may accept a 'true' climax vegetation in equilibrium with climate in the sense of Clements (1916) or in derived form 'potential natural vegetation' (Tüxen, 1956). Both concepts are misleading when describing the quaternary vegetation development (Tallis, 1991), but useful for constructing vegetation maps in cultivated landscapes.

So, 'climax communities' as well as 'more or less' stabilised intermediate successional species assemblages underlie fluctuations or cyclical micro-successions. These include 'fine scale gap and patch dynamics' caused by episodical disturbance and perturbations by grazers, storms or fire and cyclical phases from initial to mature and terminal phases of decay (*'degradative* or *heterotrophic succession'*). All mentioned fluctuations are characterised by changes in species composition and standing crop. The polyclimax and climax pattern hypothesis (Cain, 1947; Whittaker, 1953), gap, disturbance and regeneration dynamics concepts in forests including the mosaic-cycle hypothesis (Watt, 1947; Shugart, 1984; Pickett and White, 1985; Falinski, 1986; Remmert, 1991; Bergeron et al., 1998) or the hypothesis of Holling (1986) on cyclical processes in ecosystems all are based on and reflect investigated examples of vegetation and site types over contrasting soils and topographies. Grain-size in the temporal and spatial pattern of investigation varies, however. It may turn out that these approaches are covered by a broader ecosystem concept of open equilibria, which is ultimately independent from the maturity of the system. In different strata or structural components of complex ecosystems such as forests or mires, both micro-successions and fluctuations take place more or less independently from each other, in various spatial and temporal scales.

In reality, there are generally some shortcomings of these concepts. Due to a world-wide rising anthropospheric pressure, it may be difficult to differentiate in detail between primary and secondary successions and to define, whether and where a 'climax community' is present or may have been reached. Also, the distinction between pioneer and mature stages as well as between autogenic and allogenic processes is academic, for this separation becomes indistinct with an increasing time and space scale. Broad ecosystem complexes such as woodlands, dune systems and mires or secular successions involve a complex network of successional transitions which contain both autogenic and allogenic components (Tansley, 1935; Walker, 1970; Londo, 1974; Zobel, 1988).

3. Biological causes and mechanisms of successions

Apart from describing successional pathways, the biological and abiotic causes of successional development are of scientific interest. After disturbance opens up a space, plant species can become established either from an existing propagule pool or as a result of immigration from neighbouring populations (Harper, 1977; Lepart and Escarre, 1983; Hodgson and Grime, 1990; Thompson et al., 1997; Bonn and Poschlod, 1998). Diaspores may persist in seed banks for few months (transient) or up to several years (short-term or long-term persistent). Recent investigations from various grassland types indicate that most rare species have no long-term persistent seed bank. For a successful re-establishment of endangered species in fragmented ecosystems, a re-immigration or introduction of plant propagules is necessary.

After plants have become established at a given site, the availability of 'safe sites' or 'regeneration niches' rather than diaspore availability may limit the regeneration and development of the populations (Grubb, 1977, 1986; Harper 1977; Peart, 1989).

Species performance depends on environmental constraints, such as resource availability and stress, life history, ecophysiology and on interactions with other organisms, e.g., competitors and herbivores. In pioneer communities, only few species are capable of becoming established in the open space. These species can modify the environment in a way, that it facilitates the recruitment of species of later successional stages ('facilitation' in the sense of Connell and Slatyer, 1977). Many interactions are one-sided: the facilitation of one plant by another bears no advantage for the first. Examples are physical support by neighboring plants, protection against frost, abrasion, heat and transpiration, reduction of salinity or anoxic conditions, or support by exudates (Gigon and Ryser, 1986; Ryser, 1990, 1993, Gigon, 1994; Callaway, 1995).

Some facilitation processes are direct plant-to-plant or fungi-to-plant interactions, e.g. mycorrhizas, which can alter the competitive balance between species (Fitter, 1977; Grime et al., 1987; Watkinson and Freckleton, 1997). Other facilitation processes result from nitrogen enrichment in the soil by mutualistic interactions of vascular plants with N_2-fixing bacteria such as *Rhizobium* spp., *Frankia* spp.. A further example involves the attraction of pollinators or seed dispersers from which co-occurring species may benefit (Kühn et al., 1991; Read, 1993; Gigon, 1994).

Competition for resources such as photosynthetically active radiation, water and nutrients results in a mutual injury of plants growing in close proximity by shading or the reduced availability of water and nutrients in the rhizosphere. The behaviour of individuals of different plants in pure and mixed stands under defined ecological conditions has often been tested experimentally. Plants of low and slow growth are commonly displaced from their physiologically optimal habitat by more competitive species with similar preferences. The physiological and ecological response of competitive species coincides only with other competitive species, ie. their fundamental and realised niche are nearly equal (Müller-Dombois and Ellenberg, 1974; Landolt and Binz, 1989).

In addition to competition for resources, various other forms of interactions between plant species occur:

- **allelopathic effects by alkaloids, phenols or senescence hormones from root exudates, leaves or decomposed plant material which has no direct positive effect for the development of the emittant plants, but hampers the competing species (i.e. Rizvi and Rizvi 1992, Harborne 1993, Inderjit and Delmoral 1997).**
- **Furthermore, hemiparasites and parasites hamper their hosts, may reduce their fitness or even kill them.**

Independent of intra and interspecific interactions between plant species are herbivory, pollution or anthropogenous resource manipulation or pollution. They

- **influence phytomass, competition and the fitness of key species. In that way, the direction and the rapidity of successions or, on the opposite, stability or resilience of the plant communities are altered.**

The impact of herbivores and pathogens upon plant species composition and productivity cannot generally be quantified in thermodynamic terms, such as energy flow or the proportion of net primary production consumed. In particular, an emphasis upon quantity neglects that the greatest effects of herbivores on populations may occur by seed predation or the destruction of seedlings. Herbivory may influence various stages of the plant's life cycle. It may be *direct*, when a particular seedling is consumed, or *indirect*, when defoliation affects the competitive balance between plants. *"Under a particular regime of herbivory some plant genotypes may be favoured more than others. Sometimes a change in the regime may alter very rapidly the genetic composition of a plant population. ... It follows that in any community the plant species present and the genetic composition of their populations are to some extent products of past herbivory"* (Edwards and Gillman 1987).

The control of vertebrate herbivores on vegetation depends on the productivity of the sites. At low productivity levels due to low nutrient supply, low temperatures or limited moisture, vegetation seems not very attractive to herbivores. Under these conditions, productivity is frequently resource-limited. At intermediate levels of primary production, the consumption pressure is assumed to be close to the carrying capacity of the system, and large herbivores may be able to control the amount of standing crop. In eutrophic systems or those dominated by long-lived perennials, that can persist and reproduce vegetatively, productivity will be higher than the consumption by herbivores. For a long period, reproduction may not be important in maintaining the presence of the dominant species in these cases. Only herbivores which effect the performance of mature plants are likely to influence the latter's abundances. Whether and when the development of the system shifts to a 'climax stage' depends on disturbances which open the perennial canopy, facilitating the recruitment of woody plants (Edwards and Gillman, 1987; Oksanen, 1990). In detail, the story is even more complicated. For example roe deer (*Capreolus capreolus*) and red deer (*Cervus elaphus*) debark and browse highly selective young shoots and therefore at least may influence the species composition even of eutrophic systems in a fairly effective manner (Klötzli, 1965; Falinski, 1986; Ellenberg, 1988).

Plant-herbivore interactions change with successional progression. Early and mid-successional stages are often characterised by polyphagous herbivores, late successional stages preferably by oligophagous or monophagous species which vary considerably in population size. Coley et al. (1985) have tried to categorise the plant strategies to control herbivory. Their hypothesis links resource availability and growth rate of plants with the degree of investment that plants devote to herbivore protection. The authors predict that (i) herbivore damage will be less in slowly growing species, and that (ii) slow-growing species will benefit most from increased resource allocation to defence mechanisms. (iii) Quantitative defence compounds, such as tannins, are characteristic for slow growing plants in areas with low resource availability, whereas qualitative compounds like alkaloids prevail in fast growing species. Southwood et al. (1986) confirm the expectation of Coley et al. (1985), that leaf palatability and herbivore damage are inversely related.

Insect herbivores living above ground may especially affect predominant key species of various systems and in a way that determines the speed or direction of successions. Gradations of monophagous species are striking examples: the geomitride *Epirrita autumnata* attacks the climax birch-woodlands in the sub-alpine belt in boreal Scandinavia, the heather beetle *Lochmaea suturalis* raves in mesohemerobic dry heathland systems in western and central Europe. In both cases, high densities of the insect population result in severe deteriorations of the host populations. In other systems, host and herbivore interact in a more balanced way without killing the host plants, i.e. the *Larix decidua – Zeiraphera diniana* system. The rhizosphere community including insects, microfloral and microfaunal elements influences aboveground primary production and, thus, the stability of the system. An elimination of soil herbivores by insecticides can increase productivity of annual forbs in early successional stages (Brown and Gange, 1992). Therefore insect larvae living from root storage tissues may effectively control the expansion of productive host plants. In Europe, about 2/3 of the native *Lythrum salicaria* populations are attacked by the beetle *Hylobius transversovittatus*, living on ¾ of the plants. In North America, where the herbivore is missing, *Lythrum* as an alien species spreads nearly uninhibited (Blossey, 1993).

Examples for pathogens and their influence on the succession or stability of populations and communities are well known from various systems. Aggressive pathogens or pests can especially influence host population dynamics in extreme environments and mono-dominant communities, i.e. the outbreaks of *Labyrinthula macrocystis* on *Zostera marina*, and those of *Herpotrichia juniperi* on conifer needles in the sub-alpine belt in oceanic mountain ranges.

The human impact on vegetation composition results, in many cases, in regressive successions caused by an increasing intensity of land use. The gradual degradation of vegetation structure and soil processes can be expressed in hemerobic levels along the gradient ahemerobic to polyhemerobic species assemblages and systems. The abandonment of cultivated areas that were used extensively results in secondary successions. These successions may become critical from the viewpoint of landscape management for preserving or developing a high species diversity. Intensification regularly includes the drainage of wetlands, the irrigation of dry lands, the application of fertilisers and various cultivation measures in order to obtain a higher yield and higher quality of utilisable phytomass. In any case, cultivation measures try to favour stable systems on a high nutritional level, in other words to suppress successions.

4. Ecosystem mechanisms of succession

The ecophysiological behaviour of plants and animals, and their interdependencies during succession can be described, investigated and modelled on a biocoenotic level. Combined with environmental constraints, however, the biological systems are involved in a complex network of positive and negative feedbacks including energy and water exchanges, nutrient cycling, resource competition, adaptive acclimatisation and allocation strategies. For the understanding and generalisation of successional processes on this level exceeding biological interactions, an ecosystem approach is appropriate. This may be demonstrated by two simple case studies:

(i) A secondary succession from abandoned fields on sandy soils to permanent grassland in sub-oceanic temperate regions involves autogenic processes from pioneer annuals to early succession perennials and furthermore to a fairly stable grassland. The pioneer stage (therophytes predominate - ruderals according to the strategy types sensu Grime 1979) is determined by species that emerge from the persistent seed bank of the former land-use system; the vegetation covers less than 100 %. The soil system is characterised by an increasing depletion of nutrients, especially nitrogen, due to leaching from the top soil poor in humus. The following stage (the ruderal competitor *Elymus repens* dominates) is characterised by longer living species forming a closed sward, and reproducing vegetatively and sexually. The soils are nutrient deficient, but nitrogen is partly stored in the vegetation. The third, even more persistent stage is dominated by the long living *Festuca brevipila*, which reproduces preferably vegetative. The stress tolerant competitor is able to store nutrients in an efficient manner and is quite resistant to seasonal desiccation of the soil. The nutrient deficit in the soil slowly decreases due to an enrichment of humus built up by the sward (Christiansen, 1999). This development fits well with the tolerance model by Connell and Slatyer (1977). In this example, site history, seed bank, soil development, water content and nutrition, microclimate as well as autogenic processes of the vegetation interact on the ecosystem level. The plant-soil-system is partly decoupled (stage 1 and 2), but becomes closer tied again in the *Festuca* stage. Figure 1 generalises the development of the nitrogen pool of the topsoil.

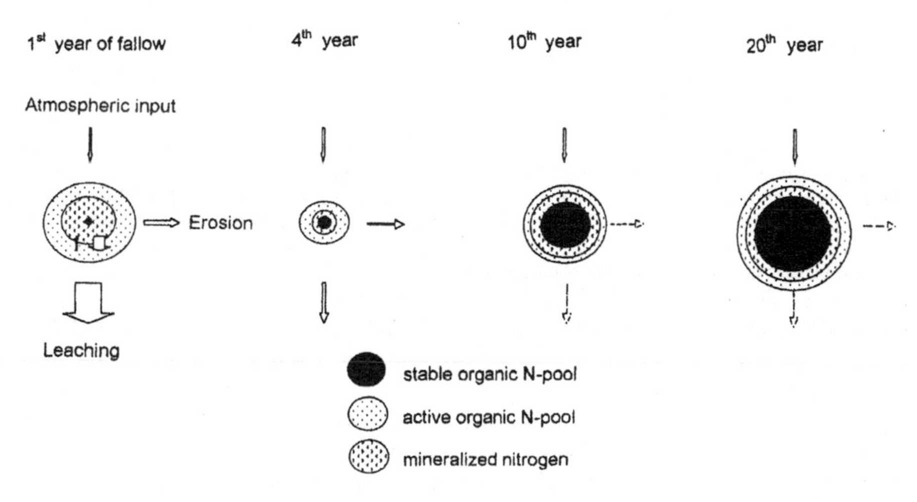

Figure 1: Generalised nitrogen storage and availability in various stages of a secondary successional from an abandoned field to low productive dry grassland (after Christiansen, 1999).

(ii) Secondary successions from the embanked Wadden sea and saltmarsh systems are characterised by desiccation of the top soil, salt leaching and an elevated nitrogen mineralisation. The composition of the halophyte vegetation of the former saltmarsh is primarily dependent on tidal inundations. The succession series reflect the degree of desiccation, desalinification, increasing nutrient availability, the accessibility of diaspores and the persistence of the former sward. Figure 2 shows the successional development derived from permanent plots. 2.1 is an example for a clearly allogenically triggered subsystem, 2.2 –2.5 indicate subsystems with a retarded response of the vegetation in relation to the decreasing soil water and salt content, triggered by autogenous feedback processes between the partly persisting former sward and slowly invading species (Neuhaus, 1997). In these cases, the developmental time of the different plots corresponds whereas the spatial succession pattern is heterogenous due to different allogenous as well as autogenous feedbacks.

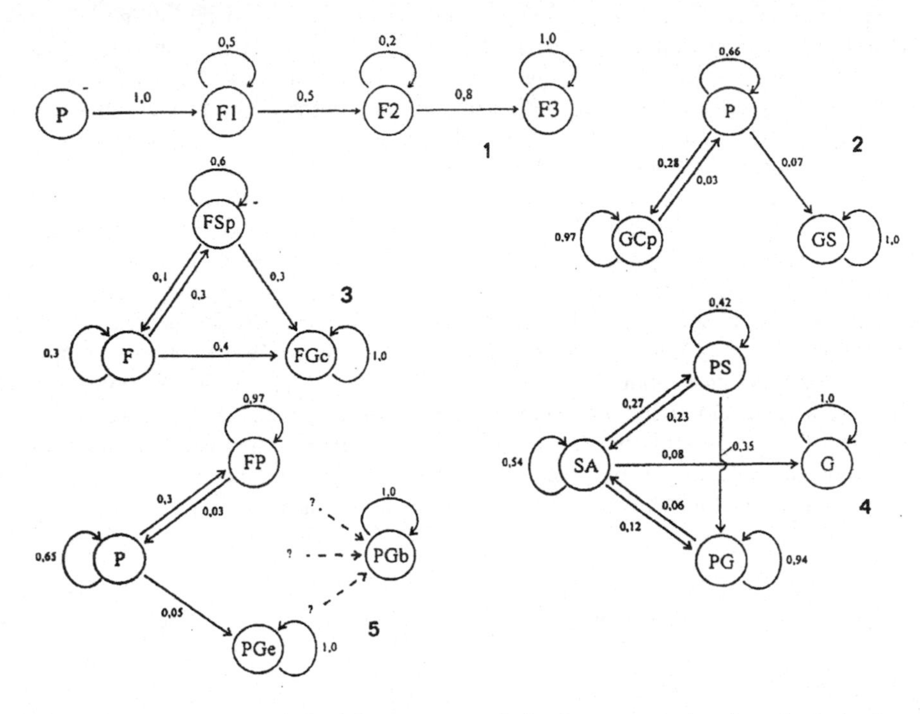

Figure 2: Successional pathways derived from cluster analysis of permanent plots in embanked saltmarshes (Beltringharder Koog, NW-Germany, 1987-1992). The arrows indicate the successional directions, the percentage values the relative amount of temporal changes. The letters characterise defined vegetation types. Abiotic changes mainly concern desalinification, leaching of calcium and drainage. Only pathway 1 is linear; 2-5 show interactions of proximate abiotic factors and ultimate interactions resulting from biological interrelationships and the site history (after Neuhaus, 1997).

From these examples one may infer, that colonising plants can behave differently within the same ecosystem complex; generalisations are insufficient unless the substrate quality is specified.

Some further examples illustrate the feedbacks between environmental constraints and the development of vegetation composition in ecosystem complexes.

Already fairly simple insect-plant interactions are by no means free from abiotic feedbacks on an ecosystem level. *Betula pubescens* agg. forms the sub-alpine timberline in western Scandinavia and is occasionally attacked by geometrids. The intensity of epidemic outbreaks of *Epirrita autumnata* in boreal birch forests depends on the ecophysiological pre-adaptation, and age structure of birch populations. It interacts with the inducible resistance triggered by moderate feeding on the leaves resulting in a lower digestibility of the phytomass and slowing down larvae development. Low winter temperatures increase the mortality of the eggs, and a high concentration of nitrogen in the leaves on more productive sites favours the development and survival of the larvae, decreasing the proportion of phenolic repellent components. Severe outbreaks, decreasing the availability of nutrients more than that of carbon, in contrast result in an increased amount of phenolic repellents. This behaviour is interpreted as a defense reaction of the plants (Haukioja et al., 1985; Tenow and Byklund, 1989; Tuomi et al., 1990; Hanhinmäki and Senn, 1992) .

In coastal dune complexes, as well as in salt marshes, succession is characterised by a synchronous progression of vegetation and soil development, especially, by the enrichment of humus and the accumulation of nitrogen in the top soil (Olff et al. 1992, Gerlach et al.

1994). Due to the leaching of cations, the dune soils behave as oligotrophic systems, and the development progresses from base rich arenosols to regosols and podzols. The degeneration of the *Ammophila arenaria* dune complexes involves at least three interacting processes: The increasing fixation of windblown sand decreases the nutrient supply in the root horizon of the pioneer plants. The age-dependent physiological senescence of the *Ammophila* plants during their life cycle reduces above and below ground phytomass, photosynthetic capacity, net carbon gain and fitness of the plants. The invasion of plant pathogen organisms, probably a combination of nematodes and soil fungi additionally weakens the populations (Wallén 1980, Van der Putten 1993).

In salt marshes, successional development is forced by an elevational gradient due to silt sedimentation and a nitrogen increase in the topsoil. Grazing intensity by herbivores (*Branta bernicla*) depends on forage quality. Cattle grazing on old salt marsh systems alter the successional development for the benefit of younger successional stages (i.e. Härdtle 1984, Olff et al. 1997).

The cyclical regeneration of heathlands suffers at least in central Europe from the withdrawal of the former management system including burning, sod-cutting and sheep grazing. The regeneration is additionally hampered by high atmospheric depositions of sulphur and nitrogen. Without phytomass export, a former nutrient source system develops in a nutrient sink system. An increasing amount of raw humus on the soil surface hampers the regeneration of the key species *Calluna vulgaris* and favours the development of the larvae of the monophagous chrysomelid heather beetle *Lochmaea suturalis*. *Lochmaea* infestation blocks water transport in the xylem vessels and affects the carbon balance of *Calluna*. Due to the increased level of air pollution, the heather beetle attacks seem to be more severe and occur more frequently nowadays. *Calluna* senescence and damages combined with the eutrophicating effect of nitrogen deposition result in a shifting of many dry inland heathlands to grassland systems dominated by *Deschampsia flexuosa*.

Scenarios taken from a simulation model indicate that a reduction of the atmospheric nitrogen deposition to 20-17 kg N ha^{-1} a^{-1} in combination with sod-cutting may stop this development (Aerts and Heil, 1993; Berdowski, 1993; Heil and Bobbink, 1993; Härdtle and Frischmuth, 1998).

The relative importance of biotic processes versus direct responses to the abiotic environment controlling both individual fitness and community structure has been the focus of controversial debate. Whether or not competition intensity increases along a gradient from pioneer to mature systems or, especially, along a productivity gradient has been tested with conentious results (Grime, 1973; Newman, 1973; Oksanen et al., 1981; Tilman 1987, 1988). If one generally accepts a high temporal and spatial variability of environmental properties for ecosystems, this implies that, both, nutrient status and stress inducing conditions may occur discontinuously (Goldberg and Novoplansky 1997, Ehrenfeld et al. 1997). In that way the debate is based in oversimplified assumptions.

5. Aspects of successions in applied landscape ecology

The decisive influence of landuse on successional trends, diversity and ecological complexity at various temporal and spatial scales (species, ecosystems, landscapes) is of increasing importance, resulting in erosion, desertification, pollution or eutrophication processes (Falkengren-Grerup, 1995; Tamm and Popovic, 1995).

Traditional land use may cause species richness, for instance, in mountain areas or dry grassland sites. Species richness declines both after abandonment and intensification of land use practices (e.g. by fertilisation). Therefore, it should be one practical aim of succession research to predict changes in the vegetation composition as affected by changes in the management regime.

Small-scale structural heterogeneity probably is an important key factor complex to maintain species richness (Van der Maarel and Sykes, 1993; Willems et al., 1993).

Abandonment leads to an accumulation of dead organic material on the soil surface, and hence, to physical and chemical changes in the uppermost layer of the soil profile. The resulting changes may occur within decades, whereas those in deeper soil layers may take thousands of years. The changes in vegetation cover and soil development influence processes, such as microclimate, energy exchange, the flow of organic and inorganic matter, primary production and decomposition.

Abandonment leads to a decrease of canopy photosynthesis as a result of low radiation near the soil surface or in some systems a decreasing nutrient availability. The changing species composition causes serious losses in the flora and vegetation of protected areas as well as an increasing proportion of photosynthetically inactive substances (attached dead plant material) (i.e. Herbich, 1986; Willems et al., 1993; Chernusca et al., 1998).

The vegetation development after abandonment of meadow ecosystems can be characterised as a sequence of different successional stages. An initial stage (I) is followed by a phase of clonal expansion of highly competitive species (II), phases of immigration and establishment of highly productive herbaceous (III) and woody (IV) species not occuring in stage I (Ekstam and Forshed, 1992; Jensen, 1997). The species contributing to successional changes can either be present in the initial aboveground vegetation, the soil seed bank or the seed rain. Seeds of some typical plants from wet and forage meadows are absent in the aboveground vegetation of later successional stages, but some may regularly persist in the seed bank. The chance of re-establishment from the seed bank, however, remains low, because the seedling-densities of meadow species decrease exponentially during succession (Jensen, 1998). Changes of species abundances during succession can be described in terms of successional categories (Figure 3). Jensen and Schrautzer (1999) distinguished four main categories (A,B,C and D). The mean cover of the species of categories A and B is highest in successional stage I and decreases during succession. While species A is missing in later successional stages, species B is still present, albeit with low cover values. Species of category C have largely the same cover during succession, while D species increase. To identify biological mechanisms underlying successional change, Jensen and Schrautzer (1999) correlated the successional categories with life history traits and ecological requirements of the species involved. They found that light competition and the limitation of sexual reproduction may play a key role in the decrease and extinction of regionally endangered meadow species in late successional stages.

6. Concepts and models describing main successional paths

Odum (1969) tried to formulate a functional word model of succession by considering how community properties may emerge during the course of succession. His definition, based on the Clementsian view, rests on three assumptions: *"(i) [Succession] is an orderly process of community development that is seasonally directional, and, therefore, predictable. (ii) It results from modification of the physical environment by the community; that is, succession is community controlled even though the physical environment determines the pattern, the rate of change, and often sets limits as to how far development can go. (iii) It culminates in a stabilized ecosystem in which maximum biomass (or high information content) and symbiotic functions between organisms are maintained per unit of energy flow, in a word, the 'strategy' of long-term evolutionary development of the biosphere - namely, increased control of, or homeostasis with, the physical environment in the sense of achieving maximum protection from its perturbations."* The description of thermodynamic orientors forms a formally concrete development of this theoretical approach (Bossel, 1998; Svirezhev, 1998; Jørgensen and Nielsen, 1998).

This organismic and teleological view, however, is idealised in respect to the predictability of successions on the object level, the homeostatic character of ecosystems and the imagination of equilibrium (climax) ecosystem.

Because structural and functional changes in successional series result from morphological, ecophysiological and population biological characteristics of the species involved, the approach of Grime (1979) has been discussed broadly. It is the central hypothesis of this approach, that land plants exhibit measurable strategies in response to habitat *stress* (environmental factors reducing photosynthetic capacity) and *disturbance* (the partial or total destruction of photosynthetically-derived phytomass). The distribution of these strategic types (competitors, ruderals, stress-tolerants and those with intermediate characters) allow a rough prognosis of the development under different nutrient, disturbance and stress regimes (e.g. Kutsch et al., 1998).

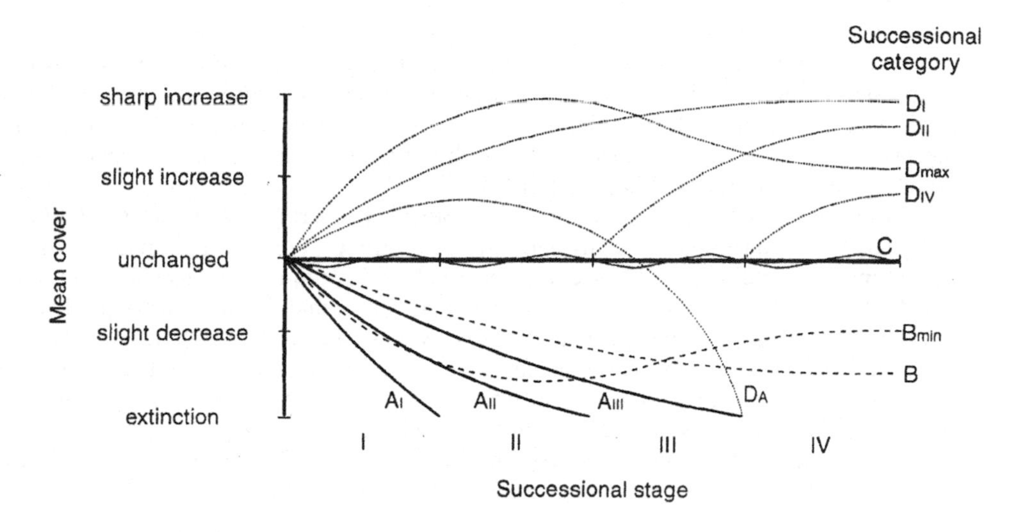

Figure 3: Schematical definition of successional categories, which represents changes of mean cover of species in the successional stages (ss) on abandoned wetland sites. AI: Highest cover in managed stands (ss I), missing in ss II, III and IV; AII: Highest cover in managed stands (ss I), decrease in ss II, missing in ss III and ss IV; AIII: Highest cover in managed stands (ss I), decrease in ss II and ss III, missing in ss IV; B: Highest cover in managed stands (ss I), decrease in ss II, III and IV; Bmin: Highest cover in managed stands (ss I), minimum in ss II or III, increase in ss IV; C: No pronounced changes of cover during succession; DA: Increase of cover with maximum in ss II, decrease in ss III, missing in ss IV; Dmax: Increase of cover with maximum in ss II; in ss III and IV higher cover than in managed stand (ss I); DI: Lowest cover in managed stands (ss I), increase in ss II, III and IV. DII: Missing in ss I, lowest cover in ss II or III, increase in ss IV; DIV: Missing in ss I, II and III, occurrence only in ss IV (after Jensen and Schrautzer 1999).

Connell and Slatyer (1977) emphasise the ability of species groups (i) to *facilitate* further colonisation for other species, (ii) to defend their captured space during colonisation against later colonists by *inhibition*, or (iii) to coexist by a differential exploitation of limited resources (*tolerance*). Other authors interpret vegetation changes during successions (or fluctuations) as probabilistic rather than deterministic. Probabilistic changes can be characterised by Markov chain models. Examples for probabilistic changes are: the tree-by-tree replacement process in forest and grassland succession (Horn, 1975, 1976; Noble and Slatyer, 1980; Usher, 1981, 1987; Baltzer, 1998), metapopulation dynamics (Valverde and Silvertown, 1997) or the 'carousel model' by Van der Maarel and Sykes (1994), characterising micro-successions and small scale species diversity in limestone grasslands.

The modeling of ecosystem processes increases in importance. Powerful tools are at hand, including numerical models at various levels of complexity. Models help to describe and to explain successional pathways, or to predict the course of particular successions either by combining different mechanisms or by specifying the relationships between mechanisms and pathways (Fresco et al., 1987; Van Andel et al., 1993). The feedback between theoretical background, sampling, experimenting and modeling can provide an impetus for further understanding and research. In the following, some examples of models with an increasing level of complexity will be given, starting with the population genetic level and ending with global climate modeling.

On the level of population genetics, gene flow by diaspore transport tends to expand the range of a species, whereas competition with already established species counteracts this tendency. The evolution of competing species with overlapping realised niches can be simulated under various assumptions for long time sequences covering many generations. The stability of subspecies may be explained by fitness reductions due to heterozygote disadvantage. If the population size of a taxon – once established – exceeds a certain minimum, it will be fairly stable against invasion by gene flow. Otherwise, extinction of a viable population due to the reduction of the minimum area below a certain threshold may be extremely slow at the beginning, but accelerates as positive feedback mechanisms are established in the final phase. This effect is characteristic of stabilised founder populations. It may help to explain the development of the recent range limits of so-called relictic species (Pils 1995), and the decrease of small populations of endangered species in the changing environment of cultural landscapes.

From the viewpoint of restoration ecology, Kleyer (1997) tried to define the potentials for functionally equivalent species group replacement under different environmental conditions. Kleyer compared field observations with a cellular-automata-based model, in with plant individuals compete in virtual habitats with gradients of disturbance intensity and resource supply. In this way, model dynamics allow insights in the functional relationships between environmental conditions and biological features. Valuable criteria were proved to be dispersal, generative reproduction, seed bank longevity, vertical and lateral expansion and vegetative regeneration.

At the landscape level, modeling can contribute to the analysis of environmental effects and environmental planning, including regional economic modules (Dabbert et al., 1999).

Modeling entire ecosystems, for example mire complexes, involves vegetation responses to elevated CO_2 concentrations on photosynthesis, nitrogen deposition and fluxes, the long term structural development of mire microsites, the peat building processes, gas fluxes, leaching and mire development in general (Wildi, 1978; Clymo, 1996; Damman, 1996; Trepel, 1999). Additionally, the coupled interactions with the radiative and moisture properties of land surface and atmosphere can be accounted for. The models indicate that global peatland dynamics may alter the future rate of increase in atmospheric CO_2 (Klinger, 1996; Franzén et al., 1996).

Global aspects of biogeophysical feedbacks concern transfer of energy and water between oceans, the soil and the air above. The key parameters governing these exchanges and their temporal variations vary with ecosystem structure. Early diagnostic biome models tried to detect the patterns of global net primary production (NPP) derived from mean annual precipitation and temperatures (Lieth, 1975). Global assessments of CO_2 and global change effects on NPP were coupled in more modern approaches (Melillo et al., 1993). They include mechanistic simulation of succession and disturbance processes (Steffen et al., 1992). Regarding climatic change processes, a hierarchy of climate models with various complexity has been developed and applied for the study of past, present and future climate (Arrhenius, 1896; Bengtsson, 1997; Manabe, 1997; Hasselmann and Hasselmann, 1998).

Keeping in mind that nature is already probabilistic from the viewpoint of quantum physics, one may ask, whether global modelling is feasible (e.g. Dürr, 1998). Anyway, because relevant anthropogenic processes interact with global ecological problems, precautionary measures and advice for policymaker by experts is necessary on the best scientific knowledge. Of course, aggregation of environmental indicators is inevitable at a global scale, but by exploiting regularities at that level, it may be possible to learn something about the general dynamics and intrinsic correlations of the systems, giving important guidance to the understanding and manipulation of the system.

7. Aspects of global change

The expansion of human economic activity, particularly, in industry and agriculture in the second part of this century has speeded up many anthropogentic interactions with terrestrial, marine and atmospheric processes, has expanded them globally and has influenced the global mega-ecosystem in quantitative and qualitative ways. Adequate explanations, how the impact of long term pollution impairs the regulatory processes of the biosphere require an understanding of how ecological systems function, and how they respond to human activities at local levels. The focus of analysis has to be linked 'upward' to the global scale of biogeochemical relationships and 'downward' to the local and regional scale of human activities. Among others Holling (1986) has outlined a conceptional framework for an understanding of resilience mechanisms and a sustainable development of ecological systems on a local and global scale. He characterises ecosystems as hierarchically organised multistable entities of spatial diffusion (Holling, 1973; Levin, 1978; Allen et al., 1982), that couple non-linear subsystems of different scales in time and space. In addition to successional processes leading to 'increasing order' from pioneer to mature systems, Holling considered periods of disorganisation by large scale discontinuous disruptions (fire, storm, pests) and local gaps (senescence). The timing of these disturbance pulses cannot be predicted precisely, because the systems in question are in a refractory stage of gradual changes for long periods, when discontinuous behaviour is inhibited. However, the increasing connection strength between elements and processes in maturing ecosystems may lead to abrupt changes, and a regeneration pulse after disturbance (compare Figure 3 in the contribution of Gunderson et al. in this volume). Derived from examples such as insect pest-host systems and forest fire dynamics, Holling (1986) developed a stimulating dynamic hypothesis which broadens the equilibrium climax concept. In detail, the hypothesis is susceptible to misleading generalisations and analogies. Nonetheless, it may help to improve the understanding between different disciplines, which are involved in the search for concepts joining economical, social and environmental approaches. The resulting key message is that (i) traditional resource-management constrains cyclical changes in ecosystems by restricting them temporally and homogenising them spatially. Internal biogeophysical relationships and feedbacks than change, leading to systems of reduced resilience, i.e., of increasing fragility. (ii) The increasing extent and intensity of modern agricultural and industrial activities have modified and accelerated many global processes, thereby changing the external variability of the ecosystems involved. As a consequence, local processes may be more frequently affected by global phenomena, and in turn can effect these interdependencies. (iii) It is necessary to develop and preserve systems which are flexible to allow recovery and renewal in the face of unexpected events. Barkmann and Windhorst (this volume) suggest the capacity of ecological systems to self-organise as a suitable target for prophylactic measures against uncertain environmental risks.

For decision-making, there is an increasing need to operationalise the notion of sustainable development (Gnauck, 1998; Werner and Bork, 1998). Even, though no general rule can be deduced for the potential biosphere-climate feedbacks by a given anthropogenic perturbation, there is no substitute for careful scientific analysis of the potential risks. Global integrated assessment models (Rotmans, 1998) may help to understand the feedbacks and interrelations between increasing pressure on the environment, the impacts on the biogeochemical cycles within and between ecosystems and the social and economic development of the human society. This includes concepts how to handle scientific, social and economic uncertainties.

8. Some conclusions

Successions are directional processes of colonisation and extinction of species in a given site, caused by biotic (autogenous) or abiotic (allogenous) forces in ecosystems. Cyclical micro-successions occur at various successional stages and spatial patterns.

Successions in reality in most cases follow multiple pathways, where more than one single force or mechanism operates. That is why it seems appropriate to leave the narrow domain of vegetation dynamics in favour of an ecosystem approach for understanding the key mechanisms.

Successions are not random, but they are heavily context-sensitive. Over-generalised hypothesis are of little use for most practical applications. History is important and spatial heterogeneity is the norm.

Apart from biocoenotic causes and mechanisms, environmental (proximal) processes drive successional pathways. Concepts and models describing and explaining vegetation and ecosystem development at various scales are of increasing importance for precautionary land use planning and as supports for decision making processes.

Acknowledgements

I would like to thank Jan Barkmann, Kai Jensen and Felix Müller for valuable comments on earlier drafts of this paper.

REFERENCES

Aerts, R., Heil, G.W., (eds.) 1993. Heatlands: patterns and processes in a changing environment. *Geobotany 20*: 223 pp., Kluwer Acad. Publ., Dordrecht.

Allen, T.F.H., Starr, T.B. 1982. Hierarchy. Perspectives of ecological complexity.- Univ. Chicago Press, Chicago.

Armesto, J.J., Pickett, S.T.A., McDonnell, M.J., 1991. Spatial heterogeneity during succession: a cyclic model of Invasion and exclusion.- In: Kolasa, J, Pickett, S.T.A. (eds.): Ecological heterogeneity. *Ecol. Stud. 86*: 256-269.

Arrhenius, S., 1896. On the influence of carbonic acid in the air upon the temperature of the ground. *Phil. Mag.* 41, 237-276.

Bakker, J.P., 1989. *Nature management by grazing and cutting*.- 400 pp., Kluwer Acad. Publ., Dordrecht.

Bakker, J.P., Olff, H., Willems, J.H., Zobel, M., 1996. Why do we need permanent plots in the study of long term dynamics? *J. Veg. Sci.* 7, 147-156.

Baltzer, H., 1998. Modellierung der Vegetationsdynamik verschiedener Pflanzengemeinschaften des Grünlandes mit Markov-Ketten. *Boden und Landschaft* 23, 152 pp.

Barkmann, J., Windhorst, W., 1999. Hedging our bets: the utility of ecological integrity. Chapter III.1.3, this volume.

Bazzaz, F.A., 1979. The physiological ecology of plant succession. *Ann. Rev. Ecol. Syst.* 10, 351-371.

Bazzaz, F.A., 1987. Experimental studies on the evolution of niche in successional plant populations. In: Gray, A.J., Crawley, M.J., Edwards, P.J. (eds.): *Colonization, succession and stability*, 245-272, Blackwell Scientific Publ., Oxford.

Beeftink, W.G.,1987. Vegetation responses to changes in tidal inundation of salt marshes. In: Van Andel, J., Bakker, J.P., Snaydon, R.W. (eds.) *Disturbance in grasslands*, 97-117, Junk, Dordrecht.

Begon, M., Harper, J.L., Townsend, C.R., 1990. Ecology – Individuals, populations and communities.- 2nd ed., 945 S., Blackwell, Oxford.

Bengtsson, L., 1997. A numerical simulation of anthropogenic climate change. *Ambio 26*, 58-65.

Berdowski, J.J.M., 1993. The effect of external stress and disturbance factors on *Calluna*-dominated heathland vegetation.- In: Aerts, R., Heil, G.W. (eds.) *Heatlands: patterns and processes in a changing environment.- Geobotany* 20, 85-124, Kluwer Acad. Publ., Dordrecht.

Bergeron, Y., Engelmark, O., Harvey, B., Morin, H., Sirois, L. (eds.), 1998. Key issues in disturbance dynamics in boreal forests. *J. Veg. Sci.* 9, 463-610.

Blossey, B., 1993. Herbivory below ground and biological weed control: life history of a root-boring weevil on purple loosetrife. *Oekologia* 94, 380-387.

Bonn, S, Poschlod, P., 1998. *Ausbreitungsbiologie der Pflanzen Mitteleuropas*.- 404 pp., Quelle and Meyer, Wiesbaden.

Bornkamm, R., 1988. Mechanisms of succession on fallow land. *Vegetatio* 77, 95-101.

Bossel, H., 1988. Ecological orientors: emergence of basic orientors in evolutionary self-organization. In: Müller, F., Leupelt, M. (eds.) *Eco targets, goal functions, and orientors*, 19-33, Springer, Berlin.

Brown, V.K., Gange, A.C., 1992. Secondary plant succession: how is it modified by insect herbivory? *Vegetatio* 101, 3-13.

Cain, S.A., 1947. Characteristics of natural areas and factors in their development. *Ecol. Monogr.* 17: 185-200.

Callaway, R.M., 1995. Positive interactions among plants. *Bot. Rev.* 61, 306-349.

Chapin III, F.S., 1993. Physiological controls over plant establishment in primary succession.- In: Miles, J., Walton, D.W.H. (eds.) *Primary succession on land*,161-178. Blackwell Scientific Publ., Oxford.

Chernusca, A., Bahn, M., Bayfield, N., Chemini, C., Fillaat, F., Graber, W., Rosset, M., Siegwolf, R., Tappeiner, U., Tasser, E., Tenhunen, J., 1998. ECOMONT: new concepts for assessing ecological effects of land use changings on terrestrial mountain ecosystems at an European scale. *Verh. GfÖ* 28, 3-11.

Christiansen, U., 1999. Zur Bedeutung der Stickstoffversorgung für den Sukzessionsverlauf von sandig-trockenen Ackerbrachen zu Sandmagerrasen.- Thesis Univ. Kiel, 139 pp.

Clements, F.E., 1916. Plant succession. An analysis of the development of vegetation. *Carnegie Inst. Washington* 242, 512 pp.

Clymo, R.S., 1996. Assessing the accumulation of carbon in peatlands. In: Laiho, R., Laine, J, Vasander, H. (eds.) *Northern peatlands in global climatic change*,207-212. Edita, Helsinki.

Coley, P.D., Bryant, J.P., Chapin III F.S., 1985. Resource availability and plant antiherbivore defense. *Science* 230, 895-899.

Connell, J.H., Slatyer, R.O., 1977. Mechanisms of succession in natural communities and their role in community stability and organization. *Am. Nat.* 111, 1119-1144.

Dabbert, S., Herrmann, S., Kaule, G., Sommer, M., (eds.) 1999. *Landschaftsmodellierung für die Umweltplanung*, 246 pp., Springer, Berlin.

Damman, A.W.H., 1996. Peat accumulation in fens and bogs: effects of hydrology and fertility. In: Laiho, R., Laine, J, Vasander, H. (eds.) *Northern peatlands in global climatic change*,213-229. Edita, Helsinki.

Daubenmire, R.F., 1968. *Plant communities: a textbook of plant synecology.* Harper and Row, New York, 300 pp.

Drake, J.A., 1985. Some theoretical and empirical exploitations of structure in food webs.- Ph.D. thesis, Purdue University, West Lafayette, Indiana.

Dröschmeister, R., Gruttke, H., (eds.) 1998. Die Bedeutung ökologischer Landzeitforschung für den Naturschutz. *SchrR. LandschPfleg. NatSchutz 58*: 435 pp.

Dürr, H.-P., 1998. Is global modelling feasible? In: Schellnhuber, H.-J., Wenzel, V. (eds.) *Earth system analysis*, 493-504, Springer, Berlin.

Edwards, P.J., Gillman, M.P., 1987. Herbivores and plant succession. In: Gray, A.J., Crawley, M.J., Edwards, P.J. (eds.): *Colonization, Succession and Stability*, 296-314, Blackwell Scientific Publ., Oxford.

Egler, F.E., 1954. Vegetation science concepts. I. Initial floristic composition, a factor in old-field vegetation development. *Vegetatio* 4: 412-417.

Egloff, T., 1998. Mittelfristige Nachwirkungen einer zweijährigen Düngung auf einer Pfeifengras-Streuwiese (Molinion) im nördlichen Schweizer Mittelland. *Telma* 28, 146-156.

Ehrenfeld, J.G., 1997. On the nature of environmental gradients: temporal and spatial variability of soils and vegetation in the New Jersey Pinelands. *J. Ecol.* 85, 785-798.

Ekstam, U., Forshed, N., 1992. *Om hävden* upphör. *Kärlväxter som indikatorarter in ängs- och hagmarker. (If grassland management ceases. Vascular plants as indicator species in meadows and pastures).* Naturvårdsverket, Solna.

Ellenberg, H., 1988. Eutrophierung – Veränderungen der Waldvegetation – Folgen für den Rehwildverbiß und dessen Rückwirkungen auf die Vegetation. *Schweiz. Z. Forstwes.* 139, 261-282.

Falinski, J.B.,1986. Vegetation dynamics in temperate lowland primeval forests. *Geobot.* 8, 537 pp., Junk Publ., Dordrecht.

Falkengren-Grerup, U., 1995. Long-term changes in flora and vegetation in deciduous forests of southern Sweden. *Ecol. Bull.* 44, 215-226.

Fitter, A.H., 1977. Influence of mycorrhizal infection on competition for phosphorus and potassium by two grasses. *New Phytologist* 79, 119-125.

Franzén, L., Chen, D., Klinger, L.F., 1997. Principles for climate regulation mechanism during the late Phanerozoic era, based on carbon fixation in peat-forming wetlands. *Ambio 25*, 435-442.

Fresco, L.F.M., Van Laarhoven, H.P.M., Loonen, M.J.J.E., Moesker, T., 1987. Ecological modeling of short term plant community dynamics under grazing with and without disturbance. In: Van Andel, J., Bakker, J.P., Snaydon, R.W. (eds.) Disturbance in grasslands, 149-165, Junk Publ., Dordrecht.

Gerlach, A., Albers, E.A., Broedlin, W., 1994. Development of the nitrogen cycle in the soils of a coastal dune succession. *Acta Bot. Neerl. 43*(2), 189-203.

Gigon, A., 1975. Über das Wirken der Standortsfaktoren; kausale und korrelative Beziehungen in jungen und reifen Stadien der Sukzession. *Mitt. Eidgen. Anst. forstl. Versuchswes. 51*, 25-35.

Gigon, A., 1994. Positive Interaktionen bei Pflanzen in Trespen-Halbtrockenrasen. *Verh. GfÖ 23*,1-6.

Gigon, A., Ryser, P., 1986. Positive Interaktionen zwischen Pflanzenarten. I. Definition und Beispiele aus Grünland-Ökosystemen. *Veröff. Geobot. Inst. Rübel 87*, 372-387.

Gleason, H.E., 1917. The structure and development of the plant association. *Bull. Torrey Bot. Club* 44, 463-481.

Gleason, H.E., 1939. The individualistic concept of the plant association. *Amer. Midl. Nat.* 21, 92-110.

Gnauck, A., 1998. Applying thermodynamic orientors: tools of orientor optimization as a basis for decision making process. In: Müller, F., Leupelt, M. (eds.) *Eco targets, goal functions, and orientors*, 511-525, Springer, Berlin.

Goldberg, D., Novoplansky, A., 1997. On the relative importance of competition in unproductive environments. *J. Ecol.* 85, 409-418.

Grime, J.P., 1973. Competitive exclusion in herbaceous vegetation. *Nature* 242, 344-347.

Grime, J.P., 1979. *Plant strategies and vegetation processes*. John Wiley and Sons, Chichester.

Grime, J.P., Mackey, J.M.L., Hillier, S.H., Read, D.J., 1987. Floristic diversity in a model system using experimental microcosms. *Nature* 328, 420-422.

Grubb, P.J., 1977. The maintenance of species-richness in plant communities: the importance of the regeneration niche. *Biol. Rev.* 52, 107-145.

Grubb, P.J., 1986. The ecology of establishment.- In: Bradshaw, A.D., Goode, D.A., Thorp, E.H.P. (eds.) *Ecology and design in landscape*, 83-97, Blackwell Scientific Publ., Oxford.

Gunderson, L.H., Holling, C.S., Peterson, G.D., 1999. Resilience in ecological systems. Chapter II.8.2, this volume.

Hanhinmäki, S., Senn, J., 1992. Sources of variation in rapidly inducible responses to leaf damage in the mountain birch-insect herbivore system. *Oecologia (Berlin)* 91, 318-331.

Härdtle, W., 1984. Vegetationskundliche Untersuchungen in Salzwiesen der ostholsteinischen Ostseeküste. *Mitt. AG Geobot. SH/HH* 34, 142 pp.

Härdtle, W., Frischmuth, M., 1998. Zur Stickstoffbilanz nordwestdeutscher Zwergstrauchheiden und ihre Störung durch atmogene Einträge (dargestellt am Beispiel des NSG Lüneburger Heide). *Jb. Naturw. Verein Fstm. Lbg.* 41, 197-204.

Harborne, J.B., 1993. *Introduction to ecological biochemie*, 4th ed., Academic Press, London.

Harper, J.L., 1977. *Population biology of plants*.- Academic press, London.

Harper, J.L., 1982. After description.- In: Newman, E.T. (ed.): *The plant community as a working mechanism*, 11-25. Blackwell scientific publ., Oxford.

Hasselmann, K., Hasselmann S., 1998. Multi-actor optimization of greenhouse gas emission paths using coupled integrated climate response and economic models.- In: Schellnhuber, H.J., Wenzel, (eds.) *Earth system analysis*, 381-413, Springer, Berlin.

Haukioja, E., Niemelä, P, Siren, S., (1985): Foliage phenols and nitrogen in relation to growht, insect damage, and ability to recover after defoliation in the mountain birch *Betula pusbescens* ssp. *tortuosa*. *Oekologia (Berlin)* 65, 214-222.

Heil, G.W., Bobbink, R., 1993. Impact of atmospheric nitrogen deposition on dry heathlands. A stochastic model simulating competition between *Calluna vulgaris* and two grass species.- In: Aerts, R., Heil, G.W. (eds.) *Heatlands: patterns and processes in a changing environment*, 181-200, Kluwer Acad. Publ., Dordrecht.

Herben, T., 1996. Permanent plots as tools for plant community ecology. *J. Veg. Sci.* 7, 195-202.

Herbich, J., 1986. Conservation and succession in the plant cover of polish reserves.- *Zesz. nauk. wydz. biol., geogr., oceanol. univ. Gdanskiego* 7, 19-29.

Hodgson, J.G., Grime, J.P., 1990. The role of dispersial mechanisms, regenerative strategies and seed banks in the vegetation dynamics of the British landscape.- In: Bunce, R.G.H., Howard, D.C. (eds.): *Species dispersal in agricultural habitats*, 65-81, Belhaven, London.

Holling, C.S., 1973. Resilience and stability of ecological systems. *Ann. Rev. Ecol. System.* 4, 1-23.

Holling, C.S., 1986. The resilience of terrestrial ecosystems: local surprise and global change.- In: Clark, W.C., Munn, R.E. (1986): *Sustainable development of the biosphere*, 292-320, Cambridge Univ. Press.

Horn, H.S., 1975. Markovian properties of forest succession.- In: Cody, M.L., Diamond, J.M. (eds.) *Ecology and evolution of communities*, 196-211, Harvard Univ. Press, Cambridge, MA.

Horn, H.S., 1976. Succession.- In: May, R.M. (ed.): *Theoretical ecology – Principles and applications*, 187-204, Blackwell Scientific Publ., Oxford.

Huisman, J., Olff, H., Fresco, L.F.M., 1993. A hierarchical set of models for species response analysis. *J. Veg. Sci.* 4, 37-46.

Inderjit, del Moral, R., 1997. Is separating resource competition from allelopathy realistic? *Bot. Rev.* 63, 221-230.

Jensen, K., 1997. Vegetationsökologische Untersuchungen auf nährstoffreichen Feuchtgrünland-Brachen - Sukzessionsverlauf und dynamisches Verhalten von Einzelarten. *Feddes Repertorium* 108, 603-625.

Jensen, K., 1998. Species composition of soil seed bank and seed rain of abandoned wet meadows and their relation to aboveground vegetation. *Flora* 193, 345-359.

Jensen, K., Schrautzer, J., 1999. Consequences of abandonment for a regional fen flora and mechanisms of successional change. *Appl. Veg. Sci.* 2, 79-88.

Jørgensen, S.E., Nielsen, S.N., 1998. Thermodynamic orientors: a review of goal functions and ecosystem orientors. In: Müller, F., Leupelt, M. (eds.) *Eco targets, goal functions, and orientors*, 123-136, Springer, Berlin.

Kiehl, K., 1997. Vegetationsmuster in Vorlandsalzwiesen in Abhängigkeit von Beweidung und abiotischen Standortfaktoren. *Mitt. AG Geobot. SH/HH 52*: 142 pp., Kiel.

Kleyer, M., 1997. Vergleichende Untersuchungen zur Ökologie von Pflanezngesellschaften – Eine Grundlage zur Beurteilung der Ersetzbarkeit in der naturschutzfachlichen Planung am Beispiel einer Agrar- und einer Stadtlandschaft. *Diss. Bot.*286, 202 pp.

Klinger, L.F., 1996. Coupling of soils and vegetation in peatland succession. *Arctic Alpine Res.28,* 380-387.

Klötzli, F., 1965. Qualität und Quantität der Rehäsung in Wald- und Grünland-Gesellschaften des nördlichen Schweizer Mittellandes.- Diss. ETH Zürich, 186 pp.

Kühn, K.-D., Weber, H.-Ch., Dehne, H.-W., Gworgwor, N.A., 1991. Distribution of vesicular-arbuscular mycorrhizal fungi on a fallow agricultural site. I. Dry habitat. *Angew. Bot* 65, 169-186.

Küppers, M., 1984. Carbon relations and competition between woody species in a central European hedgerow. I. Photosynthetic characteristics. *Oecologia* 64, 332-343.

Kutsch, W., Dilly, O., Steinborn, W., Müller, F., 1998. Quantifying ecosystem maturity - a case study. In: Müller, F., Leupelt, M. (eds.) *Eco targets, goal functions, and orientors*, 209-231, Springer, Berlin.

Lambers, H, Chapin III F.S., Pons, T.L., 1998. *Plant physiological ecology.* 540 pp., Springer, New York.

Landolt, E., Binz, H.-R., 1989. Konkurrenzuntersuchungen zwischen nah verwandten Arten von *Scabiosa columbaria* L. s.l. II. Differenzierung von Bastardpopulationen unter verschiedenen Temperatur-, Feuchtigkeits- und Nährstoffbedingungen. *Ber. Geobot. Inst. ETH Inst. Rübel* 55, 177-236.

Lawton, J.H., 1987. Are there assembly rules for successional communities? In: Gray, A.J., Crawley, M.J., Edwards, P.J. (eds.): *Colonization, Succession and stability*, 225-244, Blackwell Scientific Publ., Oxford.

Levin, S.A., 1978. Pattern formation in ecological communities. In: Steele, J.A. (ed.) *Spatial pattern in plancton communities*, 433-470. Plenum Press, New York.

Lepart, J., Escarre, J., 1983. La succession végétale, mécanisms et modèles: analyse bibliographique. *Bull. d'Écol.* 14, 133-178.

Lieth, H., 1975. Modeling the primary production of the world. In: Lieth, H., Whitakker, R.H. (eds.) *Primary production of the biosphere*, 237-263, Springer, New York.

Londo, G., 1974. Successive mapping of dune slack vegetation. *Vegetatio* 29, 51-61.

Manabe, S., 1997. Early development in the study of greenhouse warming: the emergence of climate models. *Ambio* 26, 47-51.

Melillo, J.M., McGuire, A.D., Kicklighter, D.W., Moore III, B., Vörösmarty, C.J., Schloss, A.L., 1993. Global climate change and terrestrial net primary production. *Nature* 363, 234-240.

Miles, J., 1979. *Vegetation dynamics.* Chapman and Hall, London.

Miles, J., Walton, D.W.H., (eds.) 1993. *Primary succession on land.* 309 pp., Blackwell Scientific Publ., Oxford.

Müller-Dombois, D., Ellenberg, H., 1974. *Aims and methods of vegetation ecology.* 547 pp., Wiley and Sons, New York.

Neuhaus, R., 1997. *Sukzession von Pflanzengesellschaften in eingedeichten Salzwiesen und Watten des Beltringharder Kooges.*- Diss. Bot. Inst. Univ. Kiel, 120 pp.

Newman, E.I., 1973. Competition and diversity in herbaceous vegetation. *Nature* 244, 310.

Noble, I.R., Slatyer, R.O., 1980. The use of vital attributes to predict successional changes in plant communities subject to recurrent disturbances. *Vegetatio* **43**: 5-21.

Odum, E.P., 1969. The strategy of ecosystem development. *Science* 164, 262-270.

Oksanen, L., 1990. Predation, herbivory and plant strategies along gradients of primary productivity.- In Grace, J.B., Tilmand, D. (eds.) *Perspectives of plant competition,* 447-474, Academic Press, London.

Oksanen, L, Fretwell, S.D., Arruda, J., Niemela, P., 1981. Exploitation ecosystems in gradients of primary productivity. *Amer. Naturalist* 118, 240-261.

Olff, H., de Leeuw, J., Bakker, J.P., Platering, R.J., 1992. Nitrogen accumulation, vegetation succession and geese herbivory during salt marsh formation on the Dutch island of Schiermonnikoog. In: Olff, H.: *On the mechanism of vegetation succession.*212 pp., PhD thesis, Groningen.

Olff, H., de Leeuw, J., Bakker, J.P., Platerink, R.J., van Wijnen, H.J., de Munck, W., 1997. Vegetation succession and herbivory in a salt marsh: changes induced by sea level rise and silt deposition along an elevational gradient. *J. Ecol.* 85, 799-814.

O'Neil, R.V., DeAngelis, D.L., Waide, J.B., Allen, T.F.H., 1986. A hierarchical concept of ecosystems.- Princeton Univ. Press.

Peart, D.R., 1989. Species interactions in a successional grassland. III. Effects of canopy gaps, gopher mounds and grazing on colonization. *J. Ecol.* 77, 252-266.

Peet, R.K., 1992. Community structure and ecosystem function.- In: Glenn-Levin, D.C., Peet, R.K., Veblen, Th.T. (eds.) *Plant succession*, 103-151, Chapman and Hall, London.

Pickett, S.T.A., White, P.S., 1985. *The ecology of natural disturbance and patch dynamics.* Academic Press, New York.

Pickett, S.T.A., Collins, SIL., Armesto, J.J., 1987. A hierarchical consideration of succession. *Vegetatio* 69, 109-114.

Pils, G., 1995. Die Bedeutung des Konkurrenzfaktors bei der Stabilisierung historischer Arealgrenzen. *Linzer biol. Beitr.* 27(1): 119-149.

Post, W.M., Pimm, S.L., 1983. Community assembly and food web stability. *Mathemat. Biosciences* 64, 169-192.

Prentice, I.C., 1998. Ecology and the earth system. In: Schellnhuber, H.-J., Wenzel, V. (eds.) *Earth system analysis,* 219-240, Springer, Berlin.

Read, D.J., 1993. Plant-microbe mutualism and community structure. In: Schulzc, E.D., Mooney, H.A. (eds.)

Biodiversity and ecosystem function.- Ecol. Stud. 99, 181-209, Springer, New York.
Remmert, H., (ed.) 1991. *The mosaik-cycle concept of ecosystems.* 168 pp., Springer, Berlin.
Rizvi, S.J.H., Rizvi, V., (eds.) 1992. *Allelopathy. Basic and applied aspects.* 480 pp., Chapman and Hall, London.
Rotmans, J., 1998. Global change and sustainable development: towards an integrated conceptual model. In: Schellnhuber, H.-J., Wenzel, V. (eds.) *Earth system analysis,* 421-453, Springer, Berlin.
Ryser, P., 1990. Influence of gaps and neighbouring plants on seedlings establishment in limestone grassland. Experimental field studies in northern Switzerland. *Veröff. Geobot. Inst. ETH Zürich* 104, 71pp.
Ryser, P., 1993. Influences of neighbouring plants on seedling establishment in limestone grassland. *J. Veg. Sci.* 4, 195-202.
Schmidt, W., 1981. Ungestörte und gelenkte Sukzession auf Brachäckern. *Scripta Geobot.* 15, 1-199.
Schreiber, K.-F., 1997. Sukzessionen – Eine Bilanz der Grünlandbracheversuche in Baden-Württemberg. *LFU Projekt „Angewandte Ökologie"* 23, 188 pp., Karlsruhe.
Shugart, H.H., 1984. *A theory of forest dynamics.* 278 pp. Springer, New York.
Shugart, H.H., Hett, J.M., 1973. Succession: similarities in species turnover rates. *Science* 180, 1379-1381.
Southwood, T.R.E., Brown, V.K., Redder, P.M., 1986. Leaf palatability and herbivore damage. *Oekologia* 77, 544-548.
Steffen, W.L., Walker, B.H., Ingram, J.S., Koch, G.W. (eds.), 1992. *Global change and terrestrial ecosystems. The operational plan.*Global change report 21, The royal swedish academy of Sciences. IFBP secretariat, Stockholm.
Svirezhev, Y., 1998. Thermodynamic orientors: how to use thermodynamic concepts in Ecology. In: Müller, F., Leupelt, M. (eds.) *Eco targets, goal functions, and orientors,* 102-122, Springer, Berlin.
Tallis, J.H., 1991. *Plant community history.* Chapman and Hall, London.
Tamm, C.O., Popovic, B., 1995. Long term field experiments simulating increased deposition of sulfur and nitrogen to forest plots. *Ecol. Bull.* 44, 301-321.
Tansley, A.G., 1935. The use and abuse of vegetational concepts and terms. *Ecology* 16, 284-307.
Tenow, O., Byklund, H., 1989. A survay of winter cold in the mountain birch/*Epirrita autumnata* system. *Mem. Soc. fauna flora fenn.*65, 67-72.
Thomson, K., Bakker, J., Bekker, R., 1997. The soil seed banks of North West Europe. 276 pp., Cambridge Univ. Press.
Tilman, D., 1987. On the meaning of competition and the mechanisms of competitive superiority. *Functional Ecology* 1, 304-315.
Tilman, D., 1988. *Plant strategies and the Dynamics and Structure of Plant Communities.* Princeton University Press, Princeton, NJ.
Tilman, D., 1989. Ecological experimentation: stengths and conceptual problems.- In: Likens, G.E. (ed): *Long-term studies in ecology,* 136-157, Springer, New York.
Trepel, M., 1999. Quantifizierung der Stickstoffdynamik von Ökosystemen auf Niedermoorböden mit dem Modellsystem WASMOD. Thesis Univ. Kiel, 140pp.
Tuomi, J., Niemelä, P., Sirén, S., 1990. The panglossian paradigm and delayed inducible accumulation of foliar phenolics in mountain birch. *Oikos* 59, 399-410.
Turner, M.G., Baker, W.L., Peterson, C.J., Peet, R.K., 1998. Factors influencing succession: lessons from large, infrequent natural disturbances. *Ecosystems* 1, 511-532.
Tüxen, R., 1956. Die heutige potentielle natürliche Vegetation als Gegenstand der Vegetationskartierung. *Angew. Pflanzensoz.* 13, 5-43.
Usher, M.B., 1981. Modelling ecological succession, with particular reference to Markovian models. *Vegetatio* 46, 11-18.
Usher, M.B., 1987. Modelling successional processes in ecosystems.- In: Gray, A.J., Crawley, M.J., Edwards, P.J. (eds.): *Colonization, succession and stability,* 31-56, Blackwell, Oxford.
Valsangiacomo, A., 1998. Die Natur der Ökologie.- 324 pp., *vdf Hochschulverlag AG,* ETH Zürich.
Valverde, T., Silvertown, J., 1997. A metapopulation model for *Primula vulgaris*, a temperate forest understory herb. *J. Ecol. 85:* 193-210.
Van Andel, J., Bakker, J.P., Grootjans, A.P., 1993. Mechanisms of vegetation succession: a review of concepts ands perspectives. *Acta Bot. Neerl. 42,* 413-433.
Van der Maarel, E., Sykes, M.T., 1993. Small scale species turnover in a limestone grassland: the carousel model and some comments on the nicht concept. *J. Veg. Sci. 4,* 179-188.
Van der Putten, W.H., 1993. Soil organisms in coastal foredunes involved in degeneration of *Ammophila arenaria.-* In: Miles, J, Walton, D.W.H. (eds.): *Primary succession on land:* 373-281, Blackwell Scientific Publ., Oxford.
Vitousek, P.M., Walker, L.R., 1987. Colonization, succession and resource availability: ecostystem-level interactions. In: Gray, A.J., Crawley, M.J., Edwards, P.J. (eds.) *Colonization, Succession and Stability,*207-223, Blackwell Scientific Publ., Oxford.
Walker, D., 1970. Direction and rate in some British post-glacial hydroseres.- In: Walker, D., West, R. (eds.) *Studies in the vegetation history of the Britisch Isles:* 117-137, Cambridge Univ. Press.
Wallén, B., 1980. Changes in structure and function of *Ammophila* during primary succession. *Oikos 34,* 227-238.
Watkinson, A.R., Freckleton, R.P., 1997. Quantifying the impact of arbuscular mycorrhiza on plant

competition. *J. Ecol. 85*, 541-545.
Watt, A.S., 1947. Pattern and process in the plant community. *J. Ecol. 35*,1-22.
Werner, A., Bork, H.-R., 1993. Integrating diverging orientors: sustainable agriculture: ecological targets and future land-use changes. In: Müller, F., Leupelt, M. (eds.) *Eco targets, goal functions, and orientors*, 565-584, Springer, Berlin.
Willems, J.H., Peet, R.K., Bik, L., 1993. Changes in chalk grassland structure and species richness resulting form selective nutrient addition. *J. Veg. Sci. 4,* 203-212.
Whittaker, R.H., 1953. A consideration of climax theory: the climax as a population and pattern. *Ecol. Monogr. 23,* 41-78.
Whittaker, R.H., 1975. *Communities and ecosystems.*- 2nd ed., Macmillan, New York.
Wildi, O., 1978. Simulating the development of peat bogs. *Vegetatio* 37, 1-17.
Wolfram, C., Hörcher, U., Kraus, U., Lorenzen, D., Neuhaus, R., Dierssen, K., 1998. Die Vegetation des Beltringharder Kooges 1987-1998 (Nordfriesland). *Mitt. AG Geobot. SH/HH 58*, 219 pp., Kiel.
Zobel, M., 1988. Autogenic succession in boreal mires – a review. *Fol. Geobot. Phytotax. 23*, 417-445.

II.9 Ecosystems on the Landscape: The Role of Space in Ecosystem Theory

Robert V. O'Neill

1. Introduction

The ecosystem is an enormously complex object of study. Progress in understanding the dynamics of such a complex object requires abstraction. Some aspects of the object are emphasised while other aspects are averaged or ignored. In this way, some of the complexity can be brought within the range of scientific understanding. But it is important to remember that any specific abstraction is only one of a number of alternative points of view.

Ecosystem Theory has largely developed within the Systems Analysis paradigm. The ecosystem is conceptualized as an interconnected web of biotic and abiotic components. Attention focuses on the dynamics of energy and nutrients flowing through this network. Principles of energy dissipation and mass conservation form a critical set of constraints on the positive and negative feedbacks within the system. Using this conceptualisation, the ecosystem is represented by a system of ordinary differential equations and the principles of Systems Analysis facilitate investigation of the dynamics and stability properties of the complex network.

A parallel development has occurred in Population/Community Theory. The community is conceptualized as an interconnected web of competing populations. Here, attention focuses on the number of organisms in each population and the dynamics of birth, death, and competition. The conceptualisation is implemented as a system of ordinary differential equations, again analysed for their dynamics and stability properties.

While this characterisation of theoretical ecology is a gross and an obvious simplification, it serves to emphasize an important point. The dominant paradigms have minimized the role of space. The ecosystem has been viewed as a spatially homogeneous unit, with the glacial lake and high gradient watershed as classic examples. The equations deal with the dynamics totalled or averaged over space. Space is regarded as a source of heterogeneity, a source of random deviations away from the well-mixed paradigm. The critical features involve the complex web of temporal interactions.

But interest has never been entirely lost in questions of spatial heterogeneity. Recent developments in Landscape Ecology (Urban et al. 1987) have been stimulated by the scale of land use change brought about by the human population growth. Vitousek (1994) has pointed out that "...land use change is now, and for some decades probably will remain, the single most important of the many interacting components of global change." It is impressive to realize that more than 50% of the tropical forests have been cut with the current rate of deforestation exceeding 168,000 km^2 per year (Goodland 1991). But this recent interest has simply re-stimulated an already existing interest in spatial theory for ecosystems.

The concern for ecosystem processes distributed in space has a long history. David Lack (1942) noted that remote British islands had fewer bird species with obvious implications for food web dynamics. A. S. Watt (1947) pointed out that isolated patches of vegetation on the heterogeneous landscape were fundamental to understanding community structure. Andrewartha and Birch (1954) discussed the importance of spatial relationships among 'local populations'.

Experimental work with spatial processes dates back to the classic work of Huffaker (1958, Huffaker et al. 1963). In elegantly simple experiments, he studied the dynamics of a fructivorous mite and its predator, another mite, on oranges spread in various spatial configurations on a table. He established the importance of patch structure and inter-patch distance to stability of the ecological system. Similar results were obtained for a fly-parasitoid system (Pimentel et al. 1963) and other mites (Takafugi 1977, Takafugi et al. 1983).

2. Biogeography Theory

An important line of theoretical work begins with MacArthur and Wilson's (1963, 1967, MacArthur 1972) studies on Island Biogeography. Their insights were based on bird communities of oceanic islands. Brown and Lomolino (1989) later pointed out that very much the same observations had been made earlier in an unpublished thesis (Munroe 1948) on butterflies in the West Indies. MacArthur and Wilson hypothesized that, at equilibrium, immigration, I, was a function of the distance to the 'source' mainland community and extinction, E, was a function of island size:

$$I = a(P - R)^k$$
$$E = n\,S^m \tag{1}$$

where P is the number of species in the mainland pool, R is the number of species on the island, and S is island size.

Early field studies provided empirical evidence for the theory (Diamond 1969, Simberloff and Wilson 1969, 1970) and expanded its applications to alpine 'islands' (Vuilleumier 1970) and caves (Culver 1970, Vuilleumier 1973). The theory found important applications in conservation studies by arousing interest in the size of reserves, nearness to other reserves and the relative benefits of one large versus many small reserves (Quinn and Hastings 1987, Simberloff 1988, Burkey 1989, McCoy 1983, Soule and Simberloff 1986).

Island Biogeography Theory has been subjected to a number of criticisms (Sauer 1969, Lack 1970, Carlquiest 1974, Gilbert 1980) and many modifications have been suggested. Perhaps the primary criticism has been the assumption of equilibrium (Barbour and Brown 1974, Culver et al 1973, Diamond 1972, 1973, Terborgh 1975). In many ecosystems, chronic disturbance (Villa et al 1992) or simply the lack of sufficient time (Simpson 1974) would invalidate the assumption. It has been pointed out that small islands with high extinction rates, may be 'rescued' by being close to shore and having very large immigration rates (Brown and Kodrick-Brown 1977). It is also clear that island size and distance becomes less important as dispersal ability becomes greater (Pulliam et al. 1992, Fahrig 1983, Fahrig and Paloheimo 1988, Burkey 1989, Roff 1974a,b).

In 1970, a parallel development began in population theory (Levins 1970). Levins pointed out that a population that was intrinsically unstable could still persist in a patchwork of sub-populations. His demonstration contrasted local extinction probability on a patch with the immigration probability from other patches. However, extinction was a random variable, not a function of patch size and all patches were equally connected. The extensive developments that became known as Metapopulation Theory are reviewed in Levin (1976). The parallel theory broadened the discussion from diversity of a community to persistence of individual populations on a patchy landscape and largely eliminated the assumption of a 'mainland source' for all species.

One must also be careful in applying the theory to specific ecosystems. Webb and Vermaat (1990) documented the case of isolated heathland remnants. Small islands were invaded by surrounding vegetation and had high diversity. Large islands resisted the invaders and had small diversity. In some cases, the habitat quality on the island may be more important than size (Gulve 1994, Murphy et al. 1990). In other cases, interspecific competition may be more important (Bengtsson 1989, Levins and Culver 1971, Horn and MacArthur 1972, Hanski 1981, 1983, Taylor 1990). It has also been pointed out that some island systems are dominated by catastrophic disturbances, such as hurricanes, that may cause extinction, irrespective of island size (Leigh 1981, Birch 1971, Ehrlich et al. 1980, Antolin and Strong 1987). In these cases, population size and community diversity may be dominated by reinvasion processes. In fact, when natural catastrophes are common, persistence may be higher on a landscape with several patches as this speads the risk of the entire population being wiped out (Goodman 1987). Toft and Schoener (1983) showed that spiders on Bahamian islands largely followed the theory. But if predatory lizards were present, the relationships were changed significantly.

In spite of criticisms and modifications, an impressive body of field data has accumulated to support the validity of Equation 1 and the developments of metapopulation theory. When the assumptions are met, the theory appears to capture some fundamental properties of communities and populations in space. Table 1 is provided to give the reader an impression of the scope of supporting studies that have found a significant correlation between diversity or population persistence as a function of island size and distance.

The most extensive applications of the theory have been in the field of landscape ecology. Work has focused on the problems encountered by organisms operating on a fragmented landscape. In this context, Pulliam (1988, Pulliam and Danielson 1991) have pointed out that some habitat patches, where reproduction is greater than mortality, will serve as 'sources'. Other smaller patches with sub-optimal resources will serve as 'sinks'. In some cases, a patch that is a source for one species may be a sink for other species (Danielson 1991, 1992). Although the sink patches do not ordinarily produce emigrants, their presence on the landscape maximizes population abundance. Kadmon and Schmida (1990), for example, show that desert grass produce most of their seeds in moist wadis. The grass spreads out over surrounding areas but relies on the moist areas to replace mortality.

A population may require resources from several patches (Dunning et al. 1992). Petit (1989) points out that wintering birds may use some patches for foraging and others to weather storms. The checkerspot butterfly needs cool slopes for prediapause larvae, but warmer slopes for postdiapause larvae and pupae (Weiss et al. 1988). Barred owls and piliated woodpeckers will supplement their diet from surrounding, suboptimal patches (Whitcomb et al. 1977).

In the fragmented landscape, the interisland matrix becomes important. Dispersal between patches, for example, is dependent on the quality of the matrix. Hedgerows and corridors tend to keep the patches accessible (Preston 1962, Willis 1974, Forman 1995). The matrix and the forest edges are themselves habitat for other species (McCollin 1993). Indeed, the fragmentation may actually increase total species diversity (Opdam 1991).

The edges of habitats become important as well. Forest edge birds may be nest predators and decrease the effective size of the patch for other species (Wilcove 1985, Andren 1992, Andren and Angelstam 1988, Brittingham and Temple 1983). The shape of the edge may inhibit or encourage immigration (Hardt and Forman 1989). Bach (1988a, b) showed that inhospitable species could be planted around a food crop to slow the migration of herbivores into the patch.

Table 1: Empirical studies supporting area and/or distance as significant correlates of diversity and population dynamics

Organisms	Ecosystems	Citations
Mammals	Islands	Lomolino 1982, 1984
	Talus Slopes	Smith 1980
	Forest Patches	Verboom and van Apeldoorn 1990, Gottfried 1979
	Freshwater	Lawton and Woodroffe 1991
	Habitat Patches	Weddell 1991
Birds	Islands	Hamilton et al. 1964, Diamond and Mayr 1976 Diamond et al. 1976
	Wetlands	Brown and Dinsmore 1986
	Tropical Forest	Diamond 1973, Howe et al. 1981, Newmark 1991
	Boreal Forest	Helle 1984, 1985, Martin 1983
	Temperate Forest	Opdam et al 1984, 1985, Askins et al. 1987, Van Dorp and Opdam 1987, Dickman 1987, Verboom et al. 1991, Ambuel and Temple 1983, Serrao 1985, Lynch 1987, Forman et al 1976, Askins and Philbrick 1987, Blake and Karr 1984, 1987, Howe and Jones 1977, Lynch and Whitcomb 1978,1984, Robbins 1979, 1980, Pettersson 1985a, McCollin 1993, Whitcomb et al. 1981, Whitcomb 1977, Moore and Hooper 1975, Galli et al. 1976, Ford 1987, Howe 1984
Amphibians	Ponds	Reading et al. 1991, Gulve 1994
Arthropods	Mangrove Islands	Simberloff 1976
	Habitat Islands	Kruess and Tscharntke 1994, Harrison et al 1988, Webb et al. 1984, Murphy et al. 1990, Rey and Strong 1983, Toft and Schoener 1983
Snails	Ponds	Lassen 1975, Aho 1978, Bronmark 1985
Protozoa	Microcosms	Have 1987

But even with the complexities of the terrestrial landscape, the basic theory explains a great deal of the variance in the spatial distribution of the bird community (Saunders et al. 1991). Some studies stress the important of isolation of the habitat islands (Helliwell 1976, Hayden et al 1985). Others stress island size (Williams 1964, Freemark and Merriam 1986) or 'effective' island size in the presence of nest predators (Wilcove et al. 1986, Small and Hunter 1988). But the studies generally confirm the insights expressed in Equation 1.

3. Processes in a Single Spatial Dimension: Stream Spiralling

Early ecosystem theory (Webster et al. 1975, Reichle et al. 1975) compared the response of different ecosystem types to disturbance. Using models that ignored spatial processing, they concluded that streams and rivers have little ability to resist disturbance. Contributing factors include: 1) streams cannot recycle nutrients which are swept downstream, 2) there are no large pools of organic matter to buffer the system, 3) there is no control over inputs of organic matter and nutrients, and 4) both inputs and throughput are determined by unpredictable storm events.

But at the same time, it was realized that point-space models were inappropriate for stream and river systems (Quinlin 1975). The entire river systems must be viewed as a spatial continuum (Vannote et al. 1980). Webster (1975, Webster and Patten 1979) coined the word 'spiralling' to describe the reprocessing of nutrients in space as they passed down the stream system. Limiting nutrients cannot be recycled at the same point in space, but they can be reused as they pass down the spatial continuum. Quinlin and Paynter (1976) and Wallace et al. (1977) demonstrated that nutrient spiralling could significantly alter the stability properties of the stream. O'Neill et al. (1979) showed that the spiralling system was more resistent to disturbances such as scouring of organic matter by storms.

Newbold et al. (1981) developed the concept of 'spiralling length' to describe the phenomenon. Spiralling length, S, is the distance travelled by the average molecule from the point of release to the point of resorption by the biota. It can be measured as the ratio of the downstream flux to the exchange rate or utilization rate of the biota. Extending the concept to the processing of organic matter (Newbold et al. 1982a), they showed that:

$$S = \frac{F}{R} \tag{2}$$

where F is the flux of organic matter downstream ($g\ m^{-1}\ s^{-1}$) and R is the flux of carbon lost to respiration ($g\ m^{-2}\ s^{-1}$). A shorter spiralling length corresponds, therefore, to more efficient utilization of organic matter. Biota can increase the efficiency of processing by increasing the utilisation rate R or by decreasing the downstream flux, F (Newbold et al. 1982b). This makes the important point that a measure of distance is the most important measure of processing in a spatially distributed ecosystem (Elwood et al. 1983, Newbold et al. 1983).

4. Movement in Homogeneous Space

A number of theoretical results are available on dispersal processes in space. In continuous space, the theory is primarily based on the diffusion process. If space is conceived as divided into discrete units or pixels, results are available from the physical theory of interacting particles. The theory has been applied to the dispersal of organisms and the spread of epidemics. The discussion here is limited to a few basic results that should have broad application to ecosystems processes on the landscape.

The diffusion model was largely developed by population geneticists (Fisher 1937, Dobzansky and Wright 1947) and ecological applications were contributed by Skellam (1951) and Kierstead and Slobodkin (1953). A simple and lucid presentation is provided by Andow et al. (1990). The basic equation for diffusion from a source is given by:

$$\frac{\partial N}{\partial t} = aN + D\left[\frac{\partial^2 N}{\partial x^2} + \frac{\partial^2 N}{\partial y^2}\right] \tag{3}$$

where a is the intrinsic rate of increase of the population and D is the diffusion coefficient. In this expression we have assumed a simple exponential growth function. Results for more complex functions can be found in Okubo (1980) and Levin (1986).

The most important general result of ecological diffusion theory comes from considering V, the rate of spread of the population. Considering simple exponential growth, and assuming that diffusion in the x and y directions are the same, the rate of spread asymptotes to

$$V = \sqrt{4aD} \tag{4}$$

This simple result depends on a complex web of simplifying assumptions. However, Andow et al. (1990) tested the result with data from studies on three populations. The theory was accurate in two cases. In the third case, data on movement was at a very fine scale and did not extrapolate well to the landscape. The study demonstrates that the simple expression in Eq. 4 can give a good first approximation to spread across the landscape, provided data are gathered at a sufficiently large scale.

Relatively few studies deal directly with the effect of spatial pattern on the dispersal process. O'Neill et al. (1992) adopt a basic epidemiology model (Bailey 1975) to look at the spread of a disturbance across a patterned landscape. They present equations for x, the number of spatial units or pixels that are susceptible to a disturbance like fire, y, the number of pixels actually burning, and z, pixels that have burned and are no longer susceptible:

$$\begin{aligned} \frac{dx}{dt} &= -iq\,xy \\ \frac{dy}{dt} &= iq\,xy - by \\ \frac{dz}{dt} &= by \end{aligned} \tag{5}$$

where iq is the probability of a disturbance spreading to an adjacent site and b is rate of disturbance extinction, i.e., the inverse of the length of time a site burns.

In order to introduce spatial pattern, the probability of spread is represented as the product of i, the probability of spread to an adjacent susceptible site, and q, the probability that a susceptible site is adjacent to a burning pixel. If there are N total pixels on the landscape and the proportion of those pixels that are susceptible is p, then, on a random landscape, p=q. If the susceptible sites are clumped into patches, q > p. So the parameter q permits examination of at least one aspect of pattern, the degree of clumpedness.
Analysis of Equation 5 reveals that the extent, T, of a single disturbance is given by:

$$T = 2pN - \frac{2b}{iq} \tag{6}$$

Notice that as the degree of clumping, q, increases, the total extent of the disturbance increases. If the susceptible sites are clumped, there is greater probability of reaching them. Notice that all susceptible sites will be disturbed, i.e., T = pN, if

$$i > \frac{2b}{qpN} \tag{7}$$

If the probability of spread increases beyond the right side of Equation 7, all susceptible sites will be disturbed if they can be reached. Any further increase in disturbance intensity will affect the rate of spread, but will not influence the total extent of the fire. The extent will be determined by the spatial pattern, i.e., whether a site can be reached, rather than the dynamics of the disturbance itself.

Additional theoretical results are available from the physics of interacting particles. A lucid discussion with ecological examples is given in Durrett and Levin (1994). Consider, for example, a plant population on a two-dimensional space. An individual plant occupies a unit of space and over a unit of time may die with probability γ . There is also a probability, γ , that it will produce a live offspring at an adjacent site. Obviously, for the population to survive, the birth rate must be high enough to offset the mortality. As a result, for any value of γ there is a critical value of $\gamma_c(\lambda)$. If mortality is greater than this critical value, the population dies out. If mortality is lower than this threshold value, there is a positive probability of survival. It is possible to set bounds on this critical value:

$$\frac{4\lambda}{4\lambda+1} \ge \lambda_c(\lambda) \ge \min_{n\ge1}\left(1-\left[0.82+2(1-\lambda)^{2n}\right]^{k}\right) \tag{8}$$

where: $\kappa = 1/(2n-1)$. For continuous time and large areas, mortalities less than or equal to γ_c lead to κ surviving populations with the number of occupied sites, N_i^*, close to

$$N_i^* = \frac{\lambda_i - \gamma_i}{\lambda_i} \tag{9}$$

If two populations are competing on the landscape, one population will approach N_i^* (Eq. 9)and the other population will approach:

$$N_j^* = \frac{1}{\lambda_j}\left(\frac{\lambda_j\gamma_i}{\lambda_i} - \gamma_j - (\lambda_i - \gamma_i)\right) \tag{10}$$

This permits coexistence, i.e., a positive value for N_j^*, as long as $\gamma_j + \lambda_i - \gamma_i < \lambda_j \gamma_i / \lambda_i$

5. Percolation Theory

Percolation theory (Stauffer 1985) deals with the connectedness of a landscape. Assume the landscape is a square grid with two land cover types, distributed randomly with probabilities P_A and P_B. We will consider two gridpoints of cover type A as connected, if they are adjacent to each other in any of the four cardinal directions. That is, they touch along any of their four sides. For very large random maps, there exists a critical threshold for P_A. When $P_A > 0.5928.....$, all of the gridpoints of A form a single, connected patch. The threshold value varies with different shapes for the gridpoints (e.g., hexagonal) and different rules for adjacency. The threshold becomes important, then, for predicting whether an animal will be able to move across the entire landscape and remain within habitat A (Gardner et al. 1989).

The theory provides a number of additional predictions about ecosystems on the landscape. For example, the largest number of individual patches occurs at $P_A \sim 0.3$ and the greatest variance in the size of the largest cluster occurs right at the threshold value, $P_A \sim 0.6$ (Gardner et al. 1987). Connectedness will influence the spread of disturbances such as fires. When $P_A < 0.5$, the landscape will be sensitive to disturbance frequency. When $P_A > 0.6$, the landscape will be more sensitive to disturbance intensity, and less sensitive to frequency (Turner et al. 1989). Connectedness will also affect the interplay between dispersal strategy in annual plants and the structural properties of a random landscape (Lavorel et al. 1995).

If resources are randomly scattered over the landscape, a consumer must adjust the scale of its movements in order to be able to reach all of the resources. Let us assume that the consumer can move n spatial units per unit time. When $P_A > 0.6$ and $n = 1$, the consumer should be able to find a unit of resource at each step in time. But if $P_A << 0.6$, then the resource is not continuously connected and the consumer will have to take multiple steps to find a unit of resource during each unit of time. For a large random map, we can define this *Resource Utilization Scale* (O'Neill et al. 1988a) as:

$$n = \frac{-0.89845}{\ln(1 - P_A)} \tag{11}$$

This is simply a matter of re-scaling the map so that the percolation threshold is reached. If the consumer takes this number of steps, the resource A will appear to be connected throughout the landscape. So, for example, if the consumer can take 100 steps and only needs to find a single unit of A anywhere along that path, then the critical value of P_A becomes 0.0089. This thinking then becomes key to understanding how consumer body size (and the related dispersal ability) becomes closely tied to the scale at which resources are distributed across the landscape (Holling 1992). Even if a dominant consumer removes 90% of the resource, a subdominant can adjust the scale of its movements to locate the remaining 10%.

The concept of Resource Utilization Scale can be extended to landscapes that show some degree of pattern. Consider, for example, that A is not randomly distributed but occurs in clumps or clusters. We will characterize the clumping by the conditional probability Q, equal to the probability of finding an immediately adjacent unit of A, given that one is standing on a unit of A. Then

$$n' = \frac{-0.89845 - 2\ln(1 - P_A) + \ln(1 - 2P_A + QP_A)}{\ln(1 - 2P_A + QP_A) - \ln(1 - P_A)} \tag{12}$$

If $P_A = 0.5928$, n' =1 irrespective of the value of Q. As Q approaches 1.0 and the resources are all clumped in one place, the denominator approaches 0.0 and n' approaches infinity. Since the resource is all in one place, it becomes impossible to locate additional resources by moving around on the landscape. When $Q = P_A$ the resource is randomly scattered and n' = n.

6. Quantifying Landscape Pattern

A first requirement for dealing with ecosystem processes distributed in space, is to develop some metrics of spatial pattern. Following an initial attempt to define matrices (O'Neill et al. 1988) a large number of candidates have been proposed from landscape ecology (Baker and Cai 1992) and from traditional image processing (Gonzalez and Woods 1992). Helpful reviews are available for measures based on diversity and information theory (Magurran 1988), fractal geometry (Milne 1991), and image textural methods (Musick and Grover 1991).

The available indicators emphasise the pattern of discrete patches on the landscape. Unfortunately, the number of measures is too large for many applications. Many of the descriptors are quite similar to each other and it is difficult to choose a small number of independent measures. Riitters et al. (1995) provide brief descriptions and equations for 55 candidate pattern indicators and analyze the metrics for 85 maps of land cover. Pairwise comparisons revealed that many indicators have correlation coefficients greater than ± 0.9. Eliminating the redundant measures reduced the candidates to 26.

A factor analysis was performed on the 26 measures, to identify small combinations of indices that were closely related to each other, and relatively independent of other indicators. The first 5 factors all have eigenvalues greater than 1.0 and explain about 83% of the variance. Each factor is composed of several indicators and some subjective judgment is required to choose a single metric from each group. Based on the ease of calculation and interpretation, the following indices are recommended:

(1) Average patch perimeter- area ratio:

$$P = \frac{1}{m}\sum_{k=1}^{m}\frac{E_k}{A_K} \tag{13}$$

where there are a total of m patches and E_k is the perimeter of the k'th patch and A_k is the area.

(2) Contagion:

$$C = 1 + \frac{1}{2\ln(n)}\sum_{i=1}^{n}\sum_{j=1}^{n} p_{ij}\ln(p_{ij}) \tag{14}$$

where there are a total of n land cover types and p_{ij} is the probability of type i being adjacent to type j. See Li and Reynolds (1993) and Riitters et al. (1996) for further details on this index.

(3) Relative patch area: (average ratio of patch area to the area of an enclosing circle)

$$\mathrm{R} = \frac{1}{\mathrm{m}}\sum_{\mathrm{k}=1}^{\mathrm{m}}\frac{\mathrm{A_k}}{\pi \mathrm{L_k^2}} \tag{15}$$

where L_k is one half of the longest straight line that can be drawn within the patch. Notice that the demoninator is the area of a circle, with L_k as the radius. This circle should approximately enclose the entire patch.

(4) Fractal Dimension: (Based on a patch perimeter- area relationship)

$$\mathrm{F} = 2B \tag{16}$$

where B is the slope from the regression of ln (E_k) on ln (A_k) for all patches greater than 3 pixels that do not touch the edge of the map. Very small patches and those with a straight edge along the map boundary tend to distort the estimated slope in the regression.

The final measure plays no role in describing the pattern of patches on a single landscape, but becomes important in comparing pattern across a number of different maps.

(5) Cover Types: (The total number of different land cover types on the map).

These five metrics provide relatively independent measures of spatial pattern on the landscape. They are based on empirical analysis of 85 landscapes scattered across the United States and should be valid for most applications, at least in temperate zones around the world. The biggest drawback is that they emphasise the landscape as a pattern of discrete patches of cover or vegetation. This may limit their application to problems where discrete patches are not a relevant aspect of spatial pattern.

7. Scales of Landscape Pattern

One of the most interesting developments in spatial ecosystem analysis has been the application of Hierarchy Theory (Allen and Starr 1982, O'Neill et al. 1987). The theory states that ecosystem processes are not uniformly distributed over spatial and temporal scales (Feibleman 1954). Dynamics and spatial pattern tend to be lumped into discrete scales of interaction (Rowe 1961, Simon 1962). The theory is intuitively appealing since dominant populations will structure the spatial scales around them and organisms will tend to interact only if they have the same temporal scale (Stommel 1963). Holling (1986) and Krummel et al. (1987) noted that the theory could be tested with ecosystem data.

The concept has been most carefully tested in the area of landscape pattern. McNab (1963) showed that mammal home ranges were scaled to their body size. Therefore, there should be a specific scale at which they interact with their spatial environment. Dominant organisms, therefore, would tend to structure the spatial environment into appropriate scales. The concept gained credence when Krummel et al. (1987) demonstrated that humans, as dominant organisms, typically alter the spatial hierarchy of the ecosystems about them.

One of the simplest ways to test the concept of spatial scaling was pioneered by Levin and Buttel (1986) who plotted variance as a function of scale. Variance is inversely proportional to sample size

$$S^2 \sim \frac{1}{n} \tag{17}$$

where n is the number of samples or the spatial size of the sample. Taking the log of both sides and adding a proportionality constant, a, gives

$$\ln S^2 = a - \ln n \tag{18}$$

which is a form of the equation for a straight line. Therefore, if ln S^2 is plotted as a function of ln n, we expect a straight line with a slope of -1. However, if the spatial data is organized into levels, then immediately adjacent points on this graph will be correlated. The next larger sample will not be independent of the last sample and the slope of the regression will lie between -1 and 0.

If the spatial data is organized into discrete hierarchical levels, we would expect the regression to show slopes that alternate between -1.0 (no correlation) and values much closer to 0.0 (high correlation). In other words, assume that the system is organized into patches of similar size. At scales close to this characteristic patch size, the samples would show high correlations. At smaller and larger scales, we would expect the randomness of the data to show slopes approximating -1.0. In a hierarchically structured landscape, therefore, we would expect to find ranges of scales with slopes closer to zero, representing the distinct scales of organization. These flat areas of the graph would be separated by slopes approximating -1.0. The resulting graph would look like a staircase if hierarchical organization is present on the landscape

O'Neill et al (1991) tested this concept on two grasslands and four forest landscapes. The expected 'staircase' pattern appeared on all landscapes. Multiple scales appeared on four of the landscapes and a single scale was evident on the two landscapes dominated by urban development. A similar approach has considered the slope of the log-log plots to indicate a fractal dimension (Wiens and Milne 1989). The slope should remain constant as long as the underlying process, shaping the patches, remained the same. Breaks in the line would indicate that a new process had taken over. This approach was applied by Palmer (1988) to spatial patterns in plant communities. Several other workers (Burrough 1983, Ver Hoef and Glen-Lewis 1989) applied spatial autocorrelation and also located multiple scales of pattern.

Three additional tests of the hypothesis have been conducted. O'Neill et al. (1991) applied a series of statistical tests to vegetation transects from three ecosystems. They established multiple scales of pattern on all sites. Holling (1992) started from the observation that frequency distributions of vertebrate body weights showed peaks and gaps (Brown and Nicoletto 1991). Since body weight is correlated with spatial extent of home ranges (McNab 1963), he showed that the weight distribution corresponded to distinct scales of pattern in the landscape. Kolasa (1989) correlated the abundance of certain sized organisms to the spatial pattern of resources in a stream ecosystem.

One of the most interesting applications of spatial scale has been the detailed interpretation of consumer dynamics. Kotliar and Weins (1990) pointed out that scaling requires that an insect use one set of criteria to locate a patch, a second set to choose a tree, and yet a third to select an individual leaf. Wallace et al. (1995) showed that large ungulates forage randomly within a patch of vegetation. At this scale, there is little or no pattern in forage quality. However, the grazers use a completely different set of sensory clues and are non-random in their choices as they move from one patch to another. At this larger spatial scale, there is a distinct difference in the quality of the vegetation from patch to patch.

REFERENCES

Aho, J. 1978. Freshwater snail populations and the equilibrium theory of island biogeography. *Ann. Zool. Fenn.* 15:146-154.

Allen, T. F. H. And T. B. Starr. 1982. Hierarchy: perspectives for ecological complexity. University of Chicago Press, Chicago.

Ambuel, B. and S. A. Temple. 1983, Area-dependent changes in the bird communities and vegetation of southern Wisconsin forests. *Ecology* 64:1057-1068.

Andow, D. A., P. M. Kareiva, S. A. Levin, A. Okubo. 1990. Spread of invading organisms. *Landscape Ecology* 4:177-188.

Andren, H. 1992. Corvid density and nest predation in relation to forest fragmentation: a landscape perspective. *Ecology* 73:794-804.

Andren, H. And P. Angelstam. 1988. Elevated predation rates as an edge effect in habitat islands: experimental evidence. *Ecology* 69:544-547.

Andrewartha, H. G. And L. C. Birch. 1954. The distribution and abundance of animals. University of Chicago Press, Chicago

Antolin, M. A. And D. R. Strong. 1987. Long distance dispersal by a parasitoid Anagrus delictus and its host. *Oecologia* 73:288-292.

Askins, R. A. and M. J. Philbrick. 1987. Effect of change in regional forest abundance on the decline and recovery of a forest bird community. *Wilson Bulletin* 99:7-21

Askins, R. A. M. J. Philbrick, and D. S. Sugeno. 1987. Relationship between the regional abundance of forest and the composition of forest bird communities. *Biological Conservation* 39:129-152.

Bach, C. E. 1988a. Effests of host plant patch size on herbivore density: Patterns. *Ecology* 69:1090-1102.

Bach, C. E. 1988b. Effects of host plant patch size on herbivore density. Underlying mechanisms. *Ecology* 69:1102-1117.

Bailey, N. T. J. 1975. The mathematical theory of infectious diseases and its applications. Hafner Press, New York.

Baker, W. L. and Y. Cai. 1992. The role of programs for multiscale analysis of landscape structure using the GRASS geographical information system. *Landscape Ecology* 7:291-302.

Barbour, C. D. And J. H. Brown. 1974. Fish species diversity in lakes. *American Naturalist* 108:473-478.

Bengtsson, J. 1989. Interspecific competition increases local extinction rate in a metapopulation system. *Nature* 340:713-715.

Birch, L. C. 1971. The role of environmental heterogeneity and genetical heterogeneity in determining distribution and abundance. In: P. J. De Boer and G. R. Gradwell (eds.) *Dynamics of Populations*. pp. 109-128. Pudoc, Netherlands

Blake, J. G. And J. R.Karr. 1984. Species composition of bird communities and the conservation benefit of large versus small forests. *Biological Conservation* 30:173-187.

Blake, J. G. And J. R. Karr. 1987. Breeding birds of isolated woodlots: area and habitat relationships. *Ecology* 68:1724-1734.

Brittingham, M. C. And S. A. Temple. 1983. Have cowbirds caused forest birds to decline? *Bioscience* 33:31-35.

Bronmark, C. 1985. Freshwater snail diversity: effects of pond area, habitat heterogeneity, and isolation. *Oecologia* 67:127-131.

Brown, J. H. And A. Kodric-Brown. 1977. Turnover rates in insular biogeography: effects of immigration on extinction. *Ecology* 58:445-449.

Brown, J. H. And M. V. Lomolino. 1989. Independent discovery of the equilibrium theory of island biogeography. *Ecology* 70:1954-1957.

Brown, J. H. And P. F. Nicoletto. 1991. Spatial scaling of species composition: Body masses of North American land mammals. *American Naturalist* 138:1478-1512.

Brown, M. And J. Dinsmore. 1986. Implication of marsh size and isolation for marsh bird management. *J. Wildlife Management* 50:392-397

Burkey, T. V. 1989. Extinction in nature reserves: the effect of fragmentation and the importance of migration between reserve fragments. *Oikos* 55:75-81.

Burrough, P. A. 1983. Multiscale sources of spatial variation in soil. *J. Soil Science* 34:577-597.

Carlquist, S. 1974. Island Biology. Columbia University Press, NY.

Culver, D. C. 1970. Analysis of simple cave communities. I. Caves as islands. *Evolution* 29:463-474.

Culver, D. C., J. R. Holsinger, and R. Baroody. 1973. Toward a predictive cave biogeography: the Greenbriar Valley as a case study. *Evolution* 27:689-695.

Danielson, B. J. 1991. Communities on a landscape: the influence of habitat heterogeneity on the interactions between species. *American Naturalist* 138:1105-1120.

Danielson, B. J. 1992. Habitat selection, interspecific interactions and landscape composition. Evolutionary *Ecology* 6:399-411.

Diamond, J. M. 1969. Avifauna equilibria and species turnover rates on the Channel Islands of California. *Proc. Natl. Acad. Sci. USA* 64:57-63.

Diamond, J. M. 1972. Biogeographic kinetics: estimation of relaxation times for avifauna of southwestern Pacific Islands. *Proc. National Academy of Sciences USA* 69:3199-3203.

Diamond, J. M. 1973. Distributional ecology of New Guinea birds. *Science* 179:759-769.

Diamond, J. M. And E. Mayr. 1976. Species-area relation for birds of the Solomon archipeligo. *Proc. Natl. Acad. Sci. USA* 73:262-266.

Diamond, J. M., M. E. Gilpin, and E. May. 1976. Species-distance relation for birds of the Solomon Archipeligo and the paradox of great speciators. *Proc. Natl. Acad. Sci. USA* 73:2160-2164.

Dickman, C. R. 1987. Habitat fragmentation and vertebrate species richness in an urban environment. J. *Applied Ecology* 24:337-351.

Dobzhansky, T. And S. Wright. 1947. Genetics of natural populations. XV. Rate of diffusion of a mutant gene through a population of Drosophila pseudoobscura. *Genetics* 31:303-324.

Dunning, J. B., B. J. Danielson, and H. R. Pulliam. 1992. Ecological processes that affect populations in complex landscapes. *Oikos* 65:169-175.

Durrett, R. And S. A. Levin. 1994. Stochastic spatial models: a user's guide to ecological applications. *Phil. Trans. R. Soc. London* B 343:329-350.

Ehrlich, P. R., D. D. Murphy, M. C. Singer, C. B. Sherwood, R. R. White, and I. L. Brown. 1980. Extinction, reduction, stability and increase: the responses of the checkerspot butterfly (Euphydryas) populations to the California drought. *Oecologia* 46:101-105.

Elwood, J. W., J. D. Newbold, R. V. O'Neill, and W. Van Winkle. 1983. Resource spiralling: An operational paradigm for analyzing lotic ecosystems. In: T. D. Fontaine, III and S. M. Bartell (eds.), *Dynamics of Lotic Ecosystems*, pp. 3-27. Ann Arbor Science, Ann Arbor, Michigan.

Fahrig, L. 1983. Habitat patch connectivity and population stability: a model and case study. Thesis, Carleton University, Ottawa, Ontario, Canada

Fahrig, L. And J. Paloheimo. 1988. Arrangement of habitat patches on local population size. *Ecology* 69:468-475.

Feibleman, J. K. 1954. Theory of integrative levels. *British Journal of the Philosophical Society* 5:59-66.

Fisher, R. A. 1937. The wave of advance of advantageous genes. *Ann. Eugen* (London) 7:355-369.

Ford, H. A. 1987. Bird communities on habitat islands in England. *Bird Study* 34:205-218.

Forman, R. T. T. 1995. Land Mosaics: The ecology of landscapes and regions. Cambridge University Press, Cambridge.

Forman, R. T. T., A. E. Galli, and C. F. Leck. 1976. Forest size and avian diversity in New Jersey woodlots with some landuse implications. *Oecologia* 26:1-8.

Freemark, K. E. And H. G. Merriam. 1986. Importance of area and habitat heterogeneity to bird assemblages in temperate forest fragments. *Biological Conservation* 36:115-141.

Galli, A. E., C. F. Leck, R. T. T. Forman. 1976. Avian distribution patterns in forest islands of different sizes in Central New Jersey. *Auk* 93:356-364.

Gardner, R. H., B. T. Milne, M. G. Turner, and R. V. O'Neill. 1987. Neutral models for the analysis of broad-scale landscape pattern. *Landscape Ecol.* *1*(1):19-28.

Gardner, R. H., R. V. O'Neill, M. G. Turner, V. H. Dale. 1989. Quantifying scale dependent effects with simple percolation models. *Landscape Ecology* 3:217-227.

Gilbert, F. S. 1980. The equilibrium theory of island biogeography: fact or fiction? *Journal of Biogeography* 7:209-235.

Gonzalez, R. C. and R. E. Woods. 1992. Digital Image Processing. Addison-Wesley, Reading MA.

Goodland, R. 1991. The case that the world has reached limits. In: R. Goodland, H. Daly, and S. El Serafy. *Environmentally sustainable economic development: Building on Brundtland.* pp. 5-17. The World Bank, Environment Working Paper Number 46.

Goodman, D. 1987. The demography of chance extinction. In. M. E. Soule (ed.) *Viable Populations for Conservation.* pp. 11-34. Cambridge Uiversity Press, Cambridge.

Gottfried, B. M. 1979. Small mammal populations in woodlot islands. Amer Midland Naturalist 102:105-112.

Gulve, P. S. 1994. Distribution and extinction patterns within a northern metapopulation of the pond frog, Rana lessonae. *Ecology* 75:1357-1367.

Hamilton, R. H., R. R. Barth, I. Rubinoff. 1964. The environmental control of insular variation in bird species abundance. *Proc. Nat. Acad. Sci. USA* 53:132-140.

Hanski, I. 1981. Coexistence of competitors in patchy environments with and without predation. *Oikos* 37:306-312.

Hanski, I. 1983. Coexistence of competitors in patchy environments. *Ecology* 64:493-500.

Hardt, R. A. And R. T. T. Forman. 1989. Boundary form effects on woody colonization of reclaimed surface mines. *Ecology* 70:1252-1260.

Harrison, S. D. D. Murphy, and P. R. Ehrlich. 1988. Distribution of the bay checkerspot butterfly Euphydryas editha bayensis: evidence for a metapopulation model. *Amer. Nat.* 132:360-382.

Have, A. 1987. Experimental island biogeography: immigration and extinction of ciliates in microcosms. *Oikos* 50:218-224.

Hayden, I. J., J. Faaborg, and R. L. Clawson. 1985. Estimates of minimal area requirements for Missouri forest birds. *Transactions of the Missouri Academy of Science* 19:11-22.

Helle, P. 1984. Effects of habitat area on breeding bird communities in northeastern Finland. *Ann. Zool. Fenn.* 21:421-425.

Helle, P. 1985. Effects of forest fragmentation on bird densities in northern boreal forests. *Ornis Fennica* 62:35-41.

Helliwell, D. R. 1976. The effects of size and isolation on the conservation value of wooded sites in Britain. *J. Biogeography* 3:407-416.

Holling, C. S. 1986. Resilience of ecosystems; local surprise and global change. In W. C. Clark and R. E. Munn (eds) *Sustainable development of the Biosphere.* pp. 292-317. Cambridge University Press, Cambridge.

Holling, C. S. 1992. Cross-scale morphology, geometry, and dynamics of ecosystems. *Ecological Monographs* 62:447-502.

Horn, H. S. and R. H. MacArthur 1972. Competition among fugitive species in a harlequin environment. *Ecology* 53:749-752.

Howe, R. W. 1984. Local dynamics of bird assemblages in small forest habitat islands in Australia and North America. *Ecology* 65:1585-1601.

Howe, R. W. and G. Jones. 1977. Avian utilization of small woodlots in Dane County, Wisconsin. Passenger Pigeon 39:313-319.

Howe, R. W., T. D. Howe, and H. A. Ford. 1981. Bird distributions on small rainforest remnants in New South Wales. *Australian Wild. Res.* 8:637-651.

Huffaker, C. B. 1958. Experimental studies on predation: dispersion factors and predator-prey oscillations. *Hilgardia* 27:343-383.

Huffaker, C. B., K. P. Shea, and S. G. Herman. 1963. Experimental studies on predation: complex dispersion and levels of food in an acarine predator-prey interaction. *Hilgardia* 34:305-330.

Kadmon, R. And A. Shmida 1990. Spatiotemporal demographic processes in plant populationsL an approach and a case study. *American Naturalist* 135:382-397

Kierstead, H. And L. B. Slobodkin. 1953. The size of water masses containing plankton bloom. *J. Mar. Res.* 12:141-147.

Kolasa, J. 1989. Ecological systems in hierarchical perspective: breaks in community structure and other consequences. *Ecology* 70:36-47.

Kotliar, N. B. and J. A. Wiens. 1990. Multiple scales of patchiness and patch structure - a hierarchical framework for the study of heterogeneity. *Oikos* 59:253-260.

Kruess, A. And T. Tscharntke. 1994. Habitat fragmentation, species loss, and biological control. *Science* 264:1581-1584.

Krummel, J. R., R. H. Gardner, G. Sugihara, R. V. O'Neill, P. R. Coleman. 1987. Landscape patterns in a disturbed environment. *Oikos* 48:321-324.

Lack, D. 1942. Ecological features of the bird fauna of British small islands. J. Animal Ecology 11:9-36.

Lack, D. 1970. Island Birds. *Biotropica* 2:29-31.

Lassen, H. H. 1975. The diversity of freshwater snails in view of the equilibrium theory of island biogeography. *Oecologica* 19:1-8

Lavorel, S., R. H. Gardner, and R. V. O'Neill. 1995. Dispersal of annual plants in hierarchically structured landscapes. *Landscape Ecology* 10:277-289.

Lawton, J. H. And G. L. Woodruffe. 1991. Habitat and the distribution of water voles: why are there gaps in the species range? *J. Animal Ecology* 60:79-91

Leigh, E. G. 1981. The average lifetime of a population in a varying environment. *J. Theor. Biol.* 90:213-239.

Levin, S. A. 1976. Population dynamics models in heterogeneous environments. *Ann Review of Ecology and Systematics* 7:287-310.

Levin, S. A. 1986. Random walk models of movement and their implications. In: T. G. Hallam and S. A. Levin (eds.) *Mathematical Ecology, An Introduction.* pp. 149-154. Spreinger-Verlag, Heidelberg.

Levin, S. A. And L. Buttel. 1986. Measures of patchiness in ecological systems. Ecosystem Research Center Report ERC-130. Cornell University, Ithaca, NY.

Levins, R. 1970. Extinctions. In: *Some Mathematical Questions in Biology, Lectures on Mathematics in the Life Sciences.* pp. 77-107. American Mathematical Society, Providence, RI.

Levins, R. and D. Culver 1971. Regional coexistence of species and competition between rare species. *Proc. National Academy of Sciences USA* 68:1246-1248.

Li, H. And J. F. Reynolds. 1993. A new contagion index to quantify spatial patterns of landscapes. Landscape *Ecology* 8:155-162.

Lomolino, M. V. 1982. Species-area and species-distance relationships of terrestrial mammals in the Thousand Island Region. *Oecologia* 54:72-75.

Lomolino, M. V. 1984. Mammalian island biogeography: effects of area, isolation, and vagility. *Oecologia* 61:376-382

Lynch, J. F. 1987. Responses of breeding bird communities to forest fragmentation.. In: D. A Saunders, G. W. Arnold, A. A. Burbridge, and A. J. M. Hopkins (eds.) *Nature Conservation: the role of remnants of native vegetation.* pp. 123-140. Surrey, Beatty, and Sons, Sydney, Australia.

Lynch, J. F. And R. F. Whitcomb. 1978. Effects of the insularization of the eastern deciduous forest on avifaunal diversity and turnover. In: A. Marmelsteing (ed.) *Classification, inventory, and evaluation of fish and wildlife habitat.* pp. 461-489. US Fish and Wildlife Service Publication, OBS-78176, Washington, DC.

Lynch. J. F. and R. F. Whitcomb 1984. Effects of forest fragmentation on breeding bird communities in Maryland, USA. *Biological Conservation* 28:287-324.

MacArthur, R. H. and E. O. Wilson. 1963. An equilibrium theory of insular zoogeography. *Evolution* 17:373-387.

MacArthur, R. H. and E. O. Wilson. 1967. Island Biogeography. Princeton University Press, Princeton.

MacArthur, R.H. 1972. Geograhical Ecology. Harper and Row, NY.

McCollin, D. 1993. Avian distribution patterns in a fragmented wooded landscape (North Humerside, UK): the role of between patch and within-patch structure. *Global Ecology and Biogeography Letters* 3:48-62.

McCoy, E. D. 1983. The application of island-biogeographic theory to patches of habitat: how much land is enough? *Biological Conservation* 25:53-62.

McNab, B. K. 1963. Bioenergetics and the determination of home range size. *American Naturalist* 97:133-140.

Magurran, A. E. 1988. Ecological diversity and its measurement. Princeton University Press, Princeton, NJ.

Martin, J. L. 1983. Impoverishment of island bird communities in a Finnish archipeligo. *Ornis Scand.* 14:66-77.

Milne, B. T. 1991. Lessons from applying fractal models to landscape patterns. In: M. G. Turner and R. H. Gardner (eds.) *Quantitative Methods in Landscape Ecology.* pp. 199-235. Springer-Verlag, New York, NY.

Moore, N. W. And M. D. Hooper. 1975. On the number of bird species in British woods. *Biological Conservation* 8:239-250.

Munroe, E. G. 1948. The geographical distribution of butterflies in the West Indies. Doctoral Dissertation, Cornell University, Ithaca, NY.

Murphy, D. D., K. E. Freas, and S. B. Weiss. 1990. An environement-metapopulation approach to population viability for a threatened invertebrate. *Conservation Biology* 4:41-51

Musick, H. B. And H. D. Grover. 1991. Image textual measures as indices of landscape pattern. IN M. G. Turner and R. H. Gardner (eds.). *Quantitative Methods in Landscape Ecology.* pp. 77-103. Springer-Verlag, New York, NY.

Newbold, J. D., J. W. Elwood, R. V. O'Neill, and W. Van Winkle. 1981. Measuring nutrient spiralling in streams. *Can. J. Fish. Aquat. Sci.* 38(7):860-863.

Newbold, J. D., P. J. Mulholland, J. W. Elwood, and R. V. O'Neill. 1982a. Organic carbon spiralling in stream ecosystems. *Oikos* 38:266-272.

Newbold, J. D., R. V. O'Neill, J. W. Elwood, and W. Van Winkle. 1982b. Nutrient spiralling in streams: Implications for nutrient limitation and invertebrate activity. *Am. Nat.* 120:628-652.

Newbold, J. D., J. W. Elwood, R. V. O'Neill, and A. L. Sheldon. 1983. Phosphorus dynamics in a woodland stream ecosystem: A study of nutrient spiralling. *Ecology* 64(5):1249-1265.

Newmark, W. D. 1991. Tropical forest fragmentation and the local extinction of understory birds in the Eastern Usambara Mountains, Tanzania. *Conservation Biology* 5:67-78.

Okubo, A. 1980. Diffusion and ecological problems: mathematical models. Springer-Verlag, Berlin

O'Neill, R. V., J. W. Elwood, and S. G. Hildebrand. 1979. Theoretical implications of spatial heterogeneity in stream ecosystems.. In: G. S. Innis and R. V. O'Neill (eds.) *Systems Analysis of Ecosystems.* pp. 79-101 Int Cooperative Publishing House, Fairland, MD.

O'Neill, R. V., D. L. DeAngelis, J. B. Waide, and T. F. H. Allen. 1986. A hierarchical concept of ecosystems. Princeton University Press, Princeton.

O'Neill, R. V., B. T. Milne, M. G. Turner, and R. H. Gardner. 1988a. Resource utilization scales and landscape pattern. *Landscape Ecology* 2:63-69.

O'Neill, R. V., J. R. Krummel, R. H. Gardner, G. Sugihara, B. Jackson, D. L. DeAngelis, B. T. Milne, M. G.

Turner, B. Zygmunt, S. Christensen, V. H. Dale, and R. L. Graham. 1988. Indices of landscape pattern. *Landscape Ecology* 1:153-162.

O'Neill, R. V., R. H. Gardner, B. T. Milne, M. G. Turner, and B. Jackson. 1991. Heterogeneity and spatial hierarchies. IN J. Kolasa and S. T. A. Pickett (eds.) *Ecological Heterogeneity*. pp. 85-96. Springer-Verlag, NY.

O'Neill, R. V., S. J. Turner, V. I. Cullinan, D. P. Coffin, T. Cook, W. Conley, J. Brunt, J. M. Thomas, M. R. Conley, J. Gosz. 1991. Multiple landscape scales: an intersite comparison. *Landscape Ecology* 5:137-144.

O'Neill, R. V., R. H. Gardner, M. G. Turner, W. H. Romme. 1992. Epidemiology theory and disturbance spread on landscapes. *Landscape Ecology* 7:19-26.

Opdam, P. 1991. Metapopulation theory and habitat fragmentation: a review of holarctic breeding bird studies. *Landscape Ecology* 5:93-106.

Opdam, P., G. Rijsdijk, and F. Hustings. 1985. Bird communities in small woods in an agricultural landscape: effects of area and isolation. *Biological Conservation* 34:333-352.

Opdam, P., D. Van Dorp, and C. J. F. Ter Braak. 1984. The effect of isolation on the number of woodland birds in small woods in the Netherlands. *J. Biogeography* 11:473-476.

Palmer, M. W. 1988. Fractal Geometry: a tool for describing spatial patterns of plant communities. *Vegetatio* 75:91-102.

Petit, D. R. 1989. Weather-dependent use of habitat patches by wintering birds. *J. Field Ornithology* 60:241-247.

Pettersson, B. 1985. Extinction of the Middle Spotted Woodpecker, Dendrocopos medius in Sweden and its relation to general theories on extinction. *Biological Conservation* 32:335-353.

Pimentel, D., W. P. Nagel and J. L. Madden. 1963. Space-time structure of the environment and the survival of parasite-host systems. *Amer. Nat* 97:141-166.

Preston, F. W. 1962. The canonical distribution of commoness and rarity. *Ecology* 43:185-215.

Pulliam, H. R. 1988. Sources, sinks, and population regulation. *American Naturalist* 132:652-661.

Pulliam, H. R. and B. J. Danielson. 1991. Sources, sinks, and habitat selection: a landscape perspective on population dynamics. *American Naturalist* 137:S50-S66.

Quinlin, A. V. 1975. Design and analysis of mass conservatiove models of ecodynamic systems. Doctoral dissertation, Department of Civil Engineering, Massachusetts Institute of Technology, Cambridge, Massachusetts.

Quinlin, A. V. And H. M. Paynter. 1976. Some simple nonlinear dynamic models of interacting element cycles in aquatic ecosystems. *Journal of Dynamic Systems, Measurement, and Control,* March 1976, 6-19.

Quinn, J. F. and A. Hastings. 1987. Extinction in subdivided habitats. *Conservation Biology* 1:198-208.

Reading, C. J., J. Loman, and T. Madsen. 1991. Breeding pond fidelity in the common toad, Bufo bufo. *J. Zool. London* 225:201-211.

Reichle, D. E. , R. V. O'Neill, W. F. Harris. 1975. Principles of energy and material exchange in ecosystems.. In W. H. Van Dobben and R. H. Lowe-Connell (eds.) *Unifying Concepts in Ecology*. pp. 27-43 Dr. W. Junk, The Hague.

Riitters. K.H, R.V. O'Neill, C.T.Hunsaker, J.D. Wickham, D.H. Yankee, S.P. Timmins, K. B. Jones, B.L. Jackson. 1995. A Factor Analysis of Landscape Pattern and Structure Metrics. *Landscape Ecology* 10:23-39.

Riitters, K. H., R. V. O'Neill, J. D. Wickham, and K. B. Jones. 1996. A note on Contagion indices for landscape analysis. *Landscape Ecology* 11:197-202.

Robbins, C. S. 1979. Effect of forest fragmentation on bird populations. In: R. M. DeGraaf and K. E. Evans (eds.) *Management of north central and northeastern forests for non-game birds*. pp. 198-212. US Department of Agriculture, Forest Service, General Technical Report NC-51.

Rey, J. R. And D. R. Strong. 1983. Immigration and extinction of salt marsh arthropods on islands: an experimental study. *Oikos* 41:396-401.

Robbins, C. S. 1980. Effect of forest fragmentation on breeding bird populations in the Piedmont of the mid-Atlantic region. *Atlantic Naturalist* 33:31-36.

Roff, D. A. 1974a. Spatial heterogeneity and the persistence of populations. *Oecologia* 15:245-258.

Roff, D. A. 1974b. The analysis of a population model demonstrating the importance of dispersal in a heterogeneous environment. *Oecologia* 15:259-275.

Rowe, J. S. 1961. The level-of-integration concept and ecology. *Ecology* 42:420-427.

Sauer, J. 1969. Oceanic islands and biogeographic theory: a review. *Geographical Reviews* 59:582-593.

Saunders, D., R. J. Hobbs, and C. R. Margules. 1991. Biological consequences of ecosystem fragmentation: a review. *Conservation Biology* 5:18-32.

Serrao, J. 1985. Decline of forest songbirds. *Rec. NJ Birds* 11:5-9.

Simberloff, D. 1976. Experimental zoogeography of islands: effects of island size. *Ecology* 57:629-648.

Simberloff, D. 1988. The contribution of population and community biology to conservation science. *Annual Reviews of Ecology and Systematics* 19:473-511.

Simberloff, D. And E. O. Wilson. 1969. Experimental zoogeography of islands. The colonization of empty islands. *Ecology* 50:278-296.

Simberloff, D. And E. O. Wilson. 1970. Experimental zoogeography of islands. A ten-year record of colonization. *Ecology* 50:278-316.

Simon, H. A. 1962. The architecture of complexity. *Proceedings of the American Philosophical Society* 106:467-482.

Simpson, J. B. 1974. Glacial migration of plants: island biogeography evidence. *Science* 185:698-700.

Skellam, J. G. 1951. Random dispersal in theoretical populations. *Biometrika* 38:196-218.

Small, M. F. And M. L. Hunter. 1988. Forest fragmentation and avian nest predation in forested landscapes.

Oecologia 76:62-64.

Smith, A. T. 1980. Temporal chanes in insular populations of the pika (Ochotona princeps). *Ecology* 61:8-13.

Soule, M. E. And D. Simberloff. 1986. What do genetics and ecology tell use about the design of nature reserves? *Biological Conservation* 2:75-92.

Stauffer, D. 1985. Introduction to percolation theory. Taylor and Francis, London.

Stommel, H. 1963. Varieties of oceanographic experience. *Science* 139:572-576.

Takafugi, A. 1977. The effect of the rate of successful dispersal of a phytoseiid mite, Phytoseilus persimilis (Acarinia: Phytoseiidae) on the persistence in the interactive system between the predator and its prey. *Researches on Population Ecology* 18:210-222.

Takafugi, A., Y. Tsuda, and T. Miki. 1983. Systems behavior in predator-prey interactions with special reference to acarine predator-prey systems. *Researches on Population Ecology Supplement* 3:75-92.

Taylor, A. D. 1990. Metapopulation, dispersal, and predator-prey dynamics: an overview. *Ecology* 71:429-433.

Terborgh, J. 1975. Faunal equilibria and the design of wildlife preserves. In: F. B. Golley and E. Medina (eds.) *Tropical Ecology Systems.* pp. 369-380.. Springer-Verlag, NY.

Toft, C. A. and T. W. Schoener. 1983. Abundance and diversity of orb spiders on 106 Bahamian islands: biogeography at an intermediate level. *Oikos* 41:411-426.

Turner, M.G., R.H. Gardner, V.H. Dale, R.V. O'Neill. 1989. Predicting the spread of disturbances across heterogeneous landscapes. *Oikos* 55:121-129.

Urban, D., R. V. O'Neill, and H. H. Shugart. 1987. Landscape Ecology. *Bioscience* 37:119-127.

Van Dorp, D. And P. F. M. Opdam. 1987. Effects of patch size, isolation, and regional abundance on forest bird communities. *Landscape Ecology* 1:59-73.

Vannote, R. L., G. W. Minshall, K. W. Cummins, J. R. Sedell, and C. E. Cushing. 1980. The river continuum concept. *Canadian Journal of Fisheries and Aquatic Science* 37:130-137.

Verboom, J. A. Schotman, P. Opdam, and J. A. J. Metz. 1991. European nuthatch metapopulations in a fragmented agricultural landscape. *Oikos* 61:149-156.

Verboom, B. And R. Van Apeldoorn. 1990. Effects of habitat fragmentation on the red squirrel, *Sciurus vulgaris L. Landscape Ecology* 4:171-176.

Ver Hoef, J. M. And D. C. Glenn-Lewis. 1989. Multiscale ordination: a method for detecting pattern at several scales. Vegetatio 82:59-67.

Villa, F., O. Rossi, F. Sartore. 1992. Understanding the role of chronic environmental disturbance in the context of island biogeographic theory. *Environmental Management* 16:653-666.

Vitousek, P. 1994.

Vuilleumier, F. 1970. Insular biogeography in continental regions. I. The northern Andes of South America. *Amer. Nat.* 104:373-388.

Vuilleumier, F. 1973. Insular biogeography in continental regions. II. Cave faunas from Tesin, southern Switzerland. *Syst. Zoology* 22:64-76.

Wallace, J. B., J. R. Webster, and W. R. Woodall. 1977. The role of filter feeders in flowing waters. *Arch. Hydrobiol.* 79:506-532.

Wallace, L.L., M.G. Turner, W.H. Romme, R.V. O'Neill, Y. Wu. 1995. Scale of heterogeneity of forage production and winter foraging by elk and bison. *Landscape Ecology* 10:75-83.

Watt, A. S. 1947. Pattern and process in the plant community. *J. Ecology* 35:1-22

Webb, N. R., R. T. Clarke, and J. T. Nicholas. 1984. Invertebrate diversity on fragmented Calluna-heathland: effects of surrounding vegetation. *J. Biogeography* 11:41-46.

Webb, N. R. and A. H. Vermaat. 1990. Changes in vegetational diversity on remnant heathland fragments. *Biological Conservation* 53:253-264.

Webster, J. R. 1975. Analysis of potassium and calcium dynamics in stream ecosystems on three Southern Appalachian watersheds of contrasting vegetation. Doctoral Thesis, University of Georgia, Athens, Georgia.

Webster, J. R. and B. C. Patten. 1979. Effects of watershed perturbation on stream potassium and calcium dynamics. *Ecological Monographs* 19:51-72.

Wenster, J. R. J. B. Waide, B. C. Pattern (1975) Nutrient recycling and the stability of ecosystems. In: F. G. Howell, J. B. Gentry, and M. H. Smith (eds.) *Mineral Cycling in Southeastern Ecosystems.* pp 1-27 Energy Research and Development Administration Symposium Series CONF-740513, Oak Ridge, Tennessee.

Weddell, B. J. 1991. Distribution and movements of Columbian ground squirrels (Spermophilus columbianus (Ord)): are habitat patches like islands. *J. Biogeography* 18:385-394.

Weiss, S. B., D. D. Murphy, and R. R. White. 1988. Sun, slope, and butterflies: topographic determinants of habitat quality for Euphydryas editha. *Ecology* 69:1486-1496.

Whitcomb, R. F. 1977. Island biogeography and habitat islands of eastern forests. *Amer. Birds* 31:3-5.

Whitcomb, B. L., R. F. Whitcomb, and D. Bystrak. 1977. Island biogeography and 'habitat islands' of eastern forests: III. Long-term turnover and effects of selective logging on the avifauna of forest fragments. *American Birds* 31:17-23.

Whitcomb, R. F., C. S. Robbins, J. F. Lynch, B. L. Whitcomb, M. K. Klimhiewicz, and D. Bystrak. 1981. Effects of forest fragmentation on avifauna of the eastern deciduous forest. In: R. L. Burgess and D. M. Sharpe (eds.) *Forest Island dynamics in man-dominated landscapes.* pp. 125-206. Springer-Verlag, NY.

Wiens, J. A. and B. T. Milne. 1989. Scaling of landscapes in landscape ecology. *Landscape Ecology* 3:87-96.

Wilcove, D. S. 1985. Nest predation in forest tracts and the decline of migratory songbirds. *Ecology* 60:1211-1214.

Wilcove, D. S., C. H. McLellan, and A. P. Dobson. 1986. Habitat fragmentation in the temperate zone. In: M. E.

Soule (ed.) *Conservation Biology: The science of scarcity and diversity.* pp237-256 Sinauer Assoc., Sunderland.

Williams, C. B. 1964. Patterns in the balance of nature. Academic Press, NY.

Willis, E. O. 1974. Populations and local extinction of birds on Barro Colorado Island, Panama. Ecol. Monographs 44:153-169.

II.10 Towards a Unifying Theory

Sven E. Jørgensen and Felix Müller

1. Introduction

In the introduction of Part II of this volume we have discussed why we need theories and what the advantages and necessities are of using theories in science. In this chapter, we will discuss the principal potentials to unify the various presented theories into one general integrating concept. In the previous chapters, a hierarchy of theories and sub-theories has been presented which enables us to explain many different ecological processes at various levels of observation. A unifying theory in systems ecology would of course make it possible to give one general explanation for all our observations of ecosystems. Such integration is desirable as it would be a significant simplification if we could refer all our observations to one set of hypotheses only. This could then become the basis for all scientific activity in the field including the possibility to derive new theses and conclusions about ecosystem structure and function.

Such a unified theory is the target of all scientific disciplines, and some of them may already have reached that state. Ecology has not. Thus, it is not possible at present to unify all the presented scientific approaches to ecosystem theory discussed in this volume in one general theory. Ecosystems are such complex entities and need therefore a pluralistic description covering various sides of the characteristic properties. The resulting pattern of theories explains the constraints for the behaviour of the systems under various conditions. It represents a pluralism of limitations to the development of ecosystems which to a certain extent can explain our observations.

The integration of most of the presently discussed ecosystem theories has been attempted in Jørgensen (1997) and Patten (1998) in the form of several propositions given to the characteristic properties of ecosystems. An integration will inevitably uncover the specific features of ecosystems that are elucidated by the single approaches and will show which dynamic, developmental properties the single theories describe. The presented theories have of course some overlaps and there are also a few minor disagreements. This, however, does not disturb the overall image of an interesting and stimulating pattern of approaches covering various aspects and properties of ecosystems. Thus, the multitude of theoretical concepts implies a multitude of ecosystem aspects resulting in an holistic pattern, similar to what is depicted in Figure 1.

2. The sources – what to be unified?

The various theories which are unified within the integration of approaches are listed in the outline of Part II, as one chapter is devoted to each single theory. At first glance, it seems to be an almost impossible task to produce a pattern out of approaches as varied as thermodynamics, cybernetics, information theory, network theories, utility theory, hierarchy theory, chaos and catastrophe theories, the theories about ecological stability, buffer capacities and resilience of ecosystems, and the theory of ecosystems as self-organising critical systems. To cope with these multiple concepts, it is a good starting point to consider that ecosystems are very dependent on energy inputs and energy consumptions. Therefore, from an energetic point of view the theory can be based on a few pertinent questions which determine many of the properties of ecosystems: How is the energy provided? What determines the amount of energy available for ecosystems? How is that energy used? The questions can be answered by the application of thermodynamics to ecosystems. It is a difficult task, but as presented in the first chapters of Part II, it is possible by the use of far-from-equilibrium thermodynamic and an extensive use of the concept 'exergy' which is easier to use than other thermodynamic principles because the reference state can conveniently be defined individually for each application.

Ecosystems may also be considered as information systems which contain many feed-back mechanisms. The components form a network which links all the elements and sub-systems in an extremely high interdependency. We can furthermore describe the various levels of components by a hierarchical system analysis. Ecosystems are also very dynamic systems both in time and space, and it would therefore be natural to attempt to apply chaos and catastrophe theory on this dynamic to explain some of the odd reactions of ecosystems to changed external conditions. This leads furthermore to the assumption that an ecosystem is a self-organising critical system.

Figure 1: An illustration of the well known elephant parable – four scientists are intensively analysing their objects. They are very concentrated in the details of their own observations, and only the single part is of interest to each of them. After finishing their research they will publish about tubes, carpets, walls or columns. They don't realise that these parts form an elephant because this holistic view is only possible by taking a step backwards and discussing ones own observations with those of their colleagues. Picture by J. Thomas.

The wide spectrum of properties which characterise an ecosystem requires assorted methods of descriptions to capture the high variety of features. This situation may be compared with the diverse descriptions of any complex object – for instance a ship. If we want to cover the design of a ship, we will use a physical model for instance at the scale 1:100. The cabins are covered by a special set of drawings, while the electrical circuit requires other types of construction. The system of pipelines will require yet other diagrams and a special model of the ship is needed to determine the hydrodynamic properties by physical experimentation. An ecosystem is far more complex than this ship, and it therefore will require an even greater number of methods to describe the various features. The different ecosystemic theories are to a certain extent needed to cover the many features of the object. So, the theories form a pattern in that the numerous aspects and features can be explained by various points of view. Thereby, a pattern of system properties is formed, the object is elucidated on the basis of several starting points because the features are to a certain extent connected. We can express it as follows – all the various theories are needed to cover the consequences of the most important ecological properties and their dynamics. The network theory for instance describes the consequences of the fact that ecosystems are forming networks, that the components are interrelated. Therefore, the components influence each other and a change in one element of the ecosystem will disseminate to other parts of the ecosystem and provoke minor or major changes of all the other components. Networks also ensure that cycling of the most important elements can take place which is a prerequisite for maintenance of life in ecosystems. In the same way, it is possible to explain the importance of all the other theories, covering additional characteristic features of ecosystems.

3. Temporal aspects – ecosystem development as focal level-of-integration

The diverse theories have introduced a number of what we may call ecological indicators, i.e., concepts which can be used to describe the dynamics and development of ecosystems. These indicators as a whole form a comprehensive figure of the ecosystem structure, the function, and they also represent features of the systems' performances. It is therefore pertinent to question whether or not these indicators are also forming a pattern, if yes, how should they be applied and how consistent is this application? We will try to answer these questions in the next section. In this section, we will use them to examine if, and to what degree the differing system ecological approaches are consistent. If they are, the changes of attributes stemming from theory A should have consequences for the variables that constitute theory B. Thus it will be attempted here to see to what extent the indicators can be used in parallel. The test-case to check the interdependencies of the various dynamic indicators is ecosystem succession. The variables will therefore be mentioned one by one and it will be asked how they describe the development of an ecosystem from an early stage and the further development towards mature ecological integrity.

Exergy dissipation per unit of time will increase for an ecosystem at an early stage under development in accordance with the extended version of the second law of thermodynamics (Kay, this volume). These dynamics typically takes place on the basis of rapid growth of the biological structure. The size and the structure of the biota determines the amount of exergy the system can capture from solar radiation. The increased structure means that more exergy dissipation is needed for the system's maintenance (respiration and evapotranspiration) per unit of time, but also that more exergy can be captured by the structure per unit of time. The reflected exergy which is not available for the biotic processes of the ecosystem can be reduced by an increased structure. Before all, this is only valid for a system at an early stage because when the structure captures the maximal amount of exergy which is possible from a physical point of view, the system will not be capable of further increasing the amount of exergy captured and therefore the amount of exergy dissipated can not increase when this maximal value has been reached.

Exergy storage will increase in parallel with the exergy captured, because there is a close relationship between the structure and the exergy stored in the structure (Jørgensen this volume). When the exergy storage has reached the mature stage, i.e., the level corresponding to a structure which is capturing the maximally possible from a physical point of view, it is still possible to increase the exergy storage, although there may be no more nutrients available to build up more biomass, as all the nutrients have been applied to construct the biological components. Increasing the exergy storage without increasing the exergy dissipation is possible by increasing:

1) the cycling (see for instance the chapters II. 1.3 and II. 7.3 and Jørgensen et al., 1999) as more feed-back loops are able to maintain more biomass without more dissipation,
2) the information stored in the biomass by use of more advanced genes, and
3) a better use and specialisation of the ecological niches, based upon the heterogeneity.

It is furthermore possible to increase or at least maintain the exergy storage and decrease the exergy dissipation by increasing the size of the organisms. Consequently, exergy storage can still increase in the mature stage, while exergy dissipation cannot. Particularly, the specific exergy storage defined as the exergy stored divided by the biomass seems to be a good indicator in this stage, where the biomass is maintained approximately constant and the exergy stored can still increase.

Emergy expresses the construction costs in solar energy equivalents, measured in sej = solar energy joules. The increased adaptation of an ecosystem implies an increase of feed-back mechanisms and thereby a growth of exergy and a decreased cost expressed in sej, while a man-made and controlled ecosystem – for instance a waste stabilisation pond – will often show a high cost of emergy in spite of a low exergy storage (Odum et al., this volume). The emergy/exergy ratio seems therefore to be a very useful ecological indicator which is very sensitive to the natural development of an ecosystem.

Maximum power has been introduced already by Lotka in the twenties as a possible ecological indicator. He, and also later H.T. Odum (see Hall, 1995), interpreted maximum power as the useful energy flow – which is an exergy flow since exergy is useful energy, that is energy able to do work. Increased power will imply that cycling increases as long as it involves flows of useful energy, and more cycling implies more feed-backs and more exergy storage, as discussed above (Odum et al., this volume). Maximum power is, however, a rate which means that there are limits to its increase. It is therefore probably not suitable to be used as an indicator for a mature system.

Minimum entropy production seems useful to indicate a mature system, while it can hardly be used as an indicator for an early stage under development where the exergy dissipation (entropy production) increases rapidly (Jørgensen, this volume). Minimum entropy production is crucial when energy is limiting. It means that the system has reached the level where the energy captured per unit of time is the maximally possible from a physical point of view. Under these circumstances, it is not possible to increase the entropy production further unless the energy is taken from the storage of exergy, but as mentioned above the system still attempts to increase this variable. The result is therefore that the exergy storage still is able to increase at the same (or lower) level of exergy dissipation and entropy production. The system tries to reduce the specific entropy production, i.e., the entropy production relative to the biomass or the exergy storage.

A system under development, both an early stage system and a mature system, attempts to increase the **number of feed-back** mechanisms and the storage of information in the genes and in the network. The ordered complexity (the complexity following a certain order), in contrast to the disordered or random complexity will increase parallel to the number of feed-back mechanisms, to the species richness and to the total information carried by the system. Ascendancy measures the size of the system (number of components and flows) and the information of the network (Ulanowicz, this volume). Ascendancy is well correlated to exergy storage which is not surprising from its definition (see Chapter II.7.1 and Jørgensen, 1994a), as more feed-back mechanisms and a larger network can support more organisms carrying more information.

The relationships between the components of an ecological network are reflected in the **information** carried by the network (Nielsen, this volume). It is demonstrated in Chapter II.7.2, that the development of ecosystems implies an increased network symbiosis and synergism and an enhanced dominance of the indirect effect (Patten and Fath, this volume). More cycling will furthermore involve a higher utility of the available energy. As also the number of components increases with the development of ecosystems, the control gets more distributed and more levels in the ecological hierarchy are formed. The development of the network involves an increase of the synergetic effects, as already mentioned above. This is the source of the emergent properties, i.e., the properties originated in the co-operative power of the components in the network which makes the entity more than sum of the parts.

Buffer capacity gives a good description of the ability of an ecosystem to resist changes because there are many buffer capacities corresponding to every combination of a state variable and a forcing function. There may be a relationship between the total sum of all possible buffer capacities and the exergy storage – at least there is a statistical relationship which has been found by use of models (Jørgensen and Mejer, 1977). The sum of buffer capacities will therefore increase with the development of an ecosystem, but it doesn't exclude that individual buffer capacities decrease, the potential brittleness increases and drastic changes can occur. It is dependent on the adaptability of the system to a new and maybe untested combination of forcing functions. Sometimes the drastic changes can be met by structural changes in the system whereby the consequences for the function of the system are reduced.; see Chapter II.8.3.

When a system operates to utilise the available resources at the optimum, it also becomes more vulnerable in the sense that it is closer to an over-exploitation of the resources which may cause chaotic reactions. A mature ecosystem will operate at a high and optimum exploitation of the available resources, but simultaneously at the *edge of the chaos* (see Chapter II.8.4). This explains why a mature ecosystem with a high extent of self-organisation gets close to the stage of potential creative destruction. Table 1 summarises the trends of the various ecological indicators presented in Part II for the early and mature stage of ecosystems.

The practical and direct application of the indicators is the focus of the next part of the volume and will therefore be discussed in greater detail below. The table may, however, be considered as a summary of Part II which can be used to carry us to the discussion of the direct application of indicators. The table may also be considered a part of the pattern of ecosystem theories resulting from the comparative presentation of the theories in Part II.

None of the presented theories gives the entire image of ecosystem development in all phases, but clearly they give us differing facets of the development at varied stages. Some of the theories may be used for all phases of development at least to a certain extent, but the full image of the development will still require that other facets are uncovered to obtain a complete understanding of the types of changes. Figure 2 sketches a working scheme for the extents and applications of the approaches. Here, the theories are distinguished into 3 levels: while thermodynamic features form the system of dynamic constraints, the result is self-organisation, a basic processual principal that is strictly dependent on thermodynamic rules. As a consequence of self-organising processes, emergent and collective properties arise. Their functional structures are investigated by the theories sketched on the left side of the figure while the right hand side approaches primarily try to explain the dynamics of the structures.

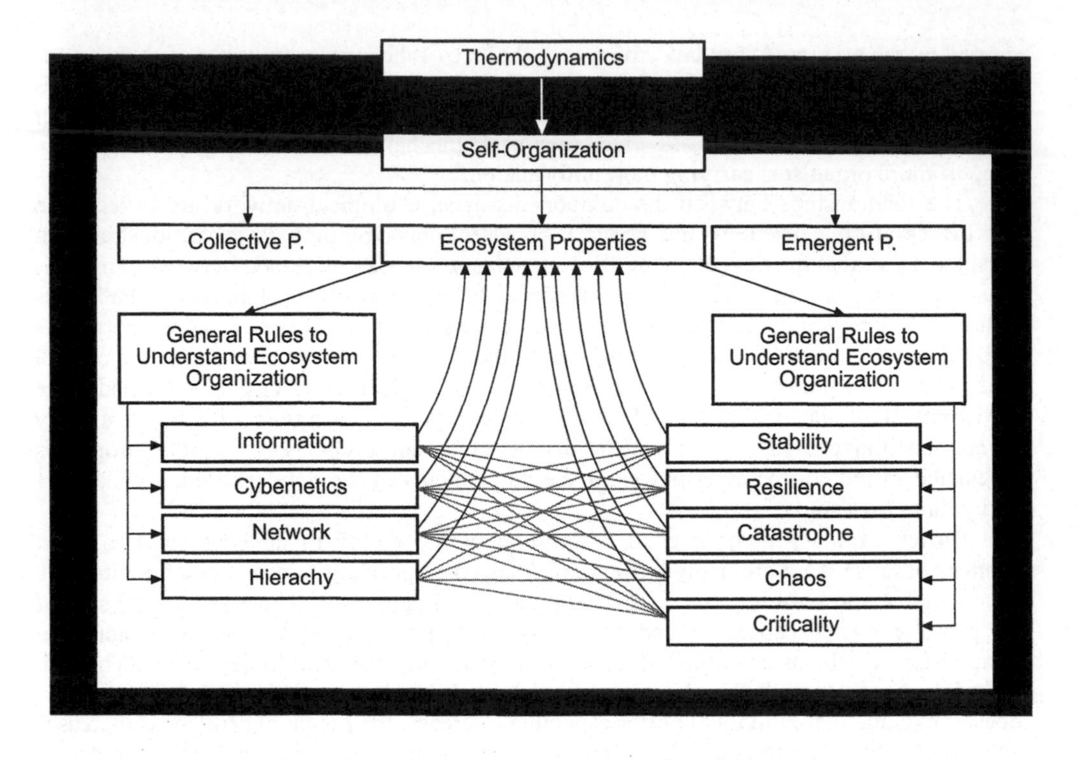

Figure 2: A pattern of ecosystem theories referring to the extents of the hypotheses and the ecological foci.

Table 1: Ecosystem properties throughout different developmental phases of ecosystem succession.

Trends in Ecosystem Development		
Indicator (property)	**System at an early stage**	**Mature system**
Feed-back loops	↑	↑
Exergy dissipation	↑	→
Exergy storage	↑	↑
Specific exergy	↑	↑↑
Emergy	↑	→
Emergy/exergy storage	↑	↓↓
Max. power	↑	↑
Global entropy	↑	→
Local entropy	→	↓
Ordered complexity	↑	↑
Structure	↑	↑
Information	↑	↑↑
Ascendancy	↑	↑↑
Indirect/direct effect	↑	↑↑
Utility	↑	↑
Heterogeneity	↑	↑
Species richness	↑	↑
Stability and resilience	↑	?
Sum of buffer capacities	↑	↑
Individual buffer capacities	?	?
Self-organising criticality	→	↑
Chaos	Come closer to the edge of chaos by a higher utilisation of the resources	

The various theories are consistent, if we apply them to describe the facets and the stages they can explain and not use them outside this range of validity. So, we can conclude that the presented theories form a pattern, but they cannot be unified into one theory because it has not been possible – and will probably never be – to cover the numerous facets in one theory due to the enormous complexity of ecosystem. To the question implied in the title of this chapter we have to answer 'no', but they form a beautiful and (almost) consistent pattern.

REFERENCES

Hall, A.S. C., 1995. Maximum Power. University Press of Colorado. 393 pp.

Jørgensen, S.E., 1994a. Fundamentals of Ecological Modelling (second edition) (Developments in Environmental Modelling, 19). Elsevier, Amsterdam, 628

Jørgensen, S.E., 1997. Integration of Ecosystem Theories: A Pattern. Kluwer Academic Publishers. Dordrecht, Boston, London. 388 pp.

Jørgensen, S.E. and Mejer H.F., 1977. Ecological Buffer Capacity. *Ecol. Modelling* 3: 39-61.

Jørgensen, S.E. , Patten, B.C. and Straskraba, M., 1999. Ecosystem Emerging: Growth. Submitted to Ecological Modelling.

Patten, B.C., 1998. Network Orientors: Steps Towards a Cosmography of Ecosystems: Orientors for Directional Development, Self Organisation, and Autoevolution, p. 137-160 In: F.Müller and M.Leupelt (eds). *Eco Targets, Goal Functions and Orientors.* Springer Verlag. Heidelberg. 1998. 616 pp

Application of Ecosystem Theoretical Aspects

Introduction

The practical application of any theory requires a certain degree of maturation of the hypotheses in which it is grounded. Sufficient time has not yet passed for ecosystem theories, as they have only developed in the last 2-3 decades. Ecosystem theory has therefore not yet been able to show its full potential as a tool to be used in a wide application of environmental management. The breadth and depth of theory utilisation will most likely increase in the coming years, as more experience with the direct use of ecosystem theoretical interrelations will be gained.

Despite this, a few applications are given in Section III, illustrating examples where ecosystem theory has had a direct influence on environmental and sustainable land use management strategies. In practice there needs to be a much wider application of ecosystem theory to gain the needed experience rapidly. Unfortunately, too few resources are allocated to these types of holistic management approaches.

Indisputably, the recent development in ecosystem theory has created an interesting challenge to the development of recipes for a sustainable management of entire ecosystems or regions. Some very first indications of such recipes are given in this section. Hopefully, they can inspire further development in this direction, by impressing on the reader, that it is possible to work with holistic ecosystem theory to solve practical environmental management problems.

Ecosystem theory is today already considered in the disciplines of conservation biology, in the conception and assessment of ecosystem health, integrity and sustainability, in ecological engineering, where engineering principles are applied on ecosystems to obtain environmentally more sound solutions to environmental problems; and in ecological economics, where it is attempted to build a hybrid of economy and ecology which can solve practical environmental management issues. Each of these possibilities will be treated in separate chapters in the following Section of the book. Finally we attempt to discuss to what extent how much ecosystem theory might be able to penetrate these practical applications by use of the ecological indicators and orientors, which were introduced in Section II and summarised in Chapter II.10.

III.1.1 Conservation Biology

Klaus Dierssen

1. Introduction

"The ecological crisis is brewing. Some day, there will be someone who wants to make his career by proving that this crisis has only been existing in people's mind." Such malicious ideas like those of the sociologist Ulrich Beck (1995) are neither pronounced by ecologists nor by conservationists. However, they may be helpful in that they probably reflect existing conceptions, which stem not only from lyrical consternation.

Two questions may be posed:

- How is it possible to provide the public with a better understanding of the problems of environment and nature conservation, and to solve these problems in a more efficient way than has been previously done? and;
- What kind of innovation could be expected in this context by natural sciences, especially by ecology and ecosystem research?

The value of ecology as 'the key science of post-modern times' is not indisputable. Mayer-Tasch (1985) for instance puts the emphasis on its limits, being too narrow while only dealing with questions concerning natural sciences, and working in a field of knowledge, which nowadays is mainly oriented towards the ecosystem concept. The centre of attention is still meant to be the effort to grasp special knowledge. He is right to stress that ecology as a natural science discipline is unable to perform a jump from scientific description to social action. Therefore, the only key science being able to guarantee a humane life worth living and viable future, would be, according to Mayer-Tasch, a human science he called political ecology.

Social decisions, simply based on common knowledge without any background on natural science (in terms of disposal knowledge), seem to be inadequate also. The following extreme statements prove that it is difficult to form a synthesis:

- the purist, self-restricting retreat of the expert scientist, which is well-known in the branch of science, to an epistemological (partial-)ethic with the ultimate aim called 'scientific knowledge' (Mohr, 1987) and
- the trans-scientific self-representation of scientists, which is not rarely demonstrated by public symposiums among laymen with a good deal of (eco-) populistic and demagogic capacity, and which cannot be efficiently controlled by the 'scientific community'.

In my opinion, the first statement contravenes with the social loyalty of the scientist in our society whereas the last point is not only 'a catastrophe for the moralistic prestige of science' (Mohr, 1977) and can be coped with, but is also able to trigger abortive developments with enormous economic consequences – for example if we consider the effectiveness of buying up whole areas by the state and of the expert planning strategies applied, the efficiency of which remains in doubt.

A first approach to find a solution could be the attempt to make the outcomes of ecological research with regard to their application more suitable for prognoses, in order to increase its place value with respect to *taking precautions for existence* (Peters 1991). This aspect will be outlined for the following complex of ecosystem research and conservation of nature.

2. Definitions

Conservation of nature is a social instruction, which is fixed by acts of nature conservation and further described by the regional regulations. The necessity of these instructions is generally accepted by the public. The financing by the state and the realisation by the government depends on changing social values and financial directories, therefore it depends on fashion and economic activity.

Conservation of nature is performed by at least three aspects: a financial-legal aspect (e.g. definitions of laws, prescriptions, guidelines), an administrative-procedural aspect (e.g. point-of-time and extent of public participation, position and density of control by the technical administration for the conservation of nature in case of a procedure for approval) and a methodological-professional aspect (e.g. inventory, assessments, definition of the targets, professional planning concerning the conservation of nature, monitoring). Especially, but by no means exceptionally the third aspect is at the moment characterised by enormous deficits of knowledge and application or realisation.

Ecosystem research fixed itself the target to gain a better understanding of structures and functions of parts of the landscape (the 'regulation of landscape'), how to simulate them by means of partial models and how to derive concepts for a sustainable cultivation, while taking care of the resources.

Consequently, ecosystem research is firstly basic research. Fundamentally, it requires, and is based on, interdisciplinary work; it takes up a great deal of time, it demands a high standard of the equipment used, it requires a complex system of evaluation, which then demands mainly well-trained specialists – in brief, it is expensive.

Taking into account the current financial situation of public budgets, we must steadily fight to get social acceptance. This is only possible with reference to the application of ecological results and the necessity of their consideration when planning based on profound ecological analyses.

The following discussion is divided into three sections with the aim of clarifying the cutting point between conservation of nature and ecosystem research:

- a short description of the objectives, protective resources/elements and instruments of nature conservation,
- an overview of the contents and the possibilities of application of the ecosystem research, and
- a characterisation of those fields of work, where proposals and decisions of conservationists in the future increasingly might be based on results of ecosystem research.

3. Objectives, protective resources or elements and instruments of the conservation of nature

The objectives of the conservation of nature are determined by nature conservation laws set by governments: the protection, maintenance and development of roughly defined protective resources/elements as well as the avoidance of any disturbances that could influence them. Such elements can be populations of wild animal and plant species, their biocoenosis, biotopes (concerning the species: habitats, concerning the biocoenosis: biotopes), further the abiotic resources of a landscape, being insufficiently described by the rough classification of water, soil and atmosphere, and finally entire ecosystems in their structure and function as well as in their visual arrangement/landscaping.

Generally speaking, nature conservation and landscape planning today doesn't just focus on isolated and 'higher-graded' ecosystems, as it is called in terms of nature conservation, but also wants to include the entire area, even though this often means working with different partial aims and priorities.

Concerning the instruments and procedures being used, the legislator has developed definite rules or guidelines. The objectives of nature conservation being fixed within the laws are meant to be transformed over the entire area in each sector by means of landscape planning. Additionally, they have to be integrated not only in the concepts of other disciplines (like water-supply, agriculture, forestry, traffic policy) but also in developmental planning, which includes socio-political targets. This should be done in a way that will make a sustainable use of landscape possible, which at the same time applies the potentials of ecosystems and won't do any harm to the environment.

If economic requirements in the area are combined with operations that do not fit with nature conservation standards, the risks should be *minimised* by checking up the environmental compatibility before taking a political or administrative decision. If this political-administrative decision is taken in favour of the operation after weighing up socio-economic needs and nature conservation necessities, the impairments of the protective resources, that occur during or after the operation will be less important and, if possible, will be minimised on the legal basis by impact regulation.

For sectorial planning in each area, an operational proceeding diagram has been developed for different integration levels (i.e. state level: landscape program; administrative district level: landscape frame schedule; municipal level: landscape plan; partial area of a municipality: greenery arrangement plan). This will be illustrated in an abbreviated form by a landscape plan: the resources and potentials for development of the area concerned will be analysed and assessed in reference to each single aim. On the basis of such an inventory, there will be definitions of aims, which step by step will be transformed from *environmental quality aims* into *environmental standards* by means of the development of regional models. Since the objectives of landscape planning will only be reached if the concerned citizens of an area are willing to accept these plans, the analysis of the current state and the definitions of aims will necessarily include a conflict analysis of competing economic requirements and disturbance variables. This assessment period of landscape planning is followed by the real procedure of planning and it ends up with a developed concept within this landscape plan, which aims at the continual application of the aims of nature conservation within a community. On the other hand, this realisation needs a certain control of the measures, always keeping in mind the final concept.

This procedure of applying the concepts of nature conservation has to be seen as a draft of an ideal situation. In reality, there are a significant number of problematic aspects, which will be roughly characterised and summed up in the following abstract.

On the analysis level, the fact of judging the state of a population or an area and its developmental capacity requires an extended knowledge about function and dynamics of each protective resource (i.e. minimal necessary size of the area, buffer capacity of partly synergistic-working disturbance variables having partly synergistic effect). For most species and systems, the knowledge is not sufficiently given and has always to be enlarged. This lack of knowledge about certain populations and areas is due to a restricted landscape inventory, which requires much time and a high methodological effort. This problem also occurs in connection with every prognosis concerning the development, which is deduced from the evaluation of a unique, relatively short-term analysis without periodical long-term studies. Inadequate data and the absence of a consistent theoretical construct of ideas normally leads to spongy prognostic statements.

In practice, conservation research is actually often limited to noncritical generalisations of hypothesis that are adopted as dogma (like the necessity of 'conjunction of biotopes', the establishment of populations with a high measured 'genetic polymorphism', or the 'diversity-stability paradigm' (Frank et al., 1994; Henle, 1994)), as well as to rough assessment schemata or to parameters which could be easily established and whose application develops concepts based on insufficient or fragmentary landscape inventory instead of differentiating and focusing on each single case. Generally speaking, there is an increasing tendency to avoid, if possible, analysis of whole areas and to take supposedly 'pragmatic' decisions on a basis that often is insufficient from the scientific point of view.

Concerning the definitions of aims, we have to keep in mind that the procedure of establishing priorities within these objectives of different protective resources is always a very subjective process. It is, in other words, influenced by sectional analysis and personal preferences which normally cannot be treated objectively (crane or orchid, conservation of species versus conservation of resources, conservation of resources versus that of processes and successions etc.). Preferences being formed on the basis of personal judgements could be understood by different persons, when the reasons of the decision are pointed out. The empirical explanation for choosing certain preferences always has to include two elements: a normative top and an empirical bottom (Schröder, 1998).

In practice, people don't always keep this in mind. Even expert/local authorities often use nebulous thought patterns which are insufficiently thought-out and whose data/information is imprecise ('conservation of rare species', 'development of diversity', 'interference of biotopes', 'natural development'), instead of defining area-related, superordinated aims (models), which will serve as orientation for every *more detailed* plan and measure in a logical and conclusive way (Jessel, 1994; Müller and Müller, 1992; Müller et al., 1997a).

As it is dissatisfying to apply concepts which are oriented at the conservation of single species – especially on small areas – to various biotypes on different levels of organisation and inundation, there is a growing tendency from the conservation of single species towards a conservation of the whole ecosystems and their underlying processes (i.e. Hansson and Larsson, 1997). The consequences of an adequate assessment of system properties and processes still remains an unsolved problem. There is a need to develop satisfying and scientific assessment frames and standards. A successful use of thermodynamic orientors as an indicator for ecosystem properties (Jørgensen and Nielsen, 1998; Svirezhew, 1998) still requires a larger data base in order to get nontrivial and useful results for the application of plans.

On the assessment level, we face a fundamental methodological problem: assessments, by nature, don't necessarily scale and judge totally equivalent (incommensurable) protective resources, for example supposed qualities, the loss of qualities of populations or biotopes and the erosion of selected functions of system compartments. The great amount of currently proposed and practised assessment procedures using a quantifying approach must be mentioned from the scientific point of view in the way that they miss to an important part the necessary logical stringency and consequently lead to different results when using the same data base (i.e. Auhagen, 1995; Scherner, 1995).

In the process of weighing up risks like it is done for the assessment of areas, often ordinal-scaled measure points are used, which are supposedly easy to obtain, whereas for system development important and cardinal quantified measure points are missing. As a result, we often find wrong or misleading assessments concerning disturbance variables and appropriate methods of control.

With regard to the application of planning concepts, we realise that nowadays, especially concerning the maintenance and development of areas, scientists often develop objectives which are not only vague/imprecise and unrealistic but also hard to translate in practice and which are expensive. Sometimes, they also help themselves with *'by chance'* free areas. On the other hand, the potential of development of these areas would be estimated rather low under the aspect of nature conservation. The result is a growing gap between *requirements* (legal task) and attainable or already reached aims.

On the monitoring level, we should admit that monitoring programs in nearly all biotopes are often too unsystematic and, either they aren't practised enough or too irregularly. Furthermore, they don't fit very well with each other on the scientific level and they aren't sufficiently evaluated (Dierssen, 1994; Hampicke, 1994; Meineke, 1994). We regret this fact, especially because either the performed translation of area-related nature conservation aims may lead to unwelcome results which were not even intended, or because some important disturbance variables could have been ignored. Only operational monitoring permits the necessary feedback in order to make changes or to optimise the use of methods even beyond the case study.

Up to now, only nature conservation problem fields have been mentioned. But there is no doubt about the fact that in everyday experience such legal and fiscal restrictions, which haven't been explained in this context, could be even more serious, because the conservation administration is usually insufficiently equipped in regard to personnel. A logical and comprehensive consequence would be the need to simplify the administrational work and on the pragmatic level to facilitate the decision-making process. Strictly speaking, an insufficiently equipped administration is normally not pushing primarily towards a high quality level of scientific analysis. They more likely prefer simple administration procedures, and they are willing to accept by force a decrease of quality concerning the content and realisation of conservation measures. It is incontestably necessary to ameliorate the structural and personnel situation in the conservation administrations. This need is not a plausible excuse for the lack of forward-looking, ameliorated conceptional approaches concerning scientific analysis, assessments and applications.

4. Contents and levels of application in ecosystem research

Ecosystems are working models of biocoenosis and biotopes which interact and which to a certain degree have the ability of self regulation (Holling, 1973, 1984; Müller, 1997; Müller et al., 1997b).

These interactions concern among others: flow of materials and metabolic circulation, energy flow, food web and ecological niche, competitive relations and dependency levels like symbiosis and parasitism, toxic effects, self regulation mechanisms and last but not least, reactions on human interventions (Fränzle et al., 1995). Some of the most important tasks and aims in ecosystem research being oriented at application will be described in the following draft:

- Analysis and interpretation of structure and dynamism of ecosystems being representative for a region, especially for its regulation of radiation, the density of material and energy fluxes, the storage capacity and consumption of material as well as of its biotic hierarchy and organisation.
- Characterisation of the interactions between diversity, productivity and stability of different types of ecosystems (Gigon, 1983; Maxwell and Costanza, 1993; Okasen, 1988; Van der Maarel and Sykes, 1993).
- Development of equipment for the simulation and prognosis of ecological processes on different scales (Gauch, 1982; Frank et al., 1994; Jørgensen, 1990; Schröder et al., 1994).
- Development of an integrative system for information and assessment using procedures of ecosystem research and environment observation as well as an efficient use of environment sample banks in order to obtain information about the balance of noxious effects and anthropogenic material fluxes (Pietsch, 1992; Seggelke, 1994).
- Elaboration of criteria for the definition of gradual sensibilities (current load and limits of capacity, potentials of release) of ecosystems, specification of developmental models and development of strategies in order to release in their potentials restricted or disturbed, with the aim of a sustainable development of relationships between human beings and the environment (SRU, 1994).

- Development of operational methods to check technical and economic plans and equipment on their ecological tolerance and to elaborate generalised groups of scenarios (alternatives in planning).
- Logistic and conceptional support concerning environment planning (concretisation of assessment scales, comparisons of the existing and the intended quality of systems, development and simulation of alternatives in planning), ecosystem management (innovative translation of operational aims, development of instruments for applied plans and decisions) and efficiency control (analysis of slow changes in the whole system due to extremely fluctuating single parameters, causal monitoring in the case of changes, the control of the management outcomes, refining the possibilities of taking prognosis).

Ecosystem research uses the following procedures and approaches to cope with these complex and extensive work fields:

- An in time and space highly integrated procedure to obtain environmental parameters, ecophysiological data of populations which are representative for certain organisms as well as for material, energy and information fluxes between habitat and biocoenosis. Data sampling is mainly done by measuring devices which register and memorise these data automatically. In fact, this demands a lot of time spent on control and maintenance.
- Centralised data banks are responsible for the evaluation and presentation of given results/data. The optimal use of software programs would make it possible, on the one hand, to evaluate the given data in a correct statistical and operational way. On the other hand, they would make it possible to illustrate for example the overlapping of spot and area data, which is necessary for the different evaluation steps ('geographical information systems', GIS) (Zölitz-Möller et al., 1993). After all, they allow different approaches for process-oriented modelling and a simulated calculation on different scale levels (organism, biocoenosis and biotope, part of a landscape) (Ebenhöh, 1993; Heil and Bobbink, 1993; Lauenroth et al., 1993; Reiche, 1995).

Adequate work procedures which follow the 'bottom-up' principle (inductive way) will make it possible to develop hypothesis about system properties and to define and test their capacity and limits of making statements for example about simulated calculations. In basic research, this procedure is first of all serving for the development/evolution of ecosystem theories which are easy to understand. The 'top-down' method (deductive way) uses these theories to generate universal planning decisions concerning landscape development, which, in fact, are geared to concrete parts of landscape and parts of biotopes and which have to be summarised to more precise and regional applicable models. In addition, this configuration of models and working methods being developed for analysis and hypothesis support can also be used by people working on practical and planning problems.

5. Characterisation of the working fields, where nature conservation may derive benefit from methods and results in ecosystem research

In the following section, different examples will be used to illustrate the tasks and objectives of ecosystem research being presented above with the focus on its potential possibilities of application to the field of nature conservation. They could help to prepare political and administrative decisions on a rational scientific basis.

During the analysis of ecosystems, it may often be possible to work out targeted prognosis about dynamic processes in ecosystems, which are the result of a combination of data dealing with the interactions of influencing factors and estimated implications of generalised statements. An example, is for instance, the structural and functional transformation of sand heath as nutrient sources during the period of extensive use, into nutrient sinks after the abandonment of usage, and a nutrient emission occurring simultaneously. This will further provide the chance to derive adequate strategies for nature conservation and to avoid inadequate ones. In unfavourable cases, it would be impossible to give a simple and definite prognosis about complex and non-linear interference between registered parameters which were considered to be important, like processes of denitrification on different kind of soils interfering with water and oxygen budget/regulation, density, temperature and acidity (e.g. Mogge, 1995). Nevertheless, long-term studies, even in these cases, at least inform about size and amplitude of the measured or registered data and can be evaluated for the estimation of risks. More detailed analysis of terrestrial systems, partly linked with stringent recommendations for further procedures, are for example available for forest, fen, field and grassland ecosystems as well as for those belonging to arctic and alpine regions (Andrén et al., 1990; Bliss et al., 1981; Ellenberg et al., 1986; Rychnowská, 1993; Sonesson, 1980).

The procedure of modelling and detailed prognosis about general development could be based on data analysis. Presently, there is a growing interest in simulations and the related simulation models used for ecosystem research and nature conservation, in order to predict for example changes in systems as a result of disturbances (Brzeziecki et al., 1995; Goodall, 1989). Dynamic developmental models register changes in vegetation by means of long-term studies. On the other hand, statistical balance models make it possible to predict the current state of vegetation at a certain point in time and at fixed/certain environmental conditions, thus, to describe scenarios (Binz and Wildi, 1988; Box, 1981; Fischer, 1994; Lindacher, 1996). Concerning the abiotic domain, models of water regulation and material flow will ameliorate the ability to provide prognosis and will simplify the registration of data; concerning the biotic domain, they will ameliorate the prediction of reaction patterns of populations (Frank et al., 1994; Wissel and Stephan 1994).

As regards the definitions of aims we have to agree, fundamentally, with the ideas of Maturana (e.g.1992) who claims that scientific knowledge isn't able to give a universally valid, independent and objective image of reality, nor it is possible to derive from it compulsory social, that is to say socially responsible or ethic norms of behaviour. However, even if the study of science and scientific knowledge doesn't provide us automatically with wisdom[1], it still doesn't exclude it. In other words, a growing knowledge of general properties, functional mechanisms and proceedings in highly complex (eco-)systems makes it easier for us to check and to enlarge our understanding of nature on the basis of a general/universal system theory and thus, to get a more rational access to problems concerning nature and environment (Henle and Kaule, 1991; Luhmann, 1990).

Having this in mind, we may confirm that our knowledge in ecosystem research makes a quite constructive contribution to the evaluation of risks and to the operationalisation of definitions concerning nature conservation, which could help to operationalise controversial opinions. This contribution of ecosystem research should for instance include, in addition to the elaboration of an extended theoretical basis, a more profound and specific data basis (data banks, geographical information systems, support of decision-taking procedures by expert systems), which could limit the risks of misjudgements on details.

Assessments have a central significance for scientific statements concerning nature, an operationalisation of assessment procedures on the basis of integral area evaluations are obligatory. Even an optimal data basis won't resolve the fundamental problem to translate ordinal scales ('good - normal – bad') into quantifying cardinal scales ('normal is 3 times better than bad'). In the meantime, modern ('fuzzy') procedures of modelling have been

[1] In other words: a striving for respect towards *the person next to you*, honour in the sense of renunciation of power, acceptance of social coexistence, sincerity and reality (Maturana, 1992).

developed, which probably open up a new perspective for resolving these problems (Salski and Sperlbaum, 1991).

Furthermore, the results of ecosystem analysis may enormously help to give a clear definition of priorities and of the scope of action after the estimation of states and risks. It is possible to define 'critical loads' concerning supported information about current nutrient inputs in oligotrophic systems like fens and heathlands combined with information about the fluxes and reservoirs (sources and sinks) of the most important nutrients in the systems themselves and their significance for vegetation distribution. In case the critical loads pass beyond their limits, this will induce radical changes in the system (Aerts and Heil, 1993; Heil, 1984; Heil and Bobbink, 1993; Lütke Twenhöven, 1992; Nilsson and Grenbelt, 1988).

Nowadays, ecosystem research is meant to occupy a central position in the field of conflict analysis resulting first of all from a direct over utilisation of the systems themselves or from remote effects (erosions, emissions). This is valid for different scales starting for example with a projection of regional nutrient outputs from plough-land in relation to the intensity of cultivation and cover crop (Schimming et al., 1995) and ending up with nation wide and global energy, nutrient and noxious fluxes (Forstliche Bundesversuchsanstalt Wien, 1992; Enquête-Kommission 'Schutz der Erdatmosphäre', 1994; Staaf and Tyler, 1995). The advantage of an ecosystem approach in comparison with sectoral approaches (for instance an isolated way of looking at single protective resources like the selective protection of special animal and plant species) is above all the possibility to clearly identify the interactions between abiotic and biotic system compartments as well as interactions within these compartments. And, referring to the 'sustainable protection of the nature regulation pathways' demanded by the legislator, to concentrate on an integral conservation of the ecosystemic integrity (Kleyer et al., 1992; Woodley et al., 1993).

It is a necessary prerequisite to do precise and subtle conflict analyses and assessments in order to realise the aims of conservation and to avoid disturbances and degradation of the system and of the conservation objects. The careful application of models being developed at selected systems, fixes the necessary frame of investigation: which data must be registered, what will be the spatial dispersal and what kind of time scale has to be chosen in order to get valid and transferable results? These kinds of models which have been developed for the causal explanation of facts cannot automatically be used for planning predictions. It is necessary as well to integrate socially accepted norms, which serve as important standards in a decision-making process, into these models (Peters, 1997).

In a further step, we have to define the marginal conditions for an intended realisation of conservation measures by means of local investigations in each specific area (in the biocoenotic field: biological conditions for population and production, in the abiotic field: nutrient potential and water regulation data (e.g. Schrautzer et al., 1996)). The development of consulting systems ('expert systems') could be seen as a supporting measure to help at least in reducing mistakes in planning and in application and to state environmental standards more precisely in case of an insufficient data basis (Asshoff, 1996). The realisation of advice, which can be derived from ecosystem research, is a continual process. This process could, under optimal conditions, already be initiated during the final period of running projects, for instance during the project of ecosystem research at Schleswig-Holsteinisches Wattenmeer (Wadden Sea, Schleswig-Holstein) (Wilhelmsen, 1994).

Finally, a monitoring of changing systems should be based on knowledge that has been acquired by scientific programs running a long time. Concerning the measuring design, the spatial and temporal effort as well as the measuring parameters important for the systems involved in the monitoring, a standardisation is useful to transfer the data and results which are scientifically safeguarded, to other regions. At the moment, the current measuring systems are not sufficiently developed to obtain fundamental disturbance variables in ecosystems like depositions of nutrient and noxious substances (Spranger, 1992). As in complex systems, it won't be possible to make as definite prognosis about future constellation and structure as about energy and material fluxes, and as a comprehensible simulation in entire models won't be carried out. Only a long-term and interdisciplinary monitoring allows the premature assessment of even slight changes in order to develop adequate landscape planning concepts deriving from this program (Hörmann, 1995).

6. Prospective

Regarding the characterisation of current nature conservation problems, different conflict fields have been mentioned which need satisfying solutions.

Discussing instruments and generalising results of ecosystem research in their relevance for nature conservation, it has been mentioned briefly in which part of ecosystem research a possible and useful support could be expected by performing more extensive problem analysis and prevention planning. The focus was placed on aspects like impact regulation, protection and development of priority areas in nature conservation, environmental compatibility check and landscape planning. Additionally, national up to global problem fields including their effects on ecosystems (for example research on forest damage, eutrophication effects, ecotoxicologic problem fields, 'global change' of radiation and temperature) can only be analysed and assessed by applying the methods of modern ecosystem research in order to derive, as much as possible, exact details for prognosis in space and time in order to develop scenarios of planning and damage defence (Hörmann, 1995; Schellnhuber and Sterr, 1993; Ulrich and Pankrath, 1983; Woodley et al., 1993).

The results of such research programs are not only relevant for the application of regional concepts for nature conservation, but they require that politicians are responsible for environmental, administrational and economical questions to find adequate measures for the development of concepts, which are suitable for a constant and ecologically practicable use of the environment, and finally to provide the necessary financial support.

7. Summary

Nature conservation is socially accepted, because it seems to be necessary. But the operational application of general objectives for nature conservation still faces a great number of administrational and scientific problems. *Applied* ecosystem research may substantially attribute to the amelioration of scientific work and partly to the simplification of the environment and nature conservation planning. It is mainly necessary to reinforce the knowledge about the functioning of ecosystems instead of their compartments by means of showing the indirect effects within working structures, illuminate the meaning/importance of processes going on within and between systems and to include them in prognosis, assessment scales and developmental concepts, and to form the integrative basis for a sectional approach of investigation, assessment and treatment.

REFERENCES

Aerts, R. and Heil, G. W. 1993. Heathland: patterns and processes in a changing environment. Dordrecht

Andrén, O.; Lindberg, T.; Paustian, K. and Rosswall, T. (Eds.) 1990. Ecology of Arable Land. Copenhagen

Asshoff, M. 1998. Expertensysteme in der biozönologischen Modellierung: Ein Beratungssystem zum Feuchtwiesenmanagement. In: *ECOSYS - Beiträge zur Ökosystemforschung* Suppl. 26

Auhagen, A. 1995. Biologische Daten zur Bewertung von Eingriffen in Natur und Landschaft und zur Bemessung einer Ausgleichsabgabe. In: *Schriftenreihe für Landschaftspflege und Naturschutz* 43: S.281-305

Beck, U. 1995. Kleine Anleitung zum ökologischen Machiavellismus. In: *Die feindlose Demokratie*. Stuttgart
Binz, H. R. and Wildi, O. 1988. Das Simulationsmodell MAB-Davos. In: *Schlußbericht Schweizer MAB-Programm 33*.Bern
Bliss, L. C.; Heal, O. W. and Moore, J. J. (Eds.) 1981. Tundra ecosystems: a comparative analysis. Cambridge
Box, E. O. 1981. Macroclimate and plant forms: an introduction to predictive modelling in phytogeography. In: *Tasks for vegetation science 1*. Den Haag
Brzeziecki, B.; Kienast, F. and Wildi, O. 1995. Modelling potential impacts of climate change on the spatial distribution of zonal forest communities in Switzerland. *Journal of Vegetation Science* 6: 257-268
Dierßen, K. 1994. Was ist Erfolg im Naturschutz? *Schriftenreihe für Landschaftspflege und Naturschutz* 40: 9-23
Ebenhöh, W. 1993 Coexistence of similar species in models with periodic environments. *Ecological Modelling* 68: 227-247
Ellenberg, H.; Meyer, R. and Schauermann, J. (Hrsg.) 1986. Ökosystemforschung. Ergebnisse des Solling-Projekts. Stuttgart
Enquête-Kommission 'Schutz der Erdatmosphäre' des Deutschen Bundestages (Hrsg.) 1994. Bd. 1, Landwirtschaft, 2 Teilbände.- Bonn
Fischer, H. S. 1994. Simulation der räumlichen Verteilung von Pflanzengesellschaften auf der Basis von Standortkarten. Dargestellt am Beispiel des MaB-Testgebietes Davos. In: Veröffentlichungen des Geobotanischen Instituts der ETH Zürich 122.- Zürich
Forstliche Bundesversuchsanstalt Wien 1992. Österreichische Waldboden-Zustandsinventur.- 2 Bde.- Wien
Fränzle, O.; Straskraba, M. and Jørgensen, S. E. 1995. Ecology and ecotoxicology. In: *Ullmann's Encyclopedia of Industrial Chemistry B7*, pp. 19-154
Frank, K.; Drechsler, M. and Wessel, C. 1994. Überleben in fragmentierten Lebensräumen - Stochastische Modelle zu Metapopulationen. *Zeitschrift für Ökologie und Naturschutz* 3: 167-178
Gauch, H. G. 1982. Multivariate analysis in community ecology. Cambridge
Gigon, A. 1983. Über das biologische Gleichgewicht und seine Beziehungen zur ökologischen Stabilität. *Berichte des Geobotanischen Institutes der ETH Zürich,* Stiftung Rübel 50: S. 149-177
Goodall, D. W. 1989. Simulation modelling for ecological applications. *Coenoses* 4(3): S. 175-180
Hampicke, U. 1994. Die Effizienz von Naturschutzmaßnahmen in ökonomischer Sicht. *Schriftenreihe für Landschaftspflege und Naturschutz* 40: 269-290
Hansson, L. and T. B. Larsson 1997. Conservation of boreal environments: a completed research program and a new paradigm. *Ecological Bulletins* 46: 9-15
Harts, J., Ottens, H.F.L., Scholten, H.J. (Eds.), 1993. Conference Proceedings, EGIS Foundation. Utrecht
Heil, G. W. 1984. Nutrients and the species composition of heathlands. Utrecht
Heil, G. W. and Bobbink, R. 1993. 'Calluna', a simulation model for evaluation of impacts of atmospheric nitrogen deposition an dry heathlands. *Ecological Modelling* 68: 161-182
Henle, K. 1994. Naturschutzpraxis, Naturschutztheorie und theoretische Ökologie. *Zeitschrift für Ökologie und Naturschutz 3*:139-153
Henle, K. and Kaule, G. (Hrsg.)1994. Arten- und Biotopschutzforschung für Deutschland. Jülich
Hörmann, G. (Red.) 1995. Auswirkungen einer Temperaturerhöhung auf die Ökosysteme der Bornhöveder Seenkette. *ECOSYS 2: 246 S.*
Holling, C. S. 1973. Resilience and stability of ecological systems. *Annual Review of Ecology and Systematics* 4: 1-23
Holling, C. S. 1983. Terrestrial Ecosystems. Local surprise and global change. In: Clark, W. C. and Munn, R. E. (Eds.) *Sustainable development of the biosphere.*Cambridge, pp 292-317
Jessel, B. 1994. Methodische Einbindung von Leitbildern und naturschutzfachlichen Zielvorstellungen im Rahmen planerischer Beurteilungen. In: Laufener Seminarbeiträge 4/94: S. 53-6
Jørgensen, S. E. 1990. Modeling in ecotoxicology. Amsterdam
Jørgensen, S. E. 1997. Thermodynamik offener Systeme. In: Fränzle, O.; Müller, F. and Schröder, W. (Hrsg.) Kap. III-1.6 *Handbuch der Ökosystemforschung*. Landsberg.
Jørgensen, S. E. and Nielsen, S. N. 1998. Thermodynamic orientors: exergy as a goal function in ecological modeling and as an ecological indicator for the description of ecosystem development. In: Müller, F. and Leupelt, M. (Eds.): *Eco-targets, goal functions and orientors.* New York, pp 62-88
Kleyer, M.; Kaule, G.; Henle, K. 1992. Landschaftsbezogene Ökosystemforschung für die Umwelt- und Landschaftsplanung. *Zeitschrift für Ökologie und Naturschutz 1:* 35-50
Lauenroth, W. K.; Urban, D. L.; Coffin, D. P.; Parton, W. J.; Shugart, H. H.; Kirchner, T. B. and Smith, T. M. 1993. Modeling vegetation structure-ecosystem process interactions across sites and ecosystems. *Ecological Modelling* 67: 49-80
Lindacher, R. 1996. Verifikation der potentiellen natürlichen Vegetation mittels Vegetationssimulation am Beispiel der TK 6434 'Hersbruck'. *Hoppea* 57: S. 5-143
Lütke Twenhöven, F. 1992. Untersuchungen zur Wirkung stickstoffhaltiger Niederschläge auf die Vegetation von Hochmooren. In: *Mitteilungen der Arbeitsgemeinschaft Geobotanik Schleswig-Holstein/Hamburg* 44. Kiel
Luhmann, N. 1990. Ökologische Kommunikation. Opladen
Maarel, E. van der and Sykes, M. 1993. Small scale species turnover in a limestone grassland: the carousel model and some comments on the niche concept. *Journal of Vegetation Science* 4: 179-188
Maturana, U. 1992 Wissenschaft und Alltagsleben: Die Ontologie der wissenschaftlichen Erklärung. In: Krohn, W. and Küppers G. (Hrsg.) *Selbstorganisation - Aspekte einer wissenschaftlichen Revolution*. Wiesbaden, S.

107-138
Maxwell, T. and Costanza, R. 1993 An approach to modelling the dynamics of evolutionary self-organization. *Ecological Modelling* 69:149-161
Mayer-Tasch, P. C. 1985. Aus dem Wörterbuch der Politischen Ökologie. München
Meineke, U. 1994. Effizienzkontrollen von Schutz- und Pflegemaßnahmen im Spannungsfeld von wissenschaftlichen Ansprüchen und administrativen Möglichkeiten anhand der Praxis in Baden-Württemberg. *Schriftenreihe für Landschaftpflege und Naturschutz 40*: S. 229-242
Mogge, B. 1995. N_2O-Emissionen und Denitrifikationsabgaben von Böden einer Jungmoränenlandschaft in Schleswig-Holstein. *ECOSYS Suppl. 9*
Mohr, H. 1977. Lectures on structure and significance of science. Berlin, Heidelberg, New York
Mohr, H. 1987. Natur und Moral - Ethik in der Biologie. Dimensionen der modernen Biologie 4. Darmstadt
Müller, C. and Müller, F. 1992. Umweltqualitätsziele als Instrumente zur Integration ökologischer Forschung und Anwendung. *Kieler Geographische Schriften* 85: 131-166
Müller, F. 1997. State-of-the-art in ecosystem theory. *Ecological Modelling* 100: 135-161
Müller, F.; Müller, C. and Dierßen, K. 1997a. Ökologische Gutachten - eine kritische Aufnahme von Problem- und Konfliktfeldern. *ECOSYS 6*: S. 103-121
Müller, F.; Breckling, B.; Bredemeier, M.; Grimm, V.; Malchow, H.; Nielsen, S. N. and Reiche, E. W. 1997b. Ökosystemare Selbstorganisation. In: Fränzle, O., Müller, F. and Schröder, W. (Hrsg.): Kap. III-2.4
Nilsson, J. and Grennbelt, P. (Eds.) 1988. Critical loads for sulphur and nitrogen. Stockholm
Okasen, L. 1988. Ecosystem organization: mutualism and cybernetics or plain darvinian struggle for existence? *American Naturalist* 131: 424-444
Peters, R. H. 1991 A critique for ecology. Cambridge
Peters, W. 1997. Zur Theorie der Modellierung von Natur und Umwelt.- Berlin
Pietsch, J. 1992. Kommunale Umweltinformationssysteme - Eine Standortbestimmung.- In: Du Bois, W. and Zimmermann, K. O. (Hrsg.) *Umweltdaten in der kommunalen Praxis*. Taunusstein
Reiche, E.-W. 1995 Ein Modellsystem zur Erstellung regionaler Wasser- und Stoffbilanzen.- In: Ostendorf, B. (Hrsg.) *Räumlich differenzierte Modellierung von Ökosystemen.* In: Bayreuther Forum Ökologie 13: S. 121-128
Rychnovská, M. (Ed.) 1993. Structure and functioning of seminatural meadows. Amsterdam
Salski, A. and Sperlbaum, C. 1991. Fuzzy logic approach to modelling in ecosystem research. In: Bouchon-Mounier, B.; Yager, R. R. and Zadeh, L. A. (Eds.) *Uncertainty in knowledge basis.* LVCS 521: pp520-527
Schellnhuber, H. J. and Sterr, H. (Hrsg.) 1993. Klimaänderung und Küste. Berlin, Heidelberg, New York
Scherner, E. R. 1995. Realität und Realsatire der 'Bewertung' von Organismen und Flächen. *Schriftenreihe für Landschaftspflege und Naturschutz 43*: S. 377-410
Schimming, C.-G.; Mette, R.; Reiche, E.-W.; Schrautzer, J. and Wetzel, H. 1995. Stickstoffflüsse in einem typischen Agrarökosystem Schleswig-Holsteins. Meßergebnisse, Bilanzen, Modellvalidierung. In: *Zeitschrift für Pflanzenernährung und Bodenkunde* 158: S. 313-322
Schrautzer, J.; Asshoff, M. and Müller, F. 1996 Restoration strategies for wet grasslands in Northern Germany. *Ecological Engineering* 7: 255-278
Schröder, W. 1998. Ökologie und Umweltrecht als Herausforderung natur- und sozialwissenschaftlicher Lehre. In: Daschkeit, A. and Schröder, W.: *Umweltforschung quergedacht. Perspektiven integrierter Umweltforschung und -lehre.* Heidelberg; S. 329-357
Schröder, W.; Vetter, L. and Fränzle, O. (Hrsg.) 1994. Neue statistische Verfahren und Modellbildung in der Geoökologie. Wiesbaden
Seggelke, J. 1994. Ganzheitlicher integrierter Modellansatz für DV-gestützte Informationssysteme am Beispiel Umweltschutz.- Polykopie
Sonesson, M. (Ed.) 1980. Ecology of a subarctic mire. *Ecological Bulletins* 30
Spranger, T. 1992. Erfassung und ökosystemare Bewertung der atmosphärischen Deposition und weiterer oberirdischer Stoffflüsse im Bereich der Bornhöveder Seenkette. *ECOSYS* Suppl. 4
SRU (Rat von Sachverständigen für Umweltfragen) 1994. Umweltgutachten 1994 für eine dauerhaft umweltgerechte Entwicklung. Stuttgart
Staaf, H. and Tyler, G. 1995. Effects of acid deposition and tropospheric ozone on forest ecosystems in Sweden. *Ecological Bulletins* 44
Svirezhev, Y. 1998. Thermodynamic orientors: how to use thermodynamic concepts in ecology. In: Müller, F. and Leupelt, M. (Eds.) *Eco-targets, goal functions and orientors.* New York, pp101-121.
Ulrich, B. and Pankrath, J. (Eds.) 1983. Effects of accumulation of air pollutants in forest ecosystems. Dordrecht.
Wilhelmsen, U. (Red) 1994. Ökosystemforschung Schleswig-Holsteinisches Wattenmeer - Eine Zwischenbilanz. *Schriftenreihe des Nationalparks Schleswig-Holsteinisches Wattenmeer* 5
Wissel, C. and Stephan, T. 1994. Bewertung des Aussterberisikos und das Minimum-Viable-Population-Konzept. *Zeitschrift für Ökologie und Naturschutz 3*: S. 155-159
Woodley, S.; Kay, J.; Francis, G. (Eds.) 1993. Ecological integrity and the management of ecosystems. Ottawa
Zölitz-Möller, R.; Schleuss, U. and Heinrich, U. 1993. Integration of GIS, data bank and simulation models for ecosystem research and environmental assessment.

III.1.2 Applications of Ecological Theory and Modelling to Assess Ecosystem Health

D. J. Rapport and R.H.H. Moll

1. Evolution of the concept of health from organism to ecosystem

The concept of health predates that of sustainability by millennia. It is a concept that has proved its worth by the test of time, not withstanding the endless quest for the true meaning and definition of health (Porn, 1984). The extension of the notion from that of the organism to the community and the whole ecosystem is of relatively recent origin – perhaps at best dating back to Hutton's efforts of approximately two hundred years ago (Calow, 1995) to characterise the dynamics of the biosphere.

In recent times, it has been practical problem solving that has driven the extension of the concept from the individual to the population and beyond. Veterinarians for example are confronted daily with questions of disease in organisms, but they soon recognised that the health of the herd (or population) was itself a legitimate concept (Ribble et al, 1997; Waltner-Toews, 1996). The concept yet extends further, and the writings of Aldo Leopold (1941) foreshadowed the explosive efforts of recent decades to examine the health of large-scale ecosystems (Rapport, Gaudet and Calow, 1995).

What does it mean to speak of the health of a community of interacting species with their environment? Are there single measures of the collective health of all the parts of the system? How can one even address the issue of health in a system in which death is a normal part – and the absence of death (of individuals) would lead to the elimination of the integrity of the system? To do this, one must let go of the tempting analogy of ecosystem as an organism – for it is clearly not that. Rather, one must examine the properties of the complex system that enables it to persist and evolve in its natural way.

Once this expansion of thinking has taken place, it is readily apparent that many, if not all the environmental problems of the day (Tolba et al, 1992) are in fact questions, if not of ecosystem health, or ecosystem pathology – which is really the other side of the coin. What presents itself today, in most regions of the globe, are dysfunctional ecosystems resulting from the intrusion of human activity. For example, in a recent special issue in *Science* (*Science*, Vol. 277, 25 July 1997) articles provide evidence that there are no places left on Earth that are not affected by human activity and that eventually all ecosystems will have to be managed to one extent or another, and to do this well, managers will need sound scientific advice.

One article by Vitousek et al (1997) demonstrates clearly the global impact of human activity on earth's biotic and abiotic processes with the conclusions that between one-third and one-half of the earth's land surface has been transformed by human action. They estimate that atmospheric carbon dioxide has increased by 30 percent since the beginning of the Industrial Revolution and that more atmospheric nitrogen is fixed by humanity than by all natural terrestrial sources combined. Furthermore, they conclude that more than half of accessible surface fresh water is being used by humans and that a quarter of the Earth's bird species have been driven to extinction.

In another article by Dobson et al (1997), a frank assessment is made of the possibilities of restoring degraded land which resulted from human conversion of natural habitats into industrial and agricultural landscapes. Using mathematical modelling, the authors demonstrate that the dynamics of habitat conversion may be represented by simple compartmental susceptible-infection-recovered (SIR) model structures used in epidemiology. Conversely, these models may be used to demonstrate the possibility of ecological restoration of currently degraded land and that it can be demonstrated that the coupling of management intervention with the power of natural processes to restore their ecological functions produces new habitats for biodiversity. For example, the restoration of 10,000 ha of barren land around the nickel smelters in Sudbury, Ontario (Winterhalder, 1996) is a case in point.

In this brief chapter, we examine the role of ecological theory and modelling in contributing to a major effort to better characterise ecosystem health, and to inform strategies for preventive health care of the earth's ecosystems.

2. Diagnosis of ecosystem functions in a health context

Rapport (1997) raised a key practical question: What are the criteria for sustainable ecosystems and landscapes to supply ecosystem services which allow society to meet its goals and aspirations? To answer this we need to integrate socio-economic, biological and health concepts and values within a holistic 'point of view' which we have called ecosystem health. This is regarded as a new direction for environmental management. This 'point of view' is illustrated using ideas developed in the last twenty or so years in forest ecosystem management and is summarised in four ecological theories proposed by Boyce (1995).

Three principles, continuity, irreversibility and structure form the basis for managing states of organisation of forested landscapes and are used to formulate the four ecological theories of Boyce (1995).

The **principle of continuity** implies that no state of organisation of a biological system is static. A biological process cannot pass from one state to the next without passing through all intermediate states that are also subject to the process of change. For example a forested landscape develops step by step from one to another state organisation. They do not randomly jump from state to state. Each succeeding state is dependent on a preceding state. A stand of old trees must pass through seedling, sapling, small tree, large tree and the intermediate states. These kinds of changes are called ecological succession, growth, or transformations (Boyce, 1995) .

The **principle of irreversibility** states that all future states of a biological system are different from past states. Natural changes in forested landscapes change biological systems state by state into states that never occurred before. Natural mortality and harvest of trees form canopy openings that irreversibly change forest landscapes to states of organisation that never before occurred. Consumption of plants and animals by predators and humans irreversibly change states of organisation.

Continuity and irreversibility explain why ecological succession is irreversible and never exactly repeated; why strip-mined lands are changed to desired states, not restored to original states; why wilderness cannot be preserved as first observed; why forested landscapes cannot be kept in one constant state of organisation (Boyce, 1995)

Finally, the **principle of structure** serves to explain the way in which parts of a biological system are linked and how energy, materials and information flow through the biological system. The flows of energy and materials determine the dynamics and behaviour of the biological system (Boyce, 1995). The structure of organisms determines whether essential variables are kept within the limits of life (Ashby, 1973). For example, natural mortality and the harvest of trees change the structure and dynamics of the stand. Death of a sunlit tree changes reception of the radiation and flows of organic compounds to soil, animals, and forest floor. Natural events and the harvest of trees shift the formation of organic compounds from a few individual trees with high canopies to many species and individuals of seedlings and plants on the forest floor. These structural changes alter the flows of carbon compounds to the soils, transfer of water nutrients from one group of species to another. These are some of the ways natural events and scheduled harvests of trees change states of organisation of forested landscapes.

Structure in a forested landscape determines aesthetics, habitats, biological diversity, and the availability of timber and fuel wood. States of organisation of forested landscapes are determined by dynamic structures, called stands, distributed in time and space.

The theories are restated from Boyce (1995) :

Theory 1. *Each living organism and its environment form an* ***individualistic system*** *with the goal of survival and dynamics of systems with negative feedback control.*

It is generally agreed that individualistic systems are cybernetic in nature in that their behaviour is governed by goal seeking negative feedback control loops. The goal directed behaviour of organisms interacting with the environment may be explained by the self regulating process of homeostasis. The population is regulated close to an equilibrium level through communication and control mechanisms between the population and the environment. Forested landscapes, stands and communities are composed of individualistic systems, each self-organising in relation to environmental changes.

For example, small trees, seedlings and other plants die in the understory of sunlit canopies. There is no evidence that the sunlit canopies (the community of individualistic systems) behave in a central organising fashion to eliminate or favour understory species. As pointed out by Boyce (1995), Sheffield and Thompson (1992) and Lorimer and Frelich, (1994), without human intervention, formations of canopy openings in forest landscapes are related to aimless, unpredictable mass flows of energy and materials.

Theory 2. *The irreversibility of biological processes results in mortality of individualistic systems whose survivors create an aimless* ***community organisation.***

Despite the attempt to extend the cybernetic paradigm used in individualistic systems to the ecosystem level by systems ecologists H.T. Odum (1960), E.P. Odum (1969), Van Dyne (1966, 1969a, 1969b) , Patten (1964), Watt (1968) , and others, it has been argued that ecosystems are not cybernetic in nature. In particular, Engelberg and Boyarsky (1979) have argued that at the ecosystem level the feedback couplings are not strong enough to support the goal directed behaviour as seen in individualistic systems. They assert that the feedbacks in ecological systems are mere brute flows of energy and are not the transmissions of information found in so-called true cybernetic regulation.

Theory 3. ***Flows of energy and materials*** *are unidirectional and have the dynamics of systems with positive feedback control.*

Forested landscapes depend on the mass flows of solar energy, carbon dioxide, oxygen and other materials. The delay and dissipation of energy and materials by organisms is not easily explained by the theories of equilibrium thermodynamics. As stated by Spanner (1964), if the universe is moving toward thermodynamic equilibrium, then as far as the biologist is concerned, that means death to all organisms. This theory states that the flows through the forest ecosystem are dissipative rather than conservative and are toward increased entropy of the universe. Such processes which do not pass through successive equilibrium states are considered in the studies of non-equilibrium thermodynamics which is also called the thermodynamics of irreversible processes (Wisniewski, Staniszewski, and Szymanik, 1976). As summarised by Boyce (1995), this theory is a statement that relates the behaviour of organisms and communities to the thermodynamic principle of increasing entropy of the universe. Organisms and communities of organisms which we call forests are irreversible systems that are in constant transformation from one non-equilibrium state to another and never return to the preceding state.

Theory 4. *States of organisation of forested landscapes determine* ***baskets of benefits*** *that satisfy human desires.*

States of organisation change over time as individualistic systems strive to keep essential variables within the limits of life and as mortality changes flows of captured energy and materials. An increase in the variety and quality of benefits results only from increasing the variety of stands classified by forest type, age, and area class. By directing the distribution of stands by forest type, age, and area classes, managers enhance aesthetics, habitats, timber, fuel wood, cash flows, and biological diversity (Boyce, 1995)

This theoretical framework will be used for discussing new directions in forest ecosystem management and how it may be applied in the assessment or diagnosis of forest ecosystem health. According to Boyce (1995) these theories provide the basis for the generalisation that a natural, unmanaged forest is an aimless system and that management can convert this system into a goal directed one. Also, they provide the foundations for simulation models in landscape forestry/forest ecosystem health. We will provide examples of these models to illustrate how they can be used to address the health status of forest ecosystems. The method can then be extended to other ecosystems. We are therefore concerned with how simulation modelling and ecological theory can be used to:

1) establish the 'norms' for operating characteristics for ecosystems, catchment areas, landscapes and regional environments,
2) ascertain what abnormalities might develop with various stresses or combination of stresses, and
3) identify causal relationships in the event of unexpected deviations.

2.1 Establishing the norms.

As pointed out by Boyce (1995) in theory 4 above, the states of organisation of a biological system at any point in time determines human access to benefits; that is, fulfils a policy which we call forest ecosystem health. The management of the biological system to fulfil a policy is the basis for ecosystem health. A healthy ecosystem is one that is in a state of organisation that approximates the state that fulfils a policy. Thus, ecosystem health is a metaphor for defined states of organisation of a biological system that satisfies human desires. This metaphor of ecosystem health defines baskets of benefits, including timber, clothing, shelter, aesthetics, habitats and endangered species for example.

Management to fill the ecosystem health basket requires a variety of states of organisation. Ecosystem health is management. It is the art and science of organising (forest) ecosystems to capture and pipe solar energy for the production of desired goods, services and effects. For example, many animals require stands of one age class or type for feeding and stands of other age classes and types for shelter, escape and reproduction. Aesthetics and wilderness values are related to patterns of stands. Scenic quality is related to order and diversity of stands dispersed over time and space. The model for landscape forestry says changes in states of organisation of the forested landscapes changes the availability of benefits.

How then can modelling and ecological theory be used to develop 'norms' for operating ecosystem characteristics? A practical use of the ecological theories proposed by Boyce (1995) is to design management schemes that direct self-organising individualistic systems to bring about desired states of forest organisation. The management models developed in Boyce (1995) are for the purpose of simulating the art of changing an initial state of organisation of a forested, step by step, toward a state which fulfils a policy which constitutes 'Ecosystem Health'.

For example, directed mortality in a forest by regulating the rates of timber harvest and the size of openings formed by harvesting result in the maintenance of a desired order in the distribution of habitats. Directed mortality is the primary method to bring about habitats for desired game animals, to increase stream flow to produce fuel wood and timber, and to change the aesthetic appearance of forests. Thinnings, changes in the composition of species, and the formation of wildlife openings all require changes in the natural rate of mortality.

Boyce (1977) illustrates this idea. Consider the habitat characteristics for the yellow-shafted flicker woodpecker (*Colaptus Aurotus, Linneus*). In the deciduous forests of Eastern North America, this bird feeds mostly on lawns, pastures and on recently harvested openings that are larger than 0.2 ha. Nesting occurs mainly in margins of forests that contain large dead trees and branches (Scott et al 1977). These characteristics may be expressed by two indexes. The feeding opportunity index (0-1) is related to the proportion of forest area in seedling habitat. For example, the feeding index varies from .05 to 1.0 corresponding to the proportion of area in seedling habitat of 1% to 6%. Similarly, the nesting opportunity index (0-1) is related to the proportion of the forest that contains mature and old-growth timber habitats interspersed among seedling, sapling, and pole stands. For example, the nesting index varies from 0.5 at 5% proportion of area in mature and old growth habitat to 1.0 at 20% proportion of old growth. Thc simulation of the forest categorised by stands according to their stage i.e. seedlings, saplings and poles can be used to find out how prospects for flicker woodpeckers might change over time as different modes of management are applied.

As the proportions of seedling, old growth, and mature timber habitats and the size of openings change over time, the food habitat and nesting habitat indexes are used to compute the change in the potential livelihood for yellow-shafted flicker woodpeckers.

Similarly, theory combined with modelling offers a guide to establish the operating range of many parameters of interest in assessing forest ecosystem health. The fundamental aspects of productivity, soil nutrients, biodiversity, can all be assessed from this perspective and the range of acceptable limits determined (Odum, 1995). Indeed, from this perspective, ecological theory provides the guidance for identification of the parameters that are most crucial to the sustainability of the ecosystem (Odum 1985; Rapport et al 1985). It was theory (combined with observation) that governed the identification of the key signs of ecosystem distress (Rapport et al 1985). These signs include aspects relating to energy flows, nutrient transport, species composition, species dominance, etc.

Where theory leaves off, and modelling takes over – is in the establishment of the 'norms' – that is quantification of the boundaries of the normal operating range for healthy systems. It is here, as we illustrated with the single example of the yellow-shafted flicker woodpecker, that models can be constructed which provide information about the expected norms under various management regimes, and under conditions with minimal human interventions (e.g., wild, or unmanaged ecosystems).

2.2 Identification of pathological states

In examining the basic features of ecosystem structure and function, guided, as suggested above, by theoretical and modelling considerations, the key question is whether there is evidence of dysfunction in the system as a whole. If so, one might suggest this is a sign of pathosis at the ecosystem level (Birkett and Rapport, 1998). Given that ecosystems are dynamic systems, it is often very difficult with spotty data – or observations over a short span of time to determine whether apparent aberrations are true deflections from the often wide boundaries of normal behaviours. This is well illustrated in the forest history of North America which Sprugel has so elegantly recounted (Sprugel, 1976, 1991). That is, the difficulty is to be sure that an observed state of the system lies outside the range of behaviour characteristic of a normal (and presumed healthy) functioning system. Generally, the extreme cases –e.g., large-scale die-offs which can be directly correlated with air pollution stress – for example, in the vicinity of the smelter operations at Sudbury, Ontario (Gunn, 1995) are readily recognised. The problem of identification comes in the vast majority of situations in which the outward signs of pathology are less clear cut, in so far as the system appears still very much alive.

For example, in a marshland in Central Ontario (Wye Marsh, in the vicinity of Midland, Ont), there is abundant vegetation, spawning and nesting habitat for fish and wildlife, and a general appearance of an almost wilderness appearance. Yet beneath the surface, there are indeed signs of pathology. One sign, for example is in the observation of aberrant flight behaviour of the Trompeter Swan – which is being introduced into the region. Another sign is an apparent unusual die-off of birds over winter – which on examination does not appear to be for lack of accessible food supplies. The problem that surfaces in this area is one of poisoning by ingestion by birds of lead shot, which is commonly used in pellets in this well established hunting area. This is not the sign of ecosystem pathology – an examination of the sediments reveals the abundance of stray pellets of lead. Further, an examination of the vegetation shows the invasion of species such as purple loosetrife – which in places has begun to displace the richness of the native vegetation.

In order to assess the health of systems like the Wye marsh, theory would suggest that reliance on a single indicator is undesirable. For example, the marsh has shown improved water quality – although it is still highly eutrophic. At the same time there are increasing problems of toxic substances, and invasions of exotic species. A multi-spectrum approach, comparing the present state of the marsh with historical states, or states of regions not so directly affected by human activities enables a determination of the degree of pathosis in these systems.

2.3 Risks of misdiagnosis

It follows from the above that if the risks of misdiagnosis are reduced, the more reliance is placed on a suite of indicators, each of which has been calibrated and validated for the particular ecosystem under investigation. To return to the example of the forests – one needs to ascertain the expected distribution of species, by major taxa – not only restricted to trees – and the expected size distributions within particular species. Further, one needs to examine soil nutrients, air quality, the presence of disease in plants and animals, to gain a full assessment of the health of forested ecosystems.

There have been spectacular cases of apparent die-offs in forested ecosystems which turn out on examination to be part of the natural dynamics of healthy systems. For example, spruce bud-worm in Eastern Canada – has a natural cycle whereby the coniferous trees largely die off at the peaks of the budworm cycle. However, one should not be hasty in declaring even this phenomena as 'healthy' until one is assured that the extent and duration of the outbreaks are within 'normal' bounds. In fact, what has happened in Canada, and we suspect elsewhere, is that the outbreaks have increased in both extent and duration, owing to the human interventions which were designed to mitigate the impacts. Pesticides used in Canada to increase the mortality of the budworm have had the effect of also reducing the budworms predators, and have contributed to the increasingly severe periodic outbreaks (Holling, 1995).

Diagnosis, at the end of the day, while governed by theory and models, is more of an art than a science. One needs to take into account as much as is known about the normal operating characteristics of these complex systems – which theory informs – and combine this with sharp intuition and observation to determine if the system appears to be in normal bounds. Often the determination is enhanced by having the 'patient' history – particularly in those ecosystems in which wide fluctuations in all the operating characteristics are the rule. In such cases, the healthy systems bounce back from perturbations, while unhealthy systems owing to their lack of organisation and ecological integrity, tend to become further disorganised over time – with the result that the ecological services (Daily, 1997) are continuously diminished.

3. Syndrome versus signs – risk analysis based on multiple versus single parameters.

As already discussed in the above section on risks of misdiagnosis, experience has shown that it is far more informative to rely on a group of signs for diagnosis rather than a specific parameter. The point is worth repeating, however, for we often find analysis of ecosystem condition based on a limited range, or even a single parameter (Kerr and Dickie, 1984).

As in medicine, it is seldom that clear cut. Rather a pathology presents itself in a seemingly disorganised fashion, with slight abnormalities in system function appearing here and there. With hindsight, as in the case of the development of anoxia in the Central Basin of Lake Erie (Rapport, 1984), it might appear that a single indicator – in this case the disappearance of the mayfly larvae, would have sufficed for a positive diagnosis of the problem. However that insight often comes too late. The preferred approach is to rely on a spectrum of indicators – which collectively is more informative than any single sign. This has been the experience in clinical medicine, in population medicine and public health, and it also holds true in the evaluation of ecosystem condition.

An example drawn from aquatic systems, illustrates the point. In the Baltic sea, the disappearance of the macrophyte, Fucus signalled to some the onset of serious degradation associated with nutrient loading. This diagnosis proved false, for a few years later, the system recovered, at least partially, without any significant reduction in nutrient input (Rapport, 1989). Had a wider spectrum of indicators been examined, including those associated with normal cycling and resuspension of bottom sediments, the temporary die-off might have been seen to be more a reflection of these processes (normal to the Baltic), rather than any sudden threshold transformation of the whole system. It may be a generalisation in both clinical medicine and ecosystem medicine, increased dimension lends more confidence to diagnosis.

4. Questions of spatial and temporal scale

The question of scale and its relationship to health assessments remains one of the key challenges for ecological theory. Can a healthy lake have an unhealthy bay? And over what period of time? When it comes to spatial and temporal dynamics, ecosystem pathology might be tracked in a manner similar to that commonly used in epidemiology. A small scale abnormality may persist for some period of time, perhaps, if isolated, with no consequence for the larger system. However, as the history of the Great Lakes has shown (Colborn et al, 1990) a number of small dysfunctional harbours or bays can result in rendering the larger system – the major basins dysfunctional. This is the case because each of the basins may serve as centres of organisation – the harbours providing habitat for breeding and foraging for both waterfowl and fish. When a significant amount of such critical habitat becomes damaged or altogether removed, the entire system may become depleted of key elements of biota, and more-over may lose the filtration services provided by wetlands – resulting in more extreme hydrological events and in increased rates and amounts of transport offshore of nutrients and contaminants. All of this, of course affects, the conditions of the larger system.

The chain or better, the web of connectivity of course doesn't stop there. Winds that pick up emissions into the bays may transport particulates (often comprising persistent organic pollutants) to regions as remote as the arctic, where they bio-accumulate within the food chain and reach significant and harmful concentrations in top predators. The result is that local dysfunctions do indeed propagate over time and space and may find their way into rendering ecosystems which are far removed also dysfunctional.

However, there may well be cases, in which local dysfunctionality is insignificant. For example, the experimental acidification of a small group of lakes in the Boreal forest has had (so far as we know) no adverse consequences for nearby lakes (Schindler, 1990). Similarly, forest fires which may totally destroy 'patches' in a large forested ecosystem, may signal death of the local areas, but in fact have no adverse impacts on the larger system – in fact the opposite, renewal by fire is part and parcel of the dynamics of many forested ecosystems.

The question then remains as a challenge to ecological theorists: under what circumstances can healthy systems have unhealthy component parts, and under what circumstances do the presence of unhealthy parts pose a threat to not only nearby systems, but even those rather remote in both space and time?

5. Questions of human values

An important element of evaluating ecosystem health can be incorporated into models. For example, a persistent concept in the literature that relates to forest ecosystem health/landscape forestry is the deeply held mental model that people must live in harmony with nature. This mental model includes perceptions of aesthetic, social, and biological values which are derived from the forested landscape. However, demands for harvest of wood for fuel, paper, houses and furniture conflict with perceptions of aesthetics, social, and biological values. As noted by Boyce (1995), out of these demands for conflicting uses of forested landscapes, landscape forestry/forest ecosystem health notions have evolved slowly over many years. How do these mental models influence ecosystem health assessments? Under what conditions are some deeply held mental models in error?

Boyce (1995), refers to the following situation when a deeply held mental model about preserving pristine conditions in old-growth forests, wilderness, and other forest reserves may be inappropriate. One mental model argues for no intervention into natural events and no actions to manage aesthetics, wildlife, habitats, and biological diversity. As pointed out by Boyce (1995) this model may be appropriate for some situations. However, in different situations the same model would limit managerial actions until failures become intolerable and corrective action is costly.

For example, a study at the Pinery Provincial Park, Ontario has identified severe grazing by a growing deer population (Crabe, 1996). Currently, the deer population is estimated at 600 by Crabe (1996) and the sustainable herd size is closer to 200 for the size and nature of the park. Over the last 15 years, park resource managers have observed a serious and steady deterioration in the quality of vegetation in the park. In particular, the understory vegetation in the oak-savannah forest is severely stressed by deer over grazing. Management intervention is inevitable to protect the ecology of the park despite reluctance by the public to culling the deer population. Recently, a joint study between the University of Guelph, Ecosystem Health and McMaster University, Environmental Health examined the relationship between impacts on the forest, forest condition, societal values, and the functions that Pinery forests provide (Patel et. al. 1999). Some conclusions from this study suggest convergence between scientific and public views of forests and forest health. In Pinery, the health of the understory vegetation, and by implication of the forest, will depend on management practices that control the deer population, continue prescribed burning and limit further trail construction. The deer management requires carefully controlled strategies that are sensitive to public views.

An extension of this work involves the construction of a computer simulation model of the relationship between the deer population levels and the understory growth dynamics of the hardwood oak-maple forest which is the food source for the deer. This modelling effort should assist in understanding the temporal interaction between the growth of the forest and the growth of the deer population and help mediate discussion between public perceptions of Pinery and management's view. We intend to calibrate this model to Pinery forest inventory data and represent the growth dynamics of the deer population so that it might lead to a more systematic approach for examining alternative policy options in the management of the Pinery.

6. Conclusions

Ecological theory provides the foundation upon which to assess ecological condition. Combined with modelling and quantification of the operating characteristics of ecosystems, the practitioner of ecosystem health is better able to assess existing situations as to evidence for pathology and better able to establish whether management practices have been effective in providing remediation to damaged systems. To do this, one must be highly selective as to the significant parameters and variables that ought to be monitored (practical limitations bear heavily on what is realistic in this domain), and one needs to establish for those selected indicators, the normal operating ranges of healthy ecosystems. Further, one needs to pay attention to the risks of misdiagnosis through reliance on too few variables, on the role of human values in establishing the operating characteristics, and on the potential (too often realised) for the transmission of ecosystem pathology far from the site of origin. Ecological theory and modelling has much to contribute to the elucidation of these questions.

REFERENCES

Ashby, W. R. 1973. An Introduction to cybernetics. London: Chapman and Hall ,University Paperbacks.

Birkett, S.H., and D.J. Rapport 1998. A framework for identifying and classifying ecosystem dysfunction. *The Environmentalist , 18(1):15-25.*

Boyce, S.G. 1977. Management of eastern hardwood forests for multiple benefits *(DYNAST-MB)* U.S. Dep. Agric. For. Serv. , Res. Pap. SE-184, 140p Southeast For. Exp. Stn. Asheville, N.C.

Boyce, S.G. 1995. Landscape Forestry. New York: John Wiley & Sons, Inc..

Calow, P. 1995. Ecosystem health: A critical analysis of concepts. In: Rapport, D.J., Gaudet, C., Calow, P. (eds.). *Evaluating and Monitoring the Health of Large-Scale Ecosystems*, 33-42. Springer-Verlag, Heidelberg.

Crabe, T. 1996. The Impact of Deer grazing on Vegetation of Pinery Provincial Park, Natural Heritage Information Centre, Ontario Ministry of Natural Resources.

Daily, G.C. 1997 Nature's Services. Societal Dependence on Natural Ecosystems. Island Press, Washington, D.C.

Dobson, A.P., A.D. Bradshaw, and A.J.M. Baker 1997. Hopes for the Future: Restoration Ecology and Conservation Biology. *Science, 277 515-522.*

Engelberg, J., and L.L. Boyarsky 1979. The noncybernetic nature of ecosystems. *Amer. Nat.* 114: 317-324.

Gunn, J.M. (Ed) 1995. Restoration and Recovery of an Industrial Region: Progress in Restoring the Smelter Damaged Landscape near Sudbury, Canada. Springer-Verlag.

Holling, C.S. 1995. Sustainability: The cross-scale dimension i. In: Munasinghe, M. & Shearer, W. (eds) *Defining and Measuring Sustainability: The Biogeophysical Foundations,* 65-75. The World Bank, Washington, DC.

Kerr, S.R., and L.M. Dickie 1984. Measuring the health of aquatic ecosystems. In Cairns, V.W., Hodson, P.V., Nriagu, J.O. (eds.) Contaminant Effects on Fisheries. J. Wiley and Sons, New York.

Leopold, A., 1941. Wilderness as a land laboratory. *Living Wilderness*, 6 (July): 3.

Lorimer, C.G., and L.E. Frelich 1994. Natural disturbance regimes in old-growth northern hardwoods. *J. Forestry.* 92(1):34-38.

Nicolis, G., and I. Prigogine 1977. *Self-organisation in non-equilibrium systems.* 491p John Wiley and Sons: New York, London, Sydney, Toronto.

Odum, E.P. 1969. The Strategy of Ecosystem Development. *Science* 164: 262-270.

Odum, E.P. 1985. Trends Expected in Stressed Ecosystems, *BioScience 35 (7): 419-422.*

Odum, E.P. 1995. Profile Analysis and Some Thoughts on the Development of the Interface Area of Environmental Health. *Ecosystem Health, 1(1):41-45.*

Odum, H.T. 1960. Ecological potential and analogue circuits for the ecosystem. *Amer. Scient.* 48 : 1-8.

Patten, B.C. 1964 The Systems Approach in Radiation Ecology. ORNL /TM-1008 Oak Ridge National Laboratory, Oak ridge, Tennessee. 19.

Patten, B.C., and S. E. Jørgensen 1995 Complex Ecology. Prentice Hall PTR, New Jersey.

Patel, A., D.J. Rapport, J. Eyles, and L. Vanderlinden 1999. Forest and human values: comparing scientific and public perception of forest health in SW Ontario. *The Environmentalist* (in press).

Porn, I. 1984. An equilibrium model of health. In: Nordenfelt, L. & Lindahl, B. (eds.). *Health, Disease and Causal Explanations in Medicine.* Reidel, Durdecht.

Rapport, D.J. 1984. State of ecosystem medicine. In: V.W. Cairns, P.V. Hodson and J.O. Nriagu (eds) *Contaminant Effects in Fisheries.* New York: John Wiley & Sons (315-324).

Rapport, D.J., H.A. Regier, and T.C. Hutchinson 1985. Ecosystem behaviour under stress, *American Naturalist*, 125, 617-640.

Rapport, D.J. 1989. Symptoms of pathology in the Gulf of Bothnia (Baltic Sea): Ecosystem response to stress from human activity. *Biological Journal of the Linnean Society* 37: 33-49.

Rapport, D.J. 1997. What is Ecosystem Health? *EcoDecision*, Winter 1997.

Rapport, D.J., C. Gaudet, and P. Calow (eds.). 1995. Evaluating and Monitoring the Health of Large-Scale Ecosystems, 33-42. Springer-Verlag, Heidelberg.

Ribble, C. et al, 1997. Ecosytem health as a clinical rotation for senior veterinary students in Canadian veterinary schools. Canadian Veterinary Journal. (38):485-490.

Scott, V. E., K.E. Evans, D.R. Patton, and C.P. Stone 1977. Cavity-nesting birds of North American forests. *U.S. Dep. Agric. For. Serv., Agric. Handbook.* 511. 112p Washington, D.C.

Schindler D.W. 1990. Experimental perturbations of whole lakes as tests of hypotheses concerning ecosystem structure and function. *Oikos* 57:25-41.

Spanner, D.C. 1964. Introduction to thermodynamics. 278p. Academic Press, London and New York.

Sheffield, R.M., and M.T. Thompson 1992. Hurricane Hugo, effects on South Carolina's forest resources. Res. Pap. SE-284. Asheville, NC: USDA Forest Service, South-eastern Forest Experimental Station.

Sprugel, D.G. 1976. Dynamic structure of wave-regenerated Abies balsamea forests in the north-east United States, *Journal of Ecology*, 64: 889-911.

Sprugel, D.G. 1991. Disturbance Equilibrium and Environmental Variability: What is 'Natural' Vegetation in a Changing Environment? *Biological Conservation*, 58: 1-14.

Tolba, M.K., O.A. El-Kholy, E. El-Hinnawi, M.W. Holdgate, D.F. McMichael, and R.E. Munn, 1992. The World Environment 1972-1992. Chapman & Hall, London.

Van Dyne, G.M. 1966 Ecosystems, Systems and Systems Ecologists. ORNL 39 3957. Oak Ridge National Laboratory, Oak Ridge, Tennessee. 40.

Van Dyne, G.M. 1969a. Grasslands Management Research, and Training Viewed in a Stems Context. Range Science Department, *Science Series, No 3.* Colorado State University, Fort Collins, Colorado 50.

Van Dyne, G.M. (ed) 1969b. The Ecosystem Concept in Natural Resource Management. Academic Press, New York. 383.

Van Dyne, G.M. 1969a. Grasslands Management, Research, and Training Viewed in a Systems Context. Range Science Department , Science Series, No 3, Colorado State University, Fort Collins, Colorado.

Vitousek, P. M., and W.A. Reiners, 1975. Ecosystem succession and nutrient retention: a hypothesis. *BioScience* 25:376-381.

Vitousek, P.M. , H.A. Mooney, J. Lubchenco, and J.M. Melillo, 1997. Human Domination of Earth's Ecosystems. *Science*, Vol 277 (25 July 1997) 494-499.

Waltner-Toews, D. 1996. Ecosystem health: A framework for implementing sustainability in agriculture. *Bioscience. 46:686-690.*

Watt, K.E.F. 1968. Ecology and Resource Management. McGraw Hill, New York 450.

Winterhalder, K. 1996. Environmental degradation and rehabilitation of the landscape around Sudbury, a major mining and smelting area. *Environ. Rev.* 4, 185.

Wisniewski, S., B. Staniszewski, and R. Szymanik, 1976. Thermodynamics of nonequilibrium processes. 274p D. Reidel Publ. Co. : Boston, Mass.

III.1.3 Hedging Our Bets: The Utility of Ecological Integrity

Jan Barkmann and Wilhelm Windhorst

1. Introduction

Ecological integrity is a successful but controversial application of ecosystem theory. As witnessed by frequent occurrence in important government documents, its struggle for influence in the environmental debate has been rather successful (e.g., U.S. Clean Water Act Amendments of 1972, Great Lakes Water Quality Agreement 1978, Agenda 21; refer to Westra 1995). Many academic ecologists remain sceptical and question the scientific existence of 'integrity' in nature. The biological reality of ecological integrity was indeed questionable if even a list of 38 phenomena of ecological systems can admittedly only 'sketch' its meaning (Regier 1993). The ensuing lack of consensus on most basic features of ecological integrity invites legitimate questions. Is ecological integrity based on a structural description relying on the 'natural composition' of the biotic community (Karr 1981) or is it a functional concept, stressing ecological process and ecosystem energetics (Schneider & Kay 1994a, 1994b)? Is ecological integrity equivalent with a 'wild' ecosystem state free from anthropogenic stress (Westra 1994:49), or is it the same as autocatalysis and self-organisation (Ulanowicz 1995)? Is ecological integrity a scientifically meaningful term at all, or is it a useless buzzword of juridical prose and environmental debate (Noss 1995)? Public appraisal for ecological integrity is clearly ahead of the theoretical fundamentals underpinning this concept at the interface of scientific ecology and environmental discourse.

Concepts at the interface of scientific ecology and environmental discourse are inevitably suspicious from a purely scientific point of view. In addition to their descriptive and empirical meaning, these concepts convey a normative (prescriptive) or an evaluative (axiological) meaning. These additions cannot be deduced from scientific ecology without recourse to pre-existing, non-scientific (= ethical) statements. Ecological integrity clearly belongs to this class of concepts because it conveys the evaluative meaning of a state of the natural environment that is in some relevant aspect *preferable* to a 'sick', 'diminished', 'compromised' or 'destroyed' state of the same environment (Woodley 1993, Kay 1993). In communicating evaluative meaning, ecological integrity is a lot more successful than ecological terms with more exclusively scientific meanings but without public appeal (Metzner 1998).

Specifically, ecological integrity belongs to a subclass of concepts at the interface of science and discourse, whose descriptive meaning changes with the evaluative preferences that are sought to be furthered in environmental debate. Apart from the most trivial cases, the 'sickness' or 'impairment' of an ecosystem cannot be measured any better than its integrity. Or as Kay & Schneider (1995) expressed it regarding the non-existing normative foundation of ecological integrity: *"Ultimately, any evaluation of the ecological acceptability of a human activity will depend on value judgements about whether the resulting changes in the affected ecosystem are acceptable to the human participants."* If we derive ecosystem attributes from evaluative preferences, and call these attributes 'ecological integrity', the result is not an element of ecological reality. Instead, we describe the ecological correlates of our preferences: we primarily chart our own *social* reality. Accordingly, ecological progress in finding the 'right' definition of ecological integrity has been slow.

In absence of *The Grand Unified Ecosystem Theory,* the right definition is not a matter of science but of societal choice[2]. Misunderstanding ecological integrity as a set of descriptive ecosystem attributes, inhibits any progress on this concept (cf. Lemons 1995).

In section 3, we employ an excursion to Sustainable Development to abstract one core value of ecological integrity. Specifically, we argue that the value at the base of ecological integrity refers to the protection of the ecological life support systems in face of ecological uncertainty. Because we identify the capacity for self-organisation as the physical correlate of ecological integrity (section 4), the need for theoretical ecology to operationalise this novel interpretation of ecological integrity is stressed (section 5). Finally, we discuss some problems of indicator development (section 6), and illustrate the 'utility of ecological integrity' with a game theoretical model (section 7).

2. Four approaches to ecological integrity

In order to capitalise on the apparent success of ecological integrity in a legitimate way, any serious interpretation of ecological integrity has to reconstruct the concept in a meta-scientifically sound way. So far, the supporters of ecological integrity have not been able to convince the sceptics that ecological integrity can stand up to the standards of inter-subjective science at all.

For opposing reasons, neither proponents of a 'holistic' approach to environmental science nor advocates of a 'reductionistic' research philosophy pay much attention to a meta-scientifically sound reconstruction of ecological integrity. Those who portray ecological integrity as a holistic concept that overcomes the 'old dichotomies of fact and value' (Westra 1994, Funtowicz & Ravetz 1995) often regard the systematic consideration of competing human goals, interests and benefits regarding ecological integrity as somewhat inappropriate. On the other hand, nomothetical purists in ecological science do not deem occupation with ecological integrity – and its 'speculative' theoretical fundamentals – a scientific enterprise at all (Kay & Schneider 1995).

In discussing the assets and deficits of different interpretations of ecological integrity in environmental science, Shrader-Frechette (1995) points to a basic dilemma of one-handed concepts at the interface of science and environmental policy making. While proponents of 'Hard Ecology' dismiss any ecological policy recommendation whose empirical foundation does not stand statistical scrutiny, supporters of 'Soft Ecology' embrace holistic, 'inspiring' conceptions of nature, such as naturalness, integrity and stability. While 'Soft Ecology' relies on largely qualitative and non-testable concepts which are compatible with any observed fact, 'Hard Ecology' cannot admit enough interesting ecological recommendations to public discourse. In most cases, our knowledge is too 'soft' due to the probabilistic character of ecological phenomena.

Furthermore, both approaches are troubled by the problem of finding a sound way to conceptualise the relation between facts and values, between ecological theory and human interests. Consider a paradigmatic situation for the ubiquitous ethical uncertainties on the preference order of societal values in environmental decision-making processes. Scientists are invited to counsel a diverse local community on a significant development project. As far as facts and knowledge are concerned, the best advice will still be based on the best ecological data and models available. Each group of stakeholders expects that the model outputs enable them to make an informed choice on the development project; this means that it must be made explicit how the project interferes with *their* interests, goals and values. Models which only address the values that the counselling scientist deems relevant (or those values that are implicitly addressed by his/her run-of-the-mill generic model) are obviously insufficient.

[2] Intrinsically good or valuable ecosystem states can only be postulated in metaphysically strong conceptions, such as in Westra's Aristotelian (= organicist) account of 'The Integrity Principle' (Westra 1994). Even if the identity of 'The Whole' and 'The Good' was granted, present ecological theory does not (yet?) identify the ontic wholes, on which the Ecosystem Good would hinge.

In this regard, Westra's Aristotelian model of ecological integrity is deficient. The delicate relationship between ecological structures on the one hand, and societal or personal goods, interests and goals on the other remain intractably entangled as she regards *The Good* and *The Whole* essentially as identical (Westra 1994, cf. also Ulanowicz 1995). Westra

a) proposes to value integrity 'on the basis of the paradigm case of organic unity',
b) defines ecosystems as unities and organismic wholes in an ontologic sense, and
c) finds the value bearing 'optimum point' or 'excellence' of ecological systems in 'pristine' ecosystems not 'impaired' or 'dismembered' by anthropogenic stress (Westra 1994:35, 43ff).

Addressing ethical uncertainty, however, requires – as clear as possible – a distinction between ecological knowledge and the subjective values necessary to judge different courses of action. We consider the separation of facts and values indispensable for any scientific enterprise, particularly, for progress on the scientific meaning of ecological integrity.

If traditional reductionist conceptions of ecological science come with a technocratic attitude, they display a complimentary blind spot. Such conceptions usually do not recognise the scientific importance of value considerations at all. In addition to inevitable methodological judgements, the ethical uncertainties rule out such a naïve conception (Funtowicz & Ravetz 1995). Ethical uncertainties require thorough reflection on personal and professional values of the individual researchers, and – most importantly – an ethically legitimising, 'extended peer community' that includes the concerned stakeholders (Funtowicz & Ravetz 1995, Barkmann 1999a). None of these are ingredients of ecological models or ecological theory.

Thus, we share Shrader-Frechette's critical assessment of the treatment of ecological integrity on many counts. There are two options for an intermediate path avoiding the extremes. Shrader-Frechette (1995) suggests a humble solution: practical ecology. Her middle ground approach recognises the urgent need of ecological policy advice even in absence of successfully tested ecological theory. Practical recommendations are to be based on 'bottom-up' ecological knowledge represented by careful case studies that use autecology, natural history information, and empirical rules of thumb. She discounts the ability of 'top-down' theoretical ecology to furnish management suggestions that stand fierce litigation[3].

But her own approach to the problems of ecosystem management is also too limited in scope. It is certainly a better idea to rely on natural-history information to manage red-cocktailed woodpeckers than to look at ecosystem thermodynamics for advice (Shrader-Frechette 1995). The empirical wisdom of practitioners is insufficient, however, when the long-term functioning of the global life support systems is the point at issue. On these matters far beyond ecological field experience and practice, it is not prudent to disregard the advice of ecosystem analysts and theoretical ecologists. As this paper argues that ecological integrity primarily deals with the long-term functioning of the global life support systems, ecosystem theory finds one if its main applications here.

Obviously, a fourth approach is required. Without neglecting advice from ecosystem theory, this approach has to avoid the ethical pitfalls of metaphysical holism and extreme reductionism. The roots of our integrity interpretation in Sustainable Development (SD) facilitate a meta-scientifically sound approach to ecological integrity that provides for these conditions. Presupposing only the most general ecological implications of SD, the norm is explicit, necessary and just, and not derived from 'value-free' science in a way that violates Hume's Law.

[3] E. g., she cites J. M. Smith (McIntosh 1985:321): 'Ecology is still a branch of science in which it is usually better to rely on the judgement of an experienced practitioner than in the predictions of a theorist.'

With its reliance on SD, our interpretation inherits the former's anthropocentric point of view, however. Because of the philosophical inability of non-anthropocentric environmental ethics to demonstrate the existence of binding obligations of humanity towards "*the sustenance of ecological and evolutionary processes, viable populations of native species, and other non-human qualities of ecosystems,* for their own sakes" we accept the SD heritage (quote from Noss 1995:64; e. g., refer to Birnbacher 1980, Pfordten 1996).

3. Being explicit in valuation space

The analysis of the main value embodied in ecological integrity benefits from an explication of the implicit values of existing definitions. The explication is not sufficient, however, for the meta-scientifically sound construction of ecological integrity. A fully rational construction of an evaluative concept can only be founded in the evaluative sources of the concept, not in existing definitions. Thus, we do not actually perform an analysis of existing definitions, but focus on the analytical procedure. A related approach had been used in a recent analysis of a similarly problematic term, forest health. Our conclusions regarding ecological integrity differ substantially, though.

The first step of the explication of the embodied values of ecological integrity identifies those particular ecosystem states, structures and processes that constitute the existing 'descriptive' integrity definitions. In a second step, these states, structures and processes are related to environmental values, e. g., aesthetic experience, material sustenance, and ethical values. Depending on the depth of the analysis sought, the main environmental values can be divided into more specific environmental needs, goals or interests. The explication of various definitions will usually yield different patterns of environmental values because different ecological phenomena are stressed by differing definitions (see, e. g., the IBI and Canadian National Park examples below). The value patterns can be mapped into an abstract, multidimensional *valuation space* (Barkmann 1999a). The environmental valuation space is constructed by a number of axes (dimensions), each representing one environmental value or its subdivisions. In valuation space, the normative loading of an integrity definition is represented by a valuation vector. If the meaning of the valuation vector is translated into non-technical language, it enables non-experts to compare the 'real-world' meanings of different integrity definitions. This explication is necessary if an informed societal choice is required on competing integrity definitions[4].

Still, the explication of the normative loading of existing definitions does not fully embrace the primacy of societal value considerations. Even vastly differing integrity definitions do not guarantee a sufficient choice among alternatives: the alternatives could be limited to integrity definitions that do not adequately fit the actually preferred valuation vector. Scientists could deliberately try to preconceive the relevant environmental value patterns, and design their integrity definitions accordingly. This ambitious procedure is equivalent with finding *the right* valuation function that transforms multidimensional valuation data to cardinal or at least ordinal valuation scores. Ecologists are not particularly well trained to accomplish this undertaking beyond traditional scientific practice.

If the valuation vector is given by law, the conditions for such an undertaking are favourable. Biologists in North America, for example, found the 'integrity' of the water bodies in the United States suddenly protected in 1972 (US Clean Water Act Amendment). The prevalent chemical and physical measures of water quality had been insufficient to trace ecosystem degradation as mandated. A genuinely biologic measure was needed to assess *biologic* integrity (Karr 1992). At this point, operationalisable methods to quantify

[4] In addition to normative and axiological concepts, ecological systems can principally also be mapped into valuation space and represented by a valuation (or assessment) vector. In this case, the vector represents the human relevance of an ecological system. This usage highlights ecosystem goods and services that can be derived from ecological systems (Barkmann 1999b).

biological integrity were not available (Karr 1992). Biologists tried to close this gap in the following years. Namely the *Indicator of Biological Integrity* (IBI), introduced by Karr and Dudley (1981), has become one of the most widely used, biologically informed instruments to enforce the management goals of the U.S. Clean Water Act Amendments. The IBI defines biological integrity as the ability to support and maintain *"a balanced, integrated adaptive community of organisms having a species composition, diversity, and functional organisation comparable to that of a natural habitat of the region"* (Karr 1992, Karr & Dudley 1981). It aims at detection of ecosystem degradation and anthropogenic stress to the water body.

The IBI example is instructive for at least two reasons. Firstly, it proves that specific, biological interpretations of ecological integrity can be scientifically operationalisable *if* the environmental values or management goals implicit in the interpretation resonate with the normative expectations of the concerned public – in this case, with the law maker, U.S. Congress –, and with the regulative body, the USEPA. More abstractly, the IBI is successful because its representation in multidimensional valuation space is roughly congruent with the multidimensional set of values that U.S. Congress intended to protect when using the integrity metaphor. Biological integrity *was meant* to correspond to some original biological condition (Karr & Chu 1995).

Secondly, the IBI example shows that a general interpretation of ecological integrity cannot simply adopt the IBI approach. The crucial SD value of the maintenance of the ecological life support systems differs from the goal mandated for the quality of the U.S. waters as the norms of SD are different from those mandated by the Clean Water Act Amendments. The same difference in scope and values is found between more general approaches to ecological integrity and 'ecological integrity as mandated by the Canadian National Parks Act of 1988'. By definition, national parks have to be in as *'natural, naturally evolving, pristine and untouched'* a condition as possible. Thus, Woodley (1993:158) can define ecological integrity in this specific case as a condition, in which *"ecosystem structures and functions are unimpaired by human-caused stresses and [in which] native species are present at viable population levels"*.

For a general notion of ecological integrity, decisive normative mandates do not exist. One of the options to deal with normative uncertainty is displayed by the report on the comprehensive assessment of the health of United States forests to the U.S. Congress (U.S. Cong. House 1997). *"The first question we asked is: what is forest health? There are 11 different definitions of forest health. And one is not more scientific than another because science isn't about definitions,"* stated Dr. Oliver, professor at the College of Forestry, University of Washington, and chair-person of the scientific committee preparing the report. *"So we looked at all of the different values people want from the forest"*, he continued. Factually, the procedure of the 'Oliver Report' defines forest health as 'the condition of the forest in reference to its ability to provide different social values'. In the terminology of this paper, the interpretation constructs the axes of valuations space relevant to forests ecosystems.

There is a serious objection to reducing forest health and ecological integrity to the *valuation space* of ecological systems: valuation space *itself* has no normative power whatsoever. Certainly, the construction of valuation space makes us aware of the multitude of aspects that potentially confer value upon ecological systems; it enables a systematic investigation of the environmental values that are associated with any descriptive management guideline. The construction of valuation space is a tool in the decision-making process, but not, however, a goal in itself. Interpreting ecological integrity as the valuation space of ecological systems remains materially empty – and even worse, it would severely threaten the public success of the integrity metaphor. Ecological integrity could be found in any ecological system no matter how severely degraded if *any* social value could be identified that is still provided by the system. One example: Even after 'ecosystem destruction' near Palmerton, Pennsylvania (USA), by zinc smelters in the early 20th

century, the affected valley system would have to be attributed some ecological integrity because it provided a *sink* for most aggressive pollutants (cf. Aber & Melillo 1991:381). Being a sink for pollutants is an important ecosystem service (Daily et al. 1997, deGroot 1992, Heydemann 1997), e. g., promoting the environmental value regarding human health, but 'ecological integrity' is clearly a misnomer of the condition of what remained of that system.

This paper proposes an approach to ecological integrity that is explicit in valuation space without resorting to descriptive integrity definitions, specific interpretations based on single purpose regulations, or without giving up all evaluative claims. We suggest to focus on *one* environmental value at the core of ecological integrity ('one-value proposition'). In normative terms, ecological integrity becomes a first order management principle to further the characteristics of ecological systems that are mapped along this axis. Before we argue for any one specific value at the core of ecological integrity, we outline a decisive advantage of this novel interpretation.

If the one-value-proposition is accepted, ecological integrity represents *one main axis in valuation space, not a multidimensional vector.* This is indeed an important advantage, because it is, at least principally, possible to identify those ecosystem states, structures and processes that correspond to this axis *without having to argue for the relative importance of this axis.* As previously discussed, any integrity definition that makes essential use of descriptive ecosystem attributes is arbitrary as long as it is not legitimised by the concerned stakeholders or their representatives. Even without a justification process, our interpretation is non-arbitrary because it is not faced with the problem to aggregate diverse multi-dimensional values. The relative importance of the environmental value of ecological integrity depends directly on the relative importance of the intergenerational long-term considerations at the core of SD detailed in the following section.

4. From Sustainable Development to the value of ecological integrity

Sustainable Development is only as good a norm for environmental conduct as the management principles for its enforcement are. Generally, there is no shortage of suggestions for management principles aiming at a more sustainable development (BMUNR 1996, BUND/MISEREOR 1996, SRU 1996, SRU 1994, Enquete-Kommission 1998[5]). The management principles focus on renewable and non-renewable resources, on acceptable pollution and discharge limits, and on the protection of human health. Contrary to its importance, only the most general suggestions are available for the protection of the ecological life support systems. Making ecological integrity an operable concept is one of the decisive steps towards such a management principle.

It may be helpful to recount some basic features of SD. SD is a procedural *norm* or a *regulative principle* (Enquete-Kommission 1998) commanding that ecological knowledge on the limitation and vulnerability of natural resources and ecosystem services be incorporated into the environmental decision making process (for an extended treatment of ecosystem services see Barkmann 1999b, Daily et al. 1997). Without neglecting the needs of the current generation, the ecological needs of future generations may not be compromised (Hauff 1987, SRU 1994). An effective application of this norm is hampered because it is impossible to know exactly how future needs and demands regarding ecological systems will change. The next few generations of the 21st century, however, will most likely continue to rely on certain 'essential' ecosystem services. A panel of senior ecologists recently concluded that *"ecosystem services operate on such a grand scale and in such intricate and little-explored ways that most could not be replaced by technology"* (Daily et al. 1997). Essential services, which are virtually impossible to substitute on a

[5] This section focuses on the European and German situation; the problems may differ in detail but not in principle under differing legal regimes.

global scale, include climatic stability and the regulation of the composition of the atmosphere, as well as the availability of fresh water and fertile soils.

SD is at odds with the *substitution paradigm* of resource economics. The substitution paradigm postulates that human-made capital can principally substitute natural resources (Hampicke 1992:110ff). Simply put, if the rate at which the economy substitutes a limited natural resource outstrips the rate at which the resource is consumed, the resource is inessential. SD fundamentally recognises the limits of substitutability: ecosystem services necessarily *complement* human-made capital in order to satisfy social and economical needs[6]. As there is no indication that the substitution of essential ecosystem services progresses at the required pace, an extension of the substitution paradigm to essential ecosystem services amounts to an unsubstantiated, dangerous act of faith in the axioms of economic models[7]. In this vein, SD asks us to treat ecological systems carefully in order to preserve their ability to provide essential goods and services for the needs of the present and of future generations – the latter beginning with our children and grandchildren. In short, SD asks us to preserve the ecological life support systems.

Likening them to the life support systems of Apollo 13, E. P. Odum has popularised the expression *natural* or *ecological life support systems* for the ecological systems that provide essential ecosystem goods and services (Odum 1991). Ecological systems, in turn, are complex structures of system elements and interactions. Accordingly, the ecological life support systems essentially rely on the network of causal relations and interactions between biotic and abiotic ecosystem elements. In German environmental law, the abstract ability of ecological systems to provide environmental goods and services is called *Leistungsfähigkeit des Naturhaushaltes* (which verbatim translates to 'efficiency of the economy of nature'; Bundesnaturschutz-Gesetz 1987). The meaning of this concept differs slightly from the meaning of *ecological life support system* because *Naturhaushalt* ('economy of nature') specifically denotes the *totality* of relations and interactions between biotic and abiotic ecosystem elements, not only potentially instrumental subsets thereof. The differentiation between ecosystem states, structures and processes that definitely furnish distinct goods and services and those states, structures and processes with uncertain function is mirrored by two kinds of environmental values embodied in SD and other environmental norms.

First, ecological systems provide immediate ecosystem goods and services. These ecosystem goods and services, such as the regulation of the atmospheric gas concentrations or the recharge of high-quality aquifers, are valuable because they directly contribute to human values (health, material subsistence, aesthetic experience, etc.). Humanity derives utility from these goods and services in a straight forward way. Each of the corresponding environmental values is represented by one axis in valuation space.

Ecological systems provide a second set of more indirect benefits that belongs to another, quite different environmental value. These indirect benefits refer to the uncertainties that the intergenerational long-term scope of SD calls to attention:

- We do not know how global change will alter the biospherical ecological baselines.
- We do not know exactly how human needs regarding the ecological life support systems will change in the future.

Furthermore, there is a third uncertainty, closely related to the first two, but of immediate importance:

- We do not even know sufficiently, how the *present* biogeochemical interactions weave the ecological fabric that holds together the *present* life support systems.

[6] To emphasise the complementary, interwoven and net-like relation of ecological systems with the social and the economical sphere, the German Council of Environmental Advisors has coined the term *retinity* ('Retinität'; SRU 1994:54).

[7] To date, there is actually no evidence that any significant substitution of these ecosystem services at *global scale* will be possible at all.

Because of this triple uncertainty, we are *prima facie* not allowed to disregard any part of an ecological system: any single component of the totality of ecosystem interactions *may* be relevant. In regard to the triple uncertainty, we have to assume that, indeed, everything is connected to everything else. We have reason to fear that a physically ever expanding human sphere, that the growing 'human appropriation of the products of photosynthesis', and that our innumerable interferences with the natural life support systems increase the risks to the long-term welfare of humanity by factually disregarding more and more ecological interactions. The environmental value dealing with the triple uncertainty seeks protection against these risks[8]:

> *The intergenerational protection against the ecological risks aggravated by a physically growing economy in a finite world is the main objective 'people want from ecological integrity'. It is the main value of ecological integrity, and at the same time it is the inadequately addressed management principle at the core of SD.*

As ecological professionals employing the integrity metaphor share the notion of a 'functional' ecosystem condition that sustainably provides essential ecosystem services, *ecological integrity* surfaces as a legitimate identifier for the appropriate first order management principle. Economically speaking, the value of ecological integrity is an 'existence value' of ecological systems (Cansier 1993, Freeman III 1979). We cannot 'prove' that this interpretation is necessarily the only admissible interpretation of ecological integrity within the SD paradigm. One could, for example, maintain that ecological integrity comprises both kinds of values, the values associated with direct ecosystem goods and services as well as the proposed environmental value in face of uncertainty. Ecological integrity again, however, became a multidimensional concept in valuation space[9]. Already this rather small extension entails the most recalcitrant problems: to yield an applicable operationalisation, the multidimensionality of that interpretation had to be aggregated again. Situating these kinds of value judgements *inside* an already ill-defined 'ecological' concept is theoretically insensible – and has definitely not turned out to yield particularly intersubjective results. For the same reason, attempts to operationalise ecological integrity validly as the "*most all-encompassing of all 'umbrella' concepts in conservation'*, comprising *'ecosystem health, biodiversity, stability [...], sustainability, naturalness wildness, and beauty"* (Noss 1995:66) are almost certainly doomed to failure.

For the sake of the argument, we were prepared to modify the 'one-value-proposition'. If required to concede that ecological integrity is a multidimensional concept, we claim that our interpretation still covers one *crucial* dimension of ecological integrity. This dimension may be called 'long-term integrity of the ecological life support systems' or 'ecological uncertainty protection'. Without this dimension, ecological integrity and SD lose an essential dimension. All following analyses apply at least to this essential dimension.

5. The self-organisation capacity of ecological systems

5.1 Justifying a top-down-approach

In this section, we apply ecosystem theory (the 'fourth approach'; see Section 2) to the problem how to operationalise the one-value interpretation of ecological integrity. Several

[8] 'Holistically inspired' authors also point to the importance of risk and uncertainty considerations regarding the value of ecological integrity (Westra 1994:39f).

[9] In an earlier stage of the discussion at the Kiel Ecology Centre, both authors did indeed subscribe to the view that ecological integrity was a management principle dealing with the *entire* ecological aspects of SD (Barkmann et al. 1998). Similar views are expressed by Müller (1998a:31) who stresses that the ecosystem goods and services within the SD paradigm *'are to be understood as a function of the realisation of integer, 'healthy' states of the ecological interaction network'*.

ecosystem scientists have employed different branches of ecological theory to deal with long-term ecosystem management problems. This paper integrates applicable findings of ecosystem theory into the novel framework of the one-value-proposition for ecological integrity. We will draw on several more recent attempts to make ecological sense from ecological integrity (e. g., Woodley 1993, Schneider & Kay 1994a, 1994b, Müller 1998a, Barkmann et al. 1998, Kutsch et al. 1999).

We had already identified the total network of ecological interactions as the relevant ecosystem correlate of the value of ecological integrity. If anything can, it is the 'functionality' of this network that protects against the triple uncertainty endangering human long-term welfare. Much of environmental science focuses on single environmental media or services, however, not on a comprehensive and integrated description of system-level interaction phenomena that is indispensable to cope with the effects of the triple uncertainty. The *total* network of ecological interactions evades description according to these traditional methods of the environmental sciences. Thus, it also evades a description of the features necessary for the long-term viability of ecological life support systems.

Because the totality of ecological interactions is not yet adequately described, a peculiar situation has arisen. In line with their importance, the ecological interactions are protected by several environmental regulations, for example, by the German Environmental Assessment Act (UVP-Gesetz 1990), which was issued to carry out the European Union directive on national Environmental Assessments laws. Other national regulations, including the Federal Nature Protection Act (Bundesnaturschutz-gesetz 1987) directly protect the 'efficiency of the economy of nature'. As 'economy of nature' means nothing else than the totality of ecological relations and interactions, both concepts basically seek to protect the same environmental value. As a result, the total network of ecological interactions is under multiple juridical protection – but without anyone knowing exactly what it is.

Similarly, SD demands the preservation of the ecological life support systems facing uncertainty, but the traditional environmental sciences do not supply any applicable operationalisation. This predicament is very similar to the situation after the U.S. Clean Water Act Amendments were enacted. Justified norms have been established, but the scientific community has yet to supply the ecological tools to operationalise them. The only difference is that neither the *general* notion of ecological integrity nor Sustainable Development are concepts enacted as enforceable public law.

The traditional environmental sciences can hardly be blamed to avoid tackling this breed of problems of ecological complexity and uncertainty. It is a hopeless enterprise to operationalise ecological integrity via an investigation of the single interactions that make up the network of ecological interactions. Ecology cannot exhaust the totality of interactions incrementally. It is the classical approach of the environmental sciences, however, to probe into ecological systems by constructing ever more detailed interaction matrices. This approach is indispensable if isolated elements or direct services of ecological systems are to be investigated. On the other hand, the approach is grossly insufficient, if criteria for the long-term protection of the natural life support systems in face of non-specific threads are needed[10]. Monitoring for the long-term protection of the natural life support systems requires a shift of attention from single interactions to manifestations of ecological interactions on a higher hierarchical level of observation.

The investigation of ecological self-organisation is a promising, alternative top-down approach on a hierarchically higher level of observation. Instead of mapping single interactions into ever more complex ecosystem models, the study of ecological self-organisation focuses on those phenomena that 'emerge' from the simultaneous interplay of all system elements on a higher level of observation.

[10] The monitoring program in the Canadian national parks recognises this difference. It complements indicators for 'thread specific' monitoring by a suite of *system level* indicators for 'Ecosystem Integrity Monitoring' to detect non-specific threads (Woodley 1993).

Ecological self-organisation is a key component of at least two explicit descriptions of ecological integrity. Kay & Schneider (1995) identify three 'organisational facets' that ecological integrity encompasses: the ability to maintain normal operations under normal environmental conditions, the ability to cope with changing environmental conditions, and the ability to continue the process of self-organisation. Müller (1998b) condenses these facets, and calls those ecological systems integer that (i) are able to maintain their organisation and steady state after small disturbances, and that (ii) have a sufficient adaptability and developmental capacity to continue self-organised development. Such interpretations of ecological self-organisation lead to indicator definitions that satisfy important requirements of the 'one-value-proposition'. As the long-term functioning of the ecological life support system depends on the long-term availability of the *conditions* for ecological self-organisation, the future-directed *potential* of an ecological system to self-organise has to be assessed. This potential depends on conditions and constraints to be detailed in the next subsection.

5.2 Thermodynamic ecosystem theory and the conditions for ecological self-organisation

The thermodynamic, albeit descriptive interpretation of ecological integrity by Schneider and Kay (1994*a*, 1994*b*) represents a natural starting point for a quantification of the factors influencing ecological self-organisation as they link thermodynamic theory with ecosystem succession via ecological self-organisation. The dissipation of thermodynamic gradients is not "*the only imperative governing life*" (Schneider & Kay 1994b), however. Consequently, one cannot base an appropriate account of the potential for ecological self-organisation exclusively on a thermodynamic analysis of ecosystems. Firstly, we outline such an exclusively thermodynamic account of ecological self-organisation. Secondly, the analysis turns to a less unidirectional application of thermodynamic principles to set up the framework for a quantification of the normatively relevant capacity of ecological systems to self-organise (section 5.3.).

Schneider & Kay (1994b) base their description of ecosystem processes on the 'restated second law' of thermodynamics:

> *"The thermodynamic principle which governs the behaviour of systems is that, as they are moved away from equilibrium, they will utilize all avenues available to counter the applied gradients. As the applied gradients increase, so does the system's ability to counter further movements from equilibrium."*

Because 'life' can be viewed as a means to dissipate the sun-induced gradients, biological systems develop in a way that increases the degradation of energy following the restated second law. Particularly, the thermodynamic hypothesis on ecosystem development states that the course of evolution and ecosystem succession is driven by the thermodynamic imperative to "*develop more complex structures with greater diversity, and more hierarchical levels to abet energy degradation*" (Schneider & Kay 1994b). Similar views on the forces driving ecosystem succession are expressed by Jørgensen (1997) who formulates a 'fourth law of thermodynamics':

> *"If a system receives a through-flow of exergy, the system will utilise this exergy to move away from thermodynamic equilibrium. If more than one pathway is offered to move away from thermodynamic equilibrium, the one yielding the most stored exergy, i.e. with the most ordered structure or the longest distance to thermodynamic equilibrium by the prevailing conditions, will be selected."*

Müller & Nielsen (1998) integrate these approaches with the classic theory of ecosystem succession (Odum 1969), network theory (e. g., Patten 1992), synergistics (Haken 1983), and information theory (Ulanowicz 1986). They propose seven principles of ecosystem succession that describe general attributes of self-organising systems, for instance, the maximisation of exergy storage and of entropy production, or the minimisation of specific entropy production. Empirical evidence (e. g., Kutsch et al. 1999; see also below) suggests that the above mentioned thermodynamic principles of ecosystem succession are best interpreted as *tendencies* in ecosystem development, not as strict laws of nature on par with the traditional laws of thermodynamics.

One reason for the absence of strict thermodynamic optimisation in ecosystem succession is closely tied to the base of non-equilibrium thermodynamics itself. By definition, non-equilibrium systems display at least one spatial, thermodynamic gradient. Initially, thermodynamic self-organisation is a *local* process, starting from random fluctuations at one point of this gradient (Schneider & Kay 1994b). The fluctuation spreads, if it provides an autocatalytic means of gradient dissipation. In systems with very large numbers of uniform elements, such as gases or liquids, a self-organising process can easily transform the entire system because the system provides a uniform 'avenue' to dissipate the gradient. Ecological systems differ from such uniform systems in three important ways:

- The structures that emerge from ecological self-organisation consist of biomass, i. e. of a source of readily utilisable energy: they create additional thermodynamic gradients.
- The spatial distribution of ecosystem biomass enhances the heterogeneity of ecological systems.
- With any change in the system, the 'avenues' of the ecological system to dissipate any gradient change in non-linear fashion.

As a result, a local fluctuation that possesses the potential to 'optimise' the thermodynamic efficiency of the system may not spread across the entire system. The local conditions in subregions of the system may not be suitable for the propagation of this particular thermodynamic process (e. g., because of lacking but essential water or nutrients).

During ecosystem development, other self-organising processes may successfully spread along the biomass gradient that emerges from the activity of the ecosystem primary producers. The emerging structures and processes can essentially modify the environment of the system, however, and can abort 'avenues' that also degrade gradients efficiently. Hot fire or gradations of phytophagous insects, for example, are very efficient facets of ecosystem self-organisation that result in rapid, local dissipations of the biomass gradient. The depressed density of photosynthetically active tissue that follows hot fires or pest outbreaks severely reduces the capacity of the ecological system to degrade the solar gradient. Even in the little disturbed instance of forest succession, net-primary productivity is not necessarily higher in 'mature' than in aggrading forest ecosystems (Bormann & Likens 1979:177ff), which also indicates a more complex relationship between thermodynamic optimisation and ecological self-organisation.

A central aspect of this complexity is captured by interpreting ecological self-organisation along two opposing, but interdependent thermodynamic attractors. The thermodynamic tendency of ecological systems to build up biomass as a means for effective gradient dissipation represents the first attractor. The tendency to dissipate the stored exergy in this biomass represents the second attractor (tendency towards thermodynamic equilibrium). The interplay of both attractors results in a state of 'self-organised criticality' (Bass 1998) in which thermodynamic maximisation is frequently – and at times unexpectedly – reversed. There are thermodynamic reasons why there is more to ecosystem succession than ecosystem thermodynamics.

Thus, we deviate slightly from the 'fourth law of thermodynamics' (Jørgensen 1997), and do not claim that only such ecosystem structures develop that maximise some thermodynamic system measure. However, there is a *tendency* in ecological systems that they organise in a manner as to counter applied energy gradients, namely, the solar gradient. In case of *decreasing* energy input, the stored energy is slowly set free. Both processes tend to keep up or enhance self-organisation in the future. These considerations justify to include observable self-organisation as a factor that favours the long-term self-organisation capability of the ecological life support systems.

This brief analysis acknowledges the heuristic usefulness of thermodynamic ecosystem theory; it also points to biological complements that are necessary to describe the *preconditions of ecological self-organisation* more comprehensively. Ecological self-organisation depends not only on an energy gradient or on present self-organisation, it also requires material substrates. Water, carbon and nutrients are the indispensable building blocks of ecological self-organisation. Investigations of ecological long-term trends in tropical plant communities, for example, point to the importance of the development of soil nutrient status for thermodynamic self-organisation (Fränzle 1994). Pedogenetic, geologic, and climatic as well as ecological processes contribute to the spatio-temporal development of these site factors. An assessment of the long-term functionality of the ecological life support systems requires an assessment of the long-term availability of the material preconditions of ecological self-organisation. Regarding the environmental value at stake, more attention has to be paid on the *material* than on the *energetical* preconditions: the solar constant changes more slowly than nutrient and water availability.

The second extension of the thermodynamic account of the conditions of ecological self-organisation refers to an aspect of heterogeneity in ecological systems: the biological system elements differ strongly in the way they contribute to ecosystem function. The genetic programme as a main instance of biological information is no spontaneous generation of random fluctuations along a thermodynamic gradient; it is a result of evolutionary history. On one hand, biological information is responsible for the differentiation between primary producers, consumers and decomposers, between different trophic guilds, and finally between single species or genetically distinct populations. This heterogeneity is responsible for 'local' non-linear interactions, for instance, between primary producers and consumers, that do not necessarily lead to an overall system optimisation. On the other hand, biological information enables the biota to 'use' the solar gradient more efficiently than simple physical or chemical self-organising systems can. In the same vein, biological information represents a source of 'time-tested' avenues to self-organisation for which no secure lower threshold can be established.

Figure 1 emphasises the role of the conditions of ecological self-organisation for a thermodynamic analysis of ecosystem development. The figure outlines our concept of a biologically informed application of thermodynamic theory to ecological systems. Energetic gradients drive the biocoenotic metabolism of ecological systems. Ecosystem metabolism is equivalent with the energetic aspect of ecological self-organisation. Depending on the availability of particular biological information, the biocoenotic metabolism provides the energy for the *construction* of ecological structures. At least on the level of the single organism or population, the biological structure is used to enhance the efficiency or effectiveness with which the organisms take part in the ecosystem's 'struggle for life'. Biological information, its environmental context, the availability of material substrates, even a part of the energetical substrates (availability of utilisable carbon and of *oxygen*) result from chance events as well as from previous turns of the ecological self-organisation loop in the centre of the Figure.

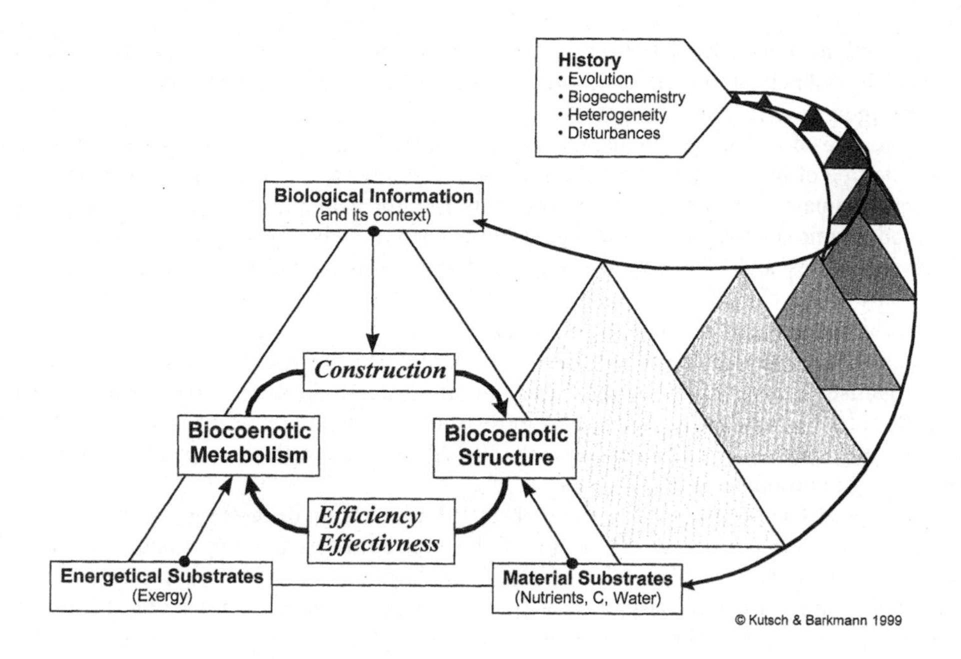

Figure 1: The conditions for ecological self-organisation (Kutsch and Barkmann 1999)

5.3 Quantifying the potential for long-term ecological self-organisation

In order to quantify the potential for long-term ecological self-organisation, biological information, long-term availability of energy and nutrients, as well as the present degree of ecological self-organisation have to be assessed and measured. Ecological self-organising capacity rests on these three conditions. It is generally agreed that quantifications of ecological integrity have to rely on sets of indicators rather than direct measurements of ecological self-organisation (Noss 1995). In this conceptual paper, we will not elaborate on indicator development in detail (see Kutsch et al. 1999). Instead, we show how the 'one-value-proposition' regarding ecological integrity solves some troubling problems of indicator development dealing with scale decisions and with human-dominated systems.

"Decisions of scale are really public policy decisions [...] while scientific information is important in resolving them, they are underdetermined by science itself" (Norton 1992:36). The one-value-proposition of ecological integrity copes ideally with Norton's insistence on the importance of normative considerations regarding scale decisions because the suggested framework easily accommodates the divergent demands of the planning process. With the exception of the entire biosphere, any planning process focuses on a specific scale that is embedded in its regional environment. Likewise, the indication of the self-organising capacity of ecological systems has to deal with the scale at issue – and with the respective regional scale. This differentiation allows for a consideration of the inherent self-organisational capabilities of the focal area, which is complemented by an assessment of cross-scale effects: As much as the self-organising capacity of the regional scale consists of the self-organising capacity at its subordinated local scales, the self-organising capacity of the regional scale provides the context for self-organisation at the focal scale. Both research or planning questions are legitimate:

- What is the contribution of the focal scale to the self-organising capacity of the regional scale? And
- how does the regional scale influence the future capacity for self-organisation at the focal scale?

The complementary use of both aspects is advised because it reduces the problem of a spuriously isolated interpretation of indicator values[11]. This procedure requires a multi-level modelling approach, however.

Because of scaling problems, mechanical interpretations of thermodynamic measures of the degree of biological self-organisation can lead to unwarranted results when applied to human-dominated systems. Corn and barley fields, for example, frequently display higher thermodynamic measures of self-organisation than a neighbouring beech forest at the Ecosystem Research Range in the Bornhöved Lakes District, Schleswig-Holstein (Germany) (Kutsch et al. 1999). Such measures have, however, been suggested to serve as appropriate indicators of ecological self-organisation, and of the 'functionality of ecological systems' (Barkmann et al. 1998, Müller 1998a). Unfortunately, the obviously 'worse', less self-organised systems reveal higher self-organisation indicators. Apart from technical considerations on the energy budget of terrestrial ecosystems (cf. Kutsch et al. 1999), the one-value-proposition yields a consistent rationale to deal with this apparent paradox by facilitating appropriate scale decisions.

Biological self-organisation can be observed not only on different geographic, but also on different conceptual scales. Not all of the conceptual scales are equally relevant in respect to the value of ecological integrity. After termination of human interferences with self-organisation at the ecosystem level (i. e., after cessation of farming), the former corn field will rapidly change in structure and species composition. While the single corn plants in the field displayed considerable *developmental* and *physiological* self-organisation, the system was overwhelmingly farmer-organised at the ecosystem scale. These short-term phenomena of biological self-organisation hardly contribute to the required long-term perspective of SD and ecological integrity. Consequently, indicators of momentary self-organisation have to be excluded or weighed down in an overall assessment of the self-organising capacity of manifestly farmer-organised systems.

The interpretation of the nutrient or biomass pool of agriculturally transformed ecosystems poses a similar problem. The application of some basic rules of soil conservation combined with the brute force of high-intensity fertilisation has resulted in increasing soil and ecosystem pools of carbon, nutrients and biomass of the arable land in many industrialised countries – in spite of high concurrent nutrient losses from the same lands. Again, the increasing pools can hardly be considered as an expression of ecological self-organisation as they depend directly on the continuance of high-input farming systems. In fertilised systems, i. e. in human-organised systems, nutrient pool size is thus not an appropriate measure of the long-term capacity for ecological self-organisation. The pool size remaining after the nutrient losses subsequent to the termination of fertilisation represents a more adequate measure. Otherwise, pool sizes are the result of an accelerated system throughput that operates on a much shorter time-scale than required for the intergenerational scope of ecological integrity and SD.

As humans interfere with biological self-organisation in a vast array of intensities, there is no clear-cut criterion which intensity precludes a mechanical application of the indicators. Based on the normative foundation of the value of ecological integrity, the necessary scale decisions can be substantiated, however: the conditions must be relevant to biological self-organisation beyond the single plant and beyond a few growing seasons. Thus, we suggest that indicators of momentary self-organisation be employed inversely proportional to human interference on the site (as measured, for example, by hemerobie classes; Jalas 1955, Sukopp 1997, Schrautzer 1988).

[11] The low species-diversity of ombrotrophic peatlands (self-organisation component 'biological information'), for example, can only be correctly evaluated in the context of the species-composition of its environment on the landscape scale.

6. The utility of ecological integrity: a risk theoretical illustration

Game theory can illustrate the benefits derived from the value of ecological integrity according to the one-value-proposition. The following excursion expands on the value of ecological integrity to introduce the notion of the 'utility of ecological integrity' in a technically unequivocal way. We employ a modified standard model in game theory, tellingly called 'Game Against Nature', which is regularly used to represent stochasticity in game theoretical models (Holler & Illing 1996). Ecologists familiar with game theory may wish to skip some of the more explanatory passages.

The two 'players' involved in our Game Against Nature are 'Nature' and 'Humanity'[12]. Both players choose from two strategies (Table. 1). Humanity can opt for two economic development options: a risky project 'A' and a less risky project 'B'. Project A is risky because it results in an economic yield of 8 units in a 'favourable' environment, but in only 2 units in an 'unfavourable' environment. B is less risky because it yields 5 units in the favourable, but still 3 units in the unfavourable case. The player 'Nature' can actualise one of two alternative options regarding the suitability of the environment in which the development projects take place. The first option brings about the favourable environment for the success of the projects, the second results in unfavourable conditions.

The ecological rationale for the set up is an extremely simplified model of the interaction of a physically growing economy in a finite world. Project A is huge, and potentially damaging to the ecological life support systems. It is optimised only in respect to economical considerations. The maximum benefits of project A are accordingly high – if it turns out that the damage to the ecological system does not counteract the economic success of the project. Put differently, none of the inevitably lessened ecosystem services turned out to be 'essential'. Project B is smaller in scale, or more 'environmentally soundly' planned. Thus, its maximum damage to the life support system is smaller, and its performance if 'Nature turns against the project' is better. On the other hand, the reduced scale (or the costs of environmental mitigation) results in a smaller maximum economic yield.

Table 1: Yield matrix of the Game Against Nature. Yields are in arbitrary units; p-values denote conditional probability estimates for 'Nature's option' (for further explanation see text)

Human Projects (X)	Yield depending on 'Nature's options'		Expected Yield
	'favourable'	'unfavourable'	of Projects E(X)[13]
A	8 (p=0.5)	2 (p=0.5)	5
B	5 (p=0.8)	3 (p=0.2)	4.6

The smaller project has a second advantage in addition to better performance in the worst case (3 vs. 2 yield units in the unfavourable environment). Because project B is potentially less damaging, it will not only disturb the ecological systems *less effectively*, it will also disturb the ecological systems *less likely* in a way that undermines its own success[14]. Because of this interaction, two different, conditional probability estimations of 'Nature's choice' are applied depending on Humanity's own option. For the sake of simplicity, we stipulate that Humanity can correctly estimate these conditional probabilities. The conditional probability estimation for a 'favourable' environment is 0.8 if

[12] The game theoretical terms 'player', 'strategy', 'choice' etc. are used in a strictly technical sense here. They are not meant to imply any 'reasoning' on the part of nature.

[13] Calculation: E(X) = yield (X; favourable) * p(X; favourable) + yield (X; unfavourable) * p(X; unfavourable)

[14] The example would also work, if the probability of 'Nature's choice' would not depend on the choice of the human 'player', or if no estimations on the probabilities were available at all (classical uncertainty calculus).

probabilities. The conditional probability estimation for a 'favourable' environment is 0.8 if project B is chosen, but it is only 0.5 if humanity chooses project A. The less damaging project has a higher probability to realise its maximum yield.

Given this model situation, it is not immediately obvious which option humanity should choose. More timid human players will certainly opt for the less risky project B; but are there strict rationality criteria that demand an option for project B? Without further information, risk theory denies this question: the choice of a certain course of action under uncertainty depends primarily on a *subjective* factor that modulates the preference order of differently risk-affected options[15]. This factor is called *risk attitude* (Holler & Illing 1998). Detailed game theoretical analyses focusing on risk attitude are used to foster an efficient management of global environmental risks (cf. Endres & Ohl 1998).

There are three types of risk attitude, ranging from *risk preference* via *risk neutrality* to *risk aversion*. Because preference orders are based on *utility* considerations, risk attitude modifies the utility that is gained from one yield unit. Risk attitude is often denoted by the identifier α, here, with $\alpha<0$: risk preference, $\alpha=0$: risk neutrality, and $\alpha>0$: risk aversion. Risk neutrality is the simplest case: The risk utility value $\Phi(X)$ of an option/project X equals its expected yield value $E(X)$[16]:

$$\Phi(X) = E(X) \qquad \text{for } \alpha = 0$$

A risk neutral decision maker would have to opt for project A because the expected yield of project A, E(A), is 5 units while E(B) amounts only to 4.6 units. In Figure 2, the expected yield of A is situated north-east of E(B) at higher yield values. For risk averse or risk preferring decision makers, the risk utility of an option differs from its expected yield. In case of risk aversion, the utility of any risk-affected project is smaller than the expected yield; in case of risk preference it is larger. In this simple 2x2-strategy Game Against Nature, the size of the modification – and accordingly the risk utility values – depend on the deviations of the two single yields of one project from their expected yields, on the probability with which Nature actualises the 'favourable' or the 'unfavourable' environment, and on the specific risk attitude.

The risk utility of the projects can be expressed formally by a *risk utility function* Φ, which integrates risk attitude, the yields of the projects and the associated probabilities of favourable environments. At risk neutrality, all projects with an identical risk utility are located on a straight line through their expected yield E(X). This line represents the *risk utility indifference curve* for $\alpha = 0$. Risk aversion results in convex risk utility indifference curves (displayed in Figure 2); risk preference leads to a concave risk utility indifference curve.

[15] We seem to confuse 'risk' with 'uncertainty' here (see also fn. 13). Regarding the illustrative purpose of this section, all considerations of risk attitude apply also to an 'uncertainty attitude' (cf. Hurwicz optimism-pessimism parameter; Cansier 1993:48). The risk utility function used below (see fn. 16) is sufficiently general to model risk as well uncertainty decision rules.

[16] For simplicity, we express risk utility in yield units here (one yield unit = one *risk utile*).

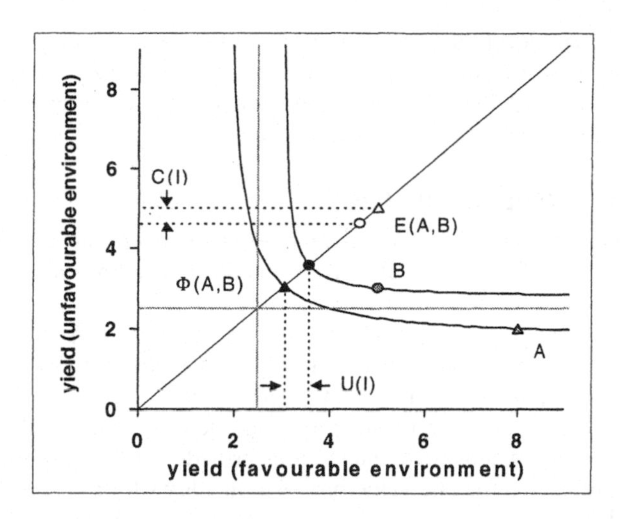

Figure 2: Risk utility indifference curves for the projects A and B at a risk preference of $\alpha = 0.5$. E(A,B): expected yield of the projects; Φ(A,B): risk utility value of the projects; U(I): utility of integrity; C(I): cost of integrity

If the analysis is accepted that SD and ecological integrity demand a reduction of the risks associated with economic development in face of ecological uncertainty, their common environmental value demands risk-averse environmental conduct. Figure 2 shows two convex risk indifference curves for an intermediate risk aversion of $\alpha = 0.5$. If, in model terms, one suspects that the yield of the risky project A under unfavourable conditions is smaller than a certain minimum value (for example, smaller than a required minimum consumption of 2.5 yield units; see Figure 2), a risk aversion of 0.5 represents a careful but plausible choice. Given the model parameters, the convexity of the risk utility indifference curve results in a higher risk utility value for project B.

We define the 'utility of ecological integrity' U(I) as the difference of the risk utility values of two alternative courses of environmental conduct (see Figure 2). Given the model parameters and the used risk utility function[17], the utility of ecological integrity U(I) equals

$$U(I) = \Phi(B) - \Phi(A) = 3.65 - 3.05 = 0.6$$

In addition to the benefits from risk adverse conduct in accordance with the value of ecological integrity, the associated costs C(I) can be expressed as the difference in the expected yield values representing the difference of the opportunity costs at $\alpha = 0$:

$$C(I) = E(A) - E(B) = 0.4$$

Both the utility of ecological integrity as well as its cost are *a priori* measures that project the utility or costs that are incurred in decision making no matter how 'Nature' acts afterwards.

Concluding this section, we stress again that the numerical values used for the above calculations are token numbers with the single purpose to illustrate the fundamental idea of

[17] The applied risk utility function can map a number of specific decision rules for risky choices as well as for choices in complete uncertainty on the probabilities involved (e. g., Bernoulli, minimax, maximin, Laplace, Hurcwicz). Specifically,

$$\Phi(X) = y + \frac{x-y}{2} - \alpha * \frac{x-y}{2} + (x-y) * (0.5 - p_y) * (1 - |\alpha|)$$

with y: 'unfavourable' yield; x: 'favourable' yield; p_y probability of the 'unfavourable' environment conditional on choosing project X; α: risk/uncertainty attitude ($-1 \leq \alpha \leq 1$)

the utility of ecological integrity. Similarly, the game theoretical model employed is reduced to its most essential features. Nevertheless, it covers one essential aspect of the developmental choices that real-world decision makers are faced with if they intend to decide in agreement with the fundamental norms of SD.

7. Summary: Hedging our bets

The one-value-proposition of ecological integrity singles out one crucial dimension of ecological integrity. This crucial dimension manifests itself as an environmental value that consists of protection against the ecological risks posed by a triple uncertainty regarding the long-term functioning of the natural life support systems. This dimension is normatively legitimised by its essential connection to the ecological long-term perspective of Sustainable Development. Provisions in favour of the 'Economy of Nature' in European environmental assessment and nature protection acts as well as other legal provisions in favour of the totality of ecosystem interactions provide similar support.

Most significantly, the one-value-proposition allows for a successful operationalisation of ecological integrity, for instance, by indicator development. The development of scientifically valid and normatively legitimised integrity indicators may turn out to be the decisive step towards an answer to the question 'How much preservation is enough?' (cf. Norton 1989). According to Norton's justice theoretical analysis of intergenerational equity considerations in the environmental decision-making process, the answer depends (i) on an assessment of the substitutability of essential ecosystem services, (ii) on an assessment of the capability of the ecological life support systems to withstand significant change (especially, species loss), and (iii) on the degree of risk aversion (Norton 1989). Because of the non-specific nature of the threads posed by the triple uncertainty associated with (i) and (ii), any antidote to the threads has to rely on the most general conditions of the functioning of the natural life support systems. We derive these features from a biologically informed, thermodynamic account of self-organisation in ecosystems. Consistent with Norton's results, the degree of risk aversion (iii) controls the magnitude of 'pre-emptive' measures and of constraints to human activity in our interpretation of ecological integrity within the SD paradigm.

The analysis of the requirements for ecological self-organisation reveals three factors that promote the long-term capability of ecological systems to self-organise: (1) availability of biological information, (2) availability of utilisable energy and of material substrates (nutrients, carbon, water), and (3) the already existing degree of ecological self-organisation. The identification of suitable indicators for these components is an ongoing process, for which criteria and first results have been presented (for the latest state of indicator development, see Kutsch et al. 1999).

Risk utility analysis demonstrates that careful, risk averse decision makers can derive *a priori* utility from protective measures no matter how robust the ecological systems *a posteriori* turn out to be. The utility gained from proactive measures in favour of the requirements of ecological self-organisation performs like the utility gained from entertaining a fire brigade: a local community profits from a fire brigade not just in the case that a house is in flames. The community is continuously gaining utility from entertaining a fire brigade because the pure possibility of raging fires is a very upsetting threat to the values of the community. With adequate preparation, however, we feel more content and secure. Ecological integrity is a first order management principle that adequately prepares against the threats highlighted by ecological uncertainty. The utility of ecological integrity is the utility of hedging the bets on the success of the global development project.

Acknowledgements

Funding was provided by the German Federal Department of Research and Technology (BMFT) for JB's doctoral work within the Bornhöved Lakes Ecosystem Research Project.

We would like to thank our collaborators at the Kiel Ecology Center, namely, F. Müller, W. Kutsch and R. Baumann, and S. Bögeholz for useful comments on the ideas expressed in the paper.

REFERENCES

Aber, J.D. and J.M. Melillo, 1991. Terrestrial Ecosystems. Saunders, Philadelphia.

Barkmann, J., 1999a. Ressourcenschutz oder Erhaltung der Biodiversität? Zu den normativen Grundlagen nachhaltigen Landschaftsmanagements. Unpublished typescript.

Barkmann, J., 1999b. Die Umweltfunktion – ein hinderliches, zielführendes oder überflüssiges Konzept im Nachhaltigkeitsdiskurs? Paper presented at the workshop 'Ökosystemare Schutzgüter, Qualitätsziele und Indikatoren', Salzau, Germany, May 18-20, 1998.

Barkmann, J., R. Baumann, B. Breckling, U. Irmler, F. Müller, C. Noell, H. Reck, E.W. Reiche, and W. Windhorst, 1998. Ökologische Integrität als Leitbild für Ökosystemschutz und nachhaltige Landschaftsentwicklung – Eine Diskussionsgrundlage. Paper presented at the workshop 'Ökosystemare Schutzgüter, Qualitätsziele und Indikatoren', Salzau, Germany, May 18-20, 1998.

Bass, B., 1998. Applying Thermodynamic Orientors: Goal Functions in the Holling Figure-Eight Model. In: Müller, F. and M. Leupelt (Eds.). *Eco Targets, Goal Functions, and Orientors*. Springer, Berlin. 193-208.

Birnbacher, D., 1980. Sind wir für die Natur verantwortlich? In: Birnbacher, D. (Ed.). *Ökologie und Ethik.* Reclam, Stuttgart. 103-139.

BMUNR, 1996. Schritte zu einer nachhaltigen, umweltgerechten Entwicklung: Umweltziele und Handlungsschwerpunkte in Deutschland – Grundlage für eine Diskussion. BMUNR, Bonn. (June 1996)

Bormann, F.H. and G.E. Likens, 1979. Pattern and Process in a Forested Ecosystem. Disturbance, Development and the Steady State Based on the Hubbard Brook Ecosystem Study. Springer, New York.

Bundesnaturschutz-Gesetz, 1987. Gesetz über Naturschutz und Landschaftspflege vom 13. März 1987. Bundesgesetzblatt I. S. 1458.

BUND/MISEREOR (Eds.), 1996. Zukunftsfähiges Deutschland – Ein Beitrag zu einer global nachhaltigen Entwicklung Erdpolitik. Eine Studie des Wuppertalinstitits für Klima, Umwelt und Energie. Birkhäuser, Basel.

Cansier, D., 1993. Umweltökonomie. UTB 1749. Gustav Fischer, Stuttgart.

Daily, G.C., S. Alexander, P. R. Ehrlich, L. Goulder, J. Lubchenco, P. A. Matson, H. A. Mooney, S. Postel, S.H. Schneider, D. Tilman, and G. M. Woodwell, 1997. Ecosystem Services: Benefits Supplied to Human Societies by Natural Ecosystems. *Issues in Ecology* 2 (Spring 1997).

DeGroot, R.S., 1992. Functions of Nature: evaluation of nature in environmental planning, management and decision making. Wolters-Noorhoff, Groningen.

Endres, A. and C. Ohl, 1998. Das Kooperationsverhalten der Staaten bei der Begrenzung globaler Umweltrisiken: Zur integration stochastischer und strategischer Unsicherheitsaspekte. Unpublished typescript.

Enquete-Kommission, 1998. Konzept Nachhaltigkeit. Vom Leitbild zur Umsetzung. Abschlußbericht der Enquete-Kommission „Schutz des Menschen und der Umwelt - Ziele und Rahmenbedingungen einer Nachhaltig zukunftsverträglichen Entwicklung". Deutscher Bundestag, Bonn.

Fränzle, O., 1994. Thermodynamic aspects of species diversity in tropical and ectropical plant communities. *Ecological Modelling* 75/76:63-70.

Freeman III, A.M., 1979.The Benefits of Environmental Improvement: Theory and Practice. Johns Hopkins, Baltimore.

Funtowicz, S.O. and J. R. Ravetz, 1995. Science for the post normal age. In: Westra, L. and J. Lemons (Eds.). *Perspectives on Ecological Integrity*. Kluwer, Dordrecht. 146-161.

Haken, H., 1983. Synergetics: an introduction: nonequilibrium phase transitions and self-organization in physics, chemistry, and biology. Springer, New York.

Hampicke, U., 1992. Ökologische Ökonomie. Individuum und Natur in der Neoklassik. Natur in der öklogischen Theorie: Teil 4. Westdeutscher Verlag , Opladen (Germany).

Hauff, V. (Ed.), 1987. Unsere gemeinsame Zukunft. Der Bericht der Weltkommission für Umwelt und Entwicklung (Brundtland-Bericht). Eggekamp, Greven (Germany).

Heydemann, B., 1997. Leistungen der Natur für die Gesellschaft. In: *BMUNR, 1997. Ökologie - Grundlage einer Nachhaltigen Entwicklung in Deutschland.* Conference papers (April 29 – 30, 1997; Wissenschaftszentrum Bonn-Bad Godesberg). BMUNR, Bonn. 21-28.

Holler, M.J. and G. Illing, 1996. Einführung in die Spieltheorie. 3rd ed. Springer, Berlin.

U.S. Cong. House, 1997. House of Representatives, Committee on Agriculture, joint with Committee on Resources, Wednesday, April 9, 1997. Washington, DC. Serial No. 105–1 (http://commdocs.house.gov/committees/resources/hiiforest.000/hiiforest_0f.htm).

Jalas, J., 1955. Hemerobe und hemerochore Pflanzenarten. Ein terminologischer Reformversuch. Acta Soc. Flora Fenn. 72(11):1-15.

Jørgensen, S.E., 1997. Thermodynamik offener Systeme. In: Fränzle, O., F. Müller, and W. Schröder (Eds.). *Handbuch der Umweltwissenschaften: Grundlagen und Anwendungen der Ökosystemforschung*. Ecomed, Landsberg (Germany). Section III-1.6.

Karr, J.R., 1981. Assessment of biotic integrity using fish communities. *Fisheries* 6:21-27.

Karr, J.R., 1992. Ecological Integrity: Protecting Earth's Life Support Systems. In: Costanza, R., B.G. Norton, and B.D. Haskell (Eds.). *Ecosystem Health – New Goals for Environmental Management*. Island Press, Washington, D.C. 223-238.

Karr, J.R. and D.R. Dudley, 1981. Ecological perspective on water quality goals. Environmental Management 5:55-68.

Karr, J.R and E.W. Chu, 1995. Ecological integrity: reclaiming lost connections. In: Westra, L. and J. Lemons (Eds.). *Perspectives on Ecological Integrity*. Kluwer, Dordrecht. 34-59.

Kay, J.J., 1993. On the Nature of Ecological Integrity: Some Closing Comments. In: Woodley, S., J. Kay, and G. Francis (Eds.). *Ecological Integrity and the Management of Ecosystems*. St. Lucie Press, Ottawa. 201-212.

Kay, J.J. and E.D. Schneider, 1995. Embracing Complexity: The Challenge of the Ecosystem Approach. In: Westra, L. and J. Lemons (Eds.). *Perspectives on Ecological Integrity*. Kluwer, Dordrecht. 49-59.

Kutsch, W. and J. Barkmann, 1999. Ecological self organisation and its constraints. Unpublished figure.

Kutsch, W., W. Steinborn, M. Herbst, R. Baumann, J. Barkmann, and L. Kappen, 1999. In search of indicators of ecological integrity from ecosystem theory. Ecosystems, *submitted*.

Lemons, J., 1995. Ecological Integrity and National Parks. In: Westra, L. and J. Lemons (Eds.). *Perspectives on Ecological Integrity*. Kluwer, Dordrecht. 177-201.

McIntosh, R.P., 1985. The Background of Ecology: Concept and Theory. Cambridge University Press, Cambridge.

Metzner, A., 1998. Constructions of Environmental Issues in Scientific and Public Discourse. In: Müller, F. and M. Leupelt (Eds.). *Eco Targets, Goal Functions, and Orientors*. Springer, Berlin. 312-333.

Müller, F., 1998a. Ableitung von integrativen Indikatoren zur Bewertung von Ökosystem-Zuständen für die Umweltökonomischen Gesamtrechnungen. Beiträge zu den Umweltökonomischen Gesamtrechnungen, vol. 2. Metzler-Poeschel, Stuttgart.

Müller, F., 1998b. Ecological components of Sustainable Development – Some Fundamentals for the Derivation of Ecosystem Indicators. Presentation at the European Workshop 'Research and Monitoring as Key Elements for Sustainable Development in the Limestone Alps', Bled, Slovenia, October 1998.

Müller, F. and S.N. Nielsen, 1996. Thermodynamische Systemauffassungen in der Ökologie. In: Mathes, K., B. Breckling, and K. Ekschmitt (Eds.). *Systemtheorie in der Ökologie*. Ecomed, Landsberg (Germany). 45-60.

Norton, B.G., 1992. A new Paradigm for Environmental Management. In: Costanza, R., B.G. Norton, and B.D. Haskell (Eds.). *Ecosystem Health – New Goals for Environmental Management*. Island Press, Washington, D.C. 23-41.

Norton, B.G., 1989. Intergenerational Equity and environmental decisions: a model using Rawls' veil of ignorance. *Ecological Economics* 1:137-159.

Noss, R.F., 1995. Ecological integrity and Sustainability: Buzzwords in conflict? In: Westra, L. and J. Lemons (Eds.). *Perspectives on Ecological Integrity*. Kluwer, Dordrecht. 60-76.

Odum, E.P., 1991. Prinzipien der Ökologie. Lebensräume, Stoffkreisläufe, Wachstumsgrenzen. Spektrum, Heidelberg.

Odum, E.P., 1969. The strategy of ecosystem development. *Science* 164:262-270.

Patten, B.C., 1992. Energy, emergy, and environs. *Ecological Modelling* 62:29-70.

Pfordten, D. v.d., 1996. Ökologische Ethik – Zur Rechtfertigung menschlichen Verhaltens gegenüber der Natur. Rowohlts Enzyklopädie. Rowohlt, Hamburg.

Regier, H.A., 1993. The Notion of Natural and Cultural Integrity. In: Woodley, S., J. Kay, and G. Francis (Eds.). *Ecological Integrity and the Management of Ecosystems*. St. Lucie Press, Ottawa. 3-18.

Schneider, E.D. and J.J. Kay, 1994a. Complexity and Thermodynamics: Towards a New Ecology. *Futures* 24:626-647.

Schneider, E.D. and J.J. Kay, 1994b. Life as a Manifestation of the Second Law of Thermodynamics. *Mathematical and Computer Modelling* 19:25-48.

Schrautzer, J., 1988. Pflanzensoziologische und standörtliche Charakteristik von Seggenriedern und Feuchtwiesen in Schleswig-Holstein. Mitteilungen der Arbeitsgemeinschaft Geobotanik in Schleswig-Holstein und Hamburg, vol. 38, Kiel.

Shrader-Frechette, K., 1995. Hard ecology, soft ecology and ecosystem integrity. In: Westra, L. and J. Lemons (Eds.). *Perspectives on Ecological Integrity*. Kluwer, Dordrecht. 125-145.

Solow, R.M., 1974. Intergenerational Equity and Exhaustible Resources. *Review of Economic Studies* 41:29-45. (cited in Hampicke 1992)

SRU, 1996. Umweltgutachten 1996. Der Rat von Sachverständigen für Umweltfragen (SRU). Metzler-Poeschel, Stuttgart.

SRU, 1994. Umweltgutachten 1994. Der Rat von Sachverständigen für Umweltfragen (SRU). Metzler-Poeschel, Stuttgart.

Sukopp, H., Indikatoren für Naturnähe. In: *BMUNR, 1997. Ökologie – Grundlage einer Nachhaltigen Entwicklung in Deutschland.* Conference papers (April 29 – 30, 1997; Wissenschaftszentrum Bonn-Bad Godesberg). BMUNR, Bonn. 71-84.

Ulanowicz, R.E., 1986. Growth and development: ecosystems phenomenology. Springer, Berlin.

Ulanowicz, R.E., 1995. Ecosystem integrity: A causal necessity. In: Westra, L. and J. Lemons (Eds.). *Perspectives on Ecological Integrity*. Kluwer, Dordrecht. 77-87.

UVP-Gesetz, 1990. Gesetz zur Umsetzung der Richtlinie des Rates vom 27. Juni 1985 über die Umweltverträglichkeitsprüfung bei bestimmten öffentlichen und privaten Projekten (85/337/EWG) vom 12. Februar 1990. Bundesgesetzblatt I. S. 205.

Westra, L., 1994. An environmental proposal for ethics: the principle of integrity. Rowman & Littlefield, Lanham (MD, USA).

Westra, L., 1995. Ecosystem integrity and sustainability: the foundational value of the wild. In: Westra, L. and J. Lemons (Eds.). *Perspectives on Ecological Integrity.* Kluwer, Dordrecht (The Netherlands). 12-32.

Woodley, S., 1993. Monitoring and Measuring Ecosystem Integrity in Canadian National Parks. In: Woodley, S., J. Kay, and G. Francis (Eds.). *Ecological Integrity and the Management of Ecosystems.* St. Lucie Press, Ottawa. 155-176.

III.1.4 Sustainability: Application of Systems Theoretical Aspects to Societal Development

Hartmut Bossel

1. Introduction

Human society is a complex system embedded in another complex system – the natural environment – on which it depends for its support. Since human activities determine developments in both the human as well as the natural system to a very significant extent, we should be well informed about the state of these complex systems. That means that we must identify and watch essential indicators; and if our goal is 'sustainable development', we must be sure that the indicator set, can inform us about the viability and sustainability of the total system and its components.

'Sustainable development' is now widely accepted as a goal for human society. Although there are different interpretations of the concept, and paths of sustainable development will differ, there is general agreement that the concept entails concern for four different aspects in particular: (1) the viability of human society, (2) the efficiency of resource use, (3) the viability of the natural system, (4) coexistence without exploitation.[1] Comprehensive attention to these aspects requires a systems approach, substantial knowledge of the systems involved and of development constraints, and the elaboration of reliable methods for assessing viability and sustainability of the systems involved.

2. Constraints of societal development

There are numerous constraints that restrict societal development. A few can be negotiated to some degree; most are unchangeable. The total range of theoretical future possibilities is reduced by these constraints, leaving only a limited, potentially accessible part of state space, the *accessibility space.*

Natural laws and processes, rules of logic: Natural laws and the laws of logic cannot be broken. Certain things are simply impossible to achieve. Examples of such restrictions are the nutrient requirements for plant growth, or the energy efficiencies of thermal processes. Natural laws, logic, and permissible physical processes therefore provide a first constraint c_1 on accessibility space.

Physical environment and its constraints: Human society evolves within, dependent on, and as part of the global environment. Its development is constrained by the conditions of the global environment: available space; waste absorption capacity of soils, rivers, oceans, atmosphere; availability of renewable and non-renewable resources; soil fertility and climate. Some of these are state limitations (depletable resources), others are rate limitations (waste absorption). Sustainable development paths must adhere to these constraints. This is a second restriction c_2 of accessibility space.

Solar energy flow, material resource stocks: There is only one permanent energy supply on earth: solar energy. All still available fossil and nuclear resources amount to a few months of global insolation – if all of the solar energy reaching the earth during those months could be captured. In a sustainability scenario, the energy limitation is therefore the rate of solar energy flux that can be captured and used (by plants and technology). The material resources are limited to the present global supplies. They have been recycling on this earth for some four billion years. Recycling is therefore also an essential requirement of sustainability. These energy and material constraints are a third restriction c_3 of accessibility space.

Carrying capacity: Organisms and ecosystems, including humans, require certain amounts of solar energy flux, nutrients, water, etc. per unit of organism supported, either directly (plants), or indirectly as food in plant or animal biomass. The consumption rate depends on the organism and its 'lifestyle'. In the long-term, it is limited by the (photosynthetic) productivity of a region, i.e. the amount of plant biomass that can be produced there per year, which is determined by the resource (nutrient, water, light) that is 'in the minimum' (Liebig's Law; limiting factor). The 'carrying capacity' is the number of organisms of a given species that can be supported by the region, given its (biomass) productivity and the demands of its organisms. The carrying capacity of a region for humans depends on their material consumption. It is not only determined by food demand, but also by the demand for other resources (water, energy, rare metals, waste absorption etc.). It would be higher in a frugal society than a wasteful one. Carrying capacity is therefore a fourth restriction c_4 of accessibility. Humans can partially, and only temporarily, overcome the carrying capacity limit of a region by bringing in critical resources from other regions. Eventually, as a resource becomes scarcer in other regions as well, this transfer would have to stop.

Human actors: Humans are self-conscious, anticipatory, imaginative, creative beings. This means that they are not restricted to act in narrowly confined ways according to fixed rules of behaviour. They can invent new solutions – or they may not even see the obvious ones. This introduces as a fifth set of constraints c_5 on accessibility space a reduction to those states that are mentally and intellectually accessible. Societies which are more innovative, have a better educated and trained population, and provide a diverse and open cultural environment, have a greater accessibility space left than others where these conditions are not found.

Human organisations, cultures, technology: For a given society, and for the world as a whole, existing human organisations, cultural and political systems, available and possible technology and its systems, with their implications for behaviour and the acceptance of change, will further constrain the accessibility space. This provides a sixth set of constraints c_6.

Role of ethics and values: Not everything that remains accessible will be tolerated by the ethical standards, or other behavioural or cultural values and norms of a given society. This introduces another set of constraints c_7.

Role of time: All dynamic processes take time. For example, building up infrastructure, or introducing a new technology, or cleaning up water in groundwater passage, or restoring soil fertility, or stopping population growth, all take time, posing severe restrictions on what can be done, how quickly, or how long it may take to change things. The characteristic time constants of essential processes therefore introduce an eighth set of constraints c_8.

Societal development – whatever its form – will be restricted to the remaining accessibility space A. Everything outside is fiction, and only confuses the discussion. However, within this accessibility space, there is still a broad spectrum of options and possible paths. This leaves choices, and it introduces subjective choice and unavoidable ethical decisions.

3. Sustainability, choice, and ethics

What do we mean by 'sustainability' and 'sustainable development'? The general idea is shared and supported by many, if not most people, but more accurate definitions differ widely. One of the most commonly cited definitions (WCED, 1987)[2], stresses the economic aspects by defining sustainable development as *"economic development that meets the needs of the present generation without compromising the ability of future generations to meet their own needs."* Another (Engel, 1990)[3], takes a broader view by defining sustainable development as *"the kind of human activity that nourishes and perpetuates the historical fulfilment of the whole community of life on earth."* 'Sustainable development' obviously implies responsible management of a complex system.

To sustain something, implies valuing it enough to put an effort into maintaining its integrity. A commitment to sustainability of human and natural systems is therefore a fundamental value decision. The sustainability postulate does not come out of thin air, but can be proposed for several good reasons: (1) We may wish to acknowledge the intrinsic value of the processes and products of natural evolution and of human cultural evolution. If we value these, we must strive for ensuring their future existence, development, and evolution, i.e. for sustainability. (2) If we take a more anthropocentric point of view, we arrive at the same conclusion: humankind is dependent on the natural systems of its environment, and its own interests compel it to be concerned for their sustainability. (3) Even if we leave human interests out of this and take a biocentric view, we may want to acknowledge that nature (including humans) is a living, evolving system, and that the products of this creative process have value in their own right. An ethical choice for 'sustainable development' can therefore be made for different reasons.

A decision for sustainable development implies a valuation of future generations and systems, and hence an ethical choice. But there are many possible paths of sustainable development, with distinctly different implications for all affected – human society in particular. A further specification of the ethical principles that guide this development is necessary. Should transnational corporations, or an eco-dictator, or a scientific elite define the rules of this society to suit their own interests, or should society aim for fair partnership of all present and future systems and actors?

Environmental sustainability coupled with a continuation of present trends, where a small minority lives in luxury, partly at the expense of an underprivileged majority, could be socially unsustainable in the long run because of the stresses caused by the institutionalised injustice. And an equitable, environmentally sustainable society that exploits the environment at the maximum sustainable rate might prove to be psychologically and culturally unsustainable.

'Sustainability' of human society therefore has environmental, material, ecological, social, cultural, and psychological dimensions that call for ethical decisions. The ethical framework we adopt determines how we deal with these different issues, and what the long-term consequences will be.

Sustainable development is a dynamic concept. It implies constant evolutionary and adaptive change. Societies and their environments change, technologies and cultures change, values and aspirations change, and a sustainable society must allow and sustain such change, i.e. it must allow continuous development under conditions of sustainability. Such change must be evolutionary and self-organising. The widest possible spectrum of adaptive responses to new challenges should compete for the 'fittest' solutions. But this means that diversity of processes and functions is one of the important prerequisites for sustainability. The greater the number of different innovative responses, the better. Monocultures of any kind carry the seeds of their own destruction, while diversity allows timely adaptation.

'Sustainable development' is possible only if subsystems as well as the total system are viable; their 'interests' must be respected. The total system of which human society is a part, and on which it depends for support, is made up of a large number of component systems. The whole cannot function properly, and is not viable and sustainable, if individual component systems cannot function properly. The ethical choice for sustainability therefore demands sufficient regard for the interests of all component systems. But what is 'sufficient regard'? Again, we face an ethical choice: We have to adopt an ethical framework that regulates the relative weights that other systems carry in our decisions and actions.[4]

'Ethical choice' means adoption of a fundamental ethical principle from which other normative standards can be derived. If we have made that fundamental choice, the bounds of normative standards follow from the fundamental ethical principle, and from the relationships in the system. The standards are no longer completely open to choice, if the normative system is to be consistent. The adoption of a particular fundamental ethical principle therefore has far-reaching consequences. It determines the future accessibility space to a great extent.

Very different paths, based on different ethics, are possible for sustainable development. They will have very different consequences for the participant systems. The 'how' of 'sustainable development' is first of all a question of ethics. Let us focus on the two ethical frameworks that are the most relevant for our future: *self-interest* and *partnership ethics* (Bossel, 1998a)[5].

3.1 Implications of self-interest ethics

Utilitarian ethics are commonly defined as ethics for assuring 'the greatest good for the greatest number'. This already implies injustice, since it means *"lesser life prospects for some for the sake of a greater sum of advantages enjoyed by others"* (Rawls, 1972).[6]

According to a widespread misinterpretation of Adam Smith's concept of the 'invisible hand', 'the greatest good for the greatest number' is best achieved by minimum moral restraint on self-interest. This then serves to justify an **ethic of self-interest** whose ethical principle can be formulated (Bossel, 1978)[7] as

"Act in such a way that the direct and indirect results of your action have a high probability of producing the greatest net benefit for you."

This implies that each human actor – individual or corporate – assigns greatest ethical weight to herself or himself. Other actors – restricted in this ethical framework to currently living humans or human organisations – are accorded ethical respect on grounds of utility only. As a consequence, the results of assessments will differ from one actor to the next; the normative conclusions of different actors will not normally agree in this ethical framework. For each, they are shaped by his or her particular relationship to the total system. Certain modes of behaviour follow as a logical consequence. Where individual normative conclusions do not agree, the only means of conflict resolution is the exertion of power. And actual or potential power therefore determines behaviour guided by self-interest ethics.

Also, under this ethical framework the decisions of individual actors cannot reflect the interests of the whole system, or of (non-human or future) component systems that may have fundamental importance but are either taken for granted (ecosystems, environment), or that cannot speak for themselves. Because of its systemic myopia, this framework has a built-in bias against the viability and sustainability of the total system.

As long as human action could not significantly affect natural systems, or threaten the total system, self-interest ethics could be tolerated by nature and human society. Since this was true until recently, and since the individual 'pursuit of happiness' has – until recently – indeed often contributed to the economic welfare of many, self-interest ethics has been elevated to the status of the 'right' ethics by neo-classical economic theoreticians. But the economic system based on this system of ethics is obviously incompatible with sustainability.

3.2 Implications of partnership ethics

John Rawls has introduced into ethics the concept of 'justice as fairness' (Rawls, 1972)[8]. This principle is the result of a hypothetical situation in which no one knows his or her (present or future) place in the total system and has to choose the ethical principle 'behind a veil of ignorance'. If this principle is extended to the co-evolution of systems, in particular in a context of sustainable development, one arrives at a partnership ethic Bossel, 1978).[9]

In a **partnership ethic**, ethical considerations are extended to all present and future component systems of the total system, human or non-human, living or non-living. The Principle of Partnership Bossel, 1978)[10] is that

"All systems that are sufficiently unique and irreplaceable have an equal right to present and future existence and development."

The principle would protect individual conscious beings, species, singular ecosystems, original works of art or cultural achievements for their uniqueness, but not, for example, the individual mosquito or chicken. A similar principle has been formulated by Johnson (Johnson, 1991):[11] *"Give due respect to all the interests of all beings that have interests, in proportion to their interests."* A system is said to have interests if it can be observed to express preferences (e.g. a plant growing towards light).

The principle of partnership is not as esoteric as it may seem. It already dominates much of our everyday behaviour in families, work teams, social groups, even traffic, i.e. situations where selfish behaviour would not be tolerated for long. An ethic based on the sustainability postulate and the partnership principle implies the following in particular:

1. With respect to the natural environment, it means acknowledging species and ecosystems as systems having their own identity and right to existence, in the present and in the future. The natural environment cannot be viewed as a (supposedly infinite) source of resources, but must be viewed as 'life space' on which our existence depends, full of systems having interests, for whose future we are responsible because of our influence on them.

2. With respect to human systems, it means respecting the right to equitable treatment for all living humans, without differentiation by region, religion, race, gender, political conviction, income, wealth, or education.

3. With respect to future systems, it means respecting the right for existence and development of future generations, species, and ecosystems.

Compared to other ethical principles, application of the partnership principle would minimise power and exploitation, and maximise diversity and hence options for sustainable futures.

3.3 The ethical framework determines societal development

We cannot objectively decide which of these (or other) frameworks is 'correct'. We must be aware however of the consequences that the different possible ethical choices have for development of human society and the natural environment, even if this development should be sustainable and we must accept the full responsibility for our decisions.

The discussion of the ethical framework is important because the choice of ethics has a significant effect on our world view, on what we care to observe, i.e. on the choice of indicators of development. What is not in our ethical horizon is of no interest to us, has no value for us. We cannot feel responsible for it. It therefore has no place on our list of indicators. A self-interest mindset will pay attention to an indicator set that is very different from a set derived with a partnership mindset.

These insights have to guide our choice of indicator sets for sustainable development. In particular, they call for proper representation of the interests of all component systems. We have to at least try to assess their role and function in the total system, now and in the future. This is the task of systems analysis. We therefore find an unexpected connection between ethics and the systems view in issues of sustainable development:

1. If we start with a systems view, trying to identify the role and importance of all component systems for the sustainable development of the total system, we shall find that ethical criteria must be developed and applied to protect the interests of the various component systems contributing to the total system.

2. If we start with the ethical choice for sustainable development and try to break it down into practical ethical criteria for decision-making, we shall find that we cannot accomplish this without fairly detailed systems studies that also take into account the dynamics of development.

4. A systems view of sustainable development

We still have to identify the components of the 'total system' for which we want to achieve 'sustainable development'. Obviously, we have to consider the huge complex of interconnected human and natural systems and subsystems.

The principles of hierarchical organisation and subsidiary require that each subsystem has a certain measure of autonomy. In its particular system environment, each subsystem must be viable. The total system can only be viable if each of the subsystems supporting it is viable. For example, a region can only be viable if its economic system is viable. This has implications for the understanding and management of 'sustainable development': We must identify the subsystems that are essential for the functioning of the total system, and must determine subsystem variables (indicators) that can provide essential information about the viability (and hence sustainability) of each subsystem.

Viability of individual subsystems is obviously not enough. There is more to a successful football team (system) than a collection of healthy football players (subsystems). Each subsystem must also contribute its characteristic share to the viability of the total system. And the viability of the total system will be reflected by indicators that may bear no relationship to the viability of the subsystems.

Note that this way of looking at complex systems is recursive: If necessary, we can apply the same system/subsystem dichotomy of viable systems again at other organisational levels. For example, a person is a subsystem of a family; a family is a subsystem of a community; a community is a subsystem of a state; a state is a subsystem of a nation, etc.

For a comprehensive assessment of societal development, we must first identify the different relevant sectors of the societal system. We must include the systems that constitute 'society' as well as the systems on which human society depends. In a systems view of sustainable development the following essential subsystems of the total system can be distinguished (Bossel, 1998a)[12]: individual development, social system, government, infrastructure, economic system, resources and environment.

These six sector subsystems correspond to 'capitals' (stocks) that must be sustainably maintained: human capital, social capital, organisational capital, infrastructure capital, production capital, natural capital. In order for the total system (human system embedded in the natural system) to be viable, each of the sector subsystems must be viable. The task will be to find relevant indicators for each subsystem. Moreover, we must identify indicators that provide information about the contribution of each subsystem to the viability of the total system.

The six sectors can be aggregated to three subsystems: 'Social system' = social system + individual development + government; 'Support system' = infrastructure + economic system; 'Natural system' = resources + environment. These three subsystems correspond to the three categories of 'capital' that are often used in analyses of the total system: human capital, structural (built) capital, natural capital.

5. Sustainability and systems viability

'Health' means 'physical and mental well-being; soundness; freedom from defect, pain, or disease; normality of mental and physical functions (Webster, 1962).'[13] And 'viable' is defined as 'able ... to live and develop; able to take root and grow (Webster, 1962).'[14] When we talk about a viable system, we mean that this system is able to survive, be healthy, and develop in its particular environment, i.e. it will be sustainable in this environment. In other words, system viability and sustainability have something to do with both the system and its properties, and with the environment and its properties. A system can only exist and prosper in its environment if its structure and functions are adapted to that environment. If a system is to be successful in its environment, the particular features of that environment must be reflected in its structure and functions. And since a system usually adapts to its environment in a process of co-evolution, we can expect that the properties of the environment will be reflected in the properties of the system. The form of a fish and its mode of motion reflect the laws of fluid dynamics of its aquatic environment, and the legal system of a society reflects the social environment in which it developed.

There is obviously an immense variety of system environments, just as there is an immense variety of systems. But could it be that all of these environments have some common general properties? If that were the case, we could expect their reflections as 'basic system needs' or 'system interests' in all systems that have been shaped by their environments. These reflections would orient not just structure and function of systems, but also their behaviour in the environment. Moreover, with proper attention to these fundamental orientations ('basic orientors') of systems towards general properties of their environment, we could design systems to be successful in a given environment (Bossel, 1977, 1989, 1994, 1998a, 1998b).[15] Indicators of viability and sustainability have to reflect how well the 'basic system needs' or 'basic orientors' are satisfied under given circumstances.

System environments are characterised by six **fundamental environmental properties** (Bossel, 1998a):[16]

Normal environmental state: The actual environmental state can vary around this state in a certain range.

Resource scarcity: Resources (energy, matter, information) required for a system's survival are not immediately available when and where needed.

Variety: Many qualitatively very different processes and patterns of environmental variables occur and appear in the environment constantly or intermittently.

Variability: The normal environmental state fluctuates in random ways, and the fluctuations may occasionally take the environment far from the normal state.

Change: In the course of time, the normal environmental state may gradually or abruptly change to a permanently different normal environmental state.

Other systems: The behaviour of other systems introduces changes into the environment of a given system.

These fundamental properties of the environment are each unique, i.e., each property cannot be expressed by any combination of other fundamental properties. If we want to describe a system's environment *fully*, we have to say something about *each* of these properties: What is the normal state of the environment? What resources are available in the environment? What is the diversity and variety of the environment? How variable is it? What are the trends of change in the environment? What other systems are present?

The specific content of these fundamental environmental properties is system-specific, however. The same physical environment presents different environmental characteristics to different systems existing in it. For example, in a meadow environment shared by cows and bees, 'resources' means grass to the cow, and nectar and pollen to the bee; 'other systems' means other cows and the farmer to the cow, and other nectar-collecting insects to the bee, etc.

The environmental properties cause distinct needs orientations or 'system interests' in systems ('basic orientors'). The term 'orientor' is used to characterise the 'dimension' of normative objects like values, norms, goals, objectives to which the system has to pay attention. Often, orientors can be arranged in an 'orientor hierarchy', where more concrete orientors specify different aspects of more general orientors (Hornung, 1988)[17]. The 'basic orientors' represent the most fundamental aspects of systems orientation. They are identical across all (self-organising) systems, irrespective of their functional type or physical nature of the system.

Basic orientors are basic system needs or basic system interests. We can identify six **basic system orientors**:

Existence: Attention to existential conditions is necessary to insure the basic compatibility and immediate survival of the system in the *normal environmental state*.

Effectiveness: The system should on balance (over the long-term) be effective (not necessarily efficient) in its efforts to secure *scarce resources* from, and to exert influence on its environment.

Freedom of action: The system must have the ability to cope in various ways with the challenges posed by *environmental variety.*

Security: The system must be able to protect itself from the detrimental effects of *environmental variability*, i.e. variable, fluctuating, and unpredictable conditions outside the normal environmental state.

Adaptability: The system should be able to change its parameters and/or structure in order to generate more appropriate responses to challenges posed by *environmental change.*

Coexistence: The system must be able to modify its behaviour to account for behaviour and orientors of *other* (actor) *systems* in its environment.

For sentient beings, ***psychological needs*** constitute an additional orientor.

Because it is better adapted to the different aspects of its environment, the system equipped for securing better overall basic orientor satisfaction will have better fitness, and will therefore have a better chance for long-term survival and sustainability (Krebs and Bossel, 1997).[18] Quantification of orientor satisfaction therefore provides a measure of system viability, sustainability, and system fitness in different environments.

The assessment of orientor satisfaction, i.e. of system viability and sustainability, can be done by identifying indicators that can provide information about how well each of the orientors is being fulfilled at a given time. In other words, the basic orientors provide us with a checklist for asking a set of questions for finding out how well a system is doing in its environment (see below).

Each of the basic orientors stands for a unique requirement. That means that a minimum of attention must be paid to each of them, and that compensation of deficits of one orientor by over-fulfilment of other basic orientors is not possible. For example, a deficit of 'freedom of action' in a society cannot be compensated by a surplus of 'security'. Note that uniqueness of each of the basic orientor *dimensions* does not imply independence of individual basic orientor *satisfactions.* For example, better satisfaction of the security orientor may require a sacrifice in freedom of action because financial resources are needed for the former, and are then unavailable for the latter. But that doesn't mean that 'freedom' can be used as a substitute for 'security'.

Viability and sustainability of a system require adequate satisfaction of each of the system's basic orientors. Planning, decisions, and actions in societal systems must therefore always reflect at least the handful of basic orientors (or derived criteria) simultaneously. In analogy to Liebig's Principle of the Minimum, which states that plant growth may be constrained by a 'limiting factor' (e.g. insufficient nitrogen or water), a system's development will be constrained by the basic orientor that is currently 'in the minimum'. Particular attention will therefore have to focus on those orientors that are currently constraining.

In the orientation of system behaviour, we deal with a two-phase assessment process where each phase is different from the other. Phase 1: First, a certain minimum satisfaction must be obtained separately for each of the basic orientors. A deficit in even one of the orientors threatens the sustainability of the whole system. The system will have to focus its attention on this deficit. Phase 2: Only if the required minimum satisfaction of *all* basic orientors is guaranteed is it permissible to try to raise system satisfaction by improving satisfaction of *individual* orientors further – if conditions, in particular other systems, will allow this.

Characteristic differences in the behaviour ('life strategies') of organisms, or of humans or human systems (organisations, political or cultural groups) can often be explained by differences in the relative importance attached to different orientors (i.e. emphasis on 'freedom', or 'security', or 'effectiveness', or 'adaptability') in Phase 2 (i.e.

after minimum requirements for all basic orientors have been satisfied in Phase 1)(cf Krebs and Bossel, 1997, Bosseel, 1998a and Bosseil, 1998b).[19]

6. Assessment of progress in sustainable development

The question of sustainable development boils down to the question of sustained viability of subsystems and their contributions to the viability and sustainability of the total system. For assessing sustainability, we have to find – for each of the subsystems of the total system – indicators in each orientor category that can answer two sets of questions: (1) 'What is the viability of each sector system?' (i.e. satisfaction of each basic orientor of that system), and (2) 'How does each sector system contribute to the viability (the basic orientors) of the total system?' The general scheme of questions to which the indicator system must provide answers is given in Table 1.

Indicator sets generated by this method focus on system viability, not on the state of individual indicators. The multidimensional pattern exhibited by the indicators reveals system viability and sustainability. The assessment requires checking every one of the orientor dimensions, and it requires a definite answer to each of the questions, but it does not prescribe the details of how this should be achieved. Very often, accurate answers can be given on the basis of available qualitative evidence or the intimate knowledge of people familiar with the (sub)system.

In some cases, it may not be possible to provide a definite answer to a basic orientor question by reference to a single indicator. It may be necessary to employ several indicators to cover different aspects of the question, and it may even be necessary to construct a hierarchy of orientors to correctly represent different aspects, and to define corresponding indicators (Hornumg, 1998).[20]

In more complex systems, or in more detailed assessments, it may be necessary to consider more than two levels of system hierarchy. In these cases, the two-level process described here can be applied recursively, i.e. by identifying the sub-subsystems of a given subsystem and applying the scheme of viability questions again at a lower level of the system hierarchy. In this way, complex hierarchies of indicators can be derived, if necessary.

Table 1: General scheme for finding indicators of sustainability

Orientor	Subsystem performance	Contribution to total system
Existence	Is the system compatible with, and able to exist in its environment?	Does the subsystem contribute its share to the existence of the total system?
Effectiveness	Is it effective and efficient in its processes and operations?	Does it contribute to the efficient and effective operation of the total system?
Freedom of Action	Does it have the necessary freedom to respond and react as needed?	Does it contribute to the freedom of action of the total system?
Security	Is it secure, safe, stable despite an unpredictable environment?	Does it contribute to the security, safety, and stability of the total system?
Adaptability	Can it adapt to new challenges from its changing environment?	Does it contribute to the flexibility and adaptability of the total system?
Coexistence	Is it compatible with interacting subsystems?	Does it contribute to the compatibility of the total system with its other systems?
Psychological Needs	Is it compatible with psychological needs and culture?	Does it contribute to the psychological well-being of people?

6.1 Assessing global sustainability

The method of sustainability assessment is demonstrated using real data from the data base of the Worldwatch Institute (Worldwatch Institute, 1998)[21]. The selection of indicators from this data base, the formulation of orientor impact assessment functions, the formal assessment process, and assessment results for the 'state of the world' from 1950 – 2000 are presented and discussed in some detail elsewhere (Bossel, 1998b)[22]. In this (pedagogic) application a reduced set of 21 indicators is used, covering the seven basic orientors and the three component systems: 'human system' (social system and human development), 'support system' (infrastructure and economy), and 'natural system' (environment and resources). The results are therefore of the nature of a 'quick check' and should not be taken too seriously. Nevertheless, they show some basic and interesting trends of global dynamics. The sustainability assessment looks at the global system as a whole.

In this exercise, the two separate assessments of Table 1 are combined into one assessment, reducing the number of indicators to 21. The relevant assessment question is now: 'What is the state of satisfaction of orientor O_i with respect to (1) the human/social aspect, (2) the infrastructure/economy aspect, (3) the environment/resources aspect of the total system?' The 21 indicators selected for this assessment are shown in Table 2. The file name and line number of the Worldwatch worksheet where the indicator is defined are also listed.

Table 2: Indicator set for sustainability assessment using Worldwatch database (January 1998)

Basic Orientor	Human system	Support system	Natural system
Existence	Grain surplus factor GRNPROD.16 / 200	Debt as share of GDP in developing countries DEBT.57	World fish catch FISH.15
Effectiveness	Unemployment in European Union INCOME.83	Gross world product per person GWP.11	Grain yield efficiency GRNPROD.16 / FERTILIZ.14
Freedom of Action	Share of population age 60 and over DEMOGRA:188	Energy productivity in industrial nation PRDUCTVT.13 (Germany)	Water use as share of total runoff WATERUSE.195
Security	Share of population in cities CITIES.14	World grain carryover stock GRAIN.126	Economic losses from weather disasters DISASTER.37
Adaptability	Persons per television set TVS.1 (1 TV per household)	Capital flow to developing countries (public funds) FINANCE.14	Carbon emissions CARBON.20
Coexistence	Income share of richest 20% of population INCOME.20	Number of armed conflicts CONFLICTS.10	Recycled content of US steel STEEL.71
Psychological Needs	Refugees per 1000 population REFUGEES.17	Immunisation of infants DISEASE.22 (DPT)	Chesapeake oyster catch RESOURCE.13

A few general comments about the choice of indicators and impact assessment functions are in order.

1. Each indicator is chosen to represent the particular aspect of orientor assessment for which it was selected, and only that aspect. It must therefore be judged under that particular aspect only, not under others (for which it may also be relevant). Example: The number of hamburgers wolfed down by a customer can be taken as an indicator of his hunger, his gourmet taste, his wealth, or his nutritional awareness. In an assessment, it must be clearly said how the indicator is interpreted. Care must be taken to avoid mixing up different concerns.

2. The restriction to one indicator for each orientor aspect and each component system is obviously a crude simplification. Each indicator must therefore be understood as representing certain general trends; it should not merely be viewed within its own limited context.

3. Some indicators represent regional, not global developments. Their use is justified for the following reason: a chain is only as strong as its weakest link, and orientor theory requires that we choose indicators representing the *weakest* features of a system.

4. The impact assessment functions focus on one particular orientor aspect. This restriction must be strictly adhered to in making the assessment. Example: a certain amount of 'income inequality' might be bad for 'coexistence', but good for 'effectiveness'. It is important to mentally separate these effects, and to resist the temptation to generate a 'balanced' assessment.

5. Despite the unavoidable subjectivity involved in indicator selection and impact function formulation, the method should not be dismissed as 'another subjective method'.

There are decisive differences between the orientor assessment method and other indicator assessments: (a) the method is based on a well-substantiated theoretical framework, and (b) all steps are formalised, documented, and reproducible (Bossel, 1997 and Bossel, 1998b).[23]

6. Note that the different indicators cannot be combined into one number describing the current state of 'sustainability'. The basic needs of systems (as represented by their basic orientors) are always multidimensional; each of the basic orientors has to be satisfied separately. It is not possible to trade or even compare, say, a lack of personal freedom with an overabundance of food.

6.2 Dynamics of global sustainability 1950 – 2000

The sustainability 'state of the world' and its dynamic development from 1950 to 2000 become obvious in orientor star diagrams for each of the component systems and the total system. These are presented separately for the years from 1950 to 2000 in 10 year intervals (Figures 1 and 2). Values of less than 3 on the graphs indicate orientor satisfaction deficits, and hence sustainability deficits of the respective system.

Human system: The system progresses from a rather unbalanced satisfaction of orientors in the earlier decades to a more balanced, but still unsustainable state in later decades – with the obvious exception of the coexistence orientor. Its inadequate state is caused by the rising income gap between the rich and the poor.

Support system: Initially extremely unbalanced, the orientor satisfactions relative to the support system (infrastructure and economic system) become more balanced in later decades, although at unsatisfactorily low levels, implying unsustainability.

Natural system: The dynamics of orientor satisfaction of the natural system reflect very clearly the continuing degradation and loss of sustainability of this system. At the beginning of the period the system appears in good shape with one exception: the coexistence orientor is already in an unacceptable state. This is due to wasteful use of resources (without any recycling attempts) threatening the sustainability of the system.

Total system: The average satisfaction ratings of the three component systems combined are shown in the orientor stars of Figure 2. They show a more balanced state of orientor satisfaction, although at low and steadily decreasing level with continuing loss of sustainability. These graphs conceal the rather more dramatic developments evident from the orientors stars of the three component systems. Referring back to the orientor stars for the three component systems, it becomes evident that any improvements appearing in the diagrams for the human and the support system are clearly coupled to corresponding sustainability losses in the natural system.

It will have become obvious that 'sustainability' and 'sustainable development' are much more than vague and intuitive concepts. They must be discussed on a deep background of systems theoretical and ethical concepts. Important conclusions are pulled together in the following (Bossel, 1996, Bossel, 1998a) .[24]

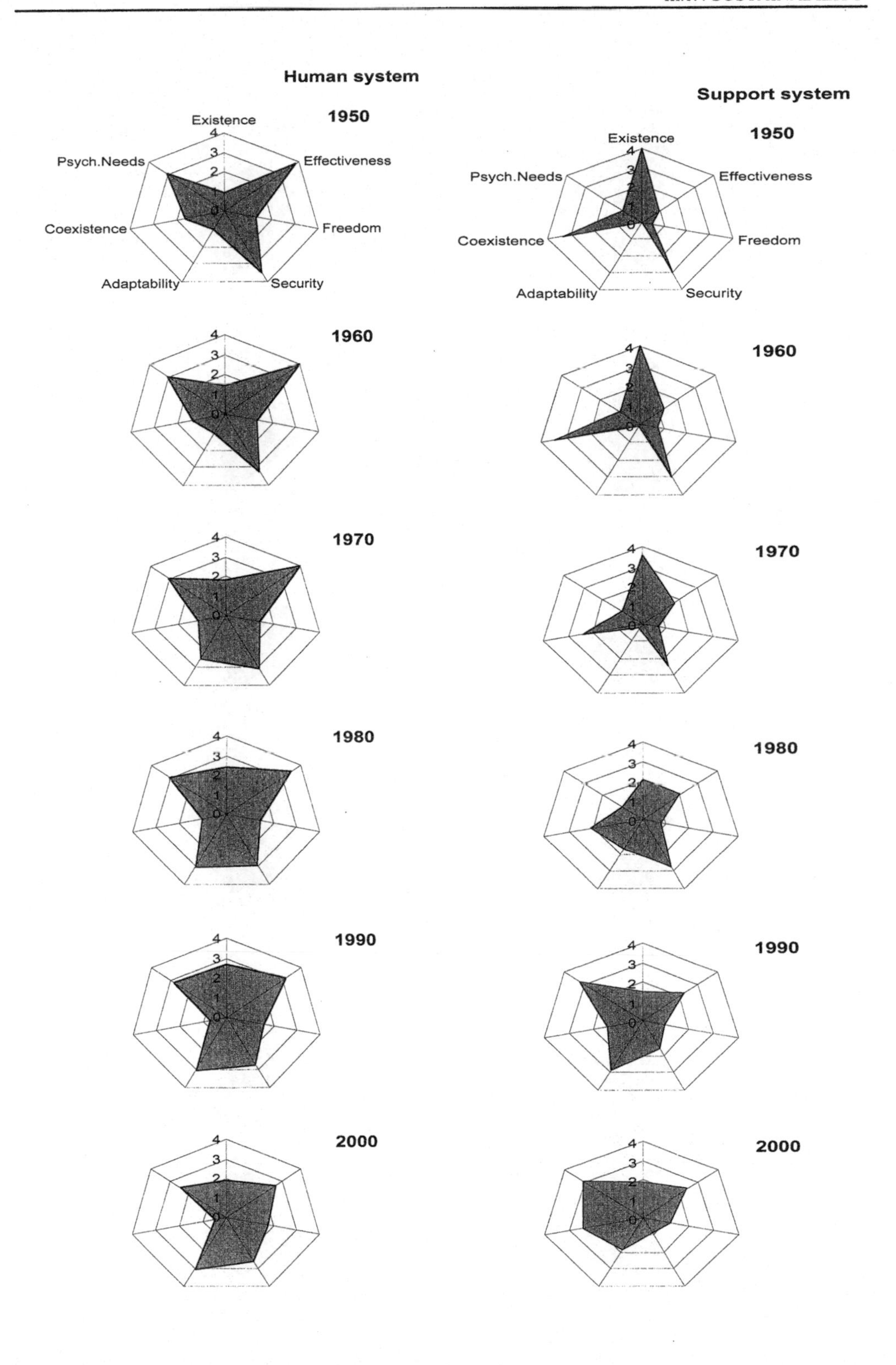

Figure 1: Basic orientor assessment of the sustainability dynamics for the global system. 1950-2000: Human system and support system.

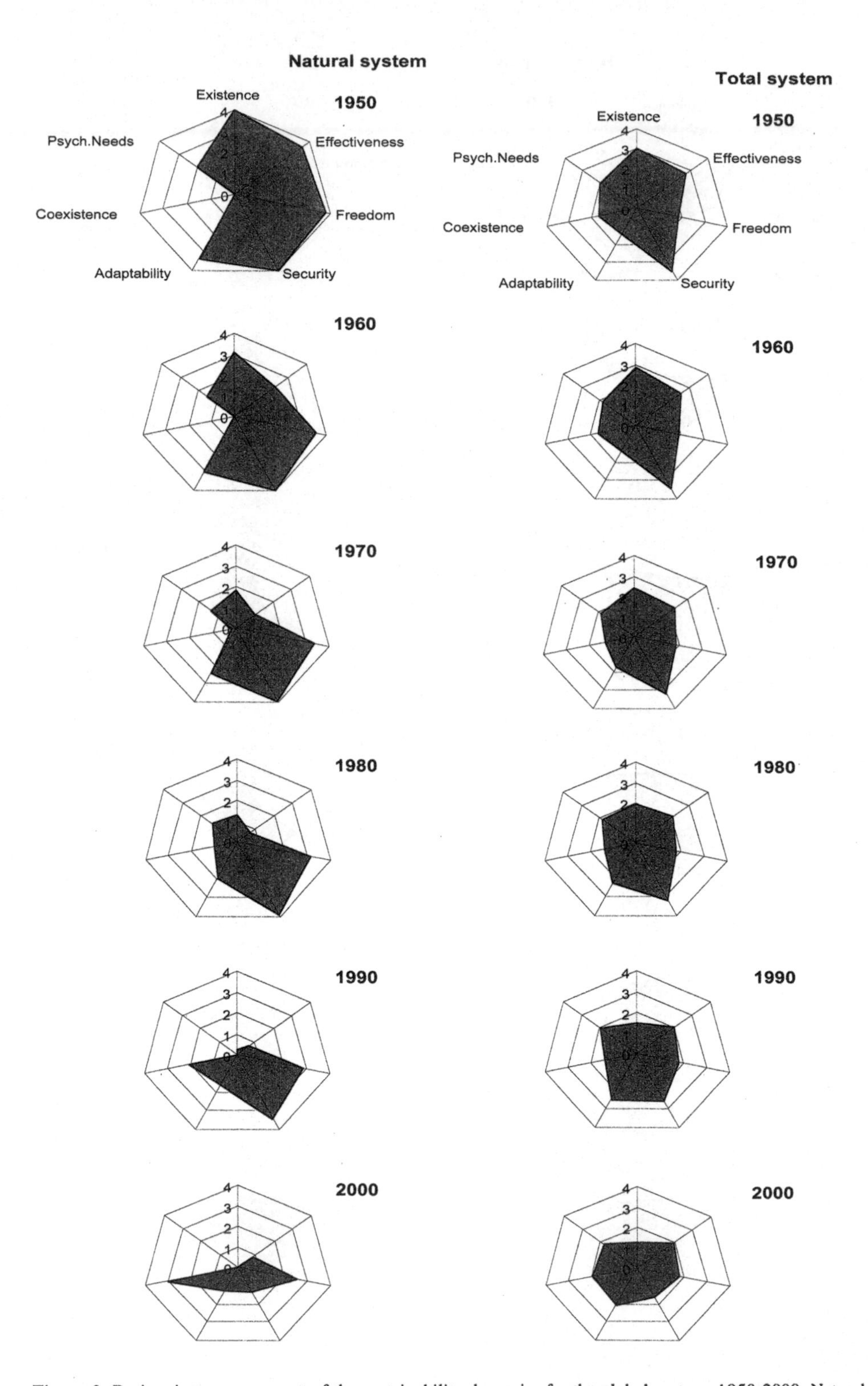

Figure 2: Basic orientor assessment of the sustainability dynamics for the global system, 1950-2000: Natural and total system.

1. Integrated view of the problem

1.1 Laws of nature and the physical constraints of planet earth put physical limits on any type of development of natural or human systems. These limits cannot be eliminated by wishful thinking or ideology.

1.2 All systems on earth, whether living or non-living, conscious or non-conscious, have to obey certain general systems laws. It is possible to analyse and understand complex system behaviour by analogy or simulation, and to use this understanding for identifying behavioural trends, for planning, and for decision-making.

1.3 The knowledge required for dealing with the issue of sustainable development has to be integrated on a general systems background. It ranges over a wide field of traditional disciplines encompassing the natural sciences, systems science, social sciences, and engineering.

1.4 In human systems, the multitude of interacting conscious actors (individuals and organisations) pursuing different goals and interests complicates things enormously. A holistic analysis of the possible complex behaviour must find ways to simplify and aggregate without losing essential features of dynamic processes.

1.5 In response to the emerging problems of the interaction of human systems and natural systems, and the realisation of the unsustainability of current developments, there is a growing perception of the need for fundamental change in human institutions, technology, behavioural principles, and ethics.

2. Principles and properties of physical systems

2.1 The First and Second Law of Thermodynamics as well as (Liebig's) Principle of the Minimum, put severe constraints on the development of systems in general, and on that of the human system embedded in the global ecosystem in particular.

2.2 Development is rate-limited, not supply-limited. The dominant rate control is affected by limited solar radiation, which drives all important geophysical and biological processes.

2.3 Material recycling, as practised in the global ecosystem, is a prerequisite for sustainable development.

2.4 The productivity of ecosystems is limited by the rate of solar energy input, physical efficiencies, availability of key materials, and waste absorption rate.

2.5 Sustainable development of the human enterprise at a high level of quality is possible in principle, but not possible at present rates of resource destruction and population increase.

3. Principles and properties of ecosystem development

3.1 The coevolution of species, ecosystems, and physical environment brings forth a growth of structural and dynamic complexity in a sustainable overall system. Certain aspects can serve as models for human development, other aspects are general system properties which also apply to human systems.

3.2 The carrying capacity of a given environment is the result of the coevolution of species populations, ecosystems, physical environment, and human activities. It is limited by the energy and material flows which can be mobilised and utilised.

3.3 Ecosystems rely almost exclusively on renewable (solar) energy and on material recycling. This has ensured sustainability of the global ecosystem over billions of years.

3.4 In the course of evolution, individual species tend to maximise exergy[25] efficiency, while ecosystems as a whole tend to maximise the use of the available exergy gradient (i.e. mainly solar radiation input) by maximum build-up of (energy-dissipating) structure. Evolution therefore implies irreversibility (loss of exergy and increase in entropy) and development (increasing organisation).

3.5 The environment not only determines the physical structure and processes of systems evolving in it, it also determines behavioural processes and cognitive structure. In order to be 'fit', a system must give simultaneous and balanced attention to different objectives: in particular, its basic orientors.

3.6 Diversity is essential for optimising the use of available gradients, and for providing the innovative potential for evolution and future progress.

4. Principles and properties of human system development

4.1 As a result of conscious, reasoned, and anticipatory decision-making, and the intellectual, technical, and organisational power of humans, the rate of development of human systems can be much faster than natural system evolution. Hence, there is little evolutionary control of human activity.

4.2 Because of the power of reasoned decision-making and the potential of putting thoughts into practice, the spectrum of possible behaviour of human systems is extremely wide.

4.3 Human action therefore requires normative guidance. It can be assumed that human behavioural norms are not completely subjective and arbitrary, but have emerged as value orientations (orientors) and ethical standards (partnership in family and community) in the course of human genetic and cultural evolution.

4.4 The evolutionary process causes the emergence of (implicit) normative standards in the behaviour of organisms, which can be interpreted as the emergence of 'evolutionary ethics' (of fair chance of competing species, etc.). This evolutionary ethic is only partially acceptable for human systems: because of their powers of imagination, anticipation, and reasoning, humans must adopt a wider ethical framework.

4.5 Maintenance of value orientations and ethical standards under changing (systems and environmental) conditions may require dynamic change of material and cognitive structure, including their 'creative' destruction and renewal.

5. Principles for sustainable society

5.1 Ecosystem principles should not be applied uncritically to human systems. Some of them are applicable, others are not.

5.2 Valuing humankind and human culture is equivalent to striving for their sustainability. Social sustainability, in particular, suggests adoption of a Principle of Partnership as ethical principle. It would extend to present and future generations, and to human and non-human systems.

5.3 Society's current ethical base, as incorporated in its institutions, is inadequate for the initialisation and protection of sustainable development. Changes in the legal system, and in the representation of the interests of human and non-human, present and future 'partners', are required.

5.4 Full representation of the interests of affected systems requires assessment of their basic orientor satisfaction in the course of time.

5.5 Population control, efficient resource use, recycling, use of renewable resources, in particular renewable energy resources, and sufficiency in consumption will be required to achieve sustainable development.

5.6 Necessary processes of control and self-organisation should be self-inducing, self-sustaining, equitable, and effective. The protection and encouragement of diversity on all system levels would provide creativity, innovation, redundancy, niches, resilience and optimum orientor satisfaction for all partner systems.

6. Required steps for the transition to sustainability

6.1 The transition to sustainability hinges on effective control of population growth and consumption growth. In view of the different time constants of these processes, the reduction of per capita consumption in the countries of the North must have absolute priority. There is an enormous potential through (a) improvements in the efficiencies of resource and energy use, and (b) social acceptance of sufficiency limits.

6.2 A truly holistic and integrated systems approach spanning the global ecosystem and the human system and their interactions is necessary. It requires new approaches in science, research, education, social and political processes, planning and decision-making.

6.3 The normative principles of society should be consistent with an overarching ethical principle. Sustainability seems to require adoption of a Principle of Partnership with other systems and organisms of the global ecosystems, today and in the future. Sustainability cannot be attained under a split ethical framework (as today, with selfish competitiveness in society at large, and partnership in the family).

6.4 In contrast to current trends for globalisation, it will be essential for sustainability to reintroduce diversity and regionalisation, or even localisation, to adapt development to regional carrying capacity, and to decouple local markets from global markets.

7. Overall conclusions

1.1 Current development of the global human system is unsustainable in more than one sense: physically, ecologically, socially.

1.2 The unsustainability is partly due to processes in human systems proper (population, economics, technologies, etc.), and partly due to destruction and change of the natural environment on which all human systems depend.

1.3 'Sustainable development' is possible, but it requires a departure from the present path. Many sustainable futures, and paths to achieve them, are possible.

1.4 The choice requires adoption of an ethical principle (Principle of Partnership).

1.5 Society (and environment) cannot afford an evolution of human systems to a sustainable path by trial and error; conscious strategic choices are required.

1.6 Strategic choices must be in agreement with general system principles, and ecosystem principles in particular, in order to be successful in the long run.

1.7 These principles should be identified, in order to provide guidance for structural and strategic changes.

1.8 The emphasis should be on changing, where necessary and possible, the rules of human-determined processes of self-organisation to ensure sustainable evolution. Omniscient design of the sustainable society is not possible. The task is rather to determine and apply functional principles which lead to self-organising evolution of sustainable human systems (of whatever shape) in a sustainable environment.

NOTES

[1] For a comprehensive survey from a systems viewpoint, and extensive bibliography, see Bossel 1996.

[2] WCED 1987

[3] Engel 1990, p. 10-11.

[4] See e.g. Fox 1996, Birnbacher 1988, Borman and Kellert 1991, Devall and Sessions 1985, Fromm 1976, Hardin 1972, Johnson 1991, Jonas 1979, Kohn 1990, Krebs 1997, Küng 1991, Potter 1971, Pojman 1989, Rawls 1972, Singer 1993, Singer 1994, Spämann 1989, Taylor 1986, Tugendhat 1993.

[5] Bossel 1998a

[6] Rawls1972, p. 14.

[7] Bossel 1978, p. 70: 'Tue alles, was heute und im Zeitraum deiner Zukunftsperspektive mit hoher Wahrscheinlichkeit zu direkten oder indirekten Folgen führt, deren Nutzen für dich größer ist als der dir entstehende Schaden.'

[8] Rawls 1972

[9] Bossel 1978

[10] Bossel 1978, p. 71: 'Alle heutigen und zukünftigen Systeme, die hinreichend einmalig und unersetzlich sind, haben gleiches Recht auf Erhaltung und Entfaltung.'

[11] Johnson 1991

[12] Bossel 1998a

[13] Webster 1962.

[14] Webster 1962.

[15] Bossel 1977, Bossel 1987, Bossel 1994, Bossel 1998a, Bossel 1998b

[16] See e.g. Bossel 1998a

[17] See e.g. Hornung 1988.

[18] This can be demonstrated in simulations of the cognitive self-organization of an artificial animal, see Krebs and Bossel 1997.

[19] Cf. Krebs and Bossel 1997, Bossel 1998a, Bossel 1998b

[20] Hornung 1988, p. 218.

[21] Worldwatch Database Disk, January 1998, data for 2000 are extrapolated. Worldwatch Institute, 1776 Massachusetts Ave., NW, Washington, DC 20036

[22] Bossel 1998b

[23] Bossel 1997, Bossel 1998b

[24] Bossel 1996, Bossel 1998a

[25] Exergy is that part of energy that can produce useful work

REFERENCES

Birnbacher, D. 1988. Verantwortung für zukünftige Generationen. Enke, Stuttgart.

Borman, F. H., and Kellert, S. R. (eds.) 1991: Ecology, Economy, Ethics: The Broken Circle. Yale University Press, New Haven CT.

Bossel, H., 1977. Orientors of nonroutine behavior. In: H. Bossel (ed.*), Concepts and Tools of Computer-assisted Policy Analysis.*: Birkhäuser, Basel 227-265

Bossel, H., 1978. Bürgerinitiativen entwerfen die Zukunft - Neue Leitbilder, neue Werte, 30 Szenarien. (NGO´s Design the Future - New Visions, New Values). Fischer, Frankfurt am Main.

Bossel, H., 1987. Viability and sustainability: Matching development goals to resource constraints. *Futures* vol. 19, no. 2, 114-128

Bossel, H., 1994: Modeling and Simulation. Wellesley MA: A K Peters, and Wiesbaden: Vieweg

Bossel, H., 1996. Ecosystems and society: Implications for sustainable development. *World Futures* Vol. 47, 143-213.

Bossel, H., 1997. Deriving indicators of sustainable Development. *Environmental Modeling and Assessment*, Vol. 1, No. 4, 193-218.

Bossel, H., 1998a. Earth at a Crossroads: Paths to a Sustainable Future. Cambridge University Press, Cambridge UK. (Globale Wende: Wege zu einem gesellschaftlichen und ökologischen Strukturwandel. Droemer-Knaur, München.)

Bossel, H., 1998b. Indicators for Sustainable Development: Theory, Method, Applications. Winnipeg, Manitoba: IISD International Institute of Sustainable Development (ISBN 1-895536-13-8).

Devall, B., and Sessions, G. 1985. Deep Ecology - Living as if Nature Mattered. Layton, UT: Peregrine Smith

Engel, J. R., 1990: Introduction: The Ethics of Sustainable Development. In: Engel, J.R., and Engel, J. G. (eds.): *Ethics of Environment and Development: Global Challenge, International Response*. Belhaven Press, London and University of Arizona Press Tucson pp1-23.

Fox, W., 1996. A critical overview of environmental ethics. *World Futures* Vol. 46, 1-21

Fromm, E. 1976. To Have or to Be? Harper and Row, New York.

Hardin, G. 1972. Exploring New Ethics for Survival. Viking Press, New York.

Hornung, B. R., 1988. Grundlagen einer problemfunktionalistischen Systemtheorie gesellschaftlicher Entwicklung. Peter Lang, Frankfurt/M.

Johnson, L. E., 1991. A Morally Deep World: An Essay on Moral Significance and Environmental Ethics. Cambridge University Press, Cambridge UK.

Jonas, H. 1979. Das Prinzip Verantwortung. Insel Verlag, Frankfurt/M.

Kohn, A., 1990. The Brighter Side of Human Nature: Altruism and Empathy in Human Nature. Basic Books, New York

Krebs, A., 1997. Naturethik – Grundtexte der gegenwärtigen tier- und ökoethischen Diskussion. Suhrkamp, Frankfurt/M.

Krebs, F., and Bossel, H., 1997. Emergent value orientation in self-organization of an animat. *Ecological Modelling* 96: 143-164.

Küng, H. 1991. Global Responsibility: In Search of a New World Ethic. New York, Crossroad Publ.

Pojman, L. J. 1989. Ethical Theory: Classical and Contemporary Readings. Wadworth, Belmont CA.

Potter, V. R. 1971. Bioethics: Bridge to the Future. Englewood Cliffs: Prentice Hall, NJ.

Rawls, J. 1972: A Theory of Justice. Oxford University Press, London.

Singer, P. (ed.) 1994. Ethics. Oxford University Press, Oxford and New York.

Singer, P. 1993. Practical Ethics, 2nd ed.: Cambridge University Press, Cambridge and New York.

Spämann, R. 1989. Basic Moral Concepts. Routledge, London and New York.

Taylor, P. W. 1986. Respect for Nature - A Theory of Environmental Ethics. Princeton University Press, Princeton NJ.

Tugendhat, E., 1993. Vorlesungen über Ethik. Suhrkamp, Frankfurt/M.

WCED (World Commission on Environment and Development), 1987. Our Common Future - The Brundtland Report. Oxford University Press, Oxford UK.

Webster's 1962: New World Dictionary of the American Language. World Publishing, Cleveland and New York.

Worldwatch Institute, 1998. Database Disk (State of the World and Vital Signs). Worldwatch Institute, Washington DC.

III.1.5 Ecological Engineering

Sven E. Jørgensen and W.J. Mitsch

1. What is ecological engineering?

H.T. Odum was among the first to define ecological engineering (Odum, 1962 and Odum et al., 1963) as the *"environmental manipulation by man using small amounts of supplementary energy to control systems in which the main energy drives are coming from natural sources."* Odum further developed the concept (Odum, 1983) of ecological engineering as follows: ecological engineering, the engineering of new ecosystems designs, is a field that uses systems that are mainly self-organising.

Straskraba (1984 and 1985) has defined ecological engineering, or as he calls it *ecotechnology* somewhat broader, as the use of technological means for ecosystem management, based on a deep ecological understanding, to minimise the costs of measures and their harm to the environment. Ecological engineering and ecotechnology are considered different by Straskraba but are considered synonymous by many others.

Mitsch and Jørgensen (1989) give a slightly different definition which, however, covers the same basic concept as the definition given by Straskraba and also encompasses the definition given by H.T. Odum. They are defining ecological engineering and ecotechnology as the design of human society with its natural environment for the benefit of both. It is engineering in the sense that it involves the design of man-made or natural ecosystems or parts of ecosystems. It is, like all engineering disciplines, based on basic science, in this case ecology and system ecology. Biological species are the components applied in ecological engineering. Ecological engineering represents, therefore, a clear application of ecosystem theory.

At a workshop on ecological engineering at the U.S. National Academy of Sciences in May 1993 (Mitch, 1996 and 1998), a light variation of the definition of ecological engineering, as originally given by Mitsch and Jørgensen (1989), was presented:

"the design of sustainable ecosystems that integrate human society with its natural environment for the benefit of both."

Ecotechnic is another often applied word, but it encompasses, in addition to ecotechnology or ecological engineering, the development of technology applied in society, based upon ecological principles, for instance all technologies based upon cycling or the use of resources in a more environmentally friendly way.

Ecological engineering should furthermore not be confused with bioengineering or biotechnology (Mitsch and Jørgensen, 1989 and Mitsch, 1993). Biotechnology involves the manipulation of the genetic structure of the cells to produce new organisms capable of carrying certain functions. Ecotechnology does not manipulate at the genetic level, but at several steps higher in the ecological hierarchy. The manipulation takes place on an assemblage of species and/or their abiotic environment as a self-designing system that can adapt to changes brought about by outside forces, controlled by humans or by natural forcing functions.

Ecological engineering is also not the same as environmental engineering which is involved in cleaning up processes in order to prevent pollution problems. It uses settling tanks, filters, scrubbers and man-made components which have nothing to do with the biological and ecological components that are applied in ecological engineering, although the use of environmental engineering aims also towards reducing man-made forcing functions on ecosystems. Ecotechnic, mentioned above, may be considered to include in addition to ecological engineering, also environmental technology based on ecological principles such as recirculation and a general better use of the resources.

The tool box of these two types of engineering are completely different. Ecological engineering uses ecosystems, communities, organisms and their immediate abiotic environment, while, as mentioned before, environmental engineering uses man-made components.

All applications of technologies are based on quantification. Ecosystems are very complex systems and the quantification of their reactions to impacts or manipulations becomes therefore complex. Fortunately, ecological modelling represents a well developed tool to survey ecosystems, their reactions and the linkage of their components. Ecological modelling is able to synthesise our knowledge about an ecosystem and makes it possible, to a certain extent, to quantify any changes in ecosystems resulting from the use of both environmental engineering and ecological engineering.

Ecological engineering may also be used directly to design constructed ecosystems. Consequently, ecological modelling and ecological engineering are two closely co-operating fields. The research of ecological engineering was originally covered by the *Journal of Ecological Modelling* which was named *Ecological Modelling – International Journal on Ecological Modelling and Engineering and Systems Ecology* to emphasise the close relationship between the three fields: ecological modelling, ecological engineering and systems ecology. *Ecological Engineering* was launched as an independent journal in 1992, and the name *Ecological Modelling* was changed to *Ecological Modelling – An International Journal on Ecological Modelling and Systems Ecology*. Meanwhile, the journal Ecological Engineering has successfully covered the field of ecological engineering which has grown rapidly during the nineties due to the increasing acknowledgement of the need for technologies other than environmental technology in our effort to solve pollution problems. This development does not imply that ecological modelling and ecological engineering are moving in different directions. On the contrary, ecological engineering is increasingly using models to perform design of constructed ecosystems or to quantify the results of application of specific ecological engineering methods for comparison with alternative methods.

2. Examples and classification of ecotechnology

Ecotechnology may be based on one or more of the following four classes of ecotechnology:

1) Ecosystems are used to reduce or solve a pollution problem that otherwise would be (more) harmful to other ecosystems. A typical example is the use of wetlands for waste water treatment.
2) Ecosystems are imitated or copied to reduce or solve a pollution problem, leading to constructed ecosystems. Examples are fishponds and constructed wetlands for treatment of waste water or diffuse pollution sources.
3) The recovery of ecosystems after significant disturbances. Examples are coal mine reclamation and restoration of lakes and rivers.
4) The use of ecosystems for the benefit of humankind without destroying the ecological balance, i.e., utilisation of ecosystems on an ecologically sound basis. Typical examples include the use of integrated agriculture and development of organic agriculture. These types of ecotechnology have wide applications in the ecological management of renewable resources.

All four classes of ecological engineering may find illustrative examples where ecological engineering is applied to replace environmental engineering because the ecological engineering methods offer an ecologically more acceptable solution and where ecological engineering is the only method that can offer a proper solution to the problem. These examples are shown in Table 1, where the alternative environmental technological solution is also indicated.

Table 1:
Ecological Engineering Examples. Alternative Environmental Engineering methods are given, if possible.

Type of Ecological Engineering	Example of Ecological Engineering Without Env. Eng. alternative	Example of Ecological Engineering with Env. Eng. alternative	Environmental Engineering alternative
1	Wetlands utilised to reduce diffuse pollution	Sludge disposal on agricultural land	Sludge incineration
2	Constructed wetland to reduce diffuse pollution	Root zone plant	Traditional wastewater treatment
3	Recovery of lakes	Recovery of contaminated land *in situ*	Transport and treatment of contaminated soil
	Agro-forestry	Ecologically sound planning of harvest rates of resources	

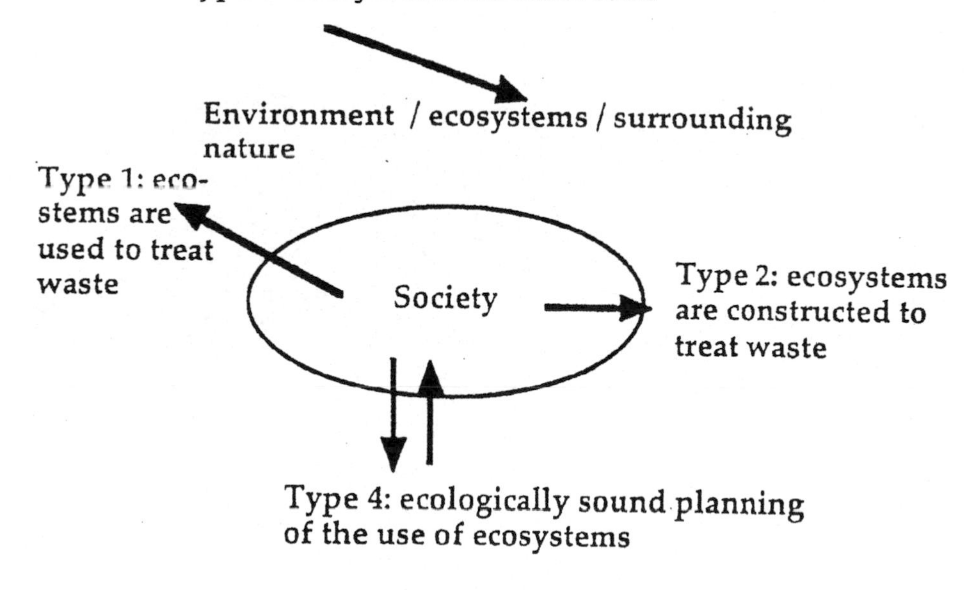

Figure 1: An illustration of the four types of engineering

It does not imply that ecological engineering consequently replace environmental engineering. On the contrary, the two technologies should work hand in hand to solve the environmental management problem better than they could do alone. This is illustrated in Figure 2, where a proper control of lake eutrophication requires both ecological engineering and environmental technology.

The type 1 ecological engineering, application of ecosystems to reduce or solve pollution problems, may be illustrated by wetlands utilised to reduce the diffuse nutrient loadings of lakes. This problem could not be solved by environmental technology. Treatment of sludge could be solved by environmental technology, namely by incineration, but the ecological engineering solution, sludge disposal on agricultural land which implies a utilisation of the organic material and nutrients in the sludge, is a considerably more sound method from an ecological perspective.

The application of constructed wetlands to cope with the diffuse pollution is a good example of ecological engineering, type 2. Again, this problem cannot be solved by environmental technology. The application of root zone plants for treatment of small amounts of waste water is an example of ecological engineering, type 2, where the environmental engineering alternative, a mechanical-biological-chemical treatment, cannot compete because it will have too high costs relative to the amount of waste water (sewage system, pumping stations and so on). A solution requiring less resources will always be ecologically more sound.

Recovery of land contaminated by toxic chemicals is possible by environmental technology, but it will require transportation of the soil to a soil treatment plant, where biological biodegradation of the contaminants takes place. Ecological engineering will propose a treatment *in situ* by adapted micro-organisms or plants. The latter method is a lot more cost moderate and the pollution related to the transport of soil will be omitted. Recovery of lakes by bio-manipulation, installation of an impoundment, by sediment removal or coverage, by siphoning off hypolimnetic water, rich in nutrients, down streams or by several other proposed ecological engineering techniques are other type 3 examples of ecological engineering. It is hardly possible to obtain the same results by environmental engineering because it requires activities in the lake and/or in the vicinity of the lake.

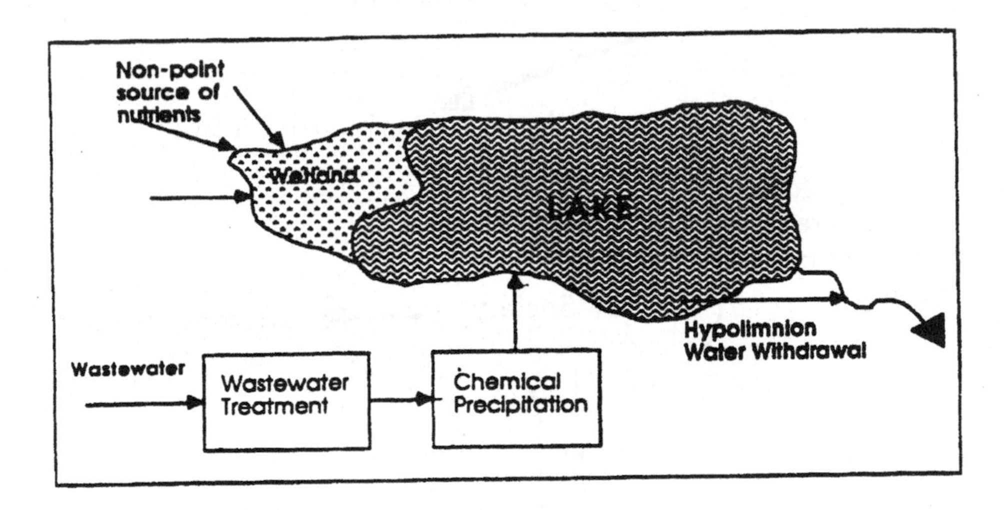

Figure 2: Control of lake eutrophication with a combination of chemical precipitation for phosphorus removal from waste water (environmental technology), a wetland to remove nutrients from the inflow (ecotechnology, type 1 or 1), and siphoning off of hypolimnetic water, rich in nutrients, down streams (ecotechnology, type 3).

The fourth type of ecological engineering is to a high extent based on prevention of pollution by utilisation of ecosystems on an ecologically sound basis. It is hardly possible to find environmental engineering alternatives in this case, but it is clear that a prudent and careful harvest rate of renewable resources, whether it be timber or fish, is the best long term strategy from an ecological and economic point of view.

3. Ecosystem theory applied in ecological engineering

Ecological engineering has been presented in the two previous sections of this chapter as a useful technological discipline. This section is devoted to drawing the relations to ecosystem theory by presentation of a number of principles applied to understand the ecological engineering methods in practice and extracted from ecosystem theory.

Mitsch and Jørgensen (1989) apply twelve system ecological principles to understand the basic concepts of ecotechnology and to ensure a proper ecologically sound application of the practical use of this approach. The principles are presented below with at least two examples as illustrations of the application and consideration of each of the twelve principles in the practical use of ecological engineering. The two examples for each principle are taken from terrestrial ecosystems, often from agriculture, and from aquatic ecosystems.

Principle 1: Ecosystem structure and functions are determined by the forcing functions of the system. Ecosystems are open systems, which implies that they exchange mass and energy with the environment. There is a close relation between the anthropogenic forcing functions and the state of the agricultural ecosystems, but due to the openness of all ecosystems, the adjacent ecosystems are also affected. Intensive agriculture leads to drainage of the surplus nutrients and pesticides to the adjacent ecosystems. This is the so-called non-point or diffuse pollution. The abatement of this source of pollution requires a wide use of ecotechnological methods.

The uses of constructed wetlands and impoundments to reduce the concentrations of nutrients in streams entering a lake ecosystem illustrate the application of this system of ecological principle in ecotechnology. The forcing functions, the nutrient loadings, are reduced and a corresponding reduction of the eutrophication should be expected.

Principle 2: Homeostasis of ecosystems requires accordance between biological function and chemical composition. The biochemical functions of living organisms define their composition, although these are not to be considered as fixed concentrations, but as ranges. The application of the principle implies that the flow of elements through agricultural systems should be according to the biochemical stoichiometry. If it is not the case the elements in surplus will be exported to the adjacent ecosystems and make an impact on the natural balances and processes there. An investigation (Skaarup and Sørensen, 1994) on a well-managed Danish farm has shown that much can be gained by a complete material flow analysis of a farm. The results will not only lead to reduction in the emission level of pollutants but will often also imply cost reductions.

Several restoration projects on lakes have considered two principles. The limiting nutrient determining the eutrophication of a lake is found and the selected restoration method will reduce the limiting nutrient further. When sediment is removed or covered in lakes, it is to prevent the otherwise limiting nutrient, often phosphorus, to reach the water phase.

Principle 3: It is necessary in environmental management to match recycling pathways and rates to ecosystems to reduce the effect of pollution. The application of sludge as a soil conditioner illustrates this principle very clearly. The recycling rate of nutrients in agriculture has to be accounted for in any use of sludge. If the sludge is applied faster than it can be utilised by the plants, a significant amount of the nutrients might contaminate the streams, lakes and/or ground water adjacent to the agricultural ecosystem. If, however, the influence of the temperature on the nitrification and denitrification processes, the hydraulic conductivity of the soil, the slope of field and the rate of the plant growth all are considered in an application plan for the manure, the loss of nutrients to the environment will be maintained on a very low and probably acceptable level. This is possible using ecological models to develop a plan for the application of the sludge.

Elements are recycled in agro-systems but to a far less extent than occurs in nature. For the last couple of decades animal husbandry has to a high extent become separated from plant production. This made the internal cycling more difficult to achieve, of less economical value and thus less attractive. We have come to accept losses of these relatively inexpensive, easy-to-apply artificial fertilisers and compensate so by increasing their application. So, the message is: know the ecological processes of farming and their rates and manage the system accordingly, i.e., recycle as much as possible in the right rates and do not use more fertilisers than can be recycled.

The recovery of a eutrofe lake by application of shading is an illustrative example of the same principle for aquatic ecosystems.

Ecosystems with pulsing patterns often have greater biological activity and chemical cycling than systems with relatively constant patterns. A specific case study will illustrate the recognition of pulsing force and how it is possible to take advantage of it in ecological engineering. Figure 3 shows a map of an estuary in Brazil, named Cannaneia. The shore of the islands and the coast are very productive mangrove wetlands and the entire estuary is an important nesting area for fish and shrimp. Channel C, (refer to the map), was built to avoid the flooding upstream, where productive agricultural land is situated. The construction of the channel has caused a conflict between farmers, who want the channel open, and fishermen, who want it closed due to its reduction of the salinity in the estuary (the right salinity is of great importance for the mangrove wetlands). The estuary is exposed to tide which is important for maintenance of a good water quality with a certain minimum of salinity. The conflict can be solved by use of an ecological engineering approach that takes advantage of the pulsing force (the tide). A sluice in the channel could be constructed to discharge the fresh water when it is most appropriate. The tide would in this case be used to transport the fresh water as rapidly as possible to the sea. The sluice should be closed when the tide is on its way into the estuary. The tidal pulse frequently is selectively filtered to produce an optimal management situation.

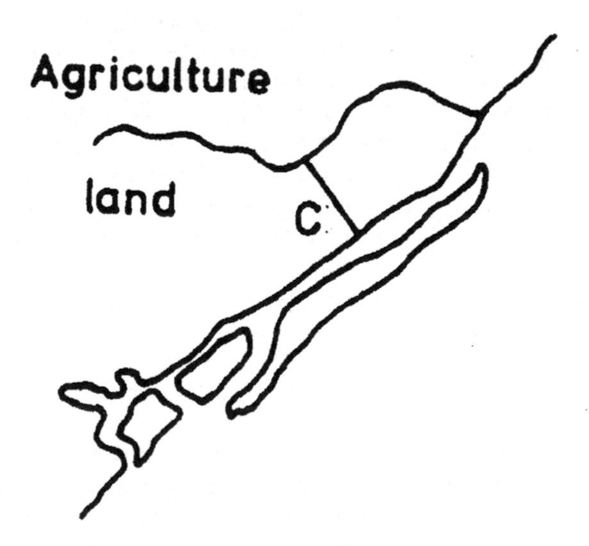

Figure 3: Map of Cannaneia Estuary in Brazil. Channel C was build to avoid upstream flooding of agricultural land.

Principle 4: Ecosystems are self-designing systems. The more one works with the self-designing ability of nature, the lower the costs of energy to maintain that system. Many of our actions are undertaken to circumvent or to counteract the process of self-design. For example, the biodiversity of agricultural fields would be significantly higher if pesticides were not used and nature let to rule on its own. While the self-designing systems are able to implement sophisticated regulations before violent fluctuations or even chaotic events occur, agriculture attempts to regulate chemically, for instance undesired organisms by the use of pesticides. This very coarse regulation sometimes causes more harm than anticipated, for instance when the insect-predators are affected more than the insects. The conclusion seems clear: do not eliminate the well-working natural regulation mechanisms, i.e., maintain a pattern of nature within the agricultural systems.

The application of green fields during the winter in Northern Europe is consistent with this principle, as the self-designing ability is maintained in this time of the year. Bare soil should generally be avoided due to possible erosion.

The closer the agricultural system is to a natural ecosystem, the more self-designing capacity the system has. Integrated agriculture is therefore less vulnerable than modern industrialised agriculture as it offers more components for self-designing regulations and it has a wider range of flows that facilitate the possibilities for recycling.

The use of constructed wetlands in lake restoration is an example on the application of this principle of self-design taken from aquatic ecosystems. If we design a wetland to remove partially the nutrients from streams entering the lake, the lake can itself do the self-design and reduce the level of eutrophication accordingly. The constructed wetland will also use self-design. The diversity (complexity) and nutrients removal efficiency will increase gradually, provided that the wetland is disturbed.

Principle 5: Processes of ecosystems have characteristic time and space scales, that should be accounted for in environmental management. The environmental management should consider the role of a certain spatial pattern for the maintenance of biodiversity. Violation of this principle by drainage of wetlands and deforestation in too large a scale has caused desertification. Wetlands and forests maintain a high humidity of the soil and regulate the precipitation. When the vegetation is removed, the soil is exposed to direct solar radiation and dries causing organic matter to be burned off. Application of too large fields prevents wild animals and plants to find their ecological niches as an important component in the pattern of agriculture and more or less untouched nature. The conclusion is to maintain ditches and hedgerows as corridors in the landscape or as ecotones between agricultural and other ecosystems. Fallow fields also should be planned as contributors to the pattern of the landscape.

The example mentioned above on the use of the tide to transport fresh water as rapidly as possible to the sea may also be used to illustrate this principle of using the right time and space scale in the application of ecological engineering.

Principle 6: Chemical and biological diversity contribute to the buffering capacity and the self-designing ability of ecosystems. A wide variety of chemical and biological components should be introduced or maintained for the ecosystem's self-designing ability to choose from. Thereby a wide spectrum of buffer capacities is available to meet the impacts from anthropogenic pollution.

Biodiversity plays an important role in the buffer capacity and the ability of the system to meet a wide range of possible disturbances by the use of the ecosystem's self designing ability. There are many different buffer capacities, corresponding to any combination of a forcing function and a state variable. It has been shown that vegetables cultivated to a high extent as mixed cultures give a higher yield and are less vulnerable to disturbances, for instance attacks by herbivorous insects. In agricultural practice it implies that it is advisable to use small fields with different crops.

This principle also implies that integrated agriculture is less vulnerable than modern industrialised agriculture simply due to its higher biological and chemical diversity.

Recovery of lakes by use of biomanipulation usually increases the biodiversity and some buffer capacities.

Principle 7: Ecotones, or transition zones, are as important for the ecosystems as the membranes are for the cells. Agricultural management should therefore consider the importance of the transition zones. Nature has developed transition zones, denoted ecotones, to make a soft transition between two ecosystems. Ecotones may be considered as buffer zones, that are able to absorb undesirable changes imposed on an ecosystem from adjacent ecosystems. We must learn from nature and use the same concepts when we design interfaces between manmade ecosystems (agriculture, human settlements) and nature. Some countries require a buffer zone between human settlements and the coast of lakes or marine ecosystems (it is for instance in Denmark 50 m).

Emissions will, without buffer zones between agriculture and natural ecosystems, be transferred directly to the ecosystem, while a buffer zone such as a wetland, would at least partly adsorb the emissions and thereby prevent their negative influences on natural ecosystems. Some countries also require a buffer zone (for instance in Denmark on 2 m) between arable land and streams or lakes.

A pattern of wetlands in the landscape will be able to remove the emission of particular nitrates from agriculture which is one of the hot issues in the environmental management of the diffuse pollution originated from agriculture.

The role of the littoral zone in lake management is another obvious example. A sound littoral zone with a dense vegetation of macrophytes will be able to absorb contamination before it reaches the lake and thereby be for a lake like a membrane for the cell.

Principle 8: The coupling between ecosystems should be utilised to the benefit of the ecosystems in the application of ecotechnology and in environmental management of agricultural systems. An ecosystem cannot be isolated – it must be an open system because it needs an input of energy to maintain the system (see also principle 1). The coupling of agricultural systems to natural systems leads to the transfer of pesticides and nutrients from agriculture to nature and measures should be taken to (almost) complete utilisation of the pesticides and the nutrients in the agricultural system, for instance by implementation of proper fertilisation plans accounting for these transfer processes. Ecological management should always consider all ecosystems as interconnected systems, not as isolated subsystems. It implies that not only local but also regional and global effects have to be considered. For instance, the methane emitted from rice fields may increase the greenhouse effect and thereby the global climate which again will feed back the cultivation of rice.

Lake management can only be successful if it is based upon this principle, as all the inputs to the open ecosystem, a lake, should be considered. Recovery of a eutrofe lake will require that all nutrient sources are quantified and that a plan which takes into account all sources is realised.

Principle 9: It is important that the application of ecotechnology and environmental management considers that the components of an ecosystem are interconnected, interrelated and form a network, which implies that direct as well as indirect effects are of importance. An ecosystem is an entity, or everything is linked to everything in the ecosystem. Any effect on any component in an ecosystem is therefore bound to have an effect on all components in the ecosystem either directly or indirectly, i.e., the entire ecosystem will be changed. It can be shown that the indirect effect is often more important than the direct one (Patten, 1991). Application of ecotechnology attempts to take the indirect effect into account, while management only considering the direct effects often fails. There are numerous example on the use of pesticides on herbivorous insects that also might have a pronounced effect on the carnivorous insects and therefore will result in the opposite effect on the herbivorous insect than intended. Therefore pesticides should not be used in a vacuum, but sufficient knowledge about the insect populations and their predators should be the basis for decisions about the application of pesticides. Preferably, a model should be developed to assure that the pesticides reach the right target organisms and not cause an inverse effect.

The use of ecotechnology in the abatement of toxic substances in aquatic ecosystems requires that this principle is considered. The biomagnification of toxic substances through the food chain is a result of the interconnectance of ecological components. Due to biomagnification, it is necessary to aim at a far lower concentration of toxic substances in aquatic ecosystems to avoid a undesirable high concentration of the toxic substance in fish for human consumption.

Principle 10: It is important to realise that an ecosystem has a history in application of ecotechnology and environmental management in general. The components of ecosystems have been selected to cope with the problems that nature has imposed on the ecosystems for million of years. The high biodiversity of old ecosystems compared with the immature ecosystems is another important realisation of this principle. The structure of mature ecosystems should therefore be imitated in the application of ecological engineering. An ecosystem with a long history is better able to cope with the emissions from its environment than an ecosystem with no history. This again emphasises the importance of establishment of a pattern of agriculture and natural terrestrial and aquatic ecosystems to ensure that the history is preserved and the right solution therefore can be offered to emerging environmental problems.

Many lakes have storage of considerable amounts of nutrients in the sediment due to a sad history of discharge of insufficiently treated waste water. Restoration of such lakes requires often removal of the sediment, an extremely expensive restoration method for large and deep lakes. Ecologically sound management will take the history into account and use prevention in proper time.

Principle 11: Ecosystems are most vulnerable at the geographical edges. Therefore ecological management should take advantage of ecosystems and their biota in their optimal geographical range. When ecological engineering involves ecosystem manipulation, the system will have enhanced buffer capacity if the species are in the middle range of their environmental tolerance. The ecosystem manipulation should therefore consider a careful selection of the involved species in accordance with this principle. For agriculture it implies that the crops should be selected following this principle. The cultivation of tomatoes and other subtropical vegetables in Northern Europe demonstrates how this principle is easily violated in modern agriculture. These products may compete in price due to good management or subsidies, but they cannot compete in quality.

Ecologically sound planning will use this principle and avoid the use of biological components which are at their geographical edges. This rule is of course important for both terrestrial and aquatic ecosystems.

Principle 12: Ecosystems are hierarchical systems and all the components forming the various levels of the hierarchy make up a structure, that is important for the function of the ecosystem. It is for instance significant to maintain the components that make up the landscape diversity such as hedges, wetlands, shorelines, ecotones, ecological niches, etc. They will all contribute to the buffer capacity of the entire landscape. Clearly, the integrated agriculture can easier follow this principle than the industrialised agriculture, because it has more components to use for construction of a hierarchical structure.

It is equally important in our management of lakes to consider the benthic zone, the littoral zone, the epilimnion and hypolimnion. All the zones require the right conditions with respect to oxygen, pH, temperature and so on to maintain the various organisms (the next lower level in the hierarchy) fitted to these zones. Selection of lake restoration methods requires consideration of this issue. What can we do to solve the problem? Which restoration methods should then be selected?

4. Concluding remarks

Ecological engineering is based on use of technological means in ecosystems to the benefit of the ecosystems and mankind. As ecosystems are extremely complex, it is of course necessary to base the use of ecological engineering on a profound understanding of ecosystems. How do they work as systems? What are their characteristic system properties? How do they react on our interactions? The use of ecological engineering as a useful technology in our effort to abate the pollution problems is therefore closely associated with our knowledge about ecosystems. Ecological engineers should, therefore, understand ecosystem theory and follow the progress in the field very carefully to be able to improve their decisions on how to solve a problem properly using ecotechnology.

REFERENCES

Mitsch, W.J., 1993. Ecological engineering - a cooperative role with the planetary life-support systems. *Environmental Science & Technology* 27: 438-445.

Mitsch, W.J., 1996. Ecological engineering: A new paradigm for engineers and ecologists. In: P.C. Schulze, (Ed.). *Engineering within Ecological Constraints.* Washington, D.C. National Academy Press, 111-128.

Mitsch, W.J., 1998. Ecological engineering - the seven-year itch. *Ecological Engineering* 10: 119-138.

Mitsch, W.J. and S.E. Jørgensen, (Eds.). 1989. *Ecological Engineering. An Introduction to Ecotechnology.* John Wiley & Sons, New York, Chichester, Brisbane, Toronto, Singapore, 430.

Odum, H.T., 1962. Man in the ecosystem. In: Proceedings Lockwood Conference on the Suburban Forest and Ecology. Bull. Conn. Agr. Station 652. Storrs, CT, 57-75.

Odum, H.T., W.L. Siler, R.J. Beyers, and N. Armstrong, 1963. Experiments with engineering of marine ecosystems. *Publ. Inst. Marine Sci. Uni. Texas* 9: 374-403.

Odum, H.T., 1983. *System Ecology.* Wiley Science, New York. 510 pp.

Patten, B.C. 1991. Network ecology: indirect determination of the life-environment relationship in ecosystems. In: M. Higashi and T.P. Burns (Eds.). Theoretical Studies of Ecosystems: The Network Perspective. Cambridge University Press, Cambridge.

Skaarup, R. and T. Sorensen, 1994. Use of green auditing in agriculture. Thesis, KVL, Copenhagen.

Straskraba, M., 1984. New ways of eutrophication abatement. In: M. Straskraba, Z. Brandl, and P. Procalova, (Eds.). *Hydrobiology and Water Quality of Reservoirs.* Acad. Sci., Ceské Budejovice, Czechoslovakia 37-45.

Straskraba, M., 1985. *Simulation Models as Tools in Ecotechnology Systems. Analysis and Simulation.* Vol. II. Academic Verlag, Berlin, 362.

Straskraba, M., 1993. Ecotechnology as a new means for environmental management. In: Ecological Engineering, Vol. 2, No. 4, 311-332.

III.1.6. Ecosystem and Economic Theories in Ecological Economics

Robert Costanza, Cutler Cleveland and Charles Perrings

1. Historical roots and motivations

Ecology and economics have developed as separate disciplines throughout their recent histories in the 20th century. While each has addressed the way in which living systems self-organise to enable individuals and communities to meet their goals, and while each has borrowed theoretical concepts from the other and shared patterns of thinking with other sciences, they began with different first principles, addressed separate issues, utilised different assumptions to reach answers, and supported different interests in the policy process. Bringing these domains of thought together and attempting to reintegrate the natural and social sciences has lead to what we call *ecological economics*. After numerous experiments with joint meetings between economists and ecologists in the 1980s (e.g. Jansson, 1984), the International Society for Ecological Economics (ISEE) was formed in 1988[18], the journal, *Ecological Economics*, was initiated and published its first issue in February of 1989 (currently publishing 12 issues per year), and major international conferences have brought together ecologists, economists and a broad range of other scientists and practitioners. Several ecological economic institutes have been formed around the world, and a significant number of books have appeared with the term *ecological economics* in their titles (e.g. Martinez Alier, 1987; Costanza, 1991; Peet, 1992; The Group of Green Economists, 1992; Jansson et al., 1994; Barbier et al., 1994).

As Martinez-Alier (1987) and Cleveland (1987) point out, ecological economics has historical roots as long and deep as any field in economics or the natural sciences, going back to at least the 17th century. Nevertheless, its immediate roots lie in work done in the 1960s and 1970s. Kenneth Boulding's classic *The economics of the coming spaceship Earth* (Boulding, 1966) set the stage for ecological economics with its description of the transition from the 'frontier economics' of the past, where growth in human welfare implied growth in material consumption, to the 'spaceship economics' of the future, where growth in welfare can no longer be fuelled by growth in material consumption. This fundamental difference in vision and world view was elaborated further by Daly (1968) in recasting economics as a life science – akin to biology and especially ecology, rather than a physical science like chemistry or physics. The importance of this shift in 'pre-analytic vision' (Schumpeter, 1950) cannot be overemphasised. It implies a fundamental change in the perception of the problems of resource allocation and how they should be addressed. More particularly, it implies that the focus of analysis should be shifted from marketed resources in the economic system to the biophysical basis of interdependent ecological and economic systems, (Clark,1973; Martinez-Alier, 1987; Cleveland, 1987 and Christensen,1989).

The broader focus of ecological economics is carried in a 'systems' framework. The systems approach, with its origins in non-linear mathematics, general systems theory, non-equilibrium thermodynamics, and ecosystem ecology, is a comparatively recent development that has opened up lines of inquiry that were off the agenda for earlier work in what Lotka termed biophysical economics (Clark, 1976; Cleveland, 1987; Martinez-Alier, 1987; Christensen, 1989; Clark and Munroe, 1994). While bioeconomic and ecological economic models both incorporate the dynamics of the natural resources under exploitation,

---

[18]For more information about ISEE, the journal, and upcoming conferences, visit the WWW home page: http://kabir.umd.edu/ISEE/ISEEhome.html, or send an email message to: button@cbl.cees.edu, or write to ISEE, PO Box 1589, Solomons, MD 20688, USA

the former tend to take a partial rather than a general equilibrium approach (van der Ploeg et al., 1987).

The core problem addressed in ecological economics is the sustainability of interactions between economic and ecological systems. Ecological economics addresses the relationships between ecosystems and economic systems in the broadest sense (Costanza, 1991). It involves issues that are fundamentally cross-scale, transcultural and transdisciplinary, and calls for innovative approaches to research, to policy and to the building of social institutions (Costanza and Daly, 1987; Common and Perrings, 1992; Holling, 1994; Berkes and Folke, 1994; d'Arge, 1994; Golley, 1994; Viederman, 1994). In this sense, ecological economics tends to be characterised by a holistic 'systems' approach that goes beyond the normal territorial boundaries of the academic disciplines.

2. Basic organising principles of ecological economics

Ecological economics is not a single new discipline based in shared assumptions and theory. It rather represents a commitment among natural and social scientists, and practitioners, to develop a new understanding of the way in which different living systems interact with one another, and to draw lessons from this for both analysis and policy. Ecological economics is conceptually pluralistic. This means that even while people writing in ecological economics were trained in a particular discipline (and may prefer that mode of thinking over others) they are open to and appreciative of other modes of thinking and actively seek a constructive dialogue among disciplines (Norgaard, 1989). There is not one *right* approach or model because, like the blind men and the elephant, the subject is just too big and complex to touch it all with one limited set of perceptual or computational tools.

Within this pluralistic paradigm, traditional disciplinary perspectives are perfectly valid as *part of the mix*. Ecological economics therefore includes some aspects of neo-classical environmental economics, traditional ecology and ecological impact studies, and several other disciplinary perspectives as components, but it also encourages completely new, hopefully more integrated, ways to think about the linkages between ecological and economic systems.

The broad spectrum of relationships between ecosystems and economic systems are the loci of many of our most pressing current problems (i.e., sustainability, acid rain, global warming, species extinction, wealth distribution) but they are not covered adequately by any existing discipline. Environmental and resource economics, as they are usually practised, are sub-disciplines of neo-classical economics focused on the efficient allocation of scarce environmental resources but generally ignoring ecosystem dynamics and scale issues, and paying only scant attention to distribution issues (Cropper and Oates, 1992). Ecology, as it is currently practised, sometimes deals with human impacts on ecosystems, but the more common tendency is to stick to 'natural' systems and exclude humans. Ecological economics aims to extend these modest areas of overlap. Its basic organising principles include the idea that ecological and economic systems are complex, adaptive, living systems that need to be studied as integrated, co-evolving systems in order to be adequately understood (Holling, 1986; Proops, 1989; Costanza et al., 1993).

Ecological economics also focuses on a broader set of goals than the traditional disciplines. Here, again, the differences are not so much the newness of the goals, but rather the attempt to integrate them. Daly (1992) lays out these goals in a hierarchical form as:

1) assessing and insuring that the scale of human activities within the biosphere is ecologically sustainable;
2) distributing resources and property rights fairly, both within the current generation of humans and between this and future generations, and also between humans and other species; and

3) efficiently allocating resources as constrained and defined by 1 and 2 above, including both marketed and non-marketed resources, especially natural capital and ecosystem services.

That these goals are interdependent and yet need to be addressed hierarchically is elaborated by Common and Perrings (1992), who differentiate between 'Solow' or economic sustainability (Solow, 1974, 1986) and 'Holling' or ecological sustainability (see Holling, 1986) and find them to be largely disjoint. The problem of ecological sustainability needs to be solved at the level of preferences or technology, not at the level of optimal prices. Only if the preferences and production possibility sets informing economic behaviour are ecologically sustainable can the corresponding set of optimal and intertemporally efficient prices be ecologically sustainable. Thus the principle of 'consumer sovereignty' on which most conventional economic solutions is based, is only acceptable to the extent that consumer interests do not threaten the overall system – and through this the welfare of future generations. This implies that if one's goals include ecological sustainability then one cannot rely on consumer sovereignty, and must allow for co-evolving preferences, technology, and ecosystems. One of the basic organising principles of ecological economics is thus a focus on this complex interrelationship between ecological sustainability (including system carrying capacity and resilience), social sustainability (including distribution of wealth and rights and co-evolving preferences) and economic sustainability (including allocative efficiency).

A major implication of this is that our ability to predict the consequences of economic behaviour is limited by our ability to predict the evolution of the biosphere. The complexity of the many interacting systems that make up the biosphere means that this involves a very high level of uncertainty. Indeed, uncertainty is a fundamental characteristic of all complex systems involving irreversible processes (Costanza and Cornwell, 1992; Ludwig et al., 1993; Costanza, 1994; Clark and Munro, 1994). It follows that ecological economics is particularly concerned with problems of uncertainty. More particularly, it is concerned with the problem of assuring sustainability under uncertainty. Instead of locking ourselves into development paths that may ultimately lead to ecological collapse, we need to maintain the resilience of ecological and socio-economic systems (Hammer et al., 1993; Holling, 1994; Jansson and Jansson, 1994; Perrings, 1994) by conserving and investing in natural assets (Costanza and Daly, 1992).

3. Material and energy flows in ecological and economic systems: theory and applications

One focus of the work on joint ecological economic systems has been material and energy flows. A dominant theme in this body of work has been the grounding of conventional economic models in the biophysical realities of the economic process. This emphasis shifts the focus from exchange to the production of wealth itself (Cleveland et al., 1984). Cleveland (1987) traces the early roots of this work dating back to the Physiocrats (Quesnay, 1758; Podilinsky, 1883; Soddy, 1922; Lotka, 1922 and Cottrell, 1955). The energy and environmental events of the 1960s and 1970s pushed work in this area to new levels. Energy and material flow analysis in recent times is rooted in the work of a number of economists, ecologists, and physicists. Economists such as Boulding (1966) and Geogescu-Roegen (1971, 1973) demonstrated the environmental and economic implications of the mass and energy balance principle. Ecologists such as Lotka (1922) and Odum (Odum and Pinkerton, 1955; Odum, 1971) pointed out the importance of energy in the structure and evolutionary dynamics of ecological and economic systems. And physicists such as Prigogine (Nicolis and Prigogine, 1977; Prigogine and Stengers, 1984) worked out the far-from-equilibrium thermodynamics of living systems.

The principle of the conservation of mass and energy has formed the basis for a number of important contributions. The assumption was first made explicit in the context of a general equilibrium model by Ayres and Kneese (1969) and subsequently by Mäler

(1974), but it also is a feature of the series of linear models developed after 1966 (Cumberland, 1966; Victor, 1972; Lipnowski, 1976; Geogescu-Roegen, 1977). All reflect the assumption that a closed physical system must satisfy the conservation of mass condition, and hence that economic growth necessarily increases both the extraction of environmental resources and the volume of waste deposited in the environment.

Perrings (1986, 1987) developed a variant of the Neumann-Leontief-Sraffa general equilibrium model in the context of a jointly determined economy-environment system subject to a conservation of mass constraint. The model demonstrates that the conservation of mass contradicts the free disposal, free gifts, and non-innovation assumptions of such models. An expanding economy causes continuous disequilibrating change in the environment. Since market prices in an interdependent economy-environment system often do not accurately reflect environmental change, such transformations of the environment often will go unanticipated.

Ayres (1978) describes some of the important implications of the laws of thermodynamics for the production process, including the limits they place on the substitution of human capital for natural capital and the ability of technical change to offset the depletion or degradation of natural capital. Although they may be substitutes in individual processes in the short term, natural capital and human-made capital ultimately are complements because both manufactured and human capital require materials and energy for their own production and maintenance (Costanza, 1980). The interpretation of traditional production functions such as the Cobb-Douglas or constant elasticity of substitution (CES) must be modified to avoid the erroneous conclusion that 'self-generating technological change' can maintain a constant output with ever-decreasing amounts of energy and materials as long as ever-increasing amounts of human capital are available.

Furthermore, there are irreducible thermodynamic minimum amounts of energy and materials required to produce a unit of output that technical change cannot alter. In sectors that are largely concerned with processing and/or fabricating materials, technical change is subject to diminishing returns as it approaches these thermodynamic minimums (Ayres, 1978). Ruth (1995) uses equilibrium and non-equilibrium thermodynamics to describe the materials-energy-information relationship in the biosphere and in economic systems. In addition to illuminating the boundaries for material and energy conversions in economic systems, thermodynamic assessments of material and energy flows, particularly in the case of effluents, can provide information about depletion and degradation that are not reflected in market price.

There is also the effect of the time rate of thermodynamic processes on their efficiency, and more importantly, their power or rate of doing useful work. Odum and Pinkerton (1955) pointed out that to achieve the thermodynamic minimum energy requirements for a process implied running the process infinitely slowly. This means at a rate of production of useful work (power) of zero. Both ecological and economic systems must do useful work in order to compete and survive and Odum and Pinkerton showed that for maximum power production an efficiency significantly worse than the thermodynamic minimum was required.

These biophysical foundations have been incorporated into models of natural resource supply and of the relationship between energy use and economic performance. Cleveland and Kaufmann (1991) developed econometric models that explicitly represent and integrate the geologic, economic, and political forces that determine the supply of oil in the United States. Those models are superior in explaining the historical record than those from any single discipline. Larsson et al. (1994) also use energy and material flows to demonstrate the dependence of a renewable resource such as commercial shrimp farming on the services generated by marine and agricultural ecosystems.

One important advance generated by this work is the economic importance of energy quality, namely, that a kcal of primary electricity can produce more output than an kcal of oil, a kcal of oil can produce more output than an kcal of coal, and so on. Odum (1971) describes how energy use in ecological and economic hierarchies tends to increase the quality of energy, and that significant amounts of energy are dissipated to produce higher quality forms that perform critical control and feedback functions which enhance the survival of the system. Cleveland et al. (1984) and Kaufmann (1992) show that much of the decline in the energy/real GDP ratio in industrial nations is due to the shift from coal to petroleum and primary electricity. Their results show that autonomous energy-saving technical change has had little, if any, effect on the energy/real GDP ratio. Stern (1993) finds that accounting for fuel quality produces an unambiguous causal connection between energy use and economic growth in the United States, confirming the unique, critical role that energy plays in the production of wealth.

The analysis of energy flows has also been used to illuminate the structure of ecosystems (e.g. Odum, 1957). Hannon (1973) applied input-output analysis (originally developed to study interdependence in economies) to the analysis of energy flow in ecosystems. This approach quantifies the direct plus indirect energy that connects an ecosystem component to the remainder of the ecosystem. Hannon demonstrates this methodology using energy flow data from the classic study of the Silver Springs, Florida food web (Odum, 1957). These approaches hold the possibility of treating ecological and economic systems in the same conceptual framework – one of the primary goals of ecological economics (Hannon et al., 1986, 1991; Costanza and Hannon, 1989).

4. Accounting for natural capital, ecological limits, and sustainable scale

Most current economic policies are largely based on the underlying assumption of continuing and unlimited material economic growth. Although this assumption is slowly beginning to change as the full implications of a commitment to sustainability sink in, it is still deeply imbedded in economic thinking as evidenced by the frequent equation of 'sustainable development' with 'sustainable growth.' The growth assumption allows problems of intergenerational, intragenerational, and interspecies equity and sustainability to be ignored (or at least postponed), since they are seen to be most easily solved by additional material growth (Arrow et al., 1995). Indeed, most conventional economists define 'health' in an economy as a stable and high rate of growth. Energy and resource depletion, pollution, and other limits to growth, according to this view, will be eliminated as they arise by clever development and deployment of new technology. This line of thinking often is called 'technological optimism (Costanza, 1989).'

An opposing line of thought (often called 'technological scepticism') assumes that technology will not be able to circumvent fundamental energy, resource, or pollution constraints and that eventually material economic growth will stop. It has usually been ecologists or other life scientists (e.g. Ehrlich, 1989; Daily and Ehrlich, 1992 - chapter 28) that take this point of view (notable exceptions among economists are Boulding, 1966 and Daly, 1968, 1977), largely because they study natural systems that invariably do stop growing when they reach fundamental resource constraints. A healthy ecosystem is one that maintains a relatively stable level. Unlimited growth is cancerous, not healthy, under this view.

Technological optimists argue that human systems are fundamentally different from other natural systems because of human intelligence and that history has shown that resource constraints can be circumvented by new ideas (Myers and Simon, 1994). Technological optimists claim that Malthus' dire predictions about population pressures have not come to pass and the 'energy crisis' of the late 70s is behind us. Technological sceptics, on the other hand, argue that many natural systems also have 'intelligence' in that they can evolve new behaviours and organisms (including humans themselves). Humans are therefore a part of nature not apart from it. Just because we have circumvented local and artificial resource constraints in the past does not mean we can circumvent the fundamental ones that we will eventually face. Malthus' predictions have not come to pass yet for the entire world, the sceptics would argue, but many parts of the world are in a Malthusian trap now, and other parts may well fall into it. This is particularly important because many industrial nations have increased their numbers and standard of living by importing carrying capacity and exporting ecological degradation to other regions.

The debate has gone on for several decades now. It began with Barnett and Morse's (1963) 'Scarcity and growth' but really got into high gear only with the publication of *The limits to growth* by Meadows et al. (1972) and the Arab oil embargo in 1973. Several thousand studies over the last fifteen years have considered aspects of our energy and resource future, and different points of view have waxed and waned. But the bottom line is that there is still considerable uncertainty about the impacts of energy and resource constraints. In the next 20 to 30 years we may begin to hit real fossil fuel supply limits. Will fusion energy or solar energy or conservation or some as yet unthought of energy source step in to save the day and keep economies growing? The technological optimists say 'yes' and the technological sceptics say 'maybe' but let's not count on it. Ultimately, no one knows.

The more specific issues of concern all revolve around the question of limits: the ability of technology to circumvent them, and the long run costs of the technological 'cures.' Do we adapt to limits with technologies that have potentially large but uncertain future environmental costs or do we limit population and per capita consumption to levels sustainable with technologies which are known to be more environmentally benign? Must we always increase supply or can we also reduce demand? Is there an optimal mix of the two?

If the 'limits' are not binding constraints on economic activity then conventional economics' relegation of energy and environmental concerns to the side of the stage is probably appropriate, and detailed energy analyses are nothing more than interesting curiosities. But if the limits are binding constraints, then energy and environmental issues are pushed much more forcefully to centre stage and the tracking of energy and resource flows through ecological and economic systems becomes much more useful and important.

Issues of sustainability are ultimately issues about limits. If material economic growth is sustainable indefinitely by technology, then all environmental problems can (in theory at least) be fixed technologically. Issues of fairness, equity, and distribution (between subgroups and generations of our species and between our species and others) are also issues of limits. We do not have to worry so much about how an expanding pie is divided, but a constant or shrinking pie presents real problems. Finally, dealing with uncertainty about limits is the fundamental issue. If we are unsure about future limits the prudent course is to assume they exist. One does not run blindly through a dark landscape that may contain crevasses. One assumes they are there and goes gingerly and with eyes wide open, at least until one can see a little better.

Vitousek et al. (1986) in an oft-cited paper estimated the percent of the earth's net primary production (NPP) which is being appropriated by humans. This was the first attempt to estimate the 'scale' or relative size of human economic activity compared to the ecological life support system. They estimated that 25% of total NPP (including the oceans) and 40% of terrestrial NPP was currently being appropriated by humans. It left open the question of how much of NPP could be appropriated by humans without damaging the life support functions of the biosphere, but it is clear that 100% is not sustainable and even the 40% of terrestrial NPP currently used may not be sustainable. Daily and Ehrlich (1992) add more depth to these arguments by considering the relationships between the size and relative impact of the human population relative to the earth's carrying capacity and the implications for sustainability. Arrow et al. (1995) add a recent interdisciplinary consensus on this relationship.

A related idea is that ecosystems represent a form of capital – defined as a stock yielding a flow of services – and that this stock of 'natural capital' needs to be maintained intact independently in order to assure ecological sustainability (El Serafy, 1991; Victor,

1991; Costanza and Daly 1992). The question of whether natural capital needs to be maintained independently ('strong sustainability') or whether only the total of all capital stocks need to be maintained ('weak sustainability') has been the subject of some debate. It hinges on the degree to which human-made capital can substitute for natural capital, and, indeed, on how one defines capital generally (Victor, 1991). In general, conventional economists have argued that there is almost perfect substitutability between natural and human-made capital (Nordhaus and Tobin, 1972), while ecological economists generally argue on both theoretical (Costanza and Daly, 1992) and empirical grounds (Kaufmann, 1995) that the possibilities for substitution are severely limited. They therefore generally favour the strong sustainability position.

Another critical set of issues revolves around the way we define economic income, economic welfare, and total human welfare. Daly and Cobb (1989) clearly distinguish these concepts, and point out that conventional GNP is a poor measure of even economic income. Yet GNP continues to be used in most policy discussions as the measure of economic health and performance, and will continue to be until viable alternatives are available. According to Hicks (1948) economic income is defined as the quantity we can consume without damaging our future consumption possibilities. This definition of income automatically embodies the idea of sustainability. GNP is a poor measure of income on a number of grounds, including the fact that it fails to account for the depletion of natural capital (Mäler 1991) and thus is not 'sustainable' income in the Hickian sense. GNP is an even poorer measure of economic welfare, since many components of welfare are not directly related to income and consumption. The Index of Sustainable Economic Welfare (ISEW) devised by Daly and Cobb (1989) is one approach to estimating economic welfare (as distinct from income) that holds significant promise. The ISEW has been calculated for several industrialised countries and shows that in all these cases, an 'economic threshold' has been passed where increasing GNP is no longer contributing to increasing welfare, and in fact in most cases is decreasing it (Max-Neef, 1995).

5. Valuation of ecological services

All decisions concerning the allocation of environmental resources imply the valuation of those resources. Ecological economics does not eschew valuation. It is recognised that the decisions we make, as a society, about ecosystems imply a valuation of those systems. We can choose to make these valuations explicit or not; we can undertake them using the best available ecological science and understanding or not; we can do them with an explicit acknowledgement of the huge uncertainties involved or not; but as long as we are forced to make choices about the use of resources we are valuing those resources. These values will reflect differences in the underlying world view and culture of which we are a part (e.g. Costanza, 1991; Berkes and Folke, 1994), just as they will reflect differences in preferences, technology, assets and income. An ecological economics approach to valuation implies an assessment of the spatial and temporal dynamics of ecosystem services, and their role in satisfying both individual and social preferences. It also implies explicit treatment of the uncertainties associated with tracking these dynamics (Costanza et al., 1993).

Ecological economics is different from environmental economics in this regard in terms of the latitude of approaches to the ecosystem valuation problem it allows. It includes more conventional willingness to pay (WTP) based approaches, but it also explores other more novel methods based on explicitly modelling the linkages between ecosystems and economic systems in the long run. Costanza et al. (1989) explore this comparison by estimating both WTP based and energy analysis based values for wetlands in coastal Louisiana and find an interesting degree of agreement. This emphasis on the direct assessment of ecosystem functions and values, independently and prior to attempting to tie it to people's perceptions of those functions and values, is extended in de Groot (1994) and Larsson et al. (1994), who enumerate these functions and estimate them for an example in Columbia, respectively. A more recent study (Costanza et al., 1997) synthesised a range of previous studies using a variety of techniques and estimated the total global value of ecosystem services at 16-54 trillion $US/yr – in the same order of magnitude as global GNP.

Spash and Hanley (1995) look at the issues of preference formation and limited information in estimating WTP based values for biodiversity preservation. They conclude on empirical grounds that a significant portion of individuals exhibit 'lexicographic' preferences – that is they refuse to make trade-offs which require the substitution of biodiversity for other goods. This places significant constraints on the use of stated preferences, as used in contingent valuation studies, for valuation of ecosystem services and decision making. It places more emphasis on the need to develop more direct methods to assess the value of these resources as a supplement to conventional WTP based methods.

Bingham et al. (1995) provide a broad interdisciplinary consensus and summary of these issues, which resulted from a U.S. EPA funded policy forum on ecosystem valuation. The forum emphasised the need to develop 'decisive information' relevant to management problems and choices.

The issue of ecosystem valuation is far from solved. In fact it is probably only in the early stages of development. Conventional WTP based approaches have severe limitations. Key directions for the future pointed to in Bingham et al. (1995) include integrated ecological economic modelling, as elaborated in the next section.

6. Integrated ecological economic modelling and assessment

The emphasis on (a) issues of scale and limits to the carrying and assimilative capacity of ecological systems, and (b) underlying dynamics of those systems both imply the need for a new approach to the modelling of joint systems. It is not surprising, therefore, that this is an active area of research in ecological economics. Indeed, it is where we most expect new advances to be made as a result of the ongoing dialogue between economists and ecologists. The seven chapters in this section share a common recognition that the valuation of ecological resources requires a deeper understanding of the ways in which economic activity depends on biogeophysical processes than is usually recognised. All of them attempt to combine (sometimes implicit) models of ecological processes and economic decision models in a way that makes the feedbacks between the two sets of processes transparent.

The range of issues that needs to be addressed in attempting to integrate economic and ecological models is explored by Braat and van Lierop (1987). While the next steps are likely to be even harder, as we shall see later, they indicate just how difficult it is to take even the first steps in bridging the modelling gap between disciplines that have long diverged both methodologically and conceptually. Costanza, Sklar and White (1990), Hall and Hall (1993), and Bockstael et al. (1995) illustrate some of the reasons why this is so, while developing new types of ecological economic models to overcome these difficulties. One reason for the difficulty in bridging the modelling gap is that economics, as a discipline, has developed almost no tools or concepts to handle spatial differentiation beyond the notions of transport cost and international trade. The spatial analysis of human activity has been seen as the domain of geographers, and has had remarkably little impact on the way that economists have analysed the allocation of resources. This makes collaboration between economists and disciplines based more directly on spatial analysis very difficult. Yet, recent work in this area shows just how important an understanding of economic and ecological landscapes is to the development of integrated models. The program of research, is having to develop new concepts as well as new models to deal with this problem. Liu, Cubbage and Pulliam (1994) offer another example of the importance of landscape in identifying the economic implications of such familiar concepts as forest rotation times. Since the development of spatially explicit integrated models is one of the areas in which ecological economics is expected to develop most rapidly in the next few years, it would seem that geographers are likely to become an increasingly important part of the research agenda in ecological-economics.

A second characteristic of ecological economic models concerns the way in which the valuation of ecological functions and processes is reflected in the model structure. The point was made in the previous section that valuation by stated preference methods (estimation of willingness to pay or accept using contingent valuation or contingent ranking) may capture the strength of people's perceptions and their level of income and endowments (their ability to pay), but it generally fails to capture the impact of a change in ecosystem functions and processes on the output of economically valued goods and services. Unless the role of non-marketed ecological functions and processes in the production of economically valued goods and services is explicitly modelled, it is hard to see how they can be properly accounted for in economic decision-making. Barbier (1994) and Ruitenbeek (1994) illustrate ways in which ecological functions and processes are embedded in decision-models, and the implications this has for valuation.

A third characteristic concerns the role of integrated modelling in strategic decision-making. One of the challenges to ecological economics has been to devise methods to address strategic 'what if' questions in a way that reflects the dynamics of the jointly determined system. This is clearly an extremely difficult task, and we indicate some of the reasons why this is so momentarily. Baker, Fennesy and Mitsch (1991) and Duchin and Lange (1994) illustrate different approaches to the task at the microeconomic and macroeconomic levels respectively. The general problem confronting anyone attempting to model long-run dynamics explicitly is that ecological economic systems are complex non-linear systems. The dynamics of economic systems are not independent of the dynamics of the ecological systems which constitute their environment, and that as economies grow relative to their environment, the dynamics of the jointly determined system can become increasingly discontinuous (Perrings,1986; Costanza et al., 1993; Arrow et al., 1995). Indeed, the development of ecological economics can be thought of as part of a widespread reappraisal of such systems.

In ecology, this reappraisal has influenced recent research on scale, complexity, stability and resilience; and is beginning to influence the theoretical treatment of the co-evolution of species and systems. The results that are most important to the development of ecological economics concern the link between the spatial and temporal structure of co-evolutionary hierarchical systems. Landscapes are conceptualised as hierarchies, each level of which involves a specific temporal and spatial scale (Holling, 1987, 1992; Costanza et al.,1993). The dynamics of each level of the structure are predictable so long as the biotic potential of the level is consistent with bounds imposed by the remaining levels in the hierarchy. Change in either the structure of environmental constraints or the biotic potential of the level may induce threshold effects that lead to complete alteration in the state of the system (O'Neill, Johnson and King, 1989).

In economics there is now considerable interest in the dynamics of complex non-linear systems (Anderson et al., 1988; Brock and Malliaris, 1989; Goodwin, 1990; Puu, 1989; Hommes 1991; Benhabib, 1992). Economists have paid less attention to spatial scale and its significance at or near system thresholds (though see Puu, 1981; Rosser, 1990), but there is now a growing body of literature with roots in geography which seeks to inject a spatial dimension into non-linear economic models (see for example White, 1990). There is also an economic analogue to the biologist's interest in evolution and the significance of co-dependence between gene landscapes. The steady accumulation of evidence that economic development is not a stationary process, that human understanding, preferences and technology all change with development and that such change is generally non-linear and discontinuous, has prompted economists to seek to endogenise technological change (Romer, 1990). Although the adaptation of this work by environmental economists has been rather disappointing, the treatment of technology and consumption preferences as endogenous to the economic process is a fundamental change that brings economics much closer to ecology.

The challenge to ecological economics in the future is to develop models that capture these features well enough to incorporate at least the major risks in economic decisions that increase the level of stress on ecological systems.

7. Summary and conclusions

This paper is a sample of the range of transdisciplinary thinking that can be put under the heading of ecological economics and the theories and models that have informed that work. While it is difficult to categorise ecological economics in the same way one would a normal academic discipline, some general characteristics can be enumerated.

- the core problem is the sustainability of interactions between economic and ecological systems.
- an explicit attempt is made at pluralistic dialogue and integration across disciplines, rather than territorial disciplinary differentiation.
- an emphasis is placed on integration of the three hierarchical goals of sustainable scale, fair distribution, and efficient allocation.
- there is a deep concern with the biophysical underpinnings of the functioning of jointly determined ecological and economic systems.
- there is a deep concern with the relationship between the scale of economic activity and the nature of change in ecological systems.
- since valuation based on stated willingness to pay reflects limitations in the valuer's knowledge of ecosystems functions, there is an emphasis on the development of valuation techniques that build on an understanding of the role of ecosystem functions in economic production.
- there is a broad focus on systems and systems dynamics, scale, and hierarchy and on integrated modelling of ecological economic systems.

These characteristics make ecological economics applicable to some of the major problems facing humanity today, which occur at the interfaces of human and natural systems, and especially to the problem of assuring humanity's health and survival within the biosphere into the indefinite future. It is not so much the individual core scientific questions that set ecological economics apart – since these questions are covered independently in other disciplines as well – but rather the treatment of these questions in an integrated, transdisciplinary way, which we feel is essential to their understanding and effective use in policy. The solutions being considered in ecological economics are deserving of increasing attention.

REFERENCES

Anderson P., K. Arrow, and D. Pines , 1988. The economy as an evolving complex system, Santa Fe Institute Studies in the Sciences of Complexity V, Redwood City CA, Addison Wesley.

Arrow, K., B. Bolin, R. Costanza, P. Dasgupta, C. Folke, C. S. Holling, B-O. Jansson, S. Levin, K-G. Mäler, C. Perrings, and D. Pimentel, 1995. Economic growth, carrying capacity, and the environment. *Science* 268: 520-521.

Ayres, R. U., 1978. Application of physical principles to economics. 37-71. In: R. U. Ayres. *Resources, Environment, and Economics: Applications of the Materials/Energy Balance Principle*. John Wiley and Sons, New York.

Ayres, R. U. and A. V. Kneese, 1969. Production, consumption, and externalities. *The American Economic Review* 59: 282-297.

Baker, K. A., M. S. Fennessy, and W. J. Mitsch, 1991. Designing wetlands for controlling mine drainage: an ecologic-economic modelling approach. *Ecological Economics* 3: 1-24.

Barbier, E. B., J. C. Burgess, and C. Folke, 1994. *Paradise lost? The ecological economics of biodiversity.* Earthscan, London. 267.

Barbier, E.B., 1994. Valuing environmental functions: tropical wetlands, *Land Economics* 70: 155-174.
Barnett, H. J. and C. Morse. 1963. Scarcity and growth: the economics of natural resource availability. Johns Hopkins, Baltimore, MD.
Benhabib J. (Ed.), 1992. Cycles and chaos in economic equilibrium, Princeton, Princeton University Press, NJ.
Berkes, F. and C. Folke, 1994. Investing in cultural capital for sustainable use of natural capital. 128-149. In: M. Jansson, M. Hammer, C. Folke, and R. Costanza, (Eds.). *Investing in natural capital: the ecological economics approach to sustainability.* Island press, Washington D.C. 504.
Bingham, G. R. Bishop, M. Brody, D. Bromley, E. Clark, W. Cooper, R. Costanza, T. Hale, G. Hayden, S. Kellert, R. Norgaard, B. Norton, J. Payne, C. Russell, and G. Suter, 1995. Issues in ecosystem evaluation: improving information for decision making *Ecological Economics* 14: 73-90.
Bockstael, N., R. Costanza, I. Strand, W. Boynton, K. Bell, and L. Wainger, 1995. Ecological economic modeling and valuation of ecosystems. Ecological Economics 14: 143-159. Boulding, K. E., 1966. The economics of the coming spaceship Earth. 3-14. In: H. Jarrett, (Ed.). *Environmental quality in a growing economy.* Resources for the Future/Johns Hopkins University Press, Baltimore, MD.
Boulding, K. E., 1966. The economics of the coming spaceship Earth. 3-14. In: H. Jarrett, (Ed.). *Environmental quality in a growing economy.* Resources for the Future/Johns Hopkins University Press, Baltimore, MD.
Braat, L. C. and W. F. J. van Lierop, 1987. Integrated economic-ecological modeling. 49-68. In: L. C. Braat and W. F. J. van Lierop, (Eds.). *Economic-ecological modeling.* North Holland, Amsterdam.
Brock W.A. and A.G. Malliaris, 1989. Differential equations, stability and chaos in dynamic economics. North Holland, Amsterdam.
Christensen, P., 1989. Historical roots for ecological economics: biophysical versus allocative approaches. *Ecological Economics* 1: 17-36.
Clark, C. W., 1973. The economics of overexploitation. *Science* 181: 630-634.
Clark, C.W., 1976. Mathematical bioeconomics: the optimal management of renewable resources. Wiley-Interscience, New York.
Clark, C. W. and G. R. Munro, 1994. Renewable resources as natural capital: the fishery. 343-361. In: A. M. Jansson, M. Hammer, C. Folke, and R. Costanza (Eds.). *Investing in natural capital: the ecological economics approach to sustainability.* Island press, Washington D.C. 504.
Cleveland, C. J., 1987. Biophysical economics: historical perspective and current research trends. *Ecological Modeling* 38: 47-74.
Cleveland, C. J. and R. K. Kaufmann, 1991. Forecasting ultimate oil recovery and its rate of production: incorporating economic forces into the models of M. King Hubbert. *The Energy Journal* 12: 17-46.
Cleveland, C. J., R. Costanza, C. A. S. Hall, and R. Kaufmann, 1984. Energy and the United States economy: a biophysical perspective. *Science* 225: 890-897.
Common M. and C. Perrings, 1992. Towards an ecological economics of sustainability. *Ecological Economics* 6: 7-34.
Costanza, R., 1980. Embodied energy and economic valuation. *Science* 210: 1219-1224.
Costanza, R., 1989. What is ecological economics? *Ecological Economics* 1: 1-7.
Costanza, R., (Ed.) 1991. Ecological economics: the science and management of sustainability. Columbia University Press, New York.
Costanza, R., 1994. Three general policies to achieve sustainability. 392-407. In: A. M. Jansson, M. Hammer, C. Folke, and R. Costanza, (Eds.). *Investing in natural capital: the ecological economics approach to sustainability.* Island press, Washington D.C. 504.
Costanza, R., R. d'Arge, R. de Groot, S. Farber, M. Grasso, B. Hannon, S. Naeem, K. Limburg, J. Paruelo, R.V. O'Neill, R. Raskin, P. Sutton, and M. van den Belt, 1997. The value of the world's ecosystem services and natural capital. *Nature* 387: 253-260.
Costanza, R. and L. Cornwell, 1992. The 4P approach to dealing with scientific uncertainty. *Environment* 34: 12–20.
Costanza, R. and H. E. Daly, 1987. Toward an ecological economics. *Ecological Modelling* 38: 1–7.
Costanza, R. and H. E. Daly, 1992. Natural capital and sustainable development. *Conservation Biology* 6: 37–46.
Costanza, R. S. C. Farber, and J. Maxwell, 1989. The valuation and management of wetland ecosystems. *Ecological Economics* 1: 335-361.
Costanza, R. and B. M. Hannon, 1989. Dealing with the "mixed units" problem in ecosystem network analysis. 90-115. In: F. Wulff, J. G. Field, and K. H. Mann, (Eds.). *Network analysis of marine ecosystems: methods and applications.* Coastal and Estuarine Studies Series, Springer-Verlag, Heidelberg. 284.
Costanza, R., F. H. Sklar, and M. L. White, 1990. Modeling coastal landscape dynamics. *BioScience* 40: 91-107.
Costanza, R., L. Wainger, C. Folke, and K-G Mäler, 1993. Modeling complex ecological economic systems: toward an evolutionary, dynamic understanding of people and nature. *BioScience* 43: 545-555.
Cottrell, W. F., 1955. Energy and society. McGraw-Hill, New York. In: Cropper, M. L. and W. E. Oates, 1992. *Environmental economics: a survey.* Journal of Economic Literature. 30: 675-740.
Cropper, M. L. and W. E. Oates, 1992. Environmental economics: a survey. *Journal of Economic Literature* 30: 675-740.
Cumberland, J.H., 1966. A regional inter-industry model for analysis of development objectives. *Regional Science Association Papers* 17: 65-94.

d'Arge, R. C., 1994. Sustenance and sustainability: how can we preserve and consume without major conflict. p. 113-127. In: A. M. Jansson, M. Hammer, C. Folke, and R. Costanza, (Eds.). *Investing in natural capital: the ecological economics approach to sustainability.* Island press, Washington D.C. 504.

Daily, G. and P. R. Ehrlich, 1992. Population, sustainability, and Earth's carrying capacity. *BioScience* 42: 761–71.

Daly, H.E., 1968. On Economics as a Life Science, *Journal of Political Economy* 76: 392-406.

Daly, H.E., 1977. Steady State Economics. The Political Economy of Bio-physical Equilibrium and Moral Growth. W.H. Freeman and Co., San Francisco.

Daly, H. E., 1992. Allocation, distribution, and scale: towards an economics that is efficient, just, and sustainable. *Ecological Economics* 6: 185-193.

Daly, H. E. and J. B. Cobb, 1989. Misplaced concreteness: measuring economic success. 62-84. In: H. E. Daly and J. B. Cobb. *For the common good: redirecting the economy toward community, the environment, and a sustainable future.* Beacon Press, Boston. 482.

de Groot, R. S., 1994. Environmental functions and the economic value of natural ecosystems. 151 168. In: A. M. Jansson, M. Hammer, C. Folke, and R. Costanza, (Eds.). *Investing in natural capital: the ecological economics approach to sustainability.* Island press, Washington D.C. 504.

Duchin, F. and G-M. Lange, 1994. Strategies for environmentally sound economic development. 250-265. In: A. M. Jansson, M. Hammer, C. Folke, and R. Costanza, (Eds.). *Investing in natural capital: the ecological economics approach to sustainability.* Island press, Washington D.C. 504.

Ehrlich, P.R., 1989. The limits to substitution: meta-resource depletion and a new economic ecological paradigm. *Ecological Economics* 1: 9-16.

El Serafy, S., 1991. The environment as capital. 168-175. In: R. Costanza. *Ecological economics: The science and management of sustainability.* Columbia University Press, New York. 525.

Georgescu-Roegen, N., 1971. The entropy law and the economic process. Harvard University Press, Cambridge, MA.

Georgescu-Roegen, N., 1973. The entropy law and the economic problem. 49-60. In: H. E. Daly. *Economics, ecology, ethics: essays toward a steady-state economy.* W. H. Freeman. San Francisco. 372.

Georgescu-Roegen, N., 1977. Matter matters, too. 293-313. In: K.D. Wilson, (Ed.). *Prospects for growth: changing expectations for the future.* Praeger, New York.

Golley, F. B., 1994. Rebuilding a humane and ethical decision system for investing in natural capital. 169-178. In: A. M. Jansson, M. Hammer, C. Folke, and R. Costanza, (Eds.). *Investing in natural capital: the ecological economics approach to sustainability.* Island press, Washington D.C. 504.

Goodwin, R.M., 1990. Chaotic economic dynamics. Clarendon, Oxford.

Hall, C. A. S. and M. H. P. Hall, 1993. The efficiency of land and energy use in tropical economies and agriculture. *Agriculture, Ecosystems, and the Environment* 46: 1-30.

Hammer, M., AM. Jansson, and B-O. Jansson, 1993. Diversity change and sustainability: implications for fisheries. *Ambio* 22: 97–105.

Hannon, B., 1973. The structure of ecosystems. *Journal of Theoretical Biology* 41: 535-46.

Hannon, B., R. Costanza, and R. A. Herendeen, 1986. Measures of energy cost and value in ecosystems. *The Journal of Environmental Economics and Management* 13: 391-401.

Hannon, B., R. Costanza, and R. Ulanowicz, 1991. A general accounting framework for ecological systems: a functional taxonomy for connectivist ecology. *Theoretical Population Biology* 40: 78-104.

Hicks, J. R., 1948. Value and capital. (2nd ed.) Clarendon, Oxford.

Holling, C. S., 1986. The resilience of terrestrial ecosystems: local surprise and global change. 292-317. In: W. C. Clark and R. E. Munn, (Eds.). *Sustainable Development of the Biosphere.* Cambridge University Press, Cambridge.

Holling C.S., 1987. Simplifying the complex: the paradigms of ecological function and structure. *European Journal of Operational Research* 30: 139-146.

Holling, C.S., 1992. Cross-scale morphology, geometry and dynamics of ecosystems. *Ecological Monographs* 62: 447-502.

Holling, C. S., 1994. New science and new investments for a sustainable biosphere. 57-73. In: A. M. Jansson, M. Hammer, C. Folke, and R. Costanza, (Eds.). *Investing in natural capital: the ecological economics approach to sustainability.* Island press, Washington D.C. 504.

Hommes C., 1991. Chaotic dynamics in economic models: some simple case studies. Wolters-Noordhoff, Groningen.

Jansson, A. M. (Ed.), 1984. Integration of economy and ecology: an outlook for the eighties. University of Stockholm Press, Stockholm.

Jansson, A. M., M. Hammer, C. Folke, and R. Costanza, (Eds.). *Investing in natural capital: the ecological economics approach to sustainability.* Island press, Washington D.C. 504.

Janson, A. M. and B. O. Jansson, 1994. Ecosystem properties as a basis for sustainability. 74-91. In: A. M. Jansson, M. Hammer, C. Folke, and R. Costanza, (Eds.). *Investing in natural capital: the ecological economics approach to sustainability.* Island press, Washington D.C. 504.

Kaufmann, R. K., 1992. A biophysical analysis of the energy/real GDP ratio: implications for substitution and technical change. *Ecological Economics* 6: 35-56.

Kaufmann, R. K., 1995. The economic multiplier of environmental life support: can capital substitute for a degraded environment? *Ecological Economics* 12: 67-79.

Larsson, J., C. Folke, and N. Kautsky, 1994. Ecological limitations and appropriation of ecosystem support by shrimp farming in Colombia. *Environmental Management.* 18: 663-676.

Lipnowski, I.F., 1976. An Input-Output Analysis of Environmental Preservation. *Journal of Environmental Economics and Management* 3: 205-214.

Liu, J. F.W. Cubbage and H.R.Pulliam, 1994. Ecological and economic effects of forest landscape structure and rotation length: simulation studies using ECOLECON. *Ecological Economics* 10: 249-263.

Lotka, A. J., 1922. Contribution to the energetics of evolution. *Proceedings of the National Academy of Science* 8: 147-155.

Ludwig, D., R. Hilborn, and C. Walters, 1993. Uncertainty, resource exploitation, and conservation: lessons from history. *Science* 260: 17-36.

Mäler, K.-G., 1974. Environmental economics - a theoretical inquiry. Johns Hopkins Press, Baltimore.

Mäler, K.G., 1991. National accounts and environmental resources. *Environmental and Resource Economics* 1: - 15.

Martinez-Alier, J., 1987. Introduction. In: J. Martinez-Alier. *Ecological economics: energy, environment, and society.* Blackwell, Cambridge, MA, 1-19

Max-Neef, M., 1995. Economic growth and quality of life: a threshold hypothesis. *Ecological Economics* 15: 115-118.

Meadows, D. H., D. L. Meadows, J. Randers, and W. W. Behrens, 1972. The limits to growth. Universe, New York.

Myers, N. and J. Simon, 1994. *Scarcity or abundance: a debate on the environment.* Norton, New York. 254.

Nicolis G. and I. Prigogine, 1977. Self-organisation in non-equilibrium systems. John Wiley, New York.

Nordhaus, W. and J. Tobin, 1972. Is growth obsolete? National Bureau of Economic Research, Columbia University Press, New York.

Norgaard, R. B., 1989. The case for methodological pluralism. *Ecological Economics* 1: 37-57.

O'Neill R.V., A.R. Johnson, and A.W. King, 1989. A hierarchical framework for the analysis of scale. *Lanscape Ecology* 3: 193-205.

Odum, H. T., 1957. Trophic structure and productivity of Silver Springs, Florida. *Ecological Monographs* 27: 55-112.

Odum, H. T., 1971. Environment, power and society. Wiley-Interscience, New York.

Odum, H. T. and R. C. Pinkerton, 1955. Time's speed regulator: the optimum efficiency for maximum power output in physical and biological systems. *American Scientist* 43: 331-343.

Peet, John, 1992. Energy and the ecological economics of sustainability. Island Press, Washington D. C.

Perrings, C., 1986. Conservation of mass and instability in a dynamic economy-environment system. *Journal of Environmental Economics and Management.* 13: 199-211.

Perrings C., 1987. Economy and environment: a theoretical essay on the interdependence of economic and environmental systems. Cambridge University Press, Cambridge.

Perrings, C., 1994. Biotic diversity, sustainable development, and natural capital. 92-112. In: A. M. Jansson, M. Hammer, C. Folke, and R. Costanza, (Eds.). *Investing in natural capital: the ecological economics approach to sustainability.* Island press, Washington D.C. 504.

Podilinsky, S., 1883. Menschliche Arbeit und Einheit der Kraft. Die Neue Zeit, März-April.

Prigogine, I. and I. Stengers,1984. Order Out of Chaos: Man's New Dialogue with Nature, New York, Bantam Books.

Proops, J. L. R. ,1989. Ecological economics: rationale and problem areas. *Ecological Economics* 1: 59-76.

Puu, T., 1981. Structural stability in geographical space, *Environment and Planning* 13: 979-989.

Puu, T., 1989. Non-linear economic dynamics. Springer, Berlin.

Quesnay, F., 1758. Tableau Economique. In: M. Kuczynski and R. L. Meek, (Editors). *Quesnay's Tableau Economique.* Macmillian, London.

Romer P., 1990. Endogenous technical change. *Journal of Political Economy* 98: S71-103.

Rosser B., 1990. From catastrophe to chaos: a general theory of economic discontinuities. Kluwer, Dordrecht.

Ruitenbeek, H.J., 1994. Modelling economy-ecology linkages in mangroves: economic evidence for promoting conservation in Bintuni Bay, Indonesia. *Ecological Economics* 10: 233-247.

Ruth, M., 1995. Information, order and knowledge in economic and ecological systems: implications from material and energy use. *Ecological Economics* 13: 99-114.

Schumpeter, J. A., 1950. Capitalism, socialism and democracy. Harper & Row, New York.

Soddy, F., 1922. Cartesian economics. Hendersons, London.

Solow, R. M., 1974. Intergenerational equity and exhaustible resources. Review of Economic Studies Symposium. 29-46.

Solow, R. M., 1986. On the intertemporal allocation of natural resources. *Scandanavian Journal of Economics.* 88: 141-149.

Spash, C. L. and N. Hanley, 1995. Preferences, information and biodiversity preservation. *Ecological Economics* 12: 191-208.

Stern, D. I., 1993. Energy use and economic growth: a multivariate approach. *Energy Economics* 15: 137-150.

The Group of Green Economists, 1992. *Ecological economics: a practical programme for global reform.* Zed Books, London. 162.

van der Ploeg S.W.F., L.C. Braat, and W.F.J. Van Lierop, 1987. Integration of resource economics and ecology. *Ecological Modelling* 38: 171-190

Victor, P., 1972. Pollution, economy and environment. George Allen and Unwin, London.

Victor, P., 1991. Indicators of sustainable development: some lessons from capital theory. *Ecological Economics* 4: 191-213.

Viederman, S., 1994. Public policy: challenge to ecological economics. 467-490. In: A. M. Jansson, M. Hammer, C. Folke, and R. Costanza, (Eds.). *Investing in natural capital: the ecological economics approach to sustainability*. Island press, Washington D.C. 504.

Vitousek, P. M., P. R. Ehrlich, A. H. Ehrlich, and P. A. Matson, 1986. Human appropriation of the products of photosynthesis. *BioScience* 36: 368–73.

Walker B.H. and Noy-Meir, I., 1982. Aspects of the stability and resilience of savanna ecosystems, 577-590. In: B.J. Huntley and B.H. Walker, (Eds.). *Ecology of tropical savannas*. Springer, Berlin.

White, R.W., 1990. Transient chaotic behaviour in a hierarchical economics system. *Environment and Planning* 22: 1309-1321.

III.2 Ecological Orientors: A Path to Environmental Applications of Ecosystem Theories

Felix Müller and Sven E. Jørgensen

1. Introduction

In the preceding chapters of this volume, many different ecosystem theories have been described and a number of applications in environmental management have been explained. In the last chapter of this book, we will attempt to integrate these different aspects into a theory based, scientific methodology which has the potential to raise new questions for the development of theory and to improve the scope and practice of environmental management. The theories will be integrated on the base of their time-related concepts, concentrating on hypotheses which especially describe the ecological dynamics of the systems. While constructing a pattern from these theoretical prognoses, some general tendencies of ecosystem development will become obvious. These patterns will be explained and described, they will be investigated from the viewpoint of the originating theory, and the orientors will be assigned to different properties of self-organisation. Their potentials in environmental management will be discussed and finally some hypotheses will be used to summarise the concept.

2.What are ecological orientors?

In the last years many fundamental aspects of the idea of ecological orientation have been developed in ecosystem theory. The concepts of self-organisation (see Müller and Nielsen, this volume) and the fundamental laws of thermodynamics (see Jørgensen, this volume), resulting from network theory (see Patten and Fath or Ulanowicz, this volume), information theory (see Nielsen, this volume), and cybernetics (see Straskraba and Jørgensen, this volume), have elucidated succession and ecosystem development from a new ecosystemic and theoretical perspective. Certain system features such as emergent or collective ecosystem properties, are regularly changed by self-organised ecological dynamics. The basic idea of the orientor approach refers to self-organising processes which build up gradients and macroscopic structures from microscopic 'disorder' without receiving directing regulations from the outside. In such dissipative structures the self-organised process sequences generate comparable series of constellations. Thus, similar changes of certain attributes can be observed in different environments. Considering these attributes, the development of the systems seems to be oriented toward specific areas in the state space (see Figure 1). The respective state variables that can elucidate these dynamics, are termed *orientors*. Their technical counterparts in modelling are called *goal functions*. They are variables or objects of optimisation procedures that are guiding the systems' functions as well as their structural developments (see Bossel, 1998a; Jørgensen et al., 1998, Nielsen et al., 1998). The points in the state space that are approached asymptotically, may be described as *attractors*. Their positions are determined by the biotic potentials of the subsystems and by the supersystem of constraints which are limiting the developmental degrees of freedom. These tendencies and attractors can be found in very different entities, but in particular they are abundant in living systems, which can be characterised by an enormous creativity referring to the evolution and succession of forms, actors, and interactions. In this situation it is feasible to follow Patten's pragmatic attitude (1998). He unifies all terms for targets, goal functions, and orientors and stresses which points these extreme variables have in common. They are '*directional criteria* accounting for systems being pushed (by forcing functions) or pulled (by attractors) toward certain configurations, either with or without goals.' As members of this class of system characteristics the three terms 'orientor,' 'goal function,' and 'attractor' should be clearly defined to facilitate a constructive discussion.

Orientors are aspects, notions, properties, or dimensions of systems which can be used as criteria to describe and evaluate a system's developmental stage (Bossel, 1992a; 1994). The degree of orientor satisfaction that is represented by the distance of an observed state from an optimum point can be taken as respective indicator. The combination of degrees of satisfaction from different orientors leads to multiple indications of systemic properties which in toto form a holistic picture of the ecosystem's state.

The underlying orientor theory (Bossel, 1992b) is closely connected with the principle of optimisation. A certain group of elements increases up to a maximum level which is determined by the constraints that regulate the respective functional unit. The idea of an optimisation process affects a similarity with technical procedures and engineering. Furthermore, the optimisation aspect brings along an economic attitude to ecosystem analysis. One of its focal points is the economy of the system's energy budget, including the minimisation of efforts and an increasing efficiency of the processes (Kull, 1995).

Another conceptual problem arises because biological evolutionary optimisation is not oriented towards future structures. It is only effective in the prevailing present situation. Considering such evolutionary processes from an ecological perspective leads to the idea of co-evolution. Here optimisation means that the single elements of a system – the organisms – become more and more adjusted to each other. This mutual adaptation affects many ecosystem properties, providing a prolonged longevity and a higher integrity of the whole (Weber et al. 1989). Most of the corresponding features have been discussed in this book, and several of them are proposed to be introduced into environmental evaluation procedures.

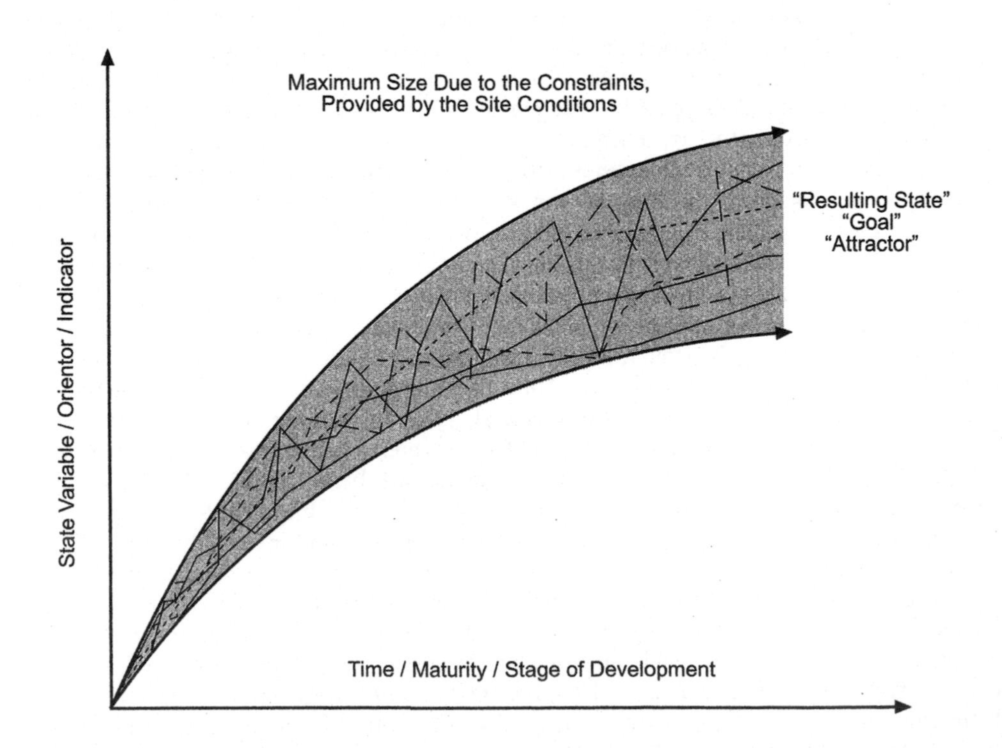

Figure 1: A generalised model of developmental trends of ecological state variables, indicators or orientors. As a consequence of conceptual uncertainty in ecological systems the tendency is directed toward a range of potential states.

Goal functions are the technical counterparts of orientors in ecological modelling. They are mathematical functions that describe the direction of a modelled ecosystem development as an outcome of its self-organising ability. This capacity enables the system to meet perturbations by directive reactions that can be described by goal functions (Jørgensen and Nielsen, 1998). Marques et al. (1998) define goal functions as follows: *"In ecological models goal functions are assumed to measure given properties or tendencies of ecosystems, emerging as a result of self-organisation processes in their development."* They are *"suitable measures of system oriented characteristics for natural tendencies of ecosystem development, and good ecological quality indicators."* In order to stress the teleological restrictions that are interrelated with this notion, Jørgensen and Nielsen (1998) formulate that *"the term goal function should solely be applied in the modelling context, while the term ecological indicator is more appropriate to discuss the propensities that characterise the development of ecosystems."* Bass (1998) introduces a third concept of directing developmental features. He defines *attractors* as relatively stable points in the state space toward which the system is drawn to. This directiveness in the trajectory of an orientor is especially evident after a perturbation.

Neither orientors nor goal functions or attractors are attained on the base of special purposes in non-human systems. Therefore, all three terms should be used without any teleological background (Barkmann et al., 1998). More importantly, the respective developments are observable consequences of self-organisation processes, reflecting the regular change of systems that are moving away from the thermodynamic equilibrium state. In this context the term 'propensity' which has been introduced into the developmental debate by Ulanowicz (1998) characterises the uncertainties that are connected with ecological dynamics: *"The system is not driven toward a certain endpoint. Rather, the probabilistic orientors guide the system in a vague direction"* (Ulanowicz, 1998). The orientation propensity describes the tendency of a certain event to occur under given circumstances, whereby the dynamics of the surroundings (e.g., the site conditions, the history of the investigated case, or the abundant species) have to be included as well as the respective probabilities. Propensities are conditional probabilities.

Both the setting of a goal and the considerations in order to reach it are intellectual activities. Natural systems can neither define a goal, consider a goal, nor reflect on it. Nevertheless, for biologists and evolutionary theoreticians it seems to be extremely tempting to impute a certain teleological purpose to natural features. The critical point about these appropriations is that they provoke the impression of an organism to be heading for a 'Lamarckian' evolutionary self-optimisation. In contrast, Darwinian theory has shown that the causality is based on mutation, competition, and selection, which may lead to new, perhaps even more efficient strategies of resource utilisation. These procedures are continuously proceeding without any conscious intention involved.

At the ecological scale, a similar teleological temptation can be found in the context of succession theory. Succeeding systems aim towards certain climax states that can be stable for long periods of time because the potentials of the sites are utilised perfectly – a stage of an optimal internal adaptation of the bioceonotic community is attained.

Thus, we have to formulate scientific hypotheses very carefully in this context. Of course, ecosystems, unlike human systems, do not have a goal. There is no intellectually fixed destination that natural development moves towards. In addition, there is no chance to formulate or realise the wish to grow into a specific direction in these systems. From a scientific perspective, there are no religious, mystic, or vital forces such as Driesch's '*vis vitalis*' or Theilhard de Chardin's final purpose 'omega.' All there is are coexisting and competing organisms that constantly interact with their environments and that therefore continuously change their abiotic life conditions. In the non-living ecosystem components the process occurs *vice versa*; the biotopes of the organisms are steadily modified, and, thus, there is a constant change, caused by internal as well as external factors. Throughout that change, the basic selection processes take place as described above.

3. Ecological orientors and ecosystem theories

All the theoretical approaches that are described in Chapter II of this book include some important hypotheses on the ecosystems' dynamic developments. In the following paragraph some of these aspects will be arranged in order to demonstrate both the diversity of the approaches and focal, common features of these concepts. Figure 2 characterises the *thermodynamic approach* to ecosystem dynamics. In general this access describes the energetic demands of a self-organised ecological development that leads from homogeneous patterns to strongly heterogeneous, gradient dominated systems (Jørgensen, 1992 and 1997; Müller and Nielsen, 1996). In a first group of parameters the energetic feature exergy is described using different criteria. Exergy is defined as the amount of work that a system can perform when it is brought into chemical equilibrium with its environment. It is the convertible, usable energy portion contained in a system with respect to its environment. In ecological systems the solar radiation functions as the primary exergy source. The imported exergy is partially transformed into biochemical energy, which is transferred and degraded into the numerous ecological reaction chains. The respective hypotheses pronounce that during their maturation, ecosystems optimise their ability to capture exergy (Schneider and Kay, 1994a,b). Thus, the photosynthetic assimilation capacity is optimised in relation to the site conditions. Along with this development, the exergy consumption increases, and the exergy flows through the system rise in their total quantity (the total energy passed through the system) and their complexity (the number and heterogeneity of individual flows). Throughout these energy transfers the initially high qualitative energy that was able to be converted into many other energy forms at the time of its import, is degraded. The convertibility is reduced and high energy fractions are lost, for example through heat and radiation exports, respiration losses, or transpiration processes.

The greater the number of elements which participate in the exergy transfer networks, the higher the demand for their maintenance, and the greater the exergy which has to be degraded. If the system receives and transforms a surplus exergy input that exceeds the demands for its maintenance, it is able to invest the additional energy into further growth (which can be indicated as an increase of biomass or the total system throughput of energy and matter) or further structurisation (which can be perceived as a rise of the structural gradients and the biocoenotical complexity; see Jørgensen, 1992). In both cases some fractions of the imported exergy are stored as a convertible reserve, either in biomass, organic matter, or in complex structures. The appropriate exergenic investments into structures can be derived on the base of information theory because the respective energy fraction is transformed into information (Jørgensen and Nielsen, 1998; Nielsen in this volume). A reference number for this energy component is the specific exergy (or structural exergy), which is the exergy per unit of biomass. The specific exergy indicates the information embedded in the biomass.

Exergy can be used to describe the 'constructive part' of the developing systems' energy budgets. On the contrary, entropy is an indicator for the cost side of the exergetic structure. To build up the complex pattern, and to enhance the distance from thermodynamic equilibrium, a high amount of energy gets lost via respiration and heat transfer. And also for the maintenance of the complex structure, a significant energy fraction has to be degraded in order to support the complex life processes. The sum of these degraded energy portions is the system's entropy production. Therefore, an optimisation of exergy degradation, gradient dissipation or exergy consumption must lead to local optima of entropy production. As ecosystems often become more and more diverse and as the ecosystem space becomes more and more filled with active biological units, the total entropy production will rise in correlation with the complexity of the system. Nevertheless, adaptational processes may lead to even better adjustments of directly linked processes or elements. This will be correlated with reduced energy and matter losses. Thus, the specific entropy production, which is restricted to one process or one transfer reaction, will decrease, although – as a function of the system's complexity – the total amount of entropy production rises throughout the development. As Svirezhev (1998) shows, this energetic adjustment is restricted to naturally evolving ecosystems. Results of anthropogenic inputs are decreasing efficiencies and therefore environmental degradation and dissipation. Also, the entropy production increases as a consequence of human disturbances.

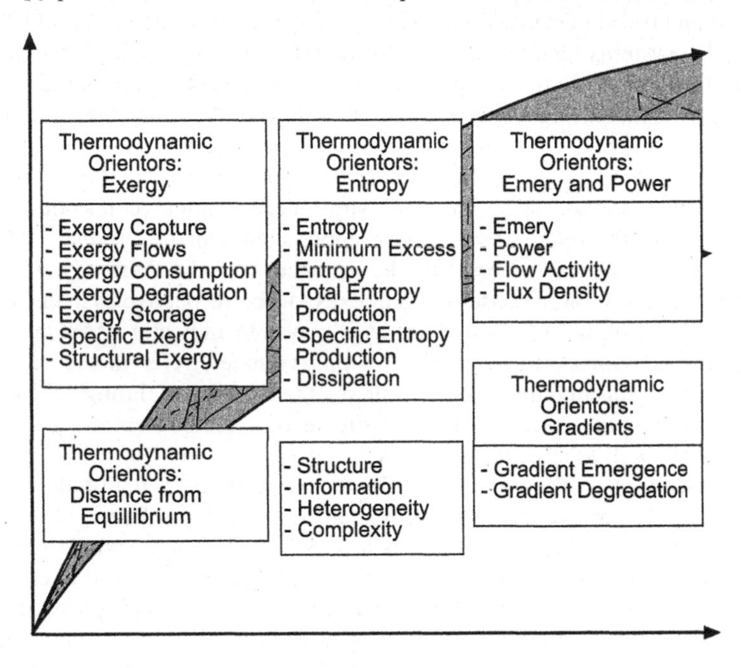

Figure 2: A list of proposed thermodynamic orientors.

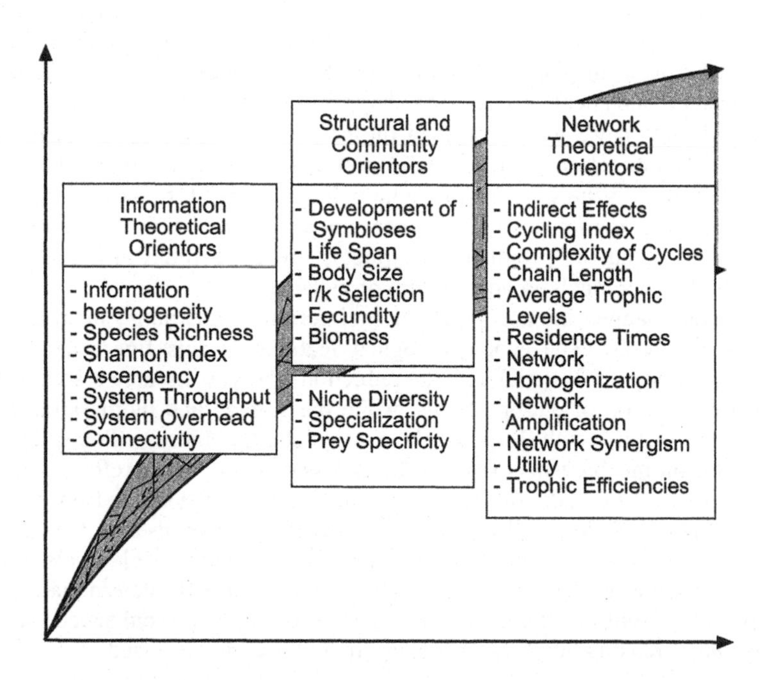

Figure 3: A list of proposed information theoretical, structural and network theoretical orientors

Similar thermodynamic ecosystem features have been developed by Lotka (1925) and Odum (1983). They are based on the increase of flux densities and activities throughout natural ecosystem developments, postulating a maximum of flows of usable energy in mature ecosystems (maximum power principle) as it has been described above. Furthermore, Odum (1983) postulates the emergy-maximisation principle (see Odum et al., in this volume) Ecosystems tend toward states where the 'energy quality,' represented by solar energy equivalents that are incorporated in the organisms, is optimised. Following these hypotheses the increasing complexity of organismic flows and the growing total amount of materials and energy flowing in the system are subjects of extreme principles.

A lot of other orientors are tightly connected with thermodynamic variables. As Nielsen (this volume) shows, there are very strict mathematical connections between thermodynamics and *information* theory. Both approaches can be used to describe the heterogeneity of a system, which represents the distance from the thermodynamic ground and which can be observed on the base of complexity of the respective systems' gradients. Besides the solely structural variables, Ulanowicz's system indicator ascendancy and its corresponding characteristics (see Ulanowicz in this volume), refer to the variability of flows as well as the total amount of energy or matter that flows through the biological networks. Ascendancy therefore is an holistic ecosystem characteristic, reflecting functional qualities as well as structural features.

Taking a closer look at the systems of flows and storages, a multitude of evolving *network properties*, such as the 20 major ecosystem features from Patten (1998) become visible. An important result of network analysis is that the utility of the whole system arises throughout undisturbed ecosystem development (Fath and Patten, 1998 and in this volume). Network synergism, which is strictly connected with the principles of network amplification and homogenisation, with increasing dominance of indirect effects and with growing residence times, offers an interesting relationship to modern, mutualistic aspects of evolutionary processes, such as the concept of Weber et al. (1989). The authors take the ecosystem level as an evolutionary unit, and they propose to integrate new organisms into this system under the precondition that they help reducing the losses of the whole and that they take part in an amplification of system-internal flows. If they do not do so, an extension of the new organism will not take place. Thus this mutualistic selection process, which can only take place in phases of dominating biological selection mechanisms and while no dominating physical events are happening, leads to the structural and thermodynamic orientation we have described on the preceding pages. As a consequence from this network theoretical concept, specialisation and niche diversity must increase in succession dynamics until a site specific, maximal value is attained. In parallel, the systems will accumulate biomass, symbioses will become more and more significant, and the life spans of the organisms can be prolonged. r-strategists will be relieved by K-strategists (Kutsch et al., 1998), and a much broader system of niches will be available.

These structural changes are accompanied by interesting *functional modifications* of the systems (see Figure 3). The ecophysiological features of this figure can be easily attributed to the thermodynamic variables. The reduction of loss (of scarce resources) refers to the relative minimisation of entropy flows, while the storage and biomass optimisation can be seen in connection with Jørgensen's exergy storage theory. Transpiration and respiration are efficient mechanisms to export the excess entropy and therefore are principle sub-processes of gradient degradation. Analysing the succeeding systems from an abstract systems-analytical point-of-view will lead to the features that have also been mentioned in the beginning: the complexity increases, the number of hierarchical levels rises, the autonomy of the systems gets higher, the systems attain a holistic determination with a maximum distributed control, the co-evolutionary processes are enhanced, and as a consequence of the increasing numbers of species, redundancy is increased.

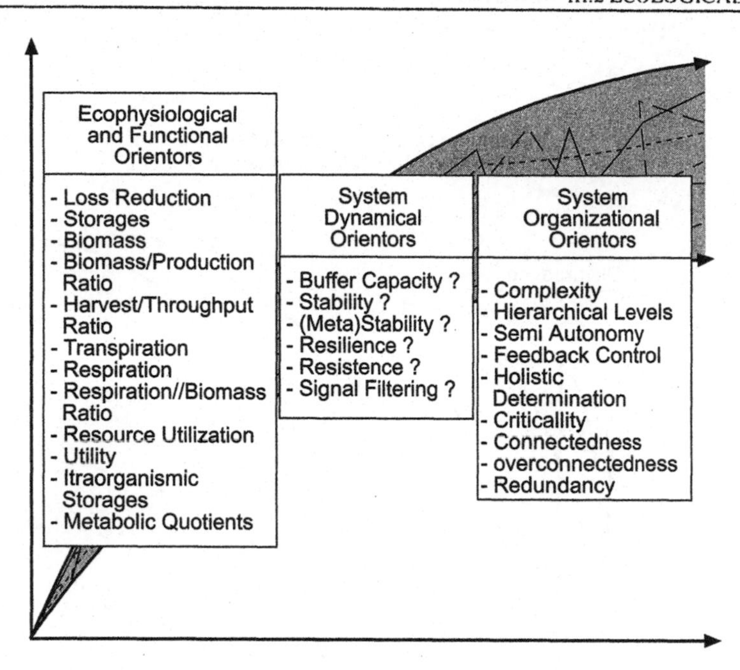

Figure 4: A list of functional, system dynamical and organisational orientors. The proposed stability attributes are indicated by question marks because the significance of their optimisation throughout succession and maturation is debatable.

On the other hand, in a climax stage of perfect maturity, the degree of the self-organised criticality is getting higher, too. The captured exergy is totally put to use for the maintenance of the existing structures. Thus, the energetic reserves for further structural development decrease. The adaptability of mature systems that have been structured for long times in similar external conditions may be low, and the high connectivity may turn into overconnectedness (Holling, 1986). Therefore, as Bass (1998) describes, the stability characteristics may not follow the orientor functions of the other parameters. Schneider and Kay (1994a,b) support this concept. They have formulated the hypothesis that the tendency to return to thermodynamic equilibrium gets higher the further a system has been moved away from thermodynamic equilibrium. Holling (1986) also points out that the longer the destructive shift has been delayed the greater the destruction will be; the longer the system remains in the critical (mature) state, the larger will be the consequences of major disturbances. The stored exergy in this case of creative destruction serves as fuel for an extensive renewal in a restarting exploitation phase. Therefore, a high maturity is always connected with a high risk.

As Bass (1998) stresses these temporal characteristics can be found at many different scales. There is a sequence of dynamics which follows the general principles of hierarchy theory (Allen and Starr, 1982; O'Neill et al., 1986; Müller, 1992, Hari and Müller, this volume). Bass summarises these aspects: *"An ecosystem is comprised of several smaller attractors, perhaps semistable attractors, of thermodynamic risk, being destabilised at different times. Although a smaller attractor may be destabilised, it is subsumed by a larger attractor which remains stable. Essentially, maintaining the system at the conservation stage changes this larger attractor into a semistable attractor, but the system cannot remain at this attractor indefinitely, thus the damage occurs on a much larger scale."*

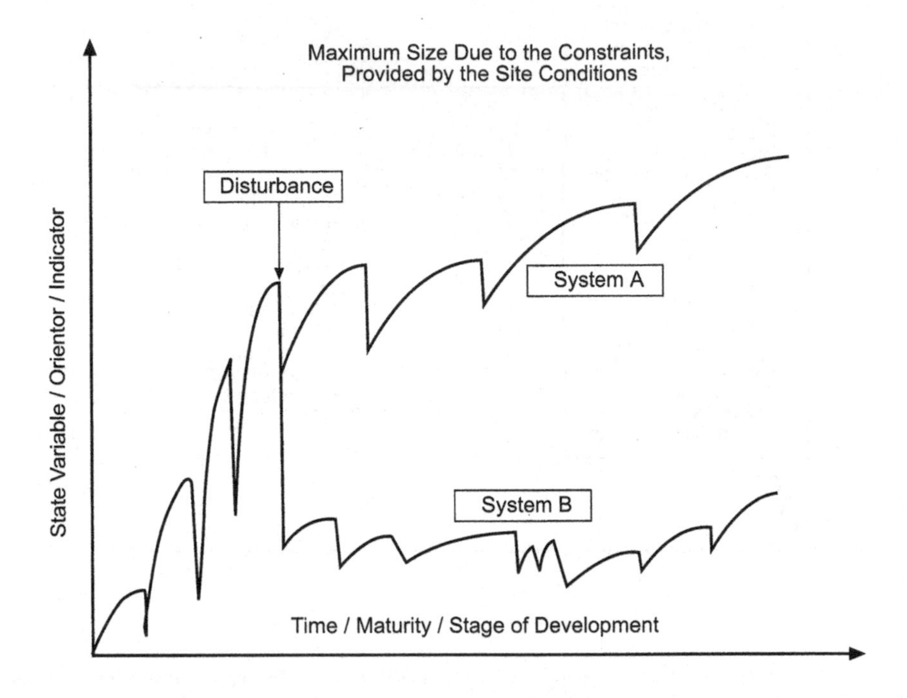

Figure 5: Sketch of a nested hierarchical organisation of orientors which are developing on different temporal scales. The resulting curves consist of many sequential 'Holling cycles' on the respective lower levels. Thus there is self-similar behaviour on different levels-of-organisation such as ecosystem dynamics, ecosystem succession, or ecosystem evolution.

If that disturbance occurs, the regulating steady state based hierarchy of signal transfers is broken, and a new orienting phase can start. The orienting functions are actualised and newly defined. If we subsume this hierarchical aspect, and if we comprehend emergence as a consequence of self-organisation (Müller, 1996a; Müller et al., 1998), it becomes obvious that the state variables that have been discussed before, function as emergent properties (if they are based on processual interactions between the parts of the system) or collective properties (if the sum of the parts is equal to the quantity of the whole) which are suitable to partially describe the system's behaviour. Therefore, the system itself constructs a regulating, autocatalytic hierarchical level. Following hierarchy theory, this step must lead to the emergence of constraints which lead system dynamics and which determine the respective orientors. The degrees of freedom are restricted self-referentially when a new self-organised level of control is built up.

All these items are linked with the concept of *ecological gradients*. The thermodynamic non-equilibrium principle (Schneider and Kay, 1994a,b) hypothesises that self-organising ecological systems build up internal, functional gradients as a consequence of the strong external exergy gradient that is provided by solar radiation. In view of the long term, the imported (captured) exergy gets transformed into a hierarchical system of interconnected gradients that imply all structural features characterising the system's distance from the thermodynamic equilibrium. Thus, all structural ecosystem properties can be comprehended as concentration gradients in space and time (Müller, 1998). They build up the potentials to carry out mechanical work, chemical reactions or biological interactions. Ecosystem function is therefore defined as the general characteristic of the system's gradient dynamics. The oriented features are valid in all dissipative systems, but living entities are able to amplify the self-organised process sequences (e.g. by their reactivity, processing of information, genetic variability, self-replication, ability for evolution, growth, development, hierarchical organisation). Therefore, the general features of orientor theory must be comprehended as a framework inside which all single attributes have to be arranged.

4. Ecological orientors and self-organisation

In many of the chapters it has been postulated that the dynamics of orientors are the result of dissipative self-organising processes, which generally can be characterised by the transformation of microscopic disorder into macroscopically ordered structures (Müller and Nielsen, this volume). For such dissipative self-organised systems some general prerequisites and features have been described (Haken, 1983; Kay, 1984; Ebeling, 1989; Schneider and Kay, 1994a; Müller, 1996a) that have to be considered with regard to the problems of orientor theory. These characteristics and their implication in reference with orientor theory are:

- *Openness of the system*: The dissipative system must be able to exchange energy, matter and information with its environment. For ecological systems this means that inputs as well as outputs of energy and nutrients are basic suppositions for any self-organised development. These essentials functions act as an external constraint for the potentials of the orientors.
- *Import of convertible energy* (exergy) *in a degradable quantity:* The dissipative developmental processes are energy-pumped. They depend on the availability of a 'high quality' energy form that can be converted into mechanical work. The quantity of this energy must fall within a window between the minimum supply, necessary for the start of the dissipative process and a maximum input which starts to turn order into turbulent structures. A respective orientation function is the capacity of exergy capture, which is optimised throughout undisturbed ecosystem development.
- *Internal energy transformations* (exergy degradation): The imported exergy is converted in dissipative processes. Throughout these reactions which are energetically connected to ramified reaction chains, the initial exergy is degraded in several succeeding steps. The number of degrading reactions and the connectivity of the relating flows are orientors and functions of the state and degree of the self-organised development.
- *Export of non-convertible, non-usable energy* (entropy): As result of the exergy degradation, non-usable energy forms (entropy) are produced. They have to be exported to the environment of the system. The ecological entropy exporting processes are heat transfers, matter transfers such as respiration, transpiration, and losses of ecosystem constituents. The degraded products are re-conducted to the next higher hierarchical level in the system's environment. The total entropy production can be taken as an orientor that provides important clues for the state of the investigated system.
- *Distance from thermodynamic equilibrium*: The self-organised processes create internal gradients. With these gradients, spatio-temporal structures and heterogeneities arise which can be passed over to functional hierarchies. The respective degree of heterogeneity and structurisation can be taken as an indicator of the distance to homogeneous distribution patterns. It is one of the most significant ecological orientors.
- *Non-linear interactions between the elements of the system*: Changes of one state variable in self-organised systems can lead to non-linear variations of other parameters. As a consequence, different stationary states can be achieved. Thus, the actual state is always dependent on the system's history. This means that all

linear descriptions of ecological interaction patterns are small-scaled sections from a non-linear reality. As a consequence, all predictions that refer to the development of ecosystems are linked with a high 'uncertainty of the detail.'

- *Strong amplifications in phase transitions*: Caused by the non-linear interactions, strong and destabilising amplitudes of state variables can appear when the system's trajectory reaches states in the proximity of structural modifications. Thus, the temporal behaviour and the variances of structuring processes can be used as ecological indicators.
- *Internal structural processes*: The self-organising processes develop in a self-referential, self-regulated manner. There is no external regulation which exceeds the limitations of the single processes' degrees of freedom that arise from hierarchical constitutions.
- *Constraints in hierarchies*: The co-ordination in self-organised systems is based on the interactions of partial processes that operate on different scales. During phases near steady states, the interrelationships are dominated by slow processes that operate within large spatial extents. The potential state space of smaller and more rapid entities is limited by these constraints. Thus the degree of hierarchical structurisation can also be used as an ecosystem indicator.
- *(Meta)stability after small impulses:* Disturbances can be buffered by self-organised systems within certain limitations.
- *Historicity and irreversibility*: Self-organisation is based upon irreversible processes. It can be understood only if the history of the system is known.

Summarising, it can be presumed that self-organisation is the basic process network for the development of ecological orientors. They therefore are capable of describing the degree and duration of self-organised processes which is important information for the scientific systems analysis as well as environmental management.

5. Potentials for environmental applications

If we utilise these orientation concepts in environmental science and management, many applications are possible. Most experience has been gained in ecological modelling. Here, certain state variables have been taken as goal functions which are observed under different model constellations, e.g., different species compositions. Assuming that the system will always develop towards a state with a higher value of the respective goal function, optimisational calculations have been carried out very successfully, for example in applying the thermodynamic variable exergy as a goal function.

As there are some basic parallels concerning the 'pushed and pulled' ecological development of orientors and the human setting and following of targets, an integration of these philosophies in environmental management seems to be a promising approach (see Müller and Leupelt, 1998). Observing this field of activities, recently some change has become obvious. A modification of the societal targets has been visible in environmental policy since the sustainability debate started (WCED, 1987; Ekins, 1992; Brand, 1997; Teichert et al., 1997). One new aspect originates in its long-term character: the temporal extents of the objectives are not restricted to the typical political election periods but to generations. With this attitude the developmental capacities of ecosystems, i.e., their potentials for future self-organisation, succession and evolution, have become emphasised targets of environmental policy. Consequently, a sustainable landscape management has to practice a holistic strategy, it has to argue about integrated ecosystems instead of structural units of the community alone, and it has to integrate ecological, social and economical goals. Therefore, the sustainability concept is an interdisciplinary strategy which brings together aspects of ecological qualities as well as the many different interpretations of human life quality. Also, the spatial extent of the debate has changed: The area that sustainability evaluations should refer to has to include neighbouring systems as well as indirectly linked zones, that may be far away from the locality of direct action. And finally, sustainable management strategies have to include indirect effects; they have to take into account non-linearities, chronic stress effects, decouplings of processual networks and combinations of these factors. In other words, they have to be based on an ecosystemic approach. Thus, the sustainability concept demands new measures, new methods, and, last but not least, new goals. Trendsetting directions and objectives have been proposed, for example with the concepts of ecosystem health (Costanza et al., 1992, Rapport in this volume) or ecological integrity (Woodley et al., 1993; Barkmann and Windhorst in this volume). Both strategies are based on an approach that integrates ecosystem structures and functions. Both approaches are orientated by the complexity of ecological systems and by the idea to support the systems' long-term developments. Constanza et al. (1992) define 'health' as an ecosystem feature which refers to the basic characteristics: vigour (metabolic activity), organisation, and resilience. 'Integrity' has been described by Kay (1993) as the ability of a system to maintain its organisation and to develop in sequences of self-organised processes, thus integrity comprises health, buffers capacity and the self-organisation capacity.

In summary, a recent pragmatic change to the idea of fundamental orientations has taken place in ecology and environmental practice. The basic ideas, models and leading principles are being revised, and thus, in this phase of change, are becoming more flexible. Also the methods of environmental evaluation and decision support are being modified. This improvement can be characterised by an increased consciousness of the significance of goals and constructions of goal hierarchies, reaching from leading environmental ideas to eco-targets, environmental quality objectives or environmental standards. This development is accompanied by a prolonged search for suitable indicators that satisfactorily represent the degree of success with respect to the environmental aims that have been, and will, be set.

6. Exemplary consequences: ecological indicators

The remaining question is: 'What will be the correct indicator set of ecosystem behaviours and at what level-of-description?' In Figures 1 to 3 many different potential orientors, goal functions, and indicators are listed. In spite of the fact that some authors believe that they have found the one parameter that can be used as the only holistic ecosystem property, we will stick to the idea of a multidimensional systems' characterisation. Of course many different arrangements of ecosystem properties are possible. For example the most basic structure can be taken from the features and prerequisites of self-organisation that have been summarised in the first half of this chapter. Although, the best answer to the question for an optimal indicator selection will be the (re)-question 'What are the specific indicanda?' Three approaches of structuring ecological orientors are demonstrated in this Chapter. At first we refer to Schneider and Kay (1994a) who take a thermodynamic, gradient orientated position and who propose the following ecosystem properties as fundamental orientors to describe the degree of organisation and maturity in ecological systems:

1) Exergy capture
2) Energy flow activity
3) Cycling of energy and materials
4) Average trophic structure
5) Respiration and transpiration
6) Biomass
7) Types of organisms

Similar high-level orientor groups have been proposed by Müller (1996b) as aggregated indications for the functionality of ecosystems. They include properties such as

1) Exergy capture
2) Energetic and material flow densities in the system
3) Cycling of energy and matter
4) Storage capacities
5) Matter retention (Loss reduction)
6) Respiration and transpiration
7) Diversity and heterogeneity
8) Organisation and hierarchy (signal filtering and buffering capacity)

A more ecophysiological set of indicators has been proposed by Kutsch et al. (i.p.). The authors have derived indicators for integrity on the basis of ecosystem research data sets, and their variables are the following

1) Intrabiotic nutrient content
2) Species number
3) Total amount of biomass
4) Production / available nutrients
5) qCO_2
6) Surface temperature
7) Rn/K* (%)
8) Transpiration / evopatranspiration

The fourth approach refers to Bossel's basic orientors (Bossel, 1992). Following this access, the viability of systems is strongly based on the properties of the system itself and the features of its environment. In viable systems, structures and functions of the systems therefore have to be adapted to the predominant environmental conditions. Bossel emphasises six classes of environmental challenges that affect the most fundamental conditions of the system's behaviour (normal environmental state, scarce resources, variety, variability, change, and other systems). Following the orientor theory (Bossel, 1992 a,b; 1994), the corresponding systems have to develop adequate responses to the dominating environmental properties. These fundamental criteria are denoted as 'basic orientors.' They are identical for all complex adaptive systems. The basic orientors are

1) Existence
2) Effectiveness
3) Freedom of action
4) Security
5) Adaptability
6) Coexistence

With these proposals for variable sets, an interesting handful of similar approaches is available to indicate the potential of self-organisation in ecosystem on the base of ecological orientors. The decision about which indicators are to be selected depends on the specific purpose of the study and on the applicability of the parameters. Future research will hopefully enhance the practical utilisation of these concepts.

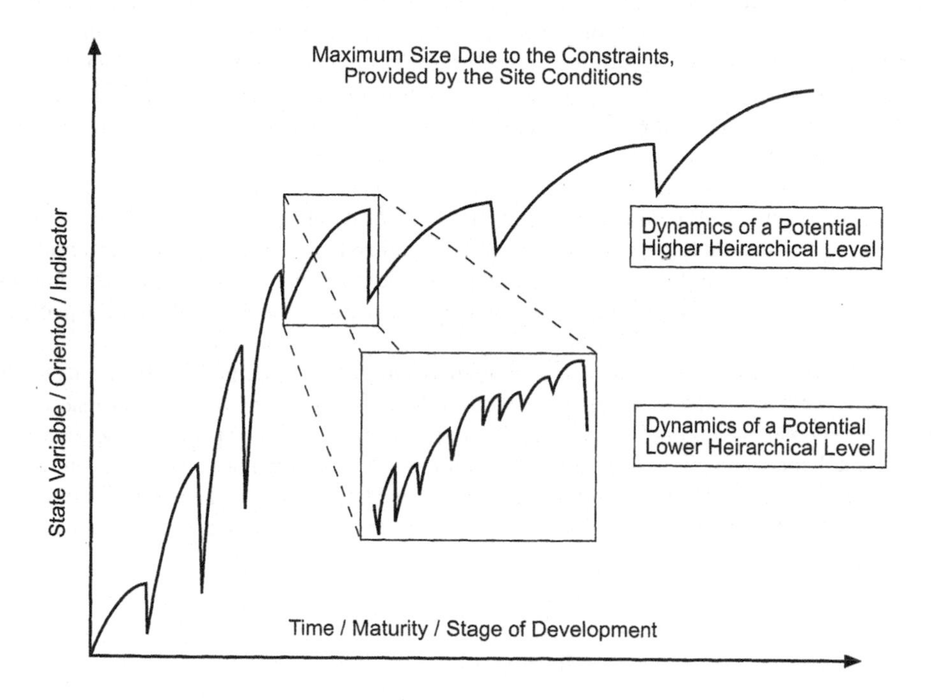

Figure 6: Two hypothetical developments of an orientor throughout systems dynamics after perturbation. While the 'healthy' ecosystem (System A) is able to develop into the maturing direction after perturbation, the stressed system (B) does not follow the natural developmental trend.

7. Summarising hypotheses and conclusions

In this chapter the concept of ecological orientors has been elucidated from many different points of view. In summary, the following hypotheses have been proved:

- During the development of ecosystems, important measurable properties are regularly optimised. These emergent or collective features are called orientors.
- Ecological orientors can be used to distinguish systems' states and to characterise different systems.
- One set of ecological orientors defines both the structural and the functional features of the investigated system. Thus, orientors are holistic ecological characteristics.
- Ecological orientors are based on thermodynamic principles. They indicate general properties of dissipative living systems.
- Therefore, they have potential for self-organisation.
- Ecological orientors also indicate the degree of naturalness in ecosystems.
- Ecological orientors are a good basis for finding usable indicators for ecosystem health, ecological integrity, and for the ecological foci of sustainability.

In fact, during the development of ecosystems there are areas of scientific research that are regularly optimised. The numerous examples, case studies, and theoretical derivations provide evident indications that the orientor principle is a valid scientific concept. Although some problems occur with regard to detailed prognoses of species abundances, the generality of the optimising principle is obvious. Case studies in the book of Müller and Leupelt (1998) show that the orientors can be used to characterise, classify and distinguish between different ecosystems. Their quantitative differences are great enough to even overview the differences between directly neighbouring ecosystems, and thus a group of orientors can in fact form a suitable tool for a holistic environmental indication. As the indicanda are holistic issues, both ecosystem functions and ecosystem structures are described. Due to the selection of indicators, the orientors describe a potential for self-organisation. They refer to developmental processes, ecological dynamics, and change. Therefore, the temporal parameter modifications should be taken into account – as a fundamental for any evaluation procedure – with greater significance than the final state that might represent a brittle configuration at the edge of creative destruction. Furthermore, there is no question, that these concepts must be introduced into modern strategies of ecosystem protection. And, of course, ecological orientors are a good basis for finding usable indicators for ecosystem health, ecological integrity and for the ecological aspects of sustainability.

In spite of these positive notions, a number of questions have not yet been answered satisfactorily. What are the prerequisites from philosophy of science for the utilisation of goal functions and orientors? How can ecological concepts be applied practically in environmental management? How can we implement ecosystem indicators in monitoring networks? How many and what indicators do we need to describe the whole ecosystem satisfactorily? How can these indicators be aggregated and on which scale do they operate?

REFERENCES

Allen, T.H.F. and T.B. Starr 1982. Hierarchy - Perspectives for ecological complexity. The University of Chicago Press.

Barkmann, J., R. Baumann, B. Breckling, U. Irmler, F. Müller, C. Noell, H. Reck, E.W. Reiche, W. Windhorst u.a. 1998. Ökologische Integrität als Beschreibungsmaßstab und Leitbild für Ökosystemschutz und nachhaltige Entwicklung.

Bass, B. 1998. Applying Thermodynamic Orientors: Goal Functions in the Holling Figure-Eight Model. In: Müller, F and M. Leupelt (eds). *Eco-targets, goal functions, and orientors*. Berlin, Heidelberg, New York pp193-208.

Bossel, H. 1992a. Real-structure process description as the basis of understanding ecosystems and their development. *Ecological Modelling* 63: 261-276.

Bossel, H. 1992b. Modellbildung und Simulation. Braunschweig.

Bossel, H.. 1994. Modelling and simulation. A. K. Peters, Wellesley MA and Vieweg, Wiesbaden.

Bossel, H. 1998a. Ecological orientors: Emergence of basic orientors in evolutionary self-organisation. In: Müller, F. and Leupelt, M. 1998. *Eco targets, goal functions and orienters*. Springer. Berlin, Heidelberg, New York, 19-33.

Costanza, R., B.G. Norton and B.D. Haskell eds., 1992. Ecosystem health. Island Press, Washington D.C.

Ebeling, W. 1989. Chaos - Ordnung - Information. Verlag Harri Deutsch, Frankfurt/Main.

Fath, B.D. and B.C. Patten 1998. Network synergism: emergence of positive relations in ecological systems. Ecological Modelling 107, 127-143.

Haken, H. 1983. Synergetics, An introduction. Berlin, Heidelberg, New York.

Holling, C.S. 1986. The resilience of terrestrial ecosystems: local surprise and global change. In: Clark, W.M. and R.E. Munn (eds.). *Sustainable development of the bioshpere*. Oxford University Press, Oxford, 292-320.

Jørgensen, S.E. 1992. Integration of ecosystem theories: A pattern. Dortrecht, Boston, London.

Jørgensen, S.E. 1997. Thermodynamik offener Systeme. In: Fränzle, O., F Müller and W. Schröder (eds) *Handbuch der Ökosystemforschung, Kapitel III-1.6.*

Jørgensen, S.E. and S. N. Nielsen 1998. Thermodynamic orientors: Exergy as a goal function in ecological modelling and as an ecological indicator for the description of ecosystem development. In: Müller, F. and Leupelt, M. (eds.). *Eco targets, goal functions, and orientors.* Berlin, Heidelberg, New York.

Jørgensen, S.E., B.C. Patten and M. Straskraba 1998. Ecosystem emerging: 3. Openness. *Ecol. Modelling.* in press.

Kay, J.J.. 1984. Self-organisation in living systems. Ph.D. Thesis, University of Waterloo.

Kay, J.J. 1993. On the nature of ecological integrity: Some closing comments. In: Woodley, S., Kay, J. and Francis, G. (eds.). *Ecological integrity and the management of ecosystems.* University of Waterloo and

Canadian Park Service, Ottawa.

Kull, U., E. Ramm and R. Reiner (eds), 1995. Evolution und Optimierung. Stuttgart.

Kutsch, W., O. Dilly, W. Steinborn and F. Müller 1998. Quantifying ecosystem maturity - A case study. In: Müller, F. and Leupelt, M. (eds.). *Ecotargets, goal functions and orienters.* Springer. Berlin, Heidelberg, New York, 209-231.

Leupelt, M. (eds.),1998. Eco-targets, goal functions, and orientors. Berlin, Heidelberg, New York.

Lotka, A.J. 1925. Elements of physical biology. Baltimore.

Müller, F. 1992. Hierarchical approaches to ecosystem theory. *Ecological Modelling* 63: 215-242.

Müller, F. 1996a. Emergent properties of ecosystems – consequences of self-organising processes? *Senckenbergiana maritima* 27 (3/6), 151-168.

Müller, F. 1996b. Ableitung von integrativen Indikatoren zur Bewertung von Ökosystem-Zuständen für die Umweltökonomischen Gesamtrechnungen. Studie für das Statistische Bundesamt, Wiesbaden.

Müller, F. and M. Leupelt (eds.).,1998. Eco-targets, goal functions, and orientors. Berlin, Heidelberg, New York.

Müller, F. and S.N. Nielsen 1996. Thermodynamische Systemauffassungen in der Ökologie. In: Mathes, K., B. Breckling and K. Ekschmitt (eds.). Systemtheorie in der Ökologie. Landsberg, 45-62.

Müller, F., B. Breckling, M.Bredemeier, V. Grimm, H. Malchow, S. N. Ulanowicz, R.E. 1998. Network orientors: Theoretical and physical considerations why ecosystems may exhibit a propensity to increase in ascendency. In: Müller, F. and Leupelt, M. 1998. *Eco targets, goal functions and orienters.* Springer. Berlin, Heidelberg.

Müller, F. 1998. Gradients in ecological systems. Proceedings of the Eco-Summit 1996 in Copenhagen, *Ecological Modelling* 108: 3-21.

Nielsen and E.W. Reiche 1997. Ökosystemare Selbstorganisation. In: Fränzle, O., F. Müller and W. Schröder (eds). *Handbuch der Ökosystemforschung, Kapitel III-2.4.*

Odum, H.T. 1983. Systems ecology. John Wiley and Sons, New York.

O'Neill, R.V., D. L. DeAngelis, J.B. Waide and T.F.H. Allen 1986. A hierarchical concept of ecosystems - Monographs in Population Biology 23, Princeton University Press, 254 pp.

Patten, B.C. 1998. Network orientors: Steps towards a cosmography of ecosystems: Orientors for directional development, self-organisation, and autoevolution. In: Müller, F. and M. Leupelt (eds.). *Eco-targets, goal functions, and orientors.* Berlin, Heidelberg, New York.

Schneider, E.D. and J. Kay 1994a. Life as a manifestation of the second law of thermodynamics. Math. *Comput. Modelling* 19: 25-48.

Schneider, E.D. and J. Kay 1994b. Complexity and thermodynamics: Towards a new ecology. *Futures* 26: 626-647

Svirezhev, Y. 1998. Thermodynamic Orientors: How to Use Thermodynamic Concepts in Ecology. In: F. Müller and M. Leupelt (eds.). *Eco Targets, Goal Functions, and Orientors.* Springer. Pp 102 122.

Ulanowicz, R.E. 1998. Network Orientors: Theoretical and Philosophical Considerations why Ecosystems may Exhibit a Propensitiy to Increase in Ascendency. Chap. 2.10 in Müller, F. and Leupelt, M., (eds.) 1998. Eco Targets, Goal Functions, and Orientors. Springer, Berlin.

Weber, B.H., D.J. Depew, C. Dyke, S. N. Salthe, E. D. Schneider, R.E. Ulanowicz, and J.S. Wicken 1989. Evolution in thermodynamic perspective: An ecological approach. *Biology and Philosophy* 4: 373-405.

WCED - World Commission on Environment and Development 1987. Our Common Future. Oxford University Press.

Woodley, S., Kay, J., and Francis, G. 1993. Ecological integrity and the management of ecosystems. University of Waterloo and Canadian Park Service, Ottawa.

Subject Index

D

E

F

N

O

P

Q

R

S